NEUROMETHODS

Series Editor
Wolfgang Walz
University of Saskatchewan
Saskatoon, SK, Canada

For other titles published in this series, go to
www.springer.com/series/7657

Animal Models of Dementia

Edited

Peter Paul De Deyn

Laboratory of Neurochemistry and Behaviour, Institute Born-Bunge, University of Antwerp, Wilrijk, Belgium

and

Debby Van Dam

Laboratory of Neurochemistry and Behaviour, Institute Born-Bunge, University of Antwerp, Wilrijk, Belgium

Humana Press

Editors
Peter Paul De Deyn, Ph.D.
Laboratory of Neurochemistry and Behaviour,
Institute Born-Bunge,
University of Antwerp,
Wilrijk,
Belgium
dedeyn@skynet.be

Debby Van Dam, Ph.D.
Laboratory of Neurochemistry and Behaviour,
Institute Born-Bunge,
University of Antwerp,
Wilrijk,
Belgium
debby.vandam@ua.ac.be

ISSN 0893-2336 e-ISSN 1940-6045
ISBN 978-1-4939-6118-4 ISBN 978-1-60761-898-0 (eBook)
DOI 10.1007/978-1-60761-898-0
Springer New York Dordrecht Heidelberg London

Softcover re-print of the Hardcover 1st edition 2011

Printed on acid-free paper

Humana Press is part of Springer Science+Business Media (www.springer.com)

Series Preface

Under the guidance of its founders Alan Boulton and Glen Baker, the Neuromethods series by Humana Press has been very successful since the first volume appeared in 1985. In about 17 years, 37 volumes have been published. In 2006, Springer Science + Business Media made a renewed commitment to this series. The new program will focus on methods that are either unique to the nervous system and excitable cells or which need special consideration to be applied to the neurosciences. The program will strike a balance between recent and exciting developments like those concerning new animal models of disease, imaging, *in vivo* methods, and more established techniques. These include immunocytochemistry and electrophysiological technologies. New trainees in neurosciences still need a sound footing in these older methods in order to apply a critical approach to their results. The careful application of methods is probably the most important step in the process of scientific inquiry. In the past, new methodologies led the way in developing new disciplines in the biological and medical sciences. For example, Physiology emerged out of Anatomy in the nineteenth century by harnessing new methods based on the newly discovered phenomenon of electricity. Nowadays, the relationships between disciplines and methods are more complex. Methods are now widely shared between disciplines and research areas. New developments in electronic publishing also make it possible for scientists to download chapters or protocols selectively within a very short time of encountering them. This new approach has been taken into account in the design of individual volumes and chapters in this series.

Wolfgang Walz

Preface

Aimed at studying human diseases, animal models emerged in the 1800s and underwent a real boom during the last decades. The discovery of new therapies for neurological disorders is especially predicated on the use of animal models both to identify new therapeutic targets and to carry out preclinical drug trials. Of primary concern to a neuroscience researcher is the selection of the most relevant animal model to achieve his or her research goals. Researchers are challenged to develop models that recapitulate the disorder in question, which is often not as straightforward as it may seem. Quite often they are confronted with the choice between models that reproduce cardinal pathological features of the disorders caused by mechanisms that may not necessarily occur in the patients versus models that are based on known etiological mechanisms that may not reproduce all clinical features. This is also the case in many neurological conditions that may be accompanied by dementia, as will be evidenced in this volume.

Dementia is defined as the loss of mental processing ability, including communication, abstract thinking, judgment, and ultimately physical abilities. Dementia typically also results in behavioral and personality changes, depending on the area(s) of the brain affected. Irrespective of etiology, dementia symptomatology significantly interferes with social and occupational functioning.

By definition, dementia refers to a syndrome consisting of a range of symptoms commonly found in people with brain diseases characterized by damage to and consequent loss of brain cells. Losing brain cells is a natural process linked to aging, but with illnesses that may lead to dementia, this loss occurs at a much faster rate resulting in abnormal brain function.

Currently, Alzheimer's disease is the most common neurological disease of adulthood, almost twice as common as stroke or epilepsy and as common as congestive heart disease. Since the elderly are the fastest growing segment of the population, the dementia epidemic poses major consequences for the health and aged care systems. It is forecasted that the worldwide number of elderly people suffering from dementia will rise to 63 million in 2030 and to 114 million by 2050. The concurrent direct (related to medical care) and indirect (reduced productivity due to illness) costs have attracted the attention of health care policy makers and motivated intensification of dementia-related research.

Preclinical research based on animal models is pivotal to our knowledge of underlying molecular mechanisms and the drug discovery pipeline for dementia aiming at the development of therapeutic strategies alleviating or preventing this devastating disorder. Part I of this volume deals with more general aspects of animal modeling as well as the related ethical issues and focuses on the dementia drug discovery pipeline. In Part II, essential methodological considerations when starting animal model research are dealt with, ranging from the choice of a certain species or strain to patenting issues. The quality and utility of any animal model should be assessed through rigorous validation; a valid model resembles the human condition in etiology, pathophysiology, symptomatology, and response to established therapeutics. Part III of this book, therefore, compiles various levels of validation including pathological, behavioral, neurochemical, pharmacological, and imaging aspects.

Neurodegeneration is the most common biological cause of dementia, with Alzheimer's disease accounting for more than 50% of the dementing subjects making it the fourth leading cause of death in Western society. Consequently, a large part of this volume (Part IV) deals with animal models of Alzheimer's disease, ranging from nonmammalian models such as fruit flies and zebra fish, over lesion and infusion-based models to the large number of available transgenic models. Given the boost transgenesis and gene targeting techniques have given to the development of valid phenocopies of the human condition, the included models are not exhaustive but give rather a representative sample of the available models. For an updated overview of available genetically modified models, we refer the readers to specialized websites, as e.g. http://www.alzforum.org. The subsequent chapters of Part V deal with other neurodegenerative conditions attended with dementia: Parkinson's disease, metachromatic leukodystrophy and adrenoleukodystrophy, amyotrophic lateral sclerosis, and frontotemporal dementia.

Dementia is not just limited to the degenerative types of dementia. Dementia after all refers to a syndrome which does not always follow the same course of development. In some cases, the person's condition may improve or remain stable over time. Types of vascular dementia, the second most common cause of dementia, are discussed in Part VI. The subsequent chapters touch upon (partially) reversible dementia types. Since various diseases or injuries may lie at the basis of these types of dementia, ranging amongst others from infections to trauma, metabolic, hormonal and toxic disorders, and tumors, only a limited selection of these types of dementia appears in this volume. Parts VII through IX reflect on animal models of, respectively, normal pressure hydrocephalus, trauma- and toxic-induced dementia.

New therapeutic avenues are opened up based on recent insights in pathophysiological mechanisms underlying dementia. This book stresses the importance of extensively validated animal models in drug discovery and development to predict clinical activity. Clinical research focuses on the diagnosis of Alzheimer's disease and related conditions in an early stage based on specific biomarkers. When conversion of mild cognitive impairment to dementia can be predicted, disease modifying treatment strategies become indispensable. Moreover, more attention is being paid to Behavioral and Psychological Signs and Symptoms of Dementia (BPSD). Animal models, therefore, should also aim at mimicking BPSD-related behaviors that will allow the evaluation of new psychopharmacological strategies.

This book will appeal to a broad readership as dementia represents an increasing socioeconomical burden in our graying population. With contributions from prominent investigators in the broad field of dementia, this book brings together a wide spectrum of expertise, from neuropathologists, pharmacists, biochemists, and biologists to clinical neurologists. As a *Neuromethods* title, this book provides a detailed, yet accessible, overview of currently available animal models in the field of dementia research, and touches, as well, upon more general areas linked to the development and use of animal models. The book will appeal to both experienced animal researchers as well as investigators on the verge of starting animal model-based dementia research.

Wilrijk, Belgium — ***Peter Paul De Deyn***
Wilrijk, Belgium — ***Debby Van Dam***

Contents

Part IV Animal Models of Alzheimer's Disease

Part V Animal Models of Non-Alzheimer Neurodegenerative Disease

PART VI ANIMAL MODELS OF VASCULAR DEMENTIA

PART VII ANIMAL MODELS OF NORMAL PRESSURE HYDROCEPHALUS

PART VIII ANIMAL MODELS OF TRAUMATIC DEMENTIA

PART IX ANIMAL MODELS OF TOXIC DEMENTIA

Contributors

MARIA ISABEL ÁLVAREZ • *Instituto Cajal, CSIC, Madrid, Spain*
FRANCESCO AMENTA • *Department of Experimental Medicine and Public Health, University of Camerino, Camerino, Italy*
PATRICK AUBOURG • *INSERM UMR745 and Paris-Descartes University, Faculté des Sciences Pharmaceutiques et Biologiques-Paris5, Paris, France*
JACOB M. BASAK • *Department of Neurology, Washington University School of Medicine, St. Louis, MO, USA*
LUC BUÉE • *Univ Lille Nord de France, Lille, France; INSERM U837, Lille, France; UDSL, Faculté de Médecine, Institut de Médecine prédictive et de Recherche thérapeutique, Centre de Recherches Jean-Pierre Aubert, Lille, France*
D. ALLAN BUTTERFIELD • *Department of Chemistry, Sanders-Brown Center on Aging, and Center of Membrane Sciences, University of Kentucky, Lexington, KY, USA*
ADRIAAN CAMPO • *Bio-Imaging Lab, Department of Biomedical Sciences, University of Antwerp, Wilrijk, Belgium*
NATHALIE CARTIER • *INSERM UMR745 and Paris-Descartes University, Faculté des Sciences Pharmaceutiques et Biologiques-Paris5, Paris, France*
SILVIA R. CASTANHEIRA PEREIRA • *Laboratório de Neurociência Comportamental e Molecular – LaNeC, Universidade Federal de Minas Gerais, Belo Horizonte, Brazil*
MARIE-FRANÇOISE CHESSELET • *Departments of Neurology and Neurobiology, The David Geffen School of Medicine at UCLA, Los Angeles, CA, USA*
RONALD A. CONLON • *Department of Genetics, Case Western Reserve University, Cleveland, OH, USA*
DAMIAN C. CROWTHER • *Department of Genetics, University of Cambridge, Cambridge, UK; Department of Medicine, University of Cambridge, Cambridge Institute for Medical Research, Cambridge, UK*
HANA N. DAWSON • *Division of Neurology, Duke University, Durham, NC, USA*
ROBERT M.J. DEACON • *Department of Experimental Psychology, University of Oxford, Oxford, UK*
PETER PAUL DE DEYN • *Laboratory of Neurochemistry and Behaviour, Institute Born-Bunge, University of Antwerp, Wilrijk, Belgium; Department of Neurology, Middelheim Hospital, ZNA, Antwerp, Belgium*
SHEILA M. FLEMING • *Department of Neurology, The David Geffen School of Medicine at UCLA, Los Angeles, CA, USA*
KARL J. FRYXELL • *Center for Biomedical Genomics & Informatics, Department of Molecular & Microbiology, George Mason University, Manassas, VA, USA*
HUGO GEERTS • *In Silico Biosciences Inc, Berwyn, PA, USA*
MARIA E. GIANNAKOU • *Department of Genetics, University of Cambridge, Cambridge, UK*

TJAKKO J. VAN HAM • *Department of Genetics, University Medical Centre Groningen, University of Groningen, Groningen, The Netherlands*
DAVID M. HOLTZMAN • *Department of Neurology, Washington University School of Medicine, St. Louis, MO, USA*
MOHAMMAD Z. HOSSAIN • *Division of BioMedical Sciences, Faculty of Medicine, Memorial University of Newfoundland, St. John's, NL, Canada*
J. DAVID JENTSCH • *Department of Psychology, The David Geffen School of Medicine at UCLA, Los Angeles, CA, USA*
MASASHI KITAZAWA • *Department of Neurobiology and Behavior and Institute for Brain Aging and Dementia, University of California, Irvine, CA, USA*
PETRA M. KLINGE • *Neurosurgical Foundation in Providence, Providence, RI, USA*
SAMIR KUMAR-SINGH • *Neurodegenerative Brain Diseases Group, VIB Department of Molecular Genetics and Laboratory of Neurogenetics, Institute Born-Bunge, University of Antwerp, Wilrijk, Belgium*
FRANK LAFERLA • *Department of Neurobiology and Behavior and Institute for Brain Aging and Dementia, University of California, Irvine, CA, USA*
DANIEL T. LASKOWITZ • *Division of Neurology, Duke University, Durham, NC, USA*
PATRICIA A. LAWLOR • *Department of Molecular Medicine & Pathology, Faculty of Medical and Health Sciences, The University of Auckland, Auckland, New Zealand*
ELLEN A.A. NOLLEN • *Department of Genetics, University Medical Centre Groningen, University of Groningen, Groningen, The Netherlands*
I. ANNA S. OLSSON • *Laboratory Animal Science, IBMC – Instituto de Biologia Molecular e Celular, Universidade do Porto, Porto, Portugal; Danish Centre for Bioethics and Risk Assessment, Faculty of Life Sciences, University of Copenhagen, Frederiksberg C, Denmark*
BEN A. OOSTRA • *Departments of Clinical Genetics and Neurology, Erasmus MC, Rotterdam, The Netherlands*
SANDRA VAN'T PADJE • *Departments of Clinical Genetics and Neurology, Erasmus MC, Rotterdam, The Netherlands*
GIANCARLO PEPEU • *Department of Preclinical and Clinical Pharmacology, University of Florence, Florence, Italy*
DANIEL PIRICI • *Neurodegenerative Brain Diseases Group, VIB Department of Molecular Genetics and Laboratory of Neurogenetics, Institute Born-Bunge, University of Antwerp, Wilrijk, Belgium; Department of Histology, University of Medicine and Pharmacy of Craiova, Craiova, Romania*
ANGELA MARIA RIBEIRO • *Laboratório de Neurociência Comportamental e Molecular – LaNeC, Universidade Federal de Minas Gerais, Belo Horizonte, Brazil*
LUIGI F. RODELLA • *Division of Human Anatomy, Department of Biomedical Sciences and Biotechnology, Faculty of Medicine, University of Brescia, Brescia, Italy*
JAMES A. ROGERS, III • *Legal Department, Mayo Clinic, Rochester MN, USA*
MARIA CRISTINA ROSI • *Department of Preclinical and Clinical Pharmacology, University of Florence, Florence, Italy*

PETER SANDØE • *Danish Centre for Bioethics and Risk Assessment, Faculty of Life Sciences, University of Copenhagen, Frederiksberg C, Denmark*
NICOLAS SERGEANT • *Univ Lille Nord de France, Lille, France; INSERM U837, Lille, France; UDSL, Faculté de Médecine, Institut de Médecine prédictive et de Recherche thérapeutique, Centre de Recherches Jean-Pierre Aubert, Lille, France*
CAROLINE SEVIN • *INSERM UMR745 and Paris-Descartes University, Faculté des Sciences Pharmaceutiques et Biologiques-Paris5, Paris, France*
JENNIFER E. SLEMMER • *Department of Biology, University of Prince Edward Island, Charlottetown, PE, Canada*
RENÃ A. SOWELL • *Department of Chemistry and Sanders-Brown Center on Aging, University of Kentucky, Lexington, KY, USA*
SUSAN L. STODDARD • *Office of Intellectual Property, Mayo Clinic, Rochester MN, USA*
PATRICK M. SULLIVAN • *Center for Aging and Division of Geriatrics, Duke University Medical Center, Durham, NC, USA*
JOHN C. VAN SWIETEN • *Departments of Clinical Genetics and Neurology, Erasmus MC, Rotterdam, The Netherlands*
MICHAEL SY • *Department of Neurobiology and Behavior and Institute for Brain Aging and Dementia, University of California, Irvine, CA, USA*
AKIHIKO TAKASHIMA • *Laboratory for Alzheimer's Disease, RIKEN Brain Science Institute, Saitama, Japan*
ADOLFO TOLEDANO • *Instituto Cajal, CSIC, Madrid, Spain*
DANIELE TOMASSONI • *Department of Experimental Medicine and Public Health, University of Camerino, Camerino, Italy*
CHRISTINE VAN BROECKHOVEN • *Neurodegenerative Brain Diseases Group, VIB Department of Molecular Genetics and Laboratory of Neurogenetics, Institute Born-Bunge, University of Antwerp, Wilrijk, Belgium*
DEBBY VAN DAM • *Laboratory of Neurochemistry and Behaviour, Institute Born-Bunge, University of Antwerp, Wilrijk, Belgium*
LUDO VAN DEN BOSCH • *Laboratory of Neurobiology, Leuven, Belgium*
ANNEMIE VAN DER LINDEN • *Bio-Imaging Lab, Department of Biomedical Sciences, University of Antwerp, Wilrijk, Belgium*
ANNEMIE VAN DIJCK • *Laboratory of Neurochemistry and Behaviour, Institute Born-Bunge, University of Antwerp, Wilrijk, Belgium*
GREET VANHOUTTE • *Bio-Imaging Lab, Department of Biomedical Sciences, University of Antwerp, Wilrijk, Belgium*
JOHN T. WEBER • *School of Pharmacy and Division of BioMedical Sciences, Faculty of Medicine, Memorial University of Newfoundland, St. John's, NL, Canada*
ROB WILLEMSEN • *Departments of Clinical Genetics and Neurology, Erasmus MC, Rotterdam, The Netherlands*
YUJI YOSHIIKE • *Laboratory for Alzheimer's Disease, RIKEN Brain Science Institute, Saitama, Japan*
DEBORAH YOUNG • *Department of Pharmacology & Clinical Pharmacology, Faculty of Medical and Health Sciences, The University of Auckland, Auckland, New Zealand*

Part I

Animal Modeling: General and Ethical Aspects and Drug Discovery Pipeline

Chapter 1

General Introduction to Animal Models of Human Conditions

Peter Paul De Deyn and Debby Van Dam

Abstract

Animal models serve to imitate (patho) physiological states and/or phenotypical characteristics known to occur in target species (usually man but sometimes other species as well). The use of animal models has had and may continue to have a tremendous impact on medical progress. Laboratory animals are now used in the study of basic (patho)physiological mechanisms, in the development, production, and evaluation of diagnostic and therapeutic agents and/or procedures, as well as in safety studies to assess carcinogenic, teratogenic, or reproductive toxicity of investigational chemical and biological agents, and in education and training. The quality or utility of a model often depends on its validity, which is highest in the so-called homologous models where the phenotypical presentation displayed as well as the cause of the condition in the animal are identical to those of the human condition. Isomorphic models display similar symptoms, but the condition is not provoked by the same events as the human condition. Partial models do not attempt to model the entire condition, but focus only on limited aspects. Models can be further classified into spontaneous, induced, negative, and "orphan" models. Uncritical extrapolation of animal findings to the human condition may lead to unreliable or even dangerous conclusions. Extrapolation tends to be most reliable when a plurispecies approach is taken, and when differences in metabolic patterns and speed, as well as several other potentially confounding variables are taken into account. Animal models have been crucial to neurological and psychiatric research, even though the search for valid models has been difficult in these fields because of the differences in brain structure and function between humans and other species.

Key words: Animal models, ethics, validity, extrapolation, generalizability

1. Introduction

Although we will presently focus on animal models, biomedical research makes and has made use of a large variety of experimental subjects and preparations ranging from human volunteers and laboratory animals through embryos, isolated organs, tissues, and cells from humans as well as animals. Experimental materials have also included bacteria, fungi, and protozoa, and artificial materials are increasingly being used in the more recent physical, chemical,

Peter Paul De Deyn and Debby Van Dam (eds.), *Animal Models of Dementia*, Neuromethods, vol. 48,
DOI 10.1007/978-1-60761-898-0_1,

and computer modeling. In biomedical science, animal subjects or materials derived from animals are commonly used to construct animal models, which serve to imitate (patho)physiological states known to occur in target species (usually man but sometimes other species as well). Animal models of such states allow the experimental study of the condition to a degree often impossible in human subjects.

The use of animal models has had a tremendous impact on medical progress. Several internationally collaborating organizations provide objective information about the use of animals in medical research: Understanding Animal Research (UK), Americans for Medical Progress (USA), National Association for Biomedical Research (USA), and European Coalition for Biomedical Research (EU). Several surveys have documented the importance of animal research in general to the advancement of human and veterinary medicine. The almost exponential growth of biomedical knowledge has coincided historically with the introduction of research on animal subjects. The website of AnimalResearchInfo.com (1) provides a nice timeline illustrating the developments in medicine since the end of the nineteenth century, transforming healthcare, extending and improving the quality of life of millions, based on vital animal (model) experimentation. Nicoll and Russell e.g. (2) found that animal experimentation contributed to almost three quarters of all important biomedical advances between 1901 and 1975. Animal research plays a crucial role in scientists' understanding of diseases and in the development of effective medical treatments. AnimalResearchInfo.com (1) lists those disorders and diseases, which are now prevented or cured with the knowledge obtained through animal experimentation. The value of animal experimentation in the advances of human health is further exemplified by the list of Nobel prizes awarded for Physiology or Medicine. Since the beginning of the twentieth century, these prizes have charted the world's greatest medical advances. Of the 98 Nobel Prizes awarded for Physiology or Medicine up to 2008, 79 were directly dependent on animal-based research, or the discovery relied on crucial data obtained from animal studies by other research groups (1,3).

2. Historical Use of Animal Models in Biomedical Science

The physiological system of the classic Greek physician Claudius Galenus (130–201) was based on a blend of Ancient Greek philosophy and some rather haphazard anatomical observation. The humoral doctrine (red blood, yellow and black bile, phleghm) was adopted from Hippocrates (460–377 BC). All matter was supposed to consist of a mixture of the elements fire, water, air, and earth, and the qualities hot, wet, cold, and dry, and came to live through the action of three varieties of pneuma (*spiritus animalis, spiritus vitalis,* and *spiritus naturalis*). Diseases and their treatments were construed according to this medical and physiological system, which endured largely unchanged for almost 15 centuries. The distinguished British physician William Harvey (1578–1657), although trained in the Galenic tradition, was forced to overthrow the system based on the dissection of human cadavers and various animals, and some simple vivisectional observations (4). His *Exercitatio anatomica de motu cordis et sanguinis in animalibus* (1628) provides the first description of blood circulation. He proved his theory on the pathway of blood flow experimentally, and showed that blood is impelled mechanically by a "pump-like" heart. Harvey opened up new avenues in scientific enquiry, and established the use of animals in physiology. Another famous pioneer, the French philosopher René Descartes (1596–1650) was the first author to attempt to explain all bodily functions according to purely mechanical laws. Descartes grounded his work on observation, including some animal experiments, but he was limited by his speculative physiological conceptions and inadequate anatomical knowledge.

The use of animal models in the experimental study of the nervous system dates back at least 300 years (5). The Dutch scientist Jan Swammerdam (1637–1680) probably conducted the first experiment in electrophysiology by showing that he could make a frog's leg wrapped in silver wire contract when it touched a copper ring. Almost 100 years later, Luigi Galvani (1737–1798) found that muscles in a frog's legs contracted in unison with intermittent production of electricity. Galvani devised many ways to generate electricity and study its effects on excitable tissues in a series of experiments leading to what then became known as the concept of animal electricity. Although the insights of later workers were required to put Galvani's observations in the right perspective, Galvani was without doubt one of the founders of electrophysiology. The frog still is an important experimental animal in the study of the functions of the nervous system.

During the end of the nineteenth and the beginning of the twentieth century, the rate of medical progress accelerated dramatically. Major discoveries of decisive significance to humanity were made in which the use of research animals again played a vital role. Louis Pasteur (1822–1895) used, of course, animals in his seminal work on pathogenic agents and vaccination. Pioneering bacteriologists like Robert Koch (1843–1910) often used mice to identify the pathogenic organisms they studied (6). The work on these animal models of infectious disease not only helped to identify the pathogenic agents of devastating diseases like anthrax, cholera, or tuberculosis, but were crucial for screening and evaluation of various antibacterial agents. Koch's former assistant Paul Ehrlich (1854–1915) and coworkers infected rabbits with syphilis, and used them to prove the efficacy of arsenical substances against this disease. However, Ehrlich's dream to find chemical substances with special affinity for pathogenic organisms, or "magic bullets" as he called them, was really made reality through the work of Gerhard Domagk (1895–1964). Domagk found that mice treated with the sulfonamide preparation *Prontosil* survived injections with hemolytic streptococci, which were shown to be lethal in control animals. Following this first demonstration of *Prontosil*'s efficacy in 1932, experiments with sulfonamide preparations were conducted everywhere, eventually leading to the therapeutic application of *Prontosil* and its derivatives, and the treatment of several previously fatal diseases. The work of Domagk marked the beginning of a new era in biomedical science. It demonstrated that diseases could be treated effectively by means of chemical compounds, but incidentally, it also showed the importance of animals in medical research. Domagk used infected mice rather than bacterial cultures to investigate the antibacterial effects of a series of substances. He thus showed that *Prontosil*, although inactive on cultured germs, did display clear-cut antibacterial effects *in vivo* through its active sulfonamide metabolite.

2.1. Contemporary Use of Laboratory Animals

Laboratory animals are now used in a variety of different applications in basic as well as applied research. Animal studies are performed in different fields of scientific enquiry ranging from psychobiology to physiology and morphology. Animal research also includes the study of animal models to increase pathophysiological knowledge or identify new therapeutic agents or procedures. Moreover, animals are successfully used for safety studies to assess carcinogenic, teratogenic, or reproductive toxicity of investigational agents. Additionally, laboratory animals are used in the production and evaluation of therapeutic and diagnostic agents (e.g., production of monoclonal antibodies, efficacy assessment of vaccines), and education (e.g., training of endoscopic or other surgical techniques).

The second report on the Statistics on the Number of Animals used for Experimental and other Scientific Purposes of the European Economic Community published in 1999 gave insight into the usage of experimental animals in 1996 (7). At that moment, a total of 11,650,000 animals were reportedly used for experimental purposes. The relative contribution of the different animal species was as follows: *Rodentia* and *Lagomorpha* (81.27%), poikilothermic animals (fish, amphibians, etc.; 12.86%), birds (4%), *Artio-* and *Perissodactyla* (pigs, goats, sheep, deer, cattle, horses, donkeys; 1.08%), carnivores (cats and dogs; 0.33%), *Prosimiae* and *Simiae* (primates; 0.09%). Looking at a similar report about a decade later summarizing the use of laboratory animals within Europe in 2005 (8), it is obvious that there are no major changes in the proportion of the different classes of animals and the concentration of biomedical research in those regions. The total number of animals in the 25 EU Member States was reported to be 12,117,583, while the 15 Member States taken up into the 1999 report (1996 data), used 11,070,299 animals in 2005, indicating a decrease of about 5% compared to 1996.

As in previous years, more than 60% of animals were used in research and development for human medicine, veterinary medicine, dentistry, and in fundamental biology studies. Production and quality control of products and devices in human medicine, veterinary medicine, and dentistry required the use of 15.3% of the total number of animals reported in 2005. Toxicological and other safety evaluation represents 8% of the total number of animals used for experimental purposes. About 1.6% of the total number of animals was used for education and training purposes (8).

In 2005, the number of animals used for the study of both animal and human diseases represented more than half (57.5%) of the total number of animals used for experimental purposes in the EU, with a mere 81% of these animals used for studies of human diseases. Of these animals specifically used for the study of human disease, over a quarter (26.7%) is employed for the study of human nervous and psychiatric disorders (8).

2.2. Types of Animals, Validity, and Extrapolation

Animal models are used as research tools in the study of certain (patho)physiological states. The quality or utility of a model often depends on its validity. Ideally, a valid model should have comparable symptomatology, etiology, and background (e.g., comparable neuropathological, neurophysiological, or electrophysiological features), and should display concordant effects of therapy as the condition it is supposed to imitate. As a consequence, validity is highest in the so-called homologous models. In homologous models, the symptoms displayed and the cause of the condition in the animal are identical to those of the human condition. Especially in neurosciences, homologous models are very rare, and mostly limited to well-defined lesion-syndromes. Isomorphic models, on the other hand, are more common but, although they display similar symptoms, the condition is not provoked by the same events as the human condition. Most common, however, are partial models, which do not attempt to model the entire condition, but focus only on limited aspects.

Models can be classified into spontaneous, induced, negative, and "orphan" models. Although spontaneous models are supposed to develop their condition without artificial manipulation, selective breeding is often necessary to establish and maintain the mutant line. Spontaneous genetic variants have been described in mouse and rat strains, but many other laboratory animals (e.g., gerbils, dogs, primates) have been used as well. Induced models, on the other hand, display the condition as a result of artificial manipulation. Such manipulations include surgical procedures (e.g., arterial ligation, nephrectomy), administration or withdrawal of biologically active substances (e.g., intracerebral injection of 6-OH-dopamine, injection of infectious agents), application of physical or physicochemical stimuli (e.g., local heat or cold application, rose-bengale illumination), and genetic manipulations (e.g., construction of transgenic animals). Negative models are models in which a specific human disorder does not develop (e.g., gonococcal infections in rabbits). Finally, models displaying a condition that has never been described in man or other target species are called orphan models.

The appraisal of work on animal models often depends on their contribution to the understanding of human (patho)physiology. In view of the phylogenetic similarity in morphology and physiology between different species, observations made in animals are extrapolated to humans based on homology. However, uncritical extrapolation of animal findings to the human condition may lead to unreliable or even dangerous conclusions. Especially when one is interested in knowing whether a condition or relationship exists in animals *to the same extent* as in humans (quantitative extrapolation), extrapolation can be extremely difficult (e.g., determining the toxic dose of a compound in humans based on results obtained in mice). As a general rule, extrapolation tends to be most reliable when a plurispecies approach is taken, and the condition is studied in a variety of relatively unrelated laboratory animals (e.g., rats, rabbits, primates). Differences in metabolic patterns and speed as well as several other confounding variables (age, diet, sex, distress, route of administration, rhythmic variations) need to be taken into account, particularly when results obtained in animals are to be extrapolated to humans in a quantitative way.

The *thalidomide* tragedy, although subject to many misinterpretations, illustrates some of the pitfalls of extrapolation. Fetal malformations caused by the administration of *thalidomide* to pregnant women form one of the most dramatic chapters of modern pharmacology. Notably, this disaster has been used for many years by activists to prove the irrelevance of medical animal experimentation as they believe it showed that even extensive animal testing failed to demonstrate the potential teratogenic nature of the drug. Actually, as Rowan (9) correctly stresses, "thalidomide was not adequately tested, and after the tragedy, drug registration authorities around the world immediately increased their animal-testing requirements." While the compound was marketed for the treatment of pregnancy-related nausea and emesis as early as 1957, reproductive toxicity studies were only performed from 1961 onwards, after suspicion of its effects on human fetuses. Human babies presented important anomalies under the form of shortening of the extremities (phocomelia). The first reproductive toxicity studies testing *thalidomide* were performed on rats and mice and demonstrated alterations in litter size as a result of fetus resorption. After administration of the compound to New Zealand white rabbits, investigators proved the induction of phocomelia, and studies in monkeys subsequently demonstrated a susceptibility to teratogenicity comparable to that found in man. The *thalidomide* drama did not demonstrate the failures of animal toxicological research, but rather it showed the dangers of insufficient testing during pharmacological development and of the ensuing premature extrapolation based on limited data. It demonstrates the necessity of toxicological research on several species before extrapolation. While researching the toxic mechanisms of action after withdrawal from the market in the 1960s, it became evident that thalidomide could have interesting positive effects in case of leprous erythema nodosum, for which clinical efficacy has been convincingly proven in the meanwhile (10). Additional research evidenced the immunomodulating and antiangiogenetic attributes (11) of thalidomide, thereby motivating its clinical evaluation in, e.g., multiple myeloma (12), graft-versus-host disease (13), Crohn's (14), and Behçet's disease (15).

3. Animal Models in Neuroscience

The two most important clinically inspired applications of animal models in neuroscience research are the development and testing of hypotheses about neurological and psychiatric disorders in general and their neural substrates in particular, and the screening and identification of new therapies (most frequently drugs). Animal models for neurological and psychiatric disorders have been developed for a variety of conditions and/or syndromes among which figure peripheral neuropathy, myasthenia, multiple sclerosis, stroke, epilepsy, amyotrophic lateral sclerosis, mental retardation, dementia, Parkinson's disease, AIDS-related dementia complex, pain, head injury, schizophrenia, affective disorders, obsessive compulsive disorders, anxiety disorders, and stress.

The question of similarities at the structural and functional level of the CNS between human and other species is often considered the most basic problem for the use of animal models in studies of human neurological and psychiatric conditions. The neuroscience community is far from reaching consensus about the level of similarity between the brains and "minds" of humans and other species. Therefore, major tasks granted to comparative neuroscience are the identification of the research areas in which animal models are likely to be useful and the appropriate species for such studies. In addition, comparative neuroscience can provide essential information for adequate comparison of both structural and functional aspects of human and nonhuman brain. When considering models of psychiatric conditions, the issue of similarities between the mental lives of animals and humans is of course fundamental. Postulations of animal consciousness or unconsciousness are difficult, if not impossible, to corroborate. However, for a model to be useful, it is not necessary for the animal mind to be as complex as the human mind. Anxiety or aggression occur both in human and nonhuman species and serve a crucial adaptive function. The differentiation between core consciousness, referring to the hypothesized level of awareness facilitated by neural structures of most – if not all – species that allows them to be aware of and react to their environment, and extended consciousness, which arise in the brain of animals with substantial capacity for memory and reason and is therefore probably limited to humans and nonhuman primates, as proposed by Damasio, may provide a useful framework (16).

Although some attributes still appear to be rather unique to the human brain, other aspects seem to be shared with more species than originally believed. For example, the complex pattern of cortical sulci and gyri is largely unparalleled in the rest of the animal kingdom. A reliable animal equivalent does not exist for many psychological functions, and consequently, human disorders of personality, affective or specific cognitive or linguistic functions are extremely difficult to model in animals. As an example, although some aspects of the cortical areas believed to mediate production and perception of speech appear to be present in man and certain nonhuman primates, the presence of unilateral, cortical areas specialized in linguistic functions, including the Broca's motor speech area, appears to be a privilege of the human brain. Animal models of traumatic aphasias are therefore unlikely ever to be developed (17,18).

Another particular aspect of the human brain involves its prefrontal cortex, which has been suggested to be the neural substrate of many higher functions, and which constitutes about one third of the total cortex in humans (19). This contrasts with findings in dogs and cats, where the prefrontal cortex accounts for only 7% and 3.5%, respectively. Even in our closest relative, the chimpanzee, the prefrontal cortex is still only 17% of the total cortex. It was therefore argued that studies on the functions of the prefrontal cortex should preferentially be conducted in primates (20). More recent studies have, however, indicated that the prefrontal cortex may be present in a broader variety of species and that the ratio between prefrontal and nonprefrontal cortical surface in itself is not an index of evolution. Detailed analyses of the anatomy of rat prefrontal cortex have modified strategies to manipulate specific cortical areas in the animal to model human conditions with prefrontal involvement, although reports delineating boundaries between prefrontal cortical subareas of the rat in comparison with the primate homologues show some dissimilarities. Nevertheless, present anatomical and functional data indicate that rats do have a prefrontal cortex, although not as differentiated as it is in primates and evolutionary later specializations are likely (21). Analogously, from an electrophysiological perspective, rat prefrontal cortex seems to combine elements of the primate anterior cingulated cortex and dorsolateral prefrontal cortex at a rudimentary level. In primates, these functions may have formed the building blocks for abstract rule encoding during the dorsolateral expansion of the cortex (22). On the other hand, it has been discussed that contemporary (functional) imaging techniques can now tell us virtually everything that we need to know about the human brain for the purposes of cognitive neuroscience, and the function of the prefrontal cortex in particular (23). Given the above-mentioned and many other difficulties, researchers have had only limited success in developing homologous and isomorphic models for human neurological and psychiatric disorders, and most models presently available are partial. Although homologous and isomorphic models may be optimal for the purpose of therapy evaluation, important aspects of a disease may even be discovered in a simple animal model that isolates a single aspect of a complex human condition, like is the case in many neurological and psychiatric conditions. Partial models may be of substantial value in the gradual process of building a more complete image of the disease and underlying pathophysiological mechanisms, whereas more complete and complex models are only possible in case of better understood diseases.

The starting point for the development of a new animal model for a specific neurological condition is often the current dominant theory about the disease. Although a logical first approach, it is essential to broaden the focus of animal models under development based on the increasing knowledge of underlying disease mechanisms. The development of animal models for Alzheimer's disease serves an excellent example of such a strategy. Early partial animal models of dementia were based on the cholinergic deficit, a known characteristic of Alzheimer's disease (24). Scopolamine-induced amnesia is the most representative model in this regard. *Scopolamine* is one of several nonselective muscarinic antagonists that have been used to induce amnesia in experimental subjects. However, scopolamine only partially mimics the memory deficits observed in patients with dementia (25, 26). Similarly, scopolamine-induced impairments in animals cannot always be unequivocally interpreted in terms of a specific learning/memory impairment (27). Despite the very partial nature of this chemical model of Alzheimer's disease, its application in experimental animals contributed to the development of the presently marketed anticholinergic agents, which have been shown to demonstrate a symptomatic effect on cognitive effects in patients. Over the years other strategies were applied, e.g., combination of anticholinergic and antiserotonergic interventions, and with the development of new techniques, in particular gene targeting techniques, many different approaches were used to model Alzheimer's dementia, as will be discussed in great detail in several chapters of this book.

Acknowledgments

This work was financed by the Fund for Scientific Research – Flanders (FWO, G.0164.09), Agreement between the University of Antwerp and the Institute Born-Bunge, Interuniversity Poles of Attraction (IUAP Network P6/43), Methusalem excellence grant of the Flemish Government, Neurosearch Antwerp, the Antwerp Medical Research Foundation, and the Thomas Riellaerts Research fund. DVD is a postdoctoral fellow of the FWO.

References

1. The AnimalResearch.info website: http://www.animalresearch.info
2. Nicoll CS, Russell SM(1991) Mozart, Alexander the Great, and the animal rights/liberation philosophy. FASEB J 5:2888–2892
3. The Nobel Foundation website: http://nobelprize.org
4. Frank RG (1980) Harvey and the Oxford physiologists. University of California Press, Berkeley, CA
5. Stevens LA (1974) Explorers of the brain. Scientific Book Club, London
6. Nobel Foundation (1970) Nobel lectures. Elsevier, Amsterdam, The Netherlands
7. Second report on the statistics on the number of animals used for experimental and other scientific purposes in the member states of the European Union. Commission of the European Communities, commission staff working document, annex to the report from the commission to the Council and the European Parliament. COM (1999) 191 final
8. Fifth report on the statistics on the number of animals used for experimental and other scientific purposes in the member states of the European Union. Commission of the European Communities, commission staff working document, annex to the report from the commission to the Council and the European Parliament. COM (2007) 675 final
9. Rowan AN (1984) The future of animals in research and training. The search for alternatives. Fundam Appl Toxicol 4:508–516
10. Walker SL, Waters MF, Lockwood DN(2007) The role of thalidomide in the management of erythema nodosum leprosum. Lepr Rev 78:197–215
11. Paravar T, Lee DJ(2008) Thalidomide: mechanisms of action. Int Rev Immunol 27: 111–135
12. Mitsiades CS, Hideshima T, Chauhan D, et al. (2009) Emerging treatments for multiple myeloma: beyond immunomodulatory drugs and bortezomib. Semin Hematol 46:166–175
13. Svennilson J (2005) Novel approaches in GVHD therapy. Bone Marrow Transplant 35(Suppl. 1):S65–S67
14. Akobeng AK, Stokkers PC (2009) Thalidomide and thalidomide analogues for maintenance of remission in Crohn's disease. Cochrane Database Syst Rev (2):CD007351
15. Gul A (2007) Standard and novel therapeutic approaches to Behçet's disease. Drugs 67:2013–2022
16. Damasio A (1999) The feeling of what happens. William Heinemann, London
17. LeMay M, Geschwind N(1975) Hemispheric differences in the brains of great apes. Brain Behav Evol 11:48–52
18. Petersen MR, Beecher MD, Zoloth SR, Moody DB, Stebbins WC(1978) Neural lateralization of species-specific vocalizations by Japanese macaques (Macaca fuscata). Science 202:324–327
19. Fuster JM (1989) The prefrontal cortex. Raven, New York

20. Fuster JM (1980) The prefrontal cortex. Anatomy, physiology, and neuropsychology of the frontal lobe. Raven, New York
21. Uylings HB, Groenewegen HJ, Kolb B (2003) Do rats have a prefrontal cortex? Behav Brain Res 146:3–17
22. Seamans JK, Lapish CC, Durstewitz D (2008) Comparing the prefrontal cortex of rats and primates: Insights from electrophysiology. Neurotox Res 14:249–262
23. Passingham R (2009) How good is the macaque monkey model of the human brain? Curr Opin Neurobiol 19: 6–11
24. Smith G (1988) Animal models of Alzheimer's disease: Experimental cholinergic denervation. Brain Res Rev 13:103–118
25. Beatty W, Butters N, Janowsky DS (1986) Patterns of memory failure after scopolamine treatment: Implications for cholinergic hypothesis of dementia. Behav Neural Biol 45:196–211
26. Kopelman MD (1986) The cholinergic neuro transmitter system in human learning and memory: A review. Q J Exp Psychol 38A: 535–573
27. Blokland A (1996) Acetylcholine: a neurotransmitter for learning and memory? Brain Res Rev 21:285–300

Chapter 2

Animal Models of Dementia: Ethical Considerations

I. Anna S. Olsson and Peter Sandøe

Abstract

This chapter aims to encourage scientists and others interested in the use of animal models of disease – specifically, in the study of dementia – to engage in ethical reflection. It opens with a general discussion of the moral acceptability of animal use in research. Three ethical approaches are here distinguished. These serve as points of orientation in the following discussion of four more specific ethical questions: Does animal species matter? How effective is disease modelling in delivering the benefits claimed for it? What can be done to minimize potential harm to animals in research? Who bears responsibility for the use of animals in disease models?

Key words: Contractarian view, utilitarian view, animal rights view, principle of 3Rs

1. Introduction

Contemporary research in the life sciences, particularly in biomedicine, involves experimentation on large numbers of live animals. Many of the animals are used for research aiming to discover new ways to prevent, cure, or alleviate human diseases. Some research animals are used as *disease models*. That is, conditions are artificially induced in them, which in some relevant respects resemble the conditions that we want to prevent, cure, or alleviate in humans.

The process of altering an animal so that it can serve as a disease model sometimes involves distressing or painful interventions; and the conditions induced in the animals may give rise to anxiety, pain, and other forms of suffering. Moreover, animals are often housed in ways that limit their freedom, and they are invariably killed when the experiment comes to an end. The overwhelming majority of animals used are vertebrates with highly developed nervous systems. They cannot, of course, consent to their own participation in research. Nor do they stand to benefit, as individuals, from such participation.

Peter Paul De Deyn and Debby Van Dam (eds.), *Animal Models of Dementia*, Neuromethods, vol. 48,
DOI 10.1007/978-1-60761-898-0_2, © Springer Science+Business Media, LLC 2011

These facts present both the scientific community and society in general with a question. In pursuit of, for the most part, admirable and understandable goals, scientists carry out experiments causing discomfort, pain, and distress to animals. Often, they limit the freedom of the animals they are dealing with, and eventually, in most cases, they kill them. Are they morally justified in acting in this way? We might also ask whether, where it exists, societal approval of this kind of scientific activity is warranted.

The answer to this question will clearly depend on one's general view of human duties to animals. In this chapter, we will not defend a single view of this kind. Rather, we shall present three ethical perspectives. We urge the reader to reflect on her own stance with these perspectives in mind. Following this, we will go through a series of further questions relating to the use of animals as models of dementia, which we think each researcher should ask herself; and we will describe what we take to be the key ethical issues raised by these questions. The questions are: Does animal species matter? How effective is disease modelling in delivering the benefits claimed for it? What can be done to minimize potential harm to animals in research? Who bears responsibility for the use of animals in disease models?

Although we do not advocate a specific ethical stance, but rather urge the researcher to make up her own mind, we do have an ethical agenda. We think that, at the very least, people who make use of animals should be prepared to devote the time and effort it takes to think through their choices from an ethical perspective. This would not only have a positive impact on the animals used in research, but also improve the credibility of animal research across society as a whole.

2. Should One Use Animals in Research?

When it comes to consideration of the right way to treat animals, there is no single, unanimous view in our societies. Even within families, for example, when the issue of using animals for research is brought up at dinner, there are strong disagreements. Some of us are outraged by the idea of exposing innocent animals to painful research. Others will take a more moderate view, arguing that as long as the research is vital and everything possible is done to protect the animals from unnecessary suffering, it is acceptable to use animals in research. Yet others will say that they do not care about the plight of rats and mice, and that we should be free to use animals to make discoveries in medicine as we wish.

The same thing happens when philosophers meet to discuss the ethics of animal use. They divide, with various groups favoring different ethical theories about human duties to animals. Such theories can be useful vehicles through which to articulate the principles underlying disagreements about animal experimentation. For a fuller outline of the ideas presented in the rest of this section, see (1).

Here, we shall present and briefly discuss three ethical views, or theoretical approaches. According to the first, the *contractarian view*, it is only one's own long-term interests that count from an ethical point of view. Since we depend for our own well-being on collaboration with other human beings, duties governing our dealings with fellow humans become established. However, no such duties exist in regard to animals (2).

According to the *utilitarian view*, on the other hand, what matters is the impact of what we do on the well-being of those affected by our actions. Here, the basic principle is that what entitles me to moral consideration, or what gives me moral status, is my capacity to suffer as a result of, or benefit from, your actions. Since not only other human beings but also other sentient animals have this capacity, we should be concerned about the welfare of both humans and animals. Of course, it is impossible to cater to the interests of every individual potentially affected by a course of conduct, so we should focus on the interests that are most dramatically served by the type of action we are considering. In essence, we should seek to produce the greatest total fulfilment of interests (3).

Finally, according to the *animal rights view*, we should distinguish between interests and rights. Rights must be respected. One should not allow interests to overrule them. In the case of human rights, this means that we do not allow an innocent being to be sacrificed for the sake of the common good. Advocates of animal rights expand this approach and apply it to all sentient animals. They therefore object to the idea of sacrificing animals for the sake of the good of others (4).

For those who adopt the *contractarian* view, the way animals are treated is not always irrelevant: once people are emotionally attached to certain kinds of animals, for example, and dislike or feel outraged by the practice of using them in painful experiments, this becomes an ethically relevant concern. For example, because most people like cats and dogs more than they like rats and mice, causing suffering to the former is likely, in the contractarian picture, to be a more serious problem than causing the same amount of suffering to the latter. Likewise, nonhuman primates will probably receive more protection than other animals, because their plight is of very considerable concern to many people.

What matters, on this view, are the feelings and beliefs of fellow humans on whose collaboration one depends to gain a licence to operate. On this approach, then, setting ethical limits to the use of animals for research boils down to the task of defining a publicly acceptable framework that allows humankind to harvest the potential benefits of animal-based research. One specific reason for looking after the welfare of animals involved in research is the avoidance, wherever possible, of experiments that are likely to cause public concern.

According to the *utilitarian* approach, the interests of every individual affected by an action deserve equal consideration. This means that for the utilitarian–unlike the contractarian–the impact of procedures, housing facilities, and the like, on the well-being of the laboratory animals must be taken into consideration in its own right. The only justification that can be given of animal use in research is that the cost to the animals used is outweighed by the benefits of the research.

On the utilitarian approach, then, ethical decisions require us to strike the most favorable balance of costs and benefits for all the sentient individuals affected by what we do. But doing the right thing, according to the utilitarian, is not just a matter of doing what is optimal. It is also essential to do something rather than nothing: if something can be done to increase well-being, we have a duty to do it. This utilitarian duty to act, proactively, so as always to bring about improvements has important consequences for society.

In the case of laboratory animals, a pragmatic utilitarian might be willing to apply something called the "Principle of the 3Rs." This principle requires researchers, where possible, to *replace* the existing live-animal experiments with alternatives, *reduce* the number of animals used, and *refine* methods so that animals are caused less suffering (5). It is not hard to see that less invasive sampling techniques, improved housing systems, and more precise models requiring fewer animals to be used are likely to be viewed as morally attractive developments within the utilitarian perspective.

Utilitarianism, as described above, suggests that animal interests can be justifiably sacrificed where that leads to the protection or satisfaction of vital human interests – as happens in much biomedical research. But is that an acceptable view? A more radical variety of utilitarianism might be worth exploring. Animal experimentation sometimes means sacrificing vital interests an animal has in continued life and the avoidance of suffering. Insisting firmly that human and animal interests deserve equal consideration, the utilitarian philosopher Peter Singer has concluded that the sacrifice of such vital animal interests is acceptable only where the benefits are extraordinarily important:

> [I]f a single experiment could cure a disease like leukemia, that experiment would be justifiable. But in actual life the benefits are always much, much more remote, and more often than not they are nonexistent (6, p. 85).

It is evident, then, that within the utilitarian approach a wide range of views are represented. Some utilitarian observers accept animal experiments when there are no alternatives and as long as we do our best to prevent and alleviate animal suffering. Others, like Singer, setting the demand for human benefit higher, would prefer to see nearly all such experiments abolished.

What all utilitarians agree on, however, is the methodological precept that ethical decisions in animal research require us to balance the harm we do to laboratory animals against the benefits we derive for humans and other animals. This precept – the notion that we can work out what is ethical by trading off one set of interests against another – is precisely what is denied by advocates of animal rights.

On the *animal rights* approach, it is always unacceptable to treat a sentient being merely as a "means to an end" – to use a sentient creature as a tool, or instrument, in pursuing one's goals. On a radical version of this view, no benefit can justify violation of the rights of an individual, whether human or animal; so, where an experiment violates an animal's rights, there is no reason to look for its expected benefits to humans or other animals. To find out whether an experiment is morally justified, we need only ask whether it respects the animal's rights and preserves its dignity. The implications of this way of looking at matters are radical. Tom Regan and many other adherents of the animal rights view argue in favor of an abolitionist position.

On this version of the animal rights view, experimentation on animals should simply stop. It matters not that an experiment will cause only minor harm to the animals it involves. It matters not that this experiment is of extraordinary importance to humanity at large. The thing that matters is that every time an animal is used for an experiment, it is treated as a mere means to an end. This being so, animal experiments are unacceptable, *period*.

It is possible to imagine a less uncompromising, more moderate advocacy of the animal rights approach. The right to life – or more accurately, the right not to be killed – is regarded as basic by some influential proponents of animal rights. But one might be doubtful about this, partly because animals have a much more limited perspective on the future than we have. What matters to animals is that, here and now, they are well off, whereas a human has aspirations and worries that reach across his or her entire potential life-span. In light of this, one might suggest that animals have something like a right to protection from suffering, or certain levels of suffering. It could then be argued, perhaps, that all animals should be protected from suffering if this involves intense or prolonged pain or distress, which the animal cannot control.

The key idea of the animal rights approach is that there are absolute, nonnegotiable limits to what can be done to animals. Certain things should not be done to animals even if this means we are prevented from doing things that would have clear benefits outweighing any pain and suffering caused along the way. If the rights approach is characterized more loosely in this way, *bans* on certain kinds of experiment–like the one introduced in Danish legislation outlawing experimentation that causes strong pain or other forms of intense suffering to animals – look like an indication that the legislators have adopted a moderate animal rights view.

So the question, raised at the start of this section, about whether animals should be used in research, has no single answer that all will agree on. There are basically different views, of which we have here distinguished three prominent types. In the rest of this chapter, we shall not attempt to adjudicate between these views, but rather discuss the issues raised by animal disease models in the light of all three.

3. Does Animal Species Matter?

Animals of very different species are used in research. The choice of animal depends on the research in question, of course. It is also affected by the experience and expertise of the researcher, the facilities of the institution, by legislation, and sometimes by public discussion in the country where the work is carried out. Let us assume that, in the case of dementia research, if the matter being investigated requires an *in vivo* approach, it will require use of an animal that has a complex enough nervous system to actually have mechanisms for learning or memory formation. Even so, there are many species that fulfil that requirement, ranging from nematodes to chimpanzees.

Does the choice of species matter when it comes to ethical evaluation of the research? It does, as we will now discuss, but quite how it matters will depend on one's ethical position. Those taking a contractarian approach will be primarily concerned with differing kinds of public sensitivity to different species. Those taking a utilitarian stance will focus on the capacity of animals of different species to suffer. But before taking a closer look at the different types of animals that can be used in dementia research (invertebrates, fish, rodents, and nonhuman primates), we need to introduce two topics that will be important in any discussion of the moral status of different species: *sentience* and the *socio-zoological scale*.

Sentience is the capacity to perceive or feel things. A sentient being is one that has its own experience of life, meaning that "there is something it is like" (7) to be that being. Scientific understanding of sentience (human and animal) is still limited. At present, neurobiology does not explain consciousness in terms of material mechanisms of the nervous system. We assume that other individuals are sentient, because they are behaviourally and physically similar to us. While this assumption is uncontroversial for adult human beings, when we extend it to nonhuman animals, the issues become more complicated, for here verbal evidence is unavailable, and the behavioral and physical similarities are smaller. Although common sense may posit sentience in many species, a scientific argument for attributing it must be based on systematically collected evidence.

Such a systematic approach is suggested by Smith and Boyd (8). To determine whether an animal has mechanisms similar to those that we know are essential for human subjective experiences, and whether an animal behaves in similar ways to sentient humans, we can consider a checklist of neuroanatomical/physiological and behavioral criteria determining the capacity to experience pain, stress, and anxiety in nonhuman animals. For any of the relevant experiences, this checklist will include the possession of higher brain centers and evidence of behavioral reaction to potentially nociceptive, anxiogenic, or stressful experiences. Further evidence will be added if these behavioral reactions are modulated by drugs, which have a known anxiolytic or analgesic effect in humans, and if there are peripheral nervous structures (including receptors, signal substances, and hormones) for each specific type of reaction, and a connection between these and the higher brain centers. The more of these criteria that are fulfilled, the stronger is the evidence that an animal is indeed sentient.

Looking at the way in which different taxonomic groups of animals fare on such a systematic analysis, there are two important lessons to be drawn. The first is that all vertebrate animals meet the criteria for sentience. When Smith and Boyd's original analysis was published, such evidence existed only for mammals and birds, but over the last decade evidence of fish sentience has accumulated (9,10). The second lesson is that, for many of the invertebrates, we still know too little to be able to give a useful answer.

However, there is another way to approach the question whether and how animals matter. Thus, there is clearly a hierarchy of animals – a moral ordering that has been called the socio-zoological scale (11). The basis of the scale is that people rate animals as morally more or less important, and therefore more or less worth protecting, according to a number of factors. These include how useful the animal is, how closely one associates with the individual animal, how cute and cuddly the animal is, how harmful the animal can be, and how "demonic" it is perceived to be.

Today, in western societies, some companion animal species, notably dogs and cats, seem to be at the top of the scale. Among other animals, large carnivores and primates are at the top end of the scale. In the middle are large farm animal species such as cattle and pigs. At the bottom of the scale are pests or vermin such as rats and mice. Fish, which are cold and slimy, also appear to be quite low down the scale. Thus, among the animals used for research, there will be a hierarchy with nonhuman primates, dogs and cats at the top, pigs (etc.) in the middle, and rodents and fish near the bottom. Below rodents and fish, one finds insects and other invertebrates.

The socio-zoological scale is in many ways based on tradition and prejudice, and its use as a basis of animal protection can be criticized from both scientifical and ethical point of view.

From both the utilitarian and the animal rights perspective, it is bound to seem unfair to discriminate animals solely on the basis of the scale – an unfairness comparable to racist treatment of humans. On the contractarian view, on the other hand, there is nothing problematic about treating animals in line with the scale, and thus giving more protection to primates and dogs than one does to rodents and fish. This is because, on the contractarian view, animals matter only to the extent that they matter to humans.

Whatever one's ethical view, it is important to be aware that the socio-zoological scale is part of social reality. This reality is, among other things, reflected in legislation that has been introduced to protect animals.

To start at the very bottom of the socio-zoological scale, the fruit fly *Drosophila melanogaster* shows a number of age-related functional types of decline that are also evident in humans, including deficiencies in learning and memory (12). Parallels with human cognitive decline related to age and oxidative stress can also be demonstrated in *Caenorhabditis elegans*, making this even simpler organism (a nematode) a potential candidate for dementia research (13). The short lifespan and the ease with which these animals can be kept, in combination with the well-developed knowledge and technology deployed in manipulating the *Drosophila* genome, means that there are clear practical advantages in using these invertebrates as research models. Their use is generally perceived as less of an ethical issue than the use of vertebrates.

In fact, from the contractarian point of view, the use of invertebrate organisms in research does not seem to be an ethical issue at all. When fruit flies appear in the kitchen most people readily kill them without thinking further; and the fact that they have been widely used in laboratories since the beginning of the twentieth century does not seem to have caused much, if any, public discussion – this, despite the fact that genetic research on *Drosophila* involves major alterations of the bodily integrity of the flies. From the utilitarian point of view, the important question is whether invertebrates such as *Drosophila* and *C. elegans* are sentient. Given the difficulty of proving sentience, the question is perhaps better expressed thus: whether these invertebrates are likely to possess the capacity for suffering and pleasure. *C elegans* has very simple nerve organization; it also lacks one of the most important components for sentience: a central nervous system.

Insects, on the other hand, have a much more complex nervous system, including a brain, and so with them it has proved difficult to give a clear-cut answer to the question of sentience. While Eismann et al. (14) have presented a list of reasons why it is unlikely that insects are able to feel pain, including their lack of a behavioral response to protect an injured limb, Lockwood (15) and Sherwin (16) have argued, appealing mainly to behavioral evidence, that we should consider extending the argument of analogy in a way that supports the conclusion that insects are sentient beings. While pain mechanisms exist in *Drosophila*, it is unclear whether pain signaling is also perceived as adverse by the insects (Fernando Casares, personal communication 2008).

While the simple organization of flies and nematodes is useful in some research, it is a disadvantage in other types of study. Biologically speaking, invertebrates are at some distance from humans, and plainly mechanisms that are specific to vertebrates can only be studied in vertebrates. Looking for a vertebrate that is smaller and easier to reproduce and manipulate genetically than the typical laboratory rodent, life science researchers are increasingly turning to zebrafish, e.g. (17). In the contractarian perspective, the use of fish in research is relatively unproblematic. Fish look very different from us; obviously they live in conditions quite unlike those we live in; and their plight matters to the average person much less than that of the domestic cat or the gorilla. For the utilitarian viewpoint, in contrast, the fact that fish are sentient makes their use in research morally significant. Consequently, from the utilitarian perspective, we are obliged to consider the harm that research may cause to fish, and to make efforts to prevent such harm. In this perspective, the perceived distance between human beings and fish may be a disadvantage for the fish, since it may make it hard for a human observer to recognize signs of distress in fish – particularly given our relative lack of knowledge about pain and fear behavior in these animals.

When they are asked about laboratory animals, most people will of course tend to think of mammals, and in particular rodents. Rats have been the lab animal of choice for experimental psychologists from Skinner and Watson onwards, and much of what we now know about learning and memory in mammals derives from research on them. Again, through the development of techniques of genetic modification, the mouse has come to play an increasingly important experimental role over the past decade. From the contractarian point of view, research with rodents used not to be problematic. Until recently, those who did not work in a research laboratory would have been most likely to view rats and mice as pests – as creatures that destroy food and property and carry serious disease.

However, the public view of rodents as pests may no longer be something we can take for granted. In an increasingly urban population, the only direct contact most people will have with animals is likely to be with a companion animal, and often that animal will be a rodent. This may well raise the moral profile of rodents as far as the contractarian is concerned.

From the utilitarian perspective, rodents, like all mammals, possess all the characteristics listed by Smith and Boyd (8) as indicative of sentience. Their use in research, then, is clearly an ethical issue – one that turns primarily on the pain and suffering these animals experience in the course of experimental work. We will come back to that question later in the chapter.

In humans, learning and memory are supported by the most complex central nervous system we know of. Nonhuman primates (NHPs) have central nervous systems that resemble that of the human being most closely, which is obviously why these animals are interesting models in dementia research. But experimental use of NHPs is also more controversial than any other research involving animals. That is, people are, in general, more concerned about research on primates than they are about similar work involving other mammals (18).

The degree of neurological resemblance between human beings and NHPs depends significantly on the species of NHP in question. Closest to humans are the great apes, of which only chimpanzees have routinely been used in research. Research involving chimpanzees remains highly controversial. It has gradually been abandoned in many countries, and the US is possibly the only country where research chimpanzees are now available (19). Chimpanzees may be remotely relevant for work on dementia, given their use in AIDS research, but the dementia researcher considering an NHP model is much more likely to encounter a macaque, such as a rhesus or cynomologous monkey, or a marmoset, e.g. (20).

Turning to utilitarianism, we need to note, first, that there is no evidence of significant differences in the levels of sentience in NHPs and other research mammals such as rodents (21). Nor is there any great difference in the way these animals are used in research: experimental uses of both NHPs and the more common rodent species will vary from noninvasive to severely invasive interventions. However, while it is not necessarily more difficult to design ethologically appropriate housing systems for NHPs than for other species, the fact that none of the NHP species are domesticated clearly distinguishes them from most laboratory animal species. Where animal housing and welfare are concerned, this is bound to be relevant to the utilitarian, because domestication involves a certain degree of genetic adaptation to the captive environment (22). Transport may also raise a more serious welfare issue vis-à-vis the NHP, since the average NHP requires lengthier transport, with more severe space restriction, than the average rodent (21). On the other hand, in handling NHPs for procedure, training through positive reinforcement is common, and once animals have learnt to collaborate, it may well be that handling stress is reduced. In contrast with this, in rodent handling, which tends to rely on physical restraint rather than training and subsequent collaboration, stress levels may not be reducible. One can also imagine that the actual and perceived closeness of the relationship between NHPs and humans may encourage researchers to treat primates well.

4. Will Research Deliver the Desired Benefits?

The ethical problem raised by animal-based research is essentially a dilemma between our interest in securing human benefits and our recognition that harm to animals ought to be avoided where possible. The 3Rs principle is a response to the second horn of this dilemma. However, the first horn can also be addressed. It can always be asked, in other words: how probable is it that this research programme will deliver the benefits to which it is expected to lead? This is an important question. It is of course impossible to make guaranteed predictions about the outcomes of a research project, and for a variety of reasons retrospective benefit reviews often examine animal procedures carried out a decade earlier. However, the difficulty of accurate prediction should not be regarded as a reason not to address the issue. Drawing on data and feedback from European ethics review committees, the Federation of European Laboratory Animal Science Associations (FELASA) working group (23) has recently described a set of key questions that ought to be asked about any research project involving animals. Here, we shall focus on just two of these, namely: the choice of animal model and scientific approach, and the validity of experimental design.

The ability of animal models to provide answers to the questions we ask about human disorders and their potential treatment is the central issue for publications such as the present volume. Critical discussion of what characterizes a good animal model is rarely engaged in by the researchers developing and using such models. Certainly it is a striking that, among scientific publications on animal models, there is a near total absence of papers identifying models as unsuitable. The assumption appears to be that as our knowledge and the technologies develop there will be a natural selection process in the course of which outdated models will be abandoned. This may ultimately be true, but it will take time – science is a more conservative field than most of us, as researchers, would like to believe. At the time of writing (May 2008), mice are still being used for the highly controversial ascites method for production of monoclonal antibodies, although it is now 10 years since an expert report concluded that "there are already several scientifically satisfactory *in vitro* methods which are both reasonably and practicably available (…) are of moderate cost, and can be shown to be either better than, or equal to, the ascites production method in terms of antibody quality" (24).

Naturally, scientists operate under practical constraints, and the fact that research is to some extent technology-driven (i.e. shaped by factors such as the models we have used before and have expertise in, the animal colony set up at high costs, and so on) may be unavoidable. However, plainly, the general aim should be to use the best model for each study in question. Critical issues to be discussed may include the question whether pharmacologically induced models are still relevant in the study of diseases with a genetic origin when genetically modified models are available; whether knock-out models should still be used if knock-in substitutes exist; whether unconditional knock-outs/knock-ins should still be used given that conditional alternatives can be employed. It is the hope of the present authors that this discussion will be developed in some of the other chapters of the book.

We now move on to the second question – about the experimental design of animal experiments – as this is an issue around which considerable and challenging evidence has accumulated over the last couple of years.

We will use the example of animal research of the kind underlying the development of treatments for stroke in humans. In this field, a number of compounds have shown neuroprotective effects in animal models, but very few have turned out to be effective in clinical trials on humans (25). This could of course be explained by the fact that animals are poor models for the human condition. However, this is not the only plausible-looking explanation. Researchers concerned over the poor translation of preclinical research results into effective human treatments carried out several systematic reviews of the earlier animal experiments and found a number of critical shortcomings in the experimental design.

In many of the animal experiments, for example, the efficacy of the prospective treatment was probably overestimated as a result of bias in the design. Often animals were not randomly allocated to treatments, and researchers who were not blinded when they administered treatment (drug or control), or assessed its outcome, may unconsciously have influenced the measurements (25,26). In addition, there were obviously significant clinical differences, in that the animals used were generally young and healthy before the experimentally induced stroke, while human stroke patients are often elderly and hypertense (26).

An additional problem in the translation of animal research into human benefits is publication bias. Publication in peer-reviewed journals is a central feature of modern academic research, and as is well known the performance of today's researchers is measured largely on the basis of the number of publications they have in influential journals. However, it is generally difficult to get negative results (no effect of treatment) published. As a direct consequence of this, publications are likely to reflect only part of the research that has been carried out in a field – the research in which differences were found between treatment groups. This has wide-ranging ethical consequences and, in particular, affects the number of animals used in research.

The consequences of using "models" of restricted validity and replicability are that results cannot be interpreted properly, that scientific progress is retarded, and that animals are used unnecessarily. In this context, the "publication bias" of journals in favor of hypothesis-confirming results – might be a reason for the slow progress in the development of new animal models and their validation. Negative results often go unpublished, and poor concepts, hypotheses, and models survive, notwithstanding a vast amount of contradictory data, merely because these data are not made available to the scientific community.

Publication of negative findings from well-conceived and performed studies can help investigators to evaluate and ultimately abandon the development of an invalid and irrelevant animal model and help reduce the unnecessary use of laboratory animals (27).

5. What Can Be Done to Minimize Potential Harm to Animals Used in Research?

Suppose that, as researchers, we have a question requiring an *in vivo* approach. Suppose further that we have selected the most appropriate model and experimental design. Assume, in short, we have done what we can to ensure that our research will deliver the expected benefits. For the research to be ethically acceptable, we will still have to make sure that the expected benefits are achieved at the cost of the lowest possible impact on animal welfare. Reducing the adverse effects of scientific protocols, or "refinement" (5), is therefore crucial in animal-based research. Major factors affecting animal welfare are the method used to induce the disorder and the way the animals are housed and handled, including experimental techniques used to administer treatment and monitor parameters.

Before the arrival of GM technologies, animal models were of two principal types: spontaneous mutations and chemically or surgically induced disorders. Although knock-out (gene removed or inactivated) and knock-in (mutated gene inserted) mice are rapidly gaining ground, some of the pre-GM techniques are still important. In the case of research into memory and dementia, this usually involves different kinds of induced lesion. Lesions are induced using stereotaxic surgery under general anesthesia. In this procedure, a part of the brain is lesioned using one of a variety of methods, such as mechanically, through injection of an excitotoxin, a neurotoxin, or a receptor-specific antibody (28). In these experimental situations, pain and pain control are major animal welfare issues.

As in any major surgical procedure involving general anesthesia, perioperative measures will have important influence on animal welfare: appropriate anesthesia and analgesia in combination with extra heat and facilitated access to food and water as the animals recover from surgery are critical (29). Here, there seems still to be considerable room for improvements in routine practice. In a review of the reported use of postoperative analgesics in the early 1990s and early 2000s, Richardson and Flecknell (30) conclude: "Although the use of analgesics has increased over the past ten years, the overall level of post-operative pain relief for laboratory rodents is still low. An additional consideration is the severity of the lesion, which may vary with the procedure used or dose administrated."

We have little reason to suppose that dementia, which develops as a result of a genetically induced disorder, causes pain – there are no signs of pain in these animals and this kind of condition is not painful in humans. Memory loss and a progressive inability to live a normal life cause negative psycho-social effects in human dementia patients, e.g. (31), and cognitive impairment is often accompanied by a series of behavioral abnormalities such as depression, apathy, anxiety, and irritability, e.g. (32). But very little is known about the way in which similar cognitive impairment in animals affects their welfare. From the growing field of veterinary geriatrics, it is known that companion dogs with age-related cognitive dysfunction become disorientated and may react with anxiety to external stimuli they are no longer able to recognize (33). Disorientation may be less of a problem in animals kept in the undemanding environment of a laboratory cage, with food and water provided within the living quarters. However, it is possible that handling and external disturbances are more stressful to animals with memory impairment and, as a result, a diminished capacity for behavioral habituation. This kind of stress should be considered at least in studies in which animals are maintained through the advanced stages of disease.

Another important aspect of refinement, particularly in experiments – common in studies of dementia – with lengthy duration, is the housing environment. By providing animals with resources that enable them to interact with, and control, features of the caged environment, and to engage in motivated behaviors, environmental enrichment improves animal well-being, e.g. (34,35). Such enrichment may also be beneficial from the scientific viewpoint, especially in the neurosciences, as it to some extent overcomes artificial aspects of deprivation resulting from more restrictive housing. This brings the animal models closer to the situation of the subjects they are intended to model of course, because human beings normally experience a wide range of physical and mental activities (36).

In experiments evaluating treatments, the method of administering the treatment will have an impact on the welfare of animals. Protocols requiring long treatment periods can be particularly harmful to the animals, especially if the administration method is invasive. Some immunization approaches recently evaluated as potential treatments for Alzheimer's disease exemplify this.

Papers report protocols of 6 months of weekly intraperitoneal injections, e.g. (37). In view of both the risk of ascites development and the stress of repeated painful manipulations, the use of osmotic minipumps has been suggested as a potential refinement (38).

In addition to minimizing the harm caused to individual animals, there is also a case for reducing the number of animals to which harm is caused. Noninvasive imaging techniques offer a novel way of pursuing this aim. For example, it is possible to monitor plaque formation *in vivo* (39), which means that the experimenter can view the same animal at several points in time rather than sacrificing a separate group of animals for each time point.

6. Who Is Responsible?

Since laboratory animals are ordered and paid for within a research project, their ownership is, in the legal sense, passed on to the research institution. But clearly this does not mean that researchers, managing their own projects, are free to do as they please with the animals (as they may do with laboratory supplies). Animals, as sentient beings, and nonliving laboratory supplies have a quite different moral status in our society. In the European Union, this distinction is reflected in one of the central legal documents, the Treaty of Amsterdam, which requires EU member states to "pay full regard to the welfare requirements of animals" (40).

Most biomedical research is funded, directly or indirectly, by public money. This is another reason why the issues raised by research on animals concern society in general rather than the research community only – why taxpayers must usually be counted among a research institution's stakeholders, to put it bluntly. Society has a number of mechanisms in place designed to guarantee that research on animals is carried out in an acceptable way. On the top level is legislation, which in terms especially of enforcement is a powerful tool. But the legislative process is sluggish, while science and technology normally develop rapidly, and this means that laws must be broad and general in application in order not to become rapidly outdated. Hence, the real decision-making about whether or not the proposed research project is to be permitted is usually delegated to the ethics, or animal care and use, committee, the remit of which is established in law.

Committees can act in a more flexible way. They also allow a dialogue with the scientist proposing an experiment to take place. In fact, the main function of these committees may rest in this dialogue, which on the one hand permits the research project to be revised, so that the scientific objective can be achieved at a mitigated cost to animals, and on the other hand poses a continuing challenge for scientists to develop their research in line with the evolving best practice. The function of the review process in influencing the culture of an institution, sometimes referred to as creating a "culture of care," is also important (23), because ultimately responsibility for the way in which animals are used rests with the researcher managing a project. This is true, not just in moral terms, but also practically, as many decisions regarding the 3Rs can only be made at the research planning stage. Again, there are cases in which the attitude of the researcher will be decisive for the choice of approach (e.g. animal or nonanimal, less severe or more severe model). In this respect, critical discussion and self-regulation within the scientific community will also be important.

Increasingly, consideration of ethics and the 3Rs is included in the review of funding applications. This occurs, for example, in European Framework programs (although with some limitations, as animal ethics review is restricted to projects with NHPs and those flagged up by the scientific review as potentially problematic ethically). In the review of manuscripts submitted for publication, in contrast, it seems that most journals continue to require merely a statement affirming that the research complies with official recommendations, or relevant legislation, or an ethics committee's decision. We think it is important for scientific journals to make better use of their potential ability to raise the ethical standards of animal use in research. Refusals to publish papers based on the ethics of the methodology applied will send a very strong signal to scientists.

This is important for several reasons. First, scientific journals are the main means of communication between scientists. When the scientist is planning to make use of a new methodology or animal model, his or her first source of information will be the leading journals in the relevant research field. In this context, it is noteworthy that information about the unexpected, adverse effects of inducing lesions are rarely reported in scientific papers, and that it is usually very difficult to find information about the relation between induction method and any impact on the animals. Thus, when reviewing animal models of Huntington's disease, we were only able to get this type of information through direct contact with researchers (41,42).

The fact that society has an interest in how animals are treated in biomedical research means that scientists are at some level accountable to society. This accountability will, of course, require communication between the scientist and society about animal experimentation. As well as meeting society's legitimate demand to be informed about such experimentation, engagement of this kind is in any case in the scientist's best interests. Previously, we have argued:

> Perhaps the main thing is to keep the channels of communications open. In the twenty-first century, transparency and accountability are watchwords. They are expected, and indeed demanded, in most areas

> of collective human endeavour. Thus, faced with questions about their work, the worst thing animal researchers can do is try to shut the enquirer out (34).

The scientist may believe that he or she is entitled to shut out ethical questions about animal research because animal rights extremists make fruitful dialogue impossible. Certainly over the last decade or so, threats, sabotage, and in rare cases even violent attacks on personnel engaged in animal experimentation have been observed in several countries. Feeling threatened, or feeling that the work one does is not publicly accepted, is an ethical issue for scientists, laboratory animal veterinarians, technicians, and caretakers. However, there is little to be said for belief that keeping quiet is the solution. In fact, the opposite has even been suggested – i.e. that there may be a link between refusing to engage in public and being exposed to animal rights activism (43).

Faced with the challenge of discussing animal experimentation in public, scientists have understandably focused on the benefits of research and the important role that animals have played in advancing medicine. In this context, brave statements have been made, attributing basically all medical advances to animal research. As was recently pointed out by Matthews (44), however, this is dishonest – no systematic review has been carried out supporting such statements. As we have explained in this chapter, there are a number of hurdles to overcome before researchers can honestly testify that animal-based research is being done in the best possible way, both in terms of potential benefit to humans and the harm caused to animals.

At this point in time, then, it would appear that the best a researcher can do is to acknowledge the dilemma of animal-based research and continue to work towards improved human benefits coupled with reduced animal harm. This is not merely an honest approach. It is also the only way to improve public confidence in animal-based research.

References

1. Sandøe P, Christiansen S (2008) Ethics of animal use. Blackwell, Oxford, UK
2. Narveson J (1983) Animal rights revisited. In: Miller HB, Williams WH (eds) Ethics and animals. Humana, Clifton, NJ
3. Singer P (1989) All animals are equal. In: Regan T, Singer P (eds) Animal rights and human obligations. Prentice Hall, Englewood Cliffs, NJ, pp 73–86
4. Regan T (1988) The case for animal rights. Routledge, London
5. Russell WMS, Burch RL (2008) The principles of humane experimental technique 1959 (Accessed June 9, 2008 at http://altweb.jhsph.edu/publications/humane_exp/het-toc.htm)
6. Singer P (1975) Animal liberation 2nd edition. Thorsons, London (1st edition 1975)
7. Nagel T (1974) What is it like to be a bat? Philos Rev 83:435–450
8. Smith JA, Boyd KM (1991) Lives in the balance: The ethics of using animals in biomedical research. Oxford University Press, Oxford, UK

9. Ashley PJ, Sneddon LU (2008) Pain and fear in fish. In: Branson EJ (ed) Fish welfare. Blackwell, UK, pp 49–77
10. Braithwaite VA, Boulcott P (2008) Can fish suffer? In: Branson EJ (ed) Fish welfare. Blackwell, UK, pp 78–92
11. Arluke A, Sanders CR (1996) Regarding animals. Temple University Press, Philadelphia, PA.
12. Grotewiel MS, Martin I, Bhandari P, et al. (2005) Functional senescence in *Drosophila melanogaster*. Ageing Res Rev 4:372–397
13. Murakami S (2007) Caenorhabditis elegans as a model system to study aging of learning and memory. Mol Neurobiol 35:85–94
14. Eismann CH, Jorgensen WK, Merritt DJ, Rice MJ, et al. (1984) Do insects feel pain? A biological view. Experientia 40:164–167
15. Lockwood JA (1987) The moral standing of insects and the ethics of extinction. Fla Entomol 70:70–89
16. Sherwin C (2001) Can invertebrates suffer? Or, how robust is argument–by analogy? Anim Welfare 10:S103–S118
17. Guo S (2004) Linking genes to brain, behavior and neurological diseases: What can we learn from zebrafish? Genes Brain Behav 3:63–74
18. Aldhous P, Coghlan A, Copley J (1999) Let the people speak. New Sci 2187:26
19. Cohen J (2007) The endangered lab chimp. Science 315:450–452
20. Tayebati SK (2006) Animal models of cognitive dysfunctions. Mech Ageing Dev 127:100–108
21. Weatherall D, Goodfellow P, Harris J (2006) The use of non-human primates in research – a working group report, London
22. Price E (1999) Behavioral development in animals undergoing domestication. Appl Anim Beh Sci 65:245–271
23. Smith JA, van den Broek FAR, Canto Martorell J, et al. (2007) Principles and practice in ethical review of animal experiments across Europe: summary of the report of a FELASA working group on ethical evaluation of animal experiments. Lab Anim 41:143–160
24. Marx U, Embleton J, Fischer R, et al. (1997) Monoclonal antibody production. The Report and Recommendations of ECVAM Workshop 231. ATLA 25:121–137
25. Van der Warp HB, de Haan P, Morrema E, et al. (2005) Methodological quality of animal studies on neuroprotection in focal cerebral ischaemia. J Neurol 252:1108–1114
26. Macleod M, Sandercock P (2005) Systematic reviews improve clinical research design – can they help improve animal experimental work? RDS News Winter issue
27. Van der Staay FJ (2006) Animal models of behavioral dysfunctions: Basic concepts and classifications, and an evaluation strategy. Brain Res Rev 52:131–159
28. Schliebs R, Rossner S, Bigl V (1996) Immunolesion by 192IgG-saporin of rat basal forebrain cholinergic system: A useful tool to produce cortical cortical cholinergic dysfunction. Prog Brain Res 109:253–264
29. Morton D (2007) Experimental procedures: general principles and recommendations. In: Kalista E (ed) The welfare of animals. Springer, Dordrecht, The Netherlands, pp 81–115
30. Richardson CA, Flecknell PA (2005) Anaesthesia and post-operative analgesia following experimental surgery in laboratory rodents: Are we making progress? ATLA 33:119–127
31. Perel VD (1998) Psychosocial impact of Alzheimer disease. J Amer Med Assoc 279:1038–1039
32. Apostolova LG, Cummings JL (2008) Neuropsychiatric manifestations in mild cognitive impairment: A systematic review of the literature. Dement Geriatr Cogn 25:115–126
33. Bradshaw J, Casey R, Blackwell E (2007) Principles of companion animal behaviour therapy. Blackwell Science, UK
34. Olsson IAS, Dahlborn K (2002) Improving housing conditions for laboratory mice: a review of 'environmental enrichment'. Lab Anim 36:243–270
35. Wolfer DP, Litvin L, Morf S, et al. (2004) Laboratory animal welfare: Cage enrichment and mouse behaviour. Nature 432:821–822
36. Nithianantharajah J, Hannan AJ (2006) Enriched environments, experience dependent plasticity and disorders of the nervous system. Nature Rev Neurosci 7:697–709
37. Bard F, Cannon C, Barbour R, et al. (2000) Peripherally administered antibodies against amyloid -peptide enter the central nervous system and reduce pathology in a mouse model of Alzheimer disease. Nat Med 6:916–919
38. Van Dam D, de Deyn PP (2006) Drug discovery in dementia: the role of rodent models. Nature Rev Drug Discov 5:956–970
39. Bacskai BJ, Kajdasz ST, Christie RH, et al. (2001) Imaging of amyloid-β deposits in brains of living mice permits direct observation of clearance of plaques with immunotherapy. Nat Med 7:369–372
40. European Union. Treaty of Amsterdam amending the Treaty on European Union, the Treaties establishing the European Communities and related acts. Official Journal C 340, 10 November 1997 (Accessed June 3, 2009 at

http://eur-lex.europa.eu/en/treaties/dat/11997D/htm/11997D.html)

41. Olsson IAS, Hansen AK, Sandøe P (2007) Ethics and refinement in animal research. Science 317:1680
42. Olsson IAS, Hansen AK, Sandøe P (2008) Animal welfare and the refinement of neuroscience research methods – a case study of Huntington's disease models. Lab Anim 42:277–283
43. Greek R (2005) Animal rights extremism revisited. The Scientist Nov 7
44. Matthews RAJ (2008) Medical progress depends on animal models – doesn't it? J R Soc Med 101:95–98

Chapter 3

The Role of Rodent Models in the Drug Discovery Pipeline for Dementia

Debby Van Dam and Peter Paul De Deyn

Abstract

Early-onset familial and late-onset dementia of the Alzheimer-type account for the major proportion of cases of dementia and of neurodegenerative diseases in general. The number of affected individuals is likely to grow in the decades to come due to demographic changes and rising life expectancy. Considerable research efforts are directed at the development and validation of animal models of AD, aiming to further unravel the underlying degenerative processes and searching for therapeutic strategies to alleviate or prevent this devastating condition. These models have contributed to a growing understanding of the molecular pathways involved in disease development and progression. The successful use of animal models in drug discovery relies on both the development of valid disease models and the availability of adequate testing paradigms for the evaluation of the effects of different therapeutic approaches aiming at symptomatic treatment of memory impairment or behavioral alterations, or at disease-modification and/or neuroprotection.

Key words: Drug discovery pipeline, animal models, validation, dementia, fenotype

1. Drug Discovery Pipeline

The general drug discovery pipeline is depicted in Fig. 1. Animal models – or laboratory animals in general – play an essential role in various stages of the drug discovery pipeline. In target discovery and validation, molecular targets pivotal to the disease process are studied. A target may be a receptor, proteins (enzymes), DNA, RNA, or ribosomes. In addition, the "drugability" of a target, which may be influenced by cellular location, development resistance, transport mechanisms, side effects, and toxicity, is assessed. Assay development encompasses the development of valid model systems, both *in vitro* and *in vivo*. This stage has known a major boost with the development of transgenesis and gene targeting techniques, but is, nevertheless, still the bottleneck of the drug discovery pipeline, since all

Peter Paul De Deyn and Debby Van Dam (eds.), *Animal Models of Dementia*, Neuromethods, vol. 48,
DOI 10.1007/978-1-60761-898-0_3,

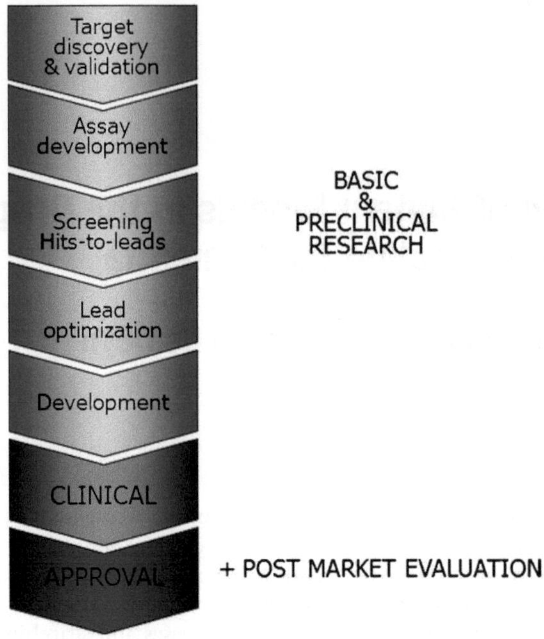

Fig. 1. Drug discovery pipeline. The drug discovery pipeline in general consists of three major stages: basic and preclinical research, clinical studies, and the (post) approval phase. Animal models – or laboratory animals in general – play essential roles in the drug discovery pipeline. See text for more detailed information on the use of animal models during the different preclinical stages.

models are partial and for some human diseases (e.g., hepatitis C infection) adequate models are still completely lacking. When converging hits-to-leads, animals are used to assess the potency (ED50) and dose-response curve of a compound. The mechanism of drug action is studied *in vitro*, on whole-cell bases systems and in animal models. During lead optimization, different species are used to study pharmacokinetics, pharmacodynamics, absorption, distribution, metabolism, and excretion, as well as to optimize formulation and delivery of the lead. All nonclinical data are compiled during the development stage. If required, some additional preclinical studies may be initiated before proceeding to the clinical phases.

2. Validation of Animal Models

Valid animal models are indispensable to the drug discovery and development pipeline. In particular, animal models have a key role to play in target discovery and validation, which require proof that a molecular target is pivotal to the disease process, and that modulation of that target has therapeutic effect. The development of relevant transgenic (knock-out/in) or chemical knockout models are vital to this stage. Additionally, the use of valid animal models for a specific human condition allows appraisal of the preclinical efficacy of candidate drugs.

The quality and utility of any animal model should be assessed through rigorous validation. A valid model resembles the human condition in etiology, pathophysiology, symptomatology, and response to therapeutic interventions. As a result, homologous models with etiology and symptomatology identical to the human condition exhibit the highest level of validity. In isomorphic models, symptomatology is not provoked by the same factors as in the human condition, leading to reduced validity (2).

The conclusions drawn from animal models largely depend on the validity of the model in representing the human condition. The following perspectives are commonly used in the characterization of a model's validity. Face validity refers to the resemblance between model and situation or process being modeled. It refers to the phenomenological similarity between the model and the human condition. Similarity of symptoms is most commonly used to assess face validity.

Predictive validity is an empirical form of validity that represents the extent to which the performance of the animal model in a test predicts the performance in the condition being modeled. This level of validation, therefore, necessitates parallel development of clinical measures for meaningful comparisons between model and man. In a more narrow sense, this term is sometimes used to indicate pharmacological isomorphism, i.e. the model's ability to identify compounds with potential therapeutic effects in the human condition. Construct validity refers to the theoretical clarification of what a test measures or a model is supposed to mimic. Because a given condition may manifest itself in different ways in different species, the behavior used in the animal model may not match that of humans, yet the test or model may still be valid. Construct validation is useful in the incessant process of further developing and refining an animal model. Etiological validity is closely related to construct validity, and refers to identical etiologies of phenomena in the model and the human condition. Models with high etiological validity are most valuable in drug development and discovery.

The more levels of validity a model satisfies, the greater its value, utility, and relevance to the human condition. A "perfect" model would account for etiology, symptomatology, treatment, and physiological basis. Animal models in general do not meet all of these criteria.

3. Validation of Animal Models of Dementia

The key etiological and symptomatic features of AD that would ideally be replicated in an animal model of the disease are outlined in the following paragraphs. In reality, most animal models are partial models, focusing only on restricted aspects of a disease, and modeling the complete condition is often not pursued.

As the prototype of cortical dementias, AD presents with prominent cognitive deficits. Initially, patients display limited forgetfulness with disruption of memory imprinting, evolving to short-term memory disruption, and with disease progression, long-term memory deficits. Most cortical dementias are characterized by the early-stage development of anomia progressing towards aphasia with perseverations and paraphasia. In addition, patients exhibit agnosia, including prosopagnosia or face blindness (representing a major burden for family members), as well as temporal and spatial disorientation. In a more advanced stage, apraxia results in executive dysfunctioning and, hence, the increasing helplessness of the AD patient (3). Besides cognitive deterioration, patients display noncognitive symptoms, such as anxiety, depression, aggression, agitation, hallucinations, circadian rhythm disturbances, and wandering. In contrast with cognitive symptomatology, these noncognitive symptoms – commonly referred to as behavioral and psychological signs and symptoms of dementia (BPSD) – do not exhibit a progressive course. The impact of BPSD is emphasized by the fact that they increase patient suffering, impose tremendous strain on family members and caregivers (often motivating institutionalization), and increase the financial burden on the family and society (3,4).

The histopathological hallmarks of AD brain are extracellular amyloid plaques and intracellular neurofibrillary tangles (NFT), accompanied by decreased synaptic density, which eventually leads to widespread neurodegeneration, loss of synapses, and failure of neurotransmitter pathways, particularly those of the cholinergic system. Clinical, epidemiological, and biochemical research over roughly the past 2 decades has significantly increased our knowledge of the pathological and molecular cascades underlying AD, uncovering potential biomarkers for disease progression or early diagnosis and revealing new drug targets. For a therapeutic intervention to slow down or halt disease progression – that is, to be disease-modifying – it must interfere with these central pathophysiological pathways. The following key pathways are being extensively scrutinized in AD rodent models.

3.1. The Amyloid Hypothesis

The hypothesis that cerebral accumulation of amyloid β (Aβ) peptides in amyloid plaques is the primary culprit driving AD pathogenesis has dominated research for 2 decades. Further disease processes are proposed to result from an imbalance between Aβ production and clearance. Cleavage of amyloid precursor protein (APP) by α-secretase precludes release of Aβ peptides (nonamyloidogenic pathway), whereas the combined effect of β- and γ-secretase cleavage releases Aβ peptides of various lengths (amyloidogenic pathway) (Fig. 2). Some early-onset AD cases showed an autosomal dominant inheritance pattern, leading to the

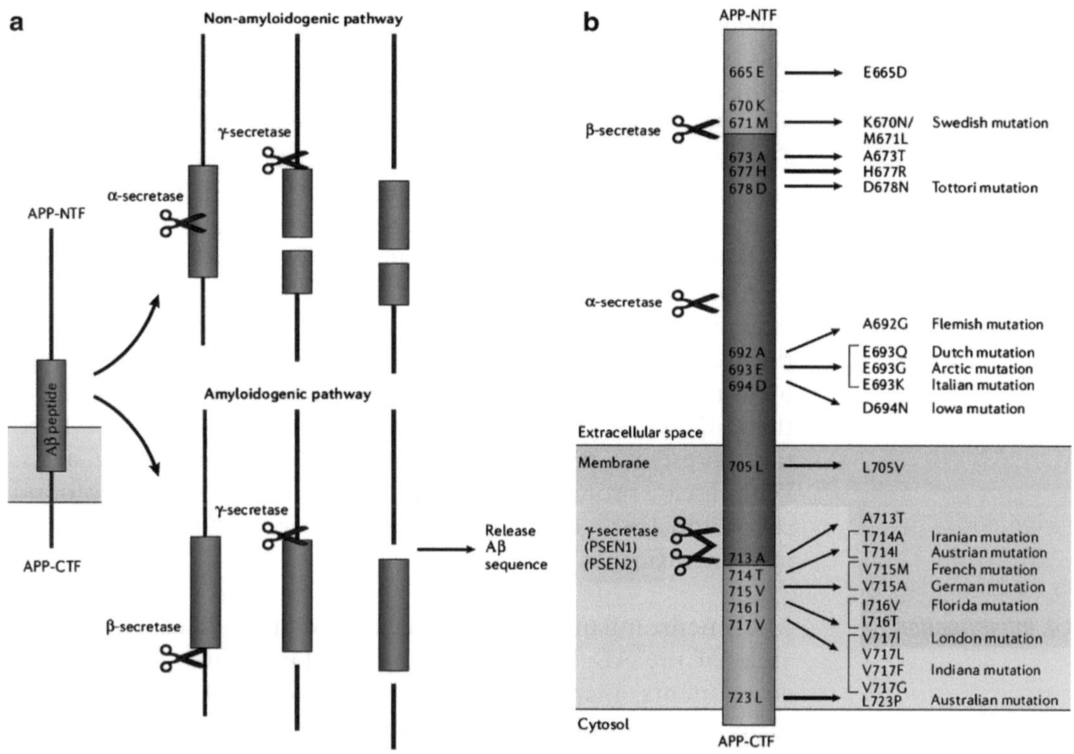

Fig. 2. APP processing and APP mutations associated with early-onset Alzheimer's disease. (**a**) Amyloid precursor protein (APP) processing involves proteolytic cleavage by several secretases. The nonamyloidogenic pathway is initiated by α-secretase cleavage occurring in the middle of the amyloid β (Aβ) sequence, and results in the release of several soluble APP fragments. The amyloidogenic pathway releases Aβ peptides through subsequent cleavage by β and γ-secretase. (**b**) Part of the APP amino acid sequence with indication of mutations associated with early-onset Alzheimer's disease. Most mutations are clustered in the close vicinity of secretase cleavage sites, thereby influencing APP processing, and are named after the localization of the first family in which a specific mutation was demonstrated. From (1). *Legend: A, alanine; Aβ, amyloid β; APP-NTF, N-terminal fragment of the amyloid precursor protein; APP-CTF, C-terminal fragment of the amyloid precursor protein; D, aspartic acid; E, glutamic acid; F, phenylalanine; G, glycine; H, histidine; I, isoleucine; K, lysine; L, leucine; M, methionine; N, asparagine; P, proline; PSEN, presenilin; Q, glutamine; R, arginine; S, serine; T, threonine; V, valine; Y, tyrosine.*

discovery of several major disease genes associated with amyloid processing (5), including mutations in APP, and in presenilin-1 (PSEN-1) and presenilin-2 (PSEN-2), proteins that form the catalytic unit of the γ-secretase protein complex (6). Another recent hypothesis attributes synaptic failure to the presence of soluble Aβ oligomers or amyloid-derived diffusible ligands (commonly abbreviated as ADDLs, pronounced "addles"), which can rapidly block long-term potentiation, and hence cause memory failure (7).

3.2. Tau-Phosphorylation, Paired Helical Filaments, and NFT

A second hallmark of the AD brain is the presence of intracellular NFT, resulting from the hyperphosphorylation and aggregation of CNS tau proteins, a group of microtubule-associated proteins that contribute to the assembly and stabilization of microtubules. Tau function is determined by the degree of phosphorylation, which is regulated by a balance between several protein kinases and phosphatases, with hyperphosphorylation preceding the formation of paired helical filaments and therefore NFT.

3.3. Oxidative Stress

Oxidative damage to different classes of biological macromolecules (sugars, lipids, proteins, and DNA) is a hallmark of both normal aging and most neurodegenerative diseases (8). Multiple lines of evidence demonstrate that oxidative stress is an early event in AD, and therefore may play a key pathogenic role. Oxidative stress may be mitigated through the use of antioxidants, and in the event of increased oxidative stress, neurons themselves upregulate antioxidant defense systems. Evidence even indicates that in the initial phase of AD, Aβ deposition and hyperphosphorylated tau represent a primary line of defense against oxidative stress. With disease progression, the brain is challenged with a profound redox imbalance, resulting in a transformation of both substances into pro-oxidants (9).

3.4. Inflammation

Many neuroinflammatory mediators are upregulated in affected areas of the AD brain, including prostaglandins, complement components, anaphylatoxins, cytokines, chemokines, proteases, protease inhibitors, adhesion molecules, and free radicals (10). Epidemiologic data suggests that prolonged use of nonsteroidal anti-inflammatory drugs (NSAIDs), which target cyclooxygenase (COX), a key mediator of the inflammatory cascade, for conditions like arthritis entails a reduced risk and delayed onset of AD. Initially, the effect of NSAID use in AD was attributed to a reduction of inflammation, but several clinical trials testing NSAIDs in AD patients yielded negative results (10 – 12) and in 2001; however, it was reported that a subset of NSAIDs reduced Aβ1–42 production in cultured cells and the mouse brain (13), effects that are independent of COX inhibition. However, the initial rationale for the use of NSAIDs in AD – that is classical anti-inflammatory mechanisms – should not be completely abandoned.

3.5. Lipid Metabolism and Apolipoprotein E

Several lines of evidence support a strong link between lipid/cholesterol metabolism and dementia. Apolipoprotein E (ApoE) is a plasma and cerebrospinal fluid protein that serves as a ligand for low-density lipoprotein receptors and is involved in the transport of cholesterol and other lipids among various cells of the body. The ApoE ε4 allele in particular is genetically associated with both vascular and Alzheimer dementia (14). Increased dietary intake of cholesterol promoted amyloid deposition in rabbit brain (15), whereas statins were able to reduce Aβ peptide levels both *in vitro* and *in vivo*, in guinea pig cerebrospinal fluid (16).

3.6. Neuronal and Synaptic Loss

The mechanisms underlying AD-related neuronal loss are not clear-cut. Apoptosis, triggered by Aβ, activation of glutamate receptors, oxidative stress, DNA damage, and elevation of intracellular calcium levels may occur, as well as other forms of programmed cell death or necrosis. Synaptic dysfunction or loss contributes to clinical symptomatology through disruption of neuronal communication and occurs early in the disease process, long before neuronal and synaptic loss, the latter currently representing the best pathological correlate of cognitive decline (17).

4. Preclinical Drug Evaluation in AD Models

In general, animal models of human disease can be classified into spontaneous, induced, negative, and orphan models, although the two latter types are not relevant when considering dementia models. Although spontaneous models are presumed to develop their condition without experimental manipulation, selective breeding is often necessary to establish and maintain the mutant line. Induced models display conditions as a result of artificial manipulation. Particularly for psychiatric and neurological conditions, few spontaneous models exist, and experimentally induced pathology is often necessary (3). Although models based on many different species, including primates and other mammals, as well as invertebrates, have contributed to the wider field of AD research, rodent models are predominant in drug discovery. The different species and types of models available for dementia-related research will be dealt with in a different chapter of this volume.

Treatment goals change with the severity of AD (3). In patients with mild to moderate disease, goals are to improve or maintain baseline performance through the administration of disease-modifying drugs targeting crucial etiological processes that thereby are neuroprotective. As the disease progresses, the goal of treatment is to slow the rate of decline in performance, mainly through highly efficacious symptomatic therapeutics improving cognitive and behavioral deficits that impair the well-being of patients and caregivers. Symptomatic therapies, however, do not address the cause of the disease. If predisposition for the development of AD should become predictable in the future, for example based on biomarker profiling in patients with mild cognitive impairment, the development of truly preventive therapies will become mandatory. Figure 3 represents the theoretical treatment outcome possibilities for different treatment options and schedules in AD, which have to be taken into account when designing a preclinical trial in an AD animal model. New putative symptomatic or preventive drugs can be tested in animal models

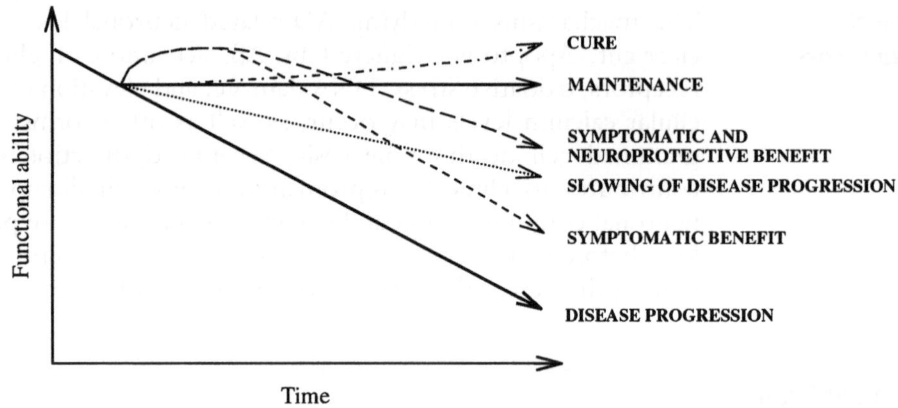

Fig. 3. Theoretical possibilities of treatment outcome in Alzheimer's disease. Treatment possibilities and goals change with the severity stage of Alzheimer's disease and can be symptomatic, curative, or disease-modifying. Curative therapy implicates complete healing and is unfortunately at present only applicable for a selected number of dementia syndromes such as normal pressure hydrocephalus. Symptomatic treatment positively influences cognitive, functional, or behavioral symptoms. Currently, for many dementia syndromes, such as Alzheimer's disease, only symptomatic treatment options exist. If administered in an early stage, some compounds may exert clinical stabilization of behavioral or functional parameters. Disease-modifying treatments can display a symptomatic action as well, but exert a positive influence on the course of the disease; they can either slow down the degenerative process (neuroprotective effect) or favorably influence the normal clinical course. From (1).

using a variety of treatment schedules and cognitive or behavioral paradigms. We will familiarize the readers with some general treatment schedules used to assess pharmacological modulation of cognitive and behavioral changes and pathophysiological mechanisms in AD models.

4.1. Evaluating Symptomatic Efficacy

As long as AD cannot be prevented, it is imperative to provide optimal symptomatic relief. The development of drugs should not only focus on cognitive decline, but also on BPSD and related disturbances, since the latter represent the major source of physical and psychological caregiver burden, often motivating institutionalization of the patient (3). Compounds with presumed symptomatic activity are mostly tested in an animal model that mimics some symptoms of AD using a single or a limited number of administrations, dependent on the duration of the behavioral paradigm that is chosen for preclinical analysis or on the time needed to reach optimal plasma and/or brain levels of the active compound.

Analysis of cognitive and behavioral alterations should be substantiated through implementation of thoroughly validated behavioral paradigms (18). Face validity alone is not a sufficient validation criterion. The predictive validity of behavioral tests can be demonstrated by showing that drugs acknowledged to work in the human setting either increase or inhibit the behavior in question. Examples include the use of anxiolytic and anxiogenic drugs to validate the elevated plus maze (19), tricyclic antidepressants to validate a learned helplessness design, and amphetamines and anticholinergic compounds to validate an operant task (20). Construct validity of a task implies solid insight into dependable variables and their effect on outcome measures; Morris water maze (MWM) performance is, for instance, influenced by lighting conditions, water temperature, trial duration, intertrial interval, and maze diameter (21–23). The development of complex disease models requires the parallel development or optimization of valid behavioral paradigms assessing complex brain-behavior relations. Besides validity of the applied paradigm, standardization at the level of experimental animals, testing procedures, and surroundings is essential to generate reliable data (24). High levels of validity and standardization can be reached only by skilled and experienced researchers.

With the development of genetically engineered mouse models with presumed CNS deficits came the use of test batteries scrutinizing a wide variety of neurological and behavioral responses. The major advantages of these test batteries are the possibility to investigate potential correlative phenotypical changes, and the fact that a CNS phenotype can be confirmed in different paradigms, thereby strengthening the reliability of the model. In addition, this approach reduces the number of mice needed. One should, however, be aware of potential training effects dependent on the order of the different behavioral paradigms and important behavioral differences between naïve and trained mice.

4.1.1. Cognitive Symptoms

Cognitive enhancement can be investigated at different stages of learning and memory processes: acquisition, consolidation and retention, using a variety of behavioral tests. To assess the effect of a compound on learning, and potentially also on early consolidation processes, administration of the test compound is usually carried out prior to the acquisition trial. To assess whether a compound enhances the consolidation process, it can be administered immediately after the acquisition trial. Whether a compound improves retrieval of learned information can be probed using an administration schedule that involves drug treatment prior to the retention or probe trial (Fig. 4).

4.1.2. Noncognitive Symptoms

Over the past decade, clinical AD-related research has been challenged with an increased interest in noncognitive symptomatology, i.e. BPSD. In accordance, major efforts have been made to mimic specific behavioral alterations in animal models and to develop useful tools to evaluate new psychopharmacological strategies to replace atypical antipsychotics or classic neuroleptics, which display only modest effect size and are frequently associated with major side-effects (1,25,26). The general preclinical treatment schedule to assess treatment efficacy on BPSD is depicted in Fig. 5.

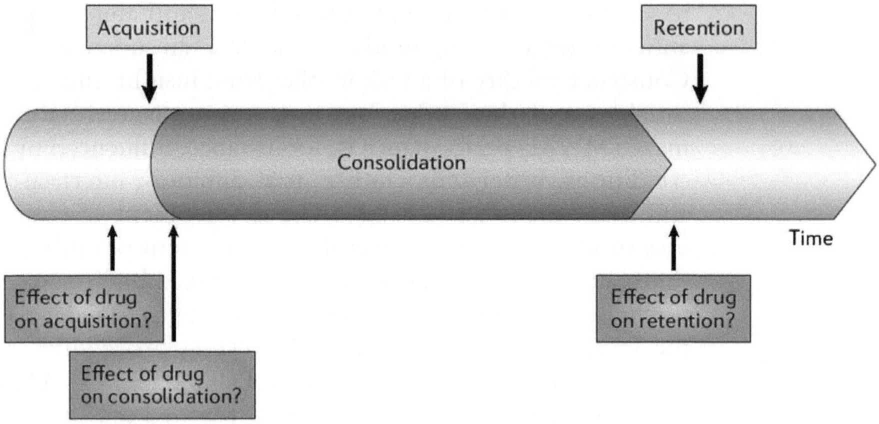

Fig. 4. Theoretical treatment schedule in preclinical evaluation targeting memory dysfunction in animal models for e.g. Alzheimer's disease. Cognitive enhancement can be investigated at different stages of learning and memory processes: acquisition, consolidation, and retention. Boxes indicate the time of drug administration used to test the effects on each of these stages. From (1).

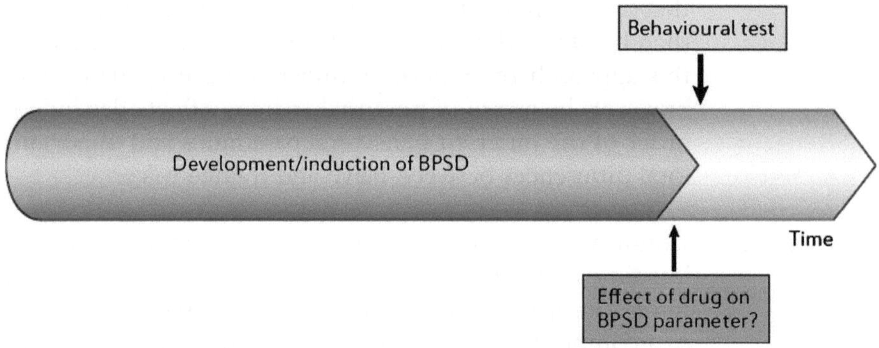

Fig. 5. Theoretical treatment schedule in preclinical evaluation when assessing treatment of BPSD in animal models for e.g. Alzheimer's disease. A compound is administered after induction or development of a BPSD-related parameter, which is subsequently quantified in a behavioral test. From (1).

A major drawback of animal models for AD is the fact that several BPSD clusters, including paranoid and delusional ideation, hallucinations, and affective disorders are difficult or even impossible to model in rodents.

5. Evaluating Neuroprotective or Disease-Modifying Efficacy

The evaluation of preventive or disease-modifying efficacy is not easily accomplished in a clinical setting. Animal models have therefore acquired a strong position in this field of research based on the rapid development of symptoms and/or pathology, availability of

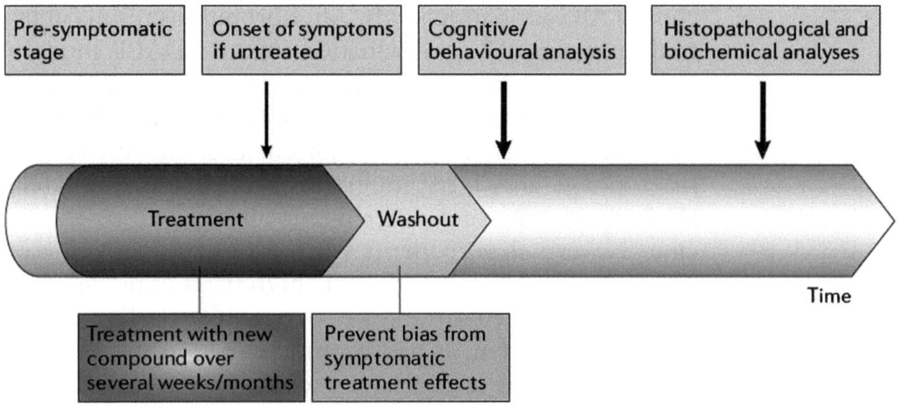

Fig. 6. Suggested treatment schedule for evaluation of disease-modifying efficacy in a model exhibiting cognitive and/or behavioral alterations. Treatment should start prior to the first presentation of the symptoms and should be followed by a wash-out period (to prevent bias from sustained symptomatic treatment effects) prior to analysis of effects on disease progression. From (1).

potentially large groups of subjects, accessibility to early-stage CNS changes, and possibility of time-linked observations. For a compound to exert neuroprotective efficacy, it must intervene in a central pathophysiological pathway of AD. Strategies for evaluating neuroprotection therefore typically involve measurement of markers of disease progression, such as levels of Aβ and plaque and tangle formation, inflammatory markers, and oxidative damage to macromolecules in animal models that develop these features over time.

The major shortcoming of many preclinical trials scrutinizing disease-modifying efficacy is the lack of a wash-out period to prevent bias from sustained symptomatic treatment effects. We have designed a preclinical treatment protocol based on clinical withdrawal designs to assess disease-modification in an AD model (27) (Fig. 6).

6. Potential Pitfalls

The reliability of preclinical pharmacological trials does not merely depend on the quality of the models itself or on the pharmacokinetics and pharmacodynamics of the compound that is administered. A large part depends on methodological aspects of the behavioral paradigm and the background strain that is chosen, as well as on external parameters characteristic for, among others, the testing environment, conditions in the animal housing facilities, and experience of experimenters.

Uncritical and premature extrapolation of animal model findings to the human condition can be unreliable and dangerous. A well-known example in dementia-related research is the first clinical Aβ vaccination study. After propitious results in preclinical trials assessing $A\beta_{1-42}$ vaccination in the PDAPP model (28), in January 2002, the first clinical trial with Aβ vaccination in AD was suspended in phase IIa due to clinical signs of CNS inflammation in some patients (29). Postmortem examination revealed CNS macrophage and T-cell infiltration, which was not detected preclinically in mice and other mammals exposed to the same vaccine. Nevertheless, patients who did not develop these adverse effects and produced sufficiently high titers of plaque-binding Aβ antibodies showed slower rates of cognitive decline and reduction of amyloid burden, as had been described in the PDAPP model. Despite the adverse effects, the search for effective immunotherapeutic strategies was not abandoned. Animal research helped uncover the causes of the vaccination-associated CNS inflammation; also in mice, vaccination with Aβ could trigger an aberrant autoimmune response associated with Aβ-specific T-cells, B-cell response, and activation of macrophage infiltration, leading to perivenular encephalomyelitis. Co-administration of pertussis toxin with the vaccine proved to be the culprit (30). After the initial gloomy future of immunotherapy for AD, animal research helped unravel innate and adaptive immune response mechanisms towards Aβ and re-opened the path for immunotherapeutic strategies in AD and other neurodegenerative diseases.

7. Conclusions

With this chapter, we have sketched the importance of rodent models for drug discovery and development in dementia, with a focus on AD. In general, high-quality and conscientious research aiming at the validation of a new model or testing a new compound requires thorough standardization of procedures, good knowledge of strains, compounds, and paradigm characteristics, and skilled personnel. Keeping in mind basic metabolic, physiological, and anatomical differences between man and other species, it is unambiguous that a plurispecies approach increases the reliability of extrapolation from animal models to man.

Animal models with a high level of validity enable us to detect and optimize therapeutically interesting drugs. However, this requirement is often considered the major bottleneck in the development of novel drug therapies because to date no animal model combines all the relevant pathophysiological and behavioral aspects of AD. Indeed, one could wonder whether it is compulsory to create a complete, and hence complex, model. It may even be more interesting to isolate a specific pathophysiological pathway or behavioral alteration of interest when attempting to intervene with new specific therapeutic approaches. On the other hand, more complex animal models manifesting reliably quantifiable pathophysiological alterations representing different etiological hypotheses may be of great value in evaluating combination therapies. This approach of targeting different pathophysiological pathways may well be the road to success in treating or preventing a complex syndrome as AD and Phase III combination therapies are currently in progress. Individual differences in response to treatment can be prompted, among others, by underlying pathogenesis, disease stage, age and nutritional status of the patient, concomitant illnesses, and accompanying treatments.

In addition to the study of the above-mentioned phenomena, genetically modified animal modeling, in particular gene knockouts, may have a great impact on the emerging field of pharmacogenomics/pharmacogenetics, which examines inherited variations in genes that dictate drug metabolism and treatment response, and explores ways in which these variations can be used to predict treatment outcome and possible adverse effects. The identification of an increasing number of genes associated with neuronal dysfunction along the human genome together with the influence of specific allelic associations and polymorphisms indicate that pharmacogenomics will become a preferential procedure for drug development in polygenic complex disorders like AD (31).

By and large, thoroughly validated animal models play a crucial role in our understanding of cellular and molecular alterations responsible for the neurodegenerative processes underlying AD, as well as in the subsequent preclinical development of new therapeutic strategies targeting these pathophysiological pathways. The future of AD-related research lies in the exploitation of new technologies such as the systems biology "omics" cluster for biomarker and genetic profiling and subtyping of the heterogeneous dementia disorders. A multidisciplinary approach combining the parallel development of valid animal models and the exploitation of new technologies improving biomarker profiling and, hence, early diagnosis of dementia subtypes and prediction of patient-specific treatment outcome (theranostics), in addition to thorough preclinical screening, combining etiopathogenic and empirical approaches, will open new routes to better treatment and ultimately prevention of AD and related disorders.

Acknowledgments

This contribution is based on (1) of which the copyright remained with the Authors, D. Van Dam D. and P.P. De Deyn, based on the Nature Publishing Group Copyright policies.

This work was financed by the Fund for Scientific Research – Flanders (FWO, G.0164.09), Agreement between the University of Antwerp and the Institute Born-Bunge, Interuniversity Poles of Attraction (IUAP Network P6/43), Methusalem excellence grant of the Flemish Government, Neurosearch Antwerp, the Antwerp Medical Research Foundation, and the Thomas Riellaerts Research fund. DVD is a postdoctoral fellow of the FWO.

References

1. Van Dam D, De Deyn PP (2006) Drug discovery in dementia: the role of rodent models. Nat Rev Drug Discov 5:956–970
2. De Deyn PP, D'Hooge R, van Zutphen LFM (2000) Animal models of human disorders – general aspects. Neurosci Res Comm 26: 141–148
3. De Deyn PP (2004) Dementie: Medisch, psychosocial, ethisch en preventief [In Dutch]. Kluwer, Mechelen, Belgium
4. Engelborghs S, Maertens K, Nagels G, et al. (2005) Neuropsychiatric symptoms of dementia: cross-sectional analysis from a prospective, longitudinal Belgian study. Int J Geriatr Psychiatry 20:1028–1037
5. St George-Hyslop PH (2000) Molecular genetics of Alzheimer's disease. Biol Psychiatry 47:183–199
6. Brunkan AL, Goate AM (2005) Presenilin function and gamma-secretase activity. J Neurochem 93:769–792
7. Lacor PN, Buniel MC, Chang L, et al. (2004) Synaptic targeting by Alzheimer's-related amyloid β oligomers. J Neurosci 24:10191–10200
8. Moreira PI, Smith MA, Zhu X, Nunomura A, Castellani RJ, Perry G (2005) Oxidative stress and neurodegeneration. Ann N Y Acad Sci 1043:545–552
9. Smith MA, Casadesus G, Joseph JA, Perry G (2002) Amyloid-beta and tau serve antioxidant functions in the aging and Alzheimer brain. Free Radic Biol Med 33:1194–1199
10. Akiyama H, Barger S, Barnum S, et al. for the Neuroinflammation Working Group (2000) Inflammation and Alzheimer's disease. Neurobiol Aging 21:383–421
11. Aisen PS, Davis KL, Berg JD, et al. (2000) A randomized controlled trial of prednisone in Alzheimer's disease. Alzheimer's disease cooperative study. Neurology 54:588–593
12. Van Gool WA, Weinstein HC, Scheltens P, Walstra GJ (2001) Effect of hydroxychloroquine on progression of dementia in early Alzheimer's disease: An 18-month randomised, double-blind, placebo-controlled study. Lancet 358:455–460
13. Weggen S (2001) A subset of NSAIDs lower amyloidogenic Abeta42 independently of cyclooxygenase activity. Nature 414:212–216
14. Poirier J (2005) Apolipoprotein E, cholesterol transport and synthesis in sporadic Alzheimer's disease. Neurobiol Aging 26:355–361
15. Sparks DL, Scheff SW, Hunsaker JC 3rd, Liu H, Landers T, Gross DR (1994) Induction of Alzheimer-like beta-immunoreactivity in the brains of rabbits with dietary cholesterol. Exp Neurol 126:88–94
16. Fassbender K, Simons M, Bergmann C, et al. (2001) Simvastatin strongly reduces levels of Alzheimer's disease beta –amyloid peptides Abeta 42 and Abeta 40 in vitro and in vivo. Proc Natl Acad Sci U S A 98:5856–5861
17. Coleman P, Federoff H, Kurlan R (2004) A focus on the synapse for neuroprotection in Alzheimer disease and other dementias. Neurology 63:1155–1162
18. Sarter M (2004) Animal cognition: Defining the issues. Neurosci Biobehav Rev 28:645–650
19. Pellow S, File SE (1986) Anxiolytic and anxiogenic drug effects on exploratory activity in an elevated plus-maze: A novel test of anxiety in the rat. Pharmacol Biochem Behav 24: 525–529
20. Bensadoun JC, Brooks SP, Dunnett SB (2004) Free operant and discrete trial performance of mice in the nine-hole box apparatus: Validation using amphetamine and scopolamine. Psychopharmacology (Berlin) 174:396–405
21. D'Hooge R, De Deyn PP (2001) Applications of the Morris water maze in learning and memory. Brain Res Rev 36:60–90
22. Klapdor K, van der Staay FJ (1996) The Morris-water maze task in mice: Strain differences and effects of intra-maze contrast and brightness. Physiol Behav 60:1247–1254
23. Van Dam D, Lenders G, De Deyn PP (2006) Effect of Morris water maze diameter on visual-spatial learning in different mouse strains. Neurobiol Learn Mem 85:164–172
24. Wahlsten D (2001) Standardizing tests of mouse behaviour: Reasons, recommendations, and reality. Physiol Behav 73:695–704

25. De Deyn PP, Rabheru K, Rasmussen A, et al. (1999) A randomized trial of risperidone, placebo, and haloperidol for behavioral symptoms of dementia. Neurology 53:899–901
26. De Deyn PP, Katz IR, Brodathy H, Lyons B, Greenspan A, Burns A (2005) Management of agitation, aggression, and psychosis associated with dementia: A pooled analysis including three randomized, placebo-controlled double-blind trials in nursing home residents treated with risperidone. Clin Neurol Neurosurg 107:497–508
27. Van Dam D, De Deyn PP (2006) Cognitive evaluation of disease-modifying efficacy of galantamine and memantine in the APP23 model. Eur Neuropsychopharmacol 16:59–69
28. Schenk D, Barbour R, Dunn W, et al. (1999) Immunization with amyloid-beta attenuates Alzheimer-disease-like pathology in the PDAPP mouse. Nature 400:173–177
29. Monsonego A, Weiner HL (2003) Immunotherapeutic approaches to Alzheimer's disease. Science 302:834–838
30. Furlan R, Brambilla E, Sanvito F, et al. (2003) Vaccination with amyloid-β peptide induces autoimmune encephalomyelitis in C57/BL6 mice. Brain 126:285–291
31. Crentsil V (2004) The pharmacogenomics of Alzheimer's disease. Ageing Res Rev 3:153–169

Part II

Methodological Considerations when Developing Animal Models of Dementia

Chapter 4

Species, Strain, and Gender Issues in the Development and Validation of Animal Models of Dementia

Annemie Van Dijck, Debby Van Dam, and Peter Paul De Deyn

Abstract

When establishing animal models of Alzheimer's disease (AD), the aim is to mimic (certain aspects of) the human condition. However, species, strain, and gender specific features interfere with this goal. Only a few species, like primates, dogs, and bears, spontaneously develop histopathological hallmarks of AD. Unfortunately, the use of these species for experimental research is limited by availability, economical and ethical reasons. Commonly used laboratory animals show age-related deterioration, but no AD-pathology. Transgenic models can give solace for this problem. Homology for AD-related genes is discussed for *Caenorhabditis elegans*, *Drosophila melanogaster*, *Danio rerio*, mice, and rats. Pros and cons of species-specific requirements for breeding and housing are argued. Effects of genetic background on pathology should be taken into consideration when studying AD in transgenic mouse models. Cognitive performance and behavior can differ between mouse strains, as well as immunology and vasculature. Strain-specific deficits in auditory or visual acuity or motor impairment can obfuscate measurements, and thus background strains should be selected cautiously. Female mice show greater variation in behavior according to estrus cycling than males, which is why males are often preferred for behavioral research. When studying gender-related differences in AD pathology or risk factors for dementia, both genders are to be compared. Gender-specific differences in learning and memory, aggression, and sexual behaviour should be borne in mind when phenotyping transgenic mice. In conclusion, this chapter provides an overview of background information to keep in mind when working with animal models of dementia.

Key words: Gene homology, breeding, housing, cognition, behavior, neuropathology

1. Species Issues

When establishing animal models of Alzheimer's disease, the aim is to reproduce as many characteristics found in humans as possible. It is obvious, however, that this goal can never entirely be reached due to species-specific characteristics. Pros and contras of different species as animal models of dementia will be discussed in this paragraph.

Peter Paul De Deyn and Debby Van Dam (eds.), *Animal Models of Dementia*, Neuromethods, vol. 48,
DOI 10.1007/978-1-60761-898-0_4,

1.1. Spontaneous Models

There are several species, which have been recognized to exhibit some characteristics of Alzheimer's disease (AD) neuropathology. Spontaneous plaque pathology was already detected in aged dogs in the 1950s (1). The majority of plaques reported in the canine brain are diffuse, whereas neuritic plaques are rare. Aged canines do not develop neurofibrillary tangles (NFTs) (2). Amyloid β (Aβ) accumulation is related to a decline in global cognitive function and specific types of behavioral tasks are affected (3–6). These features support the use of the canine as a model of the earliest stages of age-related cognitive decline and consequences of Aβ deposition in the absence of NFTs can be determined. This model is termed the canine counterpart of senile dementia of the Alzheimer's type. The neuropathology of aged cats resembles that of dogs, since Aβ is deposited as diffuse plaques and no mature NFTs are detected (7).

A wide variety of nonhuman primates has been used in aging research with the rhesus monkey as the most studied nonhuman primate (8). Senile plaques have been reported in the brains of various nonhuman primate species, including cynomolgus monkey (9), rhesus monkey (10), chimpanzee (11), marmoset (12), vervet (13), lemur (14), orang-utan (15), and cotton-top tamarins (16). In 2008, NFTs were demonstrated in chimpanzees. These NFTs consisted of dense intraneuronal bundles of paired helical filaments (PHFs) that were identical in size and helical periodicity to those in humans with AD. However, the rarity of these NFTs and the paucity of senile Aβ plaques limits the usefulness of this model (17). No ultrastructural studies have been performed on mouse lemur brains so far. Nevertheless, immunoreactive tau proteins, evidenced using monoclonal antibodies against human PHFs have been detected in the cortex of mouse lemurs. PHF-like material has also been evidenced in the mouse lemurs using immunoblotting (18).

Diffuse plaques as well as NFTs were observed in aged bear (19) and aged polar bear brains (20). In the study of an aged wolverine, senile plaques and NFTs were documented (21).

We have now seen that some mammalian species spontaneously develop plaque pathology and some species even exhibit tauopathies. Species like wolverine and polar bear resemble AD more closely than most species examined to date, but their inaccessibility to laboratory researchers limits the usefulness of these species. For nonhuman primates, long lifespan, cost, and ethical consideration are among the limiting factors for utility. The average and maximal lifespan for rhesus monkeys, for example, is about 25 and 40 years (8); Aβ plaques are known to be present over the age of 20 years (8). Mouse lemurs in contrast are considered as aged from the age of 5 years on (18). NFTs were only found in a 41-year-old chimpanzee (17). Economic factors are an important consideration when selecting a primate model, but the limited availability of research subjects is also a problem.

Additional regulations and laws concerning the use of primates that may not apply to other model systems can also shift the choice for a model organism in the direction of lower mammalian species.

Aging rodents do not develop plaques and NFT, but they do show age-related cognitive and behavioral alterations that correlate with AD-relevant neurochemical alterations, such as age-associated cholinergic hypofunction (22). In this sense, aging rodents can help to uncover the switch between normal aging and pathological states. Natural age-associated deterioration has culminated in the senescence-accelerated mouse (SAM), a model which was established through phenotypic selection of AKR/J mice (23). The SAMP8 substrain shows age-related learning and memory deficits, and is therefore useful as a model for dementia (24).

1.2. Transgenic Models

The vast majority of transgenic models have been established in the mouse, but other species can also be used in the search to unravel the pathology of AD. Both the nematode *Caenorhabditis elegans* (*C. elegans*) and the fruit fly, *Drosophila melanogaster* (*D. melanogaster*), are in the leading positions of invertebrate models of diseases. Their short reproduction time, informed genomics, variety of phenotypes, and the ability to express human genes of interest offer fast ways to screen for functional implications of human gene mutations (25). *C. elegans* was the first multicellular organism of which the genome was sequenced (26); in the meanwhile, the genome of *D. melanogaster* is also sequenced (27). The simplicity of these invertebrate models allows identifying protein interactions, which could help to unravel entire regulatory pathways. Basic forms of learning and memory and circadian rhythms can be studied in the fruit fly, but it mainly contributes to the knowledge of molecular and cellular processes. Fundamental aspects of cell biology are quite similar in man and flies, including regulation of gene expression, cell signaling, cell death, membrane trafficking, neuronal connectivity, and synaptogenesis (28). Of course, the fly is a much simpler organism with less sophisticated circulatory systems and cognitive processes. In humans, often redundancy exists, duplicated versions of genes are identified that are present in only one copy in flies. This lack of redundancy can simplify analysis of biological processes in the fly (28). Worms and fruit flies offer the opportunity to generate multiple transgenic lines simultaneously. Their short life span allows screening large numbers of gene mutations in a short period of time.

Rat models have certain advantages over mice, such as their larger size, unique genetics, and well-studied behavioral characteristics. Rats are better suited for microsurgery, cell and tissue transplantation, *in vivo* functional analyses, and studies that require multiple sampling. Stereotaxic injection is a commonly used method in neurological studies, which is easier to perform with precision in rats than in mice (29). However, less transgenic rat models have been established up to now. Compared to the mouse, rat egg microinjection is more problematic and as a consequence survival rates of microinjected rat eggs are often less than 80% (30). Recent advances in transgenesis of the rat promise the upcoming of more transgenic rat models, including AD models (29,30). It is possible that rats in general, or at least some rat strains, are more resistant to Aβ deposition than most strains of mice. Nevertheless, Flood et al. succeeded in the construction of a triple transgenic rat model in Sprague–Dawley rats with Aβ deposits similar to those observed in the transgenic mouse models of AD. Extracellular deposits were found at 7–9 months of age (31). Due to reduced fertility of inbred rat strains, generation of a transgenic model in inbred rats is more difficult to achieve than in outbred rat strains. The use of inbred rat strains, however, minimizes the individual variation among transgenic rats. Some attempts were made to produce inbred transgenic rat models. Fischer 344 rats with expression of the Swedish amyloid precursor protein (APP) mutant gene show a surprising improvement in hippocampal-dependent learning and memory tasks, and only mild increases in brain APP mRNA with no extracellular deposits (32). Another inbred rat model was established by lentiviral infection of Fischer 344 zygotes. This APP21 rat model shows three times greater cerebral APP expression compared to wild-type rats, but no extracellular plaque deposition is demonstrated either (29).

The zebrafish (*Danio rerio*) is an effective and simple model organism for studies of developmental and neurological disease processes. Since the zebrafish is a vertebrate, it is more closely related to humans than invertebrate models such as worms or flies. It is an advantageous model for genetic studies as it is genetically malleable by injection of morpholino antisense oligonucleotides, mRNA, or transgenes. These techniques make it possible to make subtle or drastic changes in gene expression and observe the influences in the developing transparent embryo (33).

1.3. Breeding and Housing

Physiological and anatomical features of primates are strongly dependent on the species. For lemurs, gestation takes 2 months, and the animals breastfeed for 6–8 weeks. In the wild, female animals usually have litters twice a year; in captivity, typically only once. One female animal has one to four babies per litter (18). Rhesus monkeys have a gestation of 5 months, in chimpanzees this can take 8 months. Latter species usually has only one baby per litter (34). Cage dimensions are dependent on the species, but the minimum floor surface can go to 2.5 m^2 and minimum height of 2 m for chimpanzees.

Dogs need relatively large cages, dependent on shoulder height. Females are ready for breeding when they are older than 12 months. Gestation lasts around 65 days and litters contain three to six animals (35). Dogs of 8 years and older can be considered middle-aged and old (6).

Breeding in mice goes relatively quickly. Mice are capable of breeding from the age of 8–10 weeks on. Female animals are polyestric, which means there are several estral cycles in a year. When groups of mice are housed without males, anestrus occurs. The introduction of a male at this time (or male pheromones) will lead to estrus synchronization, termed the Whitten effect (36). After a successful copulation, no implantation will occur if the female is introduced to another male within 24 h after mating, known as the Bruce effect (37). Until 24 h after mating, a coagulation plug can be found in the vagina. This is a very simple method to confirm mating. Pups are born blind and naked after a gestation of 18–22 days. After 3 weeks, the pups can be weaned, but survival and general health of the pups is better when weaning occurs at 4 weeks of age (38). When mating takes place during postpartum estrus, lactation and gestation can coincide. Lactation can impair the implantation of zygotes, which prolongs the gestation with 3–5 days. Mice can be housed in groups, but it is possible that males become aggressive when housed together. Individually housed animals are more aggressive than permanently group-housed mice (35).

Rats are less aggressive than mice. Males can be housed together without many problems, but large groups of adult animals may lead to territorial aggression. Too many rats in one cage can lead to death by hyperthermia. Reproduction is comparable to the mouse, although the Bruce effect does not occur in rats and estrus synchronization by pheromones is less clear. Synchronization of estrus and gestation can be induced by the administration of progesterone during 4 days (causes anestrus), followed by the administration of pregnant mare serum gonadotrophin, which make most females go to estrus. The estral cycle is light sensitive; lights on 12–16 h per 24 h results in optimal breeding results. Rats are ready for breeding when aged 12–16 weeks. Gestation takes 21–23 days and litters consist of 6–12 pups (39).

Housing requirements and recommendations for commonly used laboratory animals are described in literature in detail, e.g. (40).

C. elegans progresses from embryo to larva to fertile adult in 3 days at room temperature. A single adult can have between 300 and 1,000 progeny. *C. elegans* primarily exist as self-fertilizing hermaphrodites with the occasional occurrence of males. This is very useful for genetics: the hermaphrodites can maintain homozygous mutations without the need for mating, and males can be used for genetic crosses. *C. elegans* are grown on nematode growth medium agar plates, covered with a special strain of bacteria that provides the necessary nutrition. At any time, the animals can be cryogenically preserved (41).

The lifespan of *D. melanogaster* is about 30 days. The developmental period varies with temperature. The shortest development time is achieved at 28°C and takes 7 days. At higher temperatures, development time increases due to heat stress.

Females lay about five eggs at a time, into rotting fruit or other suitable material. The eggs hatch after 12–15 h. The resulting larvae grow for about 4 days. During this time, they feed on the micro-organisms that decompose the fruit, as well as on the sugar of the fruit itself. Then, the larvae encapsulate in the puparium and undergo a 4-day-long metamorphosis, after which the adults emerge. Females become receptive to courting males at about 8–12 h after emergence (42,43).

1.4. Gene Homology

1.4.1. App

The human APP cDNA was first cloned in 1987 (44–46) and homologous genes are being identified ever since. cDNA sequences have been published for monkey, rat, mouse, frog, chicken, polar bear, pig, rabbit, guinea pig, sheep, and other species. The homology percentages are depicted in Table 1 for the most important species in which models for AD are established. An APP-related gene, Appl, has been identified in *D. melanogaster* (47). Furthermore, in *C. Elegans*, an APP-like sequence, apl-1, has been identified (48). Both these species lack, however, a recognizable Aβ sequence. The amino acid sequence of the 43-amino acid Aβ peptide sequence is identical in dogs, polar bears, monkeys, and humans (49). Additionally, amyloid precursor-like proteins (APLPs) have also been identified; i.e., APLP1 sequences from human and mouse, as well as human, mouse, and rat APLP2 sequences. There are three conserved domains between the invertebrate and mammalian APP homologues. These domains correspond to exons 2–5 (D1), 9–13 (D2), and 17–18 (D3; cytoplasmic domain) of human APP. Obvious differences between family members are noted: the recruitment of the Kunitz Protease Inhibitor-domain or exon 8 in both APP and APLP2 members,

Table 1
Gene homology in the APP gene

Gene		Identity (%)	
Species	**Symbol**	**Protein**	**DNA**
Homo sapiens	**APP**		
vs. Pan troglodytes	APP	94.5	95.4
vs. Canis lupus familiaris	APP	91.8	87.6
vs. Mus musculus	App	97.4	89.3
vs. Rattus norvegicus	App	92.2	85.9
vs. Danio rerio	appa	73.6	70.8
vs. Caenorhabditis elegans	apl-1	35.3	46.9

inclusion of exon 9 (OX-2 homology in human APP); and the variability both in sequence and length of the protein between each of the three conserved domains (50). Deletion of Appl in *D. melanogaster* results in relatively subtle behavioral changes, such as defects in fast phototaxis, but no morphological defects. These deficiencies can be partially complemented by the expression of human APP, suggesting significant functional conservation (51). Two homologues of APP, appa and appb, have been identified in zebrafish. Both genes have approximately 70% amino acid identity to human APP-695, with 80% identity in the $A\beta_{1-42}$ region and 95% identity within the transmembrane domain (52).

1.4.2. Presenilins

The proteins encoded by the presenilin (PS) genes are 467 (PS1) and 448 (PS2) amino acids long. There is a strong sequence homology between PS1 and PS2 and between homologues of different species, as shown in Tables 2 and 3. Presenilin homologues were found in a variety of organisms, not only in mammals or vertebrates, but even in species as distantly related to humans as nematodes and fruit flies. Three presenilin homologues have been identified in *C. elegans*: Sel-12 (51), HOP (homologue of presenilin)-1 (53,54), and SPE (spermatogenesis defective)-4, a more divergent member of the presenilin family (55). Presenilins are functionally conserved between humans and *C. elegans*. The egg-laying defective phenotype of Sel-12 *C. elegans* mutants can be rescued by the expression of human PS1 (56,57) or HOP-1 (54). In contrast to the other two homologues, which are expressed somatically during all developmental stages in nearly all tissues (53,56), the third PS homologue in *C. elegans*, SPE-4, is exclusively expressed during the larval stage in spermatogenesis (55,58) and possibly plays a role in preventing spermatid

Table 2
Gene homology in the PSEN1 gene

Gene		Identity (%)	
Species	**Symbol**	**Protein**	**DNA**
Homo sapiens	**PSEN1**		
vs. Pan troglodytes	PSEN1	100.0	99.9
vs. Canis lupus familiaris	PSEN1	94.0	91.9
vs. Mus musculus	Psen1	92.7	87.2
vs. Rattus norvegicus	Psen1	92.9	88.0
vs. Danio rerio	psen1	74.5	68.0
vs. Caenorhabditis elegans	sel-12	54.5	56.1

Table 3
Gene homology in the PSEN2 gene

Gene		Identity (%)	
Species	Symbol	Protein	DNA
Homo sapiens	**PSEN2**		
vs. Pan troglodytes	PSEN2	99.8	99.3
vs. Canis lupus familiaris	PSEN2	96.0	90.7
vs. Mus musculus	Psen2	95.5	89.3
vs. Rattus norvegicus	Psen2	95.1	89.4
vs. Danio rerio	psen2	74.0	69.5
vs. Drosophila melanogaster	Psn	67.7	66.7

activation (59). *D. melanogaster* has only one PS homologue, psn (55,58). Orthologues of PS1 and PS2 have been identified in zebrafish, namely psen1 (60) and psen 2 (61). The primary structures are highly conserved, as shown by sequence alignment, though there are highly variable regions at the N-terminus and the C-terminal half of the cytoplasmic loop domains. Transcripts from zebrafish psen1 are ubiquitously expressed from fertilization, implying zebrafish psen1 has an essential function throughout embryonic development (60). Zebrafish psen2 mRNA is present from fertilization, but protein expression has only been detected from the onset of gastrulation (61).

1.4.3. Apolipoprotein E

Apolipoprotein E (ApoE) is a multifunctional lipoprotein made by the liver and by brain astrocytes, which is found in the blood and in cerebrospinal fluid. It has an important role in the transport of cholesterol to steroid-producing cells.

The primary structures of ApoE from several species, including human (62), baboon (63), cynomolgus monkey (64), rat (65), mouse (66), guinea pig (67), rabbit (68), cow (69), dog (70), and sea lion (71), has been determined. They range in length from 279 to 310 amino acid residues for guinea pig and sea lion ApoE, respectively. Overall, there is a high degree of sequence conservation across mammalian species, as can be seen in Table 4, with notable exceptions at the amino and carboxyl termini (72).

In humans, ApoE has several genetic variants, or isoforms (apoE ε2, ε3, ε4), with the ε3 allele being the most abundant, about 77% in Caucasian populations. The other apoE alleles, ε4 and ε2, comprise about 15% and 8%, respectively (73). On the contrary, primates have only one ApoE isoform that most resembles ε4 (74), like most animals (72). This suggests that the ε4 allele is the common ancestral gene, despite the high

Table 4
Gene homology in the APOE gene

Gene		Identity (%)	
Species	Symbol	Protein	DNA
Homo sapiens	APOE		
vs. Pan troglodytes	APOE	97.2	98.4
vs. Canis lupus familiaris	APOE	71.2	82.9
vs. Bos taurus	APOE	72.2	81.8
vs. Mus musculus	Apoe	73.7	78.7
vs. Rattus norvegicus	Apoe	71.3	77.7
vs. Danio rerio	apoeb	30.2	47.4

abundance of the ε3 allele in humans (73). ApoE ε4 represents the most common risk factor for AD (75) and cardiovascular disease (76). It appears that worms and flies do not have an ApoE gene (77).

1.4.4. Tau

Microtubule-associated protein (MAP) tau, MAP1 and MAP2, are the three major microtubule-associated proteins of a normal mature neuron. These MAPs apparently perform similar functions, i.e. the promotion of assembly and stability of microtubules. In AD and related disorders, tau is abnormally hyperphosphorylated and accumulates as paired helical filaments (78).

D. melanogaster models of tauopathy have been reported. The tau homolog of *D. melanogaster* has been cloned and characterized (79). Deletion of the fly homolog of tau does not result in any detectable phenotype (80). *C. elegans* possesses a structural microtubule-associate protein, termed PTL-1 (81). A tau-like protein, XTP, has also been sequenced and biochemically characterized in one amphibian, Xenopus laevis (82). MAP has been identified in numerous mammalia, including Mus musculus (83). Despite the presence of an endogenous tau in most species, investigators have opted to overexpress human tau in order to establish tauopathy models.

2. Strain Issues

Given the abundance of transgenic mouse models of dementia, this chapter will focus on strain and gender differences in this species, though some of the information given will be generally applicable and can be translated to other species, like rats or other mammalia.

2.1. Effects of Genetic Background

2.1.1. Aβ Deposition

Studies have been undertaken to examine the effects of defined genetic background on APP metabolism, Aβ metabolism, and Aβ deposition in transgenic mouse models of AD. Lehman et al. transferred the transgene with Swedish mutation into three different genetic backgrounds by repeated backcrossing into the inbred mouse strains C57Bl/6J, DBA/2J, and 129S1/SvImJ for ten generations. They show that genetic background impacts several steps leading from APP processing to Aβ metabolism and Aβ deposition. First, the choice between α- and β-secretase cleavage in APP processing was significantly different. Second, levels of $A\beta_{40}$ and $A\beta_{42}$ in brain and plasma varied among the congenic lines, with highest levels in the C57Bl/6J background, despite equivalent levels of holo-APP. Analysis of B6 × D2 revealed that C57Bl/6J alleles were dominant over the DBA/2J alleles, resulting in high $A\beta_{40}$ an $A\beta_{42}$ levels. C57Bl/6J and B6 × D2 strains exhibit enhanced Aβ deposition. These data demonstrate that significant phenotypic alterations in Aβ metabolism and deposition are conferred by genetic background (84). To map the genetic loci responsible for the observed heritable differences in brain Aβ levels, a dense whole-genome scan in the B6 × D2 background strain was performed, which revealed genomic regions associated with brain Aβ levels. Several of these regions contain biologically interesting candidate genes and are syntenic to regions with indications of potential linkage to human AD risk (85).

Overexpression of APP produces dramatically different phenotypes in transgenic mice depending on the genetic background. Concentrations of APP sufficient for amyloid plaque deposition at 1 year are lethal on susceptible backgrounds, such as C57Bl/6J and FVB/N mice. These effects cannot be subscribed to a decrease in Aβ peptide concentrations in the brain in the more resistant strains (D2 × FVB, B6 × FVB and CAST × FVB), since these were similar, thus modulating genes are to be sought (86). The effect of genetic background was proven to be of influence in the TgCRND8 model too (87). The importance of genetic backgrounds in transgenic AD models was demonstrated as well, by the APP immunoreactivity, which segregated more closely with the B6/SJL genotype than with the presence of the transgene (88).

2.1.2. Tau Models

Mutations in the Disabled-1 (Dab1) gene lead to hyperphosphorylation of endogenous mouse tau. Dab1 mutants are viable in BALB/c but nonviable in 129Sv or in 129Sv × B6 (89). Genetic background genes thus modify the survival of mutants. Some chromosome regions have been identified to affect tau hyperphosphorylation in Dab1-/- mice. In these regions, homologs of several genes associated with human AD are found (89).

2.2. Cognitive and Behavioral Performance

The phenotype of a transgenic mouse is not only the result of the targeted gene, but it also reflects interactions with background genes. Thus, genetic background should be as carefully controlled as any other experimental variable. The simplest way to do this is to derive and maintain mutations in an isogenic genetic background. However, not all isogenic backgrounds are appropriate for a given study, since the behavioral characteristics of certain isogenic strains could overshadow the effects of the introduced gene mutations. Since natural strain differences exist for behavioral traits, the genetic background of the inbred mouse strains must be carefully considered in the interpretation of behavioral phenotypes of transgenic mice (90).

Learning and memory capacity are the traits of primary interest in dementia models. Luckily, the performance of inbred mouse strains on a wide variety of paradigms is well documented. It is important to discern between true learning differences and sensory impairments that lead to poor performance. Visual acuity is necessary for spatial learning tasks (91). Albino mouse strains demonstrate poor vision under bright lights and some mouse strains have the retinal degradation gene, which leads to blindness in adult mice. Auditory function is important for paradigms involving conditioning in the presence of a tone; some strains show deafness as a function of age (92–94). Analogously, the role of strain differences in motivation needs to be dissociated from true learning and memory differences.

The Morris water maze is frequently used to examine spatial learning in rats and mice (95). Animals are trained to locate a hidden platform in a circular pool filled with opaque water, using distal room cues. Latency to locate the platform and path length are recorded in each trial. The performance in a probe trial, in which the platform is removed, is examined. C57BL/6J and C57BL/10J are good learners in contrast to 129Sv/J, DBA/2, and BALB/c (96,97). The Barnes maze is a dry land maze in which the animals escape from a brightly lit, exposed circular open field into a darkened box that is hidden beneath one of the 18 holes around the perimeter of the open field. This test for spatial learning and memory is physically less demanding and probably less stressful than the Morris water maze. There was no evidence of strain differences between 129S6, C57BL/6J, and DBA/2J mice during acquisition in the Barnes maze. During the probe trial, C57BL/6J mice made more visits to the target hole than any of the nontarget holes, thus showing the best performance on this test. 129S6 mice performed the worse and DBS/2J mice showed an intermediate profile (98). Spatial memory has also been examined in eight-way radial arm maze tasks, in which animals must remember the arm of a maze in which they previously obtained a food reward. In this paradigm, C57BL/6J and DBA/2 prove to be better learners, while BALB/c and C3H/He perform significantly poorer (99,100). Brooks et al.

have also shown that the C3H/He strain failed to acquire a visual discrimination task, but this may be due to a visual deficit in these animals (101). Contextual fear conditioning is based on the association of a shock paired with a tone. Bouts of behavioral immobility, termed "freezing" are used as a measure of performance. Again C57BL/6J and C57Bl10/J prove to be good learners, but also 129/SvJ and BALB/cByJ are among the good learners in this paradigm. FVB/NJ and DBA/2 do not show greater freezing rates in the training context versus the altered context (96). Mouse strain differences were shown as well in long-term fear memory (102), operant learning (103), lever press escape/avoidance conditioning (104), object recognition (105), and social transmission of food preference (98). The best choice of an inbred background on which to explore the impact of a mutation on learning capacities appears to be C57BL/6. These mice are moderate learners, such that either impairment or improvement could be observed.

Motor activity underlies almost every mouse behavioral paradigm. Dysfunctions in physical movement can produce false-positive and false-negative effects on behaviors of interest for transgenic mice. Automated registration of open field activity can measure the total amount of movement, rate of movement, and type of spontaneous activity. The open field test is the simplest way to evaluate emotional behavior. Decreased ambulation and increased defecation in a brightly lit open field indicate heightened emotionality (106,107). Rearing behavior decreases in an anxiogenic environment, and thigmotaxis, i.e. the proportion of time the animal remains close to the walls of the open field, increases (108). In general, the C57 inbred strains of mice, including C57BL/6, C57BL/10, C57BR, and C57L, consistently show high levels of open field locomotion and low levels of anxiety-related measures in the open field. Intermediate strains include the DBA/2, CBA, AKR, and LP. Strains typically exhibiting low locomotor activity and high levels of emotional reactivity include DBA/1, BALB/c, and A/J. Albino strains are overrepresented at the high anxiety end of the distribution. This could be partly due to the added stress of high illumination in albino mice (108).

Tests of intermale aggression in the mouse include the "resident-intruder" paradigm where males are individually housed during several weeks. For the test, an opponent male from a strain, selected for the passivity under attack, is introduced in the cage. Numbers of attacks are counted. There are robust differences among inbred mouse strains in levels of aggressive behavior: the NZB/B1NJ and AKR mice are extremely aggressive, C57BL/10 and C57BL/6 mice exhibit markedly lower proportions of attacking males, and other strains show intermediate levels of aggression (109). Other articles provide information on strain distributions and male offence (110–113). Aggression in females is discussed in the section about gender issues.

Circadian rhythm disturbances and the "sundowning syndrome" are frequently observed among AD patients (114). Likewise, in an AD mouse model, altered circadian locomotor activity was observed (115). Underlying strain differences have to be taken into account when comparing AD mouse models or when selecting a strain in which circadian rhythm disturbances are to be studied. C57BL/6J mice demonstrate two distinct peaks of activity and deep body temperature, in contrast to a single early peak for C3H/HeJ mice (116). An important chronobiological regulator is melatonin (117). Melatonin was measured in one strain of outbred mice and four strains of inbred mice. In this study, all mouse strains were able to synthesize melatonin, but the basal levels as well as the diurnal variations were very different from one strain to another. CBA and C3H strains showed a clear-cut day–night rhythm of pineal melatonin concentration. In BALB/c, the presence of a very short melatonin peak (15 min) in the middle of the dark period was confirmed. In C57BL/6 and OF1 Swiss, a very small but significant peak was observed in the middle of the darkness. In the former, another small peak was also observed at light onset (118).

Sexual behavior can be altered in the context of dementia. Underlying strain differences were examined for male copulatory mounts, intromissions, and ejaculations. The stimulus female is usually artificially brought into estrus with hormone treatment. High copulatory behaviors were reported for C57BL and C57BL/6, lower copulatory behaviors were reported for DBA/2 and AKR, whereas the lowest copulatory behaviors were seen in BALB/c and A/J (90). Parental behavior can play a role in the selection of a background mouse strain for a transgenic model, since it has important implications for survival rates of the pups. CBA/H, C4H/Ico, C57BL/6, and CBA/J are better pup retrievers than BALB/c, NZB, DBA/2, A/J, and AKR (119).

2.3. Immunology

Differences in immunologic background of mouse strains can influence the outcome of vaccination studies with Aβ. Therefore, C57BL/6 mice were compared to the offspring of a cross between C57BL/6 female and DBA 2 male (B6 × D2 F1) in the ability to generate anti-Aβ antibodies after vaccination. The B6 × D2 F1 mice made anti-Aβ antibodies earlier and in greater quantity than C57BL/6 mice (120). Even in the context of long-term immunization, differences were found between mouse background strains in antibody levels. The antibody epitopes and IgG isotype remained fairly constant from strain to strain. Additionally, B6 × D2 F1 mice had a greater Th1 response compared to C57BL/6 mice (121). These results indicate it is important to take the genetic background into account, when performing vaccination studies in dementia models.

2.4. Vasculature

When establishing a mouse model of vascular dementia, vasculature and susceptibility to ischemia of mouse strains are important features. High-resolution magnetic resonance angiography revealed highly variable arterial cerebrovascular structures in mice from different strains and within the same strain. In particular, mice from the CD1 strain showed a highly variable vascular architecture. C57BL/6 mice presented small unilateral anastomoses between the posterior cerebral and the superior cerebellar arteries. Well-developed, either unilateral or bilateral, posterior communicating arteries were detected in CBA mice. In CD1 mice, the arterial structure ranged from no detectable anastomoses to well-developed unilateral posterior communicating cerebellar arteries. 129/Sv mice showed significantly shorter middle cerebral arteries compared to the other strains, and clear bilateral anastomoses between the posterior cerebral and the superior cerebellar arteries (122). These facts might explain the strain differences in susceptibility to injury after bilateral common carotid artery or middle cerebral artery occlusion, with the C57BL/6 strain in general as the most susceptible (123–126).

A collaborative database of inbred mouse strain characteristics can be found on http://www.jax.org/phenome (127).

3. Gender

3.1. Considerations

Female hormonal influences throughout life are never static, and the variability they cause is far greater than that measured in males. In general, females used in animal research are young, have a regular reproductive cycle, and have not been pregnant. This is in contrast to the female AD patients, who are predominantly elderly women who have often given birth to one or more children. In female animals, the estrus cycle can influence the behavioral variables. Therefore, they are often thought to be too "variable" and it is considered easier and cheaper to use only males. However, gender differences in drug sensitivity have been acknowledged, and hormonal profiles have been shown to affect activity and effects of drugs (128). Epidemiological studies indicate that women have a higher risk of AD (129) even after adjusting for age (130). Hence, it is important to also study the female mice. Since the phase of the estrus cycle is of influence, it should be determined for comparative purposes. When it is important to identify a female group with stable hormone levels, e.g. for sexual behavior, gonadectomy and hormone supplementation can be used to produce artificial stability. Transgenic AD mouse models often exhibit brain pathology and behavioral alterations at reproductive ages, whereas human AD is primarily found in postmenopausal women. These are important considerations when modeling and studying gender-based differences in mice.

3.2. Variability in Female Mice

The estrus cycle influenced locomotor activity patterns in C3H/He and C57BL/6 mice, although to a lesser extent than in rats. In the young, cycling mice, the second part of the pro-estrus night was often, but not consistently, characterized by increased motor activity compared to the remaining estrus cycle nights. In the course of aging, after estrus cycling has ceased, the estrus-dependent day-to-day variability in activity was reduced (131). Reproductive stage, as well as aging, influences motor activity patterns of female mice, which can bias other behavioral measurements, such as the open field test. Palanza et al. examined mice housed individually or with siblings in a free-exploratory paradigm of anxiety (where test animals have a choice to stay in their home cage or to explore an open field). Individually housed females did not leave their home cage for long periods, explored less the unfamiliar area, and displayed higher risk assessment. This profile is suggestive of lower propensity for exploration and higher level of anxiety compared with group-housed females. Individually housed males tended to show an opposite profile. Pro-estric mice were less sensitive to the decrease of exploratory propensity induced by individually housing compared to estric and di-estric mice (132).

3.3. Behavioral and Cognitive Differences Between Genders

Behavioral phenotypes can differ between genders, depending on the strain. Behavior of male and female mice of the 129S2/SvHsd × C57BL/6JOlaHsd F1 hybrids was strikingly different in the elevated plus maze, a test of anxiety. The males displayed a more anxious behavioral profile (133). The measure of latency to startle shows sex differences in all strains tested by Tarantino et al., and at almost all decibel levels, with females showing greater latency to startle than males. DBA/2J mice showed sex differences on prepulse inhibition and startle response, whereas other strains show equal performance between sexes on these behaviors (134).

Aggression in females depends on the reproductive state, neither pregnant nor lactating, pregnant but not lactating, or lactating but not pregnant. Strain differences have been described. For example, nonpregnant and nonlactating DBA/2 and C57BL/6 females do not differ in offence against an intruder male (135), whereas C57BL/6 females are more aggressive than DBA/2 females against a lactating intruder female (90). Differences depend on life history, test situation, and type of opponent.

Sex differences in learning and memory in rodents are reviewed in a meta-analysis by Jonasson et al. (136). In the Morris water maze, male rat advantages in spatial learning were discovered, but these effects diminished with pretraining regimens. In radial mazes, male rat advantages increased when unbaited arms were included in the protocol. Mouse studies exhibited a different pattern of sex effects; small female advantages were evident in the water maze, but small male advantages were evident in the radial maze. Thus, this review shows an important species dichotomy between rats and mice.

Female and male young, middle aged, and old wild-type C57BL/6J mice were tested on memory tasks. Old males and females performed worse than young or middle-aged mice in novel location, but not novel object recognition tasks. Old mice, of both sexes, also showed impaired spatial water maze performance during training compared with young or middle-aged mice; however, only old females failed to show robust spatial bias during probe trials. While there was no age-difference in passive avoidance performance for males, females showed an age-related decline. There was no difference in cognitive performance between young and middle-aged mice of either sex on any task (137). Male C57BL/6 mice (138) and male CD1 mice (139) learn both the working and reference memory components of a water-escape motivated radial arm maze task better than females. Males were shown to outperform females on a task of simple odour discrimination learning with three repeated reversals (140).

3.4. Pathological Differences

In Tg2576 mice, the $A\beta_{1-40}$ level has been reported to be higher in females than males, but levels of $A\beta_{1-42}$ were not different at 15 months of age (141). Gender-dependent elevated plaque formation has been reported in APP23 (142) and APP/Tau double transgenic mice (143). Female APP/PS1 mice accumulate amyloid at an earlier age and they build up more amyloid deposits in the hippocampus than age-matched male mice (144). In triple transgenic AD mice, the AD-related Aβ and tau pathology has been compared between genders. Female 3xTg-AD transgenic mice have significantly more aggressive Aβ pathology. An increase in β-secretase activity and a reduction of neprilysin, an Aβ-degrading enzyme, in female mice compared to males was found. Gender did not affect levels of phosphorylated tau in 3xTg-AD mice (145). Loss of ovarian steroids at menopause may increase the susceptibility of the aging brain to AD neurodegeneration (146), and estrogen is central to the current hypothesis on the gender differences in the risk of AD. Ovariectomy increases the amount of Aβ, while estrogen replacement therapy reduces Aβ in Tg2576, PS/APP, and 3xTg-AD mice (147,148), but not in PDAPP mice (149).

4. Conclusions

Some species spontaneously exhibit pathological hallmarks of AD, like primates and dogs. Nevertheless, their use is often limited by economical or practical reasons. Other species, like rodents, develop spontaneously only a small fraction of the spectrum of symptoms of dementia. Therefore, the vast majority of animal research is conducted in induced models, with transgenic mouse models currently being the most abundant. When establishing a transgenic model, genetic background should be carefully considered, since it has great influence on the outcome of behavioral as well as molecular experiments. Inbred strains are preferred for their genetic homogeneity, but sensory and motor impairments should be avoided. In addition, one should be aware of the impact of gender. In general, one sex is chosen, and males are preferably used for behavioral studies, but comparing gender differences can be valuable to unravel the impact of hormones on pathology or predict different treatment efficacy between males and females.

Acknowledgments

This work was financed by the Fund for Scientific Research – Flanders (FWO, G.0164.09), Agreement between the University of Antwerp and the Institute Born-Bunge, Interuniversity Poles of Attraction (IUAP Network P6/43), Methusalem excellence grant of the Flemish Government, Neurosearch Antwerp, the Antwerp Medical Research Foundation, and the Thomas Riellaerts Research fund. DVD is a postdoctoral fellow of the FWO.

References

1. Braunmühl A (1956) Kongophile angiopathie und Senile Plaques bei greisen hunden. Arch Psychiatr Nervenkr 194:395–414
2. Cummings BJ, Su JH, Cotman CW, White R, Russell MJ (1993) Beta-amyloid accumulation in aged canine brain: a model of early plaque formation in Alzheimer's disease. Neurobiol Aging 14(6):547–560
3. Cummings BJ, Head E, Afagh AJ, Milgram NW, Cotman CW (1996) Beta-amyloid accumulation correlates with cognitive dysfunction in the aged canine. Neurobiol Learn Mem 66(1):11–23
4. Colle M-A, Hauw J-J, Crespeau Fet al(2000) Vascular and parenchymal Abeta deposition in the aging dog: Correlation with behavior. Neurobiol Aging 21(5):695–704
5. Pugliese M, Geloso MC, Carrasco JL, Mascort J, Michetti F, Mahy N (2006) Canine cognitive deficit correlates with diffuse plaque maturation and S100beta (-) astrocytosis but not with insulin cerebrospinal fluid level. Acta Neuropathol 111(6):519–528
6. Rofina JE, van Ederen AM, Toussaint MJ, et al. (2006) Cognitive disturbances in old dogs suffering from the canine counterpart of Alzheimer's disease. Brain Res 1069(1):216–226
7. Gunn-Moore DA, McVee J, Bradshaw JM, Pearson GR, Head E, Gunn-Moore FJ (2006) Ageing changes in cat brains demonstrated by beta-amyloid and AT8-immunoreactive phosphorylated tau deposits. J Feline Med Surg 8(4):234–242
8. Lane MA (2000) Nonhuman primate models in biogerontology. Exp Gerontol 35(5):533–541
9. Kimura N, Tanemura K, Nakamura S, et al. (2003) Age-related changes of Alzheimer's disease-associated proteins in cynomolgus monkey brains. Biochem Biophys Res Commun 310(2):303–311
10. Sloane JA, Pietropaolo MF, Rosene DL, et al. (1997) Lack of correlation between plaque burden and cognition in the aged monkey. Acta Neuropathol 94(5):471–478
11. Gearing M, Rebeck GW, Hyman BT, Tigges J, Mirra SS (1994) Neuropathology and apolipoprotein E profile of aged chimpanzees: Implications for Alzheimer disease. Proc Natl Acad Sci U S A 91(20):9382–9386
12. Geula C, Nagykery N, Wu CK (2002) Amyloid-beta deposits in the cerebral cortex of the aged common marmoset (Callithrix jacchus): Incidence and chemical composition. Acta Neuropathol 103(1):48–58

13. Lemere CA, Beierschmitt A, Iglesias M, et al. (2004) Alzheimer's disease abeta vaccine reduces central nervous system abeta levels in a non-human primate, the Caribbean vervet. Am J Pathol 165(1):283–297
14. Bons N, Mestre N, Ritchie K, Petter A, Podlisny M, Selkoe D (1994) Identification of amyloid beta protein in the brain of the small, short-lived lemurian primate Microcebus murinus. Neurobiol Aging 15(2):215–220
15. Gearing M, Tigges J, Mori H, Mirra SS (1997) beta-Amyloid (A beta) deposition in the brains of aged orangutans. Neurobiol Aging 18(2):139–146
16. Lemere CA, Oh J, Stanish HA, et al. (2008) Cerebral amyloid-beta protein accumulation with aging in cotton-top tamarins: A model of early Alzheimer's disease? Rejuvenation Res 11(2):321–332
17. Rosen RF, Farberg AS, Gearing M, et al. (2008) Tauopathy with paired helical filaments in an aged chimpanzee. J Comp Neurol 509(3): 259–270
18. Bons N, Rieger F, Prudhomme D, Fisher A, Krause KH (2006) Microcebus murinus: A useful primate model for human cerebral aging and Alzheimer's disease? Genes Brain Behav 5(2):120–130
19. Cork LC, Powers RE, Selkoe DJ, Davies P, Geyer JJ, Price DL (1988) Neurofibrillary tangles and senile plaques in aged bears. J Neuropathol Exp Neurol 47(6):629–641
20. Tekirian TL, Cole GM, Russell MJ, et al. (1996) Carboxy terminal of beta-amyloid deposits in aged human, canine, and polar bear brains. Neurobiol Aging 17(2):249–257
21. Roertgen KE, Parisi JE, Clark HB, Barnes DL, O'Brien TD, Johnson KH (1996) A beta-associated cerebral angiopathy and senile plaques with neurofibrillary tangles and cerebral hemorrhage in an aged wolverine (Gulo gulo). Neurobiol Aging 17(2):243–247
22. Sherman KA, Friedman E (1990) Pre- and post-synaptic cholinergic dysfunction in aged rodent brain regions: New findings and an interpretative review. Int J Dev Neurosci 8(6): 689–708
23. Takeda T, Hosokawa M, Higuchi K (1997) Senescence-accelerated mouse (SAM): A novel murine model of senescence. Exp Gerontol 32(1–2):105–109
24. Pallas M, Camins A, Smith MA, Perry G, Lee HG, Casadesus G (2008) From aging to Alzheimer's disease: Unveiling "the switch" with the senescence-accelerated mouse model (SAMP8). J Alzheimers Dis 15(4):615–624
25. Wu Y, Luo Y (2005) Transgenic C. elegans as a model in Alzheimer's research. Curr Alzheimer Res 2(1):37–45
26. The C. elegans Sequencing Consortium (1998) Genome sequence of the nematode C. elegans: a platform for investigating biology. Science 282(5396):2012–2018
27. Adams MD, Celniker SE, Holt RA, et al(2000) The genome sequence of Drosophila melanogaster. Science 287(5461):2185–2195
28. Sang TK, Jackson GR (2005) Drosophila models of neurodegenerative disease. NeuroRx 2(3):438–446
29. Agca C, Fritz JJ, Walker LC, et al.(2008) Development of transgenic rats producing human beta-amyloid precursor protein as a model for Alzheimer's disease: Transgene and endogenous APP genes are regulated tissue-specifically. BMC Neurosci 9:28
30. Filipiak WE, Saunders TL (2006) Advances in transgenic rat production. Transgenic Res 15(6):673–686
31. Flood DG, Lin YG, Lang DM, et al. (2007) A transgenic rat model of Alzheimer's disease with extracellular Abeta deposition. Neurobiol Aging 30:1078–1090
32. Ruiz-Opazo N, Kosik KS, Lopez LV, Bagamasbad P, Ponce LR, Herrera VL (2004) Attenuated hippocampus-dependent learning and memory decline in transgenic TgAPPswe Fischer-344 rats. Mol Med 10(1–6):36–44
33. Herrera VL, Ponce LR, Ruiz-Opazo N (2007) Genome-wide scan for interacting loci affecting human cholesteryl ester transfer protein-induced hypercholesterolemia in transgenic human cholesteryl ester transfer protein F2-intercross rats. J Hypertens 25(8):1608–1612
34. Hendrickx AG, Dukelow WR (1995) Reproductive biology. In: Bennett BT, Christian RA, Roy H (eds) Nonhuman primates in biomedical research. Academic, San Diego, CA, pp 147–191
35. Havenaar R, Ritskes-Hoitinga J, Meijer J, Zwart P (1991) Biologie en zootechniek. In: Van Zutphen L, Baumans V, Beyen A (eds) Proefdieren en dierproeven. Wetenschappelijke Uitgeverij Bunge, Utrecht, The Netherlands, pp 20–74
36. Whitten WK, Bronson FH, Greenstein JA (1968) Estrus-inducing pheromone of male mice: Transport by movement of air. Science 161(841):584–585
37. Bruce HM (1959) An exteroceptive block to pregnancy in the mouse. Nature 184:105
38. Crawley J (2000) What's wrong with my mouse? Behavioral phenotyping of transgenic

and knockout mice, 1st edn. Wiley-Liss, Wilmington, DE

39. Lohmiller JJ, Swing SP (2006) Reproduction and breeding. In: Mark AS, Steven HW, Craig LF (eds) The laboratory rat, 2nd edn. Academic, Burlington, MA, pp 147–164
40. Hau J, Van Hoosier GL (2003) Handbook of laboratory animal science, essential principles and practices, 2nd edn. CRC Press, Boca Raton, FL
41. Corsi AK (2006) A biochemist's guide to Caenorhabditis elegans. Anal Biochem 359(1):1–17
42. Ashburner M, Thompson JJ (1978) The laboratory culture of Drosophila. In: Ashburner M, Wright T (eds) The genetics and biology of Drosophila. Academic, London, pp 1–81
43. Ashburner M, Golic K, Hawley R (2005). Drosophila: A laboratory Hhndbook, 2nd edn. Cold Spring Harbor Laboratory Press, Cold Spring Harbor, NY
44. Goldgaber D, Lerman MI, McBride OW, Saffiotti U, Gajdusek DC (1987) Characterization and chromosomal localization of a cDNA encoding brain amyloid of Alzheimer's disease. Science 235(4791):877–880
45. Kang J, Lemaire HG, Unterbeck A, et al. (1987) The precursor of Alzheimer's disease amyloid A4 protein resembles a cell-surface receptor. Nature 325(6106):733–736
46. Robakis NK, Ramakrishna N, Wolfe G, Wisniewski HM (1987) Molecular cloning and characterization of a cDNA encoding the cerebrovascular and the neuritic plaque amyloid peptides. Proc Natl Acad Sci U S A 84(12):4190–4194
47. Rosen DR, Martin-Morris L, Luo LQ, White K (1989) A Drosophila gene encoding a protein resembling the human beta-amyloid protein precursor. Proc Natl Acad Sci U S A 86(7): 2478–2482
48. Daigle I, Li C (1993) apl-1, a Caenorhabditis elegans gene encoding a protein related to the human beta-amyloid protein precursor. Proc Natl Acad Sci U S A 90(24):12045–12049
49. Johnstone EM, Chaney MO, Norris FH, Pascual R, Little SP (1991) Conservation of the sequence of the Alzheimer's disease amyloid peptide in dog, polar bear and five other mammals by cross-species polymerase chain reaction analysis. Brain Res Mol Brain Res 10(4):299–305
50. Coulson EJ, Paliga K, Beyreuther K, Masters CL (2000) What the evolution of the amyloid protein precursor supergene family tells us about its function. Neurochem Int 36(3):175–184
51. Luo L, Tully T, White K (1992) Human amyloid precursor protein ameliorates behavioral deficit of flies deleted for Appl gene. Neuron 9(4): 595–605
52. Musa A, Lehrach H, Russo VA (2001) Distinct expression patterns of two zebrafish homologues of the human APP gene during embryonic development. Dev Genes Evol 211(11): 563–567
53. Levitan D, Greenwald I (1995) Facilitation of lin-12-mediated signalling by sel-12, a Caenorhabditis elegans S182 Alzheimer's disease gene. Nature 377(6547):351–354
54. Li X, Greenwald I (1997) HOP-1, a Caenorhabditis elegans presenilin, appears to be functionally redundant with SEL-12 presenilin and to facilitate LIN-12 and GLP-1 signaling. Proc Natl Acad Sci U S A 94(22):12204–12209
55. Arduengo PM, Appleberry OK, Chuang P, L'Hernault SW (1998) The presenilin protein family member SPE-4 localizes to an ER/Golgi derived organelle and is required for proper cytoplasmic partitioning during Caenorhabditis elegans spermatogenesis. J Cell Sci 111 (Pt 24):3645–3654
56. Baumeister R, Leimer U, Zweckbronner I, Jakubek C, Grunberg J, Haass C (1997) Human presenilin-1, but not familial Alzheimer's disease (FAD) mutants, facilitate Caenorhabditis elegans Notch signalling independently of proteolytic processing. Genes Funct 1(2):149–159
57. Levitan D, Doyle TG, Brousseau D, et al.(1996) Assessment of normal and mutant human presenilin function in Caenorhabditis elegans. Proc Natl Acad Sci U S A 93(25): 14940–14944
58. L'Hernault SW, Arduengo PM (1992) Mutation of a putative sperm membrane protein in Caenorhabditis elegans prevents sperm differentiation but not its associated meiotic divisions. J Cell Biol 119(1):55–68
59. Gosney R, Liau WS, Lamunyon CW (2008) A novel function for the presenilin family member spe-4: Inhibition of spermatid activation in Caenorhabditis elegans. BMC Dev Biol8:44
60. Leimer U, Lun K, Romig H, et al. (1999) Zebrafish (Danio rerio) presenilin promotes aberrant amyloid beta-peptide production and requires a critical aspartate residue for its function in amyloidogenesis. Biochemistry 38(41):13602–13609
61. Groth C, Nornes S, McCarty R, Tamme R, Lardelli M (2002) Identification of a second presenilin gene in zebrafish with similarity to the human Alzheimer's disease gene presenilin2. Dev Genes Evol 212(10):486–490

62. Rall SC, Jr., Weisgraber KH, Mahley RW (1982) Human apolipoprotein E. The complete amino acid sequence. J Biol Chem 257(8):4171–4178
63. Hixson JE, Cox LA, Borenstein S (1988) The baboon apolipoprotein E gene: Structure, expression, and linkage with the gene for apolipoprotein C-1. Genomics 2(4):315–323
64. Marotti KR, Whitted BE, Castle CK, Polites HG, Melchior GW (1989) Nucleotide sequence of the cynomolgus monkey apolipoprotein E cDNA. Nucleic Acids Res 17(4):1778
65. McLean JW, Fukazawa C, Taylor JM (1983) Rat apolipoprotein E mRNA. Cloning and sequencing of double-stranded cDNA. J Biol Chem 258(14):8993–9000
66. Rajavashisth TB, Kaptein JS, Reue KL, Lusis AJ (1985) Evolution of apolipoprotein E: Mouse sequence and evidence for an 11-nucleotide ancestral unit. Proc Natl Acad Sci U S A 82(23):8085–8089
67. Matsushima T, Getz GS, Meredith SC (1990) Primary structure of guinea pig apolipoprotein E. Nucleic Acids Res 18(1):202
68. Lee BR, Miller JM, Yang CY, et al. (1991) Amino acid sequence of rabbit apolipoprotein E. J Lipid Res 32(1):165–171
69. Yang YW, Chan L, Li WH (1991) Cloning and sequencing of bovine apolipoprotein E complementary DNA and molecular evolution of apolipoproteins E, C-I, and C-II. J Mol Evol 32(6):469–475
70. Luo CC, Li WH, Chan L (1989) Structure and expression of dog apolipoprotein A-I, E, and C-I mRNAs: Implications for the evolution and functional constraints of apolipoprotein structure. J Lipid Res 30(11): 1735–1746
71. Davis RW, Pierotti VR, Lauer SJ, et al. (1991) Lipoproteins in pinnipeds: Analysis of a high molecular weight form of apolipoprotein E. J Lipid Res 32(6):1013–1023
72. Weisgraber KH (1994) Apolipoprotein E: Structure-function relationships. Adv Protein Chem 45:249–302
73. Mahley RW, Rall SC, Jr (1999) Is epsilon4 the ancestral human apoE allele? Neurobiol Aging 20(4):429–430
74. Finch CE, Sapolsky RM (1999) The evolution of Alzheimer disease, the reproductive schedule, and apoE isoforms. Neurobiol Aging 20(4):407–428
75. Raber J, Huang Y, Ashford JW (2004) ApoE genotype accounts for the vast majority of AD risk and AD pathology. Neurobiol Aging 25(5):641–650
76. Eichner JE, Dunn ST, Perveen G, Thompson DM, Stewart KE, Stroehla BC (2002) Apolipoprotein E polymorphism and cardiovascular disease: A HuGE review. Am J Epidemiol 155(6):487–495
77. Link CD (2005). Invertebrate models of Alzheimer's disease. Genes Brain Behav 4(3): 147–156
78. Iqbal K, Alonso AC, Chen S, et al. (2005) Tau pathology in Alzheimer disease and other tauopathies. Biochim Biophys Acta 1739(2–3): 198–210
79. Heidary G, Fortini ME (2001) Identification and characterization of the Drosophila tau homolog. Mech Dev 108(1–2):171–178
80. Doerflinger H, Benton R, Shulman JM, St JD (2003) The role of PAR-1 in regulating the polarised microtubule cytoskeleton in the Drosophila follicular epithelium. Development 130(17):3965–3975
81. Goedert M, Baur CP, Ahringer J, et al. (1996) PTL-1, a microtubule-associated protein with tau-like repeats from the nematode Caenorhabditis elegans. J Cell Sci 109 (Pt 11): 2661–2672
82. Olesen OF, Kawabata-Fukui H, Yoshizato K, Noro N (2002) Molecular cloning of XTP, a tau-like microtubule-associated protein from Xenopus laevis tadpoles. Gene 283(1–2): 299–309
83. Lee G, Cowan N, Kirschner M (1988) The primary structure and heterogeneity of tau protein from mouse brain. Science 239(4837): 285–288
84. Lehman EJ, Kulnane LS, Gao Y, et al. (2003) Genetic background regulates beta-amyloid precursor protein processing and beta-amyloid deposition in the mouse. Hum Mol Genet 12(22):2949–2956
85. Ryman D, Gao Y, Lamb BT (2008) Genetic loci modulating amyloid-beta levels in a mouse model of Alzheimer's disease. Neurobiol Aging 29(8):1190–1198
86. Carlson GA, Borchelt DR, Dake A, et al. (1997) Genetic modification of the phenotypes produced by amyloid precursor protein overexpression in transgenic mice. Hum Mol Genet 6(11):1951–1959
87. Chishti MA, Yang DS, Janus C, et al. (2001) Early-onset amyloid deposition and cognitive deficits in transgenic mice expressing a double mutant form of amyloid precursor protein 695. J Biol Chem 276(24):21562–21570
88. Ali SM, Siedlak SL, Gonzalez-DeWhitt PA, et al. (1996) Artifactual strain-specific signs of incipient brain amyloidosis in APP transgenic mice. Neurobiol Aging 17(2):223–234

89. Brich J, Shie FS, Howell BW, et al. (2003) Genetic modulation of tau phosphorylation in the mouse. J Neurosci 23(1):187–192
90. Crawley JN, Belknap JK, Collins A, et al. (1997) Behavioral phenotypes of inbred mouse strains: Implications and recommendations for molecular studies. Psychopharmacology (Berl) 132(2):107–124
91. Brown RE, Wong AA (2007) The influence of visual ability on learning and memory performance in 13 strains of mice. Learn Mem 14(3):134–144
92. Johnson KR, Zheng QY, Noben-Trauth K (2006) Strain background effects and genetic modifiers of hearing in mice. Brain Res 1091(1):79–88
93. Johnson KR, Zheng QY, Erway LC (2000) A major gene affecting age-related hearing loss is common to at least ten inbred strains of mice. Genomics 70(2):171–180
94. Zheng QY, Johnson KR (2001) Hearing loss associated with the modifier of deaf waddler (mdfw) locus corresponds with age-related hearing loss in 12 inbred strains of mice. Hear Res 154(1–2):45–53
95. D'Hooge R, De Deyn PP (2001) Applications of the Morris water maze in the study of learning and memory. Brain Res Brain Res Rev 36(1):60–90
96. Owen EH, Logue SF, Rasmussen DL, Wehner JM (1997) Assessment of learning by the Morris water task and fear conditioning in inbred mouse strains and F1 hybrids: Implications of genetic background for single gene mutations and quantitative trait loci analyses. Neuroscience 80(4):1087–1099
97. Upchurch M, Wehner JM (1989) Inheritance of spatial learning ability in inbred mice: A classical genetic analysis. Behav Neurosci 103(6):1251–1258
98. Holmes A, Wrenn CC, Harris AP, Thayer KE, Crawley JN (2002) Behavioral profiles of inbred strains on novel olfactory, spatial and emotional tests for reference memory in mice. Genes Brain Behav 1(1):55–69
99. Ammassari-Teule M, Hoffmann HJ, Rossi-Arnaud C (1993) Learning in inbred mice: Strain-specific abilities across three radial maze problems. Behav Genet 23(4):405–412
100. Chapillon P, Roullet P, Lassalle JM (1995) Ontogeny of orientation and spatial learning on the radial maze in mice. Dev Psychobiol 28(8):429–442
101. Brooks SP, Pask T, Jones L, Dunnett SB (2005) Behavioural profiles of inbred mouse strains used as transgenic backgrounds. II: Cognitive tests. Genes Brain Behav 4(5):307–317
102. Balogh SA, Wehner JM (2003) Inbred mouse strain differences in the establishment of long-term fear memory. Behav Brain Res 140(1–2):97–106
103. Baron SP, Meltzer LT (2001) Mouse strains differ under a simple schedule of operant learning. Behav Brain Res 118(2):143–152
104. Brennan FX (2004) Genetic differences in leverpress escape/avoidance conditioning in seven mouse strains. Genes Brain Behav 3(2):110–114
105. Sik A, van NP, Prickaerts J, Blokland A (2003) Performance of different mouse strains in an object recognition task. Behav Brain Res 147(1–2):49–54
106. Hall CS (1934) Emotional behavior in the rat. I. Defecation and urination as measures of individual differences in emotionality. J Comp Psychol 18:385–403
107. Hall CS (1936) Emotional behavior in the rat. III. The relationship between emotionality and ambulatory activity. J Comp Psychol 22:345–352
108. Walsh RN, Cummins RA (1976) The open-field test: A critical review. Psychol Bull 83(3):482–504
109. Guillot PV, Chapouthier G (1996) Intermale aggression and dark/light preference in ten inbred mouse strains. Behav Brain Res 77(1–2):211–213
110. Jones SE, Brain PF (1987) Performances of inbred and outbred laboratory mice in putative tests of aggression. Behav Genet 17(1):87–96
111. Kulikov AV, Osipova DV, Naumenko VS, Popova NK (2005) Association between Tph2 gene polymorphism, brain tryptophan hydroxylase activity and aggressiveness in mouse strains. Genes Brain Behav 4(8):482–485
112. Guillot PV, Roubertoux PL, Crusio WE (1994) Hippocampal mossy fiber distributions and intermale aggression in seven inbred mouse strains. Brain Res 660(1):167–169
113. Miczek KA, Maxson SC, Fish EW, Faccidomo S (2001) Aggressive behavioral phenotypes in mice. Behav Brain Res 125(1–2):167–181
114. Vitiello MV, Bliwise DL, Prinz PN (1992) Sleep in Alzheimer's disease and the sundown syndrome. Neurology 42(7 Suppl 6):83–93
115. Vloeberghs E, Van Dam D, Engelborghs S, Nagels G, Staufenbiel M, De Deyn PP (2004) Altered circadian locomotor activity in APP23 mice: A model for BPSD disturbances. Eur J Neurosci 20(10):2757–2766

116. Tankersley CG, Irizarry R, Flanders S, Rabold R (2002) Circadian rhythm variation in activity, body temperature, and heart rate between C3H/HeJ and C57BL/6J inbred strains. J Appl Physiol 92(2):870–877
117. Lewy AJ (2007) Melatonin and human chronobiology. Cold Spring Harb Symp Quant Biol 72:623–636
118. Vivien-Roels B, Malan A, Rettori MC, Delagrange P, Jeanniot JP, Pevet P (1998) Daily variations in pineal melatonin concentrations in inbred and outbred mice. J Biol Rhythms 13(5):403–409
119. Carlier M, Roubertoux P, Cohen-Salmon C (1982) Differences in patterns of pup care in Mus musculus domesticus l-Comparisons between eleven inbred strains. Behav Neural Biol 35(2):205–210
120. Spooner ET, Desai RV, Mori C, Leverone JF, Lemere CA (2002) The generation and characterization of potentially therapeutic Abeta antibodies in mice: differences according to strain and immunization protocol. Vaccine 21(3–4):290–297
121. Seabrook TJ, Iglesias M, Bloom JK, Spooner ET, Lemere CA (2004) Differences in the immune response to long term Abeta vaccination in C57BL/6 and B6D2F1 mice. Vaccine 22(29–30):4075–4083
122. Beckmann N (2000) High resolution magnetic resonance angiography non-invasively reveals mouse strain differences in the cerebrovascular anatomy in vivo. Magn Reson Med 44(2):252–258
123. Yang G, Kitagawa K, Matsushita K, et al. (1997) C57BL/6 strain is most susceptible to cerebral ischemia following bilateral common carotid occlusion among seven mouse strains: Selective neuronal death in the murine transient forebrain ischemia. Brain Res 752(1–2): 209–218
124. Majid A, He YY, Gidday JM, et al. (2000) Differences in vulnerability to permanent focal cerebral ischemia among 3 common mouse strains. Stroke 31(11):2707–2714
125. Wellons JC, III, Sheng H, Laskowitz DT, Burkhard MG, Pearlstein RD, Warner DS (2000) A comparison of strain-related susceptibility in two murine recovery models of global cerebral ischemia. Brain Res 868(1): 14–21
126. Barone FC, Knudsen DJ, Nelson AH, Feuerstein GZ, Willette RN (1993) Mouse strain differences in susceptibility to cerebral ischemia are related to cerebral vascular anatomy. J Cereb Blood Flow Metab 13(4): 683–692
127. Grubb SC, Churchill GA, Bogue MA (2004) A collaborative database of inbred mouse strain characteristics. Bioinformatics 20(16): 2857–2859
128. Godfroid IO (1999) Sex differences relating to psychiatric treatment. Can J Psychiatry 44(4):362–367
129. Brookmeyer R, Gray S, Kawas C (1998) Projections of Alzheimer's disease in the United States and the public health impact of delaying disease onset. Am J Public Health 88(9):1337–1342
130. Hy LX, Keller DM (2000) Prevalence of AD among whites: A summary by levels of severity. Neurology 55(2):198–204
131. Kopp C, Ressel V, Wigger E, Tobler I (2006) Influence of estrus cycle and ageing on activity patterns in two inbred mouse strains. Behav Brain Res 167(1):165–174
132. Palanza P, Gioiosa L, Parmigiani S (2001) Social stress in mice: Gender differences and effects of estrous cycle and social dominance. Physiol Behav 73(3):411–420
133. Voikar V, Koks S, Vasar E, Rauvala H (2001) Strain and gender differences in the behavior of mouse lines commonly used in transgenic studies. Physiol Behav 72(1–2):271–281
134. Tarantino LM, Gould TJ, Druhan JP, Bucan M (2000) Behavior and mutagenesis screens: The importance of baseline analysis of inbred strains. Mamm Genome 11(7):555–564
135. Ogawa S, Makino J (1984) Aggressive behavior in inbred strains of mice during pregnancy. Behav Neural Biol 40(2):195–204
136. Jonasson Z (2005) Meta-analysis of sex differences in rodent models of learning and memory: a review of behavioral and biological data. Neurosci Biobehav Rev 28(8): 811–825
137. Benice TS, Rizk A, Kohama S, Pfankuch T, Raber J (2006) Sex-differences in age-related cognitive decline in C57BL/6J mice associated with increased brain microtubule-associated protein 2 and synaptophysin immunoreactivity. Neuroscience 137(2):413–423
138. Gresack JE, Frick KM (2003) Male mice exhibit better spatial working and reference memory than females in a water-escape radial arm maze task. Brain Res 982(1):98–107
139. LaBuda CJ, Mellgren RL, Hale RL (2002) Sex differences in the acquisition of a radial maze task in the CD-1 mouse. Physiol Behav 76(2):213–217
140. Mihalick SM, Langlois JC, Krienke JD (2000) Strain and sex differences on olfactory discrimination learning in C57BL/6J

and DBA/2J inbred mice (Mus musculus). J Comp Psychol 114(4):365–370

141. Callahan MJ, Lipinski WJ, Bian F, Durham RA, Pack A, Walker LC (2001) Augmented senile plaque load in aged female beta-amyloid precursor protein-transgenic mice. Am J Pathol 158(3):1173–1177

142. Sturchler-Pierrat C, Staufenbiel M (2000) Pathogenic mechanisms of Alzheimer's disease analyzed in the APP23 transgenic mouse model. Ann N Y Acad Sci 920:134–139

143. Lewis J, Dickson DW, Lin WL et al. (2001) Enhanced neurofibrillary degeneration in transgenic mice expressing mutant tau and APP. Science 293(5534):1487–1491

144. Wang J, Tanila H, Puolivali J, Kadish I, van GT (2003) Gender differences in the amount and deposition of amyloidbeta in APPswe and PS1 double transgenic mice. Neurobiol Dis 14(3):318–327

145. Hirata-Fukae C, Li HF, Hoe HS et al. (2008) Females exhibit more extensive amyloid, but not tau, pathology in an Alzheimer transgenic model. Brain Res 1216:92–103

146. Gandy S, Duff K(2000) Post-menopausal estrogen deprivation and Alzheimer's disease. Exp Gerontol 35(4):503–511

147. Zheng H, Xu H, Uljon SN et al. (2002) Modulation of A(beta) peptides by estrogen in mouse models. J Neurochem 80(1): 191–196

148. Carroll JC, Rosario ER, Chang L et al. (2007) Progesterone and estrogen regulate Alzheimer-like neuropathology in female 3xTg-AD mice. J Neurosci 27(48): 13357–13365

149. Green PS, Bales K, Paul S, Bu G (2005) Estrogen therapy fails to alter amyloid deposition in the PDAPP model of Alzheimer's disease. Endocrinology 146(6): 2774–2781

Chapter 5

Transgenic and Gene Targeted Models of Dementia

Ronald A. Conlon

Abstract

Animal models of disease in genetically manipulated mice are powerful tools in medical research, including the study of dementia. The time and expense required to make genetically altered mice is considerable, and the importance of this investment is amplified by the long time course of most studies of dementia. Investigators need to be able to make informed choices about the different strategies for transgenics and gene targeting in order to minimize unwanted variation, and to maximize fidelity to the disease. In recent years, large genomic fragments stably cloned in well-characterized libraries, the means to manipulate their sequence, and the ability to make transgenic mice from these clones in inbred strains have increased greatly the power of the transgenic mouse. In addition, new embryonic cell lines from the C57BL/6 inbred strain of mice have become widely adopted for gene targeting, allowing knockins, knockouts, and conditional alleles to be established on the standard C57BL/6 background much more expeditiously than in the past. These methods, the time required, and the probability of success are reviewed with respect to mouse models of dementia.

Key words: Dementia, transgenic, gene targeting, knockout, recombineering, inbred strain, mice

1. Introduction

Gene targeting and transgenic technologies have different strengths and weaknesses with respect to fidelity to the disease, minimized variation, and avoidance of unintended consequences. The generation of genetically manipulated mice for any purpose involves significant time and expense. This investment is amplified in the case of models of dementia because of the aging component in many dementia models. My goal in the following is to help the investigator make wise choices in selection of technologies such that unwanted variation is minimized and fidelity to the disease can be maximized. This is not a cookbook of how to perform gene targeting or transgenics aimed at transgenic or targeting cores, but is intended to aid the investigator in making decisions about which approaches are best suited for achieving

Peter Paul De Deyn and Debby Van Dam (eds.), *Animal Models of Dementia*, Neuromethods, vol. 48, DOI 10.1007/978-1-60761-898-0_5, © Springer Science+Business Media, LLC 2011

their experimental goals while highlighting some of the pitfalls associated with the different approaches.

2. Transgenics and Gene Targeting

In transgenics, DNA is randomly inserted into the genome by injecting the swelling sperm nucleus shortly after fertilization (1). The injected embryos are transferred into recipient female mice. About 20% of the offspring will have the injected DNA inserted into their genome. Each of the transgene-positive offspring is used to establish a line of offspring carrying that insertion of the transgene by breeding founder animals to wild-type mice of the appropriate strain.

In gene targeting, changes are introduced into an endogenous gene using homologous recombination with manipulated DNA from that gene (2,3). Gene targeting is performed in embryonic stem (ES) cells in culture. Because gene targeting is a rare event, drug selection minigenes are incorporated into the targeting DNA construct, and then bona fide gene targeted cell lines must be identified from among the cells surviving drug selection. The fraction of correctly targeted cells in the cells surviving selection can be as large as one-quarter, but can be much smaller. Cell lines are screened for homologous recombination on one side of the targeted construct, then positive cell lines are propagated and characterized in more detail to ensure that homologous recombination occurred on both sides, and all elements that should have been introduced into the gene were in fact recombined in. Correctly targeted cell lines are analyzed to ensure that they have the correct number of chromosomes, since aneuploid cell lines rarely transmit through the germ line. Euploid, correctly targeted cell lines are combined with recipient embryos, and the chimeric embryos are transferred into a host female. The ES cells and the host embryo are typically marked by different coat color genes, and typically the chimeras are mated to mice of a chosen coat color such that offspring derived from the ES cell component can be distinguished from those derived from the host embryo. Half of the ES-derived offspring should carry the targeted gene.

3. Genetic Mechanism of Disease and Choice of Technologies

Consideration must be given to the genetic mechanism of the disease to be modeled. Typical genetic mechanisms of disease include mutations that eliminate gene function, mutations that alter gene activity, and mutations that increase gene copy number.

If a disease results from complete loss of function (a null mutation), then the gene can be inactivated by gene targeting (gene knockout). Both homozygous recessive traits (both gene copies inactivated by mutation) and haploinsufficient traits (one gene copy inactivated) can be modeled by gene knockout.

If a disease results from increased or novel activity of the gene product, then either transgenes that express mutant product or targeted sequence changes (knockin) to the endogenous gene can be used. If the mutations lead to partial loss of function, then gene targeting can be used to introduce a specific compromising mutation into the endogenous gene (knockin).

If the disease is due to increases in gene copy number, then transgenes can be used to increase gene copy number.

Also, consider that while many disease-causing mutations are not complete loss-of-function mutations (null mutations), null mutations are the most informative mutation for the normal function of the gene, and can be used in conjunction with disease-causing mutations to better understand molecular mechanisms.

4. Minimizing Unwanted Variation

One strength of animal models is that variation can be minimized, such that experiments can have greater sensitivity. Minimized variation results if genetic background, environment, and epigenetic state are uniform. Of course, variation in phenotype can arise directly if different mutations at the disease-causing locus have different effects on gene activity, but this variation is apparent and in the control of the investigator.

Variation can arise through interaction of a disease-causing mutation with segregating variants elsewhere in the genome. The effects of these gene–gene interactions can be quite large (4), and these types of interactions affecting the severity of the disease phenotype have been observed for models of dementia (5–8). Many mouse strains that have no genetic variation at all are available – inbred strains – so variation due to variable gene–gene interactions can be minimized. Genetic models can be generated directly on a small number of inbred strains, or a genetic variant can be generated and crossed onto the inbred strain of choice. The characteristics of inbred strains, which ones are amenable to genetic manipulation, and practical details of breeding inbred model mice are discussed next.

Variable gene expression at the disease-causing locus can arise due to epigenetic variation. DNA methylation and chromatin structure are heritable from cell to cell, and even from generation to generation in some cases. The problem is that the epigenetic state is not completely stable and can switch stochastically, from one permissive for expression, to one that is not. Most genes in the genome probably do not have stochastic variation in epigenetic state, but it is quite common at foreign DNA experimentally inserted into the genome, particularly in intergenic regions.

It is assumed that epigenetic silencing of introduced DNA is a protective mechanism against foreign DNA, such as might arise from viral genome insertion. The likelihood of silencing is influenced by the nature of the inserted DNA and the epigenetic state of the DNA surrounding the insertion (9–11). Importantly, variation in epigenetic state is exacerbated with small transgenes inserted in large numbers at one site. Transgenes insert in direct (tandem, head-to-tail) repeats at a single site. For small transgenes, the array can contain hundreds of copies. The repetitive nature of the transgene array promotes epigenetic silencing (12–14). Epigenetic silencing at a transgene array can vary from cell to cell, can increase with age, and can vary from animal to animal (10,12,15). Transgene arrays, once silenced in the germ line, usually remain stably silenced from generation to generation (16–18). Thus, it is important to monitor gene expression from generation to generation in transgenic mice with large transgene arrays.

Modern mouse housing practices seek to eliminate environmental variation through the control of ventilation, temperature, humidity, noise, vibration, photoperiod, enrichment, infectious agents, and diet. Although housing and veterinary care of the mice typically is managed by animal care staff and veterinarians of an animal facility, investigators should take an active interest in the housing, care, and infectious disease state of their animals to ensure that stated standards are met and procedures followed.

Environmental perturbations can come from unexpected sources. Mouse housing units consist in part of plastic, which come into contact with the mice. When cages made from polycarbonate and polysulfone degrade, they release the estrogenic compound bisphenol A (19,20). Polysulfone caging is more stable than polycarbonate, releasing less bisphenol A (19). Visibly degraded or hazed plastics should be removed from use.

5. Artifacts of Genetic Manipulation

Genetic manipulations occasionally have unintended consequences. Transgene insertions can physically disrupt genes at their site of insertion. The frequency of phenotypes arising from insertion site mutation by a transgene (almost 10%) is higher than might be expected from random integration into the genome. The higher than expected mutation rate results because transgenics generated by pronuclear injection can generate large deletions and complex rearrangements at the site of DNA integration (21–31). If a transgene cannot be made homozygous, or the homozygous mice have an unexpected phenotype, then this may indicate that there is an effect from the insertion site. Moreover, if an unexpected phenotype is not seen in other lines of mice from different founder mice with the same transgene, then the phenotype may be due to the insertion site. Thus, it is prudent to generate multiple lines of transgenic mice from independent founder animals and to compare the phenotypes of these lines.

In addition to direct physical interruption of genes, integrations can also have effects on the expression of neighboring genes (32–35). This more indirect effect has been observed in a few well-characterized gene knockouts, but in principle this phenomenon could apply to transgenes as well. Where indirect effects on neighboring genes have been observed, the effect is due to a selectable marker minigene introduced into the endogenous, mutated gene. In most cases where this phenomenon has been observed, the affected neighboring genes are close to the targeted gene, and have been closely related to the targeted gene. In gene knockouts, effects on the expression of neighboring genes can be minimized by designing knockouts with the drug resistance gene cassette flanked by recognition sequences for a site-specific recombinase (such as Frt or LoxP sites for Flpe or Cre recombinase, respectively). The drug resistance minigenes can be removed after targeting by transient expression of recombinase in ES cells or by crossing the knockout to a mouse, which expresses recombinase in the germ line.

Individual transgenic lines made with characterized promoters and small transgenes are occasionally ectopically expressed. This aberrant expression may be due to influences from the site of integration, and is observed more frequently with weak promoters.

6. Inbred Strains of Mice

A large number of inbred strains of mice exist, but only a small number are commonly used to make transgenics or gene targeted mice. The C57BL/6, 129, and FVB inbred strains are commonly used to generate genetically altered mice. Inbred strains are less robust than hybrid strains in knockout and transgenic production, but if the experimental goals are best served with an inbred strain, it is wise to start on an inbred strain to avoid the almost 2½ years needed to cross the mice to a new background. The choice of strain depends on the characteristics of the strain including susceptibility to the disease and whether other genetic variants of interest are present on that strain.

Inbred strains differ in susceptibility to disease phenotype, as well as in neuroanatomy, sensory acuity, and proficiency in behavioral tests (36). The C57BL/6 inbred strain has relatively normal neuroanatomy and is susceptible to dementia phenotypes (5–8,37), and many genetic variants have been established on C57BL/6J. C57BL/6J has age-related hearing loss (38), and is susceptible to dermatitis (39). 129 mice typically have a small corpus callosum and perform poorly in learning tests (37,40), and may have a reduced severity of dementia phenotype (5). The FVB strain is blind due to retinal degeneration and thus performs poorly in behavioral assays requiring vision (41–43). Fewer genetic variants are available on 129 and FVB than on C57BL/6J. Although there are only a small number of studies thus far, C57BL/6 and FVB mice appear to be more susceptible to both Alzheimer's and Huntington's disease pathology than 129 mice.

In the past, most gene targeting was done in ES cells from 129 mice. However, recently, cell lines from C57Bl/6J and the closely related C57BL/6N have become widely used (44,45). In addition, transgenics can be made directly on C57BL/6J [46).

7. Transgenics

Three types of transgene will be considered: small, cDNA-based transgenes; large, genomic DNA-based transgenes; and dual transgene (digenic) systems.

Small cDNA-based transgenes consist of a promoter, a complete protein-coding sequence from a cDNA, and a polyadenylation signal cloned into a high-copy-number plasmid. An intron is sometimes included in the construct to increase expression (47,48). These elements are cloned into the plasmid such that the transgene can be liberated in one piece from the plasmid backbone by digestion with restriction endonuclease(s), and the backbone-free construct isolated from a gel. The backbone is not included in the injected DNA since the backbone promotes epigenetic silencing (49,50). The injected DNA makes an array of head-to-tail copies, which inserts at a single site in the genome (1). The expression levels of the transgene-encoded products typically do not correlate with the number of copies of these cDNA-based transgenes (12). Different transgenic founders can have different expression levels, and this variation can be used to investigate if the phenotype varies with transgene expression level. Expression from a transgene of this kind can be considerably higher than that from the endogenous gene.

Promoters that drive ubiquitous or tissue-specific transgene expression have been developed. A list of characterized brain-specific promoters, which have been used in dementia models, is listed in the supplementary material to Gotz and Ittner [51).

Briefly, promoters which have been used to create models of dementia include those from the Thyl (Thyl.2), Prnp (PrP), Pdgfb (Pdgf-beta, PDGF), Camk2a (CAMKII), Eno2 (NSE), and GFAP genes. These genes vary in their spatial pattern of expression and the level of expression. The Thy1.2 promoter drives strong expression in most or all neurons beginning at early postnatal stages (52,53).

The Prp promoter drives strong expression in neurons in the hippocampus, Purkinje cells, and a few other neuronal cell types (54). The PDGF promoter drives moderate expression in neurons of the cortex and hippocampus (55). The CAMKII promoter drives postnatal expression in forebrain-derived neurons (56). The NSE promoter drives strong expression in postmitotic neurons, beginning on embryonic day 9.5 (57). The GFAP promoter drives robust expression in astrocytes (58).

The coding sequence typically is derived from a full-length cDNA. Mutations or tags can be incorporated into the coding sequence by many different methods.

Polyadenylation signals are needed for transgene-encoded RNA to accumulate. The AAUAAA in the 3' untranslated region of typical mRNAs is not sufficient to promote 3' end cleavage and polyadenylation in this context. Two different polyadenylation sequences are in common use: the bovine growth hormone and SV40 polyadenylation sequences (59,60).

Given the propensity of small transgenes to be epigenetically silenced in mice, and in order to fully characterize different transgenic lines for expression, plan to monitor expression of the transgene products. Ideally, the transgene-encoded protein is monitored, rather than the RNA. This can be accomplished if the protein is from a different species and species-specific antibodies are available, if antibodies detect the transgene-encoded mutant form of the protein, or the protein is tagged with an epitope.

Recently, large genomic fragments containing entire genes have become easier to manipulate to generate transgenic mice (61,62). Large fragment genomic DNA libraries in bacterial artificial chromosome (BAC) vectors have been made from many species, including humans and many strains of mice. The average size of the cloned genomic DNA (typically 150 kb) is such that most genes and the cis-acting sequences required for their expression can be contained within a single clone (63). BAC clones are stable in their bacterial hosts, and can be isolated with commercial large plasmid purification kits (64). Some of these libraries, both human and mouse, have had many clones sequenced at both ends. The paired end sequences were used to tile the clones on the genome assemblies (65–67). The partially sequenced clones are available for purchase in their bacterial hosts. Many BAC genomic clones, when injected as transgenes into mice, recapitulate the normal temporal and spatial pattern of expression of the endogenous gene at comparable expression levels (63). A technology for introducing mutations into the BAC clones, recombineering, is available (64,68–70). If use of the recombineering technology will be limited to a small number of constructs, it may be expedient to use a commercial recombineering service.

Genomic clones in BAC vectors are typically injected as intact circular molecules including the vector. The DNA concatemerizes and a small number of intact copies integrate at a single site (71). Unlike small, cDNA-based transgenes, the expression level of these large transgenes correlates with copy number (71,72). Epigenetic silencing has not been reported to be a problem with these large transgenes, presumably because they resemble mouse genes rather than foreign DNA. Detection of expression from unmodified mouse BAC clones can be difficult because of similarity or identity to the endogenous gene, thus use of human clones, or tagging or modifying a mouse coding sequence by recombineering should be considered.

Large genomic DNA fragments cloned in yeast artificial chromosome (YAC) vectors have also been used to generate transgenic mice. Although YAC vector libraries have larger DNA fragments on average, clones are much more prone to DNA chimerism, are much more difficult to isolate and use to generate transgenics than BAC clones (73).

Temporal control of expression is possible with digenic transgene systems. The most prevalent of these are the variants of the tetracycline-regulated transcription factor (74–76). One transgene of the pair consists of a responder transgene with the target coding sequence under control of a minimal promoter and binding sites for the trans-acting factor. The activity of the trans-acting factor is regulated by binding the tetracycline analogue doxycycline, which is supplied in drinking water. The trans-acting factor is typically expressed from a tissue-specific promoter in a second, independently established transgene. The two transgenes are brought together by breeding. In this way, both timing and tissue of expression can be precisely controlled. A number of well-characterized driver genes are available.

Most but not all components of digenic systems are small transgenes, and thus are susceptible to the epigenetic silencing that affects cDNA-based transgenes. Therefore, monitoring of the expression of both transgenes of the digenic system may need to be factored into the research plan.

For the creation of models of dementia, BAC transgenes have the advantages of expression which better mimics the endogenous gene and relative resistance to epigenetic silencing. For experiments that require high-level expression of a transgene product, the plasmid-based transgene approach works well. If control over the timing of expression of the transgene product is desired, the tetracycline-regulated system would be most suitable. The genetic background of choice is the C57BL/6J inbred strain.

8. Gene Targeting Strategies

In gene targeting, an endogenous gene can be altered in a wide variety of ways: it can be rendered nonfunctional by deleting essential sequences (gene knockout), have sequences replaced or added (gene knockin), or made into a conditional mutant (e.g. a “floxed allele”).

Because the genomic DNA used to construct the targeting vector needs to be from the same strain as the cell line (77), first a decision must be made about which ES cell line to use. Historically, most gene targeting was done in lines derived from the 129 inbred strain. Cell lines derived from 129 were better able to retain their chromosomes in culture, and thus were more likely to transmit the mutations through the germ line of mice (78). Recently, a number of excellent cell lines have become available from C57BL/6N mice (45,78–80). C57BL/6N and C57BL/6J diverged in 1951. There are a handful of known genetic differences between the two strains, including a null mutation in nicotinamide nucleotide transhydrogenase (Nnt) in C57BL/6J not present in C57BL/6N (81). After targeting in C57BL/6N, the chimeras can be bred to C57BL6/J directly. If the known genetic differences are a concern, they can be eliminated in two crosses to C57BL6/J by monitoring for the variants in the offspring.

End-sequenced genomic clones in BAC vectors, which can be used for the construction of targeting vectors, are available for both 129 and C57BL/6J (65,66).

Gene targeting has great fidelity to the genetics of the disease since the endogenous gene is targeted. However, there are disadvantages to gene targeting. Production of a mouse model by gene targeting takes more time than by transgenics. Importantly, success is less assured with gene targeting than with transgenics. Occasionally, a gene targeting vector will not produce gene targeted cell lines at all. It is not clear why targeting fails in these instances, and the usual resolution involves increasing the extent of homologous DNA or choosing a different part of the gene to target. In our experience, about 80% of gene targeting vectors target successfully. In addition, only about 80% of ES cell lines with a normal number of chromosomes will transmit through the germ line. If multiple correctly targeted cell lines were generated, then this second issue can be overcome by making chimeras with two or more cell lines for each targeting experiment.

9. Timeline for Transgenics and Gene Targeting

Moving a mutation or transgene to an inbred strain requires nine consecutive crosses to mice of the target strain, including at least one cross through each sex. This process takes a minimum of slightly more than 2 years. DNA closely linked to the gene variant will remain from the original strain on which the variant was generated.

In transgenics, it is 3 weeks from DNA injection to birth of offspring, 3 weeks to weaning when the mice containing the injected DNA can be identified, and another 3–5 weeks for females and males, respectively, to reach sexual maturity. Transmission of the transgene into offspring requires 6 weeks to weaned offspring.

Thus, a minimum of 15–17 weeks is required from DNA injection until a transgene from a founder mouse is established in multiple mice, the start of a line of transgenic mice.

In gene targeting, it takes a minimum of 8 weeks to electroporate the DNA construct and do an initial characterization of cell lines to identify potential targeted clones, then a minimum of another 6 weeks to fully characterize targeted cell lines for the construction of chimeras. From injection of targeted embryonic stem cells into host embryos, it takes 12 weeks for chimeras to reach sexual maturity, and another 6 weeks for weaned offspring, which will be tested for germ line transmission. Thus, a minimum of 32 weeks is required after the targeting vector is constructed to reach heterozygous, gene targeted mice.

These timelines do not include the time required to build DNA constructs or develop assays to identify founder mice or targeted ES cells. The DNA constructs for gene targeting are more complicated to construct, and the verification of gene targeting is more involved than identifying transgenic founder mice.

10. Sources of Services, Materials, and Information

Some academic transgenic and targeting services will take orders from clients from outside their institution. Their prices are typically substantially lower than commercial services.

The UCSC genome browser (http://genome.ucsc.edu) displays end-sequenced BAC genomic clones tiled across the genome for mice and humans. These clones can be purchased from The BACPAC Resources Center (http://bacpac.chori.org/). The mouse genomic clones are from either C57BL6/J, or Mus musculus molossinus, so be certain to select the correct library. The Ensembl genome browser (http://www.ensembl.org/Mus_musculus/index.html) displays end-sequenced 129 genomic clones, which can be purchased from the Wellcome Trust.

Transgenic cores can help you to identify sources of promoters, polyadenylation signals, and selectable markers and their material transfer agreements.

The Jackson Laboratory maintains the Mouse Genome Informatics (MGI) website and database, which is an invaluable source of information about nomenclature, existing mouse mutants, strains of mice, and public repositories of mice. The MGI email discussion group is a good way to get input from mouse geneticists at other institutions. A number of transgenic cores, including our own, maintain websites with a good deal of basic information about mouse genetics, reproduction, and biology (UC Irvine, http://www.research.uci.edu/tmf/index.htm; U of Michigan, http://www.med.umich.edu/tamc/; and CWRU, http://ko.cwru.edu/).

Table 1
Strengths and weaknesses of transgenics and gene targeting

	Small insert transgene	Genomic fragment transgene	Gene targeting
Time	>17 weeks	>17 weeks	>32 weeks
Vector construction	Standard	Special technology	Most involved
Epigenetic state	Can be unstable	Stable	Stable
Off target genetic effects	~1 in 10	~1 in 10?	Rare
Spatial expression	Many available promoters	Similar to endogenous	Endogenous
Expression level	Low to very high	Endogenous to ~5x endogenous	Endogenous
Inbred strains	FVB and C57BL/6	FVB and C57BL/6	129 and C57BL/6
Likelihood of success	≤100%	≤100%	~80%

11. Summary

The advent of BAC genomic libraries and recombineering, combined with the ability to make transgenics directly on the C57BL/6J inbred strain, have led to great improvements in the generation of animal models of disease. In many cases, a dementia model can be quickly established, which has low variation and high fidelity to the genetics of the disease. On the other hand, the recent development of stable C57BL/6 ES lines has facilitated gene targeting. However, gene targeting requires more time, and is less certain to succeed. The strengths and weakness of different approaches are summarized in Table 1.

References

1. Brinster RL, Chen HY, Trumbauer M, Senear AW, Warren R, Palmiter RD (1981) Somatic expression of herpes thymidine kinase in mice following injection of a fusion gene into eggs. Cell 27:223–231
2. Thomas KR, Capecchi MR (1987) Site-directed mutagenesis by gene targeting in mouse embryo-derived stem cells. Cell 51:503–512
3. Doetschman T, Gregg RG, Maeda N, et al. (1987) Targetted correction of a mutant HPRT gene in mouse embryonic stem cells. Nature 330:576–578
4. Threadgill DW, Dlugosz AA, Hansen LA, et al. (1995) Targeted disruption of mouse EGF receptor: Effect of genetic background on mutant phenotype. Science 269:230–234
5. Lehman EJ, Kulnane LS, Gao Y, et al. (2003) Genetic background regulates beta-amyloid precursor protein processing and beta-amyloid deposition in the mouse. Hum Mol Genet 12:2949–2956
6. Lloret A, Dragileva E, Teed A, et al. (2006) Genetic background modifies nuclear mutant huntingtin accumulation and HD CAG repeat instability in Huntington's disease knock-in mice. Hum Mol Genet 15:2015–2024
7. Ryman D, Lamb BT (2006) Genetic and environmental modifiers of Alzheimer's disease phenotypes in the mouse. Curr Alzheimer Res 3:465–473
8. Van Raamsdonk JM, Metzler M, Slow E, et al. (2007) Phenotypic abnormalities in the YAC128

mouse model of Huntington disease are penetrant on multiple genetic backgrounds and modulated by strain. Neurobiol Dis 26:189–200

9. Montoliu L, Chavez S, Vidal M (2000) Variegation associated with lacZ in transgenic animals: a warning note. Transgenic Res 9:237–239
10. Robertson G, Garrick D, Wu W, Kearns M, Martin D, Whitelaw E (1995) Position-dependent variegation of globin transgene expression in mice. Proc Natl Acad Sci U S A 92:5371–5375
11. Chevalier-Mariette C, Henry I, Montfort L, et al. (2003) CpG content affects gene silencing in mice: evidence from novel transgenes. Genome Biol 4:R53
12. Garrick D, Fiering S, Martin DI, Whitelaw E (1998) Repeat-induced gene silencing in mammals. Nat Genet 18:56–59
13. McBurney MW, Mai T, Yang X, Jardine K (2002) Evidence for repeat-induced gene silencing in cultured Mammalian cells: Inactivation of tandem repeats of transfected genes. Exp Cell Res 274:1–8
14. Mehtali M, LeMeur M, Lathe R (1990) The methylation-free status of a housekeeping transgene is lost at high copy number. Gene 91:179–184
15. Robertson G, Garrick D, Wilson M, Martin DI, Whitelaw E (1996) Age-dependent silencing of globin transgenes in the mouse. Nucleic Acids Res 24:1465–1471
16. Sutherland HG, Kearns M, Morgan HD, et al. (2000) Reactivation of heritably silenced gene expression in mice. Mamm Genome 11:347–355
17. Hadchouel M, Farza H, Simon D, Tiollais P, Pourcel C (1987) Maternal inhibition of hepatitis B surface antigen gene expression in transgenic mice correlates with de novo methylation. Nature 329:454–456
18. Allen ND, Norris ML, Surani MA (1990) Epigenetic control of transgene expression and imprinting by genotype-specific modifiers. Cell 61:853–861
19. Howdeshell KL, Peterman PH, Judy BM, et al. (2003) Bisphenol A is released from used polycarbonate animal cages into water at room temperature. Environ Health Perspect 111:1180–1187
20. Hunt PA, Koehler KE, Susiarjo M, et al. (2003) Bisphenol a exposure causes meiotic aneuploidy in the female mouse. Curr Biol 13:546–553
21. McNeish JD, Scott WJ, Jr., Potter SS (1988) Legless, a novel mutation found in PHT1-1 transgenic mice. Science 241:837–839
22. Mahon KA, Overbeek PA, Westphal H (1988) Prenatal lethality in a transgenic mouse line is the result of a chromosomal translocation. Proc Natl Acad Sci U S A 85:1165–1168
23. Covarrubias L, Nishida Y, Terao M, D'Eustachio P, Mintz B (1987) Cellular DNA rearrangements and early developmental arrest caused by DNA insertion in transgenic mouse embryos. Mol Cell Biol 7:2243–2247
24. Covarrubias L, Nishida Y, Mintz B (1986) Early postimplantation embryo lethality due to DNA rearrangements in a transgenic mouse strain. Proc Natl Acad Sci U S A 83:6020–6024
25. Brown A, Copeland NG, Gilbert DJ, Jenkins NA, Rossant J, Kothary R (1994) The genomic structure of an insertional mutation in the dystonia musculorum locus. Genomics 20:371–376
26. Shawlot W, Siciliano MJ, Stallings RL, Overbeek PA (1989) Insertional inactivation of the downless gene in a family of transgenic mice. Mol Biol Med 6:299–307
27. Woychik RP, Stewart TA, Davis LG, D'Eustachio P, Leder P (1985) An inherited limb deformity created by insertional mutagenesis in a transgenic mouse. Nature 318:36–40
28. Vogt TF, Jackson-Grusby L, Wynshaw-Boris AJ, Chan DC, Leder P (1992) The same genomic region is disrupted in two transgene-induced limb deformity alleles. Mamm Genome 3:431–437
29. Perry WL, 3rd, Vasicek TJ, Lee JJ, et al. (1995) Phenotypic and molecular analysis of a transgenic insertional allele of the mouse Fused locus. Genetics 141:321–332
30. Moyer JH, Lee-Tischler MJ, Kwon HY, et al. (1994) Candidate gene associated with a mutation causing recessive polycystic kidney disease in mice. Science 264:1329–1333
31. Pohl TM, Mattei MG, Ruther U (1990) Evidence for allelism of the recessive insertional mutation add and the dominant mouse mutation extra-toes (Xt). Development 110:1153–1157
32. Yoon JK, Olson EN, Arnold HH, Wold BJ (1997) Different MRF4 knockout alleles differentially disrupt Myf-5 expression: Cis-regulatory interactions at the MRF4/Myf-5 locus. Dev Biol 188:349–362
33. Olson EN, Arnold HH, Rigby PW, Wold BJ (1996) Know your neighbors: Three phenotypes in null mutants of the myogenic bHLH gene MRF4. Cell 85:1–4
34. Ren SY, Angrand PO, Rijli FM (2002) Targeted insertion results in a rhombomere 2-specific Hoxa2 knockdown and ectopic

activation of Hoxal expression. Dev Dyn 225:305–315

35. Taveau M, Stockholm D, Marchand S, Roudaut C, Le Bert M, Richard I (2004) Bidirectional transcriptional activity of the Pgkl promoter and transmission ratio distortion in Capn3-deficient mice. Genomics 84:592–595
36. Crawley JN, Belknap JK, Collins A, et al. (1997) Behavioral phenotypes of inbred mouse strains: implications and recommendations for molecular studies. Psychopharmacology (Berl) 132:107–124
37. Chen XJ, Kovacevic N, Lobaugh NJ, Sled JG, Henkelman RM, Henderson JT (2006) Neuroanatomical differences between mouse strains as shown by high-resolution 3D MRI. Neuroimage 29:99–105
38. Jones SM, Jones TA, Johnson KR, Yu H, Erway LC, Zheng QY (2006) A comparison of vestibular and auditory phenotypes in inbred mouse strains. Brain Res 1091:40–46
39. Csiza CK, McMartin DN (1976) Apparent acaridal dermatitis in a C57BL/6 Nya mouse colony. Lab Anim Sci 26:781–787
40. Balogh SA, McDowell CS, Stavnezer AJ, Denenberg VH (1999) A behavioral and neuroanatomical assessment of an inbred substrain of 129 mice with behavioral comparisons to C57BL/6J mice. Brain Res 836:38–48
41. Gimenez E, Montoliu L (2001) A simple polymerase chain reaction assay for genotyping the retinal degeneration mutation (Pdeb(rd1)) in FVB/N-derived transgenic mice. Lab Anim 35:153–156
42. Wahlsten D, Cooper SF, Crabbe JC (2005) Different rankings of inbred mouse strains on the Morris maze and a refined 4-arm water escape task. Behav Brain Res 165:36–51
43. Wong AA, Brown RE (2006) Visual detection, pattern discrimination and visual acuity in 14 strains of mice. Genes Brain Behav 5:389–403
44. Seong E, Saunders TL, Stewart CL, Burmeister M (2004) To knockout in 129 or in C57BL/6: That is the question. Trends Genet 20:59–62
45. Hughes ED, Qu YY, Genik SJ, et al. (2007) Genetic variation in C57BL/6 ES cell lines and genetic instability in the Bruce4 C57BL/6 ES cell line. Mamm Genome 18:549–558
46. Auerbach AB, Norinsky R, Ho W, et al. (2003) Strain-dependent differences in the efficiency of transgenic mouse production. Transgenic Res 12:59–69
47. Brinster RL, Allen JM, Behringer RR, Gelinas RE, Palmiter RD (1988) Introns increase transcriptional efficiency in transgenic mice. Proc Natl Acad Sci U S A 85:836–840
48. Palmiter RD, Sandgren EP, Avarbock MR, Allen DD, Brinster RL (1991) Heterologous introns can enhance expression of transgenes in mice. Proc Natl Acad Sci U S A 88: 478–482
49. Townes TM, Lingrel JB, Chen HY, Brinster RL, Palmiter RD (1985) Erythroid-specific expression of human beta-globin genes in transgenic mice. Embo J 4:1715–1723
50. Clark AJ, Harold G, Yull FE (1997) Mammalian cDNA and prokaryotic reporter sequences silence adjacent transgenes in transgenic mice. Nucleic Acids Res 25:1009–1014.
51. Gotz J, Ittner LM (2008) Animal models of Alzheimer's disease and frontotemporal dementia. Nat Rev Neurosci 9:532–544
52. Caroni P (1997) Overexpression of growth-associated proteins in the neurons of adult transgenic mice. J Neurosci Methods 71:3–9
53. Gordon JW, Chesa PG, Nishimura H, et al. (1987) Regulation of Thy-1 gene expression in transgenic mice. Cell 50:445–452
54. Scott M, Foster D, Mirenda C, et al. (1989) Transgenic mice expressing hamster prion protein produce species-specific scrapie infectivity and amyloid plaques. Cell 59:847–857
55. Sasahara M, Fries JW, Raines EW, et al. (1991) PDGF B-chain in neurons of the central nervous system, posterior pituitary, and in a transgenic model. Cell 64:217–227
56. Mayford M, Wang J, Kandel ER, O'Dell TJ (1995) CaMKII regulates the frequency-response function of hippocampal synapses for the production of both LTD and LTP. Cell 81:891–904
57. Forss-Petter S, Danielson PE, Catsicas S, et al. (1990) Transgenic mice expressing beta-galactosidase in mature neurons under neuron-specific enolase promoter control. Neuron 5:187–197
58. Mucke L, Oldstone MB, Morris JC, Nerenberg MI (1991) Rapid activation of astrocyte-specific expression of GFAP-lacZ transgene by focal injury. New Biol 3:465–474
59. Woychik RP, Lyons RH, Post L, Rottman FM (1984) Requirement for the 3' flanking region of the bovine growth hormone gene for accurate polyadenylylation. Proc Natl Acad Sci U S A 81:3944–3948
60. Pfarr DS, Rieser LA, Woychik RP, Rottman FM, Rosenberg M, Reff ME (1986) Differential effects of polyadenylation regions on gene expression in mammalian cells. DNA 5:115–122

61. Vintersten K, Testa G, Naumann R, Anastassiadis K, Stewart AF (2008) Bacterial artificial chromosome transgenesis through pronuclear injection of fertilized mouse oocytes. Methods Mol Biol 415:83–100
62. Scott MM, Wylie CJ, Lerch JK, et al. (2005) A genetic approach to access serotonin neurons for in vivo and in vitro studies. Proc Natl Acad Sci U S A 102:16472–16477
63. Gong S, Zheng C, Doughty ML, et al. (2003) A gene expression atlas of the central nervous system based on bacterial artificial chromosomes. Nature 425:917–925
64. Gong S, Yang XW (2005) Modification of bacterial artificial chromosomes (BACs) and preparation of intact BAC DNA for generation of transgenic mice. Curr Protoc Neurosci Chapter 5:Unit 5 21
65. Gregory SG, Sekhon M, Schein J, et al. (2002) A physical map of the mouse genome. Nature 418:743–750
66. Adams DJ, Quail MA, Cox T, et al. (2005) A genome-wide, end-sequenced 129Sv BAC library resource for targeting vector construction. Genomics 86:753–758
67. McPherson JD, Marra M, Hillier L, et al. (2001) A physical map of the human genome. Nature 409:934–941
68. Warming S, Costantino N, Court DL, Jenkins NA, Copeland NG (2005) Simple and highly efficient BAC recombineering using galK selection. Nucleic Acids Res 33:e36
69. Lee EC, Yu D, Martinez de Velasco J, et al. (2001) A highly efficient Escherichia coli-based chromosome engineering system adapted for recombinogenic targeting and subcloning of BAC DNA. Genomics 73:56–65
70. Yu D, Ellis HM, Lee EC, Jenkins NA, Copeland NG, Court DL (2000) An efficient recombination system for chromosome engineering in Escherichia coli. Proc Natl Acad Sci U S A 97:5978–5983
71. Chandler KJ, Chandler RL, Broeckelmann EM, Hou Y, Southard-Smith EM, Mortlock DP (2007) Relevance of BAC transgene copy number in mice: Transgene copy number variation across multiple transgenic lines and correlations with transgene integrity and expression. Mamm Genome 18:693–708
72. Vadolas J, Wardan H, Bosmans M, et al. (2005) Transgene copy number-dependent rescue of murine beta-globin knockout mice carrying a 183 kb human beta-globin BAC genomic fragment. Biochim Biophys Acta 1728:150–162
73. Giraldo P, Montoliu L (2001) Size matters: Use of YACs, BACs and PACs in transgenic animals. Transgenic Res 10:83–103
74. Urlinger S, Baron U, Thellmann M, Hasan MT, Bujard H, Hillen W (2000) Exploring the sequence space for tetracycline-dependent transcriptional activators: Novel mutations yield expanded range and sensitivity. Proc Natl Acad Sci U S A 97:7963–7968
75. Gossen M, Bujard H (1992) Tight control of gene expression in mammalian cells by tetracycline-responsive promoters. Proc Natl Acad Sci U S A 89:5547–5551
76. Gossen M, Freundlieb S, Bender G, Muller G, Hillen W, Bujard H (1995) Transcriptional activation by tetracyclines in mammalian cells. Science 268:1766–1769
77. te Riele H, Maandag ER, Berns A (1992) Highly efficient gene targeting in embryonic stem cells through homologous recombination with isogenic DNA constructs. Proc Natl Acad Sci U S A 89:5128–5132
78. Auerbach W, Dunmore JH, Fairchild-Huntress V, et al. (2000) Establishment and chimera analysis of 129/SvEv- and C57BL/6-derived mouse embryonic stem cell lines. Biotechniques 29:1024–1028
79. Hansen GM, Markesich DC, Burnett MB, et al. (2008) Large-scale gene trapping in C57BL/6N mouse embryonic stem cells. Genome Res 18:1670–1679
80. Tanimoto Y, Iijima S, Hasegawa Y, et al. (2008) Embryonic stem cells derived from C57BL/6J and C57BL/6N mice. Comp Med 58:347–352
81. Freeman HC, Hugill A, Dear NT, Ashcroft FM, Cox RD (2006) Deletion of nicotinamide nucleotide transhydrogenase: A new quantitive trait locus accounting for glucose intolerance in C57BL/6J mice. Diabetes 55:2153–2156

Chapter 6

Transgenic Animals and Intellectual Property Concerns

Susan L. Stoddard and James A. Rogers, III

Abstract

Transgenic mouse models of neurodegenerative diseases may have considerable commercial value. In many cases, that value is linked to the intellectual property rights associated with the model. This chapter discusses the protection of intellectual property, including patent, contract, and third-party rights, related to transgenic mouse models and addresses aspects of litigation, which can occur if intellectual property rights are infringed. Various methods of commercializing transgenic mouse models are considered, including how these various methods address concerns of the National Institutes of Health regarding the use of research tools.

Key words: Transgenic mouse, patent, license, commercialization, intellectual property, research tool

Legal disclaimer: This chapter does not give legal advice, which should be obtained from an attorney familiar with the current state of the law and the specific facts applicable to the reader's situation.

Transgenic mouse models have become valuable tools for advancing discoveries related to the diagnosis and treatment of neurodegenerative diseases. Once an animal model of a disease exists, it is possible to begin to understand disease mechanisms and to develop and test therapeutic approaches. The first transgenic animal model for a neurodegenerative disease potentially has considerable scientific and commercial value. Whereas the other chapters of this volume discuss the scientific aspects of various transgenic models, this chapter will address topics relating to the patenting and commercialization of such models. Collectively, we have worked to protect and commercialize transgenic mouse models of neurodegenerative disease for over 18 years, and will use some examples from that experience.

Peter Paul De Deyn and Debby Van Dam (eds.), *Animal Models of Dementia*, Neuromethods, vol. 48, DOI 10.1007/978-1-60761-898-0_6, © Springer Science+Business Media, LLC 2011

The development of a transgenic animal raises intellectual and tangible property issues (1). These issues relate to an institution's ability to make, use, and sell such an animal. The two primary issues are: how does the institution (such as a university) protect the transgenic animal and did the university have the right to make or sell the animal in the first place? We will address each of these topics in turn.

The US government grants patents on novel, nonobvious, and useful inventions. The scope of novelty, obviousness, and usefulness, from a legal perspective, evolves over time as new cases add to the understanding of these concepts. However, it is a well-settled law that a party may obtain patent rights on a novel, nonobvious, and useful transgenic animal. Provided the invention meets these criteria, a party may also patent the gene construct, cell line, or techniques used to create the animal, in addition to the methods of using the animal, e.g., to diagnose a specified disease.

A patent allows the owner to exclude others from making, using, and selling the claimed invention. This exclusion applies to all activities, unless there is an exception under the law, of which there are very few. The Hatch–Waxman law provides an exception to acts of infringement in furtherance of seeking US Federal Drug Administration (FDA) approval for a product. However, the scope of this exception as it relates to research tools, such as transgenic animals, has not been settled. The reality is that uses of transgenic animals, including in a research setting, may constitute infringement of an issued patent claim.

A party owning a patent on a gene construct may prevent both use of that gene construct to develop a new transgenic mouse and use of the mouse if it is developed. This is true even if the mouse is used primarily for research purposes, as discussed above. There is no requirement that a party grant any right under a patent to any other party. Patents are still one of the few areas that permit a legalized monopoly, in the United States and elsewhere. In the United States, the patent term is 20 years from the date of filing, with extensions available under certain circumstances. Accordingly, a patent, which is enforced, can offer the patent owner very significant protection.

Filing for and obtaining a patent is costly. This process involves drafting a patent application with the help of a patent agent or patent attorney.[1] Once the patent application is drafted, it is filed with the US Patent and Trademark Office (PTO). The PTO examines the patent application to determine whether the application

[1] Both patent agents and patent attorneys are registered to practice before the US Patent and Trademark Office. Both can write, file, and prosecute patents. Only the patent attorney, however, can litigate patents.

meets the requirements for patentability, fails to meet the requirements, or whether there is some other deficiency in the patent application itself. The PTO then issues an office action, which requires a response from the party seeking to obtain the patent. This process, called prosecution, is usually conducted with the assistance of a patent agent or patent attorney. Typically, prosecution involves two to three rounds of office actions and responses prior to obtaining an issued patent. Once the patent is issued, the party owning the patent is required to pay periodic maintenance fees. This process from application to maintaining an issued patent in the United States may cost tens of thousands of dollars in attorney's and government fees. If the application is also filed overseas, the cost can balloon to hundreds of thousands of dollars.

To have value, the patent must cover the invention or methods of making or using the invention. If, for example, a patent claims a specific gene construct and a third party must use that gene construct to make a specific transgenic mouse, then the patent holder can exclude the third party from making, using, or selling the mouse unless the third party takes a license for the right to use the gene construct. On the other hand, if the patent claims cover, for example, only one of several different techniques that could be used to make the transgenic mouse, then ultimately the patent may be much less valuable, if not completely worthless, because a third party could practice one of the other unpatented methods.

The claims of a patent must issue for the patent owner to have enforceable rights. For example, assume that a pending patent application covers a cell line used ultimately to make a transgenic mouse. If a third party uses the cell line to help make a mouse prior to the patent claims issuing, the patent applicant cannot sue to stop use of the mouse. On the other hand, if the patent covers the mouse itself, the applicant would have a right to sue for the use of the mouse, even if the mouse was made prior to the claims issuing.

A party owning a patent must understand that the cost to enforce a patent can be significant. A patent infringement lawsuit can cost well over a million dollars in fees and there is no guarantee that the owner will recover these costs. Damages, if awarded, are calculated based on a reasonable royalty or lost profits. Further, a party may be entitled to recover attorney's fees, court costs, and have damages multiplied up to three times in the event that the court finds the infringer acted with knowledge of the patent and in bad faith.

Separate and apart from patents, a party may have rights in the underlying materials used to create the animal, such as a gene construct. These rights may be protected through a contract. For example, in order to receive a material from a third party, you may have to sign a material transfer agreement (MTA). This MTA may describe who owns the material, any materials derived from the original material, and any inventions arising from use of the material. The agreement may include restrictions on sharing the materials with others and may provide for sharing of any revenue derived from use of the materials. Finally, the MTA may include other provisions, such an indemnification of the provider of the material.

From a right-to-use perspective, a party should be concerned about third-party patent rights, contract rights, or lack of contract rights. With respect to patents, the fact an invention such as a transgenic animal or the starting material of a gene construct is disclosed in a publication does not mean that these items are dedicated to the public domain. In fact, they may be the subject of a pending published or unpublished patent application[2] or an issued patent. Further, any new invention may infringe the existing patent rights. Accordingly, if a party is concerned about the potential of infringing third-party rights, they should consult a patent attorney for advice.

Separately, given the collaborative nature of research, materials are often provided under separate agreements, including material transfer or research agreements. In some cases, materials may be provided by a party that did not have the right to provide the material. In either event, the situation is analogous to driving a stolen car, and an institution should therefore review any obligations or restrictions that may be attached to the use of any materials utilized in making a transgenic animal.

Transgenic animal models of disease are most frequently commercialized as research tools. Whereas it certainly might be possible to license a unique transgenic model exclusively for a considerable sum, such a practice would be contradictory to the US National Institutes of Health (NIH) Guidelines for the provision of research tools (2). It would also be contradictory to the principles of academic research institutions, whose mission is to provide technological advances for the betterment of humankind. A transgenic mouse model provided nonexclusively to many companies offers the licensor[3] the opportunity to receive a variety of payment streams, while broadly supplying a valuable tool. Mayo had the privilege of being able to provide the first animal model of Alzheimer's disease, the Tg2576 mouse, to companies trying to develop therapies for this disease. By putting such a valuable tool in many hands, the possibility of actually finding a therapy for Alzheimer's disease increased tremendously.

There are numerous financial structures for a nonexclusive license for a transgenic animal model. Generally, pass-through royalties are not appropriate. Pass-through royalties are royalties

---

[2] Patent applications generally publish 18 months from the earliest priority date.

[3] The licensor is the party that gives another party, the licensee, a license permitting the licensee to use the intellectual property of the licensor.

that would be paid on the sales of a drug developed using the animal model. The very logical reason given by pharmaceutical companies for not paying pass-through royalties is that if they had to pay such royalties for every research tool used in the development of a drug, the royalty burden would be too high to be commercially feasible. Additionally, the record-keeping involved in keeping track of the relevant research tools and their associated royalties would be unmanageable.

With a particularly valuable transgenic animal model, it might be possible, in certain negotiations, to obtain milestone payments for the progression of a drug, which was tested on such animals, through clinical trials. This form of financial return would likely be a single payment, and therefore more acceptable and easier to manage for a commercial licensee.

In our experience, the best model for commercialization is an upfront payment and annual license maintenance fees that are paid for as long as a company continues to use the particular transgenic model. These two components, upfront and yearly payments, can be adjusted up or down to create any number of unique financial arrangements. Some companies would rather pay a single large upfront payment and then not worry about yearly maintenance payments for several years. Other smaller companies may prefer smaller but equal payments over a number of years.

The largest financial return to a licensor of a transgenic mouse model is likely to come early in the commercialization lifetime of a particular model. This is when the model is a new and unique tool, and its value is greater while it remains unique. Over time, others develop similar, modified, or perhaps better models and the value of the first model decreases proportionally. At this point in the commercialization timeline, it makes sense to consider sublicensing the model to an animal supplier who will sell the model to all customers, while still generating modest revenues for the licensor.

It is important to keep in mind that a transgenic mouse model can be commercialized even if it is not patented. A third party will be willing to license an unpatented transgenic mouse if the mouse is difficult to engineer, if that party simply does not want to invest time and resources in developing their own model, or if the price is favorable. In fact, it may only be worth the expense of patenting a transgenic mouse in limited cases – when the model is the first for a particular neurodegenerative disease, or if the model has truly unique characteristics. Our general practice at present is not to patent transgenic mouse models, but to rely on being able to license them by saving others' time and resources.

Sublicenses to animal breeders and suppliers[4] are an excellent way to commercialize a transgenic mouse of moderate value and

[4] Taconic Farms (www.taconic.com) and Charles River Laboratories (www.criver.com) are two examples.

a follow-on way to capitalize on a particularly valuable transgenic model in the later stages of its commercial life. Other than the financial incentives, animal breeders and suppliers offer two other significant advantages to the licensor. First, the licensor is no longer responsible for supplying the mice. If a transgenic mouse model is in demand, the burdens put on the academic lab for provision of the mice, and on the technology transfer office for oversight and paperwork can be overwhelming and unmanageable. Second, a license to an animal breeder and supplier can open up a whole new market to the licensor. Typically, academic developers of transgenic mouse models provide these models to investigators at other academic and nonprofit organizations at no cost and under the terms of an MTA. With a popular model, this can become very burdensome. Additionally, considerable cost can accrue to the institution providing the transgenic model, when facility and personnel costs for breeding, housing, and shipping the animals are combined with administrative costs from the technology transfer office to execute numerous MTAs. When the model is licensed to an animal breeder and supplier, all the breeding and housing costs, and the associated paperwork, is subsumed by the licensee. Additionally, the transgenic model is now sold to everyone who wants it, academic scientists included. The return to the licensor, while likely smaller on a per-mouse basis, is paid on many more mice.

Transgenic mouse models of neurodegenerative diseases are now widely used and have become valuable research tools. They can be patented, which is an expensive proposition, and likely worth the investment only for first-in-class or truly unique models. Unpatented transgenic mouse models can be of commercial value to the developer since it is frequently easier and cheaper to purchase a validated model than to make a new model from scratch. Either patented or unpatented, transgenic mouse models are continuing to help academic and commercial scientists unravel the etiology of many neurodegenerative diseases and develop earlier diagnosis and therapy.

References

1. Abrams I, Kaiser M (2000) Licensing transgenic mice: A short tutorial. J Assoc Unit Techno Managers XII:81–100
2. Report of the National Institutes of Health (NIH) working group on research tools. National Institutes of Health, Washington, DC, 1998. (Accessed October 29, 2008, at http://www.nih.gov/news/researchtools/)

Part III

Validation of Animal Models of Dementia

Chapter 7

Pathological Validation of Animal Models of Dementia

Daniel Pirici, Christine Van Broeckhoven, and Samir Kumar-Singh

Abstract

Alzheimer's disease (AD) and frontotemporal dementia (FTD) are two most common forms of presenile dementia where insoluble protein deposits as intra- or extracellular aggregates. During the past decade, a number of mouse models have been devised based on human mutant genes associated with familial forms of disease. Partly due to such experimental models, enormous progress has been made in the understanding of mechanisms by which amyloid-β or tau protein is toxic to neurons and causes part of the cognitive/behavioral or neuropathological features characteristic of AD or FTD. This chapter enumerates transgenic mouse models commonly used in AD and FTD research and discusses how these have served as an important research tool in defining critical disease-related mechanisms. Furthermore, this chapter also summarizes how these mouse models have contributed in identification of potential drug targets or in the evaluation of novel therapeutic approaches in delaying the onset or progression of these devastating diseases.

Key words: Transgenic mouse models, Alzheimer's disease, frontotemporal dementia, frontotemporal lobar degeneration, Aβ, tau, neuropathology, behavior, therapy

1. Introduction

Alzheimer's disease (AD) is the most frequent form of neurodegenerative disease resulting in progressive loss of memory and cognitive abilities. It is responsible for the cognitive decline in up to three quarters of all dementia patients (1). It was not until the turn of the last century that this syndrome was recognized as an individual entity and its two pathological hallmarks, the amyloid plaques and the tau neurofibrillary tangles became widely accepted (2). However, the precise connections between amyloid precursor protein (APP) and tau proteinopathies as well as any relation of these proteinopathies to synaptic and neuronal loss in AD are still largely unknown. The identification of the early onset, familial

Peter Paul De Deyn and Debby Van Dam (eds.), *Animal Models of Dementia*, Neuromethods, vol. 48,
DOI 10.1007/978-1-60761-898-0_7, © Springer Science+Business Media, LLC 2011

AD (FAD)-linked APP (3), and presenilin 1 and 2 mutations (4,5) offered for the first time a possibility to model the disease, and thus opened new frontiers in AD research. Several overexpression transgenic mouse models were made that hosted different pathogenic mutations that subsequently have been extensively characterized regarding the transcriptional promoter, the earliest time at which amyloid deposition appeared, the predominant type of amyloid-β (Aβ) deposition, and the associated neuropathological, biochemical, and behavioral changes. A decade of such investigations has shown that many of these mouse models have a number of similarities with human disease, thereby helping to gain further insights into different pathological aspects of the human disease. Thus, not surprisingly, many of these mouse models are also being utilized in current preclinical drug trials.

This book chapter, after addressing to some of the salient features of AD and amyloid metabolism, will discuss some common mouse models of AD and show how such mouse models have been instrumental in understanding the etiopathogenesis of Aβ amyloidosis. Such protein misfolding pathology (or proteinopathy) is also a common feature of other dementias. Of relevance to AD is also tauopathy, a co-proteinopathy in AD but a primary proteinopathy in frontotemporal dementia (FTD) of the tau type. We will therefore also discuss mouse models of tau and see how these have helped in finding common links between Aβ and tau pathologies.

2. APP and MAPT

2.1. APP and Aβ peptide

Amyloid precursor protein is a ubiquitously expressed single-transmembrane protein with a 590–680 amino acids (aa.) long extracellular amino terminal domain and a short cytoplasmic tail, which contains intracellular trafficking signals (Fig. 1) (6). APP transcripts undergo alternative splicing to yield eight possible isoforms of which APP-695, 751, and 770 isoforms predominate in the brain, with APP-695 isoform being mainly produced by neurons (7). Thus, most of the transgenic mouse and cellular models are based on APP-695 isoform, while some of the constructs also utilize APP-770 isoform.

While APP is normally processed by α- and γ-secretases to release a secreted APPα fragment (APPsα), a 3 kD p3 peptide and a 7 kD APP intracellular domain (AICD), the alternative processing of APP by β- and γ-secretases results in full-length Aβ, an N-terminal soluble APPsβ fragment, and the C-terminal AICD tail (8). The α- and β-secretases have recently been identified as members of the ADAM and BACE family of enzymes, respectively,

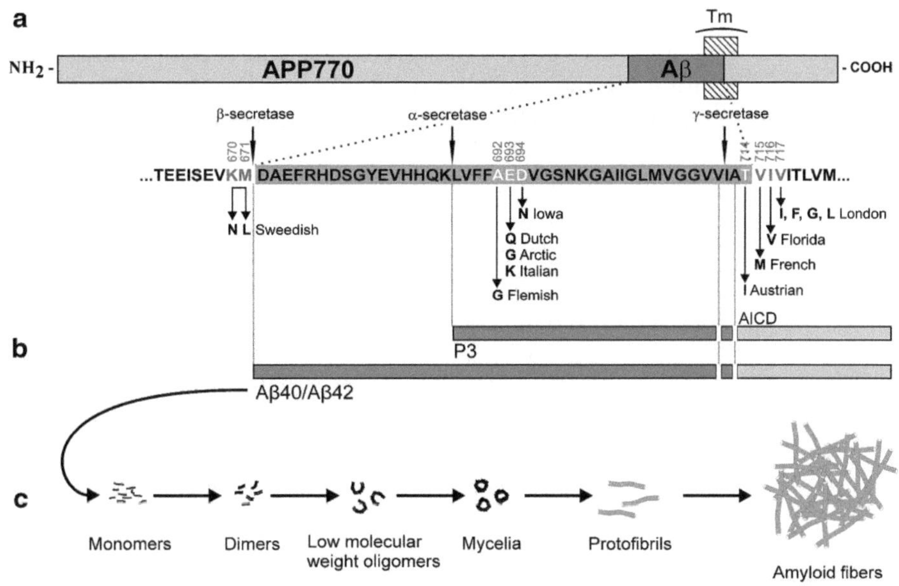

Fig. 1. Schematic diagram of the amyloid precursor protein (APP) with pathogenic mutations and metabolic processing. (**a**) The Aβ fragment is located between residues 671 and 714 and partially overlaps with the transmembrane segment (TM). The amino acid sequence of Aβ fragment (numbering accordingly with the APP770 isoform) is enlarged below and shows common pathogenic amino acid substitutions. Sites of α-, β-, and γ-secretase cleavage points are indicated by arrows. (**b**) The constitutive proteolytic cleavage by α- and γ-secretases leads to the formation of the short p3 peptide, and the alternative pathway leads to the formation of Aβ peptide, in both cases with a consecutive release of a C-terminal APP intracytoplasmic domain (AICD). The variations at the γ-secretase site cleavage lead to the formation of Aβ40/Aβ42 isoforms. (**c**) Various forms Aβ isoforms described in human pathology and transgenic mouse models.

and γ-site processing was found to be dependent on a pentameric complex in which presenilin 1 and 2 (PS) act as aspartyl transmembrane proteases (9,10). It is important to note that while these pathways were called amyloidogenic and nonamyloidogenic pathways, recent studies suggest that p3 or other N-terminally cleaved Aβ fragments are also deposited as amyloid (11) (see later). APP is processed in both the constitutive and alternative secretory pathways in the trans-Golgi network (6) and via the early endosomal-lysosomal-exosomal pathway (6,12). In neurons, APP and its catabolites are also trafficked along the microtubules within the axons via a kinesin-mediated mechanism and APP is observed to be present in synapses (13). This fits well with the arguable physiological function of APP in neurite growth and memory (14). In support of this proposed function, APP deficient mice show a moderate atrophy of corpus callosum together with astrogliosis but without any increase in morbidity or mortality (15). The anterograde transport of APP also might explain why some of the Aβ deposits in the dentate gyrus appear to originate from nerve terminals whose axons traverse the perforant pathway, and why lesions of perforant pathway in animal models of AD result in a reduction in hippocampal amyloid burden (16,17).

Aβ has two major C-terminal isoforms, Aβ40 and Aβ42. Aβ40 comprises 90–95% of secreted Aβ and is the predominant species found in the cerebrospinal fluid while less than 10% of the secreted Aβ is Aβ42 (18). The secreted fraction of Aβ presents as monomers/dimers and soluble Aβ oligomers, which are passively redistributed in brain parenchyma and cleared (19). Several Aβ clearance mechanisms have been described. First, Aβ is taken up and degraded by microglial and astroglial cells (20). In addition, Aβ is also degraded by proteases including neprilysin (NEP), insulin-degrading enzyme (IDE), endothelin-converting enzymes-1 and -2 (ECE), matrix metalloproteinase-2 and -9 (MMPs), and tissue plasminogen activator (tPA) (21). Of these, NEP, IDE, and ECE1 are active intracellularly while IDE, tPA, and MMPs act at the cell surface or are secreted and activated in the interstitial fluid (ISF) (21). Knockout models for many of these proteases show a significant elevation of murine brain Aβ, while their overexpression counterparts show a reduction in Aβ levels and retardation of plaque formation (22). Human studies also show that a local reduction in expression of some of these proteases correlates with a high plaque burden (23) (Fig. 3, upper panel). Lastly, Aβ is also drained out from the brain parenchyma by two mechanisms involving blood vessels. One of these mechanisms is a direct Aβ transport across the blood-brain barrier (BBB) mediated by LDL receptor-related proteins-1 and -2 (LRP-1, LRP-2), and regulated by α2-macroglobulin, apolipoprotein E (ApoE), and apolipoprotein J (ApoJ), respectively (24) (see later). This direct vascular clearance mechanism is highly efficient in clearing not only physiological Aβ but also pathological amounts of Aβ in transgenic mouse models (25). It is postulated that if the influx of circulating Aβ would be completely stopped, LRP-mediated transport could remove the entire Aβ pool from brain ISF within less than 1 min under physiological conditions, or in approximately 40 min at pathological levels (26). A reduction of LRP-1 expression has indeed been observed in AD patients as well as in mouse models (27). The second vessel-related Aβ clearance route utilizes the periarterial ISF drainage pathway where Aβ in ISF is collected around periarterial space (also known as the Virchow-Robbin space) that reaches the cerebrospinal fluid (CSF) in the subarachnoid space. From here, along the perivascular spaces of the circle of Willis, the fluid reaches the olfactory bulbs, then passes with perforant lymphatic channels through the cribriform plate to nasal submucosa and finally to the cervical lymph nodes (28). Tracer experiments in animal brain demonstrate that this is equivalent to lymphatics of brain (29). The ISF drainage pathway constitutes a major route of Aβ elimination in situations when BBB becomes less efficient (as in aging) or is overwhelmed (as in mouse AD models) (24,30). All mechanisms taken together, Aβ clearance from human brain normally exceeds its production with Aβ fractional production and clearance rates measured as 7.6% and 8.3% per hour, respectively (31).

Both soluble and fibrillar aggregates of Aβ are shown to be toxic and mouse models have been very useful in confirming the toxic potential of different Aβ aggregation states, either by showing inhibition of hippocampal long-term potentiation (LTP) or by demonstrating memory impairment after injecting aggregated Aβ species in rodent brain. For instance, together with human and *in vitro* studies, mouse models have been instrumental in showing that fibrillar and/or nonfibrillar forms of Aβ deposited in vessel walls are associated with degeneration of endothelial cells, smooth muscle cells (SMCs), and brain pericytes (32–34). Similarly, reduction in synaptophysin levels and disruption of neuronal calcium homeostasis has been shown in mouse models, suggesting a considerable role of soluble Aβ aggregates (35–37). Increasing data on varied models also indicate that Aβ can in turn activate microglia and thus indirectly damage neurons, contributing to the neurodegenerative process and cognitive loss (38,39). It is maybe for this reason that the "neuritic plaques" are more active as these are shown to be in histological proximity with inflammatory markers (40,41). Some of these issues will be discussed in relevant sections of this chapter.

2.2. MAPT and Tau

AD also shows tau proteinopathy with aggregated forms of hyperphosphorylated tau deposited in neurons as neurofibrillary tangles (NFTs) and neuropil threads. Hyperphosphorylated tau is also a major constituent of Pick bodies in FTD. Tau is a highly conserved protein that belongs to the family of microtubule associated proteins (MAPs) and is essential for the assembly of tubulin into microtubules (42). In human CNS, tau is abundantly expressed in the axons of mature and growing neurons; however, low levels of tau are also present in oligodendrocytes and astrocytes (43,44). The human tau gene (*MAPT*) is located on chromosome 17q21 and encodes 15 exons of which three exons (4A, 6, and 8) are never present in any mRNA expressed in human brain (reviewed in (45)). Tau mRNA is spliced in to six possible isoforms with lengths between 352 and 441 aa (46). These isoforms occur by combining two possible inserts (0N, 1N, 2N; encoded by exons 2 and 3) in the N-terminal domain, and 3 or 4 possible repeats in the C-terminal tail (3R or 4R, depending on the alternative splicing of exon 10) (45). Also, the six tau isoforms are not equally expressed in neurons, for example tau mRNA containing exon 10 is not found in granular cells of the dentate gyrus. The four repeat domains represents the microtubule-interacting domain, while the C-terminal acidic region projects between the microtubules, acting as a linker with the other components of the cytoskeleton and as a spacer for the other surrounding microtubules. In fact, tau proteins may interconnect microtubules with other cytoskeletal components such as neurofilaments and actin filaments. In mice lacking the tau gene, an increase in the microtubule associated protein 1A is observed, and this may compensate for the loss of tau functions (47). However, in these mice, axonal diameter in some neurons is particularly affected.

Phosphorylation is by far the most common posttranslational modification of tau. At least 25 different tau phosphorylation sites have been identified. Tau phosphorylation might have a physiological role, as during development fetal tau is more heavily phosphorylated than adult tau (48). Tau hyperphosphorylation and aggregation is also a major molecular change leading to accumulation of insoluble intracellular paired helical or straight filaments (PHF/SF) in a number of tau proteopathies such as AD, Pick's disease (PiD), progressive supranuclear palsy (PSP), and corticobasal degeneration (CBD) (49). In the following sections, we will discuss how Aβ and tau interact in relevant mouse models.

3. Genetics of AD and FTD

Mutations in three genes have been deemed causal for autosomal dominant forms of AD: the APP gene (3), the PS1 gene (4), and the PS2 gene (5,50). Furthermore, while advancing age is the strongest risk factor to developing AD, different ApoE alleles are also shown to modulate the disease progression (51).

3.1. APP Mutations

Characterization of vascular amyloid in AD and Down's syndrome patients led to the identification of the Aβ peptide and APP gene (52), which was followed by the identification of Aβ in parenchymal plaques (53). The first pathogenic mutation (E693Q) was also identified in a form of vascular amyloidosis called hereditary cerebral hemorrhages with amyloidosis – Dutch-type (HCHWA-D) (54,55). HCHWA-D patients are characterized by recurrent hemorrhages and extensive Aβ deposition in cerebral blood vessel walls, in the absence of senile plaques and tau-related pathology (56–58). Besides the occurrence of large lobar intracerebral hemorrhage, HCHWA-D patients may show cognitive deterioration, frequently associated with white matter abnormalities on MRI and small ischemic and hemorrhagic infarctions on pathological examination (59). At the level of APP processing, this mutation drastically increases the Aβ40/Aβ42 ratio, with decreased Aβ42 levels, both *in vivo* and *in vitro* (60,61). Moreover, the mutated Aβ Dutch peptides show an enhanced fibrillogenesis and toxicity towards vascular cells in various *in vitro* experimental setups, explaining to some extent its affinity to blood vessels (62,63).

The Flemish (A692G) APP mutation, identified subsequently, leads to both cerebral hemorrhages and AD (64). Pathologically, Flemish APP mutation carriers show senile plaques with the largest central dense-cores observed in AD and a severe degree of cerebral amyloid angiopathy (CAA) (41,65,66). Also, in contrast to HCHWA-D and consistent with a clinical diagnosis of AD, Flemish APP patients have considerable neurofibrillary pathology (41,65). Besides Dutch and Flemish APP mutations, other mutations that have been identified near the α-secretase processing site are Arctic (E693G), Italian (E693K), and Iowa (D694N) mutations (Fig. 1). These APP mutations lead to very similar phenotypes, dominated by amyloid angiopathy, diffuse amyloid deposits, and occasionally ischemic infarctions (67–69). This is explained by the fact that these α-secretase site APP mutations change the primary Aβ sequence and increase its vascular affinity and/or fibrillogenic potential including formation of Aβ oligomers.

The α-secretase site APP pathology contrasts sharply with pathology caused by mutations identified near the β- or γ-secretase sites. The only APP mutation near the β-secretase site is the double Swedish (K670N/M671L) mutation (70) that promotes β-secretase activity and increases the total Aβ production without altering the Aβ40/Aβ42 ratio (71). This is consistent with the pathological data in APP Swedish carriers, which deposit more Aβ40 compared to carriers of APP mutations near the γ-secretase site (72). The first γ-secretase site APP mutation identified was the London (V717I) APP mutation that results in an increased Aβ42/Aβ40 ratio, without altering total secreted Aβ (73,74). The enhanced secretion of more aggregatable Aβ42 directs the pathology towards a diffuse amyloid distribution predominantly constituted of Aβ42, and exceeding the burden observed in sporadic AD patients (72). Many γ-secretase site APP mutations have been identified and while almost all of these mutations decrease the total Aβ production and increase Aβ42/Aβ40 ratios (75), the pathology caused by Austrian (T714I) APP mutation is the most remarkable one. This mutation leads to a very aggressive form of AD with a very young age of onset and a diffuse, nonneuritic amyloid pathology (76). *In vitro* cell culture experiments in neuronal and nonneuronal cells showed that APP Austrian mutation led to one of the most drastic increases in the Aβ42/Aβ40 ratio in a panel of γ-secretase site mutations studied (76,77). Lastly, several independent duplications of the APP locus have been recently identified in French and Dutch families with autosomal dominant AD and/or lobar cerebral hemorrhage with prominent CAA (78,79). Similarly, Down's syndrome patients, also carrying three copies of APP, develop CAA as young as 30 years and the severity of CAA and vessel-related risks increase with age (80). APP gene dosage should be considered as a situation similar to APP Swedish mutation because it also increases total Aβ without changing the Aβ42/Aβ40 ratio. These data underpin the growing consensus that increase in absolute levels of Aβ40 leads to a predominant vascular amyloidosis (81).

3.2. Presenilin Mutations

FAD mutations in presenilin 1 and 2 are more common than APP mutations (4,5,50). Structurally, PS are multiple transmembrane domain proteins localized in the endoplasmic reticulum, Golgi apparatus, and lysosomes as a part of the γ-secretase complex (6,82). The functions of PS are not entirely clear; however, PS1 knock-out mice die *ante partum* or immediately after birth and show lissencephaly, severe altered somitogenesis, skeletal, and vascular disturbances resembling a Notch phenotype (83,84). PS mutation carriers present with an earlier age of onset than APP mutation carriers, and with abundant diffuse-like amyloid pathology, perhaps because these mutations increase the production of the more insoluble and readily depositable Aβ42 isoform (75). Interestingly, most of the PS mutations tested *in vitro* not only increase Aβ42, but also lower Aβ40 production drastically with an overall reduction of total Aβ production (75,85). This might have important implications for amyloidogenic mouse models based on PS mutations (see further). Also, interestingly, select AD-linked PS1 mutations and the Volga-German PS2 N141I mutation associate with a marked CAA (86,87). While the precise cause is again unknown, it is plausible that PS mutations associated with predominant CAA favor the production of more soluble Aβ40 that is drained and deposited along the perivascular space in association with Aβ42 nidi (81).

3.3. Tau Mutations

As previously mentioned, the deposition of hyperphosphorylated tau as insoluble filaments in the brain is a pathological hallmark of several neurodegenerative disorders, defined as tauopathies. An autosomal-dominant inherited form of frontotemporal dementia with parkinsonism (FTDP) was initially linked to chromosome 17q21-22 in 1994 (88). In the following years, 13 additional families with FTD and parkinsonism with linkage to 17q21-22 were identified (45). These patients presented with disinhibition, loss of initiative, obsessive-compulsive behavior, and/or psychosis, followed by cognitive decline. In most patients, extrapyramidal symptoms occurred only late in the clinical course, but considerable heterogeneity is observed both between mutations and within families with the same mutation (89). In 1997 at a consensus conference, the term of frontotemporal dementia with parkinsonism linked to chromosome 17 (FTDP-17) was introduced to describe these patients (89) and in the following year mutations in MAPT were reported in the majority of these families (90,91). At present, 39 mutations in MAPT have been identified in 115 families, including missense mutations, silent mutations, in-frame single codon deletions, and intronic mutations (45). Most FTDP-17 MAPT mutations result in increased 4R/3R tau ratios, and modify tau splicing pattern or cause single aa changes, which probably result in impaired microtubule binding capacity or increased aggregability (45). On gross neuropathology, FTDP-17 patients present with a severe atrophy of the frontal and temporal lobes (or blade-edge atrophy described in classical Pick's disease), although the superior part of the precentral gyrus is commonly spared. On microscopy, astrocytic gliosis and a massive neuronal loss are found in the superficial cortical layers and sometimes also in basal ganglia (92). Filamentous tau intracytoplasmic inclusions in neuronal and glial cells are the characteristic finding. Extensive mutation analysis revealed many other MAPT mutations in several additional FTDP-17 families, as well as in unrelated FTD, CBD, or PSP patients (for a complete update, visit www.molgen.ua.ac.be/FTDMutations).

4. Amyloid and Tau Deposition

4.1. Physical States of Aβ

In brain parenchyma, Aβ exist as monomers, dimers, low molecular weight oligomers, mycelia, protofibrils, and finally, mature densely compact fibrils (Fig. 1). While most of the Aβ assembly is extracellular, recent data also suggest that Aβ dimers could be constitutively secreted from neurons (93). Except the insoluble mature fibrils, all other precursor forms can be regarded as the soluble Aβ fraction, and most of these states including dimers but excluding monomers are toxic (94). While the soluble monomers and early aggregation states do not cross the threshold for immunohistochemical detection, densely packed aggregates are also less visible by immunohistochemistry due to epitope masking (see further). Thus, immunohistochemistry is most robust in identifying higher-order Aβ aggregation states including loosely arranged fibrils. Densely packed fibrils also acquire a β-pleated sheet conformation (95,96) and are therefore readily identified by silver impregnation techniques and Congo red or thioflavin (Th)-S or -T (95,96). Based on these different staining protocols, various types of extracellular and intracellular amyloid can be identified in both humans and mouse AD models.

4.2. Diffuse Deposits

Diffuse deposits are the first extracellularly visible Aβ deposits observed in AD and Down's syndrome patients (11,80,97), and are also termed preamyloid deposits or preplaques (98). In AD brains, diffuse plaques are mainly present in the entorhinal cortex, allocortex (hippocampus and olfactory cortex), followed by the superficial layers of the neocortices, especially association neocortices (99). Occasionally, a cloud of diffuse plaques are observed under the cortical pia mater or the subependymal periventricular zones (99). These plaques also involve neostriatum and the molecular layer of cerebellar cortex (100). Various morphological forms of diffuse plaques have been described. However, in most frequent forms they appear as clusters of dots and rods with irregular borders merging diffusely with the parenchyma, and ranging in size from a few microns to many hundreds of microns (Fig. 2).

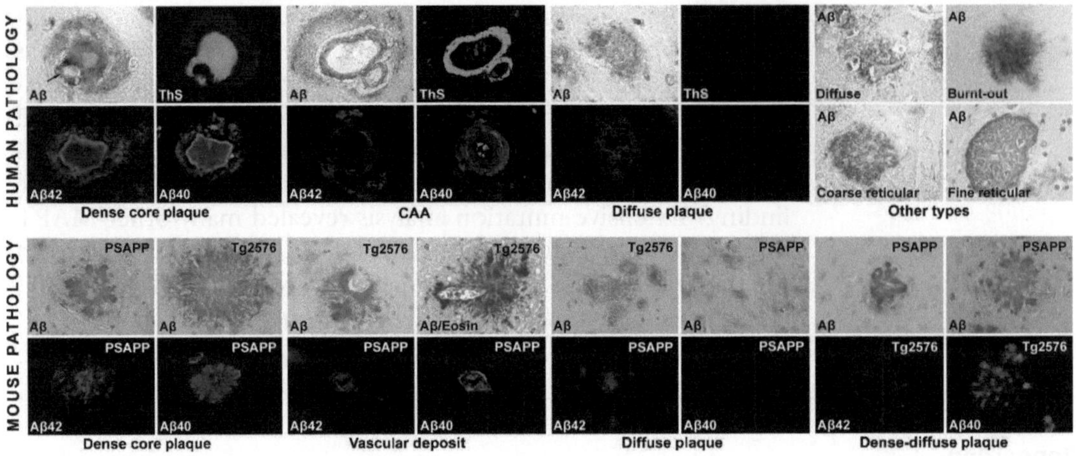

Fig. 2. Amyloid pathology in AD patients and mouse models. *Upper panel*: Spectrum of amyloid deposits in human AD pathology. Dense-core plaques typically have a ThS-positive compact core composed of Aβ40 and are typically surrounded by a corona composed chiefly of Aβ42. The center of the cores usually is faintly stained with anti-Aβ antibodies as a result of the densely packed fibrils. In the majority of the cases, a blood vessel can be identified on serial sections abutting to the dense core (*arrow*). CAA is ThS-positive, compact ring-like deposition of amyloid in the vessel wall that is chiefly composed of Aβ40 while the neighboring parenchymal diffuse amyloid is chiefly constituted of Aβ42. Diffuse-type of plaques are constituted of fine grains of nonfibrillar amyloid depositions, which is ThS-negative and mainly constituted of Aβ42. Normal appearing neurons can be sometimes discerned within diffuse plaques. Burnt-out types of plaques are represented by compact cores in a region devoid of neuronal bodies and neurites. Coarse reticular-type of plaques are composed of mesh-like dense-diffuse amyloid depositions, while a denser reticular pattern surrounded by a clear amyloid-parenchyma rim is described as fine reticular type of plaques. *Lower panel*: Spectrum of amyloid deposits encountered in mouse models. Dense-core plaques usually consist of multiple smaller cores (PSAPP model) or more organized, single cores (APP model), but are distinct from human pathology as mice lack peri-core amyloid. The plaques are positive for Aβ40 and Aβ42, with the central dense regions staining faintly due to highly aggregated amyloid. Vascular amyloid deposition is a frequent finding in AD mouse models too, appearing as dysphoric angiopathy affecting small-to-medium sized blood vessels (shown here in Tg2576 mice) with Aβ40 predominating over Aβ42 (shown here in a PSAPP mouse). Diffuse plaques in mice biochemically and morphologically resemble human diffuse plaques being also ThS-negative and composed predominantly of Aβ42. Collections of diffuse plaques frequently occupy large areas in the parenchyma in PSAPP mice but more restricted in Tg2576. Dense-diffuse plaques are intermediate forms between diffuse and dense-core pathology with multiple nidi, not bridged as dense-core plaques, and sometimes surrounded by diffuse pathology (illustrated here in PSAPP mice). They are also chiefly composed of Aβ40 perhaps representing increasingly aggregating amyloid (in Tg2576 here).

Ultrastructural studies of such deposits have revealed scattered bundles of amyloid fibrils as well as amorphous, nonfibrillar material (101). Diffuse plaques frequently surround neuronal cell bodies and neurites without implicating vascular elements (101), suggesting that such deposits might be neuronally derived. Although it was shown that the progression to the end-stage of AD is associated with a proportional increase in fibrillar plaque isoforms (102), recent studies suggest independent evolution paths for fibrillar-dense and diffuse plaques (103). Biochemically, diffuse plaques seem to be composed of full-length and N-truncated Aβ42 (100,104). Moreover, diffuse plaques neither seem to associate with any cytoskeletal neuritic alterations nor display any significant astrogliosis or microglial activation (100,105,106). Only exceptionally, in end-stage diseases or in very aggressive familial forms of AD, such as Austrian

T714I pathology, diffuse plaques are observed to be neuritic and associated with abundant glial and inflammatory pathology (76,100,107).

AD mouse models based on APP and PS mutations also develop diffuse plaques that resemble human diffuse plaques biochemically and morphologically, in that they are also ThS-negative and predominantly composed of Aβ42. In mice, diffuse deposits can exist as small patches of amyloid, but given the supraphysiological levels of Aβ production, the number of plaques increases rather rapidly to occupy large areas of brain parenchyma. All brain regions can be affected, although the first region to be involved largely depends on the specific promoter utilized. Certain brain regions such as cerebellum and basal ganglia seem to have a paucity of diffuse deposits (further discussed under section *Mouse Models*).

4.3. Compact Amyloid Deposits with Dense Cores

Dense-cored plaques were first observed in 1892 and named "senile plaques" in 1911 by Simchowicz (108). These plaques consist of a dense core of amyloid fibrils arranged radially, surrounded by an immediate clear halo and an external belt of amyloid. Sometimes, long arms of amyloid connect the surrounding corona with the inner core of the plaque. The central core regions are one of the most compact forms of amyloid observed in human brain, and are readily stained with Congo-red or ThS, appearing round, star, or cross-shaped ("Malthese cross"). With Aβ immunohistochemistry, the dense-core regions stain faintly perhaps due to the masking of Aβ epitopes. This type of dense aggregated cored plaques is the most characteristic finding in the Flemish APP pathology, with the "largest" dense-cored plaques encountered in AD (41). Dense-core plaques are also called neuritic plaques as they are frequently surrounded by dystrophic neurites with or without NFT pathology or neuropil threads and also seem to incite both astrocytosis and microgliosis (99). Occasionally, dense-core plaques seem to lack a coronal plaques region as well as dystrophic neurites and these compact cores are sometimes referred to as "burnt-out" or "end-stage" plaques with the belief that the surrounding neuronal elements are destroyed in these plaques. Biochemically, while the corona of the plaques has predominantly, if not solely, Aβ42, the central dense-core regions are rich in Aβ40. Thus, Aβ42 is present in both the core and the corona of senile plaques although in different proportions, but it seems that increased Aβ42 is a common feature for both normal aging and senile AD cored-type of deposits (109). It is possible that with increasing compactness and aggregation state of the plaque cores, an increasing ratio of Aβ40/Aβ42 occurs for the deposited amyloid.

The highest number of dense-cored plaques is found in the superior hippocampal regions (CA1–CA2), subiculum, layers II, III, and V of the entorhinal cortex, and the plaques become fewer and larger towards CA4 (110,111). Dense plaques together with reticular plaques (see later) are also observed in lower cortical layers of the association cortices, such as the temporal association cortex. Primary motor and visual cortices also deposit mostly dense-core plaques. In the basal ganglia, pallidum is usually the only striatal nucleus where dense plaques appear. In cerebellum, senile plaques are abundant in the Purkinje and granular cell layers (100) and, interestingly, the highest incidence of amyloid plaques in the cerebellum is found in presenile patients (112). Substantia nigra and other brainstem nuclei also deposit dense plaques and the deposition is proportional to the amyloid burden found in cerebellum (112).

Dense-cored plaques are also observed in mouse models, especially in the environment of a high Aβ40/Aβ42 ratio (81). Importantly, dense-cored plaques lack the characteristic surrounded corona of diffuse amyloid, and frequently show presence of multiple small cores, especially in PSAPP mice where Aβ42 levels are high. Biochemically, the dense cores resemble human dense senile cores being ThS positive and rich in Aβ40. As mentioned, peri-plaque corona is mostly absent, but when present, is predominantly of Aβ42-type as shown in AD. Interestingly, dense-core plaques in mouse models most resemble similar plaques observed in the Flemish APP pathology (34). Dense-cored plaques are observed in all regions of the mouse brain with a preference for the hippocampal region and the hippocampal sulcus (34,107).

4.4. Compact Deposits Without Amyloid Cores

Compact deposits lacking a discernable amyloid core are also called "primitive plaques" due to an arguable viewpoint that they progress in time to form dense-core plaques. Such plaques are the most common type of plaques in the cerebral cortex in most AD patients and consist of agglomerated wisps of amyloid deposits arranged in a reticular pattern (99). Typically, these plaques contain almost exclusively Aβ42. When the faint reticular pattern has no readily demarcated edge, these are called "coarse" reticular plaques, and seem to predominate in AD, especially in FAD patients such as those harboring PS1 G384A mutation (113). When the reticular pattern is clearly demarcated by a strong amyloid edge on immunohistochemistry, these are called "fine" reticular dense plaque. The latter type of plaques is the most frequent compact, noncored plaque found in the Flemish APP pathology (41) (see further).

In mice models, diffuse plaques are seen especially in very early ages, and dense-diffuse plaques are more common (34). The compacted region contains more Aβ40, and diffuse deposits are predominantly Aβ42-positive. Dense-diffuse deposits can appear throughout the mouse brain, with a relative sparing of the basal nuclei and cerebellum.

4.5. Vascular Amyloid Deposition

When Aβ is deposited in the walls of small-to-medium-sized blood vessels of the brain and of leptomeninges and incites a pathological response, it is referred to as cerebral (congophilic or ThS-positive) amyloid angiopathy of the Aβ type (114). More than 10% of the persons over 60 years of age and ≈80% of the AD patients have at least some degree of CAA (115). CAA affects most frequently the occipital, temporal, and frontal lobes, being less frequently found in hippocampus, cerebellum, and basal ganglia, and almost absent in the white matter and brain stem (116). Leptomeningeal arteries are most frequently involved, followed by cortical arteries, arterioles, and capillaries, while the venous sector is less frequently involved (116). While Aβ deposits seem to appear in the media near the smooth muscle cells in larger vessels, capillary CAA (cap-CAA) has Aβ deposited first on the abluminal side of the basement membrane (117). Interestingly, cap-CAA have frequently small perivascular caps of amyloid or perivascular amyloid fibrils radiating in the surrounding brain tissue, described as dysphoric angiopathy or "drusige Entartung" (114,118,119). In pathologies where CAA is a prevailing feature such as Flemish APP mutation, dysphoric angiopathy is also observed within larger vessels (41). Biochemically, Aβ40 is the major constituent of CAA affecting large leptomeningeal and neocortical vessels, while both Aβ40 and Aβ42 are present in CAA affecting smaller parenchymal vessels.

Vascular Aβ depositions also cause secondary changes in vessels such as hyalinization and degeneration as indicated by increased representation of collagen-IV in the basement membrane. Not infrequently, "double-barreling" or "vessel-within-vessel" configuration is also observed due to deposition of Aβ both in the vessel wall and under glia limitans, separated by the distended perivascular spaces (116,120). Thus, not surprisingly, CAA accounts for approximately 10% of all intracerebral nonfatal hemorrhages in the elderly (121). Interestingly, CAA also seems to be capable of initiating a strong local neuritic and inflammatory pathology (86) and in this respect cap-CAA, but not CAA associated with larger vessels, seems to correlate well with parenchymal pathology (119,122).

Mouse models of amyloidosis also show significant CAA pathology (123–125). As in humans, vascular Aβ in mouse models is also ThS-positive and chiefly constituted of Aβ40. Both CAA and dysphoric angiopathy can be found throughout the mouse brain including thalamus (34).

4.6. Intraneuronal Aβ

Recent data also indicate that Aβ is deposited within neurons in brains of AD and Down's syndrome patients (126,127). While this is a relatively new and understudied area of amyloidosis, it is highly likely that intraneuronal Aβ, similar to a plethora of inclusion proteinopathy diseases, may induce cytopathic effects with important roles in AD pathogenesis. Several mouse models have been established that deposit intraneuronal Aβ either solely or in early stages of diseases and shall be discussed in relevant sections.

4.7. NFT and NF Threads

Tau is the major intraneuronal proteopathy in AD inside the body of the neurons (as NFTs) or along dystrophic neurites (as neurophil threads). Early studies on tau deposits recognized phosphorylated tau as the major component of PHF (128–131). In AD, PHF consist of 10–20 nm filaments distributed especially in the entorhinal cortex, hippocampus, amygdala, and association temporal cortex (100). Staging of AD based on the prevalence of NFT pathology is also proposed, which seems to correlate better with the cognitive deficits compared to criterions based on amyloid deposits (132,133).

Tau is also the main pathological finding in FTLD-tau disorders such as Pick's disease. The histopathological features of Pick's disease are dominated by argyrophilic tau-positive inclusions (Pick bodies) and swollen achromatic cells (Pick cells). They usually occur in the granular layer of dentate gyrus, pyramidal cells of the CA1, subicular hippocampal sectors, neocortex, and in several subcortical nuclei. In the neocortex, they predominate in the layers II and VI, in contrast to the neurofibrillary tangles in AD that appear especially in layers III and V. By electron microscopy, Pick bodies contain intermediate filaments, 15 nm straight filaments, and some paired helical filaments (134). Biochemical characterization showed that the insoluble tau in Pick bodies consists of 3R-tau isoforms (135,136). However, recent studies have demonstrated much greater biochemical heterogeneity and up to 50% of patients with Pick's disease having either at least as much 4R as 3R-tau, or even a predominance of 4R-tau (49,137,138). Mouse models expressing various isoforms of tau are discussed hereunder.

5. Mouse Models of Dementia

Several mouse models of AD have been established based on FAD mutations that develop amyloid pathology. The discovery of tau mutations in FTDP-17 families also brought new perspectives with regard to common tau dysfunctions in AD and FTDP-17 (88). We have reviewed below some of the most important mouse models based on these pathogenic mutations that show how these models have expanded our molecular understanding of diseases like AD or FTD. For a complete list of AD mouse models, visit www.alzforum.org/res/com/tra.

5.1. APP Mouse Models

5.1.1. PDAPP and APP/Ld2 (APPIndiana and APPLondon) Models

The PDAPP mouse model was the first successful mouse model established in the study of AD. It carries the human APP770 with the Indiana V717F mutation under the control of human platelet-derived growth factor-β (PDGF-β) neuronal promoter and on a mixed C57BL/6, DBA, and Swiss-Webster genetic background (123). More than tenfold overexpression of the APP transgene occurred compared to the murine endogenous APP levels. The high expression level is perhaps due to the transgenic construct utilizing a splicing cassette that permits the expression of all three major APP isoforms. Also, the transgene is predominantly expressed in neurons in the cortex, hippocampus, hypothalamus, and cerebellum. Amyloid pathology is detected around 6–9 months of age, beginning in the hippocampus, corpus callosum, and cerebral cortex. The deposits range from diffuse to ThS-positive dense plaques and CAA, and the size and densities of these deposits increase with age. Dystrophic neurites are present in the vicinity of amyloid depositions; however, no NFT is identified. GFAP-positive activated astrocytes surround the plaques while isolated, activated microglia are detected throughout the neocortex. Synaptic and dendritic densities are reduced in the molecular layer of the dentate gyrus, but no significant neuronal loss is recorded here or in the neocortex. Behavioral deficits begin to appear around 3 months of age, with impairments in visuospatial reference memory (139), spatial discrimination tasks, spontaneous object-recognition, and operant learning (140). As we shall see later for other mouse models as well, behavioral deficits occur before immunohistochemically detectable amyloid depositions appear in brain (Table 1).

Four years later, a model carrying another mutation on the same codon, the V717I London APP mutation, was published as the APP/Ld2 model (141). This transgenic mouse strain carried the V642I mutation on the APP695 isoform under the regulation of the mouse thymus cell antigen 1 promoter (Thy-1) and on a FVB/N background. Human APP insert was expressed two- to fivefold higher than the APP endogenous level in hippocampus and neocortex. However, in contrast to the PDGF promoter (utilized in the PDAPP model) that drives expression predominantly in the neocortex, the murine Thy-1 promoter drives transgene expression more widely, including expression in the subcortical regions. Amyloid deposition occurred around 12 months of age as both diffuse and compact neuritic plaques were most abundant in the hippocampus and cortex, and only occasionally in thalamus and white matter. APP/Ld2 mice also develop significant vascular amyloid depositions primarily in pial, cortical, thalamic, and hippocampal vessels with predominant Aβ40 accumulation (142). Behavioral deficits begin to appear around the age of 8 weeks with alternating episodes of hyperactivity, anxiety, aggression, decreased exploration, and ambulation (141).

Table 1
Neuropathology of commonly utilized mouse models in Alzheimer's disease and other tauopathies

Protein model	Line/ Reference	Genetic background	Promoter	Associated pathology	Cognitive and behavioral changes
APP models	PDAPP (124)	Human APP770 with Indiana (V717F) mutation	PDGF-β	Amyloid depositions at 6–9 months; mostly diffuse deposits (Aβ42 rich) and CAA (Aβ40 rich). Significant loss of SMCs and hemorrhages; no neuronal loss; no NFT	Impairments in visuospatial reference memory, spatial discrimination tasks, spontaneous object-recognition and operant learning at ≈3 months
	APP/Ld2 (142)	Human APP695 with London (V717I) mutation	Thy-1	Amyloid plaques and CAA at 10–12 months. Loss of SMCs but no hemorrhages; cholinergic fiber distortions	Alternating episodes of hyperactivity, anxiety, aggression, decreased exploration, and ambulation starting at ≈8 weeks
	Tg2576 (125)	Human APP695 with Swedish double (K670N/M671L) mutation	Hamster PrP	Plaque at ≈9–12 months; mostly dense deposits and CAA (Aβ40 rich). Loss of SMCs and hemorrhages; no neuronal loss; no NFT	Spatial alternation task and longer escape latencies in the water maze at ≈9–10 months
	APP23 (126)	Human APP751 with Swedish mutation	Thy-1	Amyloid deposition at 6 months; mostly CAA and dense plaques. Loss of SMCs and hemorrhages; neurodegenerative changes with hyperphosphorylated tau, but no NFT	Severely decreased learning and training performances at 3 months
	Austrian (149)	Human APP695 with Austrian (T714I) mutation	PDGF-β	Intraneuronal Aβ in hippocampus by 6 months. Intraneuronal N-truncated Aβ42 co-localize with endosomal-lysosomal pathway markers; brain volume reduction on MRI	Subtle behavioral impairments in light–dark transition box tests at 12 months

	TgCRND8 (154)	Human APP695 with Swedish and Indiana mutations	Hamster PrP	Amyloid depositions at 6 weeks in subiculum and frontal cortex; dense cored deposits appear first in subiculum, while diffuse depositions appear later. Neuritic pathology and synaptic degeneration present	Impairment in the acquisition of spatial information during place discrimination training at ≈11 months
	APPDutch (156)	Human APP with Dutch (E693Q) mutation	Thy-1	Amyloid pathology around 22–25 months; most depositions as CAA where mutant DutchAβ40 predominates. Loss of SMCs and hemorrhages	NA
	Tg-SwDI (157)	Human APP770 with Swedish, Dutch and Iowa (D694N) mutations	Thy-1	Amyloid pathology starts at 3 months with diffuse-like plaques and vascular accumulations in thalamic region at 6 months; Decreased Aβ42/Aβ40 ratio	NA
	hAPP-Sw/Ind (159)	Human APP with Swedish and Indiana mutations	PDGF-β	Amyloid pathology at 7 months with dense-like deposits and vascular deposition. Dystrophic neurites present	NA
	hAPP-Arc/Sw/Ind (159)	Human APP with Arctic (E693G), Swedish, and Indiana mutations	PDGF-β	Amyloid deposition starts around 2.3 months and is associated with dystrophic neurites. Introducing Arctic mutation in Sw/Ind mice decreases the Aβ42/Aβ40 ratio, and favors parenchymal depositions	NA
PSAPP models	PS1 (159)	Human PS1 with M146L or M146V mutation	PDGF-β	No amyloid pathology up to 12 months; Higher Aβ42/Aβ40 ratio	No behavioral abnormalities up to 12 months

(continued)

Table 1 (continued)

Protein model	Line/ Reference	Genetic background	Promoter	Associated pathology	Cognitive and behavioral changes
	Tg2576 × PS (162)	Human APP695 with Swedish mutation Human PS1 with M156L mutation	Hamster PrP	Accelerated amyloid pathology with smaller and more numerous dense plaques, less diffuse depositions. Increased Aβ42/Aβ40 ratio	Impaired Y-maze spontaneous alternations at ≈3 months
	APPSw × PS (161)	Human APP with Swedish mutation and human PS1 with A246E mutation	Mouse PrP	Accelerated amyloid pathology with smaller and more numerous dense plaques, less diffuse depositions. Increased Aβ42/Aβ40 ratio	NA
	APP/Ld × PS1(164)	Human APP695 with London mutation and human PS1 with A246E mutation	Thy-1	Dense amyloid cores and CAA load increased more than double compared to APP/Ld mice. Increased soluble brain Aβ levels	NA
	TgCRND8 × PS (154)	Human APP with Swedish and Indiana mutations and human PS1 with M146L and L286V mutations	Hamster PrP	Accelerated amyloid plaque pathology appearing at 4–6 weeks	NA
	APPDutch × PS1 (156)	Human APP with Dutch mutation and human PS1 with G384A mutation	Thy-1	Amyloid pathology around 6 months as both diffuse and compact plaques (DutchAβ42 twice as abundant as DutchAβ40)	NA

	APP × PS1-knockin (152)	PS1 M233T and L235P knock-in and human APP751 with London and Swedish mutations	Thy-1 (PS1 knockin)	First Aβ depositions at 2.5 months that become generalized by 6 months. Amyloid deposits associated gliosis and distrophic neurites; Severe neuronal loss in hippocampus by 10 months	NA
Aβ-only expressing mice	Bri-Aβ42/ Bri-Aβ40 (166)	Fusion BRI and Aβ42 (or Aβ42) protein	Mouse PrP	Bri-Aβ42 mice start depositing compact plaques and CAA at 3 months in the cerebellum and at 6 months in the cortex. Diffuse pathology appears with increasing age. Dense plaques associated with reactive astrogliosis and rare dystrophic neurites	NA
Tau models	JNPL3 (167)	Human four repeat tau with P301L mutation	Mouse PrP	NFT in motor neurons in the spinal cord at ≈6.5 months and progress to neuronal loss in brain stem and midbrain	Absence of escape extension during tail elevation at ≈5 months, weakness in all limbs, dystonic postures, decreased vocalization, and decreased weight
	JNPL3 × Tg2576 (170)	Human four repeat tau with P301L mutation and human APP with Swedish mutation	Mouse PrP	Amyloid depositions comparable with those of Tg2576 mice. Significant increase of NFT as well as insoluble tau levels in limbic regions compared to JNPL3 mice	Same as for the JNPL3 × Tg2576 mice
	APP Sw/Tau P301L/ PS1M146V (153)	Human four repeat tau with P30lL mutation in knock-in PS1 (M146V)	Thy-1 (PS1 knockin)	Intraneuronal Aβ deposit at 3 months and extracellular plaques at 6 months. At 18 months NFT occur in the hippocampus	Retention/retrieval deficits appear at ≈4 months

APP, amyloid precursor protein; CAA, cerebral amyloid angiopathy; NFT, neurofibrillary tangles; PDGF, plateled-derivated growth factor; PrP, prion protein promoter; PS, presenilin; SMCs, smooth muscle cells; Thy-1, thymus cell antigen 1 promoter

5.1.2. Tg2576 and APP23 (APPSwedish) Models

The Tg2576 mouse model expresses human APP695 harboring the Swedish double mutation (K670N/M671L) under the control of the hamster prion protein (PrP) promoter and on a C57B6/SJL background (124). Tg2576 expresses transgenic APP sixfold over the endogenous murine APP level. Total Aβ secretion increases three- to eightfold, and although cellular secretion of Aβ42/Aβ40 ratio is not changed, these mice predominantly deposit Aβ40 (143). Amyloid pathology is not detected in animals below 3 months of age. Around 9–10 months, amyloid deposits can be identified beginning in the entorhinal and piriform cortices and continuing in the neocortex and cerebellum, in a pattern similar to that of the PDAPP model. These consist mainly of large dense-core plaques and CAA (rich in Aβ40), and occasionally of diffuse-like deposits (rich in Aβ42). Regions that deposit most of the dense deposits were neocortex and subiculum. In the thalamic region, there is an abundance of CAA and cap-CAA type of depositions (34). Compared to the PDAPP model that carries the London APP mutation and thus causes decreased Aβ40 production, this model increases the absolute amounts of Aβ40 and corroborates with a more compact amyloid pathology. Amyloid depositions are surrounded by dystrophic neurites and gliotic changes. Neither NFTs nor neuronal loss could be identified (144). Nevertheless, memory deficits such as poor performance in spatial alternation task and longer escape latencies to the hidden platform in the water maze are observed at 9–10 months (124).

Like the Tg2576 mouse model, the APP23 model also carries the Swedish double mutation, but in the APP751 human isoforms and under the control of murine Thy-1 promoter, and on a C57BL/6 background (30,125). In these mice, the transgene expression is sevenfold over murine APP levels and leads to amyloid deposits beginning with ≈6 months of age in the hippocampus, and later involving the neocortex, thalamus, and olfactory bulbs. APP23 mice also present with alterations of the cholinergic system, synaptic button loss, and neuronal loss (145). As mice age, they also deposit vascular amyloid in the majority of the pial vessels and in a significant fraction of the vessels in thalamus, cortex, and hippocampus (30). In these mice, vascular amyloid is associated with SMC degeneration leading to microhemorrhages, and they have been extensively utilized as a model for CAA with spontaneous hemorrhagic stroke (33,146). APP23 mice also show severely decreased learning and training performances by 3 months of age (147).

5.1.3. APP/Austrian and Similar Models

The APP/Austrian mouse model (APP-Au) (148) was based on identification of the APP (T714I) mutation in an Austrian family, with one of the earliest age of onsets in AD of ≈34 years, the patients having abundant intracellular and extracellular amyloid deposits in the brain (76). The latter, strikingly, was nonfibrillar diffuse amyloid, composed of N-truncated Aβ42 in the absence of Aβ40, as shown in other mouse models (149). *In vitro*, this mutation led to one of the highest Aβ42/Aβ40 ratios among all familial AD mutations studied (76,77). The APP-Au mouse model carries the T714I mutation on a human APP695 isoform, under the control of the PDGF-β promoter. Despite having tenfold lower transgene expression than endogenous murine APP, intraneuronal hippocampal Aβ deposits appear by 6 months of age. Accumulations increase with age, and parallel decreased brain sizes on volumetric MRI, compared to age-matched and similar transgene-expressing APP wild-type mice. Subtle behavioral impairments were noted in the light–dark transition box tests. In this model, the majority of the intraneuronal Aβ deposits colocalized with markers of endosomal–lysosomal pathway, supporting the hypothesis that intraneuronal accumulation of Aβ could be an important factor in AD pathogenesis. Interestingly, mice expressing both Austrian and Swedish double mutations despite having fivefold higher transgene levels comparted to APP/Au mice have significantly lower intraneuronal Aβ and lack the reduced brain volume phenotype (150). APP/Sw is processed in the secretory pathway instead of the endosomal–lysosomal pathway necessary for intraneuronal Aβ accumulation thereby suggesting that intraneuronal Aβ accumulation is related to reduced brain volumes in APP-Au mice. These data also find support in mouse models based on mutant PS1-knockin where intraneuronal N-truncated Aβ42 (but not extracellular Aβ40 or Aβ42) could correlate with neuronal loss (151,152) (see later).

5.1.4. TgCRND8 (APPSwedish/Indiana) Model

The TgCRND8 line carries the Swedish double mutation and the Indiana (V717F) mutation on a human APP695 isoform under the control of the hamster PrP promoter, and on a C3H/B6 background (153). γ-Secretase site APP mutations lead to a drastic reduction of Aβ40 or total Aβ (76,77), and Swedish APP mutation ensures a high Aβ secretion level, for instance, levels of Aβ42 in 6-month-old TgCRND8 mice are close to those seen in PDAPP mice at 16 months of age. Amyloid deposition started at ≈6 weeks of age in TgCRND8 mice, appearing first in the subiculum and frontal cortex, and later also involving corpus callosum, hippocampus, dentate gyrus, olfactory bulbs, and pial vessels. Thalamus, striatum, and cerebellum were the last regions involved where amyloid deposited at 16–35 weeks of age. The first amyloid deposits appeared as cored deposits in subiculum, changing to larger, multicored deposits with age. Diffuse pathology occurred in the cortex around 35 weeks, and by 45 weeks, diffuse amyloid depositions appeared throughout the cortex. In the olfactory bulbs, diffuse pathology occurred even earlier but remained unchanged until late ages. Dense plaques associated with dystrophic neurites, synaptic degeneration, and inflammatory response. From 11 weeks onwards, TgCRND8 mice show impairment in the acquisition of spatial information during place discrimination training (153).

5.1.5. APP/Dutch Model

The E693Q Dutch APP mutation leads to extensive CAA with recurrent cerebral hemorrhages and dementia in the HCHWA-D syndrome. Although earlier attempts with Dutch APP transgenesis did not show amyloid deposition till 18 months of age (154), a new transgenic mouse model based on this mutation has recently been generated that closely mimics human HCHWA-D pathology (155). The Dutch mice drive APP751 with the E693Q mutation under the control of murine Thy-1 promoter, and with a fivefold transgenic overexpression compared to murine APP levels. The transgenic expression, restricted to neurons and neuronal processes, was predominantly observed in the neocortex, hippocampus, and brain stem. Dutch Aβ40 was readily detected in 23-months-old mice, and vascular amyloid depositions could be also identified around this age. CAA appeared first in the leptomeningeal vessels followed by cortical vessels, while parenchymal depositions were restricted to a few diffuse accumulations. Affected vessels showed a severe loss of SMC with multiple surrounding fresh hemorrhages, and were associated with a strong perivascular microglial inflammatory reaction.

5.1.6. Tg-SwDI (APPSwedish/Dutch/Iowa) and Similar Lines

A step forward in studying the relation between Aβ and the vascular compartment was the introduction of two vasotropic APP mutations in a single construct. The Tg-SwDI model harbors human APP770 carrying the Swedish APP mutation together with the Dutch and Iowa (D694N) mutations, under the control of the Thy-1 murine promoter and on a C57Bl/6 background (156). The transgene expression was less than 50% of the level of endogenous mouse APP. Starting with 3 months of age onwards, mice had numerous amyloid deposits appearing first in subiculum, hippocampus, and cortex, and by 6 months of age also involved olfactory bulbs and thalamus. CAA involving predominantly cerebral microvessels but also large meningeal vessels appeared at about 6 months and was associated with reduced microvessel densities, endothelial loss, accumulation of inflammatory cells, and occasionally, microhemorrhages (157). Mutant Aβ40 predominated and although parenchymal accumulations presented as diffuse-like depositions, interestingly, all ThS-positive amyloid deposits colocalized with blood vessels (157).

Another recent mouse line expressing vasotropic human Aβ was APP Arctic (E693G) mutation along with APP Swedish and APP Indiana mutations, Tg-SwAI, under the control of the PDGF-β promoter and on a C57BL/6J background (158). This line had decreased Aβ42/Aβ40 ratios and induced neuritic amyloid plaques at 2.3 months, which was earlier and more extensive than in the Tg-SwDI line. In contrast to APP Dutch mice that deposits almost exclusively vascular amyloid with a predominance of Aβ40 in the parenchyma, Tg-SwAI mice also have high Aβ40 levels but develop only parenchymal plaques with no obvious increase in the CAA formation, a phenomenon that seems to indicate that fibrillogenetic properties of a specific mutant Aβ species may be a prime factor in its preferential vascular vs. parenchymal deposition.

5.2. PSAPP Models

5.2.1. PS1 Model

The first mouse model carrying a presenilin pathogenic mutation harbored the PS1 M146L or M146V mutation, under the control of the PDGF-β promoter and on a SW/B6D2F1 background (159). This initial model did not show any pathology and behavioral abnormalities up to 12 months. However, soluble Aβ42 was increased in these models and this was the basis of use of mutant presenilin transgenic mice in crossbreeding experiments with, for instance, APP transgenic mice discussed above.

5.2.2. APP/Sw × PS1 Models

Crossbreeding a model harboring a mutation that enhances the total production of soluble Aβ (Swedish APP mutation) and a mutation that preferentially increases Aβ42 secretion was an important step in the study of the role of Aβ42 in the pathogenesis of amyloid deposition (160,161). The Holcomb et al. model resulted from crossbreeding Tg2576 with PS1 M146L mice under the control of hamster PrP promoter (161). The crossbred mice had drastically accelerated amyloid pathology. For instance, amyloid pathology developed in neocortex and hippocampus at ≈3 months in these mice compared to Tg2576 where in the same regions pathology developed after 9 months. Even in older animals, amyloid plaques, although with a higher degree of compactness, were smaller but more numerous compared to the single transgenic Tg2576 model, supporting the hypothesis that the fibrillar Aβ42 is important in seeding the initial amyloid nidus, while Aβ40 is more important in the growth of the deposits (34). Associated changes included neuritic pathology, gliosis, and a modest neuronal loss (162). At about 3 months, mice also showed behavioral and cognitive deficits such as impaired Y-maze spontaneous alternations (161).

The other PSAPP mouse model carried the human APPSw together with the human PS1 A246E mutations, both under transcriptional control of mouse PrP promoter (160). The initial sites of amyloid deposition were the hippocampus, subiculum, and the neocortex, and this double model again showed at least twice accelerated amyloid pathology compared to single transgenic APPSw animals.

5.2.3. APP/Ld × PS1 and TgCRND8 × PS1 Model

APP/Ld2 mice were crossbred with PS1 A246E mice, under the control of the Thy-1 promoter and on a FVB × C57BL6 background (163). With APP/Ld mutation already increasing Aβ42/Aβ40 ratio, only a further 50% increase in Aβ42 levels occurred compared to APP/Ld2 mice. Both parenchymal and vascular depositions increased compared to APP/Ld2 mice, again underlining an important role of Aβ42 in vascular amyloid pathogenesis.

The same group that published the TgCRND8 Sw/Ind mice generated another mouse model that co-expresses 2 PS1 familial AD mutations (M146L and L286V) over the TgCRND8 line, a model that further increases Aβ42 production (153). As expected, this quadruple transgenic model deposited plaques already at the age of 30–45 days.

5.2.4. PS45 (APPDutch × PS1) Model

PS45 mice line was created by crossbreeding APPDutch mice with mice expressing the PS1 G384A mutation under the control of murine Thy-1 promoter (155). The G384A PS1 mutation is known to increase Aβ42 production (164), and consistent with these data, the level of mutant Aβ42Dutch was twice as the Aβ40Dutch in these mice (155). Crossbred PS45 mice line had consistently more parenchymal amyloid in the neocortex and hippocampus beginning with the age of 3 months, compared to APPDutch mice. By the age of 6 months, massive diffuse and compact parenchymal amyloid depositions were observed in all brain regions, with a massive decrease for the CAA type of depositions.

5.2.5. APP × PS1-Knockin Mouse Models

A number of PS1-knockin mouse models also have been made. In one such example, $APPS_{LP}S1KI$ mouse model harboring two knocked-in human PS1 mutations (M233T, L235P) and two APP mutations (APP/Swedish and APP/London) had pathology beginning at a very young age. Not only Aβ deposited as plaques, but also as intraneuronal accumulations, and interestingly intraneuronal N-truncated Aβ42 (but not extracellular Aβ40 or Aβ42) correlated with the neuronal loss observed in this model (151). Such models together with APP/Austrian mice (see above) are thus very important tools also in studying the role of intraneuronal Aβ in AD.

5.3. βA-Expressing Mice in Absence of APP/PS Mutations

5.3.1. Bri-Aβ40 and 42 Mice

Recently, transgenic mice that express Aβ40 or Aβ42 without APP overexpression have been constructed (165). In these mice, mouse PrP promoter drives expression of a fusion protein consisting of the BRI protein, involved in amyloid deposition in Familial British Dementia (FBD) (166), and either Aβ40 (BRI-Aβ40) or Aβ42 (BRI-Aβ42) peptides. The fusion transgene was expressed at endogenous mouse APP expression levels for BRI-Aβ42 mice and at approximately half this level for BRI-Aβ42 mice. While BRI-Aβ40 mice did not develop amyloid pathology at any age, BRI-Aβ42 mice deposited compact ThS-positive Aβ42 plaques beginning with 3 months of age in the molecular layer of cerebellum, while diffuse pathology appeared later and became more prominent with aging. Plaques also occurred in allocortex from 6 months of age. CAA occurred also as a prominent pathological change in BRI-Aβ42 mice with wisps of amyloid radiating from the abluminal surface of the vessels, like described for human AD. Compact plaques were associated with reactive astrogliosis and rare dystrophic neurites. Crossbreeding BRI-Aβ42 mice with Tg2576 (predominantly Aβ40 secreting) resulted in a marked augmentation of Aβ deposition throughout the cortex, hippocampus, and cerebellum. The studies on these mouse models suggest that Aβ40 also has considerable role in the progression of amyloidosis itself.

5.4. Tau and More Complex Models

5.4.1. JNPL3 (Tau P301L) Model

The JNPL3 line of transgenic mouse model harbors human 4R-tau containing the P301L mutation under the control of the mouse PrP promoter and on a SW/B6D2F1 background (167). The P301L is the most common mutation causing FTDP-17 and FTD-tau pathology. JNPL3 mice showed NFT pathology in the motor neurons of spinal cord beginning at the age of 6½ months and involving other regions such as brain stem and cerebellar nuclei as the phenotype progressed. The pretangle type of pathology was more common than NFT pathology and was observed in paleocortex, hippocampus, and basal ganglia. In pretangle pathology, neurons express abnormally phosphorylated tau that does not aggregate into fibrils, as also observed for human pathology. Furthermore, consistent with human FTDP-17 pathology, the JNPL3 mouse model demonstrated a shift from the soluble tau forms towards the insoluble forms, accompanied by neuronal loss in affected areas. For instance, spinal cord motor neurons were reduced by ≈50% accompanied by fibrillary gliosis and axonal degeneration in JNPL3 mice. Homozygous JNPL3 mice were also established that overexpress human mutated tau at ≈ twofold higher levels than hemizygous mice and, as expected, also show a more severe neuropathology. Behaviorally, absence of escape extension during tail elevation was observed around 6.5 months in hemizygotes mice, and at 4.5 months in homozygous mice. Within 2 weeks of the disease onset, weakness was observed for all limbs, accompanied by dystonic postures, decreased vocalization, and decreased weight. Within 3–4 weeks of initial signs, mice became moribund (167). These mice did not develop Aβ amyloid pathology.

5.4.2. TAPP (JNPL3 × Tg2576) Model

The first mouse to recapitulate both NFT and amyloid-related AD changes expressed both the P301L tau mutation and the Swedish APP mutation (168). TAPP mice develop amyloid plaques that are similar in distribution and morphology to Tg2576 model. Although NFT have a similar morphology with those described for JNPL3 mice, 9-months-old TAPP mice have a markedly increased NFT accumulation in limbic regions that is sevenfold higher than in JNPL3 mice of the same age. Interestingly, the limbic areas with enhanced NFT pathology of TAPP mice coincide with the areas where Aβ pathology appears first in the Tg2576 mice. Although there was no difference in Aβ42/Aβ40 ratios between Tg2576 and JNPL3 mice, or in the total soluble tau levels between TAPP and JNPL3 mice, insoluble tau species were more abundant in the TAPP mice in some areas such as cortical/limbic regions but not in the subcortical areas. However, mice show similar motor disturbances as observed for JNPL3 mice with similar age of disease onset (168). Nevertheless, these results showed for the first time a synergistic effect of APP/Aβ over tau pathology.

5.4.3. Triple Tg (APPSw × PS1 M146V × Tau P301L) Model

Finally, a triple transgenic model has been established where constructs with pathogenic mutations in APP (Swedish mutation) and human 4R-tau harboring the P30lL mutation under the control of the Thy-1 promoter are co-injected in single-cell embryos derived from PS1 (M146V) knocked-in mice (152,169). Both APP and tau transgenes express to comparable levels in the same brain regions in these mice. Intraneuronal Aβ (mostly Aβ42) is the first detectable neuropathological feature occurring at ≈3 months in the neocortex, and at 6 months in CA1 pyramidal cells. Extracellular Aβ depositions occur for the first time in the frontal cortex at approximately 6 months, followed by involvement of other cortical regions and of hippocampus by 12 months. Hyperphosphorylated tau begins to accumulate in hippocampal neurons by 12–15 months, after the appearance of plaques, and by 18 months, readily identifiable NFT start appearing in the hippocampus and neocortex. Interestingly, the triple transgenic mice also show important synaptic dysfunctions like LTP deficits that precede amyloid and NFT pathology. Cognitive impairments appear by 4 months of age as a retention/retrieval deficit, paralleling the intraneuronal Aβ accumulations in olfactory cortex (170). Importantly, clearance of the intraneuronal accumulations by immunotherapy alleviates cognitive impairments and recurrence of Aβ pathology leads to the recurrence of cognitive deficits (170). These models thus show a sequential evolution of amyloid and tau pathology and support the amyloid cascade hypothesis of AD pathogenesis, and also underline the tight link between intraneuronal Aβ and cognitive impairment in AD.

5.5. FTLD-U Mouse Models

While tau-based models are excellent tools to study FTD-tau such as FTDP-17, recent data suggest that FTD presenting with only ubiquitin-positive aggregates without tau-positive inclusions or tau gene mutations (FTLD-U) are more common (138,171,172). For instance, recent estimates suggest that FTLD-U accounts for more than 50% of all autopsy-confirmed FTLD (138). A number of FTLD-U causing genes have been identified recently such as valosin-containing protein (*VCP*) (173), charged multivesicular body protein 2B (*CHMP2B*) gene (174), and progranulin (*PGRN*) (175,176). PGRN could also be a risk factor for development of other dementias as PGRN mutations known to be causing FTLD-U have been identified in other dementias like AD (177). Interestingly, the culprit protein in ubiquitin-positive inclusions in FTLD-U and a related entity, motor neuron disease (MND) has also been identified recently as the TAR DNA-binding protein (TDP-43) (178). Thus, all necessary genetic insights are in place to identify molecular mechanisms involved in FTLD pathogenesis.

PGRN knockout mice were generated before the PGRN link with FTLD-U was known (179). PGRN knockout mice show significantly reduced transcript levels of serotonergic receptor 5-HT1A in the hippocampus. As PGRN is involved in estrogen-induced neurogenesis in the adult rat hippocampus (180), a down-regulation of 5-HT1A might be related to the significant neuronal loss and the selective high density of ubiquitin-positive inclusions observed in the dentate gyrus of FTLD-U patients. Behavioral analysis of PGRN knockout mice show decreased male sexual activity, enhanced aggressiveness, and anxiety. Interestingly, similar behavioral alterations are observed in FTLD-U patients (181) and treatment with serotonin reuptake inhibitors alleviate these symptoms (182,183). Many groups, including ours, are involved in further utilization of PGRN knockout mice as well as VCP or TDP-43 transgenic mice to elucidate FTLD-U disease mechanisms.

6. Mouse Models Shedding Light on Disease Pathogenesis

In principle, the mouse models discussed above provide support for four concepts that are important when discussing amyloid deposition versus amyloid clearance. First, mouse models have shown that Aβ has the potential to diffuse away from the site of production, as shown by transplantation experiments where non-transgenic brain tissue grafted in APP23 hosts develops both diffuse and congophilic amyloid plaques as early as 3 months after grafting (212). Second, APP23 mice crossbreed with APP knockout mice demonstrate that neuronally derived Aβ could predominantly drive Aβ pathology in both parenchymal and vascular compartments (30). Third, these mouse models also lend support to the periarterial ISF Aβ drainage hypothesis and it is now well accepted that CAA might indeed be a result of the failure of this mechanism and entrapment of Aβ in the periarterial space (29). And lastly, as we saw earlier, that transgenic AD and FTD mouse models have provided us with proof of principle concepts on Aβ and tau relationship with cognitive and behavioral abnormalities that occur in some of these mouse models. Furthermore, mouse models such as APP23 and Tg2576 have shown that vascular amyloidosis is indeed the cause of spontaneous hemorrhages as both cerebral microhemorrhages and fatal lobar hemorrhages are observed in these mouse models (34,146). In the following sections, we have discussed some mechanism(s) commonly proposed to be involved in the formation of CAA and parenchymal amyloid deposition as well as mechanism(s) proposed for Aβ and tau interaction.

6.1. Mechanisms of CAA Formation

Several hypotheses have been proposed to explain the development of cerebrovascular amyloidosis, of which the more common ones are the systemic, vascular, and drainage hypothesis. The "systemic hypothesis" proposes a hematogenous origin for Aβ and is based on the observation that most of the body cell types are able to express APP and secrete Aβ that can deposit in vessel walls. A major argument that goes against it is that ultrastructural observations of the earliest vascular Aβ deposits show Aβ deposition towards the abluminal side of BBB (between the basement membrane and the glia limitans) (184). The "vascular hypothesis" proposes a local production of Aβ from within the vessels themselves, primarily driven by the fact that essentially all elements of vessels express APP (32,117,185,186). The major argument against the vascular hypothesis is that large arteries consisting of several layers of SMCs are less severely affected by CAA than small arteries, and that CAA is also present in capillaries that lack SMC (as cap-CAA).

The more commonly accepted "drainage hypothesis" suggests that neuronally produced Aβ drains along the ISF in the perivascular spaces of parenchymal and leptomeningeal vessels, and that CAA occurs due to deposition of Aβ along these drainage pathways (29). This hypothesis has been well supported by studies in mouse models that express pathogenic APP mutations under the control of neuron-specific promoters, such as in the Tg2576 and APP23 mice, and still develop CAA (124,125). In these mouse models, perhaps the local clearance factors such as phagocytosis and protease degradation are unable to keep up with the grossly overproduced Aβ, which is further drained towards vessels. Interestingly, CNS has one of the highest water contents and injections of India ink elegantly demonstrate these perivascular clearance pathways in rodent CNS (187). Similarly, for the human sporadic variant of the disease, aging could be an important factor where decreased local degradation of Aβ would coerce more Aβ trafficking towards vessels. A saturation of vascular Aβ clearance across the BBB in transgenic mouse models or age-dependent decrease in transporters across the BBB (26) could perhaps coerce Aβ precipitation along the vessel walls (29) (Fig. 3). Interestingly, Aβ40 is the major component of CAA-Aβ deposits (188) and further supports the neuronal-vascular Aβ drainage

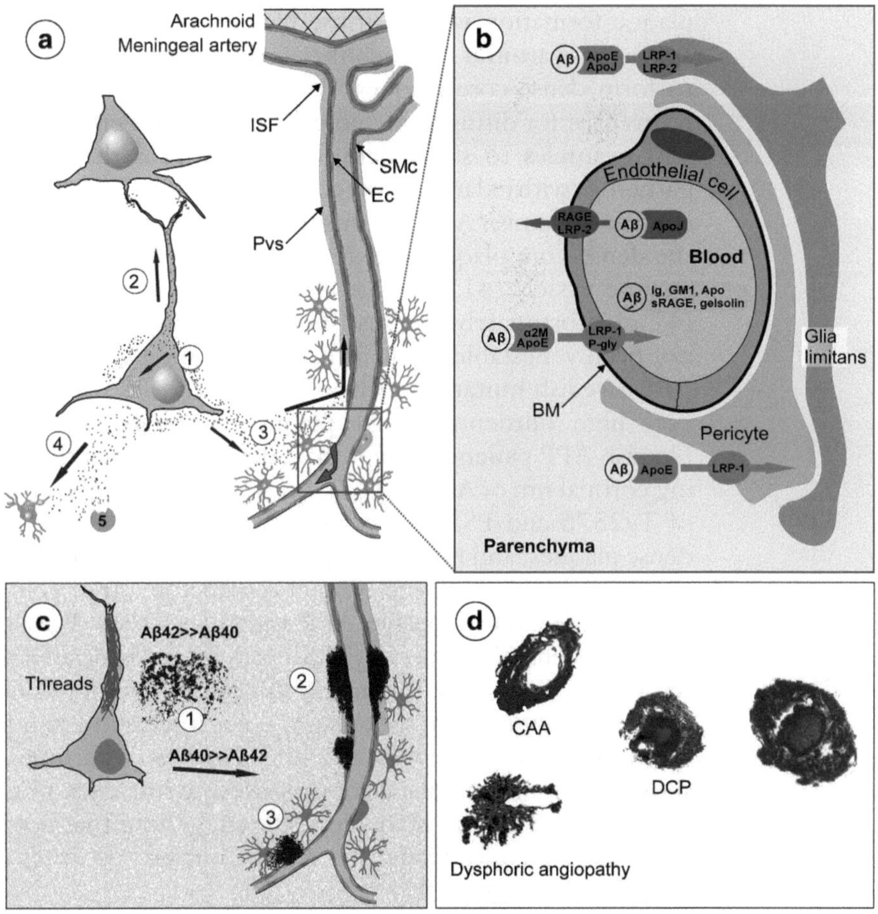

Fig. 3. Physiological and pathological Aβ metabolism. (**a**) Normal Aβ metabolism in the brain. Intraneuronally, Aβ produced by the neurons is either degradated into the lysosomes (1), transported along the axons to elicit synaptic modulation and growth (2), or is secreted in the surrounding neurophil. Parenchymal Aβ is drained around the perivascular space (3) into the interstitial fluid (ISF), or actively transported directly through the endothelium of the blood brain barrier (arrow within the box). Aβ diffusing in the parenchyma can also be degradated by a number of enzymes as well as microglia (4 and 5). (**b**) At the level of BBB, Aβ is transported into the blood flow via a transcytosis mechanism mediated by LRP-1 and P glycoprotein transporters, in association with chaperon lipoprotein molecules (α2 macroglobulin and apolipoprotein E). A reverse transport from the blood towards the parenchyma is mediated by RAGE and LRP-2 receptors in association with apolipoprotein J. Aβ might be sequestered in the blood flow by immunoglobulins, apolipoproteins, soluble RAGE receptors, and gelsolin. Astrocyte endfeet and pericytes also mediate Aβ intake by expressing LRP-1 and 2 receptors. (**c**) As the production exceeds the clearance of Aβ, it begins to deposit. Neuronal metabolism is impaired, axonal transport is blocked, and dystrophic neurites accumulate hyperphosphorylated tau. Amyloid depositions can range from diffuse deposits in the vicinity of secretion situses (1), to CAA and dense plaques around the vascular basement membranes (2 and 3).

hypothesis as a major mechanism of vascular Aβ deposition as also discussed in the next section.

6.2. Mechanism Involved in the Formation of Plaques

Two common mechanisms for the formation of plaques have been proposed – angiocentric and neurocentric development of plaques. While angiocentric theory suggests that plaques form around vessels, the neurocentric theory suggests neurons to be responsible for plaque formation where diffuse plaques of predominantly Aβ42 appear first around neurons that further coerce Aβ40 deposition to form dense-core plaques. It is possible that different mechanisms exist for diffuse and dense-core plaques. Perhaps, one of the best examples to study dense-core plaques is the Flemish APP pathology with its large dense-core deposits (41,65). Morphological studies of Flemish APP patients have identified that the majority of the dense-core plaques are directly associated with vessel walls or enclose vessels (41). Diffuse plaques had no relation to vessels (41). Unfortunately, a mouse model of Flemish APP mutation has not been yet established (154); however, mouse models based on APP Swedish mutation such as Tg2576 and PSAPP mice show a very high burden of dense-plaque pathology as described in Flemish APP patients, except that such plaques lack the surrounding coronal rim of Aβ (34,107). Interestingly, morphometric study of Tg2576 and PSAPP has also shown that up to 90% of these dense plaques, and not diffuse plaques, are centered on vessel walls (34). Similar observations were also made on a transgenic mouse model based on vasotopic APP mutant such as TgSwDI mice where all dense Aβ deposits (ThS-positive) are associated with vessel walls (157). While the jury is still out in regard to dense-core plaque development in sporadic AD, it is interesting to note that a recent study on sporadic AD and Down's syndrome patients showed a significant association between dense-core plaques and Prussian blue-labeled heme deposits, suggesting that not only are dense core plaques related to vessels, but are also sites of older microhemorrhages (189).

So how does dense-core plaque deposit in association with vessels in these transgenic mouse models and Flemish APP pathology? First, mass spectrometric and immunohistochemical analysis of Flemish APP pathology and Tg2576 and PSAPP mouse models have shown a preponderance of Aβ40 constituting both CAA and dense-core plaques (41,143). As Aβ40 is less fibrillogenic than Aβ42, these data suggest that Aβ40 could be one factor that might dictate a preferential deposition of dense-core plaques. A further reason to believe that less fibrillogenic Aβ is preferentially cleared by the vascular drainage is the observation that Aβ40 in the vessel-related deposits is always full-length (i.e., Aβ1–40), which is less fibrillogenic than N-terminally truncated Aβ (190). Conversely, situations that lead to decreased Aβ40 secretion also lead to a drastically reduced prevalence of CAA and dense-core plaques as observed in Austrian pedigree with APP T714I mutation where ≈80% reduction of Aβ40 production *in vitro* and near-absence of brain Aβ40 coincides with near-absence of CAA and dense-core plaques in the brain (76). To parallel this observation, in animal models with increasing ratios of more fibrillogenic Aβ42 either shift the pathology from the vascular to the parenchymal compartment as observed when Dutch mice are crossbred with mutant PS1 mice (155), or to smaller but higher number of plaques when Tg2576 mice are crossbred with mutant PS1 mice (34). Moreover, the *in vivo* potential of less fibrillogenic Aβ40 forming dense deposits is demonstrated by the experimental formation of congophilic, fibrillar dense deposits and electron microscopy-proven fibril formation by injection of soluble Aβ40 in rat brain, while soluble Aβ42 deposited as diffuse deposits only (191). Thus, it seems that development of diffuse plaques and dense-core plaques are distinct processes, where a predominant Aβ42 secretion drives diffuse Aβ pathology and a predominant Aβ40 secretion drives the formation of dense-core plaques. However, the role of Aβ42 in the formation of the dense plaques is still very important as it provides the initial nidus. To exemplify this, BRI-Aβ42 mice although predominantly depositing diffuse plaques, also develop CAA and, conversely, BRI-Aβ40 mice secreting similar levels of Aβ40 in the absence of Aβ42 develop neither plaque nor CAA (165).

6.3. Mechanism of Tau Formation and Relation of Tau and Amyloid

Production of tau and double tau/APP transgenic mice offered the opportunity to study more closely the pathogenic relation between tau and Aβ (167,168). Introduction of a pathogenic APP mutation that increases the overall Aβ production (APPSw) on tau P301L mice revealed enhanced deposition of NFTs compared to the single P301L line (168). Although the morphology of the NFTs between the two lines was not different, areas that first developed amyloid depositions in APP/Sw mice revealed a shift toward insoluble tau isoforms and an enhanced NFT pathology. This showed a clear interaction between pathogenic APP or Aβ leading to increased tau phosphorylation and aggregation (Fig. 3). Interestingly, amyloid deposition was not enhanced in at least two models of tau/APP mice when compared to APP mice, suggesting that the tau–Aβ interaction might be a unidirectional pathway (168,192).

Injection of fibrillar Aβ42 in 6-months-old P301L mice addressed to the relation between readily fibrilized Aβ42 and tau metabolism (193). New NFTs began to form after 18 days from the injection and still increased in number at 60 days. Overall, injected mice had five times more NFTs compared to P301L mice, proving a clear amplification effect of fibrillar Aβ42 on tau phosphorylation and aggregation. If injected into the hippocampus, amyloid depositions occurred also in the amygdala, probably by axonal transport as already mentioned (13). Moreover, injecting diluted brain extracts from Aβ-depositing APP23 mice into P301L mice showed increased tau pathology, both in the injection sites and in projectional-related areas, like in amygdala in this case, even in the absence of local amyloid deposits (192).

All together, it seems that many Aβ species present in soluble Aβ, fibrilized Aβ42 and parenchymal-derived Aβ, retain their capacity of altering tau metabolism. The work on tau mice is still very recent, but this will definitely lead to a more complete understanding of how the two proteinopathies interact and lead to a mixed amyloid/NFT pathology even in distant brain regions.

7. Mouse Models Directing Anti-Amyloid Therapeutic Strategies

Aside from being instrumental in exploring various pathogenic mechanisms, mouse models have also proved essential in testing therapeutic hypothesis based on modulating amyloid, tau, antioxidants, caloric input, metal toxicity, and other proposed pathways (194). Regarding our major focus in this chapter, amyloid pathology, mouse models are currently being utilized in testing strategies employed in reducing amyloid production (such as BACE and γ-secretase targeting), reducing Aβ aggregation (especially those involving formation of high-molecular-weight oligomers), as well as in immune- and nonimmune-mediated Aβ clearing strategies. We specifically focus here on Aβ clearance strategies on mouse models based on familial APP and PS mutations that enhance Aβ production, as these are also relevant models to understand the more common sporadic AD where an overwhelming majority of cases are thought to be due to impairment of Aβ degradation pathways. We showed that these models have provided us with proof of principle concepts in *in vivo* Aβ toxicity and of pathology that is similar to AD. For instance, some mouse models we studied have ultrastructural microvascular abnormalities such as endothelial cell loss, basement membrane thickening, and degeneration of SMCs and pericytes, changes that are typically found in AD brains. We also saw that such mouse models of amyloidosis exhibit behavioral and cognitive abnormalities similar to AD. Thus, mouse models could be considered bona fide models of AD-amyloidosis and could be utilized in prehuman clinical trials for testing treatment efficacy and for exploring possible side-effects.

Thus far, only immunotherapeutic approaches have offered one of the very few disease-modifying treatments in AD. The lead came in 1999, when studies on PDAPP mice showed that active immunization with synthetic Aβ42 peptides mounted a sustained immune response against amyloid and resulted in a decrease in the total amyloid burden in these mice (195). These results were also confirmed on TgCRND8 mice where active Aβ42 immunization reversed behavioral and cognitive abnormalities and coincided with ≈50% reduction in dense-core plaques (196). Similarly, passive immunization studies on PDAPP mouse model with an anti-Aβ monoclonal antibody (m266) also revealed a pathological and cognitive recovery (197,198). Thus, both active and passive immunization in mouse models provided an important proof of principle of reversing amyloid deposition, stopping neurodegenerative changes and improving cognitive performance.

Based on these data, the first clinical trials on AD patients started in 2001 with an active immunization procedure based on aggregated Aβ42 peptides (AN-1792 trial) (199). The clinical trial was halted in 2002, as up to 6% of the patients showed signs of aseptic meningoencephalitis, with an abundant infiltration of autoreactive T-lymphocytes into leptomeninges, CAA laden regions, and perivascular spaces (200). Nevertheless, many patients had generated anti-Aβ antibodies and showed significantly slower rates of cognitive decline and to some extent had a limited amount of plaque depositions in isocortical regions that still displayed neuritic and glial pathology, suggesting that immunization could result in amyloid burden clearance in humans (201).

At present, the precise mechanisms underlying the anti-amyloidotic effect of immune therapy are unknown and so are the mechanisms of the side-effects such as meningoencephalitis. Three major mechanisms of Aβ removal following immunization have been proposed: (i) direct solubilization of amyloid, (ii) phagocytosis by microglia, and (iii) "peripheral sink" hypothesis (202,203). The fact that like in AD pathology, studies on mouse models have shown a severe loss of BBB impermeability for serum proteins (34), and that small amount of circulating antibodies could cross even an intact BBB (204), provide a strong support for the hypothesis involving a direct anti-amyloidotic effect of antibodies in the SNC. First, this direct solubilization hypothesis is based on the fact that antibody-mediated disaggregation of Aβ fibrils has been described both *in vitro* and in peripheral-injected PDAPP mice (205,206). To some extent, it seems that antibodies raised against the N-terminal region of Aβ are more prone to inhibit and reverse fibrilization and neurotoxic effects of Aβ fibrillar forms (205). Second, strong evidence suggest that the affinity of the antibodies for Fc receptors on microglial cells is even more important compared to the affinity for Aβ itself, and that complement activation is not a prerequisite for plaque clearance (207). These studies on PDAPP mice suggest microglial Fc-mediated plaque removal as an efficient amyloid clearing and cognitive rehabilitation mechanism (207,208). Third, in a mutually nonexclusive mechanism, peripherally injected antibodies in PDAPP mice are proposed to sequester Aβ, creating a drainage effect on the brain-deposited amyloid, in what is called the "peripheral sink" hypothesis (209). To parallel this, active immunization studies inducing the formation of only large pentameric IgM antibody complexes, revealed a decrease of the total Aβ burden in Tg2576 mice (210).

Similarly, mouse models have been instrumental in understanding some of the side effects of anti-Aβ vaccinations. An important observation is breakdown of BBB, for instance in Tg2576 and PSAPP models, which have dense plaques spatially related to blood vessels and also exhibit ultrastructural evidence of BBB abnormalities (34). Moreover, in advanced stages of CAA when Aβ replaces all the vascular wall elements, any aggressive Aβ removal strategy is predicted to cause acute breakdown of BBB that could not only cause acute hemorrhages, but also auto-immune responses in an antigenically isolated organ such as the brain. In support of this, APP23 mouse model receiving passive anti-Aβ immunization showed severe CAA-associated microhemorrhages (211) and autopsy of one of the encephalitic patients from the active immunization trial revealed that antibody titers in CSF equaled those in plasma, again indicating a severe breakdown of the BBB (201).

8. Conclusions

In this chapter, we focused on AD and FTD, the two most common causes of presenile dementia and a growing cause of concern in Western societies. We especially reviewed pathological events that lead to Aβ and tau deposition as amyloid plaques and NFTs as well as to groundbreaking genetic research that also facilitated the development of AD and FTD mouse models. With a viewpoint of genotype–phenotype correlation and comparing human and mouse pathology, we discussed not only how these mouse models closely mimic the pathology and the behavioral deficits described in the human disease, but also how these studies have given us important insights into the disease mechanisms that lead to formation of different types of brain depositions. And finally, we also discussed how these mouse models are being further utilized in supporting ongoing therapeutic strategies to treat these devastating diseases.

References

1. Kawas CH (2003) Clinical practice. Early Alzheimer's disease. N Engl J Med 349(11):1056–1063.
2. Alzheimer A (1907) Über eine eigenartige erkrankung der Hinrinde. Zentralblatt für Nervenheilkunde und Psychiatrie 18:177–179.
3. Goate A, Chartier HM, Mullan M, et al. (1991) Segregation of a missense mutation in the amyloid precursor protein gene with familial Alzheimer's disease. Nature 349(6311):704–706.
4. Sherrington R, Rogaev EI, Liang Y, et al. (1995) Cloning of a gene bearing mis-sense mutations in early-onset familial Alzheimer's disease. Nature 375:754–760.
5. Levy-Lahad E, Wasco W, Poorkaj P, et al. (1995) Candidate gene for the chromosome 1 familial Alzheimer's disease locus. Science 269(5226):973–977.
6. Selkoe DJ (1998) The cell biology of ß-amyloid precursor protein and presenilin in Alzheimer's disease. Trends Cell Biol 8(11):447–453.
7. Kang J, Muller-Hill B (1990) Differential splicing of Alzheimer's disease amyloid A4 precursor RNA in rat tissues: PreA4(695) mRNA is predominantly produced in rat and human brain. Biochem Biophys Res Commun 166(3):1192–1200.

8. Cupers P, Orlans I, Craessaerts K, Annaert W, De Strooper B (2001) The amyloid precursor protein (APP)-cytoplasmic fragment generated by gamma-secretase is rapidly degraded but distributes partially in a nuclear fraction of neurones in culture. J Neurochem 78(5):1168–1178.
9. Roberts SB, Ripellino JA, Ingalls KM, Robakis NK, Felsenstein KM (1994) Non-amyloidogenic cleavage of the beta-amyloid precursor protein by an integral membrane metalloendopeptidase. J Biol Chem 269(4):3111–3116.
10. Vassar R, Bennett BD, Babu-Khan S, et al. (1999) Beta-secretase cleavage of Alzheimer's amyloid precursor protein by the transmembrane aspartic protease BACE [see comments). Science 286(5440):735–741.
11. Gowing E, Roher AE, Woods AS, et al. (1994) Chemical characterization of A beta 17–42 peptide, a component of diffuse amyloid deposits of Alzheimer disease. J Biol Chem 269(15):10987–10990.
12. Rajendran L, Honsho M, Zahn TR, et al. (2006) Alzheimer's disease beta-amyloid peptides are released in association with exosomes. Proc Natl Acad Sci U S A 103(30):11172–11177.
13. Kamal A, Almenar-Queralt A, LeBlanc JF, Roberts EA, Goldstein LS (2001) Kinesin-mediated axonal transport of a membrane compartment containing beta-secretase and presenilin-1 requires APP. Nature 414(6864):643–648.
14. Dodart JC, Mathis C, Ungerer A (2000) The beta-amyloid precursor protein and its derivatives: From biology to learning and memory processes. Rev Neurosci 11(2–3):75–93.
15. Zheng H, Jiang M, Trumbauer ME, et al. (1995) b-Amyloid precursor protein-deficient mice show reactive gliosis and decreased locomotor activity. Cell 81:525–531.
16. Lazarov O, Lee M, Peterson DA, Sisodia SS (2002) Evidence that synaptically released beta-amyloid accumulates as extracellular deposits in the hippocampus of transgenic mice. J Neurosci 22(22):9785–9793.
17. Sheng JG, Price DL, Koliatsos VE (2002) Disruption of corticocortical connections ameliorates amyloid burden in terminal fields in a transgenic model of Abeta amyloidosis. J Neurosci 22(22):9794–9799.
18. Seubert P, Vigo-Pelfrey C, Esch F, et al. (1992) Isolation and quantification of soluble Alzheimer's b-peptide from biological fluids. Nature 359:325–327.
19. Hardy J, Selkoe DJ (2002) The amyloid hypothesis of Alzheimer's disease: Progress and problems on the road to therapeutics. Science 297(5580):353–356.
20. Wyss-Coray T, Loike JD, Brionne TC, et al. (2003) Adult mouse astrocytes degrade amyloid-beta in vitro and in situ. Nat Med 9(4):453–457.
21. Saido TC, Iwata N (2006) Metabolism of amyloid beta peptide and pathogenesis of Alzheimer's disease. Towards presymptomatic diagnosis, prevention and therapy. Neurosci Res 54(4):235–253.
22. Leissring MA, Farris W, Chang AY, et al. (2003) Enhanced proteolysis of beta-amyloid in APP transgenic mice prevents plaque formation, secondary pathology, and premature death. Neuron 40(6):1087–1093.
23. Yasojima K, McGeer EG, McGeer PL (2001) Relationship between beta amyloid peptide generating molecules and neprilysin in Alzheimer disease and normal brain. Brain Res 919(1):115–121.
24. Shibata M, Yamada S, Kumar SR, et al. (2000) Clearance of Alzheimer's amyloid-ss(1–40) peptide from brain by LDL receptor-related protein-1 at the blood-brain barrier. J Clin Invest 106(12):1489–1499.
25. Zlokovic BV (2005) Neurovascular mechanisms of Alzheimer's neurodegeneration. Trends Neurosci 28(4):202–208.
26. Zlokovic BV (2004) Clearing amyloid through the blood-brain barrier. J Neurochem 89(4):807–811.
27. Bading JR, Yamada S, Mackic JB, et al. (2002) Brain clearance of Alzheimer's amyloid-beta 40 in the squirrel monkey: A SPECT study in a primate model of cerebral amyloid angiopathy. J Drug Target 10(4):359–368.
28. Kida S, Pantazis A, Weller RO (1993) CSF drains directly from the subarachnoid space into nasal lymphatics in the rat. Anatomy, histology and immunological significance. Neuropathol Appl Neurobiol 19(6):480–488.
29. Weller RO, Massey A, Newman TA, Hutchings M, Kuo YM, Roher AE (1998) Cerebral amyloid angiopathy: Amyloid beta accumulates in putative interstitial fluid drainage pathways in Alzheimer's disease. Am J Pathol 153(3):725–733.
30. Calhoun ME, Burgermeister P, Phinney AL, et al. (1999) Neuronal overexpression of mutant amyloid precursor protein results in prominent deposition of cerebrovascular amyloid. Proc Natl Acad Sci U S A 96(24):14088–14093.
31. Bateman RJ, Munsell LY, Morris JC, Swarm R, Yarasheski KE, Holtzman DM (2006) Human amyloid-beta synthesis and clearance rates as measured in cerebrospinal fluid in vivo. Nat Med 12(7):856–861.
32. Kawai M, Kalaria RN, Cras P, et al. (1993) Degeneration of vascular muscle cells in cerebral amyloid angiopathy of Alzheimer disease. Brain Res 623(1):142–146.

33. Winkler DT, Biedermann L, Tolnay M, et al. (2002) Thrombolysis induces cerebral hemorrhage in a mouse model of cerebral amyloid angiopathy. Ann Neurol 51(6):790–793.
34. Kumar-Singh S, Pirici D, McGowan E, et al. (2005) Dense core plaques in Tg2576 and PSAPP mouse models of Alzheimer's disease are centered on vessel walls. Am J Pathol 167:527–543.
35. Mucke L, Masliah E, Yu GQ, et al. (2000) High-level neuronal expression of abeta 1–42 in wild-type human amyloid protein precursor transgenic mice: Synaptotoxicity without plaque formation. J Neurosci 20(11):4050–4058.
36. Dickson DW (2004) Commentary - Building a more perfect beast - APP transgenic mice with neuronal loss. Am J Pathol 164(4):1143–1146.
37. Mattson MP (1997) Cellular actions of beta-amyloid precursor protein and its soluble and fibrillogenic derivatives. Physiol Rev 77(4):1081–1132.
38. Meda L, Cassatella MA, Szendrei GI, et al. (1995) Activation of microglial cells by beta-amyloid protein and interferon-gamma. Nature 374(6523):647–650.
39. El KJ, Hickman SE, Thomas CA, Cao L, Silverstein SC, Loike JD (1996) Scavenger receptor-mediated adhesion of microglia to beta-amyloid fibrils. Nature 382(6593):716–719.
40. Breitner JC, Gau BA, Welsh KA, et al. (1994) Inverse association of anti-inflammatory treatments and Alzheimer's disease: Initial results of a co-twin control study. Neurology 44(2):227–232.
41. Kumar-Singh S, Cras P, Wang R, et al. (2002) Dense-core senile plaques in the Flemish variant of Alzheimer's disease are vasocentric. Am J Pathol 161(2):507–520.
42. Weingarten MD, Lockwood AH, Hwo SY, Kirschner MW (1975) A protein factor essential for microtubule assembly. Proc Natl Acad Sci U S A 72(5):1858–1862.
43. LoPresti P, Szuchet S, Papasozomenos SC, Zinkowski RP, Binder LI (1995) Functional implications for the microtubule-associated protein tau: Localization in oligodendrocytes. Proc Natl Acad Sci U S A 92(22):10369–10373.
44. Gu Y, Oyama F, Ihara Y (1996) Tau is widely expressed in rat tissues. J Neurochem 67(3):1235–1244.
45. Rademakers R, Cruts M, Van Broeckhoven C (2004) The role of tau (MAPT) in frontotemporal dementia and related tauopathies. Hum Mutat 24(4):277–295.
46. Buee L, Bussiere T, Buee-Scherrer V, Delacourte A, Hof PR (2000) Tau protein isoforms, phosphorylation and role in neurodegenerative disorders. Brain Res Rev 33(1):95–130.
47. Harada A, Oguchi K, Okabe S, et al. (1994) Altered microtubule organization in small-caliber axons of mice lacking tau-protein. Nature 369(6480):488–491.
48. Kanemaru K, Takio K, Miura R, Titani K, Ihara Y (1992) Fetal-type phosphorylation of the tau in paired helical filaments. J Neurochem 58(5):1667–1675.
49. Mott RT, Dickson DW, Trojanowski JQ, et al. (2005) Neuropathologic, biochemical, and molecular characterization of the frontotemporal dementias. J Neuropathol Exp Neurol 64(5):420–428.
50. Rogaev EI, Sherrington R, Rogaeva EA, et al. (1995) Familial Alzheimer's disease in kindreds with missense mutations in a gene on chromosome 1 related to the Alzheimer's disease type 3 gene. Nature 376:775–778.
51. Roses AD (1996) Apolipoprotein E alleles as risk factors in Alzheimer's disease. Annu Rev Med 47:387–400.
52. Glenner GG, Wong CW (1984) Alzheimer's disease: Initial report of the purification and characterization of a novel cerebrovascular amyloid protein. Biochem Biophys Res Commun 122:885–890.
53. Masters CL, Simms G, Weinman NA, Multhaup G, McDonals BL, Beyreuther K (1985) Amyloid plaque core protein in Alzheimer disease and Down syndrome. Proc Natl Acad Sci USA 82:4245–4249.
54. Van Broeckhoven C, Haan J, Bakker E, et al. (1990) Amyloid beta protein precursor gene and hereditary cerebral hemorrhage with amyloidosis (Dutch). Science 248(4959):1120–1122.
55. Levy E, Carman MD, Fernandez-Madrid IJ, et al. (1990) Mutation of the Alzheimer's disease amyloid gene in hereditary cerebral hemorrhage, Dutch type. Science 248(4959):1124–1126.
56. van Duinen SG, Castaño EM, Prelli F, Bots GTAB, Luyendijk W, Frangione B (1987) Hereditary cerebral hemorrhage with amyloidosis in patients of Dutch origin is related to Alzheimer disease. Proc Natl Acad Sci USA 84:5991–5994.
57. Haan J, Hardy JA, Roos RAC (1991) Hereditary cerebral hemorrhage with amyloidosis-Dutch type: Its importance for Alzheimer research. TINS 14:231–234.
58. Maat-Schieman ML, Yamaguchi H, van Duinen SG, Natte R, Roos RA (2000) Age-related plaque morphology and C-terminal heterogeneity of amyloid beta in Dutch-type hereditary cerebral hemorrhage with amyloidosis. Acta Neuropathol (Berl) 99(4):409–419.
59. Bornebroek M, Haan J, Van Buchem MA, et al. (1996) White matter lesions and cognitive deterioration in presymptomatic carriers

of the amyloid precursor protein gene codon 693 mutation. Arch Neurol 53(1):43–48.
60. De Jonghe C, Zehr C, Yager D, et al. (1998) Flemish and Dutch mutations in amyloid beta precursor protein have different effects on amyloid beta secretion. Neurobiol Dis 5(0969–9961):281–286.
61. Bornebroek M, De JC, Haan J, et al. (2003) Hereditary cerebral hemorrhage with amyloidosis Dutch type (AbetaPP 693): Decreased plasma amyloid-beta 42 concentration. Neurobiol Dis 14(3):619–623.
62. Miravalle L, Tokuda T, Chiarle R, et al. (2000) Substitutions at codon 22 of Alzheimer's A{beta} peptide induce conformational changes and diverse apoptotic effects in human cerebral endothelial cells. J Biol Chem 275(35):27110–27116.
63. Kumar-Singh S, Julliams A, Nuyens D, et al. (2002) In vitro studies of Flemish, Dutch, and wild type Amyloid ß (Aß) provide evidence for a two-stage Aß neurotoxicity. Neurobiol Dis 11(2):300–310.
64. Hendriks L, van Duijn CM, Cras P, et al. (1992) Presenile dementia and cerebral haemorrhage linked to a mutation at condon 692 of the b-amyloid precursor protein gene. Nat Genet 1:218–221.
65. Cras P, van Harskamp F, Hendriks L, et al. (1998) Presenile Alzheimer dementia characterized by amyloid angiopathy and large amyloid core type senile plaques in the APP 692Ala Gly mutation. Acta Neuropathol 96(3):253–260.
66. Brooks WS, Kwok JB, Halliday GM, et al. (2004) Hemorrhage is uncommon in new Alzheimer family with Flemish amyloid precursor protein mutation. Neurology 63(9):1613–1617.
67. Tagliavini F, Rossi G, Padovani A, et al. (1999) A new APP mutation related to heriditary cerebral haemorrhage. Alzheimer's Rep 2:23.
68. Nilsberth C, Westlind-Danielsson A, Eckman C, et al. (2001) The 'Arctic' APP mutation (E693G) causes Alzheimer's disease by enhanced Abeta protofibril formation. Nat Neurosci 4(9):887–893.
69. Shin Y, Cho HS, Rebeck GW, Greenberg SM (2002) Vascular changes in Iowa-type hereditary cerebral amyloid angiopathy. Ann N Y Acad Sci 977:245–251.
70. Mullan M, Crawford F, Axelman K, et al. (1992) A pathogenic mutation for probable Alzheimer's disease in the APP gene at the N-terminus of b-amyloid. Nature Genet 1:345–347.
71. Citron M, Oltersdorf T, Haass C, et al. (1992) Mutation of the b-amyloid precursor protein in familial Alzheimer's disease increases b-protein production. Nature 360:672–674.
72. Mann DM, Iwatsubo T, Ihara Y, et al. (1996) Predominant deposition of amyloid-beta 42(43) in plaques in cases of Alzheimer's disease and hereditary cerebral hemorrhage associated with mutations in the amyloid precursor protein gene. Am J Pathol 148(4):1257–1266.
73. Suzuki N, Cheung TT, Cai XD, et al. (1994) An increased percentage of long amyloid beta protein secreted by familial amyloid beta protein precursor (beta APP717) mutants. Science 264(5163):1336–1340.
74. Cai XD, Golde TE, Younkin SG (1993) Release of excess amyloid beta protein from a mutant amyloid beta protein precursor. Science 259(5094):514–516.
75. Kumar-Singh S, Theuns J, Van Broeck B, et al. (2006) Mean age-of-onset of familial alzheimer disease caused by presenilin mutations correlates with both increased Abeta42 and decreased Abeta40. Hum Mutat 27(7):686–695.
76. Kumar-Singh S, De Jonghe C, Cruts M, et al. (2000) Nonfibrillar diffuse amyloid deposition due to a gamma(42)-secretase site mutation points to an essential role for N-truncated abeta(42) in Alzheimer's disease. Hum Molec Genet 9(18):2589–2598.
77. De Jonghe C, Esselens C, Kumar-Singh S, et al. (2001) Pathogenic APP mutations near the gamma-secretase cleavage site differentially affect Abeta secretion and APP C-terminal fragment stability. Hum Molec Genet 10(16):1665–1671.
78. Rovelet-Lecrux A, Hannequin D, Raux G, et al. (2006) APP locus duplication causes autosomal dominant early-onset Alzheimer disease with cerebral amyloid angiopathy. Nat Genet 38(1):24–26.
79. Sleegers K, Brouwers N, Gijselinck I, et al. (2006) APP duplication is sufficient to cause early onset Alzheimer's dementia with cerebral amyloid angiopathy. Brain 129(Pt 11):2977–2983
80. Lemere CA, Blusztajn JK, Yamaguchi H, Wisniewski T, Saido TC, Selkoe DJ (1996) Sequence of deposition of heterogeneous amyloid beta-peptides and APOE in Down syndrome: Implications for initial events in amyloid plaque formation. Neurobiol Dis 3(1):16–32.
81. Kumar-Singh S (2008) Cerebral amyloid angiopathy: Pathogenetic mechanisms and link to dense amyloid plaques. Genes Brain Behav 7(Suppl. 1):67–82.
82. De Strooper B, Saftig P, Craessaerts K, et al. (1998) Deficiency of presenilin-1 inhibits the normal cleavage of amyloid precursor protein. Nature 391(6665):387–390.
83. Shen J, Bronson RT, Chen DF, Xia W, Selkoe DJ, Tonegawa S (1997) Skeletal and CNS defects in Presenilin-1-deficient mice. Cell 89(4):629–639
84. Hartmann D, De Strooper B, Saftig P (1999) Presenilin-1 deficiency leads to loss of Cajal-Retzius neurons and cortical dyspla-

sia similar to human type 2 lissencephaly. Curr Biol 9(14):719–727.

85. Bentahir M, Nyabi O, Verhamme J, et al. (2006) Presenilin clinical mutations can affect gamma-secretase activity by different mechanisms. J Neurochem 96(3):732–742.
86. Dermaut B, Kumar-Singh S, De Jonghe C, et al. (2001) Cerebral amyloid angiopathy is a pathogenic lesion in Alzheimer's disease due to a novel presenilin 1 mutation. Brain 124(Pt 12):2383–2392.
87. Nochlin D, Bird TD, Nemens EJ, Ball MJ, Sumi SM (1998) Amyloid angiopathy in a Volga German family with Alzheimer's disease and a presenilin-2 mutation (N141I). Ann Neurol 43(1):131–135.
88. Hutton M, Lendon CL, Rizzu P, et al. (1998) Association of missense and 5'-splice-site mutations in tau with the inherited dementia FTDP-17. Nature 393(6686):702–705.
89. Foster NL, Wilhelmsen K, Sima AA, Jones MZ, D'Amato CJ, Gilman S (1997) Frontotemporal dementia and parkinsonism linked to chromosome 17: A consensus conference. Conference Participants. Ann Neurol 41(6):706–715.
90. Poorkaj P, Bird TD, Wijsman E, et al. (1998) Tau is a candidate gene for chromosome 17 frontotemporal dementia. Ann Neurol 43(6):815–825.
91. Spillantini MG, Murrell JR, Goedert M, Farlow MR, Klug A, Ghetti B (1998) Mutation in the tau gene in familial multiple system tauopathy with presenile dementia. Proc Natl Acad Sci U S A 95(13):7737–7741.
92. Dickson DW, Bergeron C, Chin SS, et al. (2002) Office of Rare Diseases neuropathologic criteria for corticobasal degeneration. J Neuropathol Exp Neurol 61(11):935–946.
93. Kawarabayashi T, Shoji M, Younkin LH, et al. (2004) Dimeric amyloid beta protein rapidly accumulates in lipid rafts followed by apolipoprotein E and phosphorylated tau accumulation in the Tg2576 mouse model of Alzheimer's disease. J Neurosci 24(15):3801–3809.
94. Shankar GM, Li S, Mehta TH, et al. (2008) Amyloid-beta protein dimers isolated directly from Alzheimer's brains impair synaptic plasticity and memory. Nat Med 14(8):837–842.
95. Divry P (1934) Etude histochimique des plaques séniles. J Neurol Psychiat 27:643–657.
96. Glenner GG (1980) Amyloid deposits and amyloidosis. N Engl J Med 302(23):1283–1292.
97. Iwatsubo T, Odaka A, Suzuki N, Mizusawa H, Nukina N, Ihara Y (1994) Visualization of Aß 42(43) and Aß 40 in senile plaques with end-specific Aß monoclonals: evidence that an initially deposited species is Aβ 42(43). Neuron 13(1):45–53.
98. Bugiani O, Tagliavini F, Giaccone G (1991) Preamyloid deposits, amyloid deposits, and senile plaques in Alzheimer's disease, Down syndrome, and aging. Ann N Y Acad Sci 640:122–128.
99. Wisniewski HM, Bancher C, Barcikowska M, Wen GY, Currie J (1989) Spectrum of morphological appearance of amyloid deposits in Alzheimer's disease. Acta Neuropathol (Berl) 78(4):337–347.
100. Dickson DW (1997) The pathogenesis of senile plaques. J Neuropathol Exp Neurol 56(4):321–339.
101. Yamaguchi H, Nakazato Y, Hirai S, Shoji M, Harigaya Y (1989) Electron micrograph of diffuse plaques. Initial stage of senile plaque formation in the Alzheimer brain. Am J Pathol 135(4):593–597.
102. Autiero M, Waltenberger J, Communi D, et al. (2003) Role of PlGF in the intra- and intermolecular cross talk between the VEGF receptors Flt1 and Flk1. Nat Med 9(7):936–943.
103. Kiemer AK, Furst R, Vollmar AM (2005) Vasoprotective actions of the atrial natriuretic peptide. Curr Med Chem Cardiovasc Hematol Agents 3(1):11–21.
104. Iwatsubo T, Saido TC, Mann DM, Lee VM, Trojanowski JQ (1996) Full-length amyloid-beta (1–42(43)) and amino-terminally modified and truncated amyloid-beta 42(43) deposit in diffuse plaques. Am J Pathol 149(6):1823–1830.
105. Storkebaum E, Lambrechts D, Dewerchin M, et al. (2005) Treatment of motoneuron degeneration by intracerebroventricular delivery of VEGF in a rat model of ALS. Nat Neurosci 8(1):85–92.
106. Dickson DW (1999) Microglia in Alzheimer's disease and transgenic models. How close the fit? Am J Pathol 154(6):1627–1631.
107. Thal DR, Capetillo-Zarate E, Del Tredici K, Braak H (2006) The development of amyloid beta protein deposits in the aged brain. Sci Aging Knowledge Environ 2006(6):re1.
108. Simchowicz T (1911) Sur la signification des plaques seniles et sur la formula senile de l'ecorde cerebrale. Rev Neurol 1:221–227.
109. Fukumoto H, Asami-Odaka A, Suzuki N, Shimada H, Ihara Y, Iwatsubo T (1996) Amyloid beta protein deposition in normal aging has the same characteristics as that in Alzheimer's disease. Predominance of A beta 42(43) and association of A beta 40 with cored plaques. Am J Pathol 148(1):259–265.

110. McKee AC, Kosik KS, Kowall NW (1991) Neuritic pathology and dementia in Alzheimer's disease. Ann Neurol 30:156–165.
111. Liau LM, Lallone RL, Seitz RS, et al. (2000) Identification of a human glioma-associated growth factor gene, granulin, using differential immuno-absorption. Cancer Res 60(5): 1353–1360.
112. Suzuki M, Nishiahara M (2002) Granulin precursor gene: A sex steroid-inducible gene involved in sexual differentiation of the rat brain. Mol Genet Metab 75(1):31–37.
113. Martin J-J, Gheuens J, Bruyland M, et al. (1991) Early-onset Alzheimer's disease in 2 large Belgian families. Neurology 41:62–68.
114. Scholtz W (1938) Studien zur Pathologie der Hirngefässe. II. Die drüsige Entartung der Hirnarterien und -capillären. Z Gesamte Neurol Psychiat 162:694–715.
115. Vinters HV (1987) Cerebral amyloid angiopathy. A critical review. Stroke 18(2):311–324.
116. Preston SD, Steart PV, Wilkinson A, Nicoll JAR, Weller RO (2003) Capillary and arterial cerebral amyloid angiopathy in Alzheimer's disease: Defining the perivascular route for the elimination of amyloid beta from the human brain. Neuropathol Appl Neurobiol 29(2):106–117.
117. Wisniewski HM, Wegiel J (1994) Beta-amyloid formation by myocytes of leptomeningeal vessels. Acta Neuropathol (Berl) 87(3):233–241.
118. Morel F, Wildi E (1952) General and cellular pathochemistry of senile and presenile alterations of the brain. Proc 1st Int Cong Neuropathol Rome 347–374.
119. Attems J, Jellinger KA (2004) Only cerebral capillary amyloid angiopathy correlates with Alzheimer pathology-a pilot study. Acta Neuropathol (Berl) 107(2):83–90.
120. Roher AE, Kuo YM, Esh C, et al. (2003) Cortical and leptomeningeal cerebrovascular amyloid and white matter pathology in Alzheimer's disease. Mol Med 9(3–4):112–122.
121. Itoh Y, Yamada M, Hayakawa M, Otomo E, Miyatake T (1993) Cerebral amyloid angiopathy: A significant cause of cerebellar as well as lobar cerebral hemorrhage in the elderly. J Neurol Sci 116(2):135–141.
122. Peers MC, Lenders MB, Defossez A, Delacourte A, Mazzuca M (1988) Cortical angiopathy in Alzheimers-disease – the formation of dystrophic perivascular neurites is related to the exudation of amyloid fibrils from the pathological vessels. Virchows Arch A Pathol Anat Histopathol 414(1):15–20.
123. Games D, Adams D, Alessandrini R, et al. (1995) Alzheimer-type neuropathology in transgenic mice overexpressing V717F beta-amyloid precursor protein. Nature 373(6514):523–527.
124. Hsiao K, Chapman P, Nilsen S, et al. (1996) Correlative memory deficits, Abeta elevation, and amyloid plaques in transgenic mice. Science 274(5284):99–102.
125. Sturchler-Pierrat C, Abramowski D, Duke M, et al. (1997) Two amyloid precursor protein transgenic mouse models with Alzheimer disease-like pathology. Proc Natl Acad Sci U S A 94(24):13287–13292.
126. Wirths O, Multhaup G, Bayer TA (2004) A modified beta-amyloid hypothesis: Intraneuronal accumulation of the beta-amyloid peptide - the first step of a fatal cascade. J Neurochem 91(3):513–520.
127. Gouras GK, Almeida CG, Takahashi RH (2005) Intraneuronal Abeta accumulation and origin of plaques in Alzheimer's disease. Neurobiol Aging 26(9):1235–1244.
128. Perry G, Rizzuto N, Autilio-Gambetti L, Gambetti P (1985) Paired helical filaments from Alzheimer disease patients contain cytoskeletal components. Proc Natl Acad Sci U S A 82(11):3916–3920.
129. Grundke-Iqbal I, Iqbal K, Tung YC, Quinlan M, Wisniewski HM, Binder LI (1986) Abnormal phosphorylation of the microtubule-associated protein tau (tau) in Alzheimer cytoskeletal pathology. Proc Natl Acad Sci U S A 83(13):4913–4917.
130. Goedert M, Wischik CM, Crowther RA, Walker JE, Klug A (1988) Cloning and sequencing of the cDNA encoding a core protein of the paired helical filament of Alzheimer disease: Identification as the microtubule-associated protein tau. Proc Natl Acad Sci U S A 85(11):4051–4055.
131. Kosik KS, Orecchio LD, Binder L, Trojanowski JQ, Lee VM, Lee G (1988) Epitopes that span the tau molecule are shared with paired helical filaments. Neuron 1(9):817–825.
132. Bierer LM, Hof PR, Purohit DP, et al. (1995) Neocortical neurofibrillary tangles correlate with dementia severity in Alzheimer's disease. Arch Neurol 52(1):81–88.
133. Braak H, Braak E (1991) Neuropathological stageing of Alzheimer-related changes. Acta Neuropathol (Berl) 82(4):239–259.
134. Dickson DW (1998) Pick's disease: A modern approach. Brain Pathol 8(2):339–354.
135. Buee L, Delacourte A (1999) Comparative biochemistry of tau in progressive supranuclear

palsy, corticobasal degeneration, FTDP-17 and Pick's disease. Brain Pathol 9(4):681–693.

136. Tolnay M, Probst A (1999) REVIEW: Tau protein pathology in Alzheimer's disease and related disorders. Neuropathol Appl Neurobiol 25(3):171–187.

137. Zhukareva V, Mann D, Pickering-Brown S, et al. (2002) Sporadic Pick's disease: A tauopathy characterized by a spectrum of pathological tau isoforms in gray and white matter. Ann Neurol 51(6):730–739.

138. Kumar-Singh S, Van Broeckhoven C (2007) Frontotemporal lobar degeneration: Current concepts and advances. Brain Res 17(1):104–114.

139. Coffey DS (1998) Self-organization, complexity and chaos: The new biology for medicine. Nat Med 4(8):882–885.

140. Dodart JC, Meziane H, Mathis C, Bales KR, Paul SM, Ungerer A (1999) Behavioral disturbances in transgenic mice overexpressing the V717F beta-amyloid precursor protein. Behav Neurosci 113(5):982–990.

141. Moechars D, Dewachter I, Lorent K, et al. (1999) Early phenotypic changes in transgenic mice that overexpress different mutants of Amyloid precursor protein in brain. J Biol Chem 10:6483–6492.

142. Van Dorpe J, Smeijers L, Dewachter I, et al. (2000) Prominent cerebral amyloid angiopathy in transgenic mice overexpressing the london mutant of human APP in neurons [In Process Citation). Am J Pathol 157(4):1283–1298.

143. McGowan E, Sanders S, Iwatsubo T, et al. (1999) Amyloid phenotype characterization of transgenic mice overexpressing both mutant amyloid precursor protein and mutant presenilin 1 transgenes. Neurobiol Dis 6(4):231–244.

144. Irizarry MC, Soriano F, McNamara M, et al. (1997) A beta deposition is associated with neuropil changes, but not with overt neuronal loss in the human amyloid precursor protein V717F (PDAPP) transgenic mouse. J Neurosci 17(18):7053–7059.

145. Calhoun ME, Wiederhold KH, Abramowski D, et al. (1998) Neuron loss in APP transgenic mice. Nature 395(6704):755–756.

146. Winkler DT, Bondolfi L, Herzig MC, et al. (2001) Spontaneous hemorrhagic stroke in a mouse model of cerebral amyloid angiopathy. J Neurosci 21(5):1619–1627.

147. Van Dam D., D'Hooge R, Staufenbiel M, Van GC, Van MF, De Deyn PP (2003) Age-dependent cognitive decline in the APP23 model precedes amyloid deposition. Eur J Neurosci 17(2):388–396.

148. Van Broeck B, Vanhoutte G, Pirici D, et al. (2008) Intraneuronal amyloid beta and reduced brain volume in a novel APP T714I mouse model for Alzheimer's disease. Neurobiol Aging 29(2):241–252.

149. Wirths O, Multhaup G, Czech C, et al. (2001) Intraneuronal Abeta accumulation precedes plaque formation in beta-amyloid precursor protein and presenilin-1 double-transgenic mice. Neurosci Lett 306(1–2):116–120.

150. Van Broeck B, Vanhoutte G, Cuijt I, et al. (2008) Reduced brain volumes in mice expressing APP-Austrian mutation but not in mice expressing APP-Swedish-Austrian mutations. Neurosci Lett 447(2–3):143–147.

151. Casas C, Sergeant N, Itier JM, et al. (2004) Massive CA1/2 neuronal loss with intraneuronal and N-terminal truncated Abeta42 accumulation in a novel Alzheimer transgenic model. Am J Pathol 165(4):1289–1300.

152. Oddo S, Caccamo A, Shepherd JD, et al. (2003) Triple-transgenic model of Alzheimer's disease with plaques and tangles: Intracellular Abeta and synaptic dysfunction. Neuron 39(3):409–421.

153. Chishti MA, Yang DS, Janus C, et al. (2001) Early-onset amyloid deposition and cognitive deficits in transgenic mice expressing a double mutant form of amyloid precursor protein 695. J Biol Chem 276(24):21562–21570.

154. Kumar-Singh S, Dewachter I, Moechars D, et al. (2000) Behavioral disturbances without amyloid deposits in mice overexpressing human amyloid precursor protein with Flemish (A692G) or Dutch (E693Q) mutation. Neurobiol Dis 7(1):9–22.

155. Herzig MC, Winkler DT, Burgermeister P, et al. (2004) A beta is targeted to the vasculature in a mouse model of hereditary cerebral hemorrhage with amyloidosis. Nat Neurosci 7(9):954–960.

156. Davis J, Xu F, Deane R, et al. (2004) Early-onset and robust cerebral microvascular accumulation of amyloid beta-protein in transgenic mice expressing low levels of a vasculotropic Dutch/Iowa mutant form of amyloid beta-protein precursor. J Biol Chem 279(19):20296–20306.

157. Miao J, Xu F, Davis J, Otte-Holler I, Verbeek MM, Van Nostrand WE (2005) Cerebral microvascular Aß protein deposition induces vascular degeneration and neuroinflammation in transgenic mice expressing human vasculotropic mutant AßPP. Am J Pathol 167(2):505–515.

158. Cheng IH, Palop JJ, Esposito LA, Bien-Ly N, Yan F, Mucke L (2004) Aggressive amyloidosis in mice expressing human amyloid peptides with the Arctic mutation. Nat Med 10(11):1190–1192.

159. Duff K, Eckman C, Zehr C, et al. (1996) Increased amyloid-beta42(43) in brains of mice expressing mutant presenilin 1. Nature 383(6602):710–713

160. Borchelt DR, Ratovitski T, vanLare J, et al. (1997) Accelerated amyloid deposition in the brains of transgenic mice coexpressing mutant presenilin 1 and amyloid precursor proteins. Neuron 19(4):939–945.
161. Holcomb L, Gordon MN, McGowan E, et al. (1998) Accelerated Alzheimer-type phenotype in transgenic mice carrying both mutant amyloid precursor protein and presenilin 1 transgenes. Nat Med 4(1):97–100.
162. Takeuchi A, Irizarry MC, Duff K, et al. (2000) Age-related amyloid beta deposition in transgenic mice overexpressing both Alzheimer mutant presenilin 1 and amyloid beta precursor protein Swedish mutant is not associated with global neuronal loss. Am J Pathol 157(1):331–339.
163. Dewachter I, Van DJ, Smeijers L, et al. (2000) Aging increased amyloid peptide and caused amyloid plaques in brain of old APP/V717I transgenic mice by a different mechanism than mutant presenilin1. J Neurosci 20(17):6452–6458.
164. De Jonghe C, Cras P, Vanderstichele H, et al. (1999) Evidence that Abeta42 plasma levels in presenilin-1 mutation carriers do not allow for prediction of their clinical phenotype. Neurobiol Dis 6(4):280–287.
165. McGowan E, Pickford F, Kim J, et al. (2005) Abeta42 is essential for parenchymal and vascular amyloid deposition in mice. Neuron 47(2):191–199.
166. Vidal R, Frangione B, Rostagno A, et al. (1999) A stop-codon mutation in the BRI gene associated with familial British dementia. Nature 399(6738):776–781
167. Lewis J, McGowan E, Rockwood J, et al. (2000) Neurofibrillary tangles, amyotrophy and progressive motor disturbance in mice expressing mutant (P301L) tau protein. Nat Genet 25(4):402–405.
168. Lewis J, Dickson DW, Lin WL, et al. (2001) Enhanced neurofibrillary degeneration in transgenic mice expressing mutant tau and APP. Science 293(5534):1487–1491
169. Guo Q, Fu W, Sopher BL, et al. (1999) Increased vulnerability of hippocampal neurons to excitotoxic necrosis in presenilin-1 mutant knock-in mice. Nat Med 5(1):101–106.
170. Billings LM, Oddo S, Green KN, McGaugh JL, LaFerla FM (2005) Intraneuronal Abeta causes the onset of early Alzheimer's disease-related cognitive deficits in transgenic mice. Neuron 45(5):675–688.
171. Kertesz A, Kawarai T, Rogaeva E, et al. (2000) Familial frontotemporal dementia with ubiquitin-positive, tau-negative inclusions. Neurology 54(4):818–827.
172. Forman MS, Farmer J, Johnson JK, et al. (2006) Frontotemporal dementia: Clinicopathological correlations. Ann Neurol 59(6):952–962.
173. Watts GD, Wymer J, Kovach MJ, et al. (2004) Inclusion body myopathy associated with Paget disease of bone and frontotemporal dementia is caused by mutant valosin-containing protein. Nat Genet 36(4):377–381.
174. Skibinski G, Parkinson NJ, Brown JM, et al. (2005) Mutations in the endosomal ESCRTIII-complex subunit CHMP2B in frontotemporal dementia. Nat Genet 37(8):806–808.
175. Baker M, Mackenzie IR, Pickering-Brown SM, et al. (2006) Mutations in progranulin cause tau-negative frontotemporal dementia linked to chromosome 17. Nature 442(7105):916–919.
176. Cruts M, Gijselinck I, van der ZJ, et al. (2006) Null mutations in progranulin cause ubiquitin-positive frontotemporal dementia linked to chromosome 17q21. Nature 442(7105):920–924.
177. Brouwers N, Sleegers K, Engelborghs S, et al. (2008) Genetic variability in progranulin contributes to risk for clinically diagnosed Alzheimer disease. Neurology 71(9):656–664.
178. Neumann M, Sampathu DM, Kwong LK, et al. (2006) Ubiquitinated TDP-43 in frontotemporal lobar degeneration and amyotrophic lateral sclerosis. Science 314(5796):130–133.
179. Kayasuga Y, Chiba S, Suzuki M, et al. (2007) Alteration of behavioural phenotype in mice by targeted disruption of the progranulin gene. Behav Brain Res 185(2):110–118.
180. Chiba S, Suzuki M, Yamanouchi K, Nishihara M (2007) Involvement of granulin in estrogen-induced neurogenesis in the adult rat hippocampus. J Reprod Dev 53(2):297–307.
181. Mendez MF, McMurtray A, Chen AK, Shapira JS, Mishkin F, Miller BL (2006) Functional neuroimaging and presenting psychiatric features in frontotemporal dementia. J Neurol Neurosurg Psychiatry 77(1):4–7.
182. Huey ED, Putnam KT, Grafman J (2006) A systematic review of neurotransmitter deficits and treatments in frontotemporal dementia. Neurology 66(1):17–22.
183. Lebert F, Stekke W, Hasenbroekx C, Pasquier F (2004) Frontotemporal dementia: A randomised, controlled trial with trazodone. Dement Geriatr Cogn Disord 17(4):355–359.
184. Yamaguchi H, Yamazaki T, Lemere CA, Frosch MP, Selkoe DJ (1992) Beta amyloid is focally deposited within the outer basement membrane in the amyloid angiopathy of Alzheimer's disease. An immunoelectron microscopic study. Am J Pathol 141(1):249–259.

185. Natte R, de Boer WI, Maat-Schieman ML, et al. (1999) Amyloid beta precursor protein-mRNA is expressed throughout cerebral vessel walls. Brain Res 828(1–2):179–183.
186. Verbeek MM, De Waal RM, Schipper JJ, Van Nostrand WE (1997) Rapid degeneration of cultured human brain pericytes by amyloid beta protein. J Neurochem 68(3):1135–1141.
187. Nicoll JAR, Yamada M, Frackowiak J, Mazur-Kolecka B, Weller RO (2004) Cerebral amyloid angiopathy plays a direct role in the pathogenesis of Alzheimer's disease Pro-CAA position statement. Neurobiol Aging 25(5):589–597.
188. Gravina SA, Ho L, Eckman CB, et al. (1995) Amyloid beta protein (A beta) in Alzheimer's disease brain. Biochemical and immunocytochemical analysis with antibodies specific for forms ending at A beta 40 or A beta 42(43). J Biol Chem 270(13):7013–7016.
189. Cullen KM, Kocsi Z, Stone J (2006) Microvascular pathology in the aging human brain: Evidence that senile plaques are sites of microhaemorrhages. Neurobiol Aging 27(12):1786–1796.
190. Pike CJ, Overman MJ, Cotman CW (1995) Amino-terminal deletions enhance aggregation of beta-amyloid peptides in vitro. J Biol Chem 270(41):23895–23898.
191. Shin RW, Ogino K, Kondo A, et al. (1997) Amyloid beta-protein (Abeta) 1–40 but not Abeta1–42 contributes to the experimental formation of Alzheimer disease amyloid fibrils in rat brain. J Neurosci 17(21):8187–8193.
192. Radde R, Bolmont T, Kaeser SA, et al. (2006) Abeta42-driven cerebral amyloidosis in transgenic mice reveals early and robust pathology. EMBO Rep 7(9):940–946.
193. Gotz J, Chen F, Van Dorpe J, Nitsch RM (2001) Formation of neurofibrillary tangles in P301l tau transgenic mice induced by Abeta 42 fibrils. Science 293(5534):1491–1495.
194. Van Broeck B, Van Broeckhoven C, Kumar-Singh S (2007) Current insights into molecular mechanisms of Alzheimer disease and their implications for therapeutic approaches. Neurodegener Dis 4(5):349–365.
195. Schenk D, Barbour R, Dunn W, et al. (1999) Immunization with amyloid-beta attenuates Alzheimer-disease-like pathology in the PDAPP mouse (see comments). Nature 400(6740):173–177.
196. Janus C, Pearson J, McLaurin J, et al. (2000) A beta peptide immunization reduces behavioural impairment and plaques in a model of Alzheimer's disease. Nature 408(6815): 979–982.
197. DeMattos RB, Bales KR, Cummins DJ, Paul SM, Holtzman DM (2002) Brain to plasma amyloid-beta efflux: A measure of brain amyloid burden in a mouse model of Alzheimer's disease. Science 295(5563):2264–2267.
198. Dodart JC, Bales KR, Gannon KS, et al. (2002) Immunization reverses memory deficits without reducing brain A beta burden in Alzheimer's disease model. Nat Neurosci 5(5):452–457.
199. Schenk D (2002) Amyloid-beta immunotherapy for Alzheimer's disease: The end of the beginning. Nat Rev Neurosci 3(10):824–828.
200. Nicoll JA, Wilkinson D, Holmes C, Steart P, Markham H, Weller RO (2003) Neuropathology of human Alzheimer disease after immunization with amyloid-beta peptide: A case report. Nat Med 9(4):448–452.
201. Hock C, Konietzko U, Streffer JR, et al. (2003) Antibodies against beta-amyloid slow cognitive decline in Alzheimer's disease. Neuron 38(4):547–554.
202. Boche D, Nicoll JA (2008) The role of the immune system in clearance of Abeta from the brain. Brain Pathol 18(2):267–278.
203. Lichtlen P, Mohajeri MH (2008) Antibody-based approaches in Alzheimer's research: Safety, pharmacokinetics, metabolism, and analytical tools. J Neurochem 104(4):859–874.
204. Banks WA, Terrell B, Farr SA, Robinson SM, Nonaka N, Morley JE (2002) Passage of amyloid beta protein antibody across the blood-brain barrier in a mouse model of Alzheimer's disease. Peptides 23(12):2223–2226.
205. Solomon B, Koppel R, Frankel D, Hanan-Aharon E (1997) Disaggregation of Alzheimer beta-amyloid by site-directed mAb. Proc Natl Acad Sci U S A 94(8):4109–4112.
206. Bard F, Cannon C, Barbour R, et al. (2000) Peripherally administered antibodies against amyloid beta-peptide enter the central nervous system and reduce pathology in a mouse model of Alzheimer disease. Nat Med 6(8):916–919.
207. Bard F, Barbour R, Cannon C, et al. (2003) Epitope and isotype specificities of antibodies to beta -amyloid peptide for protection against Alzheimer's disease-like neuropathology. Proc Natl Acad Sci U S A 100(4):2023–2028.
208. Bard F, Cannon C, Barbour R, et al. (2000) Peripherally administered antibodies against amyloid beta-peptide enter the central nervous system and reduce pathology in a mouse model of Alzheimer disease. Nat Med 6(8):916–919.
209. DeMattos RB, Bales KR, Cummins DJ, Dodart JC, Paul SM, Holtzman DM (2001)

Peripheral anti-A beta antibody alters CNS and plasma A beta clearance and decreases brain A beta burden in a mouse model of Alzheimer's disease. Proc Natl Acad Sci U S A 98(15):8850–8855.

210. Sigurdsson EM, Knudsen E, Asuni A, et al. (2004) An attenuated immune response is sufficient to enhance cognition in an Alzheimer's disease mouse model immunized with amyloid-beta derivatives. J Neurosci 24(28):6277–6282.

211. Pfeifer M, Boncristiano S, Bondolfi L, et al. (2002) Cerebral hemorrhage after passive anti-Abeta immunotherapy. Science 298(5597):1379.

212. Meyer-Luehmann M, Stalder M, Herzig MC, et al. (2003) Extracellular amyloid formation and associated pathology in neural grafts. Nat Neurosci 6(4):370–7.

Chapter 8

Behavioral Validation in Animal Models of Dementia

Debby Van Dam, Annemie Van Dijck, and Peter Paul De Deyn

Abstract

The development of complex disease models requires the parallel development or optimization of valid behavioral paradigms assessing complex brain-behavior relations. Besides validity of the applied paradigm, standardization at the level of experimental animals, testing procedures, and surroundings is essential to generate reliable data. High levels of validity and standardization can be reached only by skilled and experienced researchers. This chapter summarizes the most frequently used cognitive and behavioral paradigms in the phenotyping of rodent models of dementia.

Key words: Cognition, motor function, aggression, anxiety, activity, phenotyping

1. Introduction

The development of complex disease models as described in subsequent chapters requires the parallel development or optimization of valid behavioral paradigms assessing complex brain-behavior relations. Analysis of cognitive and behavioral alterations should be substantiated through implementation of thoroughly validated behavioral paradigms that must be correctly applied for the data to be interpretable. Face validity on itself is not a sufficient validation criterion. The predictive validity of behavioral designs can be confirmed by showing that drugs acknowledged to work in the human setting either increase or inhibit the behavior in question. Construct validity of a behavioral paradigm or protocol implies solid insight into dependable variables and their effect on outcome parameters. Besides validity of the applied paradigm, standardization at the level of experimental animals, testing procedures, and surroundings is essential to generate reliable data. High levels of validity and standardization can be reached only by skilled and experienced researchers who know for example how to handle animals to reduce potentially biasing stress.

Peter Paul De Deyn and Debby Van Dam (eds.), *Animal Models of Dementia*, Neuromethods, vol. 48,
DOI 10.1007/978-1-60761-898-0_8, © Springer Science+Business Media, LLC 2011

Along with the development of genetically engineered animal models with presumed CNS deficits came the use of test batteries scrutinizing a wide variety of neurological and behavioral responses. These test batteries allow investigating potential correlative phenotypical changes. Moreover, a CNS phenotype can be confirmed in different paradigms, thereby strengthening the reliability of the model, as well as contributing to Russell and Burch's 3 Rs with reduction of the number of experimental animals needed. One should, however, be aware of potential training effects dependent on the order of the different behavioral paradigms and potentially important behavioral differences between naïve and trained animals.

Writing a complete "how-to manual" for the behavioral phenotyping of a new dementia model would go beyond the scope of this chapter. This chapter will summarize the most frequently used behavioral paradigms and protocols to assess different aspects of altered behavior in models of dementia. For further reading, we refer the readers to excellent manuals on behavioral neuroscience as, e.g., "What's wrong with my mouse?" written by Jacqueline N. Crawley (1) and "Methods of Behavior Analysis in Neuroscience" edited by Jerry Buccafusco (2). Since the majority of models use rodents, and especially (transgenic) mice, as model organisms, this chapter focuses on behavioral studies in rodents. It serves to mention though that many behavioral paradigms originally developed in rodents have been successfully adapted to the level of nonmammalian species like fruit flies (3–5), zebrafish (6–8), or *C. elegans* (9–11).

2. Observational Test Batteries

When a new model is developed, behavioral phenotyping often starts with the application of systematized observational test batteries to mainly assess general health and reflexes. Irwin developed a screening test (12) that is still in use today in modified forms. The Irwin test is a systematic observational method to comprehensively assess and quantify the behavioral and physiological state of the mouse and its response to drugs. Different parameters corresponding to the behavioral, neurological, and autonomic state of the mice are scored, to the extent possible, through direct observation only. Observers should be extensively trained, requiring at least 2 weeks to learn and an additional 2 weeks to develop assurance and ease in observation and animal handling. Interobserver variability should be restricted to a minimum, achieved by a period of simultaneous scoring and comparison. Moser and colleagues proposed a similar screening method termed the Functional observation battery (13), which rapidly assesses the autonomic, sensory, and motor integrity of rodents. It is based on the Irwin test, but focuses less on the neurological responses and is partly performed in the home cage. A third example of a screening test is the SHIRPA primary screen, which is also an adapted Irwin test. The SHIRPA primary screen was specifically developed as a high-throughput phenotyping tool for studying mouse mutants (14). This test has been widely validated; standard operation procedures as well as databases containing the normal scores of several inbred and mutant strains are available (e.g., www.eumorphia.org).

In addition, researchers should also bear in mind to regularly include basic observations like body weight and temperature, and pay attention to body posture and gait. A healthy mouse is well groomed and displays pink coloration of ear pinna and footpads. Attention should be paid to aberrant spontaneous or stereotyped behaviors, like popping (repeatedly jumping in corner of cage), head bobbing, or circling.

3. Assessing Learning and Memory

It goes without saying that the behavioral phenotyping of a mouse model of dementia should focus on different aspects of learning and memory. A variety of paradigms and protocols have been developed to assess cognitive functions in rodents. Ideally, several paradigms requiring distinct sensory and motor abilities are chosen. Protocols distinguishing explicit versus implicit memory can be chosen, as well as designs assessing short-term versus long-term memory.

The Morris water maze (MWM) (15) is presently the most widely used paradigm for the evaluation of hippocampus-dependent visual-spatial learning and memory skills in rodents, which represents the highest cognitive level appreciable in rodents. In this test, the animals have to learn the location of a hidden platform using visual cues (Fig. 1). The task is based on the principle that rodents are reinforced by the water environment to escape from the maze via the quickest and most direct route. Spatial learning is verified in a subsequent probe trial with the platform removed; preference for the prior platform location is viewed as a measure of spatial learning. The task has built-in control parameters for potential noncognitive perturbations affecting performance; lower swim speed or the inability to reach the platform when it is marked with a clear visual cue (visible platform MWM task) may reflect sensory or motor dysfunction. Unfortunately, the MWM is among the most stressful of cognitive paradigms in rodents.

Besides the MWM, several other mazes have been developed, which are all based on the same principle; successfully passing through the maze is rewarded by escaping from the water in wet mazes, or by food as a positive reinforcer in dry-land mazes. Examples include the plus-shaped water maze, radial arm mazes, multiple Y mazes, and the Barnes maze (1).

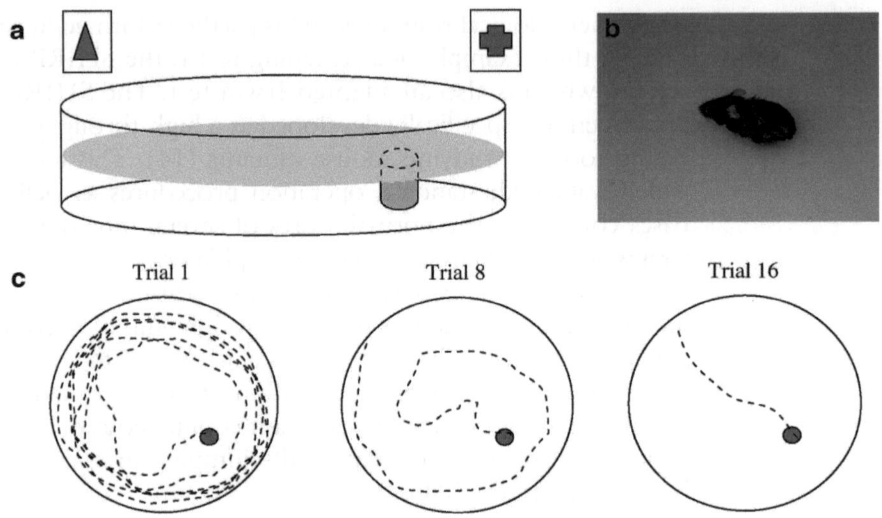

Fig. 1. Morris water maze. (**a** and **b**) In the Morris water maze test, the animals have to learn the location of a hidden platform using visual cues surrounding the maze; (**c**) With training, the animal learns the position of the platform, which is ultimately reached via the quickest and most direct route.

Because of the high stress levels associated with the MWM, and the potential bias by motor impairment, behavioral neuroscientists have developed complex nonspatial hippocampus-dependent tasks like the odor paired-associates task, analogous to the verbal paired-association task for humans. The test was originally developed for rats (16), but was adapted for mice (17,18). The test consists of several phases in which associations between different odors have to be learned in order to obtain a reward (Fig. 2). Afterwards, flexible expression of the earlier attained memories is evaluated in transitivity and symmetry tests. Provided the animals can dig, motor impairment will not influence the results, neither will impaired vision. Moreover, the results are not biased by anxiety, since the test is based on natural scavenging habits, in contrast to the MWM paradigm. On the down side, the long duration of the protocol, which encloses several months, makes it unsuitable for the assessment for memory in progressing phenotypes. The same disadvantage may arise when using schedule-induced operant tasks employing operant conditioning boxes, as originally developed by B. Skinner (19), equipped with levers, nose poke operandi, lights, buzzers, pellet dispensers, and optical lick-o-meters. Protocols often include subsequent training schedules with increasing difficulty; from fixed ratio (e.g., a fixed number of responses delivers a pellet) and fixed interval schedules (e.g., FI30 where a food pellet is delivered every 30 s on response at that specific time point), over variable ratio and variable interval schedules, to complex conditioned emotional response tasks.

Passive avoidance learning – a nonspatial form of learning – can be tested in a step-through box consisting of a small, brightly

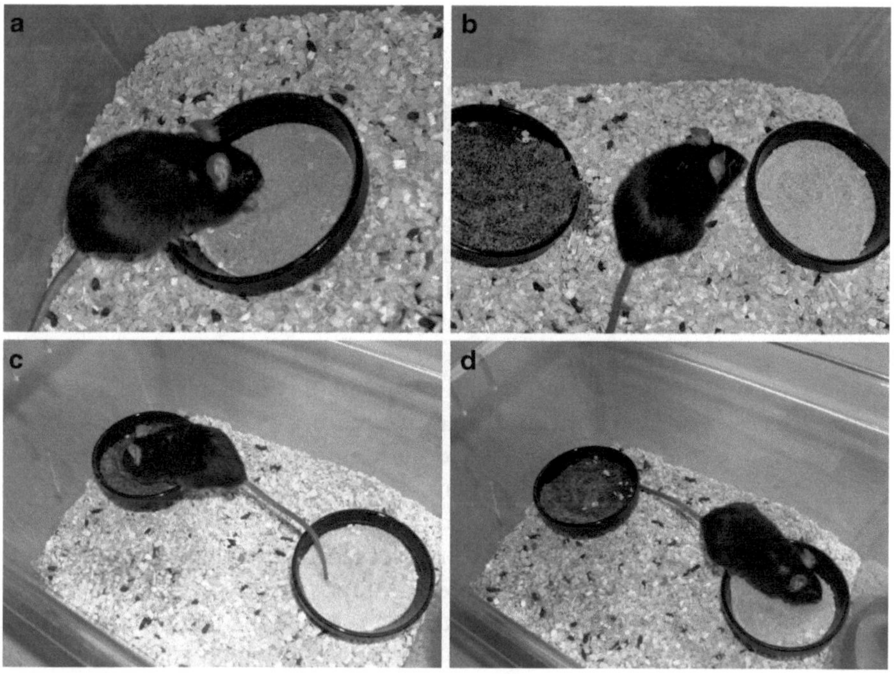

Fig. 2. Odor paired-associates task. In the odor paired-associate task, the animal has to dig up a reward from a cup filled with sand that has been scented by specific odors (e.g., coffee, curcumin, onion, nutmeg, cacao). (**a**) During the preliminary training phase, one cup with a reward (e.g., sweet cereal) is presented and the mouse has to learn to dig up the reward; (**b**) In later stages, two cups with different odors are presented. In only one cup, a reward is hidden; (**c** and **d**) The animal is allowed to dig in both cups to find the reward, while latency and the number of errors are counted.

Fig. 3. Passive avoidance learning. Passive avoidance learning is typically assessed in a step-through box consisting of a small, brightly lit compartment and a large, dark compartment connected with a sliding door. On acquisition day, the animal is placed in the first compartment of the box, and upon complete entry into a second, dark compartment, a light electric foot shock is applied through the grid floor. Exactly 24 h later, latency to re-enter the dark compartment is measured. In the photo, the cover of the dark compartment was removed to display the grid.

lit compartment and a large, dark compartment connected with a sliding door (Fig. 3). On acquisition day, the animal is placed in the first compartment of the box, and upon complete entry into a second, dark compartment, a light electric foot shock is applied through the grid floor. Exactly 24 h later, latency to re-enter the dark compartment is measured. Active avoidance (where the animal has to escape from a negative stressor in an active manner) can be tested in a two-compartment shuttle box where the animal has to jump to the other compartment to avoid an electrical shock that is predicted by a sound or light signal. Avoidance tasks are widely used to assess cognitive function in a short timeframe. However, procedural components of the task are not easily distinguishable from declarative memory components as is the case in the MWM or in cued and contextual learning. Latency to enter the dark compartment on day 1, which is based on the fact that nocturnal animals, like rats and mice, prefer dark, enclosed spaces over open, brightly lit spaces, is the single built-in control parameter. Sensory or motor dysfunction has to be ruled out by using other behavioral tasks (1,2). Avoidance tasks are comparable to cued and contextual conditioning, which is a fear condition task that measures the ability of an animal to learn and remember the association between an aversive stimulus and environmental cues (20). Associative learning is an adaptive process that allows an organism to learn to anticipate events. Cued and contextual conditioning requires a different set of sensory and motor functions, so that the procedural components do not overlap. Fear conditioning, scored as freezing responses, is among the most intuitive rodent learning and memory paradigms, and is less labor- and time-consuming than the MWM.

Other potentially useful learning and memory tasks include novel object recognition (21), conditioned taste aversion (22), social recognition (23), and discrimination learning in, e.g., a Lashley jumping stand (24).

4. Behavioral and Psychological Signs and Symptoms of Dementia (BPSD)

In accordance with the increased clinical focus on BPSD, major efforts have been made to mimic specific behavioral alterations in animal models and to develop useful tools to evaluate new psychopharmacological strategies to replace atypical antipsychotics or classic neuroleptics, which display only modest effect size and are frequently associated with significant side-effects (25).

Activity and circadian rhythm disturbances are present in normal senescence and are nonspecific symptoms of a variety of neurodegenerative conditions affecting different but often overlapping brain structures (26). In laboratory animals, they can be easily

Fig. 4. Cage activity recording. Activity and circadian rhythm disturbances can be easily screened using infrared sensors surrounding the animal's home cage. Sensors are typically linked to a microprocessor unit and/or a computer recording the number of beam interruptions over a specific time interval as a measure of spontaneous horizontal locomotor activity.

screened using infrared sensors surrounding the animal's home cage (Fig. 4). Sensors are typically linked to a microprocessor unit and/or a computer recording the number of beam interruptions over a specific time interval as measure of activity. The regular 12 h/12 h light–dark schedule should of course be strictly applied and potentially disturbing ambient sounds, e.g., by placing the cages in sound attenuated chambers, should be prevented. When food and water is provided, recordings can be run over several days to assess potential circadian rhythm disturbances comparable to actigraphic measurements in the clinical setting (27).

Emotional disturbances such as depression, anxiety, and irritability-aggressiveness are the most common noncognitive symptoms in dementia. Aggressive behavior in male rodents can be provoked using a variety of behavioral protocols based on dominance hierarchy. One method involves isolated housing lasting several weeks after which the animal is confronted with a previously group-housed male in its home cage (combined isolation-induced resident/intruder protocol) or in a new enclosure (28,29). Alternative male aggression paradigms are, e.g., the tube-dominance task, electroshock-induced aggression, or the observation of dominance in hierarchy during feeding. Agonistic behaviors generally begin with anogenital contact, tail rattling by the dominant male, defensive upright posture by the subordinate male, attacks and biting by the dominant male on the rump of the subordinate male, and running escape by the subordinate male.

Anxiety and fear-related behaviors can be assessed in rodents in various ways; both conditioned and unconditioned response tests exist. Conditioned response tests include conflict tests, as e.g., Vogel's lick-suppression test where the thirsty animal is confronted with the choice between drinking and avoiding electroshocks delivered through the spout of the drinking bottle (30). Other conditioned response tests include, among others, a conditioned emotional response recorded in Skinner boxes, fear-potentiated startle and protocols based on the active or passive avoidance of negative reinforcements (electroshock). Unconditioned anxiety tasks are based on the quantification of fear-related responses like freezing, defecation, thigmotaxis (tendency to remain near the edges of an arena, rather than moving to the center of it), and the conflict between the innate trait of nocturnal animals like mice and rats to prefer narrow, dark enclosures, and the tendency to explore new environments. Spontaneous locomotor activity, thigmotaxis, and defecation can be scored in an open field arena (31), either manually (e.g., number of squares crossed) or fully automated with video tracking software or photo cells. The elevated plus maze is based on the innate conflict between exploration of a new environment and aversion for brightly lit open areas (32). The elevated plus maze is a four-arm runway with two open brightly lit arms and two enclosed, dark arms. An anxious animal will spend most of its time in the enclosed arms and will display few to none entries in the open arms (Fig. 5). The same conflict is employed in the dark–light transition box where an animal can move from a small dark compartment to a brightly lit arena through several connecting gaps (33).

A major drawback of animal modeling is the fact that several emotional disturbances, including hallucinations, paranoid and

Fig. 5. Elevated plus maze. The elevated plus maze is based on the innate conflict between exploration of a new environment and aversion for brightly lit open areas. The elevated plus maze is a four-arm runway with two open brightly lit arms and two enclosed, dark arms. An anxious animal will spend most of its time in the enclosed arms and will display few to none entries in the open arms.

delusional ideation, and affective disorders are difficult or even impossible to model in rodents. Appraisal of depression-related symptoms in rodents is based on learned helplessness and behavioral despair phenomena, in which animals are exposed to uncontrollable and inescapable stress; e.g., Porsolt forced swim test, tail suspension test, and inescapable shock paradigms. Failure to try to escape from this type of aversive stimulus in rodents is considered to model a depression-like state (34). Anhedonia, a core symptom of clinical depression, which is defined as the loss of sensitivity to reward, can be assessed in rodents with a sucrose preference test during which the consumption of a 0.8% sucrose solution is compared to the simultaneous consumption of tap water (35).

Phenotyping can also focus on ingestive behavior. The intake of food and water can be screened using metabolic cages and weighing the amount of food consumed and spilled. More detailed analysis of food intake is possible in operant conditioning boxes equipped with pellet dispensers and optical lick-o-meters that allow appraisal of the circadian rhythm of ingestive behavior (36).

5. Motor Performance

Behavioral phenotyping of a new model generally also includes appraisal of sensorimotor function, either to exclude potentially biasing motor impairments for other cognitive and/or behavioral tasks, or to assess motor impairment in animal models for human conditions with motor impairment as one of the core symptoms, e.g., Parkinson's disease, Metachromatic leukodystrophy, Adrenoleukodystrophy, vascular dementia, and Amyotrophic lateral sclerosis. Many different types of tests, from very simple to more complex, have been developed. We will discuss the most frequently used ones here.

Spontaneous open field locomotion (31) is the most standardized general measure of motor function. A 5–10-min recording is usually sufficient to disclose gross locomotor abnormalities.

Sensorimotor coordination can be assessed with the rotarod test where a rodent is placed on a rotating rod (37). The speed of rotation is gradually increased and the rodent's ability to remain on the rotating rod is recorded during several trials (Fig. 6).

Motor coordination and equilibrium (ataxia) can also be assessed with the stationary beam task. In contrast to the rotarod, no sophisticated equipment is required. A 2-cm-diameter pole (e.g., a broomstick) is wrapped with cloth tape to improve grip and divided into several 10-cm segments. The animal is placed in the center of the horizontal beam and the number of segments crossed is counted over a 1-min period, as well as the latency to fall off the beam.

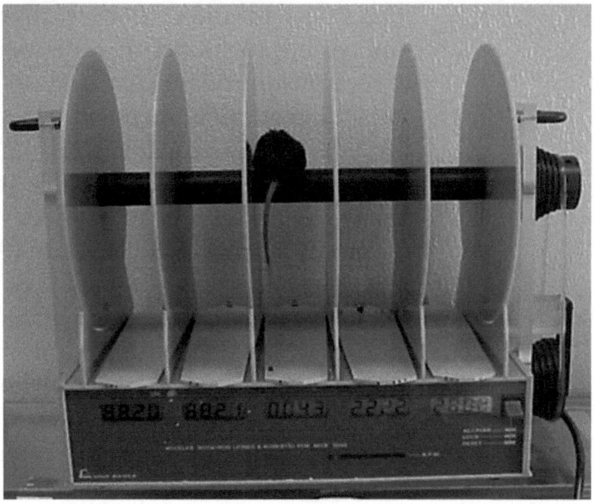

Fig. 6. Rotarod. Sensorimotor coordination can be assessed with the rotarod test where a rodent is placed on a rotating rod. The speed of rotation is gradually increased and the rodent's ability to remain on the rotating rod is recorded during several trials.

Alternatively, the same pole is used for the vertical pole test during which the pole is gradually lifted to a vertical position. Again latency to fall off the pole is recorded.

Grip strength can by quantified with a grip strength meter that senses the peak amount of force an animal applies in grasping specially designed pull bar assemblies. A metal wire horizontally stretched between two supports at a height of approximately 40 cm can be used to qualitatively measure muscle strength. The animal is suspended on the wire by its front paws and the time it can hang onto the wire is the outcome measure.

Ataxia and gait disturbances are quantified by the analysis of foot print patterns. These can be recorded by wetting the paws of the animal with ink and letting it run over a piece of paper on the bottom of a narrow aisle (38). Sophisticated fully automated systems have been developed too (e.g., the Catwalk from Noldus).

6. General Considerations

For meaningful statistical interpretations of behavioral phenotyping experiments, the absolute minimum of animals of each genotype is 10, although 12–15 or even 20 mice may serve a more optimal number based on interindividual variations in behavioral parameters. For many behavioral tasks, male mice are used because of the additional bias of hormonal fluctuations in females based on their 4–6-day estrus cycle. When studying an animal model of dementia, it goes without saying that naïve groups of different ages need to be tested in the complete test battery of choice. Adult mice aged 3–4 months are an ideal starting age; if deficits are already present, juvenile mice (<2 months) can be tested. If no deficits are present yet, older or even aged groups (>12 months) need to be screened.

Behavioral phenotyping of disease models is a highly specialized field that requires adequate training of researchers and technicians. Thorough standardization of housing, handling, and experimental procedures is essential to minimize interindividual and between-experiment variation. Various research groups have specialized in the behavioral phenotyping of rodent models and are often willing to study newly developed models on a collaborative basis. With the development of genetically modified models came the increased need for behavioral phenotyping, which has led to the establishment of several companies that offer contract-based behavioral phenotyping (e.g., NeuroDetective International, USA; Cerebricon, Finland; reMYND, Belgium; PsychoGenics, USA; PhenoPro, France).

References

1. Crawley J (2000) What's wrong with my mouse? Behavioral phenotyping of transgenic and knockout mice, 1st edn. Wiley-Liss, Wilmington, DE.
2. Buccafusco JJ (2008) Methods of behavior analysis in neuroscience, 2nd edn. CRC Press/Taylor & Francis Group, Boca Raton, FL.
3. Slawson JB, Kim EZ, Griffith LC (2009) High-resolution video tracking of locomotion in adult Drosophila melanogaster. J Vis Exp (24). pii:1096. doi: 10.3791/1096.
4. Pitman JL, DasGupta S, Krashes MJ, Leung B, Perrat PN, Waddell S (2009) There are many ways to train a fly. Fly (Austin) 3:3–9.
5. Gerber B, Stocker RF, Tanimura T, Thum AS (2009) Smelling, tasting, learning: Drosophila as a study case. Results Probl Cell Differ 47: 139–185.
6. Emran F, Rihel J, Dowling JE (2008) A behavioral assay to measure responsiveness of zebrafish to changes in light intensities. J Vis Exp (20). pii: 923. doi: 10.3791/923.
7. Kokel D, Peterson RT (2008) Chemobehavioural phenomics and behaviour-based psychiatric drug discovery in the zebrafish. Brief Funct Genomic Proteomic 7:483–490.
8. Sison M, Cawker J, Buske C, Gerlai R (2006) Fishing for genes influencing vertebrate behavior: zebrafish making headway. Lab Anim (NY) 35:33–39.
9. Giles AC, Rankin CH (2009) Behavioral and genetic characterization of habituation using Caenorhabditis elegans. Neurobiol Learn Mem 92:139–146.
10. Mori I, Sasakura H, Kuhara A (2007) Worm thermotaxis: A model system for analyzing thermosensation and neural plasticity. Curr Opin Neurobiol 17:712–719.
11. Murakami S (2007) Caenorhabditis elegans as a model system to study aging of learning and memory. Mol Neurobiol 35:85–94.
12. Irwin S (1968) Comprehensive observational assessment: Ia. A systematic, quantitative procedure for assessing the behavioral and physiologic state of the mouse. Psychopharmacologia 13:222–257.
13. Moser VC, McCormick JP, Creason JP, MacPhail RC (1988) Comparison of chlordimeform and carbaryl using a functional observational battery. Fundam Appl Toxicol 11: 189–206
14. Rogers DC, Fisher EM, Brown SD, Peters J, Hunter AJ, Martin JE (1997) Behavioral and functional analysis of mouse phenotype: SHIRPA, a proposed protocol for comprehensive phenotype assessment. Mamm Genome 8:711–713.
15. D'Hooge R, De Deyn PP (2001) Applications of the Morris water maze in the study of learning and memory. Brain Res Brain Res Rev 36:60–90.
16. Bunsey M, Eichenbaum H (1996) Conservation of hippocampal memory function in rats and humans. Nature 379:255–257.

17. Ohta A, Akiguchi I, Seriu N, et al. (2002) Deterioration in learning and memory of inferential tasks for evaluation of transitivity and symmetry in aged SAMP8 mice. Hippocampus 12:803–810.
18. Van Dijck A, Vloeberghs E, Van Dam D, Staufenbiel M, De Deyn PP (2008) Evaluation of the APP23-model for Alzheimer's disease in the odour paired-associate test for hippocampus-dependent memory. Behav Brain Res 190:147–151.
19. Skinner BF (1938) The behavior of organisms. An experimental analysis. D. Appleton-Century Company, New York.
20. Fanselow MS (1980) Conditioned and unconditional components of post-shock freezing. Pavlov J Biol Sci 15:177–182.
21. Bevins RA, Besheer J (2006) Object recognition in rats and mice: A one-trial non-matching-to-sample learning task to study 'recognition memory'. Nat Protoc 1:1306–1311.
22. Schafe GE, Sollars SI, Bernstein IL (1995) The CS-US interval and taste aversion learning: A brief look. Behav Neurosci 109:799–802.
23. Gheusi G, Bluthe RM, Goodall G, Dantzer R (1994) Ethological study of the effects of tetrahydroaminoacridine (THA) on social recognition in rats. Psychopharmacology (Berl) 114:644–650.
24. Carvell GE, Simons DJ (1990) Biometric analyses of vibrissal tactile discrimination in the rat. J Neurosci 10:2638–2648.
25. De Deyn PP, Katz IR, Brodathy H, Lyons B, Greenspan A, Burns A (2005) Management of agitation, aggression, and psychosis associated with dementia: A pooled analysis including three randomized, placebo-controlled double-blind trials in nursing home residents treated with risperidone. Clin Neurol Neurosurg 107:497–508.
26. Vitiello MV, Bliwise DL, Prinz PN (1992) Sleep in Alzheimer's disease and the sundown syndrome. Neurology 42:83–93.
27. Vloeberghs E, Van Dam D, Engelborghs S, Nagels G, Staufenbiel M, De Deyn PP (2004) Altered circadian locomotor activity in APP23 mice: A model for BPSD disturbances. Eur J Neurosci 20:2757–2766.
28. Valzelli L (1973) The "isolation syndrome" in mice. Psychopharmacologia 31:305–320.
29. Winslow JT, Miczek KA (1983) Habituation of aggression in mice: Pharmacological evidence of catecholaminergic and serotonergic mediation. Psychopharmacology (Berl) 81(4):286–291.
30. Vogel JR, Beer B, Clody DE (1971) A simple and reliable conflict procedure for testing anti-anxiety agents. Psychopharmacologia 21:1–7.
31. Blizard DA, Bailey DW (1979) Genetic correlation between open-field activity and defecation: Analysis with the CXB recombinant-inbred strains. Behav Genet 9:349–357.
32. Handley SL, Mithani S (1984) Effects of alpha-adrenoceptor agonists and antagonists in a maze-exploration model of 'fear'-motivated behaviour. Naunyn Schmiedebergs Arch Pharmacol 327:1–5.
33. Crawley J, Goodwin FK (1980) Preliminary report of a simple animal behavior model for the anxiolytic effects of benzodiazepines. Pharmacol Biochem Behav 13:167–170.
34. Chourbaji S, Zacher C, Sanchis-Segura C, Dormann C, Vollmayr B, Gass P (2005) Learned helplessness: Validity and reliability of depressive-like states in mice. Brain Res Protoc 16:70–78.
35. Sanchis-Segura C, Spanagel R, Henn FA, Vollmayr B (2005) Reduced sensitivity to sucrose in rats bred for helplessness: A study using the matching law. Behav Pharmacol 16:267–270.
36. Vloeberghs E, Van Dam D, Franck F, et al. (2008) Altered ingestive behavior, weight changes, and intact olfactory sense in an APP overexpression model. Behav Neurosci 122:491–497.
37. Jones BJ, Roberts DJ (1968) A rotarod suitable for quantitation measurements of motor incoordination in naïve mice. Naunyn Schmiedebergs Arch Pharmacol 259:211.
38. D'Hooge R, Hartmann D, Manil J, Colin F, Gieselmann V, De Deyn PP (1999) Neuromotor alterations and cerebellar deficits in aged arylsulfatase A-deficient transgenic mice. Neurosci Lett 273:93–96.

Chapter 9

Pharmacological Validation in Animal Models of Dementia

Hugo Geerts

Abstract

The field of Alzheimer's disease (AD) research has been quite fortunate – in contrast to some other neurodegenerative psychiatric diseases – in that a number of animal models have been developed based on genetic and neuropathological information. These animal models have been "validated" based on the fact that they reflect one or a few of the neuropathological features found in postmortem brain of AD patients. However, the predictive value of these models for drug discovery has been far from spectacular. Limitations of these models include the failure to capture the dynamics of the ongoing pathology in a clinical setting and to display the total neuropathology. However, animal models – because of the species barrier – have a number of additional limitations for a successful drug development program, which are not always fully appreciated. This chapter discusses (1) differences in drug affinities between human and rodent targets, (2) the absence of key human functional genotypes in rodent models, (3) the intrinsic difference in some neurotransmitter circuits, and (4) the difficulty of simulating the same amount of drug exposure as in the clinical situation. In addition, the problems associated with extrapolating cognitive tests in animal studies with actual performance of treated AD patients on clinical scales are explored. Possible solutions to this dilemma include (1) developing multitarget directed ligands where cholinesterase inhibition is combined with disease modification, (2) a better translation of clinical endophenotypes, (3) capitalizing on drug discovery efforts for cognition in other disease areas, (4) the introduction of realistic polypharmacy in early stages of preclinical tests, and (5) the systematic testing of the face-value of a specific preclinical model/readout combination using marketed drugs with documented clinical effects. Finally, Computational Neuropharmacology, a novel and highly innovative computer modeling approach of interacting brain circuits is introduced. When added to the toolbox of preclinical drug discovery, this approach intends to bridge the difference between preclinical animal models and the clinical situation and reduce the rate of attrition.

Key words: Affinity, functional genotype, neurotransmitter circuit, drug exposure, endophenotype, polypharmacy

1. Introduction

The ability of the human brain to solve higher-order problems, language, and social interaction, compared to mere food foraging and survival in the rodent has led to a vastly more complex system

Peter Paul De Deyn and Debby Van Dam (eds.), *Animal Models of Dementia*, Neuromethods, vol. 48,
DOI 10.1007/978-1-60761-898-0_9,

of neurotransmitter interactions, in addition to a more complex variety of functional genotype modifications. Some of the most stringent differences, which are often not very well appreciated in today's drug discovery environment and which are significant enough to possibly account for failures in human clinical trials, will be addressed. Too often, it is taken for granted that there is a very good correlation between the effect of drugs in so-called "validated" animal models and human clinical trials. A possible failure of a drug in clinical situations is often interpreted as the failure of the basis hypotheses on which the target for the drug was selected, rather than the failure of the animal models in which the drug was active. Such a philosophy risks of hindering the development of valuable drugs for treating Alzheimer's disease (AD). Recently, a very well-publicized Phase III trial of a disease-modifying agent, tramiprosate, a β-amyloid (Aβ) anti-aggregation drug (1), was found to be negative, despite robust effects on plaque lowering and neuroinflammation in some hAPP transgene mice. Conversely, the drugs currently marketed for AD have been identified using basic translational approaches; they did not rely upon the current "validated" AD animal models. In hindsight, these drugs show only a marginal or no effect at all in various AD animal models.

In this chapter, possible problems associated with animal models and which might hinder a successful clinical development of a candidate drug will be presented. These factors include (1) the different pharmacology of the same drug for rodent vs. human target subtypes, (2) the different wiring of specific neurotransmitter circuits in rodent vs. human brain, (3) the limited spectrum of clinically important functional genotypes in animal models, (4) the incapacity to capture the full spectrum of pathology in rodents, (5) the difference in drug metabolism, which make it difficult to simulate the human drug exposure, (6) the inability to capture the full dynamics of the pathology, and (7) the limited predictability of cognitive test outcomes in animal models in the clinical situation.

Possible – and sometimes controversial – solutions to this dilemma include (1) a better translation of clinical endophenotypes to the preclinical situation, (2) capitalizing on drug discovery efforts for cognition in other disease areas, (3) the introduction of realistic polypharmacy in early stages of preclinical tests, and (4) the systematic testing of the face-value of a specific preclinical model/readout combination using marketed drugs with documented clinical effects. In addition, examples will be given where Computational Neuropharmacology, a computer modeling approach of interacting brain circuits, is able to better predict a clinical outcome.

2. Limitations of Animal Models

2.1. The Pharmacology of Investigative Drugs Can Often Be Substantially Different Between Human and Rodent Target Subtypes

The selection of a good "Investigational New Drug" or IND candidate in drug discovery is based upon efficacy at "reasonable doses" in "validated" animal models of the disease; for instance, can a certain experimental drug improve a Morris water maze and/or reduce Aβ plaque load in a transgene amyloid model?

Using this paradigm, drugs are almost certainly optimized for their interaction with rodent receptors or targets and although the drug is sometimes tested for activity on human targets as well, the drug with the best pharmacological profile for the rodent receptors subtypes has not always the optimal profile against the human targets.

In fact, using a systematic search on publicly available data, it turns out that for certain targets, the differences in pharmacology can be quite substantial.

In one particular study, binding affinities for the same drug-tracer combinations were compared between human and rodent membrane G-protein coupled receptors for dopamine. Table 1 shows the average ratio of maximal over minimal K_i-value for four dopamine receptor subtypes. It is clear from this analysis that especially the human and rodent D1R differ significantly.

A difference of a factor of 4 can be biologically significant, especially if the affinity of the endogenous neurotransmitter is in the same range. A compound like SKF77434 has an affinity of 71 nM for the rat D1R and 12 nM for the human D1R, both measured with 3H-SCH2339 as tracer under similar conditions. With an affinity of 40 nM for the endogenous dopamine neurotransmitter (2,3), which is similar for the human and the rat receptor, it is clear that in the rat the compound SKF77434 is a rather weak antagonist, whereas in the human it is a modest to strong antagonist. Given the importance of D1R for cognitive processes (Table 1), failure to recognize this issue can lead to false negatives.

Table 1
Average ratio of maximal over minimal Ki-value for four dopamine receptor subtypes

Receptor subtype	n	Average Ki	Range Ki
D1-R	41	6.88 ± 12.3	1–58
D2-R	87	3.18 ± 3.7	1–20
D3-R	45	3.19 ± 3.31	1–45
D4-R	22	4.83 ± 7.6	1–22

Although these differences have been documented on G-protein coupled receptors (because of the availability of binding data), similar differences have been found in typical AD-related gene products. The gene sequence difference in human and mouse Aβ peptide is such that mouse amyloid has a much lower tendency to aggregate. Transgene animal models solve this by overexpressing one form or another of the human amyloid protein; however, the need for high expression levels to induce the pathology yield additional problems (see later).

2.2. Some Neurotransmitter Circuits Are Wired Differently in the Human versus the Rodent Brain

An often overlooked problem is that certain neurotransmitter systems are differently wired in human vs. rodent settings. In some instances, even certain drug receptor subtypes only exist in human brain and not in mouse brain. A typical example is the human 5HT1D-R, site of action for many marketed antimigraine drugs. However, in the mouse, the 5-HT1D receptor is downregulated at birth and no functional expression is seen *in vivo* (4). For this type of pharmacology, mice are clearly not the right animal.

Other dramatic differences can be found in actual receptor localization with Table 2 giving a number of examples.

The 5-HT6 receptor is of particular interest, because changes have been reported in the brain of AD patients (9) and a number of 5-HT6 antagonists are currently in development for cognitive deficits. The observation of a different mouse 5-HT6 receptor distribution limits dramatically the testing of 5-HT6 drugs in transgene Alzheimer mouse models.

2.3. Some Functional Human Genotypes Cannot Be Reproduced in Animal Models

Advances in genomic technology have identified key functional genotypes with clinical effects in human subjects with the APOE isoform as an obvious example in the Alzheimer field. The presence of the APOE4 gene confers a higher risk for younger age of

Table 2
Major differences in receptor localization between humans and rodents

Receptor/target	Major difference human-rodent	Reference
mGluRl in cortex	mGluRl-R found on pyramidal neurons and interneurons in cortex; in rodent brain exclusively on cortical interneurons	(5)
mGluRl in substantia nigra	Primate SNcompacta>>Primate SNreticularis Rodent SNc=Rodent SNr	(6)
5-HT3	Substantial striatal localization in human; very low striatal localization in rodents. Converse true in cortex	(7)
5-HT6	Rat: Striatum, N accumbens Human: Caudate nucleus, putamen, N accumb Mouse: very weak in these regions	(8)

onset, although the natural course as measured by the ADAS-Cog does not seem to be influenced by the presence of the APOE4 allele (10). The APOE gene in mouse does not have these different genotypes (see (11) for a discussion of the consequences in mouse models). In addition, there are important differences in the mouse vs. the human promoter (12), which would in theory necessitate the inclusion of the promoter in all transgene constructs. Despite a lot of effort in transgene animal models, the biological role of APOE4 in the human disease has not been elucidated completely.

Other examples with relevance to cognitive performance include the Val158Met SNP in the Catechol-*O*-methyl-transferase (COMT) gene, the protein product of which is involved in degradation of catecholamines. Subjects with the Met/Met allele have a less stable COMT gene and lower protein expression, which leads to higher free levels of dopamine and noradrenaline in the prefrontal cortex. A number of clinical studies have shown that subjects with the Met/Met allele perform consistently better on a variety of cognitive tasks (for a review, see (13)). The functional effect suggests that this genotype affects the clinical outcome of dopaminergic or adrenergic drugs, as shown in the effect of olanzapine on an N-back working memory test on schizophrenic patients (14).

Another example is the recent clinical observations that clinical antidepressant response of certain drugs is associated with particular genotypes of the ABCB1 drug transporter, when the drug is a substrate of this transporter (15). This suggests a more complex degree of interaction at the blood–brain barrier than can be achieved in rodent models.

Because rodents often do not have similar functional genotype, it is close to impossible to assess its effect on the functional effect of a candidate drug. Failure to recognize this issue might lead to an imbalance in the differently powered arms in clinical trials.

2.4. Difference in Metabolism Makes It Hard to Simulate Clinical Drug Exposure in Animal Models

Drug metabolism in rodents is usually much faster than in humans. Many years of expertise in pharmacokinetics has allowed to often successfully extrapolating the PK profile observed in rodent models to human situations. However, predicting the human pharmacodynamics (i.e. the pharmacological activity) has been much more difficult.

This has been illustrated in a systematic study of different antipsychotics in rats vs. humans (16). In this case, the actual brain D2R occupancy can be readily measured using PET imaging of ^{11}C-raclopride. Using *in vivo* radiography in the rat model, in order to achieve similar exposure as seen in patients, repeated drug applications or minipump solutions needed a dose at least five times higher than the optimal single-dose concentration. This is because metabolism of these drugs in rats is four to six times faster than in humans. As a consequence, many preclinical drug studies often use concentrations that are by no means relevant for the clinical situation.

The availability of an imaging tracer or a functional CNS readout, like pharmacoEEG, which links rodent plasma concentration to a functional brain readout in order to determine the active dose-range allows for a correct extrapolation to the human situation; however, such an ideal situation is seldom achieved. Furthermore, in many cases, the translational link with the clinical situation is not appreciated.

Another problem in the predictability of animal models is the possible difference in blood–brain barrier (BBB) permeability of certain drugs. While the make-up of the rodent BBB is very similar to the human, a species-specific drug–PgP interaction sometimes makes drugs behave differently. Again, having an imaging tracer or a functional CNS readout available in the clinical setting allows one to rapidly determine how readily the compound gets into the brain.

2.5. Animal Models Often Simulate Only a Fraction of the Clinical Neuropathology

Over the last years, numerous transgene animal models have been developed where the genetic make-up of the animal is modified to introduce certain wild-type or mutated human genes, involved in the familial forms of AD. Cross-breeding of different lines in theory leads to the expression of different neuropathology forms in the same animal. Examples include the triple transgene 3xTg mice with PS1, APP, and tau mutations (17), currently the model, which captures the broadest AD neuropathology. While this mouse model is an important research tool that has clarified aspects of the interaction between Aβ deposits and tangle formation, it is by no means a complete model of the human Alzheimer neuropathology. For instance, in the 3xTg mouse, an age-dependent decrease in α7 nicotinic acetylcholine receptor (nAChR) density has been reported (18), but no α4β2 nAChR loss. This is different from the clinical AD pathology, in which consistent loss of α4β2 nAChRs has been seen, even *in vivo* (19) and sometimes, but not always loss of α7 nAChRs. Some of these pathologies interact with each other; for instance, there is mounting evidence for a direct interaction between nAChRs and brain microglia that modulate the degree of brain inflammation (20). With many of the documented pathologies interacting with each other, it is clear that the limited ability to model these interactions appropriately in animals has important consequences for the predictability of clinical efficacy.

2.6. Animal Models Do Not Capture the Full Dynamics of the Human Pathology

Animal models try to capture the long-life dynamics of the pathology in a few months; this compression necessarily leads to a limited simulation of the full spectrum of the human pathology. For instance, many mice models using hAPP overexpression already report "cognitive" deficits well before plaques are present (21), suggesting a disconnection between functional deficits and amyloid aggregation. Such data have been explained by the fact that smaller nonfibrillar Aβ oligomer aggregates are toxic to neuronal synapses and have been interpreted as a pre-Alzheimer or an MCI (i.e., mild cognitive impairment) stage.

In contrast, in the human situation recent imaging studies using the amyloid agent PIB-1 have revealed that in normal volunteers and MCI, the degree of retention of the tracer (which is related to the amyloid plaque density) is well correlated with the degree of cognitive episodic memory impairment, as measured by the Rey Complex Figure test and the California Verbal learning test, both at long delays (22). There was no correlation in the case of Alzheimer patients, presumably because of a ceiling effect.

A large majority of AD animal models are based on genetic modification using human genes, which predispose to early familial type of the disease. The dynamics of this familial type are very likely different from the sporadic case, for there is between a 20 and 40 year difference in age of onset. Of note is that there are other "more sporadic" animal models of AD, such as a rabbit model fed on a diet of 2% cholesterol with traces of copper (23), which might contribute more to understanding the interplay between neuropathology, cognitive deficit, and therapeutic response for the sporadic case.

2.7. The Need for Polypharmacy in Animal Studies

Both in real-life clinical situations and in well-controlled clinical trial studies, patients are allowed to take additional medication. For instance, many AD patients are treated with acetylcholinesterase (AChE)-inhibitors or NMDA antagonists defining a placebo active comparator arm. However, some of these medications do have indirect but modest effects on some of the intracellular pathways involved in amyloid pathology.

As an example, increasing cholinergic tone by reducing ACh hydrolysis activates m1 muscarinic receptors, which have been shown in preclinical models to modulate both tau and amyloid related pathways (for a recent review see (24)). Galantamine has been shown to have pleiotropic activities *in vitro* and *in vivo*, due to its allosteric modulation of the nAChR (25). Similarly, metabotropic glutamate receptors, which are indirectly modulated by nAChR-dependent changes in presynaptic glutamate release or by antagonism at the NMDA-R, can modulate APP expression (26).

In many clinical trials, antidepressant medication is allowed and recent well-reproduced data suggest a direct effect of antidepressants on hippocampal neurogenesis (27). Also, atypical antipsychotics, which are sometimes used to address behavior problems, have been shown to be involved in neurogenesis, stimulating proliferation of neuronal stem cells and microglia (for a review, see (28)). This could have important effects on various effects of AD neuropathology. These data show that the existing medications, which are often used in combination with the investigative drug, can influence various aspects of AD pathology. Yet, in many preclinical animal models, candidate drugs are often selected based on experiments in which no real polypharmacy is simulated. Although such combination studies would delay the discovery project, assessment of possible pharmacodynamic interactions can be very important. In addition, multitarget-directed ligands, which combine symptomatic improvement (for instance using AChE inhibition) with a disease-modifying action in one molecule, will have a much larger chance of successful clinical development. However, this necessitates modifying the current mantra of "one drug, one target," currently adhered to in many medicinal chemistry departments in pharmaceutical companies.

3. Relation Between Preclinical and Clinical Readouts

3.1. Not All Clinical Scales Have the Same Sensitivity for Therapeutic Interventions

The mandatory clinical scale for FDA AD drug approval is the Alzheimer's Disease Assessment Scale – Cognitive Subscale (ADAS-Cog) (29) and all currently marketed drugs have shown statistically significant benefit on these scales. However, the ADAS-Cog does not capture all pharmacological activities, as other scales such as NTB (Neuropsychological Test Battery) have shown differential improvement in cases where the ADAS-Cog has not. For example, in the AN1792 vaccination trial, the NTB did show a small but significant improvement (0.03 vs. –0.20) in the antibody-responsive patients, whereas the ADAS-Cog, MMSE, and CGI did not (30), suggesting that these scales probe different functional dimensions.

Extensive analysis of clinical responses to AchE inhibitors has allowed to identify subdomains, such as the Clock Drawing Test, a Visual-Spatial Motor Tracking Test, and the Boston Picture Naming Test, within the ADAS-Cog who are most sensitive to cholinergic intervention (*31*). In addition, adding memantine to rivastigmine improves specific subdomains of the ADAS-Cog (*32*), such as executive/attention-mediated processes. This also suggests that different therapeutic approaches improve different subdomains of clinically well-validated cognitive scales. This has important implications for the choice of cognitive tests in animal models. A possible solution for this issue could be to (1) capture all published information on the effect of therapeutic interventions on clinical scales, (2) run various animal models with their cognitive readout in standardized conditions using the same drugs at their appropriate clinical doses, and (3) correlate the outcome of the animal models with the clinical outcomes. A good correlation would de facto give much more face-validity to the preclinical animal model.

A PubMed search at the moment of writing this chapter (2008) allows identifying 75 articles covering about 160 drug/dose combinations with documented effect on a number of cognitive tasks in healthy controls and various disease states. The drugs cover a wide variety of targets, including cholinergic, dopaminergic, serotonergic, adrenergic, and GABAergic neurotransmitter systems.

Obviously, such experiments could probably not be performed by a single laboratory; however, a central agency like the NIA could facilitate the sponsoring of a dedicated research center, much along the lines of the NIMH sponsored PsychoActive drug Screening Program at the University of North Carolina in Chapel Hill (http://pdsp.med.unc.edu/indexR.html). This particular center provides standardized testing of the binding affinity of different drugs on a battery of over 50 human receptors.

3.2. What Can Work in Other Areas, Such As Cognitive Deficit in Schizophrenia, Teach Us?

Schizophrenia patients also show cognitive deficits, although not to the same degree as Alzheimer patients and extensive studies suggest that this cognitive deficiency is a stable core deficit of the disease and does not respond well to antipsychotic medication. The areas of executive functioning, memory, and attention are particularly affected with a deficit greater than 1.5 standard deviations (33).

Recently, an industry-FDA-academia workgroup (Matrics, Measurement and Treatment Research in Cognitive deficits in Schizophrenia, http://www.matrics.ucla.edu/) has embarked on developing a framework for translational research in order to facilitate development of new therapeutic approaches for improving cognitive deficits. A clinical battery of cognitive tests has been validated and accepted by the FDA.

A number of different cognitive approaches are now in early clinical development for AD and schizophrenia, among them are a number of nicotinic receptor modulators (34), AMPAkines (35), D1-R partial agonists (36), 5-HT6 antagonists (37), and 5-HT4 partial agonists (38). Although these approaches do not directly act on the course of the neuropathology, such symptomatic therapies have an important role in the clinical management of AD patients. It is to be expected that over the course of the next years, some of these clinical experiments will be reported that can add to the face-validity database mentioned earlier.

3.3. Introduction of Clinical Endophenotypes in Early Discovery Research

Although the FDA still requires a functional assessment for clinical development of Alzheimer drugs, other biomarkers or endophenotypes can be very helpful in early stages of clinical development. These biomarkers can include biochemical measures of CSF or plasma Aβ, CSF total tau or phospho-tau, imaging with specific tracer molecules reporting on brain amyloid aggregates and pharmacoEEG of pharmacofMRI measurements. For example, Aβ peptide levels can be readily measured in the CSF of freely moving rats (39). Although the dynamics of CSF Aβ peptide changes in patients are not clear, information on the effect of an amyloid-directed therapy on CSF levels in an early stage of discovery can be very helpful.

Metabolic imaging has consistently documented the functional deficit in temporal areas of the brain and structural MRI has identified brain volume loss in well-defined areas. Various correlations between brain atrophy and functional tests such as MMSE have been identified, such as ventricular enlargement (40) or hippocampal atrophy (41). Novel techniques, such as diffusion tensor imaging, have also found a strong correlation between diffusivity and functional MMSE scores (42).

It is to be expected that the NIA funded Biomarker Consortium project for AD will provide the necessary sample size to identify more robust correlations between functional outcomes in patients and biomarkers.

Early preclinical studies with a candidate drug using these endophenotypes can help design proof-of-concept clinical trials in patients, in order to gain confidence that the drug acts in a similar way in the human as in the animal model. But more importantly, correlations between clinical scales and these biomarkers in human patients can pave the way for using these more objective endophenotypes as proxies for functional tests in preclinical animal research, rather than the more difficult to simulate cognitive scales.

It will probably be unlikely that the FDA will substitute the functional ADAS-Cog scale for a surrogate marker to prove clinical efficacy, but information on the effect of a candidate drug on these endophenotypes in an early stage of clinical development will certainly contribute to a better assessment of the success rate of the proposed therapy.

4. Introducing Computational Neuropharmacology as an Additional Tool

Recent developments in computing performance and analysis algorithms have allowed developing a more hybrid preclinical-patient in silico model. Such approaches embrace both mechanistic and physiological animal data and data on the pathology of the human patients, based on a wide array of imaging, postmortem, and functional genotype data, and as such intent to bridge the animal–human divide.

For instance, a number of neurocomputational models of working memory have explored the physiological processes involved in human cognitive tasks (for a review, see (43)). These models are detailed biophysical descriptions of different interacting cell types and are usually validated using either available animal data or even human data.

An example of the power of these techniques is based upon the observation of an inverse U-shaped dose–response after a dopamine agonist on a short-term learning task in human volunteers (44). In animal studies, D1R agonists dose-dependently improve certain forms of cognitive behavior without evidence of an inverse U-shaped dose–response (45). In contrast, computational network models of working memory, after introducing the different effects of postsynaptic D1R and presynaptic D2R stimulation on various types of ion channels and glutamate receptors, are able to explain and recapitulate the U-shaped dose–response of dopamine by the fine balance between excitatory and inhibitory circuits (46).

These cortical circuits are completed by introducing afferent serotonergic fibers from thalamic nuclei, cholinergic fibers from N. basalis of Meynert, noradrenergic fibers from Raphe Nucleus, and dopaminergic fibers from ventral tegmentum area in a human-specific fashion. The models further allow introducing receptor pharmacology according to the appropriate human distributions and Alzheimer pathology, for instance by eliminating postmortem confirmed synapse deficits in the circuits to such a degree that the functional outcome of the network corresponds to a clinically observed deficit in untreated patients.

These neuronal circuits are also optimized to generate certain types of EEG, and introduction of appropriate drug pharmacology on the receptors allow for detecting a change in EEG spectral bands, which can then be validated using clinically available data on the effect of certain drugs on the EEG spectrum. Such analysis can be helpful to identify possible pharmacoEEG markers for early proof of concept clinical trials.

While these models obviously are limited in the number of physiological processes included, in the complexity of different cell types and in the limited dynamic scales, they are able to provide additional insights, especially with regard to the effects of functional genotypes, inclusion of a more complete pathology, appropriate drug exposure, and the ability to run virtual patient trials based upon the genetic variability. It is to be expected that these models will become increasingly important as additional tools for drug discovery and drug development.

5. Conclusion and Recommendations

The prospects of a simple extrapolation of animal data to the human situation in neurodegenerative diseases seem gloomy. The human brain is way more complex than the rodent brain, because of its need to solve major problems in the language, social interaction, and higher-order cognitive area.

This overview hopefully helps to identify the major differences between animal models and human patients, which are often underappreciated in drug discovery and which can lead to failures in the clinic, disappointment among the researchers and a fatalistic attitude at drug companies, forcing them to shut down further research in this area.

Many of the limitations are inherent to the animal species used in preclinical research and need to be acknowledged as such. However, some of the other issues can be addressed readily. Here is a list of suggestions:

- Test the candidate compound on human receptors and targets.
- Explore multitarget-directed ligands in medicinal chemistry combining AChE-inhibition with disease-modifying activity in a single molecule.
- Adjust the dose for long-term studies as close to the human situation as possible.
- Simulate the clinical trial reality by introducing realistic polypharmacy in preclinical experiments.
- Use a biomarker/endophenotype with proven face-validity in the human situation and a correlation with clinical scales.
- Develop a systematic face-validity study of model/cognitive readout combinations by running the same drug/dose combinations in animal models as are reported in actual clinical situations and correlating the outcome of the two situations.
- Test the candidate compound additionally in a nontransgene animal model, maybe more representative of the sporadic AD case.

Adding aspects of computational neuropharmacology might be very useful for taking into account important clinical issues not covered by animal models, such as the effect of functional genotypes, the identification of useful biomarkers in early clinical trial, or the design of clinical trials.

Implementation of these suggestions has the potential to reduce attrition rates in drug development and to increase the chance that Alzheimer patients will get access to novel helpful medications.

References

1. Geerts H (2004) NC-531 (Neurochem). Curr Opin Investig Drugs 5:95–100.
2. Lidow MS, Roberts A, Zhang L, et al. (2001) Receptor crosstalk protein, calcyon, regulates affinity state of dopamine D1 receptors. Eur J Pharmacol 427:187–193.
3. Trantham-Davidson H, Neely LC, et al. (2004) Mechanisms underlying differential

D1 versus D2 dopamine receptor regulation of inhibition in prefrontal cortex. J Neurosci 24:10652–10659.
4. Bonnin A, Peng W, Hewlett W, et al. (2006) Expression mapping of 5-HT1 serotonin receptor subtypes during fetal and early postnatal mouse forebrain development. Neuroscience 141:781–794.
5. Muly EC, Maddox M, Smith Y (2003) Distribution of mGluR1alpha and mGluR5 immunolabeling in primate prefrontal cortex. J Comp Neurol 467:521–535.
6. Hubert GW, Paquet M, Smith Y (2001) Differential subcellular localization of mGluR1a and mGluR5 in the rat and monkey Substantia nigra. J Neurosci 21:1838–1847.
7. Marazziti D, Betti L, Giannaccini G, Rossi A, et al. (2001) Distribution of [3H]GR65630 binding in human brain postmortem. Neurochem Res 26:187–190.
8. Hirst WD, Abrahamsen B, Blaney FE, et al. (2003) Differences in the central nervous system distribution and pharmacology of the mouse 5-hydroxytryptamine-6 receptor compared with rat and human receptors investigated by radioligand binding, site-directed mutagenesis, and molecular modeling. Mol Pharmacol 64:1295–1308.
9. Garcia-Alloza M, Hirst WD, Chen CP, et al. (2004) Differential involvement of 5-HT(1B/1D) and 5-HT6 receptors in cognitive and non-cognitive symptoms in Alzheimer's disease. Neuropsychopharmacology 29:410–416.
10. Aerssens J, Parys W, Lilienfeld S, et al. (2001) "APOE genotype: No influence on galantamine treatment efficacy nor on rate of decline in AD". Dement Geriatr Cogn Disord 12:69–77.
11. Bales KR, Dodart JC, DeMattos RB, et al. (2002) Apolipoprotein E, amyloid, and Alzheimer disease. Mol Interv 2:363–375.
12. Maloney B, Ge YW, Alley GM, et al. (2007) Important differences between human and mouse APOE gene promoters: Limitation of mouse APOE model in studying Alzheimer's disease. J Neurochem 103:1237–1257.
13. Diaz-Asper CM, Weinberger DR, Goldberg TE (2006) Catechol-O-methyltransferase polymorphisms and some implications for cognitive therapeutics. NeuroRx 3:97–105.
14. Bertolino A, Caforio G, Blasi G, et al. (2004) Interaction of COMT (Val(108/158)Met) genotype and olanzapine treatment on prefrontal cortical function in patients with schizophrenia. Am J Psychiatry 161:1798–1805.
15. Uhr M, Tontsch A, Namendorf C, et al. (2008) Polymorphisms in the drug transporter gene ABCB1 predict antidepressant treatment response in depression. Neuron 57:203–209.
16. Kapur S, VanderSpek SC, Brownlee BA, et al. (2003) Antipsychotic dosing in preclinical models is often unrepresentative of the clinical condition: A suggested solution based on in vivo occupancy. J Pharmacol Exp Ther 305:625–631.
17. Oddo S, Caccamo A, Shepherd JD, et al. (2003) Triple-transgenic model of Alzheimer's disease with plaques and tangles: Intracellular Abeta and synaptic dysfunction. Neuron 39:409–421.
18. Oddo S, Caccamo A, Green KN, et al. (2005) Chronic nicotine administration exacerbates tau pathology in a transgenic model of Alzheimer's disease. Proc Natl Acad Sci U S A 102:3046–3051.
19. O'Brien JT, Colloby SJ, Pakrasi S, et al. (2007) Alpha4beta2 nicotinic receptor status in Alzheimer's disease using 123I-5IA-85380 single-photon-emission computed tomography. J Neurol Neurosurg Psychiatr 78:356–362.
20. Carnevale D, De Simone R, Minghetti L (2007) Microglia-neuron interaction in inflammatory and degenerative diseases: Role of cholinergic and noradrenergic systems. CNS Neurol Disord Drug Targets 6:388–397.
21. Kelly PH, Bondolfi L, Hunziker D, et al. (2003) Progressive age-related impairment of cognitive behavior in APP23 transgenic mice. Neurobiol Aging 24:365–378.
22. Pike KE, Savage G, Villemagne VL, et al. (2007) Beta-amyloid imaging and memory in non-demented individuals: Evidence for preclinical Alzheimer's disease. Brain 130: 2837–2844.
23. Woodruff-Pak DS, Agelan A, Del Valle L (2007) A rabbit model of Alzheimer's disease: Valid at neuropathological, cognitive, and therapeutic levels. J Alzheimers Dis 11:371–383.
24. Fisher A (2007) M1 muscarinic agonists target major hallmarks of Alzheimer's disease–an update. Curr Alzheimer Res 4:577–580.
25. Coyle JT, Geerts H, Sorra K, et al. (2007) Beyond in vitro data: A review of in vivo evidence regarding the allosteric potentiating effect of galantamine on nicotinic acetylcholine receptors in Alzheimer's neuropathology. J Alzheimers Dis 11:491–507.
26. Lee RK, Wurtman RJ, Cox AJ, et al. (1995) Amyloid precursor protein processing is stimulated by metabotropic glutamate receptors. Proc Natl Acad Sci U S A 92:8083–8087.
27. Sahay A, Hen R (2007) Adult hippocampal neurogenesis in depression. Nat Neurosci 10:1110–1115.

28. Newton SS, Duman RS (2007) Neurogenic actions of atypical antipsychotic drugs and therapeutic implications. CNS Drugs 21:715–725.
29. Peña-Casanova J (1997) Alzheimer's disease assessment scale–cognitive in clinical practice. Int Psychogeriatr 9:105–114.
30. Gilman S, Koller M, Black R, Jenkins L, et al. (2005) Clinical effects of Abeta immunization (AN1792) in patients with AD in an interrupted trial. Neurology 9:1553–1562.
31. Saumier D, Murtha S, Bergman H, et al. (2007) Cognitive predictors of donepezil therapy response in Alzheimer disease. Dement Geriatr Cogn Disord 24:28–35.
32. Riepe MW, Adler G, Ibach B, et al. (2007) Domain-specific improvement of cognition on memantine in patients with Alzheimer's disease treated with rivastigmine. Dement Geriatr Cogn Disord 23:301–306.
33. Bilder RM, Goldman RS, Robinson D, et al. (2000) Neuropsychology of first-episode schizophrenia: Initial characterization and clinical correlates. Am J Psychiatry 157:549–559.
34. Arneric S, Holladay M, Williams M (2007) Neural nicotinic receptors: A perspective on two decades of drug discovery. Biol Psych 74:1092–1102.
35. Goff DC, Lamberti JS, Leon AC, Green MF, et al. (2008) A Placebo-controlled add-on trial of the Ampakine, CX516, for cognitive deficits in schizophrenia. Neuropsychopharmacology 33:465–472.
36. El-Ghundi M, O'Dowd BF, George SR (2007) Insights into the role of dopamine receptor systems in learning and memory. Rev Neurosci 18:37–66.
37. Mitchell ES, Neumaier JF (2005) 5-HT6 receptors: A novel target for cognitive enhancement. Pharmacol Ther 108:320–333.
38. Maillet M, Robert SJ, Lezoualc'h F (2004) New insights into serotonin 5-HT4 receptors: A novel therapeutic target for Alzheimer's disease? Curr Alzheimer Res 1:79–85.
39. El Mouedden M, Haseldonckx M, Mackie C, et al. (2005) Method for the determination of the levels of beta-amyloid peptide in the CSF sampled from freely moving rats. J Pharmacol Toxicol Methods 52:229–233.
40. Ferrarini L, Palm WM, Olofsen H, et al. (2008) MMSE scores correlate with local ventricular enlargement in the spectrum from cognitively normal to Alzheimer disease. Neuroimage 39:1832–1838.
41. Yavuz BB, Ariogul S, Cankurtaran M, et al. (2007) Hippocampal atrophy correlates with the severity of cognitive decline. Int Psychogeriatr 19:767–777.
42. Bozzali M, Falini A, Franceschi M, et al. (2002) White matter damage in Alzheimer's disease assessed in vivo using diffusion tensor magnetic resonance imaging. J Neurol Neurosurg Psychiatry 72:742–746.
43. Tretter F, Albus M (2007) Computational neuropsychiatry of working memory disorders in schizophrenia: The network connectivity in prefrontal cortex - data and models. Pharmacopsychiatry 40:S2–S16.
44. Williams GV, Castner SA (2006) Under the curve: Critical issues for elucidating D1 receptor function in working memory. Neuroscience 139:263–276.
45. Di Cara B, Panayi F, Gobert A, et al. (2007) Activation of dopamine D1 receptors enhances cholinergic transmission and social cognition: A parallel dialysis and behavioural study in rats. Int J Neuropsychopharmacol 10:383–399.
46. Durstewitz D, Seamans JK, Sejnowski TJ (2000) Dopamine-mediated stabilization of delay-period activity in a network model of prefrontal cortex. J Neurophysiol 83:1733–1750.

Chapter 10

Validation of Animal Models of Dementia: Neurochemical Aspects

Giancarlo Pepeu and Maria Cristina Rosi

Abstract

The neurochemical alterations underlying the cognitive and behavioral symptoms that constitute the clinical picture of Alzheimer's disease (AD), which should be reproduced by an animal model of the disease, are briefly described. The ideal animal model of AD and related dementias should include β-amyloid deposition evolving in senile plaques (SP), tau protein hyperphosphorylation resulting in neurofibrillary tangles (NFT), an inflammatory reaction, and the degeneration of the forebrain cholinergic neurons with the ensuing cholinergic hypofunction. However, since the understanding of AD pathogenesis has grown step by step during 30 years of research, simpler models, only reproducing one or a few neurochemical changes, have been gradually developed beginning from the lesion of the cholinergic nuclei and ending with transgenic mice developing SP and NFT. The partial models still maintain their usefulness for understanding the different pathogenetic mechanisms and for developing new drugs. The animal models are validated by the demonstration, through appropriate and specific methods, that they show the neurochemical changes that they are meant to reproduce. The methods are mentioned and briefly described in this chapter, and evidence of their validity is given by quoting significant papers in which they have been used.

Key words: APP mice, β-amyloid, brain inflammation, cholinergic hypofunction, neurofibrillary tangles, tau hyperphosphorylation

1. Neurochemical Changes in Alzheimer Dementia

To validate an animal model, it is mandatory to define the pathological features of the disease that the model should mimic. Of the many forms of dementia existing, we will examine in this chapter the models of Alzheimer's disease (AD), and briefly mention the models of frontotemporal dementia (FTD). The neurochemical alterations underlying the cognitive and behavioral symptoms that constitute the clinical picture of AD, and that

Peter Paul De Deyn and Debby Van Dam (eds.), *Animal Models of Dementia*, Neuromethods, vol. 48,
DOI 10.1007/978-1-60761-898-0_10,

Table 1
The main neuropathological alterations in AD that should be mimicked by an animal model

Impairment of neurotransmitter systems:
Degeneration of forebrain cholinergic neurons (↓ ChAT; AChE, ACh)
Loss of cortical pyramidal neurons (↓ Glu)
β-Amyloid deposition (plaques, vessels, intracellular)
Tau protein hyperphosphorylation (NFT)
Inflammatory reaction

AD = Alzheimer's disease; ChAT = choline acetyltranferase; AChE = acetylcholinesterase; ACh = acetylcholine; Glu = glutamate; NFT = neurofibrillary tangles

we expect to find in a model, will be briefly described, and are summarized in Table 1.

1.1. Impairment of Neurotransmitter Systems

The first neurochemical change detected postmortem, more than 30 years ago, in the brain of patients affected by AD was a severe decrease in choline acetyltransferase (ChAT) activity in the cerebral cortex and hippocampus (1) This observation was followed by extensive investigations leading to the "cholinergic hypothesis" (2) on which the present therapy of AD was built. It was demonstrated that the decrease in ChAT activity resulted from the loss of cholinergic neurons in the forebrain cholinergic nuclei (3), which was associated with a reduction in acetylcholine (ACh) formation, observed in cortical biopsies from AD patients (4). The ChAT activity decrease correlated with the severity of cognitive impairment (5). Several reviews summarize the changes of the cholinergic systems in AD and related dementias (6–8). The changes also include a decrease in acetylcholinesterase (AChE) activity in the cerebral cortex and hippocampus, an increase in butyrylcholinesterase (BuChE) activity, a decrease in vesicular ACh transporter (VAChT) activity, a loss of α4β2 and α7 nicotinic receptors, and of M2 and M4 muscarinic receptors.

In AD, other neurotransmitter systems are also affected, as reviewed by Hardy et al. (6) and Gottfries (7), although none of them is damaged as extensively and in all patients as the cholinergic system. Loss of noradrenaline, dopamine, serotonin, and changes in their receptor-signal transduction pathways as well as their metabolites have been observed in the brains of AD patients (6, 9). Antemortem investigation of amino acid content in the neocortex of demented patients with AD showed that glutamate is the only significantly reduced amino acid (10). Finally, among the many neuropeptides investigated, a decrease in somatostatin levels was repeatedly found postmortem in the hippocampus and cortical regions of AD subjects (11).

1.2. β-Amyloid and Tau Protein

No matter how important the role of the neurotransmitter system impairment is in AD pathogenesis, it is considered secondary to the deposition of β-amyloid (Aβ) protein and the phosphorylation of tau protein, which are regarded as the biochemical events responsible for the built-up of the senile plaques (SP) and neurofibrillary tangles (NFT), respectively. These are the two histopathological features characterizing AD as described a century ago by Alois Alzheimer. The formation of Aβ from the β-amyloid protein precursor (APP), its amino acid composition, conformation, and role in physiology and pathology have been the subject of many reviews (12–14). From these studies, the "amyloid cascade hypothesis" of AD pathogenesis was formulated (15).

Two years after the discovery of the main component of SP, Grundke-Iqbal et al. (16) demonstrated that the main component of the paired helical filaments that make up the characteristic NFT is an abnormally hyperphosphorylated form of the microtubule-associated protein tau. The extensive studies leading to the understanding of tau phosphorylation and polymerization, and the identification of taupathies, including FTD, are reviewed in (17–19).

Both Aβ and tau protein can be detected in the cerebral spinal fluid (CSF) of AD patients (20) and the possibility that their changes may be used for diagnostic and prognostic studies is currently under investigation (21).

1.3. Brain Inflammation

The immunohistochemical demonstration of reactive microglia in AD brains (22) and the epidemiological observations that subjects treated with anti-inflammatory agents had a greatly reduced risk of AD (23) attracted attention towards the large inflammatory reaction associated with SP, triggered by Aβ (24). The inflammatory reaction produces cellular and biochemical changes including microglial and astroglial activation with an increase in phospho-p38 MAPK, inducible cyclooxygenase (Cox 2) and NO synthase expression resulting in the production of prostaglandins, interleukins, NO, free radicals, as summarized by several reviews (25–27). The inflammatory products play an important pathogenetic role in the neuronal neurodegeneration characterizing AD (28, 29) and are considered a potential therapeutic target.

2. Animal Models

The ideal animal model of AD and related dementias should include all the neurochemical changes described in the previous section. However, since the understanding of AD pathogenesis has grown step by step during 30 years of research, simple models, only mimicking one or a few pathogenetic mechanisms, have been gradually developed. These partial models maintain their usefulness for investigating the different pathogenetic mechanisms, and in the drug discovery pipeline for dementia.

2.1. Impairment of Neurotransmitter Systems

None of the monoaminergic, amino acidergic, and peptidergic neurotransmitter systems is so constantly and deeply affected in AD as the cholinergic system. Therefore, several models reproducing the loss of cholinergic neurons and the ensuing functional cholinergic hypofunction have been developed. The models used are: animals in which the cholinergic neurons were lesioned by excitotoxins (30), immunotoxins, such as 192 IgG-saporin (31), Aβ peptides (32), as well as AD mice in which the expression of anti-nerve growth factor (NGF) antibodies was induced (33). Other models in which a cholinergic hypofunction has been detected and investigated are, among many others, aging animals (34); rats with a diffuse brain inflammation induced by bacterial lipopolysaccharide administration (35); and transgenic mice overexpressing Aβ (36). The aims of the models were to investigate the role of cholinergic hypofunction in AD pathogenesis, particularly in the cognitive deficit, mostly to test drugs restoring the cholinergic function and treating AD memory impairment.

A detailed description of the models is given in other chapters of this book; here their neurochemical validation is presented.

2.1.1. Methods to Validate the Cholinergic Deficit

Several markers can be investigated to demonstrate and validate a cholinergic deficit:

1. *Immunohistochemical visualization and quantification of the ChAT-positive forebrain neurons by the use of polyclonal rabbit antibodies* (1:200) (Chemicon, Temecula Ca). Thirty-μm-thick coronal slices are prepared from brain removed from deeply anesthetized rats or mice perfused with a 4% paraformaldehyde solution in phosphate buffer. Free floating sections are incubated overnight at 4°C with the primary antibodies. For details of the method, see (37) and (36). Cell counting can be performed manually, using a 20x objective with a calibrated eyepiece grid. The mean area of the ChAT–positive cells can be evaluated by using, e.g., Analysis B Software (Olympus, Hamburg Germany). A local injection of the specific 192 IgG-saporin toxin in rat nucleus basalis may result in the disappearance of 92% of the cholinergic neurons (38). In aging rats (24 months and older), the loss of ChAT-positive neurons shows marked individual and strain differences (for a discussion, see (34) and (39)).
2. *Histochemical visualization of AChE-positive neurons and nervous fibers.* This is a valid alternative to the visualization of ChAT-positive neurons, which allows not only to count the AChE-positive neurons but also to examine the network of AChE-positive fibers in the neocortex and hippocampus.

Modifications of the classical method of Koelle (40), based on the use of acetylthiocholine as a substrate and its visualization with copper sulfate (see (41) for details), are used either on sections from brain fixed with 4% paraformaldehyde solution in phosphate buffer (42) or prepared with a freezing microtome (43). Alternatively, the direct-coloring Karnovsky–Roots method can be used (44). AChE-positive fiber density was estimated by McGaughy and Sarter (43) by means of a Vanox Olympus Research microscope with a 25x magnification. After intracortical infusions of 192 IgG-saporin toxin, the fiber loss ranged between 40% and 60% of total cortical cholinergic fibers.

3. *Determination of ChAT activity.* The usefulness of ChAT determination for assessing the cholinergic dysfunction in animal models of neurodegenerative diseases was recently reviewed by Contestabile (45). ChAT activity is measured by the conversion of 1-[^{14}C]acetyl-coenzymeA to [^{14}C]acetylcholine according to the method of Fonnum (46). [^{3}H] acetyl-coenzymeA has also been used (47). In the standard procedure, brain tissue is homogenized in 20 volumes of 10 mM EDTA buffer (pH 7.4) and 2% (v/v) Triton X-100. Incubation time is 15 min. Ballmaier and coworkers (38) found a ChAT activity of 2.42 ± 0.26 and 2.65 ± 0.29 μmol/h/100 mg protein in the frontal and parietal cortex, respectively, with a decrease ranging between 44% and 60% after 192 IgG-saporin toxin injection in rat nucleus basalis. Comparable levels of ChAT activity and similar post-lesion decreases were observed in the cortex by Robinson and colleagues (48). According to these authors, ChAT activity in the hippocampus was 8.57 ± 0.39 μmol/h/100 mg protein with a 52% decrease after 192 IgG-saporin toxin injection in the medial septum.

   In aging rats, investigations on cortical ChAT activity did not reveal consistent changes in the cortex (39), as it should be expected from the absence of a clear cut effect of age on the number of cholinergic neurons in the basal forebrain.

4. *Determination of AChE and BuChE activity in discrete brain areas.* The determination of AChE activity is an easy and inexpensive method for detecting and evaluating cholinergic deficits (49). Brain samples of about 50–70 mg are homogenized in 50 mM Na-phosphate buffer and incubated with acetylthiocholine as a substrate according to the method of Ellman (50). AChE activity is expressed as mmol acetylthiocholine iodide hydrolyzed per min/mg protein, measured spectrophotometrically at 412 nm. The radiometric method (51, 52) is an alternative based on the enzymatic hydrolysis of [^{3}H]acetylcholine. The [^{3}H] acetic acid formed is extracted

and the radioactivity counted. The correspondence between the two methods was checked by Cerbai et al. (52). For the measurement of BuChE activity, the method of Ellman et al. (50), with modifications suggested by Lockridge (52), has been used. Butyrylthiocholine is added as the substrate, instead of acetylthiocholine, to the homogenate, in the presence of a selective inhibitor of AChE, such as BW 248-C51 (52).

Examples of the sensitivity of AChE and ChAT determinations can be found in the papers of Harkany et al. (53) and Van Dam et al. (54). In the first paper, a 26% decrease in cortical ChAT and 28% in AChE activity was detected 14 days after injection of $A\beta_{1-42}$ peptide in the rat nucleus basalis. Van Dam et al. (54) found a 30% decrease in ChAT and a 40% decrease in AChE activities in the basal forebrain nuclei of 7–8-month-old APP/SWE (APP23) mice. Gil-Bea et al. (49) found a significant correlation between the decrease in ChAT and AChE activities in the frontal cortex after 192 IgG-saporin injection in the nucleus basalis of rats. Taken together, these findings demonstrate that ChAT and AChE methods are equally useful for evaluating the cholinergic deficit.

5. *Determination of high-affinity choline uptake (HACU).* HACU is considered a biochemical marker of the localization and integrity of the cholinergic nerve endings, like ChAT activity (55). However, HACU can also be taken as a measure of the activity of the cholinergic neurons (56). For these reasons, HACU has been used for validating the functional cholinergic impairment induced by lesions of the forebrain nuclei (57). In the first days after the lesion, ChAT and HACU decreases are comparable (47), but after a few weeks the decrease in HACU becomes smaller than that of ChAT due to the compensatory increase in activity of the remaining cholinergic neurons. No decrease in cortical HACU activity has been detected in aging rats (39). In aging transgenic Tg2576 mice expressing the Swedish mutation of human APP, no changes have been detected in ChAT and AChE activities; only an HACU decrease (58) was found, a finding indicating that in these mice there was a functional cholinergic impairment without a morphological damage.

   HACU is measured by incubating the tissue homogenate with [^{3}H]choline for 5 min at 37°C followed by centrifugation and counting of the amount of [^{3}H]choline accumulated in the pellet, according to Rossner et al. (47) or Atweh et al. (56). An alternative to HACU is the determination of [^{3}H] hemicholinium-3 (HC-3) binding to choline transport sites, as described by Wenk et al. (59), who observed a 70–80% decrease in [^{3}H]C-3 cortical binding a month after ibotenic or quisqualic lesions of the forebrain neurons.

6. *Measurement of ACh extracellular levels in discrete brain areas of freely moving animals.* ACh extracellular levels are the expression of the activity-dependent ACh release from the cholinergic nerve endings. For this reason, the changes in extracellular ACh levels, measured by microdialysis, are considered the most dynamic and sensitive indication of functional integrity of the cholinergic neurons. Therefore, the microdialysis technique coupled with a high-sensitivity HPLC detection and quantification method has been often used for assessing the cholinergic impairment and its recovery in many AD models including aging rats, rats with lesions of the forebrain cholinergic nuclei, and transgenic mice overexpressing Aβ. The procedure for measuring extracellular ACh levels in discrete brain regions in rats or mice is more complex, time–consuming, and expensive than the procedure for measuring ChAT and AChE activity and requires some training. It has been described in detail in the Handbook of Microdialysis edited by Westerink and Cremers (60). In the same handbook, the methodological issues specifically concerning ACh release are discussed by Pepeu and Giovannini (61). The microdialysis probes (commonly CMA, Carnegie Medicine or Hospal) are stereotaxically implanted, either vertically or transversally, under anesthesia. The probes are perfused with artificial cerebrospinal fluid beginning 24 h after implantation. Calcium concentration in the fluid is critical and is usually 1.2 mM. AChE inhibitors – usually neostigmine 0.1–0.5 μM – are added in order to protect ACh from hydrolysis and obtain ACh levels more easily detectable with the HPLC methods commonly available. However, if the sensitivity of the methods is increased, AChE inhibitors can be avoided (52).

In aging rats, the cholinergic hypofunction may result in a 35–40% decrease in ACh release from the cerebral cortex and hippocampus (62, 63). This age-associated hypofunction can be corrected by the administration of AChE inhibitors, such as donepezil, rivastigmine, and metrifonate (64). However, according to Sarter and Bruno (39), aging rats do not always show a decrease in basal ACh release, in comparison to young rats; they do so only when the cholinergic forebrain neurons are partially lesioned and activated by drugs or attention-requiring behaviors. Rat strains, perfusion fluid composition, and different placement of the dialysis probe may explain the variability in detecting the age-dependent decrease in ACh extracellular levels.

A decrease in ACh release from the cerebral cortex and hippocampus has been detected following $A\beta_{1-40}$ or $A\beta_{1-25}$ peptides injection in the cholinergic forebrain nuclei (53, 65). These experiments validate Aβ peptide intracerebral injections as a useful experimental tool for investigating the mechanisms of their toxicity *in vivo*.

In rats in which the forebrain cholinergic nuclei have been destroyed by 192 IgG-saporin toxin, Gil-Bea et al. (66) found a 35% decrease in cortical ACh release and a correlation between decrease in ACh release and the decrease in AChE but not in ChAT activity (49).

A cholinergic hypofunction in the brain of transgenic mice, overexpressing APP, has been demonstrated as well. In 7-month-old TgCRND8 mice, a 36% decrease in the basal ACh efflux and a suppression of the increase in ACh release induced by potassium depolarization and scopolamine was shown by Bellucci et al. (36) in comparison with WT controls. Similar results were obtained by Bales et al. (67) in PDAPP mice; Watanabe et al. (68) found no decrease in basal ACh release in Tg2576 mice tested at any age but observed a marked decrease in potassium-evoked release in 9–11 month-old mice.

In conclusion, the examples described in this section demonstrate that of the many methods available to validate the cholinergic hypofunction in AD models, ChAT and AChE determinations are the simplest and most convenient. However, a decrease in these enzymatic activities is detected only when a consistent loss of cholinergic neurons has occurred, whereas the measure of HACU and ACh extracellular levels detect the cholinergic hypofunction in an earlier stage and make it possible to observe its recovery.

As mentioned above (8), in AD patients changes in muscarinic and nicotinic receptors associated with the cholinergic hypofunction take place. Changes in muscarinic receptors have also been detected in rats with a destruction of the cholinergic neurons (47), but there is not enough information to use this feature for validation of a cholinergic hypofunction.

2.2. AD Animal Models with β-amyloid (Aβ) Plaques

Since the definition of the nature of the SP and the formulation of the "amyloid cascade hypothesis" of AD pathogenesis (15), animal models were sought in which spontaneous Aβ plaque formation occurred or SP formation could be induced. Aging monkeys (69) and dogs (70) are animal species in which SP can be found. Transgenic mice have been generated in which APP overexpression leads to Aβ deposits and SP formation (71). Aβ peptides have been injected intracerebrally to form artificial plaques and investigate their toxicity (37).

The validation of these animal models requires the demonstration of the presence of plaques and their quantification. This can be done by using several histological, immunohistochemical, and biochemical methods.

2.2.1. Congo Red Staining

Congo Red is a commonly used histological dye for amyloid detection (72) since the work of Divry (73). It has been used for detecting Aβ deposits in aging animals (74) and in transgenic mice (75) and for visualizing intracerebrally injected Aβ peptide (37). Amyloid in Congo Red-stained fixed paraffin sections reveals the typical pale pinkish to pink-reddish coloration under normal light microscopy, whereas the dye-stained amyloid gives an apple-green birefringence when viewed under polarized light. At closer inspection under electron microscopy, it reveals the fibril nature of its protein constituents. The original method of Congo Red staining has undergone several modifications to improve its sensitivity, specificity, and reliability. The most common modification is the alkaline Congo Red method described by Puchtler (76). Specificity is improved by using freshly prepared stain and a staining solution fully saturated with sodium chloride. For more details on the procedure, see (77–80).

2.2.2. Fluorescent Staining: Thioflavin S

The thioflavin-S histochemical method is used more frequently than the Congo Red method for Aβ deposits visualization in AD animal models. See Selkoe et al. (81) for aging monkeys, Games et al. (76), Bellucci et al. (36) for transgenic mice. The method allows visualization of insoluble proteinaceous material in a pleated β-sheet conformation. For this reason, it is not specific for amyloid deposits only, but detects insoluble aggregates of hyperphosphorylated protein tau as well (82). Five-μm-thick, paraffin-embedded slices are used. The description of the method can be found in Yamamoto and Hirano (83) and Bellucci et al. (84). The staining (green) is observed using a fluorescence microscope. After photography using imaging software, within 24 h of staining, the total number of plaques in the cerebral cortex and hippocampus can be counted and averaged to quantify the plaque load (85).

2.2.3. Immunohistochemistry

The most specific method for visualizing and quantifying Aβ deposits is their detection by means of polyclonal antibodies raised against the human amyloid peptide followed by quantitative analysis of size, number, and total area occupied by plaques in distinct brain regions by means of imaging analysis software. This method is frequently used for assessing the Aβ load in AD animal models and to investigate environmental conditions and pharmacological treatments that may affect it. A few examples are the studies on the effects on Aβ load of immunization with Aβ in mice (86) and dogs (87), environmental enrichment (88), drugs (85), and nutrients (89). In these papers, the standard method is described. Briefly, free floating cortical and hippocampal sections are incubated overnight with the primary antibody, a polyclonal antibody raised against $A\beta_{1-42}$. On the second day, the sections are incubated with an anti-mouse secondary antibody and the immunostaining is visualized using the avidin–biotin system and a diaminobenzidine substrate kit. To quantify Aβ plaque burden, the stained sections are digitized under constant light and filter setting. Plaque number, size (maximum area), and total area are determined automatically. Data from four to five sections are then summed to derive representative values of the total plaque area for each animal.

2.2.4. Aβ ELISA

Although the histochemical and immunohistochemical visualization of Aβ deposits is sufficient for validating an animal model of AD, a more precise and reliable method to evaluate the Aβ load, including the intraneuronal Aβ deposits, and the nature of the peptides consists in the application of highly specific Enzyme Linked-Immuno-Sorbent Assay (ELISA) systems for quantification of $A\beta_{1-40}$ and $A\beta_{1-42}$. According to standard protocols, brains are rapidly harvested and snap-frozen until biochemical assessment of Aβ concentrations. At the time of use, whole brains or specific brain regions are homogenized or sonicated, and cerebral Aβ is solubilized and extracted. After centrifugation, the supernatant is carefully decanted, eventually diluted if the Aβ content is too high, and finally pipetted into the wells of the ELISA plate. Detailed information on the experimental procedure can be found in (85, 87, 90, 91). Each brain sample is usually analyzed in duplicate or triplicate, with the average value reported for each brain sample. The assay provides sensitive and reproducible detection of Aβ peptides through the development of a colorimetric reaction, whose intensity is directly proportional to the concentration of human $A\beta_{1-40}$ or $A\beta_{1-42}$ present in the original specimen.

2.3. Animal Models with Neurofibrillary Tangles

Together with Aβ deposition, the presence of NFT, resulting from intraneuronal aggregation of hyperphosphorylated tau protein, represents the other neuropathological hallmark that characterizes the AD brain. Therefore, an ideal animal model, fully mimicking the neurochemical picture of AD, should develop also NFTs and its validation requires a demonstration of their presence. Transgenic mouse models overexpressing mutated human APP with or without additional expression of mutated PS1/2 (please visit http://www.alzforum.org/res/com/tra/default.asp for a complete overview on the genetically engineered AD mouse models generated so far) closely recapitulate many features of the human pathology, including diffuse and neuritic amyloid deposits, dystrophic neurites and synapses, amyloid-associated neuroinflammation, impaired neurotransmitter system functionality, and cognitive deficits (92, 93). On the other hand, a major limitation consisting in the lack of NFT formation reduces their reliability (94, 95). Although abnormal patterns of tau hyperphosphorylation can be observed (84, 96), no typical flame-shaped NFT are generally detected in mice carrying mutated human APP or mutated PS1/2. On the contrary, tangle-like structures occur in the brains of the anti-NGF mice (33) indicating that, as with rodent amyloid, rodent tau is able to form pathological structures itself. The development of several (mutated) tau models (97), the crossing of tau and amyloidosis models (98, 99), and the recent generation of triple models, which featured enhanced amyloid deposition, tau phosphorylation, NFT-like formation, and overt neuronal loss in AD-relevant brain regions (100, 101), seem to overcome this hurdle.

Alterations of tau protein, without Aβ deposition, also occur in a heterogeneous group of neurodegenerative diseases called taupathies, of which FTD is the most frequent. Animal models of FTD have been generated and have become a widely used tool to study disease-related pathogenic mechanisms (97, 102). The first tau transgenic mouse model only showed pretangle formation and tau hyperphosphorylation (103). Later, the expression of human FTD mutant P301L tau reproduced aggregation and NFT-formation in mice (104).

In both AD and FTD transgenic mouse models, the visualization and quantification of hyperphosphorylated tau are obtained by immunohistochemical and/or immunoblotting methods, which employ phospho-epitope specific antibodies (AT8 (105), PHF-1 (106), AT100, CP13 (84), and others), specifically designed to recognize tau in correspondence of those Ser/Thr sites which become abnormally phosphorylated in the multistep process leading to NFT formation. Hyperphosphorylation is believed to be an early event in the pathway that leads from soluble to insoluble and filamentous tau (107). Early deposits of tau, normally referred to as "pretangles," cannot be detected by β-sheet-specific dyes (i.e. Congo Red or thioflavine-S), indicating that these intermediate forms of aggregated tau do not exhibit the pleated β-sheet structure typically found in amyloid aggregates (108). A structural transition leads to these more organized aggregates and the eventual development of NFT, which can be ultrastructurally analyzed under electron microscopy (109, 110). Aggregated tau can be isolated and quantified performing the extraction of Sarkosyl-insoluble tau from total brain homogenates, according to the method originally described by Greenberg and Davies (111). All the methods that are mentioned here are routinely used to investigate and characterize the nature of tau protein in tissues from human brain biopsies, as well as in brain samples from animal models: for details on the experimental procedures, see (84, 110, 112).

2.4. Brain Inflammation in AD Models

As described in a previous section, Aβ plaques are surrounded by inflammatory cells such as astrocytes and microglia. The inflammatory reaction is always present in AD brains (22), in the brain of AD animal models such as APP mice (96), and rats in which preaggregated Aβ peptides have been injected intracerebrally (24). The latter is a useful model for investigating the mechanisms through which Aβ induces the inflammatory response and the formation and release of inflammatory mediators including cytokines, chemokines, nitric oxide, and others (26). Microglia and astrocyte activation and inducible nitric oxide synthase immunoreactivity were observed in TgCRND8 mice (36). A review comparing the inflammatory reactions in different APP mice strains can be found in Howlett and Richardson (96). It is useful to validate an AD animal model also for its inflammatory component. This can be easily done by visualizing the activated microglia and astrocytes by standard immunohistochemical methods. Activated microglial cells are detected by means of the monoclonal mouse antibody OX-6 against MHC class II complex. Astrocytes are visualized by means of monoclonal mouse antibodies against glial fibrillary acidic protein (GFAP). A detailed description of the procedures can be found in (24). The formation of interleukin-1β and prostaglandin-E2 can be quantified by spectrophotometer in brain homogenates using commercially available specific ELISA kits and EIA kits, respectively (113).

3. Conclusions

It should be kept in mind that by definition, a model of a disease, no matter how similar to the disease it is, will always remain a model. The validation will tell its limits and usefulness. The progress in understanding the pathogenetic mechanisms of AD made the generation of accurate and verisimilar models possible. In turn, the models helped and are helping to unravel the neuro chemical mechanisms of the disease. The models of cholinergic hypofunction only mimic a component of AD pathogenesis. Nevertheless, they have been useful in understanding the role of the brain cholinergic system and for testing in animals the drugs that are still the mainstay of AD treatment and developing new drugs. The APP transgenic mice in which the validation demonstrates Aβ deposition with the formation of mature plaques such as the Tg2576, APP751 mice, and others (114) have been and still are useful tools for investigating the mechanisms of Aβ synthesis and disposition and to test drugs that may lower Aβ plaque load. Even more useful are models like the TgCRND8 (36, 84) and the 3xTg mice (100, 101), in which tau hyperphosphorylation leading to NFTs, inflammatory response, cholinergic hypofunction, and cognitive impairment have been reproduced. They may help in detecting new therapeutic targets such as glycogen synthase kinase 3 (GSK3) and its role in tau phosphorylation (115). However, the doubt casted by Howlett and Richardson (96) should be always present: "APP transgenic mice: a model of AD or simply overexpression of APP?"

References

1. Davies P, Maloney AJR (1976) Selective loss of central cholinergic neurons in Alzheimer's disease. Lancet 2:1403–1405.
2. Bartus RT, Dean RL, Beer B et al (1982) The cholinergic hypothesis of geriatric memory dysfunction. Science 217:408–417.
3. Whitehouse PJ, Price DL, Clark AW et al (1981) Alzheimer's disease:evidence for selective loss of cholinergic neurons in the nucleus basalis. Ann Neurol 10:122–126.
4. Sims NR, Bowen DM, Smith CC et al (1980) Glucose metabolism and acetylcholine synthesis in relation to neuronal activity in Alzheimer's disease. Lancet 1:333–336.
5. Wilcock GK, Esiri MM, Bowen DM et al (1982) Alzheimer's disease. Correlation of cortical choline acetyltransferase activity with the severity of dementia and histological abnormalities. J Neurol Sci 57:407–417.
6. Hardy J, Adolfsson J, Alafuzoff I et al (1985) Transmitter deficits in Alzheimer's disease. Neurochem Int 7:345–363.
7. Gottfries CG (1985) Alzheimer's disease and senile dementia: biochemical characteristics and aspect of treatment. Psychopharmacology 86:245–252.
8. Francis PT, Perry EK (2006) Neurochemical pathology of cholinergic systems in neurodegenerative and neurodevelopmenta disorders. In: Giacobini E, Pepeu G. (eds) The brain cholinergic system in health and disease. Informa, Abington, England, pp 59–74.
9. Fowler CJ, O'Neill C, Winblad B et al (1992) Neurotransmitter, receptor and signal transduction disturbances in Alzheimer's disease. Acta Neurol Scand Suppl 139:59–62.
10. Lowe SL, Bowen DM, Francis PT et al (1990) Ante mortem cerebral amino acid concentrations indicate selective degeneration of glutamate-enriched neurons in Alzheimer's disease. Neuroscience 38:571–577.
11. Bissette G (1997) Neuropeptides and Alzheimer's disease pathology. Ann NY Acad Sci 814:17–29.
12. Selkoe DJ (1994) Normal and abnormal biology of the beta-amyloid precursor protein. Ann Rev Neurosci 17:489–517.
13. Selkoe DJ (2002) Deciphering the genesis and fate of amyloid beta-protein yields novel therapies for Alzheimer disease. J Clin Invest 110:1375–1381.
14. Masters LC, Beyreuther K (2006) Pathways to the discovery of the Abeta amyloid of Alzheimer's disease. In: Perry G, Avila J, Kinoshita J et al (eds) Alzheimer's disease: a century of scientific and clinical research. IOS Press, Amsterdam, pp 155–161.
15. Hardy J, Allsop D (1991) Amyloid deposition as the central event in the aetiology of Alzheimer's disease. Trends Pharmacol Sci 12:383–388.
16. Grundke-Iqbal I, Iqbal K, Quinlan M et al (1986) Microtubule-associated protein tau. A component of Alzheimer paired helical filaments. J Biol Chem 261:6084–6089.
17. Goedert M (1993) Tau protein and the neurofibrillary pathology of Alzheimer's disease. Trends Neurosci 16:460–465.
18. Yancopoulou D, Spillantini MG (2003) Tau protein in familial and sporadic diseases. Neuromol Med 4:37–48.
19. Hernandez F, Avila J (2007) Tauopathies. Cell Mol Life Sci 64:2219–2233.
20. Blennow K, Cowburn RF (1996) The neurochemistry of Alzheimer's disease. Acta Neurol Scand Suppl 168:77–86.
21. Aluise CD, Sowell RA, Butterfield DA (2008) Peptides and proteins in plasma and cerebrospinal fluid as biomarkers for the prediction, diagnosis, and monitoring of therapeutic efficacy of Alzheimer's disease. Biochim Biophys Acta 1782:549–558.
22. McGeer PL, Itagaki S, Tago H et al (1987) Reactive microglia in patients with senile dementia of the Alzheimer type are positive for the histocompatibility glycoprotein HLA-DR. Neurosci Lett 79:195–200.
23. McGeer PL, Shulzer M, McGeer EG (1996) Arthritis and anti-inflammatory agents as possible protective factors for Alzheimer's disease: a review of 17 epidemiologic studies. Neurology 47:425–432.
24. Giovannini MG, Scali C, Prosperi C et al (2002) Beta-amyloid-induced inflammation and cholinergic hypofunction in the rat brain in vivo: involvement of the p38MAPK pathway. Neurobiol Dis 11:257–274.
25. Griffin WST, Sheng JG, Royston MC et al (1998) Glial-neuronal interactions in Alzheimer'disease: the potential role of a 'Cytokine Cycle'in disease progression. Brain Pathol 8:65–72.
26. Giovannini MG, Scali C, Prosperi C et al (2003) Experimental brain inflammation and neurodegeneration as model of Alzheimer's disease: protective effects of selective COX-2 inhibitors. Int J Immunopathol Pharmacol 16:31–40.
27. Eikelenboom P, Veerhuis R, Familian A et al (2008) Neuroinflammation in plaque and

vascular beta-amyloid disorders: clinical and therapeutic implications. Neurodegener Dis 5:190–193.

28. Sheng JG, Jones RA, Zhou XQ et al (2001) Interleukin-1 promotion of MAPK-p38 overexpression in experimental animals and in Alzheimer's disease: potential significance for tau protein phosphorylation. Neurochem Int 39:341–348.
29. Zipp F, Aktas O (2006) The brain as a target of inflammation: common pathways link inflammatory and neurodegenerative diseases. Trends Neurosci 29:518–527.
30. Olton DS, Wenk GL (1987) Dementia: animal models of the cognitive impairments produced by degeneration of the forebrain cholinergic system. In: Meltzer HY (ed) Psychopharmacology: the third generation of progress. Raven Press, New York, pp 941–953.
31. Wiley RG, Oeltmann TN, Lappi DA (1991) Immunolesioning: selective destruction of neurons using immunotoxin to rat NGF receptor. Brain Res 562:149–153.
32. Pepeu G (2001) Overview and perspective on the therapy of Alzheimer's disease from a preclinical viewpoint. Prog Neuropsychopharmacol Biol Psychiatry 25:193–209.
33. Capsoni S, Ugolini G, Comparini A et al (2000) Alzheimer-like neurodegeneration in aged antinerve growth factor transgenic mice. Proc Natl Acad Sci USA 97:6826–6831.
34. Pepeu G, Giovannelli L (1994) The central cholinergic system during aging. In: Bloom FE (ed) Neuroscience: from the molecular to the cognitive. Progress in Brain Research, vol 100. Elsevier, Amsterdam, pp 67–71.
35. Willard LB, Hauss-Wegrzyniak B, Wenk GL (1999) Pathological and biochemical consequences of acute and chronic neuroinflammation within the basal forebrain cholinergic system of rats. Neuroscience 88:193–200.
36. Bellucci A, Luccarini I., Scali C. et al (2006) Cholinergic dysfunction, neuronal damage and axonal loss in TgCRND8 mice. Neurobiol Dis 165:1643–1652.
37. Giovannelli L, Scali C, Faussone-Pellegrini MS et al (1998) Long-term changes in the aggregation state and toxic effects of beta-amyloid injected into the rat brain. Neuroscience 87:349–357.
38. Ballmaier M, Casamenti F, Scali C et al (2002) Rivastigmine antagonizes deficits in prepulse inhibition induced by selective immunolesioning of cholinergic neurons in nucleus basalis magnocellularis. Neuroscience 114:91–98.
39. Sarter M, Bruno JP (1998) Age-related changes in rodent cortical acetylcholine and cognition: main effects of age versus age as an intervening variable. Brain Res Brain Res Rev 27:143–156.
40. Koelle GB (1955) The histochemical identification of acetylcholinesterase in cholinergic, adrenergic and sensory neurons. J Pharmacol Exp Ther 114:167–184.
41. McGaughy J, Kaiser T, Sarter M (1996) Behavioral vigilance following infusions of 192 IgG-saporin into the basal forebrain: selectivity of the behavioral impairment and relation to cortical AChE-positive fiber density. Behav Neurosci 110:247–265.
42. Torres EM, Perry TA, Blokland A et al (1994) Behavioural, histochemical and biochemical consequences of selective immunolesions in discrete regions of the basal forebrain cholinergic system. Neuroscience 63:95–122.
43. McGaughy J, Sarter M (1998) Sustained attention performance in rats with intracortical infusions of 192 IgG-saporin-induced cortical cholinergic deafferentation: effects of physostigmine and FG 7142. Behav Neurosci 112:1519–1525.
44. Karnovsky MJ, Roots L (1964) A "direct-coloring" thiocholine method for cholinesterases. J Histochem Cytochem 12:219–221.
45. Contestabile A, Ciani E, Contestabile A (2008) The place of choline acetyltransferase activity measurement in the "cholinergic hypothesis" of neurodegenerative diseases. Neurochem Res 33:318–327.
46. Fonnum F (1975) A rapid radiochemical method for the determination of choline acetyltransferase. J Neurochem 24:407–409.
47. Rossner S, Schliebs R, Hartig W et al (1995) 192 IgG-saporin-induced selective lesion of cholinergic basal forebrain system: neurochemical effects on cholinergic neurotransmission in rat cerebral cortex and hippocampus. Brain Res Bull 38:371–381.
48. Robinson JK, Wenk GL, Wiley RG et al (1996) 192 IgG-saporin immunotoxin and ibotenic acid lesions of nucleus basalis and medial septum produce comparable deficits on delayed nonmatching to position in rats. Psychobiology 23:179–186.
49. Gil-Bea FJ, Garcia-Alloza M, Dominguez J et al (2005) Evaluation of cholinergic markers in Alzheimer's disease and in a model of cholinergic deficit. Neurosci Lett 375:37–41.
50. Ellman GL, Courtney KD, Andres V Jr et al (1961) A new and rapid colorimetric determination of acetylcholinesterase activity. Biochem Pharmacol 7:88–95.

51. Thiermann H, Eyer P, Worek F et al (2005) Effects of oximes on muscle force and acetylcholinesterase activity in isolated mouse hemidiaphragms exposed to paraoxon. Toxicology 214:190–197.
52. Cerbai F, Giovannini MG, Melani C et al (2007) N1phenethyl-norcymserine, a selective butyrylcholinesterase inhibitor, increases acetylcholine release in rat cerebral cortex: a comparison with donepezil and rivastigmine. Eur J Pharmacol 572:142–150.
53. Harkany T, Lengyel Z, Soos K et al (1995) Cholinotoxic effects of beta-amyloid (1–42) peptide on cortical projections of the rat nucleus basalis magnocellularis. Brain Res 695:71–75.
54. Van Dam D, Marescau B, Engelborghs S et al (2005) Analysis of cholinergic markers, biogenic amines, and amino acids in the CNS of two APP overexpression mouse models. Neurochem Int 46:409–422.
55. Kuhar MJ (1976) The anatomy of cholinergic neurons. In: Goldberg AM, Hanin I (eds) Biology of cholinergic function. Raven Press, New York, pp 3–27.
56. Atweh S, Simon JR, Kuhar MJ (1975) Utilization of sodium-dependent high affinity choline uptake in vitro as a measure of the activity of cholinergic neurons in vivo. Life Sci 17:1535–1544.
57. Pedata F, Lo Conte G, Sorbi S et al (1982) Changes in high affinity choline uptake in rat cortex following lesions of the magnocellular forebrain nuclei. Brain Res 233:359–367.
58. Apelt J, Kumar A, Schliebs R (2002) Impairment of cholinergic neurotransmission in adult and aged transgenic Tg2576 mouse brain expressing the Swedish mutation of human beta-amyloid precursor protein. Brain Res 953:17–30.
59. Wenk GL, Harrington CA, Tucker DA et al (1992) Basal forebrain neurons and memory: a biochemical, histological, and behavioral study of differential vulnerability to ibotenate and quisqualate. Behav Neurosci 106: 909–923.
60. Handbook of Microdialysis (2007) Methods, applications and perspectives. Elsevier, Amsterdam.
61. Pepeu G, Giovannini MG (2007) Changes in acetylcholine extracellular levels during cognitive processes. In: Westerink BH, Cremers TI (eds) Handbook of microdialysis. Methods, applications and perspectives. Elsevier, Amsterdam, pp 377–396.
62. Wu CW, Bertorelli R, Sacconi M et al (1988) Decrease of brain acetylcholine release in aging freely-moving rats detected by microdialysis. J Neurobiol 9:357–361.
63. Scali C, Vannucchi MG, Pepeu G et al (1995) Peripherally injected scopolamine differentially modulates acetylcholine release in vivo in the young and aged rats. Neurosci Lett 197:171–174.
64. Scali C, Casamenti F, Bellucci A et al (2002) Effect of subchronic administration of metrifonate, rivastigmine and donepezil on brain acetylcholine in aged F344 rats. J Neural Transm 109:1067–1080.
65. Abe E, Casamenti F, Giovannelli L et al (1994) Administration of amyloid beta-peptides into the medial septum of rats decreases acetylcholine release from hippocampus "in vivo". Brain Res 636:162–164.
66. Gil-Bea FJ, Dominguez J, Garcia-Alloza M et al (2004) Facilitation of cholinergic transmission by combined treatment of ondansetron with flumazenil after cortical cholinergic deafferentation. Neuropharmacology 47: 225–232.
67. Bales KR, Tzavara ET, Wu S et al (2006) Cholinergic dysfunction in a mouse model of Alzheimer disease is reversed by an anti-A beta antibody. J Clin Invest 116:825–832.
68. Watanabe T, Yamagata N, Takasaki K et al (2009) Decreased acetylcholine release is correlated to memory impairment in the Tg2576 transgenic mouse model of Alzheimer's disease. Brain Res 1249: 222–228.
69. Price DL, Martin LJ, Sisodia SS et al (1991) Aged non-human primates: an animal model of age-associated neurodegenerative disease. Brain Pathol 1:287–296.
70. Cummings BJ, Head E, Ruehl W et al (1996) The canine as an animal model of human aging and dementia. Neurobiol Aging 17:259–268.
71. Emilien G, Maloteaux JM, Beyreuther K et al (2000) Alzheimer disease: mouse models pave the way for therapeutic opportunities. Arch Neurol 57:176–181.
72. Frid P, Anisimov SV, Popovic N (2007) Congo red and protein aggregation in neurodegenerative diseases. Brain Res Rev 53:135–160.
73. Divry P, Florkin MP (1927) Présentée par J. Firket. Sur les propriétes optiques de l'amyloide. C R Soc Biol (Paris) 97:1808–1810.
74. Rofina JE, van Ederen AM, Toussaint MJ et al (2006) Cognitive disturbances in old dogs suffering from the canine counterpart of Alzheimer's disease. Brain Res 1069:216–226.

75. Games D, Adams D, Alessandrini R et al (1995) Alzheimer-type neuropathology in transgenic mice overexpressing V717F beta-amyloid precursor protein. Nature 373:523–527.
76. Puchtler H, Sweat F, Levine M (1962) On the binding of Congo red by amyloid. J Histochem Cytochem 10:355–364.
77. Puchtler H, Sweat F (1965) Congo red as a stain for fluorescence microscopy of amyloid. J Histochem Cytochem 13:693–694.
78. Elghetany MT, Saleem A (1988) Methods for staining amyloid in tissues: a review. Stain Technol 63:201–212.
79. Elghetany MT, Saleem A,Barr K (1989) The Congo red stain revisited. Ann Clin Lab Sci 19:190–195.
80. Bely M, Makovitzky J (2006) Sensitivity and specificity of Congo red staining according to Romhanyi. Comparison with Puchtler's or Bennhold's methods. Acta Histochem 108: 175–180.
81. Selkoe DJ, Bell DS, Podlisny MB et al (1987) Conservation of brain amyloid proteins in aged mammals and humans with Alzheimer's disease. Science 235:873–877.
82. Sun A, Nguyen XV, Bing G (2002) Comparative analysis of an improved thioflavin-s stain, Gallyas silver stain, and immunohistochemistry for neurofibrillary tangle demonstration on the same sections. J Histochem Cytochem 50:463–472.
83. Yamamoto T, Hirano A (1986) A comparative study of modified Bielschowsky, Bodian and thioflavin S stains on Alzheimer's neurofibrillary tangles. Neuropathol Appl Neurobiol 12:3–9.
84. Bellucci A, Rosi MC, Grossi C et al (2007) Abnormal processing of tau in the brain of aged TgCRND8 mice. Neurobiol Dis 27:328–338.
85. Klausner AP, Sharma S, Fletcher S et al (2008) Does oxybutynin alter plaques, amyloid beta peptides and behavior in a mouse model of Alzheimer's disease? J Urol 179:1173–1177.
86. Schenk D, Barbour R, Dunn W et al (1999) Immunization with amyloid-b attenuates Alzheimer-disease-like pathology in the PDAPP mouse. Nature 400:173–177
87. Head E, Pop V, Vasilevko V et al (2008) A two-year study with fibrillar beta-amyloid (Abeta) immunization in aged canines: effects on cognitive function and brain Abeta. J Neurosci 28:3555–3566.
88. Lazarov O, Robinson J, Tang YP et al (2005) Environmental enrichment reduces Abeta levels and amyloid deposition in transgenic mice. Cell 120:701–713.
89. Wang J, Ho L, Zhao Z et al (2006) Moderate consumption of Cabernet Sauvignon attenuates Abeta neuropathology in a mouse model of Alzheimer's disease. FASEB J 20: 2313–2320.
90. Jankowsky JL, Slunt HH, Gonzales V et al (2005) Persistent amyloidosis following suppression of Abeta production in a transgenic model of Alzheimer disease. PLoS Med 2:e355.
91. Abramowski D, Wiederhold KH, Furrer U et al (2008) Dynamics of Abeta turnover and deposition in different beta-amyloid precursor protein transgenic mouse models following gamma-secretase inhibition. J Pharmacol Exp Ther 327:411–424.
92. Morrissette DA, Parachikova A, Green KN et al (2009) Relevance of transgenic mouse models to human Alzheimer disease. J Biol Chem 284:6033–6037.
93. Codita A, Winblad B, Mohammed AH (2006) Of mice and men: more neurobiology in dementia. Curr Opin Psychiatry 19:555–563.
94. Van Dam D, De Deyn PP (2006) Drug discovery in dementia: the role of rodent models. Nat Rev Drug Discov 5:956–970.
95. Radde R, Duma C, Goedert M et al (2008) The value of incomplete mouse models of Alzheimer's disease. Eur J Nucl Med Mol Imaging 35:S70–S74.
96. Howlett DR, Richardson JC (2009) The pathology of APP transgenic mice: a model of Alzheimer's disease or simply overexpression of APP? Histol Histopathol 24:83–100.
97. Gotz J, Deters N, Doldissen A et al (2007) A decade of tau transgenic animal models and beyond. Brain Pathol 17:91–103.
98. Gotz J, Schild A, Hoerndli F et al (2004) Amyloid-induced neurofibrillary tangle formation in Alzheimer's disease: insight from transgenic mouse and tissue-culture models. Int J Dev Neurosci 22:453–465.
99. Ribe EM, Perez M, Puig B et al (2005) Accelerated amyloid deposition, neurofibrillary degeneration and neuronal loss in double mutant APP/tau transgenic mice. Neurobiol Dis 20:814–822.
100. Oddo S, Caccamo A, Shepherd JD et al (2003) Triple-transgenic model of Alzheimer's disease with plaques and tangles: intracellular Abeta and synaptic dysfunction. Neuron 39:409–421.
101. Oddo S, Caccamo A, Kitazawa M et al (2003) Amyloid deposition precedes tangle formation in a triple transgenic model of Alzheimer's disease. Neurobiol Aging 24:1063–1070.
102. Gotz J, Ittner LM (2008) Animal models of Alzheimer's disease and frontotemporal dementia. Nat Rev Neurosci 9:532–544.
103. Gotz J, Probst A, Spillantini MG et al (1995) Somatodendritic localization and hyperphos-

phorylation of tau protein in transgenic mice expressing the longest human brain tau isoform. EMBO J 14:1304–1313.

104. Lewis J, McGowan E, Rockwood J et al (2000) Neurofibrillary tangles, amyotrophy and progressive motor disturbance in mice expressing mutant (P301L) tau protein. Nat Genet 25:402–405.

105. Goedert M, Jakes R, Vanmechelen E (1995) Monoclonal antibody AT8 recognises tau protein phosphorylated at both serine 202 and threonine 205. Neurosci Lett 189:167–169.

106. Otvos L Jr, Feiner L, Lang E et al (1994) Monoclonal antibody PHF-1 recognizes tau protein phosphorylated at serine residues 396 and 404. J Neurosci Res 39:669–673.

107. Brandt R, Hundelt M, Shahani N (2005) Tau alteration and neuronal degeneration in tauopathies: mechanisms and models. Biochim Biophys Acta 1739:331–354.

108. Ballatore C, Lee VM, Trojanowski JQ (2007) Tau-mediated neurodegeneration in Alzheimer's disease and related disorders. Nat Rev Neurosci 8:663–672.

109. Goedert M, Spillantini MG, Jakes R et al (1995) Molecular dissection of the paired helical filament. Neurobiol Aging 16:325–334.

110. Delobel P, Lavenir I, Fraser G et al (2008) Analysis of tau phosphorylation and truncation in a mouse model of human tauopathy. Am J Pathol 172:123–131.

111. Greenberg SG, Davies P (1990) A preparation of Alzheimer paired helical filaments that displays distinct tau proteins by polyacrylamide gel electrophoresis. Proc Natl Acad Sci USA 87(15):5827–5831.

112. Ksiezak-Reding H, Tracz E, Yang LS et al (1996) Ultrastructural instability of paired helical filaments from corticobasal degeneration as examined by scanning transmission electron microscopy. Am J Pathol 149:639–651.

113. Scali C, Prosperi C, Vannucchi MG et al (2000) Brain inflammatory reaction in an animal model of neuronal degeneration and its modulation by an anti-inflammatory drug: implication in Alzheimer's disease. Eur J Neurosci 12:1900–1912.

114. Yu P, Oberto G (2000) Alzheimer's disease: transgenic mouse models and drug assessment. Pharmacol Res 42:107–114.

115. Pei JJ, Sjogren M, Winblad B (2008) Neurofibrillary degeneration in Alzheimer's disease: from molecular mechanisms to identification of drug targets. Curr Opin Psychiatry 21:555–561.

Chapter 11

Validation of Dementia Models Employing Neuroimaging Techniques

Greet Vanhoutte, Adriaan Campo, and Annemie Van der Linden

Abstract

Dementia is a clinical diagnosis; however, none of the clinical scales guarantee high sensitivity or specificity. Therefore, neuroimaging is often crucial for proper assessment. The most typical neurological symptoms of dementia are often discerned using computed X-ray tomography (CT), magnetic resonance imaging (MRI), magnetic resonance spectroscopy (MRS), single-photon emission computerized tomography (SPECT), and positron-emission tomography (PET) techniques. In general, these different imaging modalities provide different information and can be considered as being complementary rather than competitive. Nowadays, structural neuroimaging is a routine component of the diagnostic evaluation of dementia. Neuroimaging offers promise as a surrogate marker for clinical trials, and new technologies have been developed to provide more molecular and physiological biomarkers such as markers for amyloid plaques, one of the neuropathological hallmarks of AD. The combined use of imaging technology and experimental *in vivo* animal models for human diseases provides a unique platform to study pathological mechanisms in longitudinal studies, develop accurate and early translational diagnostic tools, and evaluate therapeutic strategies. In this chapter, an overview of how brain imaging can be used to detect both early and late stages of dementia in small animal models is explained. The early stage offers a unique therapeutic window and is diagnostically most challenging.

Key words: Small animal, brain, early diagnosis, noninvasive

1. Introduction to *In Vivo* Small Animal Neuroimaging

The technology of neuroimaging and the development of a variety of small animal models for neurodegenerative diseases and dementia are two fields that have made rapid progress over the last years. The combined use of imaging technology and experimental *in vivo* animal models for human diseases provides a unique platform to study pathological mechanisms in longitudinal studies, develop accurate and early translational diagnostic tools, and evaluate therapeutic strategies. Technical advances in imaging modalities have

Peter Paul De Deyn and Debby Van Dam (eds.), *Animal Models of Dementia*, Neuromethods, vol. 48,
DOI 10.1007/978-1-60761-898-0_11, © Springer Science+Business Media, LLC 2011

allowed the evaluation of functional, neurochemical, anatomical, and even molecular changes in brains of small animals.

Inherent to neurodegeneration and subsequent dementia is neuronal loss, which can be detected with neuroimaging tools at terminal stages. However, monitoring of altered biological processes that might precede this stage, including changes in neurotransmitter turnover, cerebral blood flow (CBF), brain activity, and connectivity, is possible through noninvasive imaging. Molecular imaging is a rather new discipline that focuses particularly on early diagnostics combining molecular biology and *in vivo* imaging tools. It enables the visualization of cellular function and the follow up of molecular processes in living organisms. It differs from traditional imaging by probing disease-specific biomarkers to help imaging various targets or pathways, in particular, at the basis of disease rather than imaging the end-stage effects of these molecular alterations.

Clinically established brain imaging methods include X-ray computerized tomography (CT), MRI, and nuclear imaging, i.e., PET and SPECT. Microversions of these imaging systems are available for use in small animals. In addition to this, optical imaging methods such as fluorescence and bioluminescence – based on a light generating principle – are two other imaging methods restricted to experimental animal use.

Success in animal models, as supported by positive imaging results, does not guarantee success in clinical trials, as was demonstrated by investigation of treatment for Parkinson's disease (PD) (1). Relying solely on imaging technology is probably not appropriate enough, but combination with other research for ratification is necessary. Until now, imaging technology has advanced our understanding of animal models of dementia and plays a critical role in the development of valid animal models at a basic science level and in preclinical evaluation of potential therapies for human disorders. Valid animal models must satisfy the criteria of having good face, construct, and predictive validity, i.e., animals have to show behavior that resembles the clinical condition, to have similar etiology as the human disease, and treatments that have beneficial effects in humans should induce also positive changes in the animal model (2). One has to keep in mind that preclinical research is only as good as the animal model being employed.

This introduction gives an overview of the different imaging tools applied within the context of small animal neurological research. Traditionally, neuroimaging techniques have been categorized as either structural or functional, according to the primary information they provide, nowadays, initial structural techniques can be altered to observe functional information (e.g., functional MRI (fMRI)) as well. Similarly, traditional functional methods, such as PET, can also be used to view structures (e.g., amyloid plaques) (3–5). In general, different imaging modalities provide different information and can be considered as being complementary rather than competitive. Therefore, the combination of techniques such as coregistration of structural information with functional and molecular information is of high interest and receives a lot of attention.

1.1. MRI and MRS

MRI uses radio waves in the presence of a powerful magnetic field to align the nuclear magnetization of (usually) hydrogen atoms in water in the body. Atomic nuclei absorb and reemit electromagnetic waves at a resonant frequency, which falls in the radiofrequency range. Originally, MRI was used for producing anatomical images (6) but developed to a technique with a wealth of information contained in the signal regarding the physicochemical state of tissues, flow, diffusion, motion, and more recently, molecular targets. MR has much greater soft tissue contrast than CT, making it especially useful in neurological imaging.

For most medical brain examinations in general, both T_1- and T_2-weighted images are acquired for observation of white-matter and gray-matter brain structures and dementia-related morphological changes, such as changes in brain volume and size of the ventricle and specific substructures. Other specific MRI parameters correlate with the diffusion of water molecules (i.e., Brownian motion) in biological tissues (7) and are quite valuable in studying neuropathology-related changes in intra/extracellular water balance and subcellular structural changes (8). The recent development of diffusion tensor imaging (DTI) enables diffusion measurement simultaneously in different directions from which tissue anisotropy can be calculated per voxel (9). From these images, brain maps of fiber directions can be reconstructed to study brain connectivity, a feature affected in several neuropathologies (10). fMRI is capable of mapping functional regions of the cortex in real time during specific task activation. This is achieved using magnetic susceptibility image contrast, which detects the local increase in CBF due to activation. Deoxyhemoglobin is paramagnetic and interacts with the magnetic field, whereas oxyhemoglobin is not. As a consequence, a higher influx of oxygenated blood during activation increases local T_2 and T_2* values. Such signal intensity change is termed blood oxygenation level-dependent (BOLD) contrast (11). MRI is currently still a new modality in the fields of *in vivo* molecular and cellular imaging compared with the established methods such as PET and optical imaging. This delayed development is probably inherent to the low sensitivity of MRI. However, MRI is gaining headway, since its main contribution to molecular imaging is its high spatial resolution compared to other methods. MRI can achieve a microscopic resolution of 100 μm in all dimensions on a routine basis in living animals, and 10 μm is feasible in fixed specimens (12). This almost represents conventional histology, while allowing the object to be viewed interactively in any plane. Merely prolonging scan time to increase spatial resolution is not a viable option in fragile, anesthetized transgenic animals. However, with the development of higher field strength magnets and the use of small cryogenically operated coils (13), spatial resolution is still improving because of the increase in signal-to-noise ratio. Some *in vivo* MRI studies of mouse brain reported so far cover a wide range of voxel sizes and field strengths, e.g., $98 \times 156 \times 1{,}500\ \mu m^3$ at 11.7 T (14), $140 \times 140 \times 1{,}200\ \mu m^3$ at 4.7 T (15), $156 \times 156 \times 1{,}000\ \mu m^3$ at 14 T (16), and $78 \times 156 \times 343\ \mu m^3$ at 7 T (17).

Advances in MRI have expanded its use in neuroimaging by applying exogenous contrast agents. The range of applications of these compounds is determined largely by their biodistribution and pharmacokinetics. Contrast agents are typically injected intravenously and, depending on their chemical structure, may remain in the vasculature (iron oxide particles), enter the interstitial space (Gadolinium chelates), or be taken up by cells (manganese). Macromolecular contrast agents are intravascular agents and are well suited to quantitative imaging of perfusion and vascular volume in brain activity studies (18). Freely diffusible contrast agents such as Gadolinium chelates are used in tissue permeability studies such as blood–brain barrier (BBB) leakage, as none of the contrast agents can pass the BBB. For molecular imaging applications, "targeted" and "smart" contrast agents were developed (19,20). Targeted agents incorporate ligands, such as antibodies that bind to specific molecular markers in the tissue, whereas smart agents are activated by the presence of specific ions or enzymes. These agents allow *in vivo* visualization of gene expression, an emerging field within molecular imaging (21). Moreover, MRI is also one of the completely safe imaging tools since this technique does not make use of ionizing or nuclear compound injections.

MRS is a magnetic resonance technique that can measure brain biochemistry. This technique together with other MRI applications provides metabolic information such as anatomical, physiological, functional, or molecular imaging. There is increasing evidence that impairment of energy metabolism, mitochondrial dysfunction, and altered glutamatergic neurotransmission are critical factors in the development and progression of neurodegenerative diseases. Therefore, *in vivo* MRS provides a useful tool in the evaluation of the underlying processes, as many of the observable metabolites prove to be relevant in the assessment of neurochemical changes associated with neurodegeneration and dementia.

In localized ^{1}H-MRS, the target is the proton with a natural abundance of almost 100%, ultimately allowing the quantification of more than 18 ^{1}H containing metabolites *in vivo* (22). Localized MRS techniques allow assessment of metabolites resorting from a well-defined volume, e.g., an affected brain region, and can return single- or multivolume information with microliter resolution.

Multivolume methods are known as chemical shift imaging (CSI) or spectroscopical imaging (SI), which give a voxelwise metabolite distribution, i.e., a metabolite image (23,24) (Fig. 5).

Metabolites relevant for neurodegeneration can be used as fingerprints for the diseased brain. The ^{1}H-MRS spectrum of the normal brain is dominated by the large *N*-acetyl aspartate (NAA) resonance at 2.01 ppm. The exact role of NAA remains unknown, but it has been used extensively as a marker for neuronal density, thereby giving a measure for neuron survival and death (25). Myo-inositol (mI) is a cyclic sugar alcohol present in high concentrations. The precise function of mI is not known, but it has been dedicated as being an osmolyte, a glial marker, or a breakdown product of myelin. Typical changes of mI over time are seen in AD, cognitive impairment, and brain injury (26,27). Taurine (Tau) is another putative osmolyte of the brain. It has also been implicated as a modulator of neurotransmitter action. Tau is present in all cell types of the CNS, being a prominent metabolite in rodents rather than in humans (28). Glutamate (Glu), the major excitatory neurotransmitter and main precursor of γ-aminobutyric acid (GABA) and gluthation (GSH) (29), can be detected with MRS together with glutamine (Gln). Lactate (Lac) is normally present at low concentrations in the brain and is elevated in cases such as stroke, ischemia, or tumors (30–32). One of the most apparent resonances in a ^{1}H-MRS spectrum of the brain is the combined resonance of both creatine (Cr) and phosphocreatine (PCr) at 3.03 ppm, also referred to as total creatine (tCr). The tCr signal is relatively constant with only negligible changes reported with age or disease. This is the reason why it is often used as an internal reference for quantification (33). Choline (Cho)-containing compounds form a prominent resonance in ^{1}H-MRS spectra of the brain at 3.2 ppm. Since there are contributions of glycerophosphocholine (GPC), Cho, and phosphocholine (PCho), it is often referred to as total choline (tCho). Together with mI, tCho is often considered as a glial marker, and thus a marker for gliosis, since concentrations of these compounds are typically much higher in glia than in neurons (34,35).

1.2. Computerized Tomography (CT)

The techniques of CT and MRI are commonly used as instrumental aids for structural imaging in the diagnosis of dementia (21,36,37). However, new imaging tools are emerging and might significantly change the way CT and MRI are used nowadays in clinical settings. CT is a medical imaging method employing tomography, which means that digital geometry processing is used to generate a 3D image of the inside of an object from a large series of 2D X-ray images taken around a single axis of rotation. CT visualizes differences in the physical densities of biological tissues through X-ray absorption. The main issue within radiology today is how to reduce the radiation dose during CT examinations without compromising image quality.

For animal research, μCT systems are available. They are fivefold cheaper than μMRI systems and are straightforward to site in a laboratory. 3D μCT has proven to be a powerful technique for imaging and analysis of bone structure and density in small animals (38). Although initial μCT imaging was performed ex vivo, recent advances allow for similar imaging protocols to be performed *in vivo* (39). Unlike ex vivo μCT, which provides very fine spatial resolution (typically less than 10 μm), the spatial resolution of *in vivo* μCT studies suffers from physiologic (e.g., cardiac and respiratory) motion. High-resolution *in vivo* μCT has become important for examining morphology in small animal models for tumor disease (40,41).

Experimental CT imaging in animal models for dementia is very limited since soft tissue imaging using CT requires the use of X-ray absorbing contrast agents due to the lack of native CT contrast in the brain. Especially in the brain, distinction between gray and white matter (in quantitative terms, gray matter 20–35 HU (Hounsfield units), white matter 30–40 HU) needs addition of a radio-opaque component (iodine anion). There is, however, a CT study in a mouse model for dementia wherein postmortem material amyloid plaques could be discerned (42).

Upon application of intravenous injection of a CT contrast agent, investigation of cerebral vascular system impairments is made possible (43). The occurrence of cerebral vascular abnormalities is not only linked to vascular dementia but is also an issue often discussed concerning all types of dementia.

1.3. Nuclear Imaging Techniques (PET and SPECT)

Together with MRI, PET and SPECT are the most commonly used imaging techniques in animal models. PET uses biologically active molecules in μmolar or ηmolar concentrations that have been labeled with short-lived positron-emitting isotopes. Opposed to PET, the imaging agent used in SPECT emits gamma rays.

Both PET and SPECT can be used to measure regional CBF, which is physiologically coupled to brain metabolism making its use relevant in dementia studies. The radiotracer is assumed to accumulate in areas of the brain, in proportion to the rate of delivery of nutrients to that volume of brain tissue.

PET also provides measures of regional cerebral glucose metabolism (rCMRglc) by using a positron-emitting glucose analog, fluorodeoxyglucose (FDG). FDG-PET has become a standard technique in oncology and is gaining headway in dementia research in animal models.

Limitations to the widespread use of PET arise from the high costs of cyclotrons needed to produce the short-lived radionuclides for PET scanning and the need for specially adapted on-site chemical synthesis apparatus to produce the radiopharmaceuticals. SPECT scans using ^{99m}Tc-HMPAO (hexamethylpropylene amine oxime) competes with FDG-PET scanning of the brain to provide very similar information about local brain damage from many processes. SPECT is more widely available because the radioisotope generation technology is longer lasting and far less expensive. The same is true for the gamma scanning equipment.

A combined or multimodality scanner, such as SPECT/CT or PET/CT, can acquire coregistered structure and function in a single study. The data are complementary, allowing CT to accurately localize functional abnormalities and SPECT or PET to highlight areas of abnormal metabolism. Although technically more challenging, the simultaneous acquisition of MR and PET has also been demonstrated, initially for preclinical imaging of small animals, but more recently for the human brain (44).

PET is an extremely sensitive technique allowing for the identification of subtle changes in brain neurochemistry by labeling the molecules with high specificity. Selective tracers for many neurotransmitter receptors and other cellular proteins or tracers that act as an enzyme substrate, which allow the detection of a specific neuronal function, are available. Novel disease-specific probes are being developed very rapidly. New perspectives are opened by tracers for imaging amyloid plaques, tracers for measuring local acetylcholinesterase (AChE) activity and the binding capacity of nicotinic and serotonergic receptors, which address neurotransmitter deficits in dementia.

Nuclear medicine is a part of molecular imaging because it produces images that reflect biological processes taking place at the cellular and subcellular level. Nuclear imaging techniques, however, suffer from poor spatial resolution, with the highest resolution scanners today giving at best 4-mm resolution and for animal systems up to 1.5 mm. Therefore, PET and SPECT scans are increasingly read alongside CT or MRI scans, taking advantage of the coregistration of both anatomic and metabolic information to estimate brain regions of altered radioligand uptake. For the same reason, and very recently, small animal SPECT/MRI dual hybrid device combining pinhole SPECT imager adjacent to a small low-field MR imager has been developed (45).

1.4. Optical Imaging (BLI and FLI)

BLI and FLI enable visualization of genetic expression and physiological processes at the molecular level in living tissues. Reporter genes that encode either fluorescent or bioluminescent proteins have been used to create internal biological light in the study of animal models for human biology and are commonly used in tumor research (46). Longitudinal recordings of living animal models with these modalities have shown to provide clues for the development of clinical applications. Because of the presence of the skull, optical methods for brain applications are more restricted, but nevertheless abundant.

Near-infrared fluorescent (NIRF) probes are typically small fluorescent molecule dyes designed to absorb and emit light in the near-infrared (650–950 nm), where tissue scattering is lowest. The simple synthesis, the low cost, and the long life of NIRF probes present an attractive alternative for MRI and PET techniques in the field of molecular imaging.

BLI does not rely on an external light as for FLI, instead uses luminescence sourcing (e.g., luciferase) from, e.g., the firefly. Sensitive charged-coupled device (CCD) cameras are used to capture the bioluminescent data, which are superimposed on the photographic image for interpretation. Because there is no competing background signal, those techniques can detect low levels of light emission. The DNA encoding the luminescent protein is incorporated into the laboratory animal either via a viral vector or by creating a transgenic animal. Within these experimental animals, luciferine is injected and its oxidation by luciferase provides bioluminescence. Because luciferase imaging requires genetic transfection, it is very unlikely to ever find an application for this technique in human studies. Nevertheless, it can be very useful in predicting human response to therapy in the early stages of drug development.

In dementia research, progression is made in designing smart optical probes that emit a characteristic fluorescence signal only when bound to the studied target, i.e., a biomarker for the disease. The first NIRF probes for use in dementia were designed in 2001 and targeted amyloid in the brain of mouse models for AD (47). Until now, targeting amyloid is the most common application in dementia research.

The potential of BLI in dementia research finds its application in stem-cell-based therapies using bioluminescent stem cells and BLI as a superior viability marker. Stem-cell-based therapies may offer a potentially attractive and unlimited source of cells for the repair of damaged tissue in CNS disease. Experimental CNS transplantation studies have largely evaluated the *in vivo* survival, proliferation, and migrational capacity of neural grafts within the host tissue with histology at several different time points. Suitable noninvasive imaging technologies may serve as tools for probing neural graft behavior in the living brain. While other imaging approaches, such as MRI or PET, might have been chosen for this purpose, the limitations associated with these modalities outweigh their benefits. Particularly in MRI, which can monitor stem-cell populations that are iron-labeled, the *in vivo* distinction between viable and nonviable cells is not possible. Moreover, iron uptake can be confounded by other resident cells such as macrophages.

2. Small Animal Brain Atlases Using Imaging Methods

Localization of the major findings in the brain is an essential parameter both for the interpretation in an anatomical and functional context and for comparing findings across specimens and between healthy and diseased brains. As the importance of digital atlases, spatial indexes, and management systems for distribution data is increasingly recognized worldwide, several parallel advanced atlas systems emerge, providing both overlapping and complementary resources (48–54) (BIRN; www.nbirn.net).

Tomographic neuroimaging techniques allow visualization of functionally and structurally specific signals in the mouse and rat brain. The interpretation of the image data relies on accurate determination of anatomical location. However, lack of structural background information in the surrounding tissue limits the identification of anatomical boundaries. The primary benefit of CT is its ability to view the brain within the skull with a high spatial resolution. The acquisition of CT brain images has shown to be beneficial in the reconstruction of a reference brain atlas for rodents since skull landmarks (directly detectable with CT) could be directly mapped to the existing atlas coordinate space (55,56).

Side-by-side comparison of image data with conventional atlas diagram is hampered by the 2D format of the atlases and by the lack of an analytical environment for accumulation of data and integrative analyses. Hjornevik et al. (57) presented a method for reconstructing 3D atlases from digital 2D atlas diagrams and exemplified 3D atlas-based analysis of PET and MRI data. Using the *in vivo* pathway tracer $MnCl_2$, Hjornevik et al. revealed the topographical organization of the corticothalamic structures. After injection of $MnCl_2$ in the somatosensory cortex, the size and location of the $MnCl_2$ injections were analyzed by 3D combined visualization of the MR-based representation of the brain surface and the atlas representations of different cerebrocortical areas. The position of the manganese-enhanced signal within the thalamus was analyzed by slicing the combined image and atlas volumes in 200-μm-thick slices using a multiplatform tool (Fig. 1).

Recently, an MRI-based small animal PET template was reconstructed for FDG metabolic activity and the dopamine transporter (58). The 3D MRI T2-weighted images were oriented according to the rat brain Paxinos atlas (56), which facilitated the accurate assessment and spatial localization of ligand-specific PET information. Currently, 30 mouse brain structures and 60 rat brain structures have been reconstructed. To exploit the 3D atlas models, they have developed a multiplatform atlas tool (available via The Rodent Workbench, http://rbwb.org), which allows combined visualization of experimental image data within the 3D atlas space together with 3D viewing and user-defined

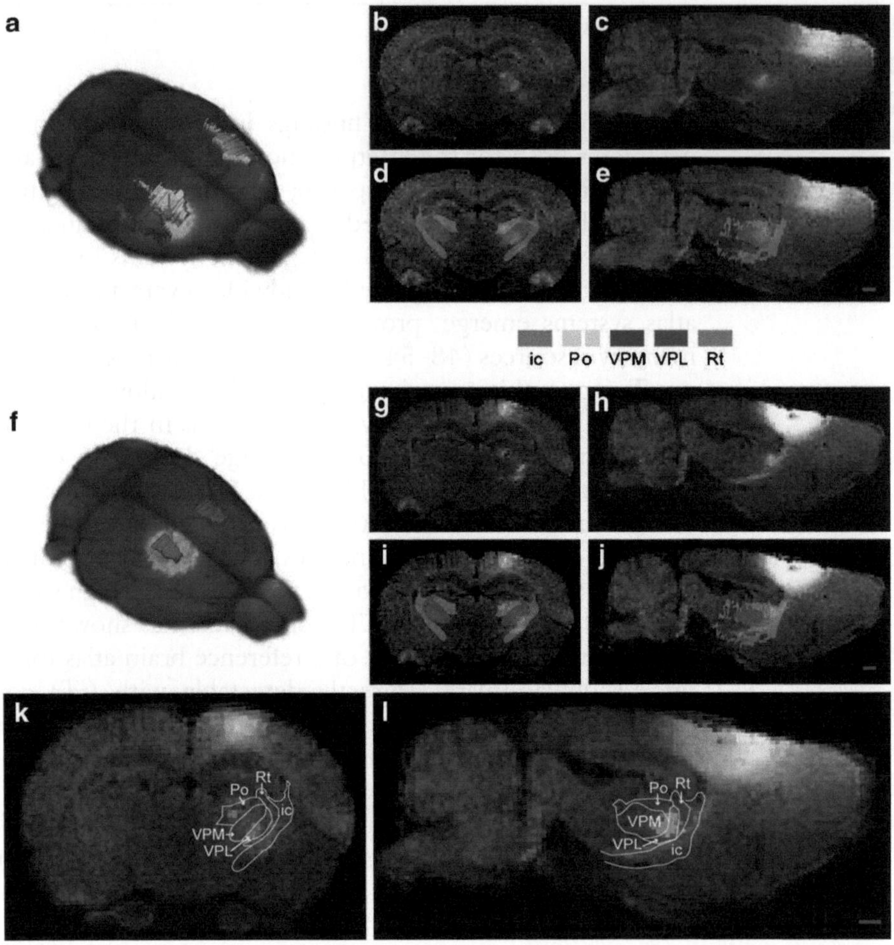

Fig. 1. Corticothalamic topographical organization revealed by $MnCl_2$/MRI tracing and 3D atlasing. Two cases are shown, case 1 (**a–e**), in which $MnCl_2$ was injected in the dysgranular zone adjacent to the SI forelimb representation (**b**), and case 2 (**f–j**), in which $MnCl_2$ was injected in the hind limb representation of SI. (**b**, **c** and **g**, **h**) Raw coronal and sagittal T_1-weighted MRI images obtained 9 h after injection. (**d**, **e** and **i**, **j**) Corresponding images with overlay of selected reconstructed atlas structures. (**k**, **l**) Superimposed, pseudocolored images (case 1, yellow; case 2, blue) with an overlay of atlas-derived boundary lines, demonstrating a somatotopical organization of the different signals within the Po and VPM. Bar, 1 mm (Adapted from (57)) *ic, internal capsule; Po, posterior thalamic nucleus; VPM, ventral posterolateral thalamic nucleus; VPL, ventral posteromedial thalamic nucleus; Rt, reticular thalamic nucleus.*

slicing of selected atlas structures. The presented tool facilitates assignment of location and comparative analysis of signal location in tomographic images with low structural contrast. The possibility of multimodal imaging responds to the desire of integrating information obtained by various imaging devices to unravel pathological mechanisms underlying neurological diseases. This means that the frequently used reference atlas space (59) for use in human brain imaging data has been made available in animals in many different formats to have a common reference space in the scientific community.

3. Imaging in Neurodegeneration and Dementia

We will focus on how brain imaging can be used to detect both early and late stages of dementia. The early stage offers a unique therapeutic window and is diagnostically most challenging. However, one problem that investigators still face is how to distinguish primary from secondary events.

Accumulating evidence suggests that chronic neurodegenerative disorders such as AD, PD, Huntington's disease (HD), and amyotrophic lateral sclerosis (ALS) are caused by a combination of events that impair normal neuronal function. Clinical signs are evident prior to neuronal loss. Therefore, new efforts have focused on identifying various crucial changes that hamper normal neuronal function.

3.1. Imaging Early Events

The etiologies of neurodegenerative diseases may be diverse; however, a common pathological denominator is the formation of aberrant or misfolded proteins. There is an ongoing debate about the nature of the harmful proteins and how toxic conformations selectively can damage neuronal populations. Soluble oligomers are associated with early pathological alterations, and strikingly, oligomeric assemblies of different disease-associated proteins may share common structural features. A major step toward the understanding of mechanisms implicated in neuronal degeneration is the identification of related genes. Studies based on these disease-associated genes illuminated the two faces of protein misfolding in neurodegeneration: a gain of toxic function and a loss of physiological function, which can even occur in combination (60).

Here, we review the occurrence of specific protein accumulations in several dementia disorders and summarize how these early-stage accumulations can be detected through neuroimaging methods in animal models.

3.1.1. Probing Misfolded Proteins

HD is caused by expansion of a CAG repeat coding for polyglutamine in the N-terminus of the Huntington protein (61). In AD, extraneuronal aggregates are mainly composed of amyloid β (Aβ) peptide, whereas intraneuronal aggregates are made up of the microtubule-associated tau-protein and also amyloid in more diffuse plaques (62,63). Transgenic mice overexpressing superoxide dismutase (SOD) have cytoplasmic inclusions containing aggregates of SOD protein (64), similar to what is found in human brain samples. Prion disease can be caused by mutations in the prion gene, leading to alterations of the prion protein (PrP). In all cases, prion disease is caused by abnormally folded PrP, and aggregation can take place both extracellularly and intracellularly (65). The pathological hallmark of PD is the Lewy body, an inclusion found in the cytoplasm of neurons containing aggregates of the synuclein protein (66).

To localize and quantify the deposition of misfolded proteins with high specificity, researchers aim at probing the proteins, i.e., a molecular approach. Expectations are high for the development of techniques to detect biomarkers that will significantly improve diagnosis and therapy follow-up.

Such imaging efforts have been made mostly in mouse models for AD trying to identify extraneuronal amyloid. In contrast to research for other forms of dementia in which the cloning of novel causative genes led to the development of biochemical hypotheses, AD research has largely developed in the reverse direction. The identification of the proteins that make up the classic amyloid plaques suggested their respective genes as sites to search for pathogenic mutations in AD.

A plaque-depositing transgenic mouse model is an obvious model system for the development of PET tracers aimed at imaging amyloid. The development of amyloid-specific ligands for the application of *in vivo* imaging started in 1992 (67). A number of potential amyloid PET ligands have been developed with most of them in clinical context (68,69). The first PET study in mice was published in 2002 (70); unfortunately, failure to detect a significant difference in [11C)PIB retention in the brains of double mutant presenilin PS1/amyloid precursor protein (APP) mice compared with PS1 littermate controls was reported. Marked reductions in brain uptake of the radioligand in transgenic mice enlightened the difficulties in PET imaging due to high requirements of the radioligands. Zhang et al. (71) therefore suggested to lower the lipophilicity and to improve bioavailability. It was also suggested that the lower uptake could be linked to a reduced CBF relative to that in wild-type mice (5). Human PET studies attempting to label amyloid were more successful. Translation of animal results to human application is not always straightforward since in humans AD pathology develops over one or more decades, whereas mouse pathology develops over months. The long time frame of deposition in human brain allows for many posttranslational and postdeposition modifications, including oxidation, crosslinking, and N- and C-terminal truncation (3).

Novel fluorescent compounds, including styryl-fluorobenzoxazole derivatives, were proposed as new PET probes for imaging of neuritic and diffuse amyloid deposits.

These compounds showed a high binding affinity for both synthetic $A\beta_{1-40}$ and $A\beta_{1-42}$ aggregates. Some of these compounds also displayed distinct staining of neuritic and diffuse amyloid plaques in AD brain sections. A biodistribution study of styryl-fluorobenzoxazole derivatives in normal mice exhibited excellent brain uptakes (4.5–5.5% injected dose/g at 2 min postinjection). Furthermore, intravenous administration of BF-145, a styryl-fluorobenzoxazole derivative, demonstrated specific *in vivo* labeling of compact and diffuse amyloid deposits in APP23 transgenic mouse brain, in contrast to no accumulation in wild-type mouse brain. These findings suggest that BF-145 is a potential candidate as a probe for imaging early brain pathology in AD patients (72).

In MRI, specificity is low because of small intrinsic contrast between protein accumulations and the surrounding brain tissue. However, in many cases, MRI has surpassed the capabilities of PET and SPECT, offering a combination of physiological sensitivity and high spatial resolution. Typically, in human AD subjects and in most of the transgenic mouse models, plaques diameter ranges from 50 to 200 μm. This is beyond the spatial resolution of PET and SPECT, and, therefore, MR microscopy with a resolution of up to 40 μm harbors the realistic potential of imaging amyloid, at least in animal models.

Many different intrinsic MR contrast mechanisms have low specificity. Although it makes use of the inherent presence of iron in amyloid plaques in AD, it was feasible to visualize the amyloid plaques *in vivo* in the brains of mouse models using MRI T2 and T2*-weighted contrast (73–78). In the last decade, its repertoire has been extended with targeted contrast agents improving the sensitivity of detection. In this alternative concept, three properties to enhance the plaque-related signal are essential: (1) transport across the BBB following intravenous injection, (2) binding to plaques with molecular specificity, and (3) MRI contrast. The magnetic label carried by the amyloid peptide is constructed of either paramagnetic Gd-chelates or superparamagnetic monocrystalline iron oxide nanoparticles. Transient osmotic disruption of the BBB to allow the reporter molecules to penetrate the brain parenchyma may be necessary (79). The design of BBB-penetrating smart molecular probes is still ongoing, and MR images showed an excellent correlation with immunohistochemistry. Amyloid plaques in APP-overexpressing Tg2576 and double mutant PS/APP mice were probed *in vivo* by putrescine-gadolinium-amyloid-β peptide (PUT-Gd-Aβ), which, because of the presence of the putrescine moiety, is transported across the BBB following intravenous injection, thereby causing specific plaque enhancement in T1-weigthed images (80,81).

In regard of the BBB as being an obstacle, approaches that use plaque-specific ligands for PET or fluorescent dyes (82,83) are more advantageous. NIRF probes for amyloid are an exciting option for molecular imaging in AD research and may translate to clinical diagnostics. The design of smart optical probes emitting characteristic fluorescence signal only when bound to amyloid (84) results in tremendous improvements. An attractive NIRF ligand, the oxazine derivative AOI987, showed specific interaction with amyloid plaques in APP23 transgenic mice as illustrated in Fig. 2 (83).

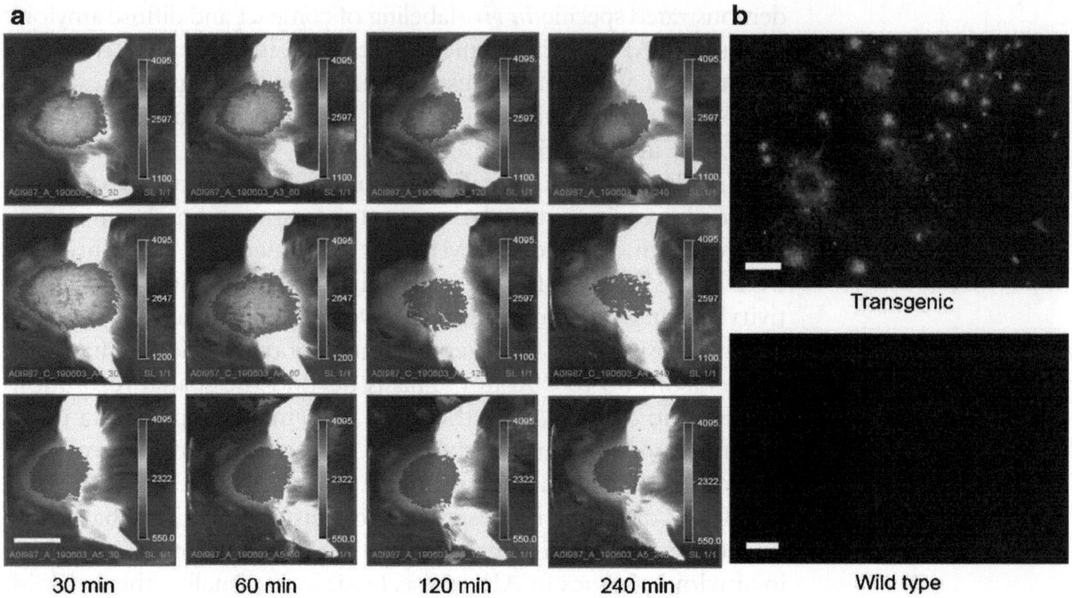

Fig. 2. (**a**) Representative images of 17-month-old APP23 transgenic (*top row*) and wild-type (*middle row*) female mice injected i.v. with 0.1 mg/kg AOI98. The images were recorded 30, 60, 120, and 240 min after injection of the fluorescent dye. In the bottom row, corresponding images of a 17-month-old transgenic APP23 female mouse treated with 0.9% saline only are shown. Scale bar, 1 cm; color scale bars in arbitrary units. (**b**) NIRF microscopy of air-dried cryotome sections (20-μm thickness) of 16-month-old female mice that have been dosed with 0.1 mg/kg AOI987, i.v. The brains were excised and fixed 4 h after dye administration; left, APP23 transgenic; right: wild-type mouse. Scale bar, 100 μm (Adapted from (83)).

Unless CT has low soft contrast, neuritic plaques in PS/APP mice could be demonstrated making use of phase-contrast X-ray combined with 3D CT (42). High-density aggregates being neuritic plaques closely associated with neuritic dystrophy were identified ex vivo.

3.1.2. Probing Early Morphological Changes

If the detection of protein accumulations cannot be tackled directly because, for instance, the accumulations are present intraneuronally, or because of lack of specific probing ligands, research can also focus on the early loss of physiological function associated with the occurrence of protein aggregates. First, subtle functional changes in the brain can be reflected by altered brain perfusion, diffusion, or neuronal connectivity.

Proteins will deposit either extraneuronally of intraneuronally, thereby altering the diffusion properties of the brain tissue. Mueggler et al. (85) used diffusion-weighted MRI and demonstrated that extraneuronal amyloid deposits in APP23 mice are accompanied by a reduction of the diffusion coefficient. Reduced diffusivity within the interstitial space may afflict on its turn the transport of endogenous signaling molecules during synaptic or extrasynaptic transmission and constitute an early sign of cognitive impairment in AD.

With DTI, the early development of AD pathology can be studied in the white matter by quantifying axial and radial diffusivities. This was reported by Song et al. (86) showing that increased levels of amyloid deposition in the brains of PDAPP mice lead to white matter injury.

However, intraneuronal protein aggregates are intrinsically located in the gray matter, and because DTI correlates with tissue microstructure, it should find its application in the gray matter too, e.g., in cortical layers.

It is obvious that in neurons displaying aggregations, critical processes might be affected and the viability of the cell is in danger. Brain perfusion is a parameter of tissue viability and can be measured by MRI, PET, and SPECT. Detection of decreased brain perfusion might not be early enough to serve as diagnostic biomarker (see "Imaging functional changes" section). Therefore, one has to search for more subtle changes, such as brain vulnerability. It was hypothesized that pathological metabolism of APP enhances inflammation and thereby sensitizes the brain to ischemic insults (87). Unlike other mutant AD mouse models, APP751 mice do not develop plaques, and it is yet to be shown that they are highly vulnerable to a focal ischemic insult, while basal CBF was normal (as measured with MRI).

Another approach are studies conceived to elucidate the extent of deformations and eliminations that alter the morphology and architecture of the (micro)vasculature in AD mouse models (88,89). In APP23 mice, Meyer et al. clearly showed that before amyloid plaques and other hallmarks of the disease appear, APP23 mice already have altered cerebrovasculature. Analogously, Beckmann et al. (88) reported flow disturbances in large arteries of APP23 mice in the absence of plaques using MRI angiography. Absence of amyloid deposits in large arteries in the region of the circle of Willis suggests that soluble amyloid may exert deleterious effects on the vasculature, thereby supporting the idea that evolving progressively cerebral circulatory abnormalities could contribute to AD pathogenesis, and that soluble amyloid may exert deleterious effects on the vasculature. Other groups have used MRI tools to investigate the effects of dietary lipids on amyloid deposition and blood circulation in the brains of 18-month-old PS1/APP mice. Relative cerebral blood volume (CBV) and flow (CBF) were determined with proton MRS and gradient-echo contrast-enhanced MRI and support the involvement of hemodynamic changes in the development of AD (90).

3.1.3. Probing Axonal Transport Rates

Manganese has recently emerged as an easily accessible and promising positive contrast agent for MRI that covers a broad spectrum of functional and anatomical applications in small animal models. This application of manganese became known as manganese-enhanced MRI (MEMRI) (91).

The capacity of the Mn^{2+} ion to substitute Ca^{2+} in excitable tissues, such as muscle and nervous tissue, combined with the ability to obtain highly localized MRI contrast changes by using high-resolution scanning techniques, provides a new route to map cellular activity or malfunctions with high accuracy. During cellular activations of excitable tissues, Mn^{2+} can enter through voltage-gated Ca^{2+} channels and an acute increase of extracellular Mn^{2+} concentrations together with the activation of specific brain regions has been shown to result in an increased local MRI signal intensity. After uptake in the neuronal body, Mn^{2+} is transported to the synapse via fast anterograde axonal transport (92). Finally, Mn^{2+} can travel across the synapse and thus can enter a subsequent step of the neuronal network. These properties can be extremely useful for *in vivo* imaging of anatomy, physiology, and activity of neuronal networks. Moreover, because of the way Mn^{2+} is transported, time-related (dynamic) changes in MRI signal intensity in specific brain regions can be used to quantify axonal transport rates between the injected and the projecting area, thereby providing an accurate *in vivo* tool to monitor axonal transport rates.

Axonopathy is a pronounced attribute of many neurodegenerative diseases. In AD, axonal swelling and degeneration are prevalent and may contribute to the symptoms of AD senile dementia. Current limitations in identifying the contribution of axonal damage to AD include the inability to detect when this damage occurs in relation to other identifiers of AD because of the invasiveness of the existing methods. However, MEMRI after Mn^{2+} administration to the nostrils of the mouse was recently used to noninvasively assess *in vivo* axonal transport rates in wild-type mice and the Tg2576 mouse model of AD (93). These authors discerned that *in vivo* axonal transport rates decrease before plaque formation in the Tg2576 mouse model and illustrated the power of this *in vivo* methodology to assess axonopathy.

3.2. Imaging Brain Connectivity Changes

MEMRI can also be used to study brain connectivity and potential changes upon neurodegeneration. In that case, specific brain nuclei of the affected circuit are injected (targeted) with manganese, and subsequently, the signal changes in the projection areas reveal the integrity and the activity of the connections. Pelled et al. (94) performed such a study in healthy and in unilateral 6-hydroxydopamine (6-OHDA) Parkinson disease rat model to test the basal ganglia interhemispheric connectivity after injection of $MnCl_2$ into the entopeduncular nucleus, substantia nigra (SN), and the habenula. They discovered modulations in the effective intra- and interhemispheric basal ganglia connectivity in the unilateral 6-OHDA rat model, which suggests a linkage between the dopaminergic and serotonergic systems in PD, in line with clinical symptoms. The use of manganese has been, however, limited due to concerns about toxicity, specifically acute cardiovascular depression, as it acts as a transient competitive antagonist to calcium activity, reducing contractility and systemic blood pressure. To date, it has been approved for clinical imaging only in a chelated form, i.e., manganese dipyridoxyl diphosphate (Mn-DPDP). While chelation markedly improves the acute safety profile of manganese, it does so at the expense of many of the properties that make it desirable for neuroimaging.

In humans, changes in functional connectivity in specific regions of the brain affected by dementia are investigated with DTI, resting-state, and activation fMRI studies. A multitude of studies exist in patients with AD but also in patients with HD (95,96). However, no such studies were performed on animal models, but some recent papers have exploited the technique in anesthetized control rats (97).

On the other hand, DTI studies to detect white matter changes associated with dementia were done in both patients (98) and in animal models of dementia. DTI studies in animals offer the enormous benefit of providing insight into the underlying mechanism of the observed changes in the DTI parameters being the radial and the axial component. This allows discriminating connectivity loss due to demyelinization or axonal loss (99,100). DTI studies in AD (86) and PD models (101) have been helpful in discerning the specific degenerating fibers in each disease.

3.3. Imaging Structural and Functional Changes Related with Neurogenesis at the Level of the Subventricular Zone or the Dentate Gyrus

The mammalian subventricular zone (SVZ) is the largest germinative zone of the adult brain, which contains a well-characterized stem cell niche with neuronal progenitor cells (NPCs). While most studies highlight the neurogenic potential of SVZ progenitors, recent data indicate that SVZ cells become reactivated in response to different pathological cues, such as trauma, ischemia, inflammation, demyelination, and also neurodegeneration like in PD (102) and HD (103). Similar multiple studies exist and what they have in common is that they assess cell proliferation and differentiation with the gold standard being 5-bromo-2'-deoxyuridine (BrdU) labeling, and subsequent, histology and immunohistochemistry for cell type-specific markers.

However, recently, several *in vivo* visualization methods were developed, which would ultimately permit to follow neuronal recruitment at the rostral migratory stream (RMS) and olfactory bulb (OB) as a read-out for neurogenesis and/or cell survival in animal models for different brain pathologies. NPCs were labeled, *in vivo*, *in situ* in adult rats with fluorescent, micron-sized iron oxide particles (MPIOs), and using MRI, the migrating neural precursors carrying MPIOs were discerned along the RMS to the OB. Immunohistochemistry and electron microscopy indicated that particles were located inside NPCs positive for glial fibrillary acid protein (GFAP+) in the SVZ, inside migrating NPCs positive for polysialylated neural cell adhesion molecule (PSA-NCAM+), in doublecortin positive NPCs within the RMS and OB, and in mature neurons in the OB staining positive for neural nucleus (Neu-N+) (104). This work demonstrates that *in vivo* cell labeling of progenitor cells for MRI is possible and enables serial, non-invasive visualization of endogenous progenitor/precursor cell migration. Undoubtedly, this method will be used in the near future to phenotype neurodegenerative models and to study the efficacy of endogenous progenitor cells as an attractive alternative to the existing therapeutic options for treating neurodegenerative diseases.

The group of Aigner L. et al. (105) developed a new tool based on the doublecortin promoter, driving the expression of the luciferase reporter gene (DCX-promo-luciferase) in transgenic mice to perform *in vivo* imaging of neurogenesis. Indeed, the DCX-promo-luciferase mice allowed optical *in vivo* imaging of the onset of and increase in neurogenesis in developing fetal brains, as well as imaging of neurogenesis in the intact adult mouse central nervous system. This imaging approach allows longitudinal study of neurogenesis in intact animals without the requirement of cellular prelabeling. Moreover, it guarantees that detection is specific for neuronal precursors and restricted to viable cells (106). The group of Baekelandt finally extended this to a truly quantitative method (107,108), as illustrated in mice in Fig. 3.

The dentate gyrus is the second germinative zone in the mammalian brain that continuously provides the adult hippocampus with new neurons. Because the germinative layer is in close vicinity to the target region, the above-mentioned methods are not applicable. However, in dementia such as in AD, the neuroproliferative (subgranular) zone of the dentate gyrus seems to be affected, while the physiological destination of these neurons (granule cell layer), and the CA1 region of Ammon's horn, is the principal site of hippocampal pathology. Jin et al. (109) reported that neurogenesis is enhanced in hippocampus of patients with AD, and they reproduced their findings in a transgenic mouse model, PDGF-APP(Sw, Ind) mice, which express the Swedish and Indiana APP mutations (109). Others have shown impairment in hippocampal neurogenesis in mice carrying targeted mutations in APP, PS1, or both APP and PS1 (110).

A recent study (111) used *in vivo* MRS to determine the load of NPCs in both human and rat dentate gyrus. Also, this might become an attractive tool for further investigating alterations in neurogenesis capacity in neurodegenerative animal models.

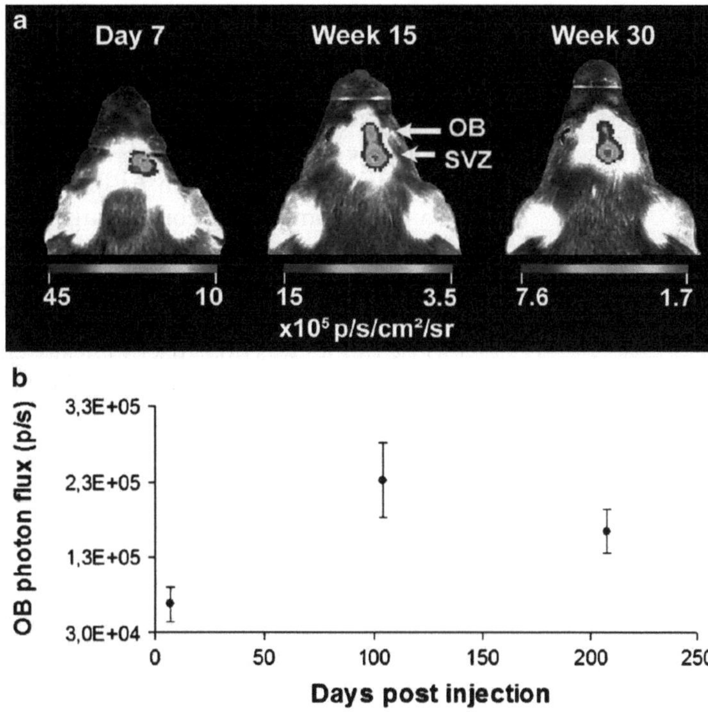

Fig. 3. (**a**) At day 7 postinjection (pi), there is a detectable BLI signal only at the site of injection (SVZ). At 15 and 30 weeks pi, an additional focus is detected at the bulbus olfactorius (OB) projection site as well as the original focus at the SVZ. A representative mouse is shown. (**b**) The graph shows the quantification of the *in vivo* OB BLI signal from all the animals followed for 30 weeks (n = 14). The OB photon flux at week 15 and week 30 is significantly higher than on day 7 (p: 0.002 and p: 0.045) (Adapted from (108)).

3.4. Imaging Functional Changes (Neuronal Performance)

3.4.1. Probing Brain Functioning with fMRI

fMRI probes secondary changes associated with neuronal activation, such as regional changes in CBV, CBF, and blood oxygenation (BOLD contrast). Such fMRI methods have emerged as a powerful tool for investigating the functional organization in the mammalian brain and are widely applied for mapping human brain function both under normal and pathological conditions. In parallel, experimental biomedical applications of fMRI techniques have been developed at a rapid pace providing relevant information on neuronal function in animal models of human disease covering a wide range of species from mice to primates (112).

It is assumed that functional imaging, including CBV, CBF, and BOLD measurements, might detect subtle alterations of neuronal function before gross anatomical abnormalities become detectable in structural images. In the APP23 transgenic AD mouse model that overexpresses human APP with the Swedish mutation, elevated brain levels and deposits of Aβ occur in an age-dependent manner and are associated with behavioral and learning deficits. Quantitative fMRI was applied to characterize brain function in this mouse model (113). Electrical stimulation of the paw led to CBV increases in the contralateral somatosensory cortex. In APP23 mice, this hemodynamic response decreased with increasing age of the animal and with increasing stimulus amplitude when compared with wild-type animals. The age-dependent dysfunction in APP23 mice may be attributed in part to a compromised cerebrovascular reactivity. These data indicate that quantitative functional brain mapping that uses standardized sensory inputs could allow for assessment of disease progression and therapy response in mice and also in patients. The same mouse model was pharmacologically challenged with the GABA-a receptor antagonist bicuculline, thereby significantly reducing the cerebral hemodynamic response. This is in part attributable to a compromised cerebrovascular reactivity, as revealed by the reduced responsiveness to vasodilatory stimulation by acetazolamide (114).

fMRI could also assist in finding abnormalities, such as reduction of olfactory capacities, in AD mice models. The tool has not yet been used for AD mice, but a study from Xu et al. (115) discerned neuronal activity in the OB using BOLD-fMRI in urethane-anesthetized mice allowing the visualization and quantification of activity patterns in response to specific olfactory stimulation. This study illustrates the potential of noninvasive fMRI to test the activation of the neural substrate of different odorants and to spot potential deviations in mouse models for neurodegeneration.

3.5. Imaging Functional and Metabolic Changes

3.5.1. Probing Brain Function and Metabolism with PET

Several FDG autoradiography studies in AD mice models confirm a pattern of reduced activity in several brain areas previously demonstrated in FDG-PET studies of AD patients (116–118).

With the achievements in detector technology, spatial resolution of PET has been considerably improved (1–2 mm), enabling for the first time investigations in small experimental animals such as mice. With the developments in radiochemistry and tracer technology, a variety of endogenously expressed and exogenously introduced genes can be analyzed by PET. This opens up the exciting and rapidly evolving field of molecular imaging, aiming toward the noninvasive localization of a biological process of interest in normal and diseased cells, both in animal models and humans *in vivo*. To date, cerebral glucose metabolism as measured by [18F] FDG-PET is a sensitive noninvasive surrogate marker for the diagnosis of AD. Moreover, degeneration of cholinergic neurons located in basal ganglia, particularly in the nucleus basalis of Meynert and subsequent alteration of cortical AChE activity, is one of the central features in AD and Lewy body dementia, which can be assessed by [11C]MP4A-PET. After passing the BBB, [11C] MP4A is hydrolyzed by AChE and trapped within the brain proportional to regional AChE activity. Currently, the value of [11C]MP4A-PET is being investigated in the differential diagnosis of AD from other forms of dementia. [11C] flumazenil (FMZ) binds to central benzodiazepine receptors, and, as measured by PET, it is an early indicator of preserved cortical neuronal integrity. [18F]FDG together with [11C]FMZ was shown to be also helpful in the diagnosis of other forms of dementia, including variants of Creutzfeld–Jakob disease (119).

A multitracer PET imaging study in the APP23 mouse model investigated the effect of locus ceruleus (LC) degeneration [by induction with *N*-(2-chloroethyl)-*N*-ethyl-bromo-benzylamine (dsp4)] and its contribution to AD pathogenesis by assessing cerebral glucose metabolism, AChE activity, and FMZ binding (120). Additional biochemical, histological, and behavioral findings demonstrated loss of LC neurons, depletion of noradrenaline (NA), increase in microglial and astroglial activation, and an increase in amyloid plaques in specific brain regions, as well as increased cognitive deficits after dsp4 treatment. Twelve weeks after LC degeneration, significant differences for all three radiotracers ([18F] FDG, [11C]MP4A, and [11C]FMZ) could be detected, indicating that LC degeneration and inflammatory reaction contributes significantly to AD pathogenesis. These results, as illustrated in Fig. 4, show the usefulness of radiotracers originally applied to show a characteristic pattern of altered brain glucose metabolism, AChE activity, or neuronal integrity in patients with AD, also in multimodal μPET imaging for characterization and noninvasive phenotyping of mouse models of AD. However, it should be pointed out that μPET imaging in mice still has major limitations with regard to spatial resolution of currently available μPET scanners, attenuation correction, and correct quantification, as well as anesthesia-induced changes of radiotracer uptake.

3.5.2. Probing Brain Metabolism with MRS

HD patients show several signs of impaired metabolism such as weight loss, alterations in cerebral glucose use and cerebral Lac levels and altered activities of mitochondrial enzymes involved in glycolysis (121–125). Because of the potential of MRS to probe energy metabolism, it has become a popular tool in the study of HD both in patients and animal models. In animals, the NMDA agonist quinolinic acid (QA) can accurately model HD because it relatively spares NADPH-diaphorase neurons, which also survive in HD (126,127). In patients, putative defects in energy metabolism lead to a continuous activation of the NMDA receptors, causing base levels of Glu to be lethal for the cell. MRS studies in the striatum of QA-treated rats show an increase in Lac and a decrease in NAA, indicating energetic failure and decreased neural health (128,129). It is hypothesized that in the QA model, consecutive activation of

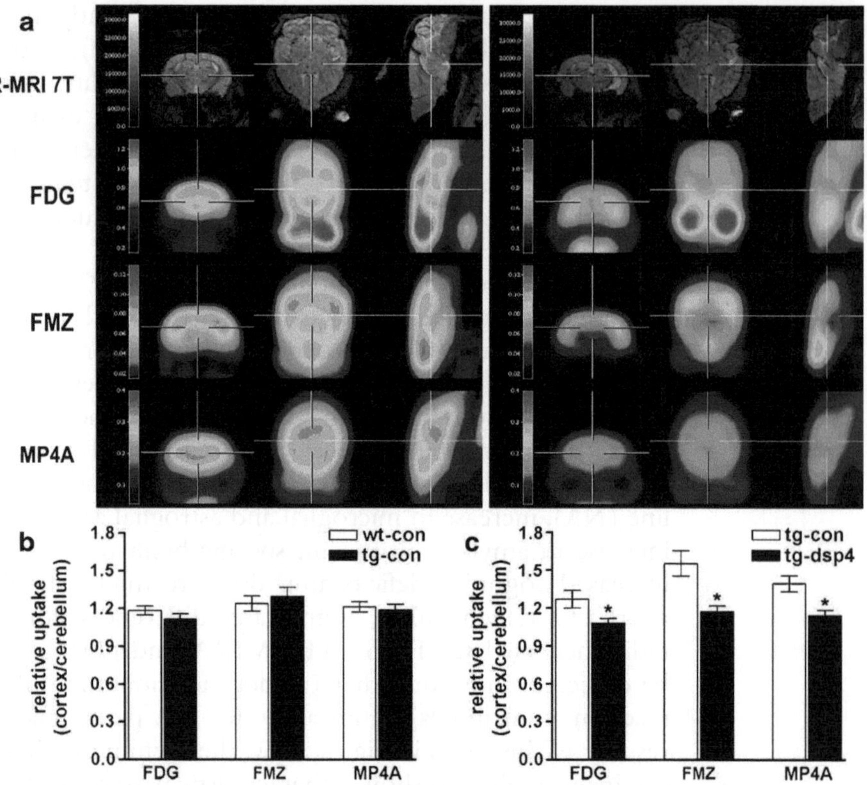

Fig. 4. Altered cerebral glucose metabolism, neuronal integrity, and cholinergic function detected *in vivo* after noradrenergic depletion of APP23. (**a**) Representative high-resolution magnetic resonance images (*first row*) and matched representative [^{18}F]FDG, [^{11}C]FMZ, and [^{11}C]MP4A micro-PET images (*second–fourth rows*, coronal is *left*; transaxial, *middle*; sagittal, *right*) through the brain of saline-treated (*left panel*) and dsp4-treated APP23 (*right panel*). (**b**) Quantification of [^{18}F]FDG, [^{11}C]FMZ, and [^{11}C]MP4A uptake in saline-treated wild-type (wt-con) and saline-treated APP23 (tg-con) mice at 13 months of age. No significant differences were detected. (**c**) Quantification of [^{18}F]FDG, [^{11}C]FMZ, and [^{11}C]MP4A uptake in saline-injected (tg-con) and dsp4-treated (tg-dsp4) APP23 transgenic mice at the same age revealed a decrease in all parameters after LC degeneration (mean ± SEM; n = 4 animals per group; Student's t test; *$p < 0.05$) (Adapted from (120)).

the NMDA receptor causes a constant depolarization of the cell, leading to energy depletion (Lac raises), oxidative stress, and consecutively, neural deterioration (NAA drops), partly mimicking the events in real patients (122). Another toxic model for HD can be induced with mitochondrial toxins such as 3-nitropropionic acid (3-NP) and malonate. These toxins induce age-dependent striatal lesions in humans, rodents, and primates that closely resemble those in HD (130–133), leading to similar changes in NAA and Lac as seen in QA-treated animals (e.g., Fig. 5) (134–139). As a rule, toxin-induced metabolic changes can be reverted by diminishing glutamatergic input, e.g., with NMDA antagonists, such as MK-801 (130,136,137,140,141) or by repleting striatal energy levels (136,140,142). All provide additional evidence to the excitotoxicity hypothesis. MRS research in transgenic models overall led to

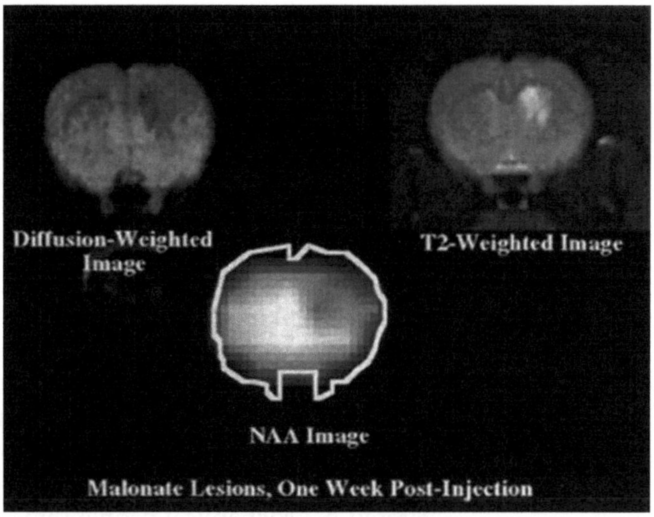

Fig. 5. Diffusion-weighted, T_2-weighted, and *N*-acetylaspartate (NAA) images of a rat brain 1 week after stereotaxic intrastriatal injection with malonate. Field of view for the NAA image was 35 mm, 2 × 2 mm in plane. Note low NAA intensity at the lesion site and high NAA intensity in the lateral septum, which is also well demarcated in the diffusion-weighted but not the T2 image (Adapted from (136)).

similar conclusions (143–145). Of course, toxic models only provide part of the picture, as they only depict more terminal stages of the disease. There exist a number of different transgenic mouse models with varying numbers of CAG repeats. By comparing these different transgenic mouse models, it was possible to investigate the effects of CAG repeat length on the phenotype. It seems that R6/2, the model with the highest CAG repeat length, displays the most severe pathology, with NAA loss both in the striatum and in the cortex. Furthermore, the neural damage of mice with shorter CAG repeats was more concentrated to the striatum, with a linear decrease in NAA over time, rather than an exponential loss as noted in the R6/2 mice (146,147). It should be noted that recent findings from Tkac et al. also indicate a change in Cr over time in R6/2 mice as measured against the unsuppressed water signal assuming 80% water content in the brain (148). The reader should keep this in mind when scrolling through current MRS literature, since most metabolite changes are referred against Cr signal intensity. Although Tkac et al. find the traditional decrease in NAA, their article conflicts with several earlier reports considering the other measured metabolites (129,134). A possible explanation is that the R6/2 model is a severe model for juvenile HD, while QA- and 3-NP-induced striatal damage mimic more terminal events in the disease progression.

Concerning AD, most mouse models being developed are based on known familial mutations in genes encoding APP, PS1, and PS2 (149). In this aspect, MRS is mainly used to validate the paradigm. In the cortex, the APP model and the APP+PS2 model show an age-dependent drop of NAA that correlates well with β-amyloid plaque deposition. Also, there is a raise of Tau. Interestingly, mI, a compound that is typically raised in human AD brain, is not increased in these mice (150–152). More in line with observations in patients, the APP+PS1 model does show a progressive increase in mI, together with an age-dependent drop in NAA (151). Why there should be this difference between different mouse models in concentrations of putative brain osmolytes is not known. The apparent divergent evolution of NAA levels when compared with the concentrations of other metabolites may either suggest neural degradation or an altered volume ratio of neurons to other cells, which is known to be decreased in the presence of amyloid plaques and related neuroinflammation. Correlation between the severity of amyloid deposition and altered neurochemical profile remains to be studied, both in animals and in human subjects. Nevertheless, the altered neurochemical profile may be a valuable marker to test therapeutics in animal models since most of the metabolite changes appear long before the onset of the disease.

PD is a neurological disorder characterized by progressive degeneration of dopaminergic neurons in the SN and concomitant loss of dopamine (DA) in the striatum (153). The etiology of PD is still elusive, but there exists a valuable model, based on the toxin 1-methyl-4-phenyl-1,2,3,6-tetrahydropyridine (MPTP). In astrocytes, MPTP is metabolized to MPP+, which is selectively taken up by dopaminergic (DAergic), causing specific decimation of the targeted neurons (154). This toxin is able to replicate almost all the pathological features of PD in rodents (155,156), cats (157) and nonhuman primates (158). Other well-known models are the rotenone model and the 6-OHDA model (159). Since the pathology of PD is suspected to be at least partly due to energetic defects, MRS has been used to evaluate the available animal models extensively (160–163). Unfortunately, these studies tend to deliver rather conflicting results. For example, studies in MPTP intoxicated cats, rodents, and monkeys showed decreased NAA levels in the striatum (136,160,161,164), while several studies using MPTP-treated mice did not reveal a significant decline in NAA concentration (162,163). The reasons for this variability are multiple, but important factors are the small size of the SN, its high iron content, and interspecies differences in MPTP sensitivity (165). Moreover, the population of DAergic neurons is relatively small, which means that their decline in health or even death may remain unnoticed with MRS, as DAergic nerve terminals affected by MPTP represent only 9% of all synapses in the striatum (166). However, several independent studies agree in the finding of a temporal rising in Lac levels after MPTP treatment, which reduces to baseline eventually (136,163), reflecting possible defects in energy impairment. In contrast, it is often seen that striatal NAA levels (160,161) are permanently reduced after MPTP treatment indicating a drop in neural health. Moreover, these changes can be attenuated by decreasing the glutamatergic input (136,167). In addition, several models show high levels of Glu and Gln in the striatum after induction of parkinsonism (164,168,169). These and other findings suggest that at least part of the pathology of PD is caused by an impairment in energy metabolism, leading eventually to a selective loss of the DAergic nigro-striatal connection. It is assumed that this loss of DAergic input in the striatum leads on its turn to hyperactivity of the corticostriatal pathway, as reflected by the high Glu and Gln levels.

3.6. Imaging Late-Stage and Irreversible Neuronal Changes

Neurological diseases can be classified by the assessment of their molecular characteristics, which is also the way to find new targets for intervention, which could lead to novel therapies. In the end, all the molecular pathways of dementia lead to neurodegeneration, i.e., loss of both structure and function of neurons. At this stage, the damage in the nervous system is irreversible and is no longer a target for therapies.

Neuronal loss manifests itself as reduction in brain volume and a passive dilatation of the ventricles within the atrophic brain tissue. High-resolution 3D MRI is by far the best suited *in vivo* technique for the accurate determination of brain volumes. Moreover, as long as the offered resolution and contrast-to-noise permits to distinguish between different structures, subregions of the brain can also be delineated for volume quantification.

Kooy et al. (170) reviewed the potential of MRI to image the brain of (transgenic) mouse models for neurological diseases and introduced these exciting new technological developments to the nonspecialist reader. MR volumetric measurements are merely performed in mouse models for AD. Redwine et al. (171) hypothesized that a hippocampal volume deficit would be detectable using MRI in aged mice corresponding to the time point when plaque accumulations and behavioral deficits were maximally evident. Volume reductions were predominantly localized in the dentate gyrus. Reduced hippocampal volume together with reduced corpus callosum length could be detected in APP/V717F mouse models for AD at different time points. Results of reduced hippocampal volume in murine models for AD build upon the spatial learning deficits that have already been reported for these mice. The most recent article (172) measuring brain volumes with MRI demonstrated cerebral volume reduction in a APP(T714I) mouse model for AD bearing intraneuronal amyloid.

Another mental retardation disorder is hydrocephalus. In 1998, a knockout (L1CAM) mouse model was generated, and after neuropathological and behavioral studies, an MRI was also performed (173). The ventricular system was shown to be abnormal with dilatation of the lateral ventricles and the fourth ventricle, and an altered shape of the Sylvius aquaduct. Ventricle volume and shape changes were thought to be a general and similar effect associated with all cases of brain atrophy. However, results on ventricle shape changes alters this perception and it was shown that specific shape variations can be solely linked to AD and not for instance to frontal lobe dementia (174). Since the contrast between grey and white matter reduces with aging, this new approach of ventricle shape analysis might become very valuable in diagnosing AD. Moreover, this tool can be applied to volumetric 3D MRI data of mouse models too.

Next to volume measurements, neurodegeneration can also be assessed through T_2-weighted imaging since hyperintensity signals on T_2 indicate vacuolization of the neurons. This was demonstrated in a mouse model for ALS overexpressing mutated SOD1. Hyperintensities in well-defined motor nuclei of the brain stem paralleled with a decreased number of neurons (175,176).

In neurodegenerative diseases, the pattern of atrophy in different brain regions could be more informative in the differential diagnostic process than the total amount of atrophy in a single brain region. Increasing evidence demonstrates that there is not only marked damage in the grey matter but also in the white matter. Synaptic loss provides indeed an excellent correlation with cognitive ability in dementia in PD (177) in AD (178) and Prion disease (179). The loss of synaptic connectivity may represent a loss of brain plasticity. *In vivo* imaging of rCMRGlc (using PET) was used in AD to identify potentially reversible and irreversible stages of brain metabolic failure in patients due to regional brain synaptic activity (180,181).

In animal models, DTI is exploited for the observation of axonal damage. Song et al. (100) have demonstrated that the radial diffusivity changes were specific to the time course of changes in myelin integrity. An *ex vivo* study using cuprizone-treated mouse brain as an experimental model of demyelinization, showed an increase in radial diffusivity with demyelinization. This application of DTI holds promise as an attractive alternative to detect and quantify white matter injury noninvasively in human demyelinating diseases.

Song et al. (86) showed in PDAPP mouse models that as amyloid deposits increase over time, the profile of DTI parameters changes significantly. The affected white matter displayed decreased relative anisotropy and a significantly elevated radial diffusivity in old PDAPP mice versus nontransgenic controls in several white matters structures (corpus callosum, external capsula, fornix, and cerebral peduncle).

4. Conclusions

In general, animal models are extensively used to understand the etiology and pathophysiology of human disorders of the central nervous system. In recent years, technical advances in imaging modalities have allowed the use of these techniques for the evaluation of functional, neurochemical, and anatomical changes in the brains of animals. Combining animal models of neurodegenerative disorders with neuroimaging are not only essential components in the follow up of the disease progression but also in the development of therapeutic interventions to derive knowledge that will ultimately inform our clinical decisions. The biggest advantages of *in vivo* neuroimaging deal with its translational character allowing experimental studies to develop further in the clinic or vice versa and, on the other hand, the possibility to use clinical imaging knowledge to validate rodent models before further exploration of underlying pathological mechanisms. Another major bonus of *in vivo* animal model imaging is that it allows presymptomatic investigations and thus enabling improvements to be made on the early diagnostics while in the clinical only post-symptomatic patients come forward as imaging subjects.

References

1. Lindvall O, Bjorklund A (2004) Cell therapy in Parkinson's disease. NeuroRx 1:382–393
2. Willner P (2008) Methods for assessing the validity of animal models of human psychopathology. In: Boulton AA, Baker GB, Martin-Iverson MT (eds) Neuromethods, animal models in psychiatry I, 18th edn. Humana, Clifton, NJ
3. Klunk WE, Lopresti BJ, Ikonomovic MD, et al. (2005) Binding of the positron emission tomography tracer Pittsburgh compound-B reflects the amount of amyloid-beta in Alzheimer's disease brain but not in transgenic mouse brain. J Neurosci 25: 10598–10606.
4. Suhara T, Higuchi M, Miyoshi M (2008) Neuroimaging in dementia: In vivo amyloid imaging. Tohoku J Exp Med 215:119–124
5. Toyama H, Ye D, Ichise M, et al. (2005) PET imaging of brain with the beta-amyloid probe, [11C]6-OH-BTA-1, in a transgenic mouse model of Alzheimer's disease. Eur J Nucl Med Mol Imaging 32(5):593–600
6. Lauterbur PC (1980) Progress in n.m.r. zeugmatography imaging. Philos Trans R Soc Lond B Biol Sci 289:483–487
7. Le Bihan D, Breton E, Lallemand D, Grenier P, Cabanis E, Laval-Jeantet M (1986) MR imaging of intravoxel incoherent motions: Application to diffusion and perfusion in neurologic disorders. Radiology 161:401–407
8. Gass A, Niendorf T, Hirsch JG (2001) Acute and chronic changes of the apparent diffusion coefficient in neurological disorders-biophysical mechanisms and possible underlying histopathology. J Neurol Sci 186:S15–S23
9. Basser PJ (1995) Inferring microstructural features and the physiological state of tissues from diffusion-weighted images. NMR Biomed 8:333–344
10. Assaf Y, Pasternak O (2008) Diffusion tensor imaging (DTI)-based white matter mapping in brain research: A review. J Mol Neurosci 34(1):51–61
11. Ogawa S, Lee TM, Kay AR, Tank DW (1990) Brain magnetic resonance imaging with contrast dependent on blood oxygenation. Proc Natl Acad Sci U S A 87(24):9868–9872
12. Benveniste H, Einstein G, Kim KR, Hulette C, Johnson GA (1999) Detection of neuritic plaques in Alzheimer's disease by magnetic resonance microscopy. Proc Natl Acad Sci U S A 96(24):14079–14084
13. Ginefri JC, Poirier-Quinot M, Girard O, Darrasse L (2007) Technical aspects: Development, manufacture and installation of a cryo-cooled HTS coil system for high-resolution

in-vivo imaging of the mouse at 1.5 T. Methods 43(1):54–67

14. Dubowitz DJ, Tyszka JM, Sewry CA, Moats RA, Scadeng M, Dubowitz V (2000) High resolution magnetic resonance imaging of the brain in the dy/dy mouse with merosin-deficient congenital muscular dystrophy. Neuromuscul Disord 10(4–5):292–298
15. Hesselbarth D, Franke C, Hata R, Brinker G, Hoehn-Berlage M (1998) High resolution MRI and MRS: A feasibility study for the investigation of focal cerebral ischemia in mice. NMR Biomed 11(8):423–429
16. Yang QX, Smith MB, Briggs RW, Rycyna RE (1999) Microimaging at 14 tesla using GESEPI for removal of magnetic susceptibility artifacts in T(2)(*)-weighted image contrast. J Magn Reson 141(1):1–6
17. Kooy RF, Reyniers E, Verhoye M, et al. (1999) Neuroanatomy of the fragile X knockout mouse brain studied using in vivo high resolution magnetic resonance imaging. Eur J Hum Genet 7(5):526–532
18. Van Camp N., Peeters RR, Van der LA (2005) A comparison between blood oxygenation level-dependent and cerebral blood volume contrast in the rat cerebral and cerebellar somatosensoric cortex during electrical paw stimulation. J Magn Reson Imaging 22:483–491
19. Artemov D (2003) Molecular magnetic resonance imaging with targeted contrast agents. J Cell Biochem 90:518–524
20. Meade TJ, Taylor AK, Bull SR (2003) New magnetic resonance contrast agents as biochemical reporters. Curr Opin Neurobiol 13:597–602
21. Dzik-Jurasz AS (2003) Molecular imaging in vivo: An introduction. Br J Radiol 76:S98–S109
22. Pfeuffer J, Tkac I, Provencher SW, Gruetter R (1999) Toward an in vivo neurochemical profile: Quantification of 18 metabolites in short-echo-time (1)H NMR spectra of the rat brain. J Magn Reson 141:104–120
23. Brown TR, Kincaid BM, Ugurbil K (1982) NMR chemical shift imaging in three dimensions. Proc Natl Acad Sci U S A 79:3523–3526
24. Maudsley AA, Hilal SK, Perman WH, Simon HE (1983) Spatially resolved high resolution spectroscopy by 'four dimensional ' NMR. J Magn Reson 51:147–152
25. Moffett JR, Ross B, Arun P, Madhavarao CN, Namboodiri AM (2007) N-Acetylaspartate in the CNS: From neurodiagnostics to neurobiology. Prog Neurobiol 81:89–131
26. Kantarci K, Jack CR, Jr., Xu YC, et al. (2000) Regional metabolic patterns in mild cognitive impairment and Alzheimer's disease: A 1H MRS study. Neurology 55:210–217
27. Miller BL, Moats RA, Shonk T, Ernst T, Woolley S, Ross BD (1993) Alzheimer disease: Depiction of increased cerebral myo-inositol with proton MR spectroscopy. Radiology 187(2):433–437
28. Govindaraju V, Young K, Maudsley AA, Proton NMR (2000) Chemical shifts and coupling constants for brain metabolites. NMR Biomed 13(3):129–153
29. Erecinska M, Silver IA (1990) Metabolism and role of glutamate in mammalian brain. Prog Neurobiol 35(4):245–296
30. Behar KL, den Hollander JA, Stromski ME, et al. (1983) High-resolution 1H nuclear magnetic resonance study of cerebral hypoxia in vivo. Proc Natl Acad Sci U S A 80(16):4945–4948
31. Bruhn H, Frahm J, Gyngell ML, Merboldt KD, Hanicke W, Sauter R (1989) Cerebral metabolism in man after acute stroke: New observations using localized proton NMR spectroscopy. Magn Reson Med 9(1):126–131
32. Bruhn H, Frahm J, Gyngell ML, et al. (1989) Noninvasive differentiation of tumors with use of localized H-1 MR spectroscopy in vivo: Initial experience in patients with cerebral tumors. Radiology 172(2):541–548
33. Jansen JF, Backes WH, Nicolay K, Kooi ME (2006) 1H MR spectroscopy of the brain: Absolute quantification of metabolites. Radiology 240(2):318–332
34. Brand A, Richter-Landsberg C, Leibfritz D (1993) Multinuclear NMR studies on the energy metabolism of glial and neuronal cells. Dev Neurosci 15(3–5):289–298
35. Urenjak J, Williams SR, Gadian DG, Noble M (1993) Proton nuclear magnetic resonance spectroscopy unambiguously identifies different neural cell types. J Neurosci 13(3):981–989
36. Frisoni GB (2001) Structural imaging in the clinical diagnosis of Alzheimer's disease: Problems and tools. J Neurol Neurosurg Psychiatr 70:711–718
37. Rose M, Scharf S (2008) Is there any role for computed tomography measurements of medial temporal lobe atrophy in dementia? A review of the literature and case series from a memory clinic. Intern Med J 38:136–139
38. Muller R, Gerber SC, Hayes WC (1998) Micro-compression: A novel technique for the nondestructive assessment of local bone failure. Technol Health Care 6:433–444
39. Barck KH, Lee WP, Diehl LJ, et al. (2004) Quantification of cortical bone loss and repair for therapeutic evaluation in collagen-induced arthritis, by micro-computed tomography and

automated image analysis. Arthritis Rheum 50:3377–3386
40. Kennel SJ, Davis IA, Branning J, Pan H, Kabalka GW, Paulus MJ (2000) High resolution computed tomography and MRI for monitoring lung tumor growth in mice undergoing radioimmunotherapy: Correlation with histology. Med Phys 27:1101–1107
41. Paulus MJ, Gleason SS, Kennel SJ, Hunsicker PR, Johnson DK (2000) High resolution X-ray computed tomography: An emerging tool for small animal cancer research. Neoplasia 2:62–70
42. Noda-Saita K, Yoneyama A, Shitaka Y, et al. (2006) Quantitative analysis of amyloid plaques in a mouse model of Alzheimer's disease by phase-contrast X-ray computed tomography. Neuroscience 138:1205–1213
43. Seo Y, Hashimoto T, Nuki Y, Hasegawa BH (2008) In vivo microCT imaging of rodent cerebral vasculature. Phys Med Biol 53:N99–N107
44. Pichler BJ, Judenhofer MS, Wehrl HF (2008) PET/MRI hybrid imaging: Devices and initial results. Eur Radiol 18(6):1077–1086
45. Goetz C, Breton E, Choquet P, Israel-Jost V, Constantinesco A (2008) SPECT low-field MRI system for small-animal imaging. J Nucl Med 49(1):88–93
46. Klerk CP, Overmeer RM, Niers TM, et al. (2007) Validity of bioluminescence measurements for noninvasive in vivo imaging of tumor load in small animals. Biotechniques 43(1 Suppl):7–13, 30
47. Bacskai BJ, Kajdasz ST, Christie RH, et al. (2001) Imaging of amyloid-beta deposits in brains of living mice permits direct observation of clearance of plaques with immunotherapy. Nat Med 7(3):369–372
48. Carson JP, Ju T, Thaller C, et al. (2004) Automated characterization of gene expression patterns with an atlas of the mouse brain. Conf Proc IEEE Eng Med Biol Soc 4:2917–2920
49. Dhenain M, Ruffins SW, Jacobs RE (2001) Three-dimensional digital mouse atlas using high-resolution MRI. Dev Biol 232(2):458–470
50. Kenzie-Graham A, Jones ES, Shattuck DW, Dinov ID, Bota M, Toga AW (2003) The informatics of a C57BL/6J mouse brain atlas. Neuroinformatics 1(4):397–410
51. Kenzie-Graham A, Boline J, Toga AW (2007) Brain atlases and neuroanatomic imaging. Methods Mol Biol 401:183–194
52. Kovacevic N, Henderson JT, Chan Eet al., (2005) A three-dimensional MRI atlas of the mouse brain with estimates of the average and variability. Cereb Cortex 15(5):639–645
53. Lein ES, Hawrylycz MJ, Ao N, et al. (2007) Genome-wide atlas of gene expression in the adult mouse brain. Nature 445(7124): 168–176
54. Ma Y, Hof PR, Grant SC, et al. (2005) A three-dimensional digital atlas database of the adult C57BL/6J mouse brain by magnetic resonance microscopy. Neuroscience 135:1203–1215
55. Paxinos GaFKB (2001) The mouse brain in stereotaxic coordinates. Elsevier, San Diego, CA
56. Paxinos GaWC (2005) The rat brain in stereotaxic coordinates. Elsevier, San Diego, CA
57. Hjornevik T, Leergaard T, Dariune D, et al. (2007) Three-dimensional atlas system for mouse and rat brain imaging data. Front NeuroInformatics1:1–4
58. Casteels C, Vermaelen P, Nuyts J, et al. (2006) Construction and evaluation of multitracer small-animal PET probabilistic atlases for voxel-based functional mapping of the rat brain. J Nucl Med 47:1858–1866
59. Talairach JaTP (1988) Co-planar stereotaxic atlas of the human brain. Thieme Medical Publishers, New York
60. Winklhofer KF, Tatzelt J, Haass C (2008) The two faces of protein misfolding: Gain- and loss-of-function in neurodegenerative diseases. EMBO J 27:336–349
61. Ho LW, Carmichael J, Swartz J, Wyttenbach A, Rankin J, Rubinsztein DC (2001) The molecular biology of Huntington's disease. Psychol Med 31:3–14
62. Hardy J, Selkoe DJ (2002) The amyloid hypothesis of Alzheimer's disease: Progress and problems on the road to therapeutics. Science 297:353–356
63. Oakley H, Cole SL, Logan S, et al. (2006) Intraneuronal beta-amyloid aggregates, neurodegeneration, and neuron loss in transgenic mice with five familial Alzheimer's disease mutations: Potential factors in amyloid plaque formation. J Neurosci 26:10129–10140
64. Bruijn LI, Houseweart MK, Kato S, et al. (1998) Aggregation and motor neuron toxicity of an ALS-linked SOD1 mutant independent from wild-type SOD1. Science 281(5384):1851–1854
65. Ma J, Wollmann R, Lindquist S (2002) Neurotoxicity and neurodegeneration when PrP accumulates in the cytosol. Science 298(5599):1781–1785
66. Dawson TM, Dawson VL (2003) Rare genetic mutations shed light on the pathogenesis of Parkinson disease. J Clin Invest 111(2):145–151
67. Majocha RE, Reno JM, Friedland RP, VanHaight C, Lyle LR, Marotta CA (1992)

Development of a monoclonal antibody specific for beta/A4 amyloid in Alzheimer's disease brain for application to in vivo imaging of amyloid angiopathy. J Nucl Med 33(12):2184–2189
68. Agdeppa ED, Kepe V, Liu J, et al. (2003) 2-Dialkylamino-6-acylmalononitrile substituted naphthalenes (DDNP analogs): Novel diagnostic and therapeutic tools in Alzheimer's disease. Mol Imaging Biol 5(6):404–417
69. Mathis CA, Wang Y, Holt DP, Huang GF, Debnath ML, Klunk WE (2003) Synthesis and evaluation of 11C-labeled 6-substituted 2-arylbenzothiazoles as amyloid imaging agents. J Med Chem 46(13):2740–2754
70. Wang Y, Klunk WE, Huang GF, Debnath ML, Holt DP, Mathis CA (2002) Synthesis and evaluation of 2-(3'-iodo-4'-aminophenyl)-6-hydroxybenzothiazole for in vivo quantitation of amyloid deposits in Alzheimer's disease. J Mol Neurosci 19(1–2):11–16
71. Zhang W, Oya S, Kung MP, Hou C, Maier DL, Kung HF (2005) F-18 Polyethyleneglycol stilbenes as PET imaging agents targeting Abeta aggregates in the brain. Nucl Med Biol 32(8):799–809
72. Okamura N, Suemoto T, Shiomitsu T, et al. (2004) A novel imaging probe for in vivo detection of neuritic and diffuse amyloid plaques in the brain. J Mol Neurosci 24:247–255
73. El Tannir El TN, Delatour B, Le CC, Guegan M, Volk A, Dhenain M (2006) Age-related evolution of amyloid burden, iron load, and MR relaxation times in a transgenic mouse model of Alzheimer's disease. Neurobiol Dis 22:199–208
74. Falangola MF, Lee SP, Nixon RA, Duff K, Helpern JA (2005) Histological co-localization of iron in Abeta plaques of PS/APP transgenic mice. Neurochem Res 30:201–205
75. Falangola MF, Dyakin VV, Lee SP, et al. (2007) Quantitative MRI reveals aging-associated T2 changes in mouse models of Alzheimer's disease. NMR Biomed 20:343–351
76. Jack CR Jr, Garwood M, Wengenack TM, et al. (2004) In vivo visualization of Alzheimer's amyloid plaques by magnetic resonance imaging in transgenic mice without a contrast agent. Magn Reson Med 52:1263–1271
77. Vanhoutte G, Dewachter I, Borghgraef P, Van LF, Van der LA (2005) Noninvasive in vivo MRI detection of neuritic plaques associated with iron in APP[V717I] transgenic mice, a model for Alzheimer's disease. Magn Reson Med 53:607–613
78. Zhang J, Yarowsky P, Gordon MN, et al. (2004) Detection of amyloid plaques in mouse models of Alzheimer's disease by magnetic resonance imaging. Magn Reson Med 51:452–457
79. Wadghiri YZ, Sigurdsson EM, Sadowski M, et al. (2003) Detection of Alzheimer's amyloid in transgenic mice using magnetic resonance microimaging. Magn Reson Med 50:293–302
80. Poduslo JF, Wengenack TM, Curran GL, et al. (2002) Molecular targeting of Alzheimer's amyloid plaques for contrast-enhanced magnetic resonance imaging. Neurobiol Dis 11:315–329
81. Poduslo JF, Ramakrishnan M, Holasek SS, et al. (2007) In vivo targeting of antibody fragments to the nervous system for Alzheimer's disease immunotherapy and molecular imaging of amyloid plaques. J Neurochem 102:420–433
82. Bacskai BJ, Skoch J, Hickey GA, Allen R, Hyman BT (2003) Fluorescence resonance energy transfer determinations using multiphoton fluorescence lifetime imaging microscopy to characterize amyloid-beta plaques. J Biomed Opt 8:368–375
83. Hintersteiner M, Enz A, Frey P, et al. (2005) In vivo detection of amyloid-beta deposits by near-infrared imaging using an oxazine-derivative probe. Nat Biotechnol 23:577–583
84. Raymond SB, Skoch J, Hills ID, Nesterov EE, Swager TM, Bacskai BJ (2008) Smart optical probes for near-infrared fluorescence imaging of Alzheimer's disease pathology. Eur J Nucl Med Mol Imaging 35:S93–S98
85. Mueggler T, Meyer-Luehmann M, Rausch M, Staufenbiel M, Jucker M, Rudin M (2004) Restricted diffusion in the brain of transgenic mice with cerebral amyloidosis. Eur J Neurosci 20:811–817
86. Song SK, Kim JH, Lin SJ, Brendza RP, Holtzman DM (2004) Diffusion tensor imaging detects age-dependent white matter changes in a transgenic mouse model with amyloid deposition. Neurobiol Dis 15:640–647
87. Koistinaho M, Kettunen MI, Goldsteins G, et al. (2002) Beta-amyloid precursor protein transgenic mice that harbor diffuse A beta deposits but do not form plaques show increased ischemic vulnerability: Role of inflammation. Proc Natl Acad Sci U S A 99:1610–1615
88. Beckmann N, Schuler A, Mueggler T, et al. (2003) Age-dependent cerebrovascular abnormalities and blood flow disturbances in APP23 mice modeling Alzheimer's disease. J Neurosci 23:8453–8459

89. Meyer EP, Ulmann-Schuler A, Staufenbiel M, Krucker T (2008) Altered morphology and 3D architecture of brain vasculature in a mouse model for Alzheimer's disease. Proc Natl Acad Sci U S A 105:3587–3592
90. Hooijmans CR, Rutters F, Dederen PJ, et al. (2007) Changes in cerebral blood volume and amyloid pathology in aged Alzheimer APP/PS1 mice on a docosahexaenoic acid (DHA) diet or cholesterol enriched Typical Western Diet (TWD). Neurobiol Dis 28:16–29
91. Koretsky AP, Silva AC (2004) Manganese-enhanced magnetic resonance imaging (MEMRI). NMR Biomed 17:527–531
92. Pautler RG, Silva AC, Koretsky AP (1998) In vivo neuronal tract tracing using manganese-enhanced magnetic resonance imaging. Magn Reson Med 40:740–748
93. Smith KD, Kallhoff V, Zheng H, Pautler RG (2007) In vivo axonal transport rates decrease in a mouse model of Alzheimer's disease. Neuroimage 35:1401–1408
94. Pelled G, Bergman H, Ben-Hur T, Goelman G (2007) Manganese-enhanced MRI in a rat model of Parkinson's disease. J Magn Reson Imaging 26:863–870
95. Thiruvady DR, Georgiou-Karistianis N, Egan GF, et al. (2007) Functional connectivity of the prefrontal cortex in Huntington's disease. J Neurol Neurosurg Psychiatr 78(2): 127–133
96. Wang K, Liang M, Wang L, et al. (2007) Altered functional connectivity in early Alzheimer's disease: A resting-state fMRI study. Hum Brain Mapp 28(10):967–978
97. Zhao F, Zhao T, Zhou L, Wu Q, Hu X (2008) BOLD study of stimulation-induced neural activity and resting-state connectivity in medetomidine-sedated rat. Neuroimage 39(1):248–260
98. Huang J, Friedland RP, Auchus AP (2007) Diffusion tensor imaging of normal-appearing white matter in mild cognitive impairment and early Alzheimer disease: Preliminary evidence of axonal degeneration in the temporal lobe. AJNR Am J Neuroradiol 28(10):1943–1948
99. Song SK, Sun SW, Ramsbottom MJ, Chang C, Russell J, Cross AH (2002) Dysmyelination revealed through MRI as increased radial (but unchanged axial) diffusion of water. Neuroimage 17(3):1429–1436
100. Song SK, Yoshino J, Le TQ, et al. (2005) Demyelination increases radial diffusivity in corpus callosum of mouse brain. Neuroimage 26(1):132–140
101. Boska MD, Hasan KM, Kibuule D, et al. (2007) Quantitative diffusion tensor imaging detects dopaminergic neuronal degeneration in a murine model of Parkinson's disease. Neurobiol Dis 26(3):590–596
102. Aponso PM, Faull RL, Connor B (2008) Increased progenitor cell proliferation and astrogenesis in the partial progressive 6-hydroxydopamine model of Parkinson's disease. Neuroscience 151(4):1142–1153
103. Tattersfield AS, Croon RJ, Liu YW, Kells AP, Faull RL, Connor B (2004) Neurogenesis in the striatum of the quinolinic acid lesion model of Huntington's disease. Neuroscience 127:319–332
104. Shapiro EM, Gonzalez-Perez O, Manuel Garcia-Verdugo J, varez-Buylla A, Koretsky AP (2006) Magnetic resonance imaging of the migration of neuronal precursors generated in the adult rodent brain. Neuroimage 32:1150–1157
105. Aigner L, Bogdahn U (2008) TGF-beta in neural stem cells and in tumors of the central nervous system. Cell Tissue Res 331:225–241
106. Couillard-Despres S, Finkl R, Winner B, et al. (2008) In vivo optical imaging of neurogenesis: Watching new neurons in the intact brain. Mol Imaging 7:28–34
107. Geraerts M, Eggermont K, Hernandez-Acosta P, Garcia-Verdugo JM, Baekelandt V, Debyser Z (2006) Lentiviral vectors mediate efficient and stable gene transfer in adult neural stem cells in vivo. Hum Gene Ther 17:635–650
108. Reumers V, Deroose CM, Krylyshkina O, et al. (2008) Non-invasive and quantitative monitoring of adult neuronal stem cell migration in mouse brain using bioluminescence imaging. Stem Cells 26:2382–2390
109. Jin K, Peel AL, Mao XO, et al. (2004) Increased hippocampal neurogenesis in Alzheimer's disease. Proc Natl Acad Sci U S A 101:343–347
110. Zhang C, McNeil E, Dressler L, Siman R (2007) Long-lasting impairment in hippocampal neurogenesis associated with amyloid deposition in a knock-in mouse model of familial Alzheimer's disease. Exp Neurol 204:77–87
111. Louis N, Manganas LN., Zhang X, et al. (2007) Magnetic resonance spectroscopy identifies neural progenitor cells in the live human brain. Science 318:980–985
112. Van der Linden A., Van CN, Ramos-Cabrer P, Hoehn M (2007) Current status of functional MRI on small animals: Application to

physiology, pathophysiology, and cognition. NMR Biomed 20:522–545

113. Mueggler T, Baumann D, Rausch M, Staufenbiel M, Rudin M (2003) Age-dependent impairment of somatosensory response in the amyloid precursor protein 23 transgenic mouse model of Alzheimer's disease. J Neurosci 23:8231–8236

114. Mueggler T, Baumann D, Rausch M, Rudin M (2001) Bicuculline-induced brain activation in mice detected by functional magnetic resonance imaging. Magn Reson Med 46:292–298

115. Xu F, Liu N, Kida I, Rothman DL, Hyder F, Shepherd GM (2003) Odor maps of aldehydes and esters revealed by functional MRI in the glomerular layer of the mouse olfactory bulb. Proc Natl Acad Sci U S A 100:11029–11034

116. Liang WS, Reiman EM, Valla J, et al. (2008) Alzheimer's disease is associated with reduced expression of energy metabolism genes in posterior cingulate neurons. Proc Natl Acad Sci U S A 105:4441–4446

117. Valla J, Berndt JD, Gonzalez-Lima F (2001) Energy hypometabolism in posterior cingulate cortex of Alzheimer's patients: Superficial laminar cytochrome oxidase associated with disease duration. J Neurosci 21: 4923–4930

118. Valla J, Chen K, Berndt JD, et al. (2002) Effects of image resolution on autoradiographic measurements of posterior cingulate activity in PDAPP mice: Implications for functional brain imaging studies of transgenic mouse models of Alzheimer's Disease. Neuroimage 16:1–6

119. Winkeler A, Waerzeggers Y, Klose A, et al. (2008) Imaging noradrenergic influence on amyloid pathology in mouse models of Alzheimer's disease. Eur J Nucl Med Mol Imaging 35:S107–S113

120. Heneka MT, Ramanathan M, Jacobs AH (2006) Locus ceruleus degeneration promotes Alzheimer pathogenesis in amyloid precursor protein 23 transgenic mice. J Neurosci 26:1343–1354

121. Browne SE, Bowling AC, MacGarvey U, et al. (1997) Oxidative damage and metabolic dysfunction in Huntington's disease: Selective vulnerability of the basal ganglia. Ann Neurol 41:646–653

122. Browne SE, Beal MF (2004) The energetics of Huntington's disease. Neurochem Res 29:531–546

123. Gu M, Gash MT, Mann VM, Javoy-Agid F, Cooper JM, Schapira AH (1996) Mitochondrial defect in Huntington's disease caudate nucleus. Ann Neurol 39:385–389

124. Jenkins BG, Koroshetz WJ, Beal MF, Rosen BR (1993) Evidence for impairment of energy metabolism in vivo in Huntington's disease using localized 1H NMR spectroscopy. Neurology 43:2689–2695

125. Koroshetz WJ, Jenkins BG, Rosen BR, Beal MF (1997) Energy metabolism defects in Huntington's disease and effects of coenzyme Q10. Ann Neurol 41:160–165

126. Beal MF, Kowall NW, Ellison DW, Mazurek MF, Swartz KJ, Martin JB (1986) Replication of the neurochemical characteristics of Huntington's disease by quinolinic acid. Nature 321:168–1671

127. Ferrante RJ, Kowall NW, Cipolloni PB, Storey E, Beal MF (1993) Excitotoxin lesions in primates as a model for Huntington's disease: Histopathologic and neurochemical characterization. Exp Neurol 119:46–71

128. Strauss I, Williamson JM, Bertram EH, Lothman EW, Fernandez EJ (1997) Histological and 1H magnetic resonance spectroscopic imaging analysis of quinolinic acid-induced damage to the rat striatum. Magn Reson Med 37:24–33

129. Tkac I, Keene CD, Pfeuffer J, Low WC, Gruetter R (2001) Metabolic changes in quinolinic acid-lesioned rat striatum detected non-invasively by in vivo (1)H NMR spectroscopy. J Neurosci Res 66:891–898

130. Beal MF, Brouillet E, Jenkins B, Henshaw R, Rosen B, Hyman BT (1993) Age-dependent striatal excitotoxic lesions produced by the endogenous mitochondrial inhibitor malonate. J Neurochem 61:1147–1150

131. Beal MF, Brouillet E, Jenkins BG, et al. (1993) Neurochemical and histologic characterization of striatal excitotoxic lesions produced by the mitochondrial toxin 3-nitropropionic acid. J Neurosci 13:4181–4192

132. Brouillet E, Hantraye P, Ferrante RJ, et al. (1995) Chronic mitochondrial energy impairment produces selective striatal degeneration and abnormal choreiform movements in primates. Proc Natl Acad Sci U S A 92:7105–7109

133. Greene JG, Porter RH, Eller RV, Greenamyre JT (1993) Inhibition of succinate dehydrogenase by malonic acid produces an "excitotoxic" lesion in rat striatum. J Neurochem 61:1151–1154

134. Dautry C, Conde F, Brouillet E, et al. (1999) Serial 1H-NMR spectroscopy study of metabolic impairment in primates chronically treated with the succinate dehydrogenase inhibitor 3-nitropropionic acid. Neurobiol Dis 6:259–268

135. Henry PG (2001) In vivo NMR measurement of TCA cycle rate alteration following

3NP intoxication. Proc Intl Soc Mag Reson Med 9

136. Jenkins BG, Brouillet E, Chen YC, et al. (1996) Non-invasive neurochemical analysis of focal excitotoxic lesions in models of neurodegenerative illness using spectroscopic imaging. J Cereb Blood Flow Metab 16:450–461
137. Lee WT, Shen YZ, Chang C (2000) Neuroprotective effect of lamotrigine and MK-801 on rat brain lesions induced by 3-nitropropionic acid: Evaluation by magnetic resonance imaging and in vivo proton magnetic resonance spectroscopy. Neuroscience 95:89–95
138. Lee WT, Lee CS, Pan YL, Chang C (2000) Temporal changes of cerebral metabolites and striatal lesions in acute 3-nitropropionic acid intoxication in the rat. Magn Reson Med 44:29–34
139. Lee WT, Chang C (2004) Magnetic resonance imaging and spectroscopy in assessing 3-nitropropionic acid-induced brain lesions: An animal model of Huntington's disease. Prog Neurobiol 72:87–110
140. Beal MF, Henshaw DR, Jenkins BG, Rosen BR, Schulz JB (1994) Coenzyme Q10 and nicotinamide block striatal lesions produced by the mitochondrial toxin malonate. Ann Neurol 36:882–888
141. Henshaw R, Jenkins BG, Schulz JB, et al. (1994) Malonate produces striatal lesions by indirect NMDA receptor activation. Brain Res 647:161–166
142. Matthews RT, Yang L, Jenkins BG, et al. (1998) Neuroprotective effects of creatine and cyclocreatine in animal models of Huntington's disease. J Neurosci 18: 156–163
143. Chou SY, Lee YC, Chen HM et al. (2005) CGS21680 attenuates symptoms of Huntington's disease in a transgenic mouse model. J Neurochem 93:310–320
144. Ferrante RJ, Andreassen OA, Jenkins BG, et al. (2000) Neuroprotective effects of creatine in a transgenic mouse model of Huntington's disease. J Neurosci 20:4389–4397
145. Ferrante RJ, Andreassen OA, Dedeoglu A, et al. (2002) Therapeutic effects of coenzyme Q10 and remacemide in transgenic mouse models of Huntington's disease. J Neurosci 22(5):1592–1599
146. Jenkins BG, Klivenyi P, Kustermann E, et al. (2000) Nonlinear decrease over time in N-acetyl aspartate levels in the absence of neuronal loss and increases in glutamine and glucose in transgenic Huntington's disease mice. J Neurochem 74(5):2108–2119
147. Jenkins BG, Andreassen OA, Dedeoglu A, et al. (2005) Effects of CAG repeat length, HTT protein length and protein context on cerebral metabolism measured using magnetic resonance spectroscopy in transgenic mouse models of Huntington's disease. J Neurochem 95(2):553–562
148. Tkac I, Dubinsky JM, Keene CD, Gruetter R, Low WC (2007) Neurochemical changes in Huntington R6/2 mouse striatum detected by in vivo 1H NMR spectroscopy. J Neurochem 100(5):1397–1406
149. McGowan E, Eriksen J, Hutton M (2006) A decade of modeling Alzheimer's disease in transgenic mice. Trends Genet 22(5):281–289
150. Dedeoglu A, Choi JK, Cormier K, Kowall NW, Jenkins BG (2004) Magnetic resonance spectroscopic analysis of Alzheimer's disease mouse brain that express mutant human APP shows altered neurochemical profile. Brain Res 1012(1–2):60–65
151. Marjanska M, Curran GL, Wengenack TM, et al. (2005) Monitoring disease progression in transgenic mouse models of Alzheimer's disease with proton magnetic resonance spectroscopy. Proc Natl Acad Sci U S A 102(33):11906–11910
152. von Kienlin M, Kunnecke B, Metzger F, et al. (2005) Altered metabolic profile in the frontal cortex of PS2APP transgenic mice, monitored throughout their life span. Neurobiol Dis 18:32–39
153. Marsden CD (1994) Parkinson's disease. J Neurol Neurosurg Psychiatr 57:672–681
154. Tipton KF, Singer TP (1993) Advances in our understanding of the mechanisms of the neurotoxicity of MPTP and related compounds. J Neurochem 61:1191–1206
155. Hadjiconstantinou M, Cavalla D, Anthoupoulou E, Laird HE, Neff NH (1985) N-Methyl-4-phenyl-1,2,3,6-tetrahydropyridine increases acetylcholine and decreases dopamine in mouse striatum: Both responses are blocked by anticholinergic drugs. J Neurochem 45:1957–1959
156. Heikkila RE, Hess A, Duvoisin RC (1984) Dopaminergic neurotoxicity of 1-methyl-4-phenyl-1,2,5,6-tetrahydropyridine in mice. Science 224:1451–1453
157. Schneider JS, Markham CH (1986) Neurotoxic effects of N-methyl-4-phenyl-1,2,3,6-tetrahydropyridine (MPTP) in the cat. Tyrosine hydroxylase immunohistochemistry. Brain Res 373:258–267
158. Burns RS, Chiueh CC, Markey SP, Ebert MH, Jacobowitz DM, Kopin IJ (1983) A primate model of parkinsonism: Selective destruction of dopaminergic neurons in the pars compacta of the substantia nigra by N-methyl-4-phenyl-1,2,3,6-tetrahydropyridine. Proc Natl Acad Sci U S A 80:4546–4550

159. Betarbet R, Sherer TB, Di Monte DA, Greenamyre JT (2002) Mechanistic approaches to Parkinson's disease pathogenesis. Brain Pathol 12:499–510
160. Boska MD, Lewis TB, Destache CJ, et al. (2005) Quantitative 1H magnetic resonance spectroscopic imaging determines therapeutic immunization efficacy in an animal model of Parkinson's disease. J Neurosci 25(7): 1691–1700
161. Brownell AL, Jenkins BG, Elmaleh DR, Deacon TW, Spealman RD, Isacson O (1998) Combined PET/MRS brain studies show dynamic and long-term physiological changes in a primate model of Parkinson disease. Nat Med 4(11):1308–1312
162. Chassain C, Bielicki G, Durand E, et al. (2008) Metabolic changes detected by proton magnetic resonance spectroscopy in vivo and in vitro in a murin model of Parkinson's disease, the MPTP-intoxicated mouse. J Neurochem 105(3):874–882
163. Koga K, Mori A, Ohashi S, et al. (2006) H MRS identifies lactate rise in the striatum of MPTP-treated C57BL/6 mice. Eur J Neurosci 23(4):1077–1081
164. Podell M, Hadjiconstantinou M, Smith MA, Neff NH (2003) Proton magnetic resonance imaging and spectroscopy identify metabolic changes in the striatum in the MPTP feline model of parkinsonism. Exp Neurol 179(2):159–166
165. Choi JK, Dedeoglu A, Jenkins BG (2007) Application of MRS to mouse models of neurodegenerative illness. NMR Biomed 20(3):216–237
166. Groves PM, Linder JC (1994) Young SJ. 5-hydroxydopamine-labeled dopaminergic axons: three-dimensional reconstructions of axons, synapses and postsynaptic targets in rat neostriatum. Neuroscience 58(3):593–604
167. Storey E, Hyman BT, Jenkins B, et al. (1992) 1-Methyl-4-phenylpyridinium produces excitotoxic lesions in rat striatum as a result of impairment of oxidative metabolism. J Neurochem 58(5):1975–1978
168. Chassain C, Bielicki G, Donnat JP, Renou JP, Eschalier A, Durif F (2005) Cerebral glutamate metabolism in Parkinson's disease: An in vivo dynamic (13)C NMS study in the rat. Exp Neurol 191:276–284
169. Moussa CE, Rusnak M, Hailu A, Sidhu A, Fricke ST (2008) Alterations of striatal glutamate transmission in rotenone-treated mice: MRI/MRS in vivo studies. Exp Neurol 209:224–233
170. Kooy RF, Verhoye M, Lemmon V, Van der LA (2001) Brain studies of mouse models for neurogenetic disorders using in vivo magnetic resonance imaging (MRI). Eur J Hum Genet 9:153–159
171. Redwine JM, Kosofsky B, Jacobs RE, et al. (2003) Dentate gyrus volume is reduced before onset of plaque formation in PDAPP mice: A magnetic resonance microscopy and stereologic analysis. Proc Natl Acad Sci U S A 100:1381–1386
172. Van Broeck B., Vanhoutte G., Pirici D, et al. (2008) Intraneuronal amyloid beta and reduced brain volume in a novel APP T714I mouse model for Alzheimer's disease. Neurobiol Aging 29:241–252
173. Fransen E, D'Hooge R, Van CG, et al. (1998) L1 knockout mice show dilated ventricles, vermis hypoplasia and impaired exploration patterns. Hum Mol Genet 7:999–1009
174. Ferrarini L, Palm WM, Olofsen H, et al. (2008) Ventricular shape biomarkers for Alzheimer's disease in clinical MR images. Magn Reson Med 59:260–267
175. Angenstein F, Niessen HG, Goldschmidt J, Vielhaber S, Ludolph AC, Scheich H (2004) Age-dependent changes in MRI of motor brain stem nuclei in a mouse model of ALS. Neuroreport 15:2271–2274
176. Niessen HG, Angenstein F, Sander K, et al. (2006) In vivo quantification of spinal and bulbar motor neuron degeneration in the G93A-SOD1 transgenic mouse model of ALS by T2 relaxation time and apparent diffusion coefficient. Exp Neurol 201:293–300
177. Revuelta GJ, Rosso A, Lippa CF (2008) Neuritic pathology as a correlate of synaptic loss in dementia with lewy bodies. Am J Alzheimers Dis Other Demen 23:97–102
178. Sorrentino G, Bonavita V (2007) Neurodegeneration and Alzheimer's disease: The lesson from tauopathies. Neurol Sci 28:63–71
179. Ferrer I, Puig B, Blanco R, Marti E (2000) Prion protein deposition and abnormal synaptic protein expression in the cerebellum in Creutzfeldt-Jakob disease. Neuroscience 97: 715–726
180. Rapoport SI (1999) In vivo PET imaging and postmortem studies suggest potentially reversible and irreversible stages of brain metabolic failure in Alzheimer's disease. Eur Arch Psychiatry Clin Neurosci 249:46–55
181. Rapoport SI (2005) In vivo imaging for evaluating synaptic integrity in Alzheimer disease. Psychol Neuropsychiatr Vieil 3:97–106

Part IV

Animal Models of Alzheimer's Disease

Chapter 12

Drosophila Melanogaster as a Model Organism for Dementia

Maria E. Giannakou and Damian C. Crowther

Abstract

In the quest for understanding human neurodegenerative disorders, a variety of organisms have been used to create disease models. Because of its many advantages, *Drosophila melanogaster* is currently being used to model many human conditions including poly Q expansion disorders such as Huntington's disease, Parkinson's disease, Alzheimer's disease (AD), and other dementias. AD is characterized by two pathologies; the first, extracellular amyloid β (Aβ) plaques, consist mainly of toxic $A\beta_{42}$ peptide, and the second, intracellular neurofibrillary tangles, consists mainly of hyperphosphorylated *tau* protein. *Drosophila* AD models have focused either on replicating the amyloid precursor protein (APP) processing model (Aβ is a proteolytic product of APP) or on expression of simpler secreted Aβ peptides in the fly nervous system. These models replicate many of the features of human AD, including Aβ deposition, neuronal loss and neurodegeneration, and behavioral phenotypes such as impaired learning and memory. In recent years, following the characterization of these models, focus has shifted to utilizing the genetic power of *Drosophila* and screens are being conducted for identifying modifiers of AD pathology.

Key words: Alzheimer's disease, *Drosophila melanogaster*, fruit fly, dementia, survival, locomotor, genetic screens, modifiers

1. Introduction

1.1. Why Use the Fruit Fly to Model Neurodegenerative Diseases?

The devastating symptoms of neurodegenerative diseases, such as movement disorders, ataxia, tremor, and cognitive and memory loss, are caused by neuronal dysfunction and loss. Many of these diseases are caused by aberrant processing and accumulation of proteins in the brain. To model these diseases in the fruit fly *Drosophila melanogaster*, we require some pre-existing knowledge about the pathology of the disease. In many cases, this understanding has come from studying families in which a normally sporadic disease is inherited as a dominant trait (1). The findings from studying such families have uncovered pathological events

Peter Paul De Deyn and Debby Van Dam (eds.), *Animal Models of Dementia*, Neuromethods, vol. 48, DOI 10.1007/978-1-60761-898-0_12,

that are sufficient to cause the disease. In Alzheimer's disease (AD) for instance, where only 10% of cases are familial, all the mutations uncovered (2) involve the proteolytic processing of the amyloid precursor protein (APP) and the generation of the amyloid-β peptide (Aβ). Since the symptoms of patients with familial and sporadic AD are very similar (3), it makes sense to use the insights from studying familial AD mutations in order to recapitulate AD in model organisms.

For an organism to be a faithful model of disease, it is necessary that the basic aspects of cell biology in this organism are conserved in higher-order metazoans (i.e. humans). Indeed, it has been shown that 75% of human disease-linked loci have homologues in *Drosophila* (4). It is also essential that the basic developmental programs are conserved between fruit flies and humans. In fact, many signaling pathways conserved among metazoans such as the *wnt* pathway were first identified and characterized in *Drosophila* (5), and subsequently shown to be crucial in humans as well (6). The *Drosophila* brain, despite being small, is amenable to study, as it comprises more than 300,000 neurons and is divided into compartments with specialized functions such as learning, memory, olfaction, and vision. In addition, fruit fly researchers have made available a plethora of genetic tools, including large collections of mutants (7) and transposon-based methods for generation of transgenics, which when coupled with a well-annotated genome and limited genetic redundancy allow for large genetic screens to identify genetic interactors (8). Finally, compared to mice, *Drosophila* are small, fecund, cheap-to-maintain organisms, with a short generation time, which allows us to quickly generate results that can then be exploited for experimental design in higher organisms.

1.2. Fly Models of Neurodegenerative Diseases

Drosophila can be used to model human diseases in many different ways. In some cases, researchers have investigated the *Drosophila* homologue of a human disease, in order to examine the normal function of the disease-linked protein. This approach has been used in examining, for instance, the role of the *Drosophila* homologues of the AD-associated genes, APP (*APPL* in the fly) (9,10) and presenilins 1 and 2 (PS1 and PS2, DPsn in the fly) (11), uncovering roles for APPL and DPsn in axonal transport and notch signaling, respectively. Alternatively, since most human neurodegenerative diseases are caused by autosomal dominant mutations or the abnormal accumulation of a protein, a model system may be generated by overexpressing the human disease gene (or a disease-causing mutant) or its *Drosophila* homologue. Such an approach has been used in modeling diseases caused by *tau* phosphorylation (12), α-synuclein (13), and various polyQ-containing proteins (14,15) that are involved in tauopathies, Parkinson's disease (PD), and polyQ-diseases such as Huntington's disease (HD), respectively. Neurodegenerative diseases can also be studied in fruit flies by unbiased, forward genetics approaches, where flies are mutagenized and then screened for certain phenotypes such as behavioral defects or decreased longevity. Such screens have identified interesting mutants such as *bubblegum* (16), *eggroll*, and *spongecake* (17), and have identified genes and processes potentially involved in human diseases with similar pathology (for review see (18)).

The first fly model of neurodegenerative disease was a model of spinocerebellar ataxia type 3 (SCA3, one of the coding and noncoding CAG expansion (polyQ) diseases) (14). Various faithful models for many polyQ diseases have since been generated by expression of human polyQ-containing proteins, with varying numbers of CAG repeats. These models for both ataxins (causing different SCAs) and for Huntingtin have shown progressive neurodegeneration and phenotype severity that correlates with the number of CAG repeats ((14,15), for review see (19,20)). The fly models for HD and the various ataxias have been utilized for investigating disease modifiers both by searching for modifiers using genetic screens (21,22), and by candidate approaches (23). Such studies have highlighted the importance of chaperones (22), protein degradation pathways (21), and signaling pathways (24) in polyQ disease process.

Various models have also been developed to mimic Parkinson's disease and related dementias by expressing α-synuclein in the fly (13). A-synuclein is the protein that is found in the Lewy body (LB) inclusions in patients with Parkinson's disease and Lewy body dementia (25). Other studies have focused on *Drosophila* mutants for parkin, a ubiquitin ligase linked to autosomal recessive juvenile onset parkinsonism (AR-JP) (26–28). The α-synuclein fly model has been shown to recapitulate features of PD including loss of dopaminergic neurons, neuronal LB-like accumulations, and locomotor deficits (13), and a role for chaperones has been uncovered in this model (29). However, there are discrepancies in the literature about the faithfulness of the α-synuclein model as various groups have failed to find any locomotor defects in these flies and also the histopathological features of this model have not proved reproducible between labs (30).

Tauopathies, characterized by the presence of neurofibrillary tangles (NFTs) (for review see (31)), have also been modeled in *Drosophila* (12). Expression of wild-type or a mutant *tau* associated with a familial tauopathy leads to adult onset, progressive neurodegeneration, accompanied by reduced longevity, *tau* accumulation, and increased toxicity of mutant *tau* (12). However, this model is characterized by lack of the NFTs present in human patients (12). Further studies have used this fly model to investigate the effects of kinases, such as the *wingless* pathway kinase glycogen synthase kinase (GSK)-3β (in flies, *shaggy*) (32,33), and also in screens looking for modifiers of *tau* pathology (34). Interestingly, fibrillar *tau* pathology was observed when wild-type human *tau* was co-expressed with *shaggy*, the main kinase responsible for *tau* phosphorylation in humans (32). Other studies have utilized the *tau* model to investigate potential roles of *tau*, thereby showing a link between human *tau* overexpression and defects in synaptic transmission in larval neurons (35).

AD is one of the more recent diseases to be modeled in the fruit fly. The rest of this chapter will focus on different *Drosophila* AD models developed in the last 5 years, how these models have been useful in identifying modifiers of AD, and how they are likely to be used in the future.

2. Making a Model of Alzheimer's Disease

2.1. Components of AD Pathology

AD was first described by Alois Alzheimer in 1906 as a rare form of presenile dementia; however, it is in fact the most common form of dementia and shows a dramatic increase in incidence with increasing age. It is thought that up to 20 million people worldwide suffer from AD, and the number is projected to double in the next 30 years. The main diagnostic lesions in AD patients are extracellular amyloid plaques and intracellular NFTs. The main component of the senile plaques is a peptide of about 4 kDa, called Aβ peptide, first identified in the plaques of brains of Down's syndrome patients and subsequently shown to be found in AD brains (36). The Aβ peptide has been shown to be a proteolytic product of the amyloid precursor protein, APP (40). The role of both *tau* and APP have been investigated in *Drosophila* with evidence for an axonal transport role for *tau* and a possible cell–cell adhesion or axonal transport role for APP (10,33,41,42).

The APP holoprotein is a transmembrane protein that becomes metabolized into a variety of peptides, by either the amyloidogenic or the nonamyloidogenic pathway. The catabolism of APP is shown in Fig. 1. In the amyloidogenic pathway, APP is sequentially cleaved by the β- and γ-secretases. The β-secretase activity has been identified to be the beta-site APP cleaving enzyme (BACE) (43,44), whereas the γ-secretase activity appears to arise from a complex of at least four proteins including presenilins 1 and 2, and nicastrin (for review see (45)). The precise subcellular location of proteolytic events is still unclear: cleavage by α-secretase has been shown to occur at the plasma membrane (46), but the β- and γ-secretases are thought to generate Aβ after re-internalization of APP, along the endosomal/lysosomal pathway (47,48). The subcellular location of the β- and γ-cleavage of APP may be significant in disease pathogenesis, as the low pH of the endosomal pathway has been shown to increase Aβ aggregation (49).

2.2. Aβ Is Primary for Neurotoxicity

Various lines of evidence suggest that $A\beta_{42}$ generation is critical to AD toxicity. First, all FAD mutations are localized in either APP or the presenilin genes (3). Mutations in APP cause FAD (50),

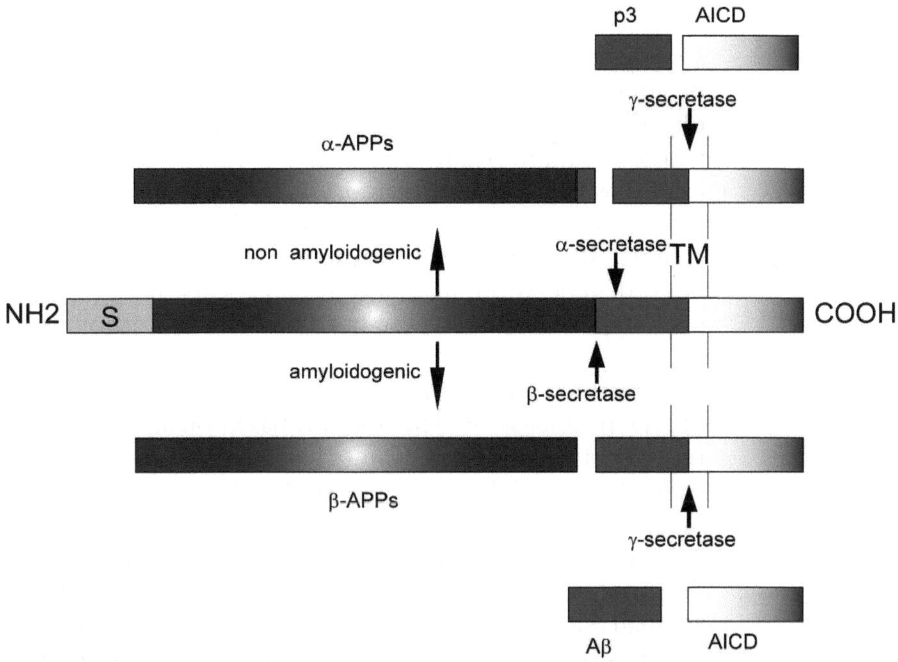

Fig. 1. Schematic diagram of β-amyloid precursor protein (APP) and its metabolic derivatives. The APP holoprotein is shown in the middle of the diagram, the signal peptide is marked with S, and the Aβ sequence is shown in dark gray. The Aβ sequence resides partially in the plasma membrane (marked TM). Cleavage of APP by the secretases follows either a nonamyloidogenic pathway or an amyloidogenic pathway (generation of Aβ peptides). Cleavage of APP by α-secretase leads to production of the α-APP soluble fragment and C83, which is further cleaved by γ-secretase to produce p3 and the APP intracellular domain (AICD). Cleavage of APP via β-secretase leads to the production of β-APP soluble fragment and C99, which is further cleaved by γ-secretase to produce Aβ and AICD. Depending on the precise site of cleavage by γ-secretase, the size of Aβ, Aβ peptides of varying lengths are produced, namely 40 or 42 residues long.

either by increasing the aggregation properties of the generated $A\beta_{42}$ peptide (51), or by increasing the ratio of $A\beta_{42}$ to $A\beta_{40}$ (52). In patients with Down's syndrome (trisomy 21), FAD is thought to be the result of an extra copy of wild-type APP (APP is on chromosome 21) (53), resulting in increased AD prevalence and reduced age of onset (54). FAD mutations in the components of γ-secretase, presenilins 1 and 2, affect processing of APP and these mutations lead to an increase in $A\beta_{42}$ production or to an increased ratio of $A\beta_{42}$ peptide compared to the less aggregatory $A\beta_{40}$ peptide (55,56). Mutations in the microtubule-associated *tau* protein have not been shown to cause FAD; however, *tau* variants are sufficient to cause related neurodegenerative disorders called familial tauopathies (57) that often manifest clinically as frontotemporal dementia.

Despite $A\beta_{42}$ being the main component of Aβ plaques (58), the number of plaques in a postmortem brain does not correlate well with the severity of AD symptoms in life (60). However, an important step forward in our understanding of the role of Aβ came with the observation that total Aβ levels, which includes soluble forms of the peptide, were found to correlate with disease progression (61). Other evidence points to a role for soluble $A\beta_{42}$ oligomers in causing toxicity *in vivo* (66,67), backed by studies where immunization with Aβ oligomer-specific antibodies improves learning and memory in AD transgenic mice (68,69). Taken together, the studies mentioned above point to a role for soluble Aβ aggregates in causing toxicity and being critical for neurodegeneration; furthermore, it appears that Aβ generation is probably sufficient to cause both *tau* and Aβ pathology.

2.3. Drosophila Models of AD

Two related strategies have been employed in generating models of AD in *Drosophila*. The first was based on an attempt to closely replicate the APP processing that is found in human patients. Indeed, *Drosophila* have much of the molecular machinery that is involved in human APP processing, specifically an orthologue of APP, called APPL. Interestingly, the homology to human APP does not stretch to the region that generates Aβ peptides (70). Human APP can rescue APPL mutants (9), and APP overexpression has subsequently uncovered endogenous γ-secretase activity in *Drosophila* (42). When human β-secretase was cloned (43), the path opened for a complete APP processing *Drosophila* model. In this model, expression of APP and BACE in the fly eye led to deposition of Aβ plaques and age-dependent neurodegeneration (71). Ubiquitous APP and BACE expression also led to shortened life span and defects in wing vein development (71).

We and others have focused on simpler models, directly expressing various Aβ peptides as secreted peptides in the fly nervous system or in the developing eye (72–74). These models have successfully shown progressive intracellular Aβ accumulation, extracellular Aβ plaque deposition, and neurodegeneration accompanied by olfactory memory defects, reduced longevity, and defective locomotor behavior (72–74). In addition, these models have demonstrated that the phenotypes caused by Aβ production were age- and dosage-dependent, with correlation both between Aβ levels and neurodegeneration (72–74), and also between propensity of Aβ peptide to aggregate into oligomers and the severity of the phenotypes (72,74). Expression of the less aggregatory $A\beta_{40}$ peptide does not cause neurodegeneration apart from an age-dependent defect in olfactory learning and memory (72). However, a mutation in the $A\beta_{42}$ peptide (E22G substitution, arctic mutation (51)), which increases its aggregation propensity, leads to more severe, earlier onset of neurodegeneration, locomotor defects, and a markedly reduced survival in comparison with the wild-type $A\beta_{42\text{-arctic}}$ peptide (74).

Both types of models have successfully recapitulated many aspects of human AD and each has benefits for studying different aspects of AD neurotoxicity. The APP-based models are good for screening secretase inhibitors (71), and potential modifiers of the effects of secretases on Aβ_{42} production. The simpler secreted Aβ models are useful for looking at the toxicity of different Aβ peptide variants (75), and for modifiers that affect Aβ metabolism and toxicity. While there appears to be good concordance between the types of phenotypes observed in these two complementary approaches to modeling Aβ toxicity in the fly, there may be some differences in the subcellular localization of the peptide. Specifically, in the normal processing of APP, it is thought that Aβ is generated in an endosomal compartment and may subsequently be released to the extracellular space. In our model system, the peptide enters the secretory pathway from the endoplasmic reticulum (47,48).

3. Characterizing Drosophila AD Model Phenotypes

Some of the phenotypes observed in the *Drosophila* AD models are analogous to those observed clinically in human patients, such as learning and memory defects, and histopathological phenotypes, such as Aβ plaques and neuronal loss. However, other, easy-to-score phenotypes in the flies are used as surrogate markers for neurodegeneration, such as reduced longevity, locomotor defects, and rough-eye phenotypes.

3.1. Histology

The two main pathologies observed in human AD patients are amyloid plaques and NFTs. In the *Drosophila* APP and Aβ models, Aβ accumulates intracellularly in the nervous system of the flies, in a manner similar to Aβ accumulation in human AD patient brains. Flies expressing either Aβ peptides or APP/BACE under control of either the neuronal *elav*^{c155}-Gal4 or the eye-specific *gmr*-Gal4 driver are maintained at either normal fly culture temperature (25°C), higher (29°C), or lower (18°C) temperature, in order to accentuate or slow phenotype progression. In order to analyze the progressive Aβ accumulation and neurodegeneration, flies are decapitated, fixed, wax-embedded, and sectioned at different ages and disease stages. For most applications, wax embedding and sectioning are sufficient to obtain material for immunohistochemical analysis (Fig. 2).

Aβ peptides accumulate in an age-dependent manner in the flies' brains, especially in areas of the brain containing neuronal cell bodies (74). Some deposits are found in aging fly brains, which stain for Aβ; however, they appear not to be mature amyloid deposits as they do not stain with Congo red (74). In other studies, thioflavin S-positive deposits have been identified, and these are comparative to diffuse plaques (71,72). The presence of diffuse nonfibrillar Aβ deposits is supported by positive staining using an oligomer-specific antibody (74,76). Moreover, the Aβ deposition in the flies' brains

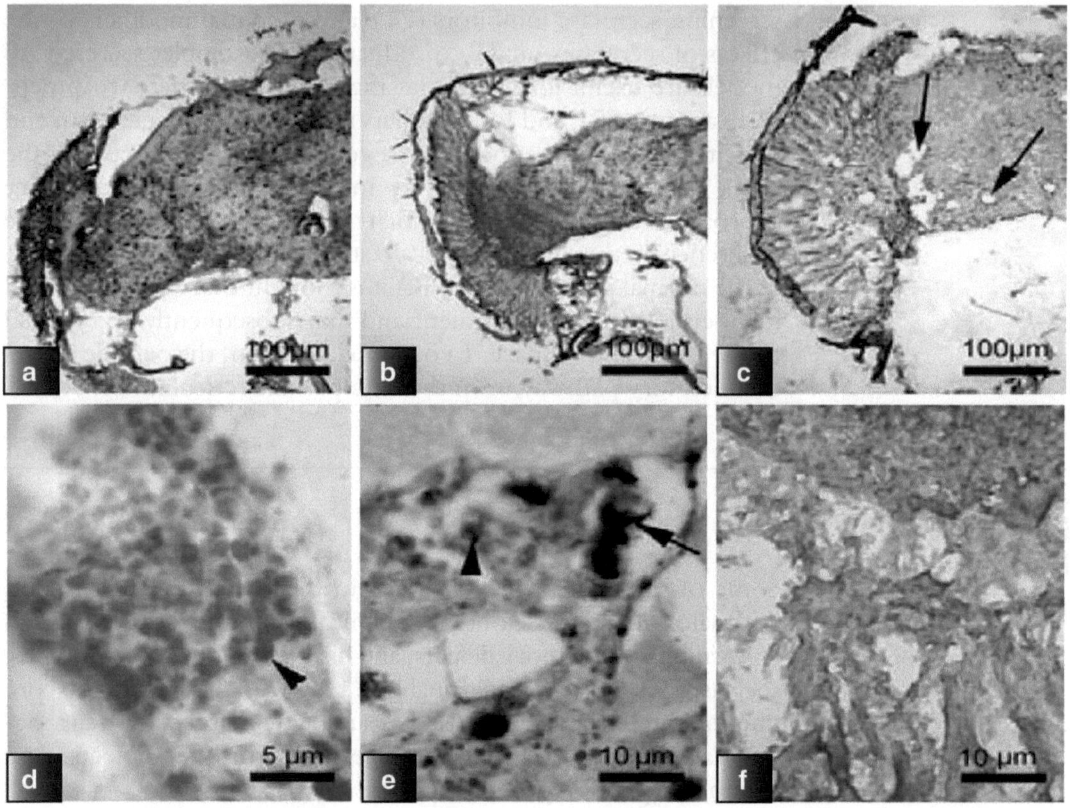

Fig. 2. Progressive Aβ accumulation, aggregate deposition, and neurodegeneration in the brain and retina of flies expressing $A\beta_{42\text{-arctic}}$ under control of Gal4-elav^{c155}. (**a–c**) Scanning electron microscopy was used to examine the appearance of the *Drosophila* eyes (scale bar =100 µm). (**d–f**) Immunohistochemistry with the 4G8 monoclonal antibody of the retina and brain of flies. On day 0 (**a**, **c**), there was widespread, intense intracellular accumulation (*arrowhead*). Intracellular staining was still present at day 5 (**b**, **e**); however, extracellular deposits (*arrow*) were now visible. By day 9, there was widespread vacuolation (*arrows*) with concomitant reduction in staining intensity (**c**, **f**) (Reproduced from (74)).

is also accompanied by age-dependent neuronal loss (72,74), demonstrating that amyloid plaques are not required for Aβ to cause neuronal toxicity and neurodegeneration.

3.2. Pavlovian Olfactory Learning Assays

The Pavlovian olfactory associative learning abilities of flies have been used to train them to associate an electric shock with a particular odor. Iijima and colleagues used olfactory learning assays to investigate learning and memory in flies expressing secreted Aβ peptides. The loss of this associative learning capacity, or the acceleration of forgetting, may be considered a good analogue of the memory impairment seen in AD patients. They first trained flies by exposure to an electric shock paired with one odor (octanol or methylcyclohexanol) for 60 s and subsequent exposure to a second odor without electroshock for 60 s (72). Immediately after training, learning was measured by allowing flies to choose between the two odors for 120 s. The performance index was calculated by subtracting the number of flies making the incorrect choice from those making the correct one, dividing by the total number of flies, and multiplying by 100. Iijima and colleagues have shown that Pavlovian olfactory associative learning is impaired in flies expressing Aβ peptides that are older than 6 days. Moreover, the degree of impairment is greater in flies expressing $A\beta_{42}$ than $A\beta_{40}$ (72).

3.3. Longevity

It is widely accepted that the insulin/insulin-like growth factor (IGF)-like signaling pathway regulates life span across metazoans and much of this evidence has come from studying longevity in invertebrates, such as *Drosophila melanogaster* (77). However, the precise cause of fly death in general, and in *Drosophila* AD models in particular, is still unclear, although it is thought that locomotor defects are associated with concomitant feeding deficits. There is a strong correlation between the locomotor ability of the fly and its life span, suggesting that longevity can be used as a surrogate marker for AD as it provides a measure of the health of a group of flies. A main advantage of a longevity assay as a marker is its clear endpoint, as a fly at any time point is either alive or dead. Kaplan–Meyer statistics are used to analyze survival assays of flies and can also take into account censored flies due to escape.

As with many disease-related phenotypes, longevity can be measured at temperatures between 18°C and 29°C, with shorter life spans at higher temperatures. All the groups of flies to be compared in a longevity assay are reared under identical conditions and are placed in groups of ten flies/vial, to a total number of at least 100 flies/genotype. Deaths are scored and flies are transferred to new food medium three times a week. In such assays, flies expressing the nontoxic $A\beta_{40}$ peptide are used as a control for neuronal peptide expression, and their life span is indistinguishable from that of *w1118* control flies (Fig. 3 (72,74)). Flies expressing the toxic $A\beta_{42}$ peptide have reduced life span compared to *elav*^{c155}-Gal4/$A\beta_{40}$ flies and the effect of $A\beta_{42}$ expression is dose-dependent, as two copies of the $A\beta_{42}$ transgene result in a further reduction of life span (Fig. 4 (74)). Expression of the $A\beta_{42\text{-arctic}}$ mutant peptide results in a marked reduction in survival compared to expression of wild-type $A\beta_{42}$ peptide (Fig. 3 (74)).

3.4. Locomotor Ability

In addition to reduced survival, flies expressing $A\beta_{42}$ peptides show an age-related decline in locomotor function (72,74), which can also be used as a surrogate marker of AD-related neuronal dysfunction. Climbing assays, such as negative geotaxis assays, have been performed on flies at either 25°C or 29°C, and have shown a progressive decline in climbing ability in *elav*^{c155}-Gal4/$A\beta_{42}$ flies age compared to *elav*^{c155}-Gal4/$A\beta_{40}$ flies and control flies (72,74). *elav*^{c155}-Gal4/$A\beta_{42\text{-arctic}}$ expressing flies show an accelerated decline in climbing compared to wild-type $A\beta_{42}$ flies (74).

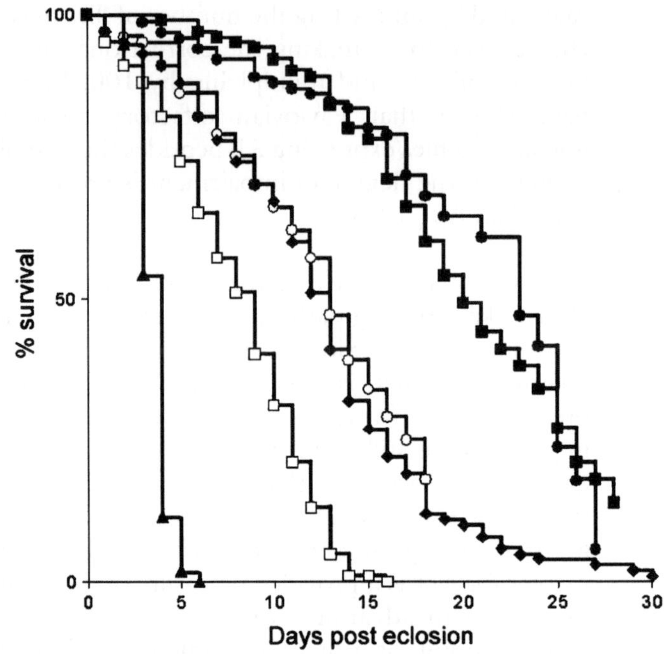

Fig. 3. Reduced survival of flies expressing Aβ peptides. The longevity of control *w1118* flies (*filled squares*) was compared with flies expressing one (*diamonds and empty circles*, two independent insertions) or two copies (*empty squares*) of the $A\beta_{42}$ transgene, one copy (*triangles*) of the $A\beta_{42}$-arctic or one copy (*filled circles*) of $A\beta_{40}$ under the control of the *Gal4-elav*^{c155} driver (Reproduced from (74)).

Negative geotaxis assays were conducted as described in Rival and colleagues (78) as a measure of fly locomotor performance. For assaying climbing abilities of AD flies at different ages, three sets of 15 flies per genotype were placed in 25-mL plastic pipettes and left to acclimatize for 20 min before starting the assays. For each experiment, flies were tapped down to the bottom of the pipette and allowed to climb for 1 min. The numbers of flies at the top and at the bottom of the pipette were recorded and expressed as a percentage of the total. The assays were repeated three times for each pipette generating nine data sets from 15 flies per genotype. The mean number of flies at the top (m_{top}) and at the bottom (m_{bot}), expressed as percentages of the total number of flies (n_T), were recorded as graphs with error bars indicating the mean ± standard deviation of the scores obtained in the three independent experiments. A climbing performance index was calculated as defined:

$$\text{Performance Index} = 1\ /12\ -2\frac{\left(n_T + m_{top} - m_{bot}\right)}{n_T}$$

Student's *t*-tests of the performance index were performed to test for significant differences in climbing performance between two genotypes.

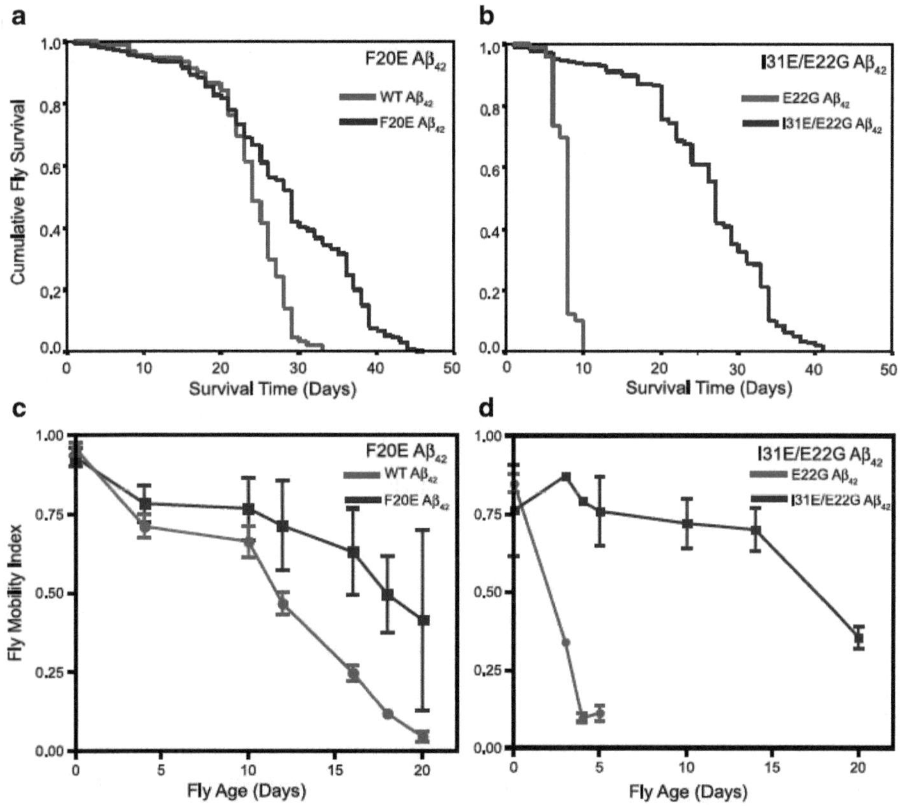

Fig. 4. Survival and locomotor analysis of various $A\beta_{42}$ mutants including rational mutations based on computational predictions. (**a**) Flies expressing F20E $A\beta_{42}$ live significantly longer (median survival 29 ± 1 d, n = 400, $p < 0.0001$) than flies expressing WT $A\beta_{42}$ (median survival 24 ± 1 d, n = 100). (**b**) Flies expressing I31E/E22G $A\beta_{42}$ show a dramatic increase in longevity (median survival = 27 ± 1 d, n = 600, $p < 0.0001$) compared to flies expressing E22G $A\beta_{42}$ (median survival = 8 ± 1 d, n = 100). (**c**) Flies expressing the F20E $A\beta_{42}$ peptide have significantly improved locomotor ability ($p < 0.001$, $n = 90$ observations per line per time point) compared with flies expressing the WT $A\beta_{42}$ peptide. (**d**) Flies expressing the I31E/E22G $A\beta_{42}$ peptide have significantly improved locomotor ability ($p < 0.001$, $n = 90$ observations per line per time point) compared to flies expressing E22G Aβ42 (Reproduced from (75))

Locomotor dysfunction can also be assessed in the *Drosophila* larval stages, looking at the velocity of crawling and time taken for larvae to right themselves when turned over. These assays have been used primarily for determining the effects of *tau* expression in larval motor neurons (33).

3.5. Rough-Eye/Pseudo-Pupil Phenotype

In many of the *Drosophila* AD models, expression of APP/BACE and $A\beta_{42}$ peptides has been directed to the developing eye, using the *gmr*-Gal4 driver (71,73,74). Main advantages of studying the eye include the easy detection of adult eye phenotypes, and that the eye is tolerant to mutations that disrupt its basic biological processes (20). The eye is also dispensable for survival under laboratory conditions, making it an ideal organ for analyzing the effects of highly toxic transgenes. Expression of peptides that affect eye structure can be investigated using two assays: the qualitative rough-eye phenotype and the quantitative pseudo-pupil assay. Normally, the *Drosophila* compound eye consists of approximately 800 units known as ommatidia; expression of toxic transgenes can produce a series of phenotypes in the eye, ranging from slight misalignments of ommatidia to fusion of ommatidia, and in severe cases to almost a complete loss of the eye.

3.5.1. Qualitative Assessment of the Rough-Eye Phenotype

The assessment of rough-eye phenotypes is qualitative and so it is important for experiments to be carried out blind, by a single observer. The data are presented as the proportion of flies with a particular genotype that have a rough eye. When there is a range in the severity of the rough-eye phenotype, it is possible to grade the phenotype and so demonstrate shifts in the spectrum of severity. Moderate rough-eye phenotype is observed in flies expressing one copy of the $A\beta_{42}$ peptide, which is further enhanced by expression of a second $A\beta_{42}$ copy (73,74), or by expression of the more toxic $A\beta_{42\text{-arctic}}$ peptide (74). In experiments using the *gmr*-gal4 driver, it is imperative to also analyze the phenotype of control *gmr*-Gal4 driver flies, as it is known to cause a mild rough-eye phenotype on its own, especially at 29°C.

3.5.2. Counting Rhabdomeres

The pseudo-pupil assay (74,79) provides a sensitive and quantitative method for assessing neurodegeneration as it allows for observation of the progressive neuronal loss in the eye during adult life (15). Ommatidia in the fly compound eye are composed of a regular trapezoidal arrangement of seven photoreceptors, called rhabdomeres. The regular packing of the rhabdomeres may be disturbed, or there may be loss of rhabdomeres, when toxic transgenes are expressed in the eye. The rhabdomeres are observed using back-illumination of the eyes on a light microscope. The average number of rhabdomeres per ommatidium are then compared, across a time course for control and AD flies, and this allows for a quantitative demonstration of neuronal loss. It is also important for these assays to include appropriate control flies to control for mild toxic effects of overexpressing Gal4 drivers in the retina in the absence of transgenes.

4. Using Drosophila Models of Alzheimer's Disease

4.1. Study of Aβ Peptide Variant Toxicity Supports Current Theories of Aβ Pathogenesis

The secreted Aβ model has recently been used to investigate the *in vivo* properties of Aβ peptides designed by rational mutagenesis using a computational approach (75). Luheshi and colleagues used an algorithm to compute the intrinsic aggregation propensities (Z_{agg}) of all 798 possible single point mutants in both the wild-type $A\beta_{42}$ peptide and the more toxic $A\beta_{42\text{-arctic}}$ (E22G) peptide. They, subsequently, expressed 17 variants of these peptides

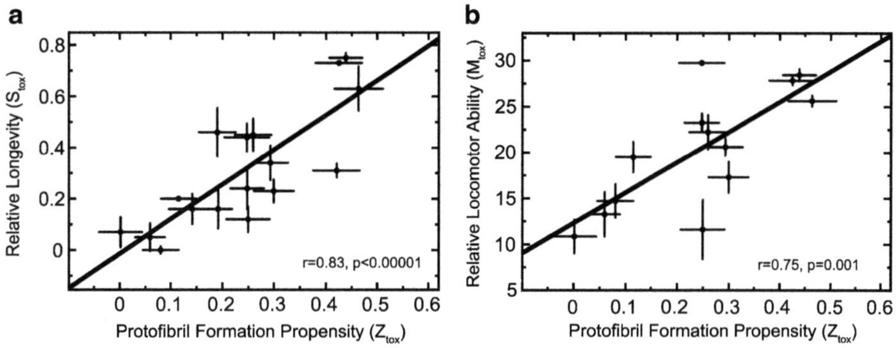

Fig. 5. Propensity of peptides to form protofibrillar aggregates (Z_{tox}) is a good predictor of the effects of Aβ_{42} in flies. (**a**) Z_{tox} predicts more accurately than Z_{agg} (aggregation propensity) the relative longevity (S_{tox}) of flies expressing different Aβ_{42} variants. (**b**) Z_{tox} predicts more accurately than Z_{agg} the relative locomotor ability (M_{tox}) of flies expressing different Aβ_{42} variants (Reproduced from (75)).

in the *Drosophila* nervous system (75). The longevity of four to six independent lines for each variant was compared to that of flies expressing wild-type Aβ_{42} and A$\beta_{42\text{-arctic}}$. Examples of locomotor ability assessment of these lines are shown in Fig. 4 (75). By investigating the *in vivo* phenotypes and *in vitro* characteristics of these peptides, and in particular those for which the predicted aggregation propensity and the *in vivo* toxicity did not correlate, a new algorithm was then developed, taking into account the protofibril formation propensity for each peptide. The toxicity of these peptides *in vivo*, as measured by analyzing the locomotor and life span phenotypes of flies expressing Aβ_{42} variants, showed a strong correlation with the propensity of the peptides to form protofibrils *in vitro* (Fig. 5 (75)). This study demonstrates that propensity of peptides to aggregate correlates with toxicity and therefore supports the current theories of Aβ_{42} pathogenesis.

4.2. Genetic Screens for Modifiers of Aβ Toxicity

One of the main advantages of *Drosophila* disease models is that they can be used in genetic screens for identifying disease process modifiers. In second-site modifier screens, a collection of lines (often P element lines) are combined with flies, which have an easily scored disease phenotype and are screened for their ability to enhance or suppress that phenotype. Phenotypes that have been used in *Drosophila* Aβ models genetic screens include the longevity and the rough-eye phenotype.

In our lab, a genetic modifier screen was conducted with longevity as the primary observation (Rival and Crowther et al., 2009 (77)) . We screened for modifiers from a library of 3,000 unique insertions of a second-generation EP element (the Gene Search, or GS, element; (80)). Using female flies expressing A$\beta_{42\text{-arctic}}$ in the brain, we were able to observe changes in life span caused by the EP element as compared to the control population of flies that did not possess the modifier. The median survival of the flies expressing A$\beta_{42\text{-arctic}}$ and each of the GS-element in the library was calculated, and the lines that had significantly prolonged survival were further investigated.

Another screen for modifiers of Aβ metabolism and toxicity was conducted using the rough-eye phenotype as a marker. Toxic Aβ_{42} expression was under control of the *gmr* promoter element and expression of a collection of 1963 EP strains was also directed to the eye using the *ey*-Gal4 driver (73,81). EP lines that modified the *gmr*-Aβ42 rough-eye phenotype were further characterized. This screen identified various pathways that affected Aβ metabolism and toxicity, such as the protease neprilysin (73), secretory pathway enzymes, and other pathways affecting cholesterol homeostasis and, unexpectedly, regulation of chromatin structure and function (81). Many of the genes identified to alter Aβ toxicity when overexpressed or disrupted did not affect total Aβ levels, but did affect the ratio of soluble/insoluble Aβ, suggesting that this effect is likely to be instrumental in determining the severity of Aβ toxicity (81). Cao and colleagues also investigated whether these modifiers had any effects on expression of *tau* and polyQ expanded huntingtin, and identified some overlap between these diverse neurodegenerative diseases (81).

4.3. Drug Screens Using Drosophila AD Models

Another potential use for a characterized *Drosophila* disease model is its use for novel drug screening. The secreted Aβ peptide fly model was verified as a platform for drug discovery by testing the efficacy of a drug that has been used in treating human AD patients and shown to slow progression of AD (74). The drug memantine, a noncompetitive glutamate antagonist, is effective in slowing progression of human AD (82). In addition, the life span of flies expressing two copies of Aβ_{42} or one copy of A$\beta_{42\text{-arctic}}$ was increased when flies were treated with MK-801, an inhibitor of the excitatory action of glutamate on the NMDA receptor (74). A therapeutic intervention that is effective in human AD patients is, therefore, also effective in the fly, making the fly AD model a good model for testing novel human drugs. Congo red, which binds to Aβ and has been shown to reduce neurodegeneration in a fly model of polyQ disease (83) and a mouse model of Huntington's disease (84), has also been shown to rescue the reduced life span of flies expressing two copies of Aβ_{42} or one copy of A$\beta_{42\text{-arctic}}$ (74).

The APP processing model in *Drosophila* has also been used in drug validation studies. Ubiquitous expression of APP, BACE, and DPsn resulted in reduced survival and a visible wing phenotype, which were used for screening β- and γ-secretase inhibitors (71). Feeding flies either β- or γ-secretase inhibitors resulted in increased survival of APP/ BACE/ DPSn expressing transgenic flies, making this fly model a good model for investigating drugs that modulate APP processing and have the potential to decrease Aβ-induced cellular degeneration (71).

5. Clinical Relevance

Various *Drosophila* models of AD have now been characterized and are currently being used in genetic screens for identifying modifiers of Aβ production, metabolism, and neurotoxicity. *Drosophila* is an ideal organism for such screens as it combines both faithful models of AD (which replicate various human AD pathologies), and also because of the tools available in this organism for such screens, such as extensive mutant collections. Various pathways are emerging from these screens, some predicted and some novel, that appear to alter Aβ metabolism and toxicity, illustrating the power of these models. Aβ modifiers identified in such a manner can be used for validation in murine models of AD and as potential therapeutic drug targets. The *Drosophila* models have also been validated as a platform for drug screening as studies have confirmed the efficacy of the classes of compounds used in treating AD patients in reducing neurodegeneration in fly Aβ models (71,74). Such screens would be much cheaper and faster to conduct in flies compared to mouse AD models, where AD-associated phenotypes take months to occur and are laborious to characterize. It is clear therefore that *Drosophila* AD models have an important role to play in further understanding AD and designing novel successful therapies.

References

1. Crowther DC (2002) Familial conformational diseases and dementias. Hum Mutat 20:1–14
2. Van Broeckhoven CL (1995) Molecular genetics of Alzheimer disease: Identification of genes and gene mutations. Eur Neurol 35:8–19
3. Selkoe DJ (2001) Alzheimer's disease: Genes, proteins, and therapy. Physiol Rev 81:741–766
4. Reiter LT, Potocki L, Chien S, Gribskov M, Bier E (2001) A systematic analysis of human disease-associated gene sequences in Drosophila melanogaster. Genome Res 11:1114–1125
5. Sharma RP, Chopra VL (1976) Effect of the Wingless (wgl) mutation on wing and haltere development in Drosophila melanogaster. Dev Biol 48:461–465
6. Siegfried E, Perrimon N (1994) Drosophila wingless: A paradigm for the function and mechanism of Wnt signaling. Bioessays 16:395–404
7. Bellen HJ, Levis RW, Liao G, et al. (2004) The BDGP gene disruption project: Single transposon insertions associated with 40% of Drosophila genes. Genetics 167:761–781
8. Bier E (2005) Drosophila, the golden bug, emerges as a tool for human genetics. Nat Rev Genet 6:9–23
9. Luo L, Tully T, White K (1992) Human amyloid precursor protein ameliorates behavioral deficit of flies deleted for Appl gene. Neuron 9:595–605
10. Torroja L, Chu H, Kotovsky I, White K (1999) Neuronal overexpression of APPL, the Drosophila homologue of the amyloid precursor protein (APP), disrupts axonal transport. Curr Biol 9:489–492
11. Ye Y, Lukinova N, Fortini ME (1999) Neurogenic phenotypes and altered Notch processing in Drosophila Presenilin mutants. Nature 398:525–529
12. Wittmann CW, Wszolek MF, Shulman JM, et al. (2001) Tauopathy in Drosophila: Neurodegeneration without neurofibrillary tangles. Science 293:711–714
13. Feany MB, Bender WW (2000) A Drosophila model of Parkinson's disease. Nature 404: 394–398
14. Warrick JM, Paulson HL, Gray-Board GL, et al. (1998) Expanded polyglutamine protein forms nuclear inclusions and causes neural degeneration in Drosophila. Cell 93:939–949
15. Jackson GR, Salecker I, Dong X, et al. (1998) Polyglutamine-expanded human huntingtin

transgenes induce degeneration of Drosophila photoreceptor neurons. Neuron 21:633–642
16. Min KT, Benzer S (1999) Preventing neurodegeneration in the Drosophila mutant bubblegum. Science 284:1985–1988
17. Min KT, Benzer S (1997) Spongecake and eggroll: Two hereditary diseases in Drosophila resemble patterns of human brain degeneration. Curr Biol 7:885–888
18. Cauchi RJ, van den Heuvel M (2006) The fly as a model for neurodegenerative diseases: is it worth the jump? Neurodegener Dis 3:338–356
19. Bilen J, Bonini NM (2005) Drosophila as a model for human neurodegenerative disease. Annu Rev Genet 39:153–171
20. Sang TK, Jackson GR (2005) Drosophila models of neurodegenerative disease. NeuroRx 2:438–446
21. Fernandez-Funez P, Nino-Rosales ML, de Gouyon B, et al. (2000) Identification of genes that modify ataxin-1-induced neurodegeneration. Nature 408:101–106
22. Kazemi-Esfarjani P, Benzer S (2000) Genetic suppression of polyglutamine toxicity in Drosophila. Science 287:1837–1840
23. Warrick JM, Chan HY, Gray-Board GL, Chai Y, Paulson HL, Bonini NM (1999) Suppression of polyglutamine-mediated neurodegeneration in Drosophila by the molecular chaperone HSP70. Nat Genet 23:425–428
24. Ravikumar B, Vacher C, Berger Z, et al. (2004) Inhibition of mTOR induces autophagy and reduces toxicity of polyglutamine expansions in fly and mouse models of Huntington disease. Nat Genet 36:585–595
25. Polymeropoulos MH, Lavedan C, Leroy E, et al. (1997) Mutations in the α-synuclein gene identified in families with Parkinson's disease. Science 276:2045–2047
26. Kitada T, Asakawa S, Hattori N, et al. (1998) Mutations in the parkin gene cause autosomal recessive juvenile parkinsonism. Nature 392: 605–608
27. Greene JC, Whitworth AJ, Kuo I, Andrews LA, Feany MB, Pallanck LJ (2003) Mitochondrial pathology and apoptotic muscle degeneration in Drosophila parkin mutants. Proc Natl Acad Sci U S A 100:4078–4083
28. Pesah Y, Pham T, Burgess H, et al. (2004) Drosophila parkin mutants have decreased mass and cell size and increased sensitivity to oxygen radical stress. Development 131:2183–2194
29. Auluck PK, Chan HY, Trojanowski JQ, Lee VM, Bonini NM (2002) Chaperone suppression of alpha-synuclein toxicity in a Drosophila model for Parkinson's disease. Science 295:865–868
30. Pesah Y, Burgess H, Middlebrooks B, et al. (2005) Whole-mount analysis reveals normal numbers of dopaminergic neurons following misexpression of alpha-Synuclein in Drosophila. Genesis 41:154–159
31. Gotz J, Deters N, Doldissen A, et al. (2007) A decade of tau transgenic animal models and beyond. Brain Pathol 17:91–103
32. Jackson GR, Wiedau-Pazos M, Sang TK, et al. (2002) Human wild-type tau interacts with wingless pathway components and produces neurofibrillary pathology in Drosophila. Neuron 34:509–519
33. Mudher A, Shepherd D, Newman TA, et al. (2004) GSK-3beta inhibition reverses axonal transport defects and behavioural phenotypes in Drosophila. Mol Psychiatr 9:522–530
34. Shulman JM, Feany MB (2003) Genetic modifiers of tauopathy in Drosophila. Genetics 165:1233–1242
35. Chee FC, Mudher A, Cuttle MF, et al. (2005) Over-expression of tau results in defective synaptic transmission in Drosophila neuromuscular junctions. Neurobiol Dis 20:918–928
36. Glenner GG, Wong CW (1984) Alzheimer's disease and Down's syndrome: Sharing of a unique cerebrovascular amyloid fibril protein. Biochem Biophys Res Commun 122: 1131–1135
37. Kang J, Lemaire HG, Unterbeck A, et al. (1987) The precursor of Alzheimer's disease amyloid A4 protein resembles a cell-surface receptor. Nature 325:733–736
38. Gunawardena S, Goldstein LS (2001) Disruption of axonal transport and neuronal viability by amyloid precursor protein mutations in Drosophila. Neuron 32:389–401
39. Fossgreen A, Brückner B, Czech C, Masters CL, Beyreuther K, Paro R (1998) Transgenic Drosophila expressing human amyloid precursor protein show γ-secretase activity and blistered wing phenotype. Proc Natl Acad Sci USA 95:13703–13708
40. Vassar R, Bennett BD, Babu-Khan S, et al. (1999) Beta-secretase cleavage of Alzheimer's amyloid precursor protein by the transmembrane aspartic protease BACE. Science 286: 735–741
41. Luo Y, Bolon B, Kahn S, et al. (2001) Mice deficient in BACE1, the Alzheimer's beta-secretase, have normal phenotype and abolished beta-amyloid generation. Nat Neurosci 4:231–232
42. Wolfe MS (2006) The gamma-secretase complex: Membrane-embedded proteolytic ensemble. Biochemistry 45:7931–7939
43. Parvathy S, Hussain I, Karran EH, Turner AJ, Hooper NM (1999) Cleavage of Alzheimer's

amyloid precursor protein by alpha-secretase occurs at the surface of neuronal cells. Biochemistry 38:9728–9734

44. Perez RG, Soriano S, Hayes JD, et al. (1999) Mutagenesis identifies new signals for beta-amyloid precursor protein endocytosis, turnover, and the generation of secreted fragments, including Abeta42. J Biol Chem 274: 18851–18856

44. Pasternak SH, Callahan JW, Mahuran DJ (2004) The role of the endosomal/lysosomal system in amyloid-beta production and the pathophysiology of Alzheimer's disease: Reexamining the spatial paradox from a lysosomal perspective. J Alzheimers Dis 6:53–65

46. Carrotta R, Manno M, Bulone D, Martorana V, San Biagio PL (2005) Protofibril formation of amyloid beta-protein at low pH via a non-cooperative elongation mechanism. J Biol Chem 280:30001–30008

47. Chartier-Harlin MC, Crawford F, Houlden H, et al. (1991) Early-onset Alzheimer's disease caused by mutations at codon 717 of the beta-amyloid precursor protein gene. Nature 353:844–846

48. Nilsberth C, Westlind-Danielsson A, Eckman CB, et al. (2001) The 'Arctic' APP mutation (E693G) causes Alzheimer's disease by enhanced Abeta protofibril formation. Nat Neurosci 4:887–893

49. Haass C, Lemere CA, Capell A, et al. (1995) The Swedish mutation causes early-onset Alzheimer's disease by beta-secretase cleavage within the secretory pathway. Nat Med 1:1291–1296

50. Tokuda T, Fukushima T, Ikeda S, et al. (1997) Plasma levels of amyloid beta proteins Abeta1–40 and Abeta1–42(43) are elevated in Down's syndrome. Ann Neurol 41:271–273

51. Lemere CA, Blusztajn JK, Yamaguchi H, Wisniewski T, Saido TC, Selkoe DJ (1996) Sequence of deposition of heterogeneous amyloid beta-peptides and APO E in Down syndrome: Implications for initial events in amyloid plaque formation. Neurobiol Dis 3:16–32

52. Suzuki N, Cheung TT, Cai X-D, et al. (1994) An increased percentage of long amyloid β protein secreted by familial amyloid β protein precursor (βAPP717) mutants. Science 264:1336–1340

53. Scheuner D, Eckman C, Jensen M, et al. (1996) Secreted amyloid beta-protein similar to that in the senile plaques of Alzheimer's disease is increased in vivo by the presenilin 1 and 2 and APP mutations linked to familial Alzheimer's disease. Nat Med 2:864–870

54. Hutton M, Lendon CL, Rizzu P, et al. (1998) Association of missense and 5'-splice-site mutations in tau with the inherited dementia FTDP-17. Nature 393:702–705

55. Tamaoka A, Sawamura N, Odaka A, et al. (1995) Amyloid beta protein 1–42/43 (A beta 1–42/43) in cerebellar diffuse plaques: Enzyme-linked immunosorbent assay and immunocytochemical study. Brain Res 679:151–156

56. Katzman R, Terry R, DeTeresa R, et al. (1988) Clinical, pathological, and neurochemical changes in dementia: A subgroup with preserved mental status and numerous neocortical plaques. Ann Neurol 23:138–144

57. Naslund J, Haroutunian V, Mohs R, et al. (2000) Correlation between elevated levels of amyloid beta-peptide in the brain and cognitive decline. Jama 283:1571–1577

58. Klein WL, Krafft GA, Finch CE (2001) Targeting small Abeta oligomers: The solution to an Alzheimer's disease conundrum? Trends Neurosci 24:219–224

59. Walsh DM, Klyubin I, Fadeeva JV, Rowan MJ, Selkoe DJ (2002) Amyloid-beta oligomers: their production, toxicity and therapeutic inhibition. Biochem Soc Trans 30:552–557

60. Dodart JC, Bales KR, Gannon KS, et al. (2002) Immunization reverses memory deficits without reducing brain Abeta burden in Alzheimer's disease model. Nat Neurosci 5:452–457

61. Lee EB, Leng LZ, Zhang B, et al. (2006) Targeting amyloid-beta peptide (Abeta) oligomers by passive immunization with a conformation-selective monoclonal antibody improves learning and memory in Abeta precursor protein (APP) transgenic mice. J Biol Chem 281:4292–4299

62. Rosen DR, Martin-Morris L, Luo LQ, White K (1989) A Drosophila gene encoding a protein resembling the human beta-amyloid protein precursor. Proc Natl Acad Sci U S A 86: 2478–2482

63. Greeve I, Kretzschmar D, Tschape JA, et al. (2004) Age-dependent neurodegeneration and Alzheimer-amyloid plaque formation in transgenic Drosophila. J Neurosci 24: 3899–3906

64. Iijima K, Liu HP, Chiang AS, Hearn SA, Konsolaki M, Zhong Y (2004) Dissecting the pathological effects of human A{beta}40 and A{beta}42 in Drosophila: A potential model for Alzheimer's disease. Proc Natl Acad Sci U S A 101:6623–6628

65. Finelli A, Kelkar A, Song HJ, Yang H, Konsolaki M (2004) A model for studying Alzheimer's Abeta42-induced toxicity in Drosophila melanogaster. Mol Cell Neurosci 26:365–375

66. Crowther DC, Kinghorn KJ, Miranda E, et al. (2005) Intraneuronal Aβ, non-amyloid aggregates and neurodegeneration in a Drosophila model of Alzheimer's disease. Neuroscience 132:123–135
67. Luheshi LM, Tartaglia GG, Brorsson AC, et al. (2007) Systematic in vivo analysis of the intrinsic determinants of amyloid beta pathogenicity. PLoS Biol 5:e290
68. Kayed R, Head E, Thompson JL, et al. (2003) Common structure of soluble amyloid oligomers implies common mechanism of pathogenesis. Science 300:486–489
69. Piper M, Selman C, McElwee JJ, Partridge L (2005) Models of insulin signalling and longevity. Drug Discov Today Dis Models 2:249–256
70. Rival T, Soustelle L, Strambi C, Besson MT, Iche M, Birman S (2004) Decreasing glutamate buffering capacity triggers oxidative stress and neuropil degeneration in the Drosophila brain. Curr Biol 14:599–605
71. Franceschini N (1972) Pupil and pseudopupil in the compound eye of Drosophila. In: Wehner R (ed) Information processing in the visual system of Drosophila. Springer, Berlin, pp 75–82
72. Toba G, Ohsako T, Miyata N, Ohtsuka T, Seong KH, Aigaki T (1999) The gene search system. A method for efficient detection and rapid molecular identification of genes in Drosophila melanogaster. Genetics 151: 725–737
73. Cao W, Song HJ, Gangi T, et al. (2008) Identification of novel genes that modify phenotypes induced by Alzheimer's beta amyloid overexpression in Drosophila. Genetics 178:1457–1471
74. Reisberg B, Doody R, Stoffler A, Schmitt F, Ferris S, Mobius HJ (2003) Memantine in moderate-to-severe Alzheimer's disease. N Engl J Med 348:1333–1341
75. Apostol BL, Kazantsev A, Raffioni S, et al. (2003) A cell-based assay for aggregation inhibitors as therapeutics of polyglutamine-repeat disease and validation in Drosophila. Proc Natl Acad Sci U S A 100: 5950–5955
76. Sanchez I, Mahlke C, Yuan J (2003) Pivotal role of oligomerization in expanded polyglutamine neurodegenerative disorders. Nature 421:373–379
77. Rival T, Page RM, Chandraratna DS, Sendall TJ, Ryder E, Liu B, Lewis H, Rosahl T, Hider R, Camargo LM, Shearman MS, Crowther DC, Lomas DA. Eur J Neurosci. 2009 Apr;29(7):1335–47. Epub 2009 Mar 23

Chapter 13

Caenorhabditis elegans as a Model Organism for Dementia

Tjakko J. Van Ham and Ellen A.A. Nollen

Abstract

The free living nematode worm *Caenorhabditis elegans* (*C. elegans*) has been extensively studied by biological, agricultural, and medical scientists for over 40 years. The animal has several characteristics that make it useful as a model organism. For example, the nematodes are transparent, which allows study of embryonic development and gene expression in living animals under the microscope. They also have a very short life cycle of about 3 days and a relatively short lifespan of about 3 weeks, which allow genetic dissection of the mechanisms that affect aging and ultimately determine lifespan. In addition, the mechanism of gene silencing by RNA interference has been discovered in *C. elegans* and has been developed into a potent reverse genetic tool.

Because of the strong conservation of molecular genetic pathways between *C. elegans* and mammals, it represents a powerful addition to the small animal model repertoire. Genetic mechanisms in human disease, such as Alzheimer's disease, have been elucidated in *C. elegans*, indicating its potential as a model for human dementia. Here, we will discuss the existing models, and what they have revealed about the genetic pathways and pathogenesis of different forms of dementia. We will also describe how to set up forward and reverse genetic screens in *C. elegans*, which can be used to identify additional genes and processes involved in dementia.

Key words: *C. elegans*, genome-wide RNAi, protein aggregation, α-synuclein, Lewy bodies

1. Introduction

The free living nematode worm *Caenorhabditis elegans* (*C. elegans*) has been studied extensively as an experimental model organism by scientists from many fields of biological, agricultural, and medical research for over 40 years. The animal, introduced into molecular biology research by Sydney Brenner in 1963, lives in soil, where it feeds mostly on bacteria (1). It has a very short life cycle of about 3 days and a relatively short lifespan of about 2–3 weeks, which facilitates genetic dissection of the mechanisms that determine lifespan. Because of its anatomical and genetic simplicity, it has proven to be an excellent model organism. In fact, it is the most

Peter Paul De Deyn and Debby Van Dam (eds.), *Animal Models of Dementia*, Neuromethods, vol. 48,
DOI 10.1007/978-1-60761-898-0_13, © Springer Science+Business Media, LLC 2011

completely understood metazoan organism in terms of development, genetics, and nervous system (1–5).

The *C. elegans* nervous system represents the animal's largest organ in cell numbers, being composed of 302 neurons. The morphology of the neurons is relatively simple and unbranched, most neurons having two or less processes. Owing to this simplicity, it has been possible to deduct their anatomical location and connectivity within the animal from electron microscopic images. This has enabled researchers to map the full circuitry of neuronal transmission in *C. elegans*, making it an unprecedented model for metazoan neurobiology. Although the *C. elegans* nervous system is fairly complex, it does not constitute an actual brain, merely a local concentration of neurons in the head region known as the nerve ring. Subtypes of neurons such as dopaminergic, GABAergic, and cholinergic neurons are present that are homologous to classes of neurons in vertebrates (6).

Reproduction of *C. elegans* is generally hermaphroditic by self-fertilization of eggs, which hatch outside of the animal. Larvae hatch from the eggs after all cell divisions are complete and go through four larval stages before entering the adult reproductive age after about 3 days. The soma consists of a constant 959 cells, of which each cell ends up in predetermined positions in the organs and tissues that make up the animal. The complex pattern of cell divisions and anatomical position is known as the *C. elegans* cell lineage (4,7). Much of what is currently known on the genetics of apoptosis has been elucidated in *C. elegans* by genetic studies of this cell lineage (7).

The basic mechanism of RNA interference (RNAi) was identified in *C. elegans* in 1998 by Andrew Fire and Craig Mello (8). Soon thereafter, the process of RNAi was extensively applied to silence expression of target genes, first in *C. elegans*, but also in other animal models and cells. The application of RNAi in *C. elegans* has greatly facilitated genetic research.

Initial RNAi experiments in *C. elegans* used injections of double-stranded RNA (dsRNA) in the gonad, leading to robust knockdown of mRNA levels in various cell types, thereby suggesting that RNAi was able to penetrate cell and tissue membranes. This indeed appeared to be the case when it was shown that RNAi can be obtained by soaking worms in dsRNA solution or feeding them on *Escherichia coli* (*E. coli*) expressing dsRNA (9–11). Feeding of *E. coli* expressing target gene dsRNA is now the most commonly used way of knocking down genes systemically in *C. elegans*. Such feeding has been used for rapid genome-wide RNAi screening, which has led to the discovery of comprehensive sets of genes involved in many genetic, cellular, and developmental processes, some related to human disease. The application of RNAi and its high-throughput capacity in genome-wide screens have revolutionized genetics research in *C. elegans* (10–13).

All the above has led to a thorough understanding of the animal's behavior, genetics, molecular, cell, and neurobiology, and even molecular pathways that can increase and decrease the animal's aging and lifespan. In addition, genetic mechanisms underlying human disease have been elucidated in *C. elegans* indicating its value as a model for human diseases, including age-related dementias. Here, we discuss the existing *C. elegans* models for dementia and provide examples of how these models can be used to screen for genes involved in the disease process, either by classical genetics or genome-wide RNAi.

2. Dementia

As described in earlier chapters, dementia is part of the pathologic spectrum of many sporadic as well as familial neurodegenerative diseases, such as Alzheimer's disease (AD), frontotemporal lobar degeneration (FTLD), and dementia with Lewy bodies (DLB). *C. elegans* models for these diseases can be subdivided into at least two categories, including mutant strains to study endogenous pathways associated with disease and transgenic strains to study human disease-related genes. These models can be used for large-scale phenotypic, genetic, and drug screens to find genetic or chemical modifiers of pathways and phenotypes related to disease. These studies could, therefore, lead to a better understanding of disease pathology, the development of clinical diagnostic tools, and uncover therapeutic targets for treatment of dementia in humans.

In this chapter, we discuss the existing *C. elegans* models for dementias and large-scale screens that can be used to identify novel genetic and chemical components that modify disease phenotypes. We explain what these models have taught us about disease mechanisms, and what they have revealed from therapeutic and diagnostic perspectives.

2.1. C. elegans Counterparts of Human Dementia-Related Genes

Familial early-onset forms of AD are often caused by mutations in presenilin-1 and 2 (PSEN1 and PSEN2). These presenilin mutations are thought to change the ratio of the highly amyloidogenic 42-amino acid-long amyloid β (Aβ) peptide to the less toxic Aβ40 amino acid peptide variant of this disease-related protein (for review see (14)). In *C. elegans*, an ortholog for PSEN1 has been found in a screen for suppressors of the egg-laying defective phenotype in *lin-12* (*sel-12*) gain-of-function worms (15). The sel-12 gene has a function similar to that in vertebrates, functioning mostly during embryonic development to facilitate Notch/*lin-12* signaling. Apart from *sel-12*, the *C. elegans* genome encodes two other orthologs to human presenilins: a homolog of presenilin-1, *hop-1*, which is in fact more homologous to human PSEN2, and an ortholog *spe-4*, which has no obvious human counterpart (16). *Sel-12* and *hop-1* are highly similar to PSEN1 and PSEN2 in structure as well as in function (16,17).

Interestingly, genetic deletion mutants for *sel-12* and *hop-1* are viable, which allows for modifier screens to find components of the *C. elegans* PSEN1 and PSEN2 pathways. For example, mutations in multiple genes were found in forward mutagenesis screens for suppression of the *sel-12*/PSEN1 egg-laying phenotype (18–21). Two of them, *spr-3* and *spr-4*, encode C(2)H(2) zinc-finger proteins that are similar to the human REST/NRSF (Rel silencing transcription factor/neural-restrictive silencing factor) transcriptional repressors, involved in neuronal differentiation (18). Two others, *spr-1* and *spr-5*, encode an ortholog of human CoREST, a co-repressor of REST, which can be substituted by human CoREST, and a homolog of p110b, which is another member of the CoREST co-repressor complex (19). In all, studies in *C. elegans* have provided important clues to the genetic context and regulation of presenilins.

In addition to these AD-related pathways, many other pathways have been investigated in *C. elegans*. For example, an ortholog of Aβ (*apl-1*), and at least three orthologs of genes encoded in loci associated with familial Parkinson's disease (PD), PARK loci, are present and under investigation in *C. elegans*, including LRRK2/*lrk-1*, PINK1/*pink-1*, and Parkin/*prk-1* (20–24).

2.2. Aggregation of Human Dementia-Associated Proteins

2.2.1. Amyloid β

In AD, Aβ forms amyloid fibrils, which accumulate in deposits known as plaques. To study aggregation of Aβ, the human Aβ peptide was overexpressed in *C. elegans*. By using the *C. elegans* Myosin/*unc-54* muscle-specific promoter to drive expression, robust expression levels were obtained. A 189-base pair fragment was overexpressed coding for amino acids 1–42 of Aβ fused to a synthetic signal peptide driven by the *unc-54* promoter. Transgenic worms expressing this transgene showed formation of aggregates with important characteristic of amyloid fibrils, as shown by Thioflavin S binding, an amyloid-specific dye, which has altered fluorescent properties on binding. In addition, paralysis of the nematodes, which can be easily distinguished by dissection microscopy, occurred indicating a specific toxicity of Aβ to the muscle cells (25). Interestingly, oligomeric species of Aβ were detected in these strains that might be similar to the neurotoxic Aβ-derived diffusible ligand (ADDL) (24).

The insulin/insulin growth factor (IGF)-1 like signaling (IIS) pathway strongly affects lifespan and stress resistance in *C. elegans*. Inhibition of *daf-2*/IGF-1 receptor by RNAi increases lifespan and stress resistance in nematodes. Mice deficient in the IGF-1 receptor also show increased lifespan, indicating the conservation of this mechanism (26). The effect of *daf-2* partially depends on heat shock factor-1 (*hsf-1*), an important transcriptional regulator of heat shock genes. Interestingly, knockdown of the *daf-2* gene in Aβ transgenic nematodes results in decreased paralysis, whereas *hsf-1* knockdown increases paralysis, which is accompanied by alterations in the aggregation phenotype (27). It will be interesting to find out if these genetic pathways also affect Aβ aggregation and toxicity in mice and humans, which would directly link aging with the prevalence of disease related to Aβ aggregation.

2.2.2. α-Synuclein

Another protein that is associated with forms of dementia is the presynaptic protein α-synuclein. This protein provides a remarkable link between several neurodegenerative diseases, most of which involve dementia. It is the main constituent of protein inclusions in the brains of PD patients that represent the pathological hallmark of this disease. Such inclusions (known as Lewy bodies) also occur in a subtype of dementia called DLB, which is currently the second most prevalent cause of dementia. Importantly, DLB often has an overlap with Alzheimer's pathology, and is sometimes difficult to distinguish from either AD or dementia related to PD.

Interestingly, α-synuclein mutations and multiplications of the α-synuclein locus have been found to cause familial forms of PD. One familial PD mutation, E46K, is also known to be associated with dementia (28). Some aspect of aggregation of α-synuclein is most likely directly involved in toxicity with the cells that degenerate in these diseases. The current hypothesis is that multimers (multiple associated molecules) of α-synuclein are somehow able to damage cells leading to degeneration. Although it is very likely that α-synuclein is directly involved in the pathogenesis, the mode of toxicity remains unknown. Physiologically, α-synuclein appears to regulate docking events of synaptic vesicles, likely by influencing the fusion of membranes (28–32).

To study expression and accumulation of a disease-related protein such as α-synuclein in a living animal, one can make the accumulation visible under a fluorescence microscope by fusion to a fluorescent molecule such as Yellow Fluorescent Protein (YFP). The *C. elegans* body wall muscle provides a system in which one can track accumulation of various disease-related aggregating proteins during aging (33,34). The muscle cells are easily seen under a microscope, and are amenable to RNAi by feeding (Fig. 1). In addition, the robust expression levels obtained by the muscle-specific *unc-54* promoter yield high protein levels, which facilitate biochemical and cell biological analysis of protein expression.

Transgenic strains expressing human α-synuclein fused to YFP in the body wall muscle show accumulation of the fusion protein during aging (35). Interestingly, the inclusions formed are mobile at young age, but show properties of immobilized aggregated protein at old age. To find genes involved in the formation of

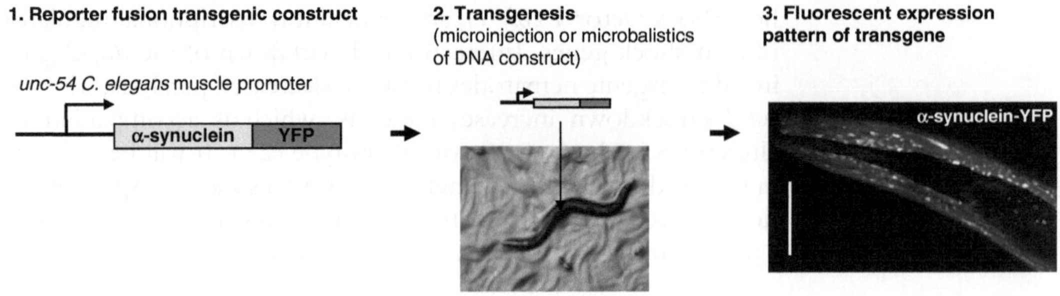

Fig. 1. Tissue-specific transgene expression in transgenic *C. elegans*. (1) A transgene is created using a known *C. elegans* tissue-specific promoter followed by the (human) gene of interest. By fusing the transgene to a fluorescent protein such as YFP (yellow fluorescent protein), its expression can be visualized. (2) Transgenic animals can be created by microinjection into the *C. elegans* gonad together with a marker or by microparticle bombardment. (3) Confocal laser scanning image showing head region of an adult *C. elegans* expressing α-synuclein fused to YFP in the body wall muscle using the *unc-54* promoter element.

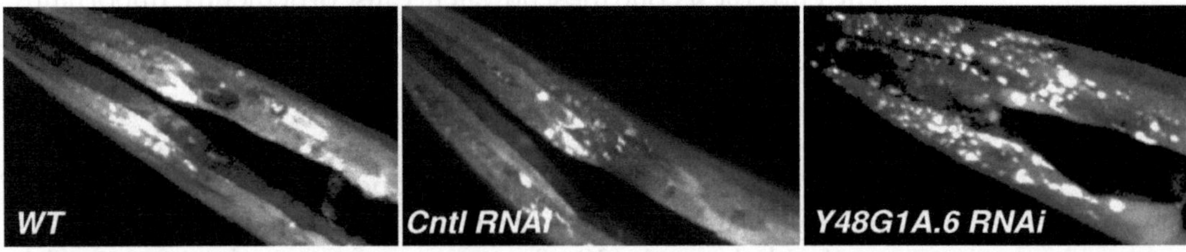

Fig. 2. RNA interference of modifiers of α-synuclein inclusion. Confocal images showing yellow fluorescent protein (YFP) fluorescence in the head region of a nematode expressing α-synuclein-YFP transgene in the body wall muscle. Worms are grown on regular food (wt, wild-type), HT115 bacteria containing empty RNAi vector (L4440) and bacteria containing *Y48G1A.6* RNAi, to knock down one of the genes found in a genome-wide RNAi screen for modifiers of α-synuclein inclusion formation.

α-synuclein inclusion, a genome-wide screen was performed for genes that result in increased aggregation when knocked down by RNAi. Of the 80 genes found, genes that function in endomembrane-related compartments of the cell are overrepresented, which indicates a role for the endomembrane system in age-related synucleinopathies such as DLB (Fig. 2).

2.2.3. Tau (τ)

Tau (MAPT) is a microtubule-associated protein (MAP) mainly found in neuronal axons where it binds and stabilizes microtubules. In sporadic AD and familial frontotemporal dementia with Parkinsonism-17 (FTDP-17), a form of presenile dementia affecting the frontal and temporal cortex and some subcortical nuclei, hyperphosphorylated tau protein forms deposits in the brain. To model disease related to tau expression and deposition in *C. elegans*, Kraemer and colleagues have expressed both wild-type and mutated (301L and 337M) tau protein in *C. elegans* neurons (36). These nematodes developed a progressive phenotype of defective motility known as "uncoordinated phenotype" that can be distinguished by dissection microscopy, which was more apparent in the mutants. Interestingly, another characteristic of disease was captured in this model, since hyperphosphorylation of tau also appeared to occur in the transgenic lines (36). To identify modifiers of this toxicity, a genome-wide RNAi screen for enhancement of tau-related toxicity was performed (37). Some of the genes found are known to be involved in tau-related pathology, such as Glycogen synthase kinase 3β (GSK-3β) that can phosphorylate tau. New components have also been found, such as homologs of carboxyl terminus of Hsp 70-interacting protein (CHIP) and Heat shock cognate 70 (Hsc70) that cooperate in ubiquitination and proteasomal degradation. Thus, such genes might normally be involved in protection against tau-mediated toxicity.

Future work will show what modifier genes found in these genome-wide screens can tell us about the disease mechanism, and whether they represent diagnostic or even therapeutic targets to treat disease.

3. Genetics and Genomics

Although *C. elegans* are invertebrates, genetically they are similar to humans, and many disease-related genes and pathways found in humans have orthologs in *C. elegans.* These orthologous genetic pathways can be studied by making transgenics by microinjection or microparticle bombardment of transgenic constructs, or by conventional forward and reverse mutagenesis methods. For example, human disease-related genes or specific mutants, such as wild-type or mutant tau, Aβ, and α-synuclein, as described above, can be overexpressed in a tissue-specific manner using worm-specific plasmids created by Andrew Fire's lab (Stanford University Medical Center, CA, USA) to study disease-related phenotypic effects on those cells.

To create transgenic animals by microinjection, plasmid DNA is injected into the gonad with a marker to select for transgene-expressing animals. These carry the transgene extrachromosomally. By using UV or γ irradiation that causes double-stranded breaks in the DNA, these extrachromosomal plasmids can be integrated into the genome to create stable transgenic strains. Another way to create transgenic animals is to use microparticle bombardment (biolistic transformation), in which small, 1-μm gold particles are shot into the gonads (38,39). The efficiency of bombardment has been greatly increased by using a specific paralyzed (uncoordinated) mutant background and co-bombarding of a plasmid that rescues this paralysis phenotype (38).

Alternatively, endogenous genetic pathways related to disease in humans can be studied by forward genetics, i.e. screening for

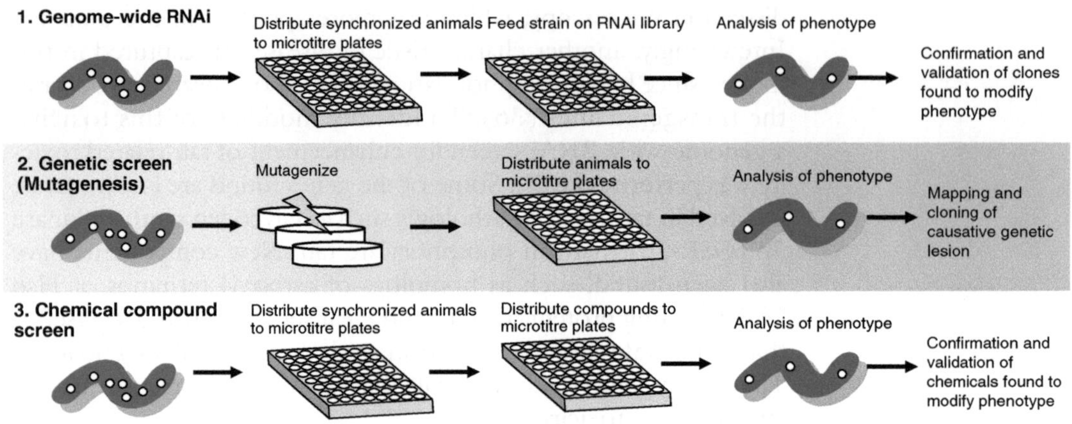

Fig. 3. *C. elegans* high-throughput screening. Scheme depicting three basic screening paradigms used in *C. elegans*: genome-wide RNAi screening, forward and reverse mutagenesis genetic screening, and chemical library and drug screening.

suppressors and enhancers of the disease-related phenotype, and reverse genetics, which includes knocking down disease-associated genes and searching for a specific phenotype. A variety of approaches can be employed for forward as well as reverse genetics, and depending on the process to be studied, one of these approaches can be chosen (see review (40); Fig. 3).

The first forward screens, F_2 screens, were done by Sydney Brenner in 1973. A chemical mutagen, ethyl methane sulfonate (EMS), is used to induce mutations in the sperm and oocytes of wild-type hermaphrodites, and the mutagenized animals are placed on dishes and grown for two generations to produce homozygous mutants (1). Worms that show a mutant phenotype are then transferred individually to new Petri dishes to determine whether the phenotype has been passed on to the next generation. Mapping of the point mutations induced by EMS has been made easier by using the great diversity in single nucleotide polymorphisms (SNP) diversity between the most often used *C. elegans* strain, namely Bristol N2, and the Hawaiian *C. elegans* strain CB1584 (41). This approach can be used to find modifier genes of protein aggregation or suppression of any specific phenotype.

Alternatively, this approach can be used for reverse genetics by target-selected mutagenesis. EMS has a mutation frequency of 5×10^{-4} per gene, of which 13% are deletions, while the mutagen trimethylpsoralen (TMP) in combination with UV irradiation has a mutation frequency of 3×10^{-5}, of which 50% are deletions. By combining EMS or TMP/UV mutagenesis with visualization of deletions by PCR, specific genes can be inactivated by deletion (42).

Another strategy for target-selected deletion is the use of the *C. elegans* transposon Tc1 (43). The Tc1 transposon DNA element randomly integrates into the genome, often inducing deletions in flanking DNA inserts on excision. By establishing a frozen library of Tc1 mutagenized animals, insertion mutants of targeted genes can be obtained. By screening with PCR, mutants can then be detected in which Tcl and 1,000–2,000 base pairs of flanking DNA are deleted, thereby inactivating the targeted gene.

A relatively new method is the application of chimeric Zinc finger nucleases consisting of a DNA-binding domain and a nuclease that can create specific double-stranded DNA breaks that could abrogate a specific allele (44,45).

The application of RNAi by dsRNA-mediated gene silencing will be described in the next two sections.

3.1. Genome-Wide RNAi Screening

The discovery of dsRNA-mediated RNAi quickly led to its use as a means for knocking down gene expression and it has revolutionized genetic studies in *C. elegans*, as well as in other model organisms. In an attempt to study the phenomenon of RNA-mediated silencing of gene expression, Andy Fire and Craig Mello (Carnegie Institution of Washington, Baltimore, MD, USA), have discovered that silencing could be triggered by dsRNA in a mRNA-specific manner (8). They have also shown that silencing is transferred to cells and tissues after injection. RNAi spreads through cells and tissues after soaking with dsRNA or intestinal exposure after feeding *E. coli* expressing dsRNA complementary to a specific messenger, suggesting that this could be used as a whole animal gene knockdown approach (9,10).

For bacterial feeding of RNAi, cDNA of the corresponding gene is cloned into a vector (L4440) containing two bacteriophage T7 RNA polymerase promoters. The two T7 promoter sites flank both sides of a multiple cloning site, which is used to clone in the cDNA fragment. The resulting vector is transformed into *E. coli* strain HT115(ED3) that is deficient in bacterial RNA polymerase III and produces bacteriophage T7 polymerase on induction by isopropyl β-D-1-thiogalactopyranoside (10). T7 polymerases traveling in opposite directions synthesize two complementary RNA strands that form the duplex RNA, which mediates RNAi. By feeding *C. elegans* on such engineered bacteria that have been induced to produce dsRNA expression, the corresponding gene can be silenced.

Although this RNAi by feeding has greatly sped up genetic research in *C. elegans*, it is nonetheless quite laborious to knock down a specific gene. First, cDNA of a specific target needs to be generated and cloned into the RNAi vector that, in turn, needs to be transformed into dsRNA expressing *E. coli*. In Julie Ahringer's lab at the Gurdon Institute in Cambridge, UK, a strategy was designed to rapidly clone cDNAs into a bacterial feeding strain, to generate a genome-wide library of bacteria containing genomic cDNA clones (46,13). The library generated contains clones for knocking down 17,575 genes of the *C. elegans* genome, corresponding to approximately 87% of all *C. elegans* genes

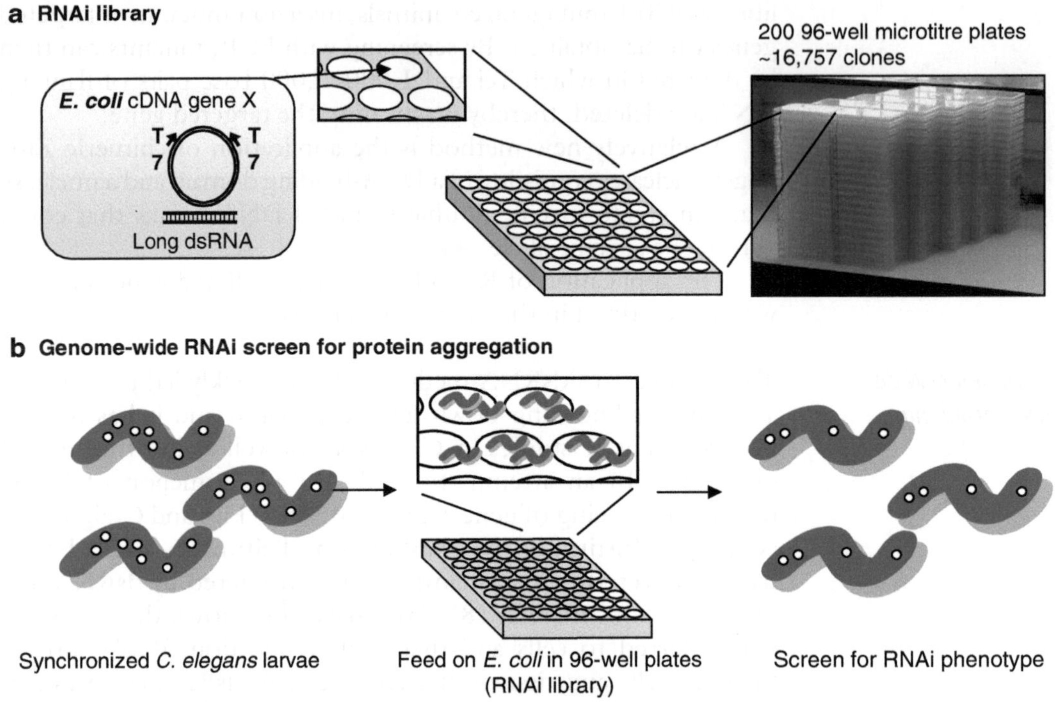

Fig. 4. RNAi library and liquid culture RNAi screening. (**a**) RNAi feeding library. Target gene cDNA is cloned into a T7 RNA polymerase promoter plasmid (L4440). This plasmid is transformed into HT115 *E. coli* strain that expresses the corresponding dsRNA on induction by isopropyl β-D-1-thiogalactopyranoside (IPTG). The library created by Ahringer's lab (Cambridge, UK) contains bacterial clones for knocking down 17,575 genes of the *C. elegans* genome, corresponding to ~87% of all *C. elegans* genes, in about 200 96-well microtitre plates. (**b**) Genome-wide liquid culture RNAi screen. Nematodes can be synchronized by hypochlorite bleaching and cultured in microtitre plates that contain the different bacterial clones induced to express dsRNA. After approximately 3–4 days, the animals can be scored for a specific knockdown phenotype, in this case reduction of aggregation.

(Fig. 4a). Another RNAi library consisting of 11,800 open reading frame RNAi clones was created by Mark Vidal's lab (Dana-Farber Cancer Institute, Boston, Ma) (47). Currently, an effort is underway to knockdown the remaining genes (48). By culturing worms in liquid culture in 96-well microtiter plates, genome-wide screens can be performed in a matter of weeks (49,50) (Fig. 4b).

3.2. Setting Up RNAi Screens

There are several important factors that determine the success of a genome-wide RNAi screen (also see (48)). The first most critical aspect is that a clear phenotype is needed that can potentially be modified by knocking down genetic components involved in the process. For example, if a genetic pathway is known to be involved in a very specific, well-characterized phenotype, such as the *sel-12* phenotype, an RNAi screen can reveal novel unknown components in that specific corresponding pathway (18,19,51,52). Another possibility is the case in which a genetic lesion does not cause a specific phenotype, and there might be a redundancy of components in the pathway. In a so-called synthetic screen, such redundant genes can be picked up by screening in a background genetic lesion without a phenotype (53).

Depending on the phenotype scored for, nematodes can be synchronized by hypochlorite bleaching, by which only fertilized eggs remain, 1 day before starting the actual experiments in hatched larvae (L1) to study the consequence of knocking down genes in one generation (49,50). The advantage is that essential genes can also be picked up in this way, because such an approach circumvents fertility and embryonic development. The clones identified are verified by sequencing.

In conclusion, there are several *C. elegans* models available for studying different molecular biological aspects of dementia, and new models can be created that mimic other aspects of dementia. New genetic pathways and modifier chemicals can be found by using these models in genetic and chemical screenings. Such pathways, genes, and compounds teach us about the mechanism of disease. In addition, they may provide direct or indirect therapeutic perspectives to halt, treat, or even cure disease.

4. Web Resources

1. http://elegans.swmed.edu/
2. http://www.wormatlas.org/index.htm
3. http://www.wormbase.org/
4. http://www.cbs.umn.edu/CGC/
5. http://www.wormbook.org/

Acknowledgments

The authors thank J. Senior for editing the manuscript and ZonMw (NWO), Research Institute for Diseases in the Elderly (RIDE), for the funding.

References

1. Brenner S (v1974) The genetics of Caenorhabditis elegans. Genetics 77(1):71–94
2. Byerly L, Cassada RC, Russell RL (1976) The life cycle of the nematode Caenorhabditis elegans. I. Wild-type growth and reproduction. Dev Biol 51(1):23–33
3. Lewis JA, Fleming JT (1995) Basic culture methods. Methods Cell Biol 48:3–29
4. Sulston JE, Schierenberg E, White JG, Thomson JN (1983) The embryonic cell lineage of the nematode Caenorhabditis elegans. Dev Biol 100(1):64–119

5. Wood WB (1988) Determination of pattern and fate in early embryos of Caenorhabditis elegans. Dev Biol (N Y 1985) 5:57–78
6. White JG, Southgate E, Thomson JN, Brenner S (1986)The structure of the nervous systemn of the nematode Caenorhabditis elegans. Phil Trans Royal Soc London Series B, Biol Sci 314:1–340
7. Sulston JE, Horvitz HR (1977) Post-embryonic cell lineages of the nematode, Caenorhabditis elegans. Dev Biol 56(1): 110–156
8. Fire A, Xu S, Montgomery MK, Kostas SA, Driver SE, Mello CC (1998) Potent and specific genetic interference by double-stranded RNA in Caenorhabditis elegans. Nature 391(6669):806–811
9. Tabara H, Grishok A, Mello CC (1998) RNAi in C. elegans: Soaking in the genome sequence. Science 282(5388):430–431
10. Timmons L, Fire A (1998) Specific interference by ingested dsRNA. Nature 395(6705):854
11. Timmons L, Court DL, Fire A (2001) Ingestion of bacterially expressed dsRNAs can produce specific and potent genetic interference in Caenorhabditis elegans. Gene 263(1–2):103–112
12. Kamath RS, Ahringer J (2003) Genome-wide RNAi screening in Caenorhabditis elegans. Methods 30(4):313–321
13. Kamath RS, Fraser AG, Dong Y, et al. (2003) Systematic functional analysis of the Caenorhabditis elegans genome using RNAi. Nature 421(6920):231–237
14. Haass C, Selkoe DJ (2007) Soluble protein oligomers in neurodegeneration: Lessons from the Alzheimer's amyloid beta-peptide. Nat Rev Mol Cell Biol 8(2):101–112
15. Levitan D, Greenwald I (1995) Facilitation of lin-12-mediated signalling by sel-12, a Caenorhabditis elegans S182 Alzheimer's disease gene. Nature 377(6547):351–354
16. Li X, Greenwald I (1997) HOP-1, a Caenorhabditis elegans presenilin, appears to be functionally redundant with SEL-12 presenilin and to facilitate LIN-12 and GLP-1 signaling. Proc Natl Acad Sci U S A 94(22): 12204–12209
17. Levitan D, Doyle TG, Brousseau D, et al. (1996) Assessment of normal and mutant human presenilin function in Caenorhabditis elegans. Proc Natl Acad Sci U S A 93(25): 14940–14944
18. Lakowski B, Eimer S, Gobel C, Bottcher A, Wagler B, Baumeister R (2003) Two suppressors of sel-12 encode C2H2 zinc-finger proteins that regulate presenilin transcription in Caenorhabditis elegans. Development 130(10): 2117–2128
19. Jarriault S, Greenwald I (2002) Suppressors of the egg-laying defective phenotype of sel-12 presenilin mutants implicate the CoREST corepressor complex in LIN-12/Notch signaling in C. elegans. Genes Dev 16(20):2713–2728
20. Hornsten A, Lieberthal J, Fadia S, et al. (2007) APL-1, a Caenorhabditis elegans protein related to the human beta-amyloid precursor protein, is essential for viability. Proc Natl Acad Sci U S A 104(6):1971–1976
21. Sakaguchi-Nakashima A, Meir JY, Jin Y, Matsumoto K, Hisamoto N (2007) LRK-1, a C. elegans PARK8-related kinase, regulates axonal-dendritic polarity of SV proteins. Curr Biol 17(7):592–598
22. Schmidt E, Seifert M, Baumeister R (2007) Caenorhabditis elegans as a model system for Parkinson's disease. Neurodegener Dis 4(2–3):199–217
23. Springer W, Hoppe T, Schmidt E, Baumeister R (2005) A Caenorhabditis elegans Parkin mutant with altered solubility couples alpha-synuclein aggregation to proteotoxic stress. Hum Mol Genet 14(22):3407–3423
24. Wu Y, Wu Z, Butko P, et al. (2006) Amyloid-beta-induced pathological behaviors are suppressed by Ginkgo biloba extract EGb 761 and ginkgolides in transgenic Caenorhabditis elegans. J Neurosci 26(50):13102–13113
25. Link CD (1995) Expression of human beta-amyloid peptide in transgenic Caenorhabditis elegans. Proc Natl Acad Sci U S A 92(20):9368–9372
26. Holzenberger M, Dupont J, Ducos B, et al. (2003) IGF-1 receptor regulates lifespan and resistance to oxidative stress in mice. Nature 421(6919):182–187
27. Cohen E, Bieschke J, Perciavalle RM, Kelly JW, Dillin A (2006) Opposing activities protect against age-onset proteotoxicity. Science 313(5793):1604–1610
28. Zarranz JJ, Alegre J, Gomez-Esteban JC, et al. (2004) The new mutation, E46K, of alpha-synuclein causes Parkinson and Lewy body dementia. Ann Neurol 55(2):164–173
29. Cabin DE, Shimazu K, Murphy D, et al. (2002) Synaptic vesicle depletion correlates with attenuated synaptic responses to prolonged repetitive stimulation in mice lacking alpha-synuclein. J Neurosci 22(20):8797–8807
30. Chandra S, Gallardo G, Fernandez-Chacon R, Schluter OM, Sudhof TC (2005) Alpha-synuclein cooperates with CSPalpha in preventing neurodegeneration. Cell 123(3): 383–396

31. Larsen KE, Schmitz Y, Troyer MD, et al. (2006) Alpha-synuclein overexpression in PC12 and chromaffin cells impairs catecholamine release by interfering with a late step in exocytosis. J Neurosci 26(46):11915–11922
32. Murphy DD, Rueter SM, Trojanowski JQ, Lee VM (2000) Synucleins are developmentally expressed, and alpha-synuclein regulates the size of the presynaptic vesicular pool in primary hippocampal neurons. J Neurosci 20(9):3214–3220
33. Morley JF, Brignull HR, Weyers JJ, Morimoto RI (2002) The threshold for polyglutamine-expansion protein aggregation and cellular toxicity is dynamic and influenced by aging in Caenorhabditis elegans. Proc Natl Acad Sci U S A 99(16):10417–10422
34. Satyal SH, Schmidt E, Kitagawa K, et al. (2000) Polyglutamine aggregates alter protein folding homeostasis in Caenorhabditis elegans. Proc Natl Acad Sci U S A 97(11):5750–5755
35. van Ham TJ, Thijssen KL, Breitling R, Hofstra RM, Plasterk RH, Nollen EA (2008) C. elegans model identifies genetic modifiers of alpha-synuclein inclusion formation during aging. PLoS Genet 4(3):e1000027
36. Kraemer BC, Zhang B, Leverenz JB, Thomas JH, Trojanowski JQ, Schellenberg GD (2003) Neurodegeneration and defective neurotransmission in a Caenorhabditis elegans model of tauopathy. Proc Natl Acad Sci U S A 100(17):9980–9985
37. Kraemer BC, Burgess JK, Chen JH, Thomas JH, Schellenberg GD (2006) Molecular pathways that influence human tau-induced pathology in Caenorhabditis elegans. Hum Mol Genet 15(9):1483–1496
38. Praitis V, Casey E, Collar D, Austin J (2001) Creation of low-copy integrated transgenic lines in Caenorhabditis elegans. Genetics 157(3):1217–1226
39. Wilm T, Demel P, Koop HU, Schnabel H, Schnabel R (1999) Ballistic transformation of Caenorhabditis elegans. Gene 229(1–2): 31–35
40. Jorgensen EM, Mango SE (2002) The art and design of genetic screens: Caenorhabditis elegans. Nat Rev Genet 3(5):356–369
41. Koch R, van Luenen HG, van der HM, Thijssen KL, Plasterk RH (2000) Single nucleotide polymorphisms in wild isolates of Caenorhabditis elegans. Genome Res 10(11):1690–1696
42. Jansen G, Hazendonk E, Thijssen KL, Plasterk RH (1997) Reverse genetics by chemical mutagenesis in Caenorhabditis elegans. Nat Genet 17(1):119–121
43. Zwaal RR, Broeks A, van MJ, Groenen JT, Plasterk RH (1993) Target-selected gene inactivation in Caenorhabditis elegans by using a frozen transposon insertion mutant bank. Proc Natl Acad Sci U S A 90(16): 7431–7435
44. Carroll D, Beumer KJ, Morton JJ, Bozas A, Trautman JK (2008) Gene targeting in Drosophila and Caenorhabditis elegans with zinc-finger nucleases. Methods Mol Biol 435:63–77
45. Morton J, Davis MW, Jorgensen EM, Carroll D (2006) Induction and repair of zinc-finger nuclease-targeted double-strand breaks in Caenorhabditis elegans somatic cells. Proc Natl Acad Sci U S A 103(44):16370–16375
46. Ashrafi K, Chang FY, Watts JL, et al. (2003) Genome-wide RNAi analysis of Caenorhabditis elegans fat regulatory genes. Nature 421(6920):268–272
47. Rual JF, Ceron J, Koreth J, et al. (2004) Toward improving Caenorhabditis elegans phenome mapping with an ORFeome-based RNAi library. Genome Res 14(10B):2162–2168
48. Boutros M, Ahringer J (2008) The art and design of genetic screens: RNA interference. Nat Rev Genet 9(7):554–566
49. Nollen EA, Garcia SM, van HG, et al. (2004) Genome-wide RNA interference screen identifies previously undescribed regulators of polyglutamine aggregation. Proc Natl Acad Sci U S A 101(17):6403–6408
50. van Haaften G., Vastenhouw NL, Nollen EA, Plasterk RH, Tijsterman M (2004) Gene interactions in the DNA damage-response pathway identified by genome-wide RNA-interference analysis of synthetic lethality. Proc Natl Acad Sci U S A 101(35):12992–12996
51. Eimer S, Lakowski B, Donhauser R, Baumeister R (2002) Loss of spr-5 bypasses the requirement for the C.elegans presenilin sel-12 by derepressing hop-1. EMBO J 21(21):5787–5796
52. Wen C, Levitan D, Li X, Greenwald I (2000) spr-2, a suppressor of the egg-laying defect caused by loss of sel-12 presenilin in Caenorhabditis elegans, is a member of the SET protein subfamily. Proc Natl Acad Sci U S A 97(26):14524–14529
53. van Haaften G, Romeijn R, Pothof J, et al. (2006) Identification of conserved pathways of DNA-damage response and radiation protection by genome-wide RNAi. Curr Biol 16(13):1344–1350

Chapter 14

Zebrafish (Danio rerio) as a Model Organism for Dementia

Rob Willemsen, Sandra van't Padje, John C. van Swieten, and Ben A. Oostra

Abstract

Zebrafish, a freshwater tropical fish, is a premiere model organism to study vertebrate development. Fast external development and transparency during embryogenesis allow for visual screening at the macroscopical and microscopical level, including visualization of organogenesis. High fecundity and short generation times facilitate genetic analyses. Zebrafish may be a particular powerful model for the study of human disease because many cellular processes are conserved throughout vertebrate evolution, including the corresponding disease genes. Finally, the ability to manipulate gene expression has broad usefulness in the study of modeling human disease, including dementia.

Key words: Zebrafish, Morpholino, knock down, transgenic, TILLING, dementia

1. Introduction

The Latin name for zebrafish is *Danio* (formerly *Brachydanio*) *rerio*, which originates from the river Ganges in India and is a common aquarium fish throughout the world (1). Zebrafish belong to the cyprinid family of teleost fish (Fig. 1). In 1981, George Streisinger introduced the freshwater tropical zebrafish as a genetic model to study vertebrate development. Because of its transparent embryo that develops outside the mother's body, the zebrafish represents an ideal vertebrate model system to study embryonic development. All developmental stages, including organogenesis, are clearly visible within the embryo and are described in detail by Kimmel et al. and Haffter et al. (2,3).

Zebrafish is a relatively simple vertebrate and there is considerable conservation of pathways across species. This means that despite the divergence of teleost fish more than 400 million years

Peter Paul De Deyn and Debby Van Dam (eds.), *Animal Models of Dementia*, Neuromethods, vol. 48,
DOI 10.1007/978-1-60761-898-0_14,

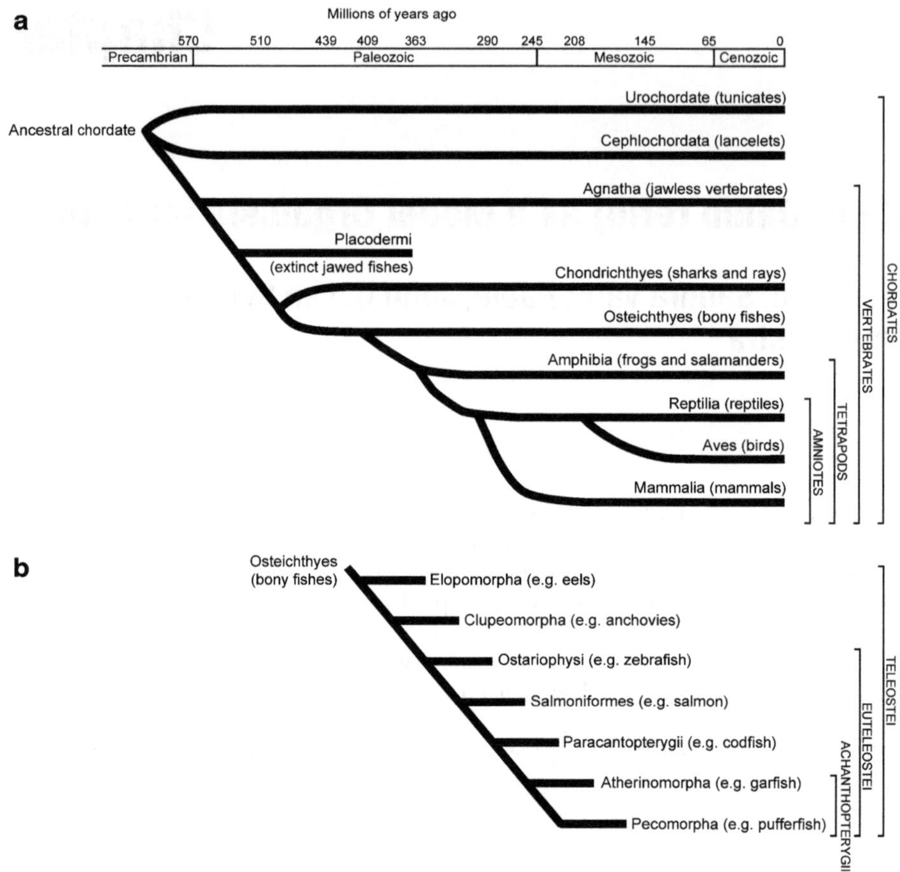

Fig. 1. Phylogenetic trees of vertebrates and fish. (**a**) Evolutionary tree of the chordates and vertebrates. (**b**) Evolutionary tree of the bony fish (Osteichthyes). Zebrafish belongs to the euteleosti and ostariophysi.

ago, they are more closely related to humans than invertebrates and, thereby, offer numerous advantages to researchers interested in many aspects of embryonic development, physiology, and disease. By virtue of their being more closely related to humans, they have many biological functions comparable to human and, hence, many cellular processes have been conserved. Furthermore, it is becoming more and more clear that all vertebrates follow an evolutionary conserved developmental program.

The small size of the zebrafish (3–5 cm), their ability to be kept in large numbers, and the ease of breeding make them easy to maintain. Breeding and collecting eggs from the zebrafish is relatively easy. Since zebrafish are photoperiodic in their breeding, they are kept at a day–night cycle with an automatic timer (14 h light/10 h dark). The day before fertilization, male and female fish are maintained separate using a special breeding tank with two separate compartments divided by a removable partition. By simply removing the partition in the morning, shortly after sunrise, embryos will be produced. As mentioned above, the fertilization of eggs occurs externally and females are very fecund generating hundreds of eggs on a weekly basis.

The embryos develop quickly from a single cell to something that is recognizable as a fish after 24 h of development. Between 4 and 8 h post fertilization (hpf), several characteristic processes occur, including epiboly (migration of cells over the yolk sac), involution, and convergent extension. These processes start with the migration of the cells (approximately 1,200 cells) over the yolk followed by extensive rearrangements. Subsequently, in the next 3 h, the three primary embryonic germ layers are formed through cell movements, a process called gastrulation. After gastrulation is complete, at around 11 hpf, the basic vertebrate body plan has developed and the formation of the first individual somites will begin. Somitogenesis starts anterior (close to the head) and sequentially moves towards the tail of the embryo (posterior). At 18 hpf, 18 somite pairs are formed and the total number that eventually forms is variable, from 30 to 34 pairs. At 24 hpf, a heartbeat and associated blood flow can be recognized. Within 48 hpf, all common vertebrate-specific body features can be seen. Larvae hatch and are able to swim and search for food within 5 days. The generation time of the zebrafish is 2–3 months. For a schematic presentation, see Fig. 2. The external fertilization and the development outside the mother make it easy to access the embryos and to manipulate them, including exposure of larvae/fish to water-soluble chemicals and drugs. The transparency of the embryo enables us to follow the development *in vivo*. The accessibility and transparency in combination with a fluorescent marker make it possible to visualize detailed developmental cellular movements using real-time imaging. In contrast, the embryonic development of a mouse takes 21 days. To study mouse embryos, the mother has to be sacrificed to get at the embryos, which sacrifices them as well. Owing to these features, the zebrafish has become an important model organism to determine the *in vivo* function of a gene (functional genomics) during embryonic development.

Two general approaches are used to reveal the function of a gene: forward and reverse genetics. The process of forward genetics starts with a mutant phenotype and moves towards the gene; this was the first approach in functional genomics. The attractive features of zebrafish, as mentioned above, led to large-scale mutagenesis screenings in zebrafish (3–5). It is easy to chemically induce mutations in the germ cells in the testis of adult zebrafish, which may give a phenotype in their offspring. The process of reverse genetics starts with a particular gene and assays the effect of its disruption, that is, studying the phenotype associated with the mutant gene. This approach usually focuses on the inactivation of a gene and requires a way to selectively mutate a chosen gene. In general, the inactivation of the chosen gene is accomplished by homologous gene targeting, resulting in a complete (or sometimes

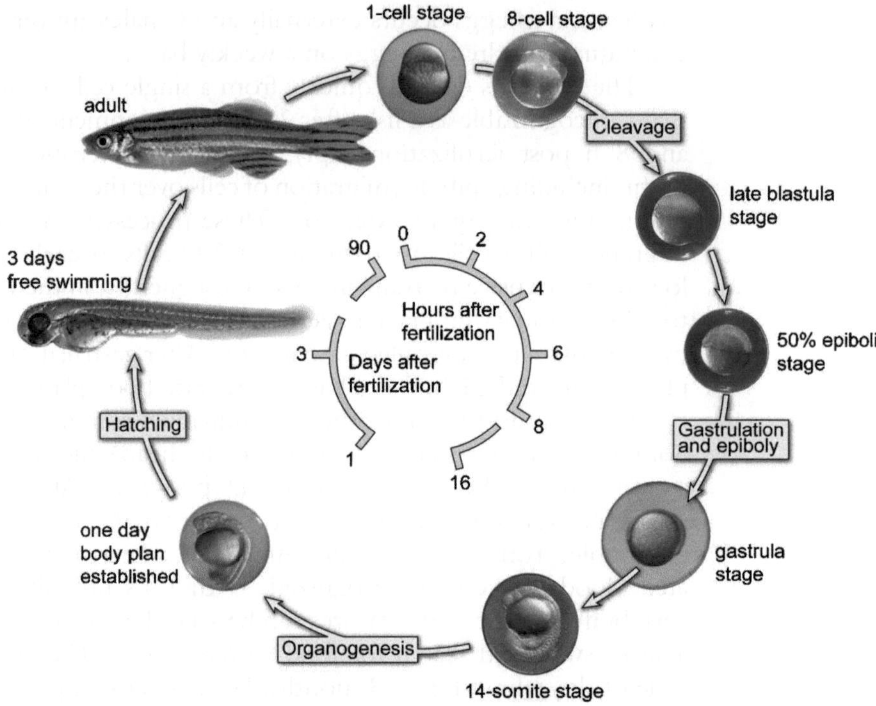

Fig. 2. A schematic representation of the embryonic developmental stages of zebrafish. The cycle starts at the top with a fertilized single stage. The embryos develop quickly to a 14-somite stage within 16 h. After 24 h, the complete body plan of the embryo has been established meaning that all organs are present. Two days after fertilization, the embryos hatch and become free swimming. Zebrafish are adult at 3 months of age and fertile.

conditional) knockout of the gene. Tools for reverse genetics typically include: gene knockouts/knock-ins using homologous recombination in embryonic stem (ES) cells, gene knock down using morpholino antisense technology (MO; see section gene knock down technologies for detailed information) or RNA Interference (RNAi), and targeted induced local lesions in genomes (TILLING) technology. Some of these applications in zebrafish research will be discussed later in more detail.

Zebrafish models generated by both forward and reverse genetics are not only used as a genetic or developmental model system. In the last decade, zebrafish has been successfully applied as an organism to elucidate the etiology of human disease. Zebrafish models of human disease are widely used in many different fields of medical research, like cancer, infectious diseases, cardiovascular disease, kidney disease, diabetes, blindness, deafness, digestive diseases, hematopoiesis, muscle disorders, and neural disorders (5,6).

In summary, the zebrafish has become a well-established model organism, making important contributions to the identification and characterization of genes and pathways involved in development, organ function, and behavior. Additionally, it has become a valuable resource for identifying genes involved in human disease.

2. The Zebrafish Genome

The zebrafish is a vertebrate with a diploid genome consisting of 25 chromosomes (1n). Although the exact number of genes in zebrafish is currently unknown, estimates about the number of base pairs point to approximately 1.7×10^9 base pairs for the haploid genome. The zebrafish is expected to have at least the same number of genes as the humans. Almost all human genes can be found in the zebrafish and approximately 20% of the human genes have two orthologs in zebrafish. This suggests genome duplication shortly before the teleost radiation, which was either partial or was followed by rapid gene loss. Importantly, if two orthologs are present, they often show different expression patterns (both spatial and temporal). This suggests that the function of the ancestral gene has been divided up between two orthologs with more restricted (less complex) functions (7,8).

Importantly, orthologs of the genes involved in familial Alzheimer's disease and frontotemporal dementia have been identified in zebrafish as well, including presenilin 1, presenilin 2, *APP*, Progranulin (*GranulinA* and *GranulinB*), and *TDP-43* (TAR DNA binding protein of 43 kDa) (9–13).

In February 2001, the Sanger Institute started sequencing the genome of the zebrafish (www.sanger.ac.uk/Projects/D_rerio/). This whole genome-sequencing project leads to the identification of genes of which the *in vivo* function is unknown. The assembled genomic zebrafish sequences are publicly available (http://www.ensembl.org/Danio_rerio/; http://vega.sanger.ac.uk/Danio_rerio/).

3. Mutagenesis Screens

The zebrafish is very well suitable for large-scale forward genetic screens in which phenotypic defects are identified before the identification of the gene causing these defects. This is due to its large quantity of eggs, short generation time, and the external development of the transparent embryos. In addition, an important practical advantage is that zebrafish sperm can be frozen for future studies. The three techniques to induce mutations in zebrafish include chemical mutagenesis, gamma irradiation, and insertional mutagenesis (5).

Chemical mutagenesis is the most favorable and efficient method applied thus far (3,14). The chemical mutagenesis by exposing adult zebrafish to *N*-ethyl-*N*-nitrosourea (ENU) is used for these screens. ENU, an alkylating agent, generates point mutations throughout the entire genome in premeiotic germ cells by transferring its ethyl group to individual bases of the DNA, which are misread by DNA polymerase in subsequent replications. However, most induced mutations are recessive and must be rendered to homozygosity to reveal a phenotype. This is accomplished by a multigeneration backcross model (3). Mutagenized adult male zebrafish are crossed with wild-type females. The F1 offspring are heterozygous for individual mutations and are once more crossed with wild-type females. The resulting F2 generation is intercrossed randomly to produce F3 families in which homozygous mutations occur. In the F1 and F2 generation, rare dominant mutations might occur, but most ENU-induced mutations are recessive. Finally, the F3 embryos are further selected on defects in organogenesis using microscopic examination between 1 and 5 days post fertilization (dpf). Positional cloning can then identify the affected gene in isolated mutants. Large numbers of mutations that disrupt embryonic development have now been isolated in the zebrafish, many of which may serve as models for human diseases or syndromes. Further characterization of these mutants (~2,000) and identification of the genetic defect will advance our knowledge of the pathogenesis of the corresponding human disease. It will advance our understanding of the underlying molecular basis of the disease and ultimately may lead to the development of drugs aimed at treatment of the disease.

The second approach to induce mutations disrupting developmental processes in zebrafish is radiation, mainly gamma. Gamma-ray mutagenesis produces a very high locus mutation rate of approximately 1:100 and has mainly been used in screens for morphological defects (2). In contrast to chemical mutagenesis, gamma rays induce translocations and large deletions at high frequency in the zebrafish genome and thus chemical mutagenesis is the method of first choice (15).

The third alternative approach to induce mutations in zebrafish is insertional mutagenesis, which can be established by injection of plasmid DNA, a mouse pseudotyped retrovirus or using a P-element transposon as insertional mutagens (16). For retroviruses, a molecular tag at the site of the mutagenic lesion enables detection of the mutated gene. Although the efficiency of the mutagenesis is less than with ENU mutagenesis, the detection is seven to eight times higher than for ENU-induced mutations (17,18). These genetic screens allow the identification of novel genes and mutants for specific organs or processes. After examining the phenotypes by random mutagenesis, the mutation responsible for the specific defect has to be found using positional cloning. Major drawback of forward genetics is that it is slow and laborious due to positional cloning methods.

Mutagenic screening technology using a reverse genetic approach has been established as well. In 2002, owing to the lack of a working protocol to produce an ES-cell-based knockout (or targeted gene expression), the TILLING technique was developed (19–21). TILLING involves randomly induced mutations by ENU and subsequent screening for mutations in target genes. This screen is an enzyme-mediated (CEL-I endonuclease) mismatch recognition procedure to detect heterozygous germline mutations in the F1 generation. Further generation of embryos with mutant phenotypes is similar to the breeding scheme described above. TILLING can be performed in a high-throughput setup. The only disadvantage is that the mutations are randomly introduced. The TILLING method would lead to more null mutants in zebrafish than achieved by homologous recombination in mice. Initially, TILLING was developed to reduce the time and costs of mutation detection using DNA sequencing. However, recent advances in DNA sequencing technology make this method equal to TILLING.

4. Transgenic and Knockout Zebrafish

Genetically modified animal models are widely used to characterize the function of many newly identified (disease) genes. Transgenic techniques in the mouse to generate transgenic and loss-of-function mutations are well established and have significantly improved our understanding of the roles of specific gene products. Genetically modified mice also serve as valuable models to study the pathogenesis of human disease and to test or develop experimental treatment regimes. However, with the zebrafish emerging as an important model organism to study human disease, the development of similar or additional genetic techniques specifically focused on zebrafish was needed. Methods for generating a transgenic zebrafish are pseudotyped retrovirus infection (17,22,23), transposons (24–26), transfection of sperm nuclei (27), and DNA microinjection. The latter is the most frequently used method for generating transgenic lines expressing a gene of interest. DNA microinjection can be achieved by injection of plasmid DNA or bacterial artificial clones (BACs) into the cytoplasm of a 1-cell stage embryo. The frequency of DNA integration into the germline by microinjection in zebrafish is 1–30%, which is comparable to mouse (28). Coinjection of I-SceI meganuclease and a construct flanked by meganuclease recognition sites has been shown to improve the integration in fish (29). The gene of interest, often cloned in a fusion vector containing a cDNA encoding Enhanced Green Fluorescent Protein (EGFP), is randomly integrated or under the control of a general or tissue-specific promoter. This approach has been successfully applied in the generation of a transgenic zebrafish model for dementia (see later).

Another approach to investigate the function of a (disease) gene in an animal model is to inactivate or disrupt the gene of interest. A major advancement in the ability to generate mouse disease models was the development of technology that makes it possible to introduce loss-of-function mutations into endogenous genes and then transmit these through the mouse germline (30). The desired null mutations are first created via homologous recombination in ES cells, which contribute to all cell lineages when injected into blastocysts. However, mouse developmental genetics is impeded by the high cost of maintenance of animals and by the intrauterine mode of development. Because of the expense and effort required to produce a genetically modified mouse, and the inaccessibility of the embryos inside the mother, the zebrafish might be the vertebrate model to allow these genetic techniques.

The strategy to generate knockout zebrafish by the germline transmission of targeted loss-of-function alleles using ES cells, as described for mice, has not yet been achieved. The only method to produce a knockout zebrafish was mutagenesis followed by targeted screening for point mutations as described earlier (TILLING) (19). Pluripotent zebrafish ES cell lines have been established (31). Recently, targeted incorporation of plasmid DNA into these cells by homologous recombination followed by *in vitro* drug selection was successful (32). In addition, the authors were able to introduce these ES cells, expressing a marker gene such as EGFP, into host embryos using microinjection techniques and achieved contribution to the germline. Although the frequency of germline chimera production was 2–4%, the availability of large quantities of fertilized eggs makes it potentially feasible to establish a knockout fish line in the near future (32).

5. Gene Knock Down Technologies

Zebrafish reverse genetics is slowly catching up with *Drosophila* and/or mouse, as the techniques to perform gene-specific knock downs, target-selected mutagenesis, and transgenesis in zebrafish are quickly developing. Next to the ability to make a knockout (see above), the generation of a morpholino-mediated knock down zebrafish was the favorite technology thus far to study gene function in zebrafish (6,33). Currently, the use of antisense modified oligonucleotides is still widely applicable due to their ease and quick results. Although morpholinos have been tested in different species, including sea urchin, ascidian, frog, chicken, and mouse, the most favorable model organism to test morpholinos has been carried out on zebrafish embryos. MOs are synthetic oligonucleotides of 25 bases, which hybridize specifically to complementary sequences of mRNA, thereby disrupting translation initiation or pre-mRNA splicing. The backbone of MOs is similar to the backbone of DNA or RNA, but with some changes. In MOs, the ribose or deoxyribose sugar molecules that link the bases of the DNA or RNA are replaced by morpholino rings (hence the name). Anionic phosphates of bases replace nonionic phosphorodiamidate linkages. Because of this modified backbone, MOs are uncharged, very stable, and cannot be degraded by nucleases (for more details about MOs, see www.gene-tools.com). The reduction of translation will never be 100%, but can be up to 90%, and is therefore called a knock down (Fig. 3). Zebrafish embryos displaying a phenotype as a consequence of ectopic MO administration are called morphants (34).

The MOs can either be microinjected in the cytoplasm of a one-cell stage embryo or into the yolk of a one- to four-cell stage embryo. MOs can diffuse into the cell until the 16-cell stage when a membrane is formed between the yolk and the cells that will form the actual embryo. MOs can be dissolved in Danieau buffer, which is thought to reduce lethality. As an injection tracer, 0.05%

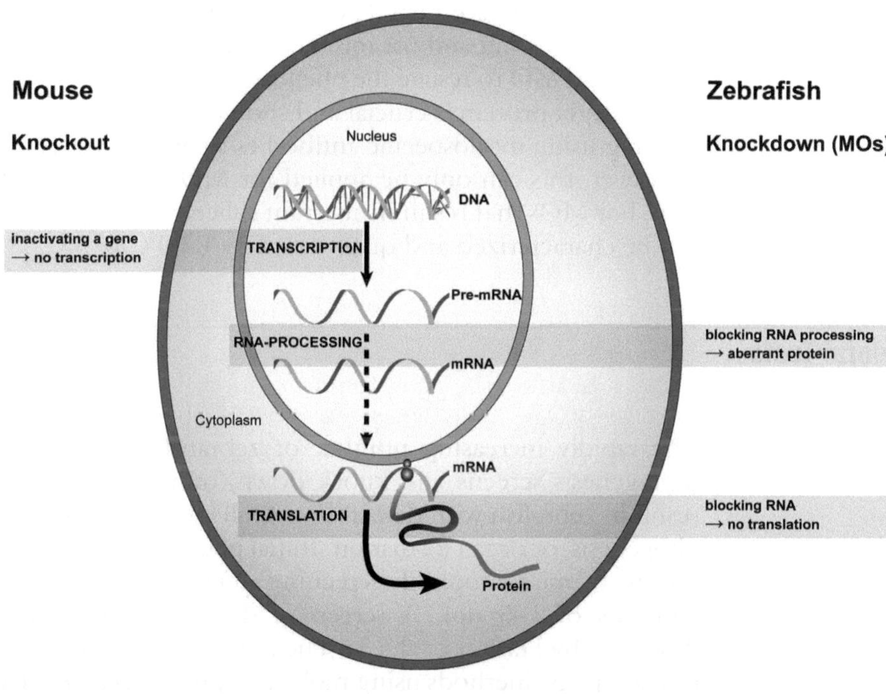

Fig. 3. Differences between the gene knockout technologies, often used in mice, versus the gene knock down technology in zebrafish. In case of knockout, the gene becomes fully inactivated at the level of transcription and subsequent total lack of protein. Knock down by morpholinos occurs either on the level of RNA processing or translation. The injected morpholino technology can be designed against either a splice site or start site of the target mRNA. When a morpholino is targeted against a splice site, an exon could be spliced out or an intron could be spliced in during pre-mRNA processing. This incorrect splicing of the mRNA leads to aberrant or truncated protein. If the morpholino is targeted against the start site, the translation initiation complex is blocked and results in reduced protein expression (up to 90%). Another important difference between knockout and knock down is the effect, which is 100% for knockout and maximum 90% for knock down.

Phenol Red can be added, which also works as a pH indicator and will turn yellow if the MO solution is acidic. MOs will directly bind to endogenous (also maternal) mRNAs at the translational start site and disrupt translation initiation. However, MOs can also be designed against a splice junction, thereby preventing correct splicing into a mature mRNA (20,35). This aberrant splicing includes exon deletion or intron insertion, which might lead to premature stop or a nonfunctional protein. Knock down of gene function is transient and effective until 3–5 days of development due to dilution of the MOs. MOs exert their effect throughout embryogenesis in a dose-dependent fashion, thereby allowing the identification of morphants that might be masked or lethal in case of null mutations (34). Interestingly, MOs against different genes of interest can be coinjected simultaneously to either study/ exclude redundant function or assess interactions between gene products within a pathway. Large quantities of eggs can be microinjected in one experiment and due to the fast development of the embryos, a phenotype can often be observed just after 24 h. MO microinjection might cause nonspecific side–effects; thus control experiments are necessary. These controls might include microinjection of a second and/or mismatch MO or coinjection of target mRNA and MO to rescue the phenotype. The efficacy of depletion of the target protein is crucial and should be analyzed by Western blotting using monospecific antibodies against the target protein. However, this can only be applied for MOs that disrupt translation. For MOs that result in aberrant spliced mRNA, all outcomes can be characterized and quantitated by RT-PCR (35).

6. Behavioral Tests

The rapidly increasing number of zebrafish mutants from the mutagenesis screens and knock down/transgenic strategies will result in zebrafish with defects in overall embryonic pattern, morphogenesis, or organ formation. Initial phenotypic characterization is done by microscopically screening all morphological features of embryos of 1–5 dpf. A screen of defects in organogenesis is described by Haffter et al. (3). The next step is to characterize the phenotype by methods using molecular probes like *in situ* hybridization and/or (immuno) histochemical techniques or biochemical tests. Other important aspects of characterization of the phenotype are behavioral studies. For mice, these tests are well defined; however, with respect to zebrafish behavior little is known.

Zebrafish is a typical diurnal schooling fish that prefers light over dark during the day. However, in response to danger, they hide in the dark. Males exhibit territoriality, including dancing movements and agonistic behavior. Zebrafish embryos (0–5 dpf) already exhibit simple sensory and locomotor abilities, whereas larvae (5–14 dpf) possess many patterns of behavior. Some simple behavioral tests concerning locomotion of the embryos/larvae have been described, including rhythmic tail movements, the escape response, equilibrium control, and the touching assay. More sophisticated assays to identify defects in optokinetic and phototactic behaviors have been described as well (36–40). Recently, a system to monitor behavior in zebrafish has been established. Ethovision from Noldus is a computerized video tracking system that enables recording movement of animals including swimming patterns of small-sized fish. The fish can be monitored in an open tank as well as a 96-well plate for high-throughput screens. The coordinates of the swimming performance can be stored and used for further software analysis. Many behavioral tests can be performed due to automated locomotion recording and data analysis.

Learning and memory have been studied in many model organisms, each requiring the use of specific behavioral paradigms. Also, for zebrafish a simple spatial alternation paradigm for evaluation of spatial learning and memory function has been developed (41). This paradigm is based on an aquatic version of an alternation task (T-maze) commonly used for rats and mice. The fish receive a food reward after choosing the correct arm when they observe a light tap on the tank. A similar approach has been applied for the use of different colors to show visual discrimination learning. The aquatic T-maze can be combined with an automated tracking and analysis system (e.g. Ethovision) to track zebrafish reliably at a high sample rate. In addition, a simple associative learning test in zebrafish has recently been described based on learning paradigms related to identification of conspecifics (42,43).

These newly developed behavioral assays for zebrafish will help to characterize (new) zebrafish mutants produced by large-scale forward genetic screens and reverse-genetics approaches.

7. Zebrafish and Dementia

In general, zebrafish has shown great utility as a model organism for studying the underlying molecular mechanisms of human neurodegenerative disorders, including different types of dementia. In this chapter we focus on two familial forms of dementia; Alzheimer's disease (AD) and frontotemporal lobar degeneration (FTLD).

For familial AD, mutations in three genes have been identified: amyloid precursor protein (*APP*), presenilin 1 (*PSEN1*), and presenilin 2 (*PSEN2*). Importantly, orthologs of the genes involved in familial AD have been identified in zebrafish (9,11,12,44). Transcripts from *psen1* are maternally expressed, whereas *psen2* transcripts are expressed after zygotic transcription. Compelling evidence shows that presenilin interacts with the *Notch* signaling pathway, which is also present in zebrafish. Recent studies have demonstrated a direct intersection between early development and neurodegeneration later in life, because both Notch and *APP* are substrates for presenilin. All the molecular components to process *APP* are conserved in zebrafish, which allows studies to investigate presenilin function and regulation *in vivo*. Indeed, an MO strategy has been applied to study presenilin function in zebrafish. Zebrafish embryos microinjected with MOs to inhibit translation of *psen1* mRNA exhibit somite formation defects (11). Similar defects are seen in *Psen1* knockout mice. Elucidating the molecular mechanisms underlying the processing of the substrates of presenilin, like membrane proteins *Notch*, and *APP*, will reveal knowledge about the signal transduction pathways involved.

Another strategy is the generation of transgenic zebrafish that express one of the genes with an AD mutation. Importantly, these transgenic fish should be adult viable and with a clear phenotype. A first step to study the effect of mutant human APP expression on the development of AD was achieved by the generation of transgenic zebrafish expressing EGFP under control of zebrafish *app* gene regulatory elements (45). EGFP expression was found to be present in subregions of brain and spinal cord. Expression in the brain started during the first day of development and then increased in intensity during later development. Strikingly, EGFP expression was also present in the developing and adult vasculature. Of course, the next step is to apply this vector to clone a PCR product containing mutant human *APP*. Expression of mutant human *APP* may lead to the formation of Aβ plaques and depending on the level of expression at earlier ages. In addition, in this way, effects of environmental toxicants and natural and synthetic products on AD susceptibility and progression can be assessed in a high-throughput fashion by exposure to these compounds at embryonic stages or later stages.

Also, the genes involved in FTLD have been identified in zebrafish. It remains unclear how mutations in the gene encoding microtubule-associated Tau (*MAPT*) are related to the neuronal and glial tau pathology. To study the functional consequences and transport kinetics of human *MAPT* containing mutations associated with hereditary FTLD, transgenic zebrafish lines have been generated (46,47). For this, four repeat human MAPT-EGFP constructs under control of either a neural-specific GATA-2 promoter or enolase-2 promoter were microinjected into zebrafish embryos at the one to two-cell stage. Transgenic zebrafish expressed four repeat large human Tau isoforms at high levels compared with normal human brain throughout the zebrafish brain, which resulted in the disruption of cytoskeletal filaments, as well as the presence of neurofibrillary tangles (NFT) that closely resembled the NFT in human disease (47). However, to establish a zebrafish tauopathy model, studies should be focused on whether zebrafish develop the full spectrum of phenotypic characteristics similar to human tauopathies. If so, these transgenic lines can be used to test several hypotheses to clarify the pathogenesis of tau-FTLD, including cytoskeletal organization, transport kinetics of mutant tau, and formation of NFT. In addition, the model should be applied to introduce into living neurons other compounds involved in the pathogenesis of neurodegenerative disorders.

Another form of hereditary FTLD, namely FTLD with ubiquitin-immunoreactive inclusions (FTLD-U), is caused by mutations in the progranulin gene (*PGRN*). All identified mutations result in null alleles and lead to 50% loss of functional *PGRN*. Recently, TDP-43 has been identified as a component of the ubiquitin-positive inclusions. In zebrafish, two orthologs are present of human PGRN and TDP-43, named *granulinA* and *granulinB*, and *tardbp* and *tardbpl*, respectively (13). *GranulinB* is expressed in zebrafish brain, while *granulinA* expression is restricted to hematopoetic lineage. In contrast, *tardbp* and *tardbpl* are co-expressed in zebrafish brain. To investigate a possible relationship between reduced expression of *granulinB* and both *tardbp/tardbpl* relocalization and deposition, knock down studies have been performed using antisense gripNAs against an exon–intron junction between exon 6 and intron 6–7. Effective knock down of *granulinB* did not result in a change of localization or marked reduction of nuclear staining (13).

In conclusion, zebrafish is a valid model organism in which to study the genes involved in hereditary dementias for their interactions, regulation, and function. The knowledge gathered by these studies will contribute to our understanding of the pathogenesis of sporadic forms of dementia as well.

References

1. Wullimann MF, RB RH (1996) Neuroanatomy of the zebrafish brain; A topological atlas. Birkhauser Verlag, Basel, Switzerland
2. Kimmel CB, Ballard WW, Kimmel SR, Ullmann B, Schilling TF (1995) Stages of embryonic development of the zebrafish. Dev Dyn 203:253–310
3. Haffter P, Granato M, Brand M, et al. (1996) The identification of genes with unique and essential functions in the development of the zebrafish, Danio rerio. Development 123:1–36
4. Driever W, Solnica-Krezel L, Schier AF, et al. (1996) A genetic screen for mutations affecting embryogenesis in zebrafish. Development 123:37–46
5. Amsterdam A, Hopkins N (2006) Mutagenesis strategies in zebrafish for identifying genes involved in development and disease. Trends Genet 22:473–478
6. Lieschke GJ, Currie PD (2007) Animal models of human disease: Zebrafish swim into view. Nat Rev Genet 8:353–367
7. Amores A, Force A, Yan YL, et al. (1998) Zebrafish hox clusters and vertebrate genome evolution. Science 282:1711–1714
8. Postlethwait JH, Yan YL, Gates MA, et al. (1998) Vertebrate genome evolution and the zebrafish gene map. Nat Genet 18:345–349
9. Groth C, Nornes S, McCarty R, Tamme R, Lardelli M (2002) Identification of a second presenilin gene in zebrafish with similarity to

the human Alzheimer's disease gene presenilin2. Dev Genes Evol 212:486–490

10. Leimer U, Lun K, Romig H, et al. (1999) Zebrafish (Danio rerio) presenilin promotes aberrant amyloid beta-peptide production and requires a critical aspartate residue for its function in amyloidogenesis. Biochemistry 38:13602–13609
11. Nornes S, Groth C, Camp E, Ey P, Lardelli M (2003) Developmental control of Presenilin1 expression, endoproteolysis, and interaction in zebrafish embryos. Exp Cell Res 289:124–132
12. Musa A, Lehrach H, Russo VA (2001) Distinct expression patterns of two zebrafish homologues of the human APP gene during embryonic development. Dev Genes Evol 211:563–567
13. Shankaran SS, Capell A, Hruscha AT, et al. (2008) Missense mutations in the progranulin gene linked to frontotemporal lobar degeneration with ubiquitin-immunoreactive inclusions reduce progranulin production and secretion. J Biol Chem 283:1744–1753
14. Driever W, Fishman MC (1996) The zebrafish: Heritable disorders in transparent embryos. J Clin Invest 97:1788–1794
15. Fritz A, Rozowski M, Walker C, Westerfield M (1996) Identification of selected gamma-ray induced deficiencies in zebrafish using multiplex polymerase chain reaction. Genetics 144:1735–1745
16. Amsterdam A, Nissen RM, Sun Z, Swindell EC, Farrington S, Hopkins N (2004) Identification of 315 genes essential for early zebrafish development. Proc Natl Acad Sci U S A 101:12792–12797
17. Chen W, Burgess S, Golling G, Amsterdam A, Hopkins N (2002) High-throughput selection of retrovirus producer cell lines leads to markedly improved efficiency of germ line-transmissible insertions in zebra fish. J Virol 76:2192–2198
18. Amsterdam A, Burgess S, Golling G, et al. (1999) A large-scale insertional mutagenesis screen in zebrafish. Genes Dev 13:2713–2724
19. Wienholds E, Plasterk RH (2004) Target-selected gene inactivation in zebrafish. Methods Cell Biol 77:69–90
20. Draper BW, Morcos PA, Kimmel CB (2001) Inhibition of zebrafish fgf8 pre-mRNA splicing with morpholino oligos: A quantifiable method for gene knockdown. Genesis 30:154–156
21. Stemple DL (2004) TILLING-a high-throughput harvest for functional genomics. Nat Rev Genet 5:145–150
22. Gaiano N, Allende M, Amsterdam A, Kawakami K, Hopkins N (1996) Highly efficient germ-line transmission of proviral insertions in zebrafish. Proc Natl Acad Sci U S A 93:7777–7782
23. Linney E, Hardison NL, Lonze BE, Lyons S, DiNapoli L (1999) Transgene expression in zebrafish: A comparison of retroviral-vector and DNA-injection approaches. Dev Biol 213:207–216
24. Fadool JM, Hartl DL, Dowling JE (1998) Transposition of the mariner element from Drosophila mauritiana in zebrafish. Proc Natl Acad Sci U S A 95:5182–5186
25. Kawakami K, Amsterdam A, Shimoda N, et al. (2000) Proviral insertions in the zebrafish hagoromo gene, encoding an F-box/WD40-repeat protein, cause stripe pattern anomalies. Curr Biol 10:463–466
26. Raz E, van Luenen HG, Schaerringer B, Plasterk RH, Driever W (1998) Transposition of the nematode Caenorhabditis elegans Tc3 element in the zebrafish Danio rerio. Curr Biol 8:82–88
27. Jesuthasan S, Subburaju S (2002) Gene transfer into zebrafish by sperm nuclear transplantation. Dev Biol 242:88–95
28. Udvadia AJ, Linney E (2003) Windows into development: Historic, current, and future perspectives on transgenic zebrafish. Dev Biol 256:1–17
29. Thermes V, Grabher C, Ristoratore F, et al. (2002) I-SceI meganuclease mediates highly efficient transgenesis in fish. Mech Dev 118:91–98
30. Hickman-Davis JM, Davis IC. Transgenic mice (2006) Paediatr Respir Rev 7:49–53
31. Fan L, Crodian J, Collodi P (2004) Production of zebrafish germline chimeras by using cultured embryonic stem (ES) cells. Methods Cell Biol 77:113–119
32. Fan L, Moon J, Crodian J, Collodi P (2006) Homologous recombination in zebrafish ES cells. Transgenic Res 15:21–30
33. Heasman J (2002) Morpholino oligos: Making sense of antisense? Dev Biol 243:209–214
34. Nasevicius A, Ekker SC (2000) Effective targeted gene 'knockdown' in zebrafish. Nat Genet 26:216–220
35. Sumanas S, Larson JD (2002) Morpholino phosphorodiamidate oligonucleotides in zebrafish: A recipe for functional genomics? Brief Funct Genomic Proteomic 1:239–256
36. Granato M, van Eeden FJ, Schach U, et al. (1996) Genes controlling and mediating locomotion behavior of the zebrafish embryo and larva. Development 123:399–413
37. Brockerhoff SE, Hurley JB, Janssen-Bienhold U, Neuhauss SC, Driever W, Dowling JE (1995) A behavioral screen for isolating

zebrafish mutants with visual system defects. Proc Natl Acad Sci U S A 92:10545–10549

38. Brockerhoff SE, Hurley JB, Niemi GA, Dowling JE (1997) A new form of inherited red-blindness identified in zebrafish. J Neurosci 17:4236–4242
39. Burgess HA, Granato M (2007) Modulation of locomotor activity in larval zebrafish during light adaptation. J Exp Biol 210:2526–2539
40. Giacomini NJ, Rose B, Kobayashi K, Guo S (2006) Antipsychotics produce locomotor impairment in larval zebrafish. Neurotoxicol Teratol 28:245–250
41. Williams FE, White D, Messer WS (2002) A simple spatial alternation task for assessing memory function in zebrafish. Behav Processes 58:125–132
42. Saverino C, Gerlai R (2008) The social zebrafish: Behavioral responses to conspecific, heterospecific, and computer animated fish. Behav Brain Res 191:77–87
43. Al-Imari L, Gerlai R (2008) Sight of conspecifics as reward in associative learning in zebrafish (Danio rerio). Behav Brain Res 189:216–219
44. Nornes S, Newman M, Verdile G, et al. (2008) Interference with splicing of Presenilin transcripts has potent dominant negative effects on Presenilin activity. Hum Mol Genet 17:402–412
45. Lee JA, Cole GJ (2007) Generation of transgenic zebrafish expressing green fluorescent protein under control of zebrafish amyloid precursor protein gene regulatory elements. Zebrafish 4:277–286
46. Tomasiewicz HG, Flaherty DB, Soria JP, Wood JG (2002) Transgenic zebrafish model of neurodegeneration. J Neurosci Res 70:734–745
47. Bai Q, Garver JA, Hukriede NA, Burton EA (2007) Generation of a transgenic zebrafish model of Tauopathy using a novel promoter element derived from the zebrafish eno2 gene. Nucleic Acids Res 35:6501–6516

Chapter 15

Spontaneous Vertebrate Models of Alzheimer Dementia: Selectively Bred Strains (SAM Strains)

Renã A. Sowell and D. Allan Butterfield

Abstract

The senescence-accelerated mouse (SAM) strains, consisting of nine SAM-prone (SAMP) mice strains and three SAM-resistant (SAMR) strains, have been used extensively as models for various age-related disorders. SAMP mice undergo accelerated aging while SAMR mice undergo normal aging processes. One of the most employed SAM strains is SAMP8, which has deficits in learning and memory. Coupled to age-dependent deposition of amyloid β-peptide, such deficits allow it to serve as a good model of dementia-related disorders such as Alzheimer's disease (AD). Many studies have characterized the behavioral, pathological, genetic, and protein abnormalities of SAMP8 mice. Interestingly, genes and proteins that undergo significant alterations in SAMP8 brains are related to the following functional categories: neuroprotection, signal transduction, immune response, energy metabolism, mitochondrion, protein folding and degradation, reactive oxygen species production, cytoskeleton and transport, lipid abnormalities, and cholinergic dysfunction. This chapter provides a summary of these findings with regard to better understanding of AD pathogenesis.

Key words: Senescence-accelerated mouse strain (SAM), Dementia, Alzheimer's disease, Murine models, SAMP8

1. Introduction

The senescence-accelerated mouse (SAM) strains are one of the many murine models used to better understand the molecular basis of aging and age-related disorders such as Alzheimer's disease (AD). In 1975, Takeda et al. noticed litters of AKR/J mice that either exhibited early age senility and shortened lifespan or normal aging, and through a selective inbreeding process, they developed SAM populations that were either prone (SAMP) or resistant (SAMR) to accelerated aging (1,2). There are eight major SAMP strains and three SAMR strains (generally used as

Peter Paul De Deyn and Debby Van Dam (eds.), *Animal Models of Dementia*, Neuromethods, vol. 48,
DOI 10.1007/978-1-60761-898-0_15, © Springer Science+Business Media, LLC 2011

controls) that are commonly used to investigate various age-related disorders, such as senile amyloidosis, immune dysfunction, cataracts, lymphoma, sarcomas, colitis, osteoporosis, and kidney disorders (3). Table 1 provides a list of these strains and their accompanying age-related phenotypes.

Most relevant to AD research is the SAMP8 strain, which exhibits senile amyloidosis and displays age-related progressive deficits in learning and memory (3,4). Specifically, SAMP8 mice exhibit significant impairments in the passive avoidance task (4,5) due to severe acquisition disability (6). These animals have a minor spatial learning deficit (5), reduced anxiety-like behavior (7), and disordered circadian rhythms (6). Senile dementia patients and elderly aged individuals have disordered circadian rhythms that cause reduced night-time sleeping (8). In addition to these changes, SAMP8 also have impaired immune function (9).

Table 1
Phenotypes associated with SAM strains

Strain	Phenotypes
SAMP1	Senile amyloidosis, contracted kidney, impaired immune response, impaired auditory system, retinal atrophy, hypertensive vascular disease, pulmonary hyperinflation, deficits in learning and memory
SAMP2	Senile and secondary amyloidosis, impaired immune response, contracted kidney, cataracts, alveolar bone loss
SAMP3	Degenerative arthrosis
SAMP6	Senile osteoporosis, secondary amyloidosis
SAMP7	Senile amyloidosis, thymic lymphoblastic lymphoma
SAMP8	Deficits in learning and memory, impaired immune response, emotional disorder, abnormal circadian rhythms, age-dependent deposition of amyloid β-peptide
SAMP9	Cataracts, thymic lymphoblastic lymphoma, senile amyloidosis
SAMP10	Deficits in learning and memory, brain atrophy, emotional disorder abnormal circadian rhythms,
SAMP11	Senile amyloidosis, contracted kidney, age-related thickening of tunic media of thoracic aorta
SAMR1	Normal aging, nonthymic lymphoma, histiocytic sarcoma, ovarian cysts
SAMR4	Normal aging, nonthymic lymphoma, histiocytic sarcoma
SAMR5	Normal aging, colitis

SAMP8 animals have life spans that generally range from 10 to 17 months, while their SAMR1 counterparts have longer life spans of 19–21 months (10). Based on its shorter life span, age-associated pathologies, defined environmental conditions, and genetic characteristics, SAMP8 is a widely accepted spontaneous mammalian model for accelerated aging (11) and cognitive impairment.

SAMP8 mice have been widely used to understand biological processes associated with aging and AD. Interestingly enough, SAMP8 mice do not have the characteristic hallmarks of AD, which includes extracellular senile plaques (SP) and intracellular neurofibrillary tangles (NFT). However, SAMP8 mice do deposit amyloid β-peptide (Aβ) with increasing age. Moreover, memory loss and learning are behaviors characteristic of AD, and thus this model can still provide valuable information about aging processes, dementia, and disease pathogenesis. This review will describe in some detail the pathological alterations of SAMP8 animals that make it a suitable model for dementia and AD-related research. In addition, several reports have examined changes in gene (12–19) and/or protein (20–32) expression in SAMP8 mice with respect to SAMR1 mice or as a function of age. Such studies have provided insights into pathways associated with accelerated aging and the cognitive deficits underlying AD. Butterfield and Poon recently provided a mini-review of both genetic and protein alterations of SAMP8 mice (33) by focusing on specific functionally related groups of genes and proteins. In this chapter, an updated synopsis of gene and protein expression changes in SAMP8 brains with respect to AD will also be provided.

2. Pathological Alterations in SAMP8 Mice

Recent reviews have detailed the pathological alterations associated with SAMP8 mice (34,35); however, here only a brief synopsis of SAMP8 pathology is provided. Various regions of SAMP8 brains show 1.5–2.5-μm-sized, irregular-shaped Aβ protein-like immunoreactive granular structures (β-LIGS) (34). Aβ peptides are derived from the amyloid precursor protein (APP), and are the main constituents of SP in AD brain. The number of β-LIGS and the level of Aβ (and APP) increase with age in these animals and provide some insight into the processing of APP.

SAMP8 brains show age-related increases in spheroidal axonal dystrophy in the dorsal column nuclei (34). Brain atrophy is a prominent feature of SAMP10 animals (see Table 1), which also show age-related memory deficits. Studies of spheroidal axonal dystrophy in SAMP8 animals suggest that axons and terminals are susceptible targets of aging (36). Spine density decreases in dendrites of hippocampal neurons in SAMP8 mice relative to SAMR1 mice and may cause neuronal membrane deterioration (36,37). Several neurotransmitters and receptors, including *N*-methyl-D-aspartate, acetylcholine, serotonin, dopamine, opiods, γ-aminobutyric acid, and α-adrenoreceptors, also decrease with age in SAMP8 brains (38–44). Other neuronal changes in this strain include age-related shrinkage of cholinergic neurons (45,46), early age (i.e., 4–8 weeks old) increases in glucose metabolism (47), and contrasting, older age reduction in glucose metabolism (48). Glucose is the primary source of energy in the brain, and perhaps relatedly, AD brains have altered glucose metabolism (49–51).

Other neuronal changes in SAMP8 brains involve glial alterations. For example, variously sized vacuoles representative of spongy degeneration reach a maximum size in 4–8-month-old mice, while none are present in SAMR1 animals (4,52). The severity of spongy degeneration generally correlates with memory and learning deficits (34), and results in the presence of activated microglia (53), the proliferation of glial fibrillary acidic protein-positive astrocytes (34), and the appearance of periodic acid Schiff-positive granular substances (54). These changes are indicative of an inflammatory or immune response in the vacuole regions of SAMP8 brains that are associated with spongy degeneration (34). Finally, blood–brain barrier dysfunction related to macromolecules is prominent in SAMP8 (55,56).

Accompanying all the pathological alterations discussed here, SAMP8 also undergo increased oxidative stress with age (57) that correlates with learning and memory deficits (22). Antioxidant treatment in SAMP8 animals has been shown to decrease oxidative stress levels and improve performance on learning and memory tasks (22,27,57,58), thus providing evidence in support of the oxidative stress/Harman's free radical theory of aging (59). Similar benefits are also observed when SAMP8 are intracerebroventricularly injected with antisense oligonucleotides directed at the Aβ region of the *APP* gene (29,60,61) or with Aβ antibody (62,63). These studies show that increased expression of Aβ is related to learning and memory deficits and oxidative stress in SAMP8, hence, making it a good model to study Aβ-related effects in AD.

3. Gene Expression Changes in SAMP8 Mice

AD patients suffer from either familial or sporadic forms of the disease, with a majority of the cases arising from the latter form. Familial AD is associated with mutations in the following genes: *APP* (64), *presenilin-1* and *presenilin-2* (*PS2*) genes (65), *apolipoprotein E* (*apoE*) (66), *endothelial nitric oxide synthase-3* gene (67), and *α-2-macroglobulin* (68). Substantial evidence links Aβ (1–42), derived from *APP*, to the pathogenesis of AD (69,70). In SAMP8 mice, the *APP* gene differs from human *APP* by an amino acid substitution of alanine for valine at residue 300 (71). Despite this similarity, *APP* cDNA shows no familial AD gene mutations in SAMP8 mice (71), although increases in *APP* mRNA are related to the learning and memory deficits observed in this strain (63). Furthermore, subsequent decreases in *APP* expression by antisense oligonucleotide brain injection improves cognitive deficits in these mice (60,61,71), thereby providing additional evidence to support the importance of *APP* in cognitive deficiency in AD patients and SAMP8 mice.

The SAMP8 genome is unique from other SAM strains and the parental AKR/J strain (3). Various SAMP8 and CD-1 crossbred strains revealed that the number of SAMP8 genes is proportional to the degree of learning and memory impairment (72). Recent reports have examined gene expression changes in the brains of SAMP8 animals relative to SAMR1 controls (12–17), and as a function of age in SAMP8 mice (12,18,19). Table 2 provides a comprehensive list of more than 140 genes with known functions whose expression changes in the brains of aged (i.e., 11–16 months) SAMP8 mice relative to SAMR1 controls or as a function of age in the SAMP8 strain (e.g., 3 vs. 16 months). While a detailed discussion of individual genes may prove beneficial to understanding of network mechanisms in AD pathogenesis, it is beyond the scope of this chapter to pursue this line of discussion. The altered SAMP8 brain genes listed in Table 2 can be grouped into the following general biological categories: neuroprotection, mitochondrion-related, immune response, signal transduction, energy metabolism, cytoskeleton and transport, protein folding and degradation, nucleic acid and protein synthesis, and others.

3.1. Neuroprotection, Signal Transduction, Protein Folding, and Degradation-Related Genes

As shown in Fig. 1, more than half of the altered genes in SAMP8 brain are related to pathways involving neuroprotection, signal transduction, or protein folding and degradation. The best-known neuroprotection-related genes in Table 2 are APP, ApoE, and the microtubule-associated protein, tau. APP is upregulated in SAMP8 mice, which results in increased Aβ deposition (15). The extracellular senile plaques found in AD brains contain Aβ oligomers, protofilaments, and amyloid-derived diffusible ligands (73). While SAMP8 animals have no senile plaque formation, increased APP can also contribute to increased amounts of β-LIGS (as described earlier). ApoE was found to decrease in SAMP8 brain, which can affect neuronal plasticity and the fibril formation of Aβ (19). Tau is also increased in SAMP8 brain (12). Tau is implicated in AD pathology because hyperphosphorylation of tau

Table 2
Functional characterization of altered genes in SAMP8 mice brain

Gene	Gene symbol	Expression change	Reference
Neuroprotection			
Amyloid precursor protein	APP	↑	(71)
Apolipoprotein E	ApoE	↓1.20	(18)
Cytochrome P450 IIB9	Cyp2b9	↓2.10	(15)
Cytochrome P450 IIIA	Cyp3a	↑2.86	(15)
Glial cell line-derived neurotrophic factor	Gdnf	↓	(17)
Glutathione S-transferase α 4	Gsta4	↑2.3	(12)
Heat shock protein 2	Hspa2	↑2.0	(12)
Heat shock protein 4-like	Hspa4l	↓1.8	(12)
Heat shock 70kD protein 5	Hspa5	↓1.8	(12)
Heme oxygenase 1	Hmox1	↓1.26	(15)
Metallothionein 2	Mt2	↑1.8	(12)
Microsomal UDP-glucorosyltransferase 1-1 precursor		↓1.46	(15)
Nerve growth factor	Ngf	↓	(14)
Neurotrophin-3	Ntf3	↓	(14)
Protein disulfide isomerase-related protein	P5	↓1.7	(12)
Quinone oxidoreductase; NADPH	Nqo2	↑8.74	(15)
Tau	Mapt	↑1.34, ↑1.6[a]	(18), (12)
Thioredoxin interacting protein	Txnip	↑1.7	(12)
Transthyretin	Ttr	↑4.3,[b] ↑4.5[a]	(12)
Signal transduction			
Adrenergic receptor kinase, β 1	ADRBK1	↑2.09	(13)
Calcium/calmodulin-dependent protein kinase type II α chain	CAMK2A	↓24.51	(13)
Calmodulin binding protein 4	Strn4	↓2.07	(13)
Calsyntenin-1	Cstn1	↑2.26	(13)
Cyclic AMP phosphoprotein 19 kDa	Arrpp19	↓8.6	(12)
Dickkopf homolog 3 (Xenopus laevis)	Dkk3	↑2.0	(12)
Dual specificity protein phosphatase containing protein	DUSP12	↓1.84	(13)
Endothelial differentiation, lysophosphatidic acid G-protein coupled receptor 2	Edg2	↓2.1	(12)

(continued)

Table 2 (continued)

Gene	Gene symbol	Expression change	Reference
Eph receptor B6	Ephb6	↑2.06	(13)
Eukaryotic translation initiation factor 2B subunit 1	EIF-2B	↑	(19)
Fibroblast growth factor inducible 14	Fin14	↑1.97	(13)
Folate receptor 1 (adult)	Folr1	↑2.8	(12)
FYVE coiled-coil domain containing 1	Fyco1	↓1.80	(13)
GABA-A receptor α 2	Gabra2	↓2.3	(12)
GABA-A receptor β 3	Gabrb3	↓2.0	(12)
GABA-A receptor, subunit δ	Gabrd	↓3.0	(12)
JNK stress-activated protein kinase		↓3.07	(15)
Karyopherin β3	Kpnb3	↑1.72	(13)
KDEL endoplasmic reticulum protein retention receptor 1	KDELR1	↑1.92	(13)
Mineralocorticoid receptor	MR	↓1.57	(18)
Mitogen-activated protein kinase 4	MAPKK4	↑1.12	(15)
Signal transduction			
Mitogen-activated protein kinase 6	Mapk6	↓2.1	(12)
Mitogen-activated protein kinase kinase kinase kinase 6	Map4k6-pending	↓1.96	(13)
Mitogen-actived protein kinase p38	MAPK p38	↓3.29	(15)
N-ethylmaleimide sensitive fusion protein	Nsf	↑1.8	(12)
Neuronal pentraxin 1	Nptx1	↑2.3	(12)
Neurotrophic tyrosine kinase receptor type 2	Ntrk2	↑1.92	(13)
Peroxisome proliferator activated receptor β		↓1.27	(15)
Phosphodiesterase 1B	Pde1b	↓3.8	(12)
Phospholipase D	Pld	↑	(19)
Preproenkephalin 1	Penk1	↓11.6	(12)
Protein kinase C-γ	Prkcc	↑2.2	(12)
Protein kinase raf1		↑2.22	(13)
Protein phosphatase 1-regulatory subunit 1A	Ppp1r1a	↓7.1	(12)
Rab26	RAB26	↓1.91	(13)
RAS-guanyl releasing protein 2	Rasgrp2	↓5.6	(12)

(continued)

Table 2 (continued)

Gene	Gene symbol	Expression change	Reference
Regulator of G-protein signaling 2	Rgs2	↓3.7	(12)
Reticulolalbin 2		↑2.81	(13)
Rho-associated coiled-coil forming kinase 1	Rock1	↑1.71	(13)
STIP1 homology and U-Box containing protein 1	STUB1	↓3.14	(13)
Transforming protein RhoB	RhoB	↓1.25	(15)
Tetratricopeptide repeat domain 3	Ttc3	↓1.91	(13)
Tubby	Tub	↓6.5	(12)
Nucleic acid and protein synthesis			
40S ribosomal protein S21	Rps21	↑2.97	(13)
Aryl hydrocarbon receptor nuclear translocator 2	ARNT2	↑3.24	(13)
CUG triplet repeat-RNA binding protein 2	Cugbp2	↓1.7	(12)
D site albumin promoter binding protein	Dbp	↑2.5	(12)
Germinal center associated nuclear protein	GANP	↓1.66	(13)
Neurogenic differentiation 6	Neurod6	↓1.6	(12)
Nuclear receptor subfamily 4, A1	Nr4a1	↓2.5	(12)
Pre B-cell leukemia transcription factor 3	Pbx3	↓6.0	(12)
Rnase H		↓2.74	(13)
RNA binding motif protein 3	Rbm3	↓1.9	(12)
SWI/SNF-related, matrix-associated, actin-dependent, regulation of chromatin, c1	Smarcc1	↑2.0	(12)
Upstream transcription factor 1	USF1	↓1.9	(12)
WW domain binding protein 11	Wbp11	↑2.2	(12)
Zinc finger protein 238	Zfp238	↑2.14	(13)
Zinc finger protein 36-C3H type-like 2	Zfp36l2	↓1.6	(12)
Protein folding and degradation			
Calnexin precursor		↓1.15	(15)
Endoplasmin precursor		↑1.88	(15)
Phospholipase C-alpha	Pdia3	↓2.61	(15)
Presenilin 2	PS-2	↑2.13	(19)
Sarcolemmal-associated protein	Slmap	↓1.8	(12)

(continued)

Table 2
(continued)

Gene	Gene symbol	Expression change	Reference
T-complex protein 1 δ subunit	Tcp1δ	↓6.69	(15)
T-complex protein 1 ε subunit	Tcp1ε	↓1.75	(15)
T-complex protein 1 η subunit	Tcp1η	↑10.38	(15)
T-complex protein 1 γ subunit	Tcp1γ	↓1.92	(15)
T-complex protein 1 θ subunit	Tcp1θ	↑1.44	(15)
T-complex protein 1 ζ subunit	Tcp1ζ	↓5.46	(15)
T-complex proteins α and β subunits	Tcpα&β	↑3.51	(15)
Ubiquitin-like protein NEDD8		↑2.11	(15)
Energy metabolism			
Glycogen-debranching enzyme isoform		↑	(19)
Insulin-like growth factor binding protein 2	Igfbp2	↑2.0	(12)
Δ 9-desaturase		↓2.00	(16)
Mitochondrion-related			
ATP-binding cassette, D3	Abcd3	↑2.6	(12)
Cytochrome c oxidase subunit I	MTCO1	↑2.28	(13)
Cytochrome c oxidase subunit III	MTCO3	↑1.75	(13)
Fragile histidine triad	Fhit	↓2.03	(13)
Glycerol phosphate dehydrogenase 1, mitochondrial	Gdm1	↑1.8	(12)
GrpE-like 2-mitochondrial	Grpel2	↑2.7	(12)
Ubiquinol-cytochrome c reductase subunit	UQCRFS1	↑2.32	(13)
Immune response			
Glucocorticoid receptor-alpha	GR alpha	↓2.64	(19)
Chemokine (C-C motif) ligand 19	CCL19	↑8.2	(12)
Chemokine (C-C motif) ligand 27	CCL27	↑1.6	(12)
Complement component 4	C4	↓2.6,[b] ↑ 2.2[a]	(12)
Histocompatibility 2, D region	H2-D	↑1.8	(12)
Ig superfamily containing leucine-rich repeat	Islr	↑1.9	(12)
Leukocyte receptor cluster (LRC) member 8	Leng8	↓3.1	(12)
NFK light chain gene enhancer in B-cells inhibitor, α	Nfkbia	↑2.8	(12)
Recombination activating gene 1	RAG-1	↑3.98	(15)

(continued)

Table 2 (continued)

Gene	Gene symbol	Expression change	Reference
Cytoskeleton and transport			
Activated leukocyte cell adhesion molecule	Alcam	↓2.7	(12)
Apolipoprotein D	Apod	↑1.9	(12)
Bullous pemphigoid antigen		↓	(19)
Ca channel voltage dependent-a2/d subunit 3	Cacna2d3	↓1.9	(12)
Complexin 2	Cplx2	↓1.9	(12)
Decorin	Dcn	↑4.7	(12)
Cytoskeleton and transport			
Dynactin 3	Dctn3	↑2.9	(12)
Dynein cytoplasmic heavy chain 1	Dnchc1	↑8.51	(13)
Ectonucleotide pyrophosphatase/ phosphodiesterase 2	Enpp2	↑2.2	(12)
Fatty acid binding protein 7, brain	Fabp7	↓2.1	(12)
Hemoglobin, β adult minor chain	Hbb-b2	↑1.7	(12)
K+ voltage gated channel-shaker related B1	Kcnab1	↓2.2	(12)
Kinectin	Ktn1	↑2.21	(13)
Kinesin family member 5B	Kif5b	↓3.6	(12)
Kinesin family member 5C	Kif5c	↓1.7	(12)
Microtubule-associated protein 2	Mtap2	↓1.8	(12)
Myosin 1b	Myo1b	↓2.8	(12)
Neurofilament protein L	NF-L	↑2.11	(13)
Opioid receptor σ 1	Oprs1	↑2.2	(12)
Procollagen-type VI α 1	Col6a1	↓1.9	(12)
Ryanodine receptor 1-skeletal muscle	Ryr1	↓3.0	(12)
Ubiquitous kinesin heavy chain		↓1.12	(15)
Other			
Cold inducible RNA-binding protein	Cirbp	↑2.4	(12)
Cysteine and histidine-rich domain (CHORD)-containingzinc-binding protein 1	Chordc1	↓2.0	(12)
Ecotropic viral integration site 2	Evi2	↓1.7	(12)

(continued)

Table 2 (continued)

Gene	Gene symbol	Expression change	Reference
Erythroid differentiation regulator	edr	↓3.0	(12)
Hepatoma-derived growth factor	Hdgf	↓2.1	(12)
Inhibitor of DNA binding 4	Idb4	↓1.8	(12)
Intracisternal A particles	Iap	↑1.9	(12)
Keratin complex 1-acidic-gene 10	Krt1-10	↓2.5	(12)
Melanoma antigen-80 kDa	Mela	↑26.9,[b] ↑3.2[a]	(12)
Nasal embryonic LHRH factor	NELF	↑2.8	(12)
Paternally expressed 3	Peg3	↓4.1	(12)
Spindlin	Spin1	↓2.6	(12)
Serine protease inhibitor 2-2	Serpina3n	↑4.1	(12)

[a]Gene expression change observed in a comparison of young vs. old SAMP8 brain in (12).
[b]Gene expression change observed in old SAMP8 brain comparison relative to age-matched SAMR1 brains in (12).

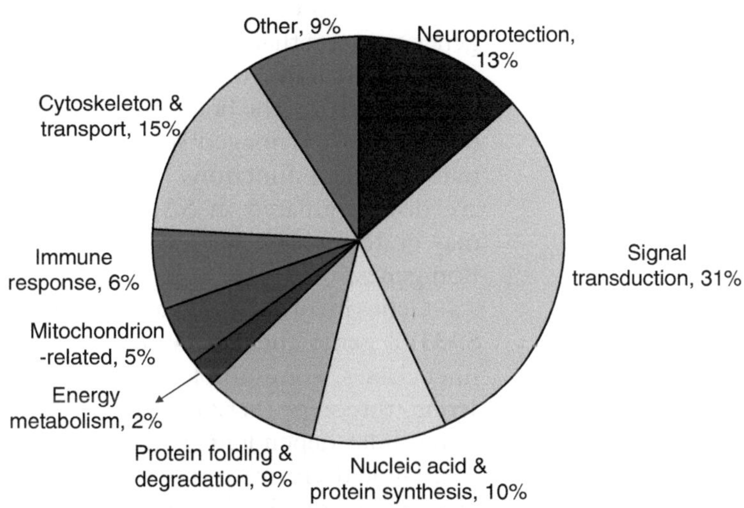

Fig. 1. A pie-chart representation of altered genes in SAMP8 brains (144 total genes, see Table 2) related to biological functions.

protein leads to NFT formation, another pathological hallmark of AD. The slight increase in tau expression in SAMP8 mice relative to SAMR1 mice (1.34-fold increase), and as a function of age (1.6-fold increase) (12), suggests that if protein expression follows gene expression, tau may undergo increased phosphorylation and be indirectly related to the cognitive deficits of SAMP8 mice.

Other altered neuroprotective genes include stress response-related proteins such as heat shock proteins (Hsps) and redox regulatory and oxidative stress-related genes. For example, Hspa4-like (1.8-fold decrease), Hspa5 (1.8-fold decrease), heme oxygenase 1 (1.26-fold decrease), and Hspa2 (2.0-fold increase) belong to the Hsp family that functions as protein chaperones, thus regulating the proper folding/unfolding of proteins. Overall, Hsps are downregulated in SAMP8 mice relative to SAMR1 mice, and as a function of age (12). These observations suggest that Hsp alteration can lead to a decrease in protein chaperoning ability and influence the age-related pathology observed in SAMP8 mice. Furthermore, heme oxygenase 1 has been reported to have altered expression in human AD (74–76), and in neuronal cultures treated with Aβ (1–42) (77). Likewise, redox regulatory and detoxification genes were altered in SAMP8 mice. These include glutathione S-transferase α 4 (2.3-fold increase), metallothionein 2 (1.8-fold increase), quinone oxidoreductase; nicotinamide adenine dinucleotide phosphate (NADPH) (8.74-fold increase), and thioredoxin interacting protein (1.7-fold increase). Both heme oxygenase 1 and thioredoxin are known to be associated with the so-called class of *vitagenes* that offer neuroprotection in AD (78). Upregulation of redox regulatory and detoxification genes suggests that SAMP8 animals may be undergoing oxidative attack by reactive oxygen species (ROS) and are in a state of oxidative stress. The brain's response to this attack would normally be to remove damaged species (i.e., proteins, lipids, DNA) partially by Hsps functions. However, because several Hsp genes are downregulated in SAMP8 brains, another primary action may be to increase the levels of redox regulatory and detoxification genes.

Genes involved in signal transduction represent 31% of altered SAMP8 genes and can be further broken down into several distinct classes. For example, altered SAMP8 genes belonging to the serine/threonine (Ser/Thr) protein kinase family include calcium/calmodulin-dependent protein kinase type II α chain (CAMPK2A, 24.51-fold decrease), c-Jun N-terminal kinase (JNK) stress-activated protein kinase (3.07-fold decrease), mitogen-activated protein kinase 4 (MAPK4, 1.12-fold increase), MAPK6 (2.1-fold decrease), MAPK p38 (3.29-fold decrease), MAPK kinase kinase kinase 6 (MAPK6-pending, 1.96-fold decrease), adrenergic receptor kinase, β1 (ADRBK1, 2.09-fold increase), protein kinase C-γ (PKC-γ, 2.2-fold increase), protein kinase raf 1 (2.22-fold increase), and Rho-associated coiled-coil forming kinase 1 (Rock1, 1.71-fold increase) (12,13,15). The most substantial change observed in Ser/Thr protein kinase altered genes is the downregulation of CAMPK2A. CAMPK2A is rich in the central nervous system and plays roles in neurotransmitter synthesis and release, synaptic plasticity formation, long-term potentiation, and spatial learning that are all regulated by phosphorylation mechanisms (79).

Downregulation of CAMPK2A can reduce the secondary messenger mechanisms of Ca^{2+} required for normal neuronal transmission (80) and, subsequently, result in the observed learning impairments in SAMP8 mice. Second, messenger-independent protein kinases such as JNK stress-activated protein kinase, MAPK6, MAPK p38, and MAPK6-pending, are all downregulated in SAMP8 brain relative to SAMR1 brain (12,13,15) with the exception of MAPK4, which is upregulated (15). On the other hand, PKC-γ, protein kinase raf 1, and Rock1 are all upregulated in SAMP8 mice (12,13). It is apparent that alteration of Ser/Thr protein kinase-related genes results in disturbances to signal transduction pathways in SAMP8 brains. These changes are most likely detrimental to SAMP8 mice because more than 99% of protein phosphorylation occurs on Ser and Thr residues.

Other alterations in SAMP8 hippocampal brain regions include genes in the Tyr protein kinase family such as Eph receptor B6 (2.06-fold increase) and neurotrophic tyrosine kinase receptor type 2 (NTRK2, 1.92-fold increase). NTRK2 is a receptor for the most abundant neurotrophin in the central nervous system and especially in hippocampus, namely neurotrophin-3 (Ntf3), which plays roles in basal forebrain cholinergic neurons. Cholinergic loss is another known pathological hallmark of AD (81,82). Interestingly, both NTRK and Ntf3 are downregulated in SAMP8 animals, suggesting that deficiencies in the phosphorylation signaling events in cholinergic neurons may contribute to the cognitive deficits in SAMP8 mice. The Tyr and Ser phosphatase families are also altered in SAMP8 brains and include downregulation of dual specificity protein phosphatase containing protein (1.84-fold decrease) (13) and protein phosphatase 1-regulatory subunit 1A (7.1-fold decrease) (12). Tyrosine phosphorylation is involved in most neuron functions such as cell survival and differentiation, axon extension, synaptogenesis, and synaptic transmission. Thus, alterations of genes in the Ser/Thr and Tyr kinase and phosphatase families suggest that dysregulation of phosphorylation events plays a key role in the cognitive deficits observed in SAMP8 mice.

Thirteen altered genes (9%) in SAMP8 brains were associated with protein folding and degradation. PS2 protein is involved in the proteolysis of APP and Notch-1. It is intuitive that increased expression (2.13-fold increase) (19) of PS2 is necessary to maintain proteolytic balance of increased APP levels (see Table 2). As described earlier, PS2 has known genetic mutations that are related to familial AD. Most of the protein folding/degradation genes, however, belong to the T-complex protein (Tcp) family and exhibit variable expression changes. For example, T-complex proteins Tcp1η (10.38-fold increase), Tcp1θ (1.44-fold increase), and Tcpα&β (3.51-fold increase) are upregulated in SAMP8 brains relative to SAMR1, while Tcp1δ (6.69-fold decrease), Tcp1ε (1.75-fold decrease), Tcp1γ (1.92-fold decrease), and Tcp1ζ (5.46-fold decrease) are downregulated in SAMP8 brains.

Tcp genes, after translation, result in chaperoning proteins that play roles in protein folding and stress response. Altered expression of protein folding and degradation-related genes may contribute to the learning and memory impairments observed in SAMP8 mice.

3.2. Energy Metabolism, Mitochondrion, Nucleic Acid, and Protein Synthesis-Related Genes

Genes related to energy metabolism and mitochondria were also altered in SAMP8 mice (see Table 2). Cytochrome c oxidase subunit I (MTCO1, 2.28-fold increase), MTCO3 (1.75-fold increase), glycerol phosphate dehydrogenase 1, mitochondrial (Gdm1, 1.8-fold increase), and ubiquinol-cytochrome c reductase subunit (UQCRFS1, 2.32-fold increase) are example genes that were upregulated in SAMP8 mice. These genes are involved in the respiratory chain processes and glycolysis. For example, MTCO1 is an enzyme that catalyzes the reduction of oxygen to water, while UQCRFS1 belongs to the proton pump involved in ATP synthesis (13). In the respiratory chain, electrons flow from ubiquinol to cytochrome *c*; thus, an increase in UQCRFS1 would require elevated expression of MTCO1 in order to maintain proper respiration. Gdm1 plays a role in the formation of ubiquinol due to its high-potential electrons, and is also involved in gluconeogenesis. Alterations in energy metabolism and mitochondrial-related hypotheses are implicated in AD. Thus, the observed alterations of energy metabolism and mitochondrial-related genes in SAMP8 brains suggest that these pathways play a role in neurodegeneration and cognitive deficits in SAMP8. In addition, Fig. 1 shows that 10% of altered SAMP8 genes were associated with nucleic acid and protein synthesis pathways. As can be seen in Table 2, overall, these altered genes, including transcription factors, do not display specific patterns of regulation. However, disturbances to nucleic acid and/or protein synthesis pathways could result in dysregulation of other crucial genes or downstream proteins necessary for proper cellular communication in the brains of SAMP8 mice.

3.3. Immune Response, Cytoskeleton, Transport-Related, and Other Genes

Immune response and cytoskeleton and transport related-genes represent more than 20% of altered genes with known functions in SAMP8 brains. Proinflammatory genes, such as chemokine (C-C motif) ligand 19 (CCL19, 8.2-fold increase) and CCL27 (1.6-fold increase) are upregulated in SAMP8 brains relative to SAMR1 (12). Variable expression of complement component 4, involved in innate immune response activation, was observed in SAMP8 brains (i.e., 2.6-fold decrease relative to SAMR1 while 2.2-fold increase in SAMP8 as a function of age). Alterations in immune response-related genes imply that SAMP8 brains respond to the presence of pathogens and/or build-up of irregular biomolecules. Such alterations could play a role in SAMP8 pathology such as the presence of β-LIGS.

Cytoskeletal structural integrity and the ability to correctly transport molecules (e.g., proteins, neurotransmitters) are crucial for proper neural transmission and cellular communication.

Several cytoskeleton structural-related genes that are involved in axonal and dendritic transport were downregulated in SAMP8 mice (i.e., kinesin family member 5B, 3.6-fold decrease; kinesin family member 5C, 1.7-fold decrease; microtubule-associated protein 2, 1.8-fold decrease; and myosin 1b, 2.8-fold decrease). On the other hand, microtubule-associated motor proteins involved in the retrograde transport of various biomolecules were upregulated in SAMP8 brains (i.e., dynein cytoplasmic heavy chain 1, 8.51-fold increase; and dynactin 3, 2.9-fold increase). Disturbances to cytoskeletal and transport genes in SAMP8 mice could affect axonal transport, dendritic outgrowth and extension, structural integrity, and proper neural communication in the brain. Furthermore, these alterations may contribute to the observed cognitive deficits and behavioral changes observed in SAMP8 mice. Finally, Table 2 also includes altered genes with various cellular functions that are disrupted in SAMP8 brain and may be related to age-related pathology in this strain.

4. Protein Expression Changes in SAMP8 Mice

The proteins listed in Table 3 represent proteins whose expression level, enzymatic activity, or oxidation level has been observed to change in SAMP8 mice. Similar to the biological functions associated with altered genes in SAMP8 brains, altered proteins can be grouped into the following categories: neuroprotection, signal transduction, energy metabolism, ROS production, protein folding and degradation, cytoskeleton and transport, and lipid abnormalities and cholinergic dysfunction. Overall, Table 3 includes 26 reported altered proteins in SAMP8 brain.

4.1. Energy Metabolism/ Cytoskeleton and Transport-Related Proteins

As was observed for altered genes in SAMP8 brains, altered proteins in SAMP8 mice are also involved in energy metabolism. Our laboratory has identified oxidized proteins in SAMP8 brains as a function of age that are related to energy metabolism, such as lactate dehydrogenase (LDH), creatine kinase, and aldolase 3 (30). Oxidation of proteins can lead to a reduction in activity and the aggregation of proteins (83–87). Aldolase and LDH are involved in glycolysis (i.e., LDH catalyzes the oxidation of pyruvate to generate lactate under anaerobic conditions), and creatine kinase is involved in the regeneration of ATP when ATP levels become low in various tissues (e.g., in brain and muscle). The levels of LDH and triosephosphate isomerase, another glycolytic enzyme, were also reported as increased in aged SAMP8 brain (30). Decreased activity of hexokinase (16), a glycolytic associated protein, and Δ 9-desaturase, a protein potentially involved in altered fatty acid oxidation, altered membrane fluidity, and altered synapse signaling, (87) was observed in SAMP8 brain.

Table 3
Functional characterization of altered proteins in SAMP8 mice brain

Gene	Expression change[a]	Reference
Neuroprotection		
Manganese superoxide dismutase	Activity ↓	(87)
Glutamine synthase	Activity ↓	(22)
		(31)
		(23)
Glutathione peroxidase	Activity ↓	(27)
Catalase	Activity ↓	(31)
Peroxiredoxin 2	Oxidation ↑	(29)
Signal transduction		
Calbindin	Level ↓	(21)
Protein kinase C γ	Level ↓	(21)
Synaptotagmin 1	Level ↑	(88)
Energy metabolism		
Lactate dehydrogenase 2	Level ↓	(29)
	Oxidation ↑	
Triosephosphate isomerase	Level ↑	(29)
Creatine kinase	Oxidation ↑	(29)
Aldolase 3	Oxidation ↑	(29)
Δ 9-desturase	Activity ↓	(29)
Hexokinase	Activity ↓	(87)
ROS production		
Aβ	Level ↑	(29)
Nitric oxide synthase	Activity ↑	(24)
Acyl-CoA oxidase	Activity ↑	(31)
Protein folding and degradation		
Cathepsin D	Level ↑	(20)
Cathepsin E	Level ↑	(20)
Heat shock protein 86	Level ↓	(29)
	Oxidation ↑	
Ubiquitin carboxyl-terminal hydrolase L3	Level ↓	(32)

(continued)

Table 3 (continued)

Gene	Expression change[a]	Reference
Cytoskeleton and transport		
Dihydropyrimidinase-like protein 2	Oxidation ↑	(29)
Neurofilament triplet L protein	Level ↓	(29)
α-Spectrin	Level ↓	(29)
Coronin 1a	Oxidation ↑	(29)
Lipid abnormalities and cholinergic dysfunction		
Hippocampal cholinergic neurostimulating peptide precursor protein	Oxidation ↑	(26)

[a]Protein expression change observed in aged SAMP8 brain relative to age-matched SAMR1 brain or in aged SAMP8 brain relative to young SAMP8 brain. See references for details.

Altered glucose metabolism is observed in AD patients using positron emission tomography (49–51).

Cytoskeleton and transport proteins were also altered in the brains of SAMP8 mice. For example, dihydropyrimidinase-like protein 2 (DRP2), involved in neuronal repair and axon outgrowth, and coronin 1a, involved in cell locomotion and cytokinesis, have increased oxidation in SAMP8 brain as a function of age (29). Neurofilament triplet L protein, which is abundant in neurons, and α-spectrin, which plays important roles in the maintenance of plasma membrane cytoskeletal integrity, have age-related decreases in expression in SAMP8 brain (29). Overall, these reported age-related changes suggest that alterations to metabolic, cytoskeletal, and transport proteins play roles in the cognitive deficits observed in SAMP8 mice.

4.2. Signal Transduction/Protein Folding and Degradation/Lipid Abnormalities and Cholinergic Dysfunction Related Proteins

As listed in Table 3, the signal transduction proteins, calbindin and protein kinase C-γ (PKC-γ), have age-related increased expression levels in SAMP8 brains (29), while synaptotagmin 1 has decreased expression (88). Calbindin is a Ca^{2+} binding protein that helps in the regulation of intracellular Ca^{2+} concentration, and synaptotagmin 1 is an integral membrane protein found in synaptic vesicles and may participate in stimulus-coupled fast chemical transmission by acting as a Ca^{2+} sensor (89). PKC-γ belongs to the family of Ser/Thr protein kinases and is involved in acquisition and retention processes by phosphorylation mechanisms. Alteration of calbindin, PKC-γ, and synaptotagmin levels may be responsible for Ca^{2+} dysregulation that would lead to apoptosis and neuronal death, and to the cognitive deficits observed in SAMP8 mice.

Cathepsin D and E are aspartic proteases involved in the breakdown of other proteins and are reported to have age-related increases in expression levels in SAMP8 brains (20). On the other hand, two other proteins involved in folding and degradation, namely heat shock protein 86 (hsp86) and ubiquitin carboxyl terminal hydrolase L3 (UCHL3), have decreased expression as a function of age (30) and relative to SAMR1 mice (32), respectively. Hsp86 was also reported to show increased oxidation in SAMP8 mice as a function of age (28), which may lead to reduced activity. UCHL3 is part of the ubiquitin-proteasome system. Alterations to proteins involved in folding and degradation could lead to protein aggregation from improper clearance due to proteasome dysfunction and may contribute to the observed pathological and cognitive deficits in SAMP8 mice. Finally, hippocampal cholinergic neurostimulating peptide precursor protein (HCNP-pp), a multifunctional protein involved in lipid abnormalities and cholinergic dysfunction was recently reported to have increased oxidation in SAMP8 mice when compared to SAMR1 mice (26). HCNP is involved in the regulation of choline acetyltransferase and in signal transduction pathways. Choline acetyltransferase loss results in lowered levels of the neurotransmitter, acetylcholine (90), and cholinergic neuronal loss has been reported in AD (81,82).

5. Conclusion and Future Outlook

This review has summarized the to-date findings of behavioral, pathological, genetic, and protein alterations in SAMP8 brain as a function of age and/or in comparison to SAMR1 control mice. Pathological alterations in SAMP8 brain are related to β-LIGS, neuronal shrinkage, neuronal and synaptic loss, axonal dystrophy, spongiform degeneration, granular inclusions, astrogliosis, glial activation, and blood-brain barrier (BBB) dysfunction. Genetic and protein alterations discussed are generally related to but not limited to the following pathways: neuroprotection, ROS production, signal transduction, energy metabolism, mitochondrion, nucleic acid and protein synthesis, immune response, protein folding and degradation, cytoskeleton and transport, and lipid abnormalities and cholinergic dysfunction. The observed genetic and protein changes are consistent with the pathological alterations, behavioral abnormalities and neurochemical changes observed in SAMP8 mice (3,6,34,91). It is evident that impairments to signal transduction at the genetic and protein level are partially responsible for the cognitive deficits observed in SAMP8 mice because they constitute a large number of gene and protein changes. Synaptic dysfunction has been previously observed in SAMP8 brain (33). Changes to ROS producing proteins and impairments in neuroprotective proteins and genes may contribute to the observed oxidative stress and cognitive deficits reported in SAMP8 mice (22,27,57,58,92). Dysregulation of metabolic and mitochondrion-related genes and proteins in SAMP8 mice is consistent with the known decreases in energy metabolism in AD, and thus may be responsible for learning and memory impairments observed in SAMP8 mice. Impairment of genes and proteins involved in protein folding and degradation, cytoskeleton, and transport pathways may lead to poor neuronal communication and protein aggregation and the subsequent cognitive deficits observed in SAMP8 mice. Overall, the pathological, genetic and protein alterations that have been reported in SAMP8 mice has provided substantial insight into the underlying mechanisms of learning and memory impairment in this murine model of dementia-related disorders.

The SAMP8 strain has proven to be a relevant model for AD and further studies may lead to the development of drug targets for disease diagnosis and prevention. For example, antioxidant sulfur-containing compounds found in onions such as di-n-propyl trisulfide have been shown to reverse memory impairments in SAMP8 mice (93). Treatment with lipoic acid, another antioxidant sulfur-containing compound, has been shown to decrease oxidative stress, lead to loss of oxidation of specific brain proteins, and lead to improvement of cognitive functions in SAMP8 mice (22,29). Our laboratory and others have shown that intracerebroventricular injection of antisense oligonucleotides, directed at the Aβ region of the *APP* gene (60,61) or with Aβ antibody (62,63,94), also decreased oxidative stress and restored cognitive function in SAMP8 mice. These results are consistent with the notion that SAMP8 cognitive changes are associated with Aβ-associated oxidative stress. More recent studies have shown that chronic melatonin treatment in SAMP8 decreases oxidative stress, provides neuroprotection (95), and reduces age-dependent inflammation (96). The peptide hormone, leptin, which is permeable across the BBB, improves memory retention in SAMP8 mice (97). The findings reported in this review provide substantial support of the suitability of the SAMP8 strain as an animal model to investigate the underlying mechanisms associated with age-related impairments to learning and memory as is necessary to develop treatments relevant to dementia-related disorders such as AD.

Acknowledgments

This work was supported in part by NIH grants to D.A.B. [AG-10836; AG-05119; AG-029839].

References

1. Takeda T, Hosokawa M, Higuchi K (1991) Senescence-accelerated mouse (SAM): A novel murine model of accelerated senescence. J Am Geriatr Soc 39:911–919.
2. Takeda T, Hosokawa M, Takeshita S, et al. (1981) A new murine model of accelerated senescence. Mech Ageing Dev 17:183–194.
3. Takeda T (1999) Senescence-accelerated mouse (SAM): A biogerontological resource in aging research. Neurobiol Aging 20:105–110.
4. Yagi H, Katoh S, Akiguchi I, Takeda T (1988) Age-related deterioration of ability of acquisition in memory and learning in senescence accelerated mouse: SAM-P/8 as an animal model of disturbances in recent memory. Brain Res 474:86–93.
5. Miyamoto M, Kiyota Y, Yamazaki N, et al. (1986) Age-related changes in learning and memory in the senescence-accelerated mouse (SAM). Physiol Behav 38:399–406.
6. Miyamoto M (1997) Characteristics of age-related behavioral changes in senescence-accelerated mouse SAMP8 and SAMP10. Exp Gerontol 32:139–148.
7. Miyamoto M, Kiyota Y, Nishiyama M, Nagaoka A (1992) Senescence-accelerated mouse (SAM): Age-related reduced anxiety-like behavior in the SAM-P/8 strain. Physiol Behav 51:979–985.
8. Skene DJ, Vivien-Roels B, Sparks DL, et al. (1990) Daily variation in the concentration of melatonin and 5-methoxytryptophol in the human pineal gland: Effect of age and Alzheimer's disease. Brain Res 528:170–174.
9. Abe Y, Yuasa M, Kajiwara Y, Hosono M (1994) Defects of immune cells in the senescence-accelerated mouse: A model for learning and memory deficits in the aged. Cell Immunol 157:59–69.
10. Flood JF, Morley JE (1998) Learning and memory in the SAMP8 mouse. Neurosci Biobehav Rev 22:1–20.
11. Masoro EJ (ed) Animal models in aging research, 3rd edn. Academic, San Diego, CA.
12. Carter TA, Greenhall JA, Yoshida S, et al. (2005) Mechanisms of aging in senescence-accelerated mice. Genome Biol 6:R48.
13. Cheng XR, Zhou WX, Zhang YX, Zhou DS, Yang RF, Chen LF (2007) Differential gene expression profiles in the hippocampus of senescence-accelerated mouse. Neurobiol Aging 28:497–506.
14. Kaisho Y, Miyamoto M, Shiho O, Onoue H, Kitamura Y, Nomura S (1994) Expression of neurotrophin genes in the brain of senescence-accelerated mouse (SAM) during postnatal development. Brain Res 647:139–144.
15. Kumar VB, Franko MW, Farr SA, Armbrecht HJ, Morley JE (2000) Identification of age-dependent changes in expression of senescence-accelerated mouse (SAMP8) hippocampal proteins by expression array analysis. Biochem Biophys Res Commun 272:657–661.
16. Kumar VB, Vyas K, Buddhiraju M, Alshaher M, Flood JF, Morley JE (1999) Changes in membrane fatty acids and delta-9 desaturase in senescence accelerated (SAMP8) mouse hippocampus with aging. Life Sci 65:1657–1662.
17. Miyazaki H, Okuma Y, Nomura J, Nagashima K, Nomura Y (2003) Age-related alterations in the expression of glial cell line-derived neurotrophic factor in the senescence-accelerated mouse brain. J Pharmacol Sci 92:28–34.
18. Wei X, Zhang Y, Zhou J (1999) Differential display and cloning of the hippocampal gene mRNAs in senescence accelerated mouse. Neurosci Lett 275:17–20.
19. Wei X, Zhang Y, Zhou J (1999) Alzheimer's disease-related gene expression in the brain of senescence accelerated mouse. Neurosci Lett 268:139–142.
20. Amano T, Nakanishi H, Oka M, Yamamoto K (1995) Increased expression of cathepsins E and D in reactive microglial cells associated with spongiform degeneration in the brain stem of senescence-accelerated mouse. Exp Neurol 136:171–182.
21. Armbrecht HJ, Boltz MA, Kumar VB, Flood JF, Morley JE (1999) Effect of age on calcium-dependent proteins in hippocampus of senescence-accelerated mice. Brain Res 842:287–293.
22. Farr SA, Poon HF, Dogrukol-Ak D, et al. (2003) The antioxidants alpha-lipoic acid and N-acetylcysteine reverse memory impairment and brain oxidative stress in aged SAMP8 mice. J Neurochem 84:1173–1183.
23. Howard BJ, Yatin S, Hensley K, et al. (1996) Prevention of hyperoxia-induced alterations in synaptosomal membrane-associated proteins by N-tert-butyl-alpha-phenylnitrone and 4-hydroxy-2,2,6,6-tetramethylpiperidin-1-oxyl (Tempol). J Neurochem 67:2045–2050.
24. Inada K, Yokoi I, Kabuto H, Habu H, Mori A, Ogawa N (1996) Age-related increase in nitric oxide synthase activity in senescence accelerated mouse brain and the effect of long-term

administration of superoxide radical scavenger. Mech Ageing Dev 89:95–102.

25. Kurokawa T, Asada S, Nishitani S, Hazeki O (2001) Age-related changes in manganese superoxide dismutase activity in the cerebral cortex of senescence-accelerated prone and resistant mouse. Neurosci Lett 298:135–138.
26. Nabeshi H, Oikawa S, Inoue S, Nishino K, Kawanishi S (2006) Proteomic analysis for protein carbonyl as an indicator of oxidative damage in senescence-accelerated mice. Free Radic Res 40:1173–1181.
27. Okatani Y, Wakatsuki A, Reiter RJ, Miyahara Y (2002) Melatonin reduces oxidative damage of neural lipids and proteins in senescence-accelerated mouse. Neurobiol Aging 23:639–644.
28. Poon HF, Castegna A, Farr SA, et al. (2004) Quantitativc proteomics analysis of specific protein expression and oxidative modification in aged senescence-accelerated-prone 8 mice brain. Neuroscience 126:915–926.
29. Poon HF, Farr SA, Banks WA, et al. (2005) Proteomic identification of less oxidized brain proteins in aged senescence-accelerated mice following administration of antisense oligonucleotide directed at the Abeta region of amyloid precursor protein. Brain Res Mol Brain Res 138:8–16.
30. Poon HF, Farr SA, Thongboonkerd V, et al. (2005) Proteomic analysis of specific brain proteins in aged SAMP8 mice treated with alpha-lipoic acid: Implications for aging and age-related neurodegenerative disorders. Neurochem Int 46:159–168.
31. Sato E, Oda N, Ozaki N, Hashimoto S, Kurokawa T, Ishibashi S (1996) Early and transient increase in oxidative stress in the cerebral cortex of senescence-accelerated mouse. Mech Ageing Dev 86:105–114.
32. Wang Q, Liu Y, Zou X, et al. (2008) The hippocampal proteomic analysis of senescence-accelerated mouse: Implications of Uchl3 and Mitofilin in cognitive disorder and mitochondria dysfunction in SAMP8. Neurochem Res 33:1776–1782.
33. Butterfield DA, Poon HF (2005) The senescence-accelerated prone mouse (SAMP8): A model of age-related cognitive decline with relevance to alterations of the gene expression and protein abnormalities in Alzheimer's disease. Exp Gerontol 40:774–783.
34. Kawamata T, Akiguchi I, Yagi H, et al. (1997) Neuropathological studies on strains of senescence-accelerated mice (SAM) with age-related deficits in learning and memory. Exp Gerontol 32:161–169.
35. Takeda T, Matsushita T, Kurozumi M, Takemura K, Higuchi K, Hosokawa M (1997) Pathobiology of the senescence-accelerated mouse (SAM). Exp Gerontol 32:117–127.
36. Kawamata T, Nakamura S, Akiguchi I, et al. (1994) Dystrophic changes in axons acccumulating nitric oxide synthase are accelerated with age in dorsal column nuclei of senescence accelerated mice (SAMP8). In: Takeda T (ed) The SAM model of senescence. Excerpta Medica, Elsevier Science B. V., Amersterdam, The Netherlands, pp 347–350.
37. Sugiyama H, Akiyama H, Akiguchi I, Kameyama M, Takeda T (1987) Loss of dendritic spines in hippocampal CA1 pyramidal cells in senescence accelerated mouse (SAM) –a quantitative Golgi study. Rinsho Shinkeigaku 27:841–845.
38. Flood JF, Morley JE, La Reginna M (1993) Age-related changes in the pharmacological improvement of retention in senescence accelerated mouse (SAM). Neurobiol Aging 14:159–166.
39. Flood JF, Morley JE, Lanthorn TH (1992) Effect on memory processing by D-cycloserine, an agonist of the NMDA/glycine receptor. Eur J Pharmacol 221:249–254.
40. Kitamura Y, Zhao XH, Ohnuki T, Nomura Y (1989) Ligand-binding characteristics of [3H] QNB, [3H]prazosin, [3H]rauwolscine, [3H] TCP and [3H]nitrendipine to cerebral cortical and hippocampal membranes of senescence accelerated mouse. Neurosci Lett 106:334–338.
41. Kitamura Y, Zhao XH, Ohnuki T, Takei M, Nomura Y (1992) Age-related changes in transmitter glutamate and NMDA receptor/channels in the brain of senescence-accelerated mouse. Neurosci Lett 137:169–172.
42. Liu J, Mori A (1993) Age-associated changes in superoxide dismutase activity, thiobarbituric acid reactivity and reduced glutathione level in the brain and liver in senescence accelerated mice (SAM): A comparison with ddY mice. Mech Ageing Dev 71:23–30.
43. Zhao XH, Kitamura Y, Nomura Y (1992) Age-related changes in NMDA-induced [3H] acetylcholine release from brain slices of senescence-accelerated mouse. Int J Dev Neurosci 10:121–129.
44. Zhao XH, Nomura Y (1990) Age-related changes in uptake and release on L-[3H]noradrenaline in brain slices of senescence accelerated mouse. Int J Dev Neurosci 8:267–272.
45. Kawamata T, Nakamura S, Akiguchi I, et al. (1990) Effect of aging on NADPH-diaphorase neurons in laterodorsal tegmental nucleus and striatum of mice. Neurobiol Aging 11:185–192.

46. Kawamata T, Nakamura S, Akiguchi I, et al. (1990) Effect on aging on NADPH-diaphorase neurons in laterodorsal tegmental nucleus and striatum of senescence accelerated mouse (SAM). In: Basic, clinical, and therapeutic aspects of Alzheimer's and Parkinson's diseases. Plenum, New York, pp 705–709.
47. Sato E, Inoue A, Kurokawa T, Ishibashi S (1994) Early changes in glucose metabolism in the cerebrum of senescence accelerated mouse: Involvement of glucose transporter. Brain Res 637:133–138.
48. Fujibayashi Y, Waki A, Wada K, et al. (1994) Differential aging pattern of cerebral accumulation of radiolabeled glucose and amino acid in the senescence accelerated mouse (SAM), a new model for the study of memory impairment. Biol Pharm Bull 17:102–105.
49. Erecinska M, Silver IA (1989) ATP and brain function. J Cereb Blood Flow Metab 9:2–19.
50. Hoyer S (2004) Causes and consequences of disturbances of cerebral glucose metabolism in sporadic Alzheimer disease: Therapeutic implications. Adv Exp Med Biol 541:135–152.
51. Rapoport SI (1999) In vivo PET imaging and postmortem studies suggest potentially reversible and irreversible stages of brain metabolic failure in Alzheimer's disease. Eur Arch Psychiatry Clin Neurosci 249:46–55.
52. Yagi H, Irino M, Matsushita T, et al. (1989) Spontaneous spongy degeneration of the brain stem in SAM-P/8 mice, a newly developed memory-deficient strain. J Neuropathol Exp Neurol 48:577–590.
53. Kitabayashi T, Tomimoto H, Akiyama H, Akiguchi I, Kimura J, Takeda T (1993) Reactive microglia in the brain of senescence-accelerated mouse (SAM): An immunohistochemical study. Can J Neurol Sci 20:146.
54. Akiyama H, Kameyama M, Akiguchi I, et al. (1986) Periodic acid-Schiff (PAS)-positive, granular structures increase in the brain of senescence accelerated mouse (SAM). Acta Neuropathol 72:124–129.
55. Ueno M, Akiguchi I, Naiki H, et al. (1991) The persistence of high uptake of serum albumin in the olfactory bulbs of mice throughout their adult lives. Arch Gerontol Geriatr 13:201–209.
56. Ueno M, Akiguchi I, Yagi H, et al. (1993) Age-related changes in barrier function in mouse brain I. Accelerated age-related increase of brain transfer of serum albumin in accelerated senescence prone SAM-P/8 mice with deficits in learning and memory. Arch Gerontol Geriatr 16:233–248.
57. Butterfield DA, Howard BJ, Yatin S, Allen KL, Carney JM (1997) Free radical oxidation of brain proteins in accelerated senescence and its modulation by N-tert-butyl-alpha-phenylnitrone. Proc Natl Acad Sci U S A 94:674–678.
58. Yasui F, Matsugo S, Ishibashi M, et al. (2002) Effects of chronic acetyl-L-carnitine treatment on brain lipid hydroperoxide level and passive avoidance learning in senescence-accelerated mice. Neurosci Lett 334:177–180.
59. Harman D (1956) Aging: A theory based on free radical and radiation chemistry. J Gerontol 11:298–300.
60. Kumar VB, Farr SA, Flood JF, et al. (2000) Site-directed antisense oligonucleotide decreases the expression of amyloid precursor protein and reverses deficits in learning and memory in aged SAMP8 mice. Peptides 21:1769–1775.
61. Poon HF, Joshi G, Sultana R, et al. (2004) Antisense directed at the Abeta region of APP decreases brain oxidative markers in aged senescence accelerated mice. Brain Res 1018:86–96.
62. Morley JE, Farr SA, Flood JF (2002) Antibody to amyloid beta protein alleviates impaired acquisition, retention, and memory processing in SAMP8 mice. Neurobiol Learn Mem 78:125–138.
63. Morley JE, Kumar VB, Bernardo AE, et al. (2000) Beta-amyloid precursor polypeptide in SAMP8 mice affects learning and memory. Peptides 21:1761–1767.
64. Goate A, Chartier-Harlin MC, Mullan M, et al. (1991) Segregation of a missense mutation in the amyloid precursor protein gene with familial Alzheimer's disease. Nature 349:704–706.
65. Cruts M, van Duijn CM, Backhovens H, et al. (1998) Estimation of the genetic contribution of presenilin-1 and -2 mutations in a population-based study of presenile Alzheimer disease. Hum Mol Genet 7:43–51.
66. Levy-Lahad E, Lahad A, Wijsman EM, Bird TD, Schellenberg GD (1995) Apolipoprotein E genotypes and age of onset in early-onset familial Alzheimer's disease. Ann Neurol 38:678–680.
67. Dahiyat M, Cumming A, Harrington C, et al. (1999) Association between Alzheimer's disease and the NOS3 gene. Ann Neurol 46:664–667.
68. Blacker D, Wilcox MA, Laird NM, et al. (1998) Alpha-2 macroglobulin is genetically associated with Alzheimer disease. Nat Genet 19:357–360.
69. Selkoe DJ (1996) Amyloid beta-protein and the genetics of Alzheimer's disease. J Biol Chem 271:18295–18298.
70. Selkoe DJ (2001) Alzheimer's disease results from the cerebral accumulation and cytotoxicity of amyloid beta-protein. J Alzheimers Dis 3:75–80.

71. Kumar VB, Vyas K, Franko M, et al. (2001) Molecular cloning, expression, and regulation of hippocampal amyloid precursor protein of senescence accelerated mouse (SAMP8). Biochem Cell Biol 79:57–67.
72. Morley JE, Farr SA, Flood JF (2001) Alzheimer's disease through the eye of a mouse. Acceptance lecture for the 2001 Gayle A. Olson and Richard D. Olson prize. Peptides 23:589–599.
73. Fawzi NL, Kohlstedt KL, Okabe Y, Head-Gordon T (2007) Protofibril assemblies of the Arctic, Dutch and Flemish mutants of the Alzheimer's A{beta}1–40 peptide. Biophys J 94:2007–2016.
74. Premkumar DR, Smith MA, Richey PL, et al. (1995) Induction of heme oxygenase-1 mRNA and protein in neocortex and cerebral vessels in Alzheimer's disease. J Neurochem 65:1399–1402.
75. Schipper HM (2000) Heme oxygenase-1: Role in brain aging and neurodegeneration. Exp Gerontol 35:821–830.
76. Takeda A, Perry G, Abraham NG, et al. (2000) Overexpression of heme oxygenase in neuronal cells, the possible interaction with Tau. J Biol Chem 275:5395–5399.
77. Sultana R, Ravagna A, Mohmmad-Abdul H, Calabrese V, Butterfield DA (2995) Ferulic acid ethyl ester protects neurons against amyloid beta- peptide(1–42)-induced oxidative stress and neurotoxicity: Relationship to antioxidant activity. J Neurochem 92:749–758.
78. Calabrese V, Guagliano E, Sapienza M, et al. (2007) Redox regulation of cellular stress response in aging and neurodegenerative disorders: Role of vitagenes. Neurochem Res 32:757–773.
79. Fink CC, Meyer T (2002) Molecular mechanisms of CaMKII activation in neuronal plasticity. Curr Opin Neurobiol 12:293–299.
80. Schulman H (1995) Protein phosphorylation in neuronal plasticity and gene expression. Curr Opin Neurobiol 5:375–381.
81. Coyle JT, Price DL, DeLong MR (1983) Alzheimer's disease: A disorder of cortical cholinergic innervation. Science (New York) 219:1184–1190.
82. Wevers A, Witter B, Moser N, et al. (2000) Classical Alzheimer features and cholinergic dysfunction: Towards a unifying hypothesis? Acta Neurol Scand 176:42–48.
83. Butterfield DA, Lauderback CM (2002) Lipid peroxidation and protein oxidation in Alzheimer's disease brain: Potential causes and consequences involving amyloid beta-peptide-associated free radical oxidative stress. Free Radic Biol Med 32:1050–1060.
84. Butterfield DA, Stadtman ER (1997) Protein oxidation processes in aging brain. Adv Cell Aging Gerontol 2:161–191.
85. Poon HF, Calabrese V, Scapagnini G, Butterfield DA (2004) Free radicals and brain aging. Clin Geriatr Med 20:329–359.
86. Poon HF, Calabrese V, Scapagnini G, Butterfield DA (2004) Free radicals: Key to brain aging and heme oxygenase as a cellular response to oxidative stress. J Gerontol A Biol Sci Med Sci 59:478–493.
87. Kurokawa T, Sato E, Inoue A, Ishibashi S (1995) Evidence that glucose metabolism is decreased in the cerebrum of aged female senescence-accelerated mouse; possible involvement of a low hexokinase activity. Neurosci Lett 214:45–48.
88. Chen GH, Wang YJ, Qin S, Yang QG, Zhou JN, Liu RY (2007) Age-related spatial cognitive impairment is correlated with increase of synaptotagmin 1 in dorsal hippocampus in SAMP8 mice. Neurobiol Aging 28:611–618.
89. Kidokoro Y (2003) Roles of SNARE proteins and synaptotagmin I in synaptic transmission: Studies at the Drosophila neuromuscular synapse. Neurosignals 12:13–30.
90. Ojika K (1998) Hippocampal cholinergic neurostimulating peptide. Seikagaku 70:1175–1180.
91. Nomura Y, Yamanaka Y, Kitamura Y, et al. (1996) Senescence-accelerated mouse. Neurochemical studies on aging. Ann N Y Acad Sci 786:410–418.
92. Alvarez-Garcia O, Vega-Naredo I, Sierra V, et al. (2006) Elevated oxidative stress in the brain of senescence-accelerated mice at 5 months of age. Biogerontology 7:43–52.
93. Nishimura H, Higuchi O, Tateshita K, Tomobe K, Okuma Y, Nomura Y (2006) Antioxidative activity and ameliorative effects of memory impairment of sulfur-containing compounds in Allium species. Biofactors 26:135–146.
94. Banks WA, Farr SA, Morley JE, Wolf KM, Geylis V, Steinitz M (2007) Anti-amyloid beta protein antibody passage across the blood-brain barrier in the SAMP8 mouse model of Alzheimer's disease: An age-related selective uptake with reversal of learning impairment. Exp Neurol 206:248–256.
95. Gutierrez-Cuesta J, Sureda FX, Romeu M, et al. (2007) Chronic administration of melatonin reduces cerebral injury biomarkers in SAMP8. J Pineal Res 42:394–402.
96. Rodriguez MI, Escames G, Lopez LC, et al. (2007) Chronic melatonin treatment reduces the age-dependent inflammatory process in senescence-accelerated mice. J Pineal Res 42:272–279.
97. Farr SA, Banks WA, Morley JE (2006) Effects of leptin on memory processing. Peptides 27:1420–1425.

Chapter 16

Lesion-Induced Vertebrate Models of Alzheimer Dementia

Adolfo Toledano and Maria Isabel Álvarez

Abstract

No animal spontaneously suffers from Alzheimer's disease (AD), nor can it be experimentally induced. However, there is a huge research need for models of AD. Lesion-induced vertebrate models of this disease have been, and indeed remain, extremely important in the study of AD pathogenesis and possible treatment. The cholinergic hypothesis of AD has led to the development of a number of animal models for studying the pathogeny of cortical cholinergic involution *in vivo*. Focal lesions in the cholinergic centers of the basal forebrain, especially of the nucleus basalis magnocellularis of rodents, as well as more general lesions of all the cholinergic neurons of the basal forebrain, have been the most used methods for obtaining these models. Surgical procedures and intraparenchymal or intracerebroventricular (i.c.v.) microinjections of toxic substances, such as quinolic, kainic, *N*-methyl-D-aspartic, ibotenic and quisqualic acids, the specific cholinotoxin AF64, the immunotoxin 192 IgG-saporin, and amyloid, have all been used to produce such lesions. They have also been used to produce lesion-induced cortical models of AD. Other models have been developed that involve brain implants or microinjections of amyloid as well as the administration, via different routes, of toxic agents such as aluminum, okadaic acid, or alcohol. The research carried out with these models has helped our understanding of AD, especially in the role of cerebral cholinergic innervation in cognitive disorders and their treatments. However, conflicting results are also obtained, and much controversy has developed concerning the role of the cholinergic system and the suitability of these models. It is very important to take into account a wide variety of factors (including e.g., the model protocol, the lesion-inducing agent, the type and concentration of toxin used, and even the morphohistochemical, biochemical, and cognitive methods used to evaluate the changes induced) that could condition the appreciated features of the lesion and its neuropathological and cognitive consequences. This chapter covers the theoretical and practical use of lesion-induced models and examines the main advantages and disadvantages associated with their use. The further use of classic lesion-induced models, and of new models developed in other rodent subspecies, should lead to important advances in our knowledge and treatment of AD and related disorders.

Key words: Alzheimer's disease (AD), Lesion-induced vertebrate AD models, Cholinergic models, Basal forebrain, Nucleus basalis, Cortical cholinergic innervations, Excitotoxins, Cholinotoxins, AD treatments

Peter Paul De Deyn and Debby Van Dam (eds.), *Animal Models of Dementia*, Neuromethods, vol. 48,
DOI 10.1007/978-1-60761-898-0_16, © Springer Science+Business Media, LLC 2011

1. Introduction

Alzheimer's disease (AD) is a devastating neurodegenerative disorder. Over the last 40 years, there has been a huge research effort to determine its cause, understand its pathogenic pathways, and establish potential therapeutic targets to delay its associated neurodegeneration or indeed prevent it. Unfortunately, the successes achieved have been less than hoped for. Experimental models of AD, in particular lesion-induced vertebrate models, have played an important role in many of the advances made. This chapter is devoted to their description, the analysis of their characteristics, and a discussion on their advantages and disadvantages.

As for the majority of diseases, research on AD has been undertaken in two complementary fields. The first has involved the clinical study of patients and the morphological and biochemical analysis of postmortem brains; the second has involved the use of experimental animal models in which it is more feasible to analyze the progress of the disease and its possible modification. Clinical studies began by Alois Alzheimer himself, but experienced an important impulse in the 1960s. Gradually, clinical AD research has revealed the enormous complexity of the alterations in brain function, in the structure and function of the neurons involved, in neuronal circuits and glial cells, and in the relationship between the CNS and the rest of the body (it should not be forgotten that AD is a systemic disease). The mental and behavioral changes seen in most patients with AD dementia are clinical manifestations exclusive to humans – the higher brain functions of man have no equal among animals. The symptoms of brain dysfunction, spontaneous or induced, observed in other mammals in pathological settings supposedly similar to AD cannot be directly equated with human dementia.

The morphological, biochemical, and molecular abnormalities seen in the brains of people with AD are complex; an array of constitutive elements and functional systems involving neurons and glial cells has been implicated, and the number of related pathological abnormalities continues to grow, particularly at the molecular level and with respect to intracellular functional systems. Some factors appear to facilitate the onset of AD (although how is not always clear), such as the possession of certain alleles (e.g., apolipoprotein E4). Certain functional anomalies at critical points in cell systems seem to be closely related to the neurodegeneration that characterizes this disease (e.g., the anomalous processing of amyloid precursor protein (APP) via the amyloidogenic route, the formation of aberrant intracellular deposits of phosphorylated tau protein, the involution of basalocortical cholinergic neurons and cortical regions, and changes in neuroprotective systems) (1–5). Other changes, in contrast, appear to be more generic and are involved in many neurodegenerative diseases, such as gliosis (astrogliosis and microgliosis), increased oxidative stress, and apoptosis (1–5). Each of these complex changes are pieces of a puzzle that is so hard to put together that up-to-date we are still unable to describe a scenario in which one or more cascades of pathological events lead to AD.

Nevertheless, the constant confirmation of abnormalities, sometimes precisely located in particular brain regions, neuronal circuits or even cells, has led to different theories that – at least partially – explain AD pathogeny (1,2). The cholinergic theory (which involves the involution of the basalocortical cholinergic system) (1–3), the amyloid cascade theory (1,4), and the theory that suggests AD to be a variant of the tau diseases (5) are the most important ones. All have been instrumental in major advances in AD-related research made so far. They have also been important in the development of models for the study of AD.

The development of AD models has been somewhat peculiar, since no animal naturally suffers from AD; nor can it be induced by any endogenous or exogenous agent. The typical set of neuropathological alterations seen in human AD simply does not appear in other aging mammals. Only in a few species (e.g., dogs, polar bears, the Argentine rabbit, and primates), similarities with human AD are seen (1). However, the incidence and prevalence of disease in these animals is so low, and the clinical characteristics so different from human AD (the neuropathological symptoms seen in AD are not all present) that no good model is afforded. Clinically, the abnormalities seen in the brain functions of "pathologically aged" members of the above-mentioned species are not fully comparable to the mental and behavioral symptoms suffered by humans with AD. However, while this prevented early researchers using the classic model animals employed to investigate other diseases, the 1970s saw the development of a number of animal models designed to allow the detailed study (over time) of particular AD abnormalities. These also allowed some of the pathogenic theories regarding AD to be tested. Before the development of transgenic animals, the models that allowed the greatest steps forward involved experimentally induced lesions designed to produce cortical and subcortical abnormalities believed important in AD (e.g., involution of the basalocortical cholinergic system, alterations to the cortical and hippocampal circuits, the presence of amyloid protein, and neurofibrillary abnormalities) (6). Although theoretically simple in their design, many of these lesion-induced vertebrate models are still highly valuable. For example, models of this sort have been used to study local alterations in markers of oxidative stress, apoptosis, and synaptic function caused by amyloid β (Aβ) deposits after Aβ protein was injected into the cortex, hippocampus, or nucleus basalis of rats (7–10). Other situations require a more sophisticated design and an in-depth study of the modifications caused by the induced lesion in the different parts of the brain, as well as of its repercussions on different brain functions. For example, the study of hippocampal cholinergic denervation in rats with a selective septohippocampal cholinergic lesion produced by injecting a specific cholinotoxin (192 IgG-saporin) into the septum requires morphological, histochemical, and biochemical studies of the septum, hippocampus, and other regions be performed, as well as behavioral studies involving validated techniques (11–14). The use of these more complex models requires their prior, ample characterization if the results are to be correctly interpreted (6). The results obtained with the techniques for quantifying changes in morphohistochemical and biochemical, neuronal and glial variables in simple lesion models are not usually associated with interpretive difficulties when comparing studies or when extrapolating to humans. However, problems do arise with the more sophisticated models when the intention is to compare changes in mental functions and the behavior recorded when using different types, or when comparisons are made with altered higher functions in human AD brains (6,15–19). Several authors indicate that all tests assessing learning and memory functions, such as the conditioned eye blink reflex in the rabbit (20) and the passive and active avoidance, T-maze, and Morris water maze tests for rats (15,16,19) should be very well characterized, controlled, and validated (15).

This chapter examines the different lesion-induced, vertebrate models of Alzheimer dementia and analyses the characteristics that make them useful in the study of the pathogenic mechanisms of AD, and in verifying the efficacy of possible medications. The characteristics that make them diverge from the disease they try to represent are also discussed.

2. Types of Lesion-Induced Alzheimer Models

2.1. Types of Models

Many types of lesion models exist they can be classified depending on the region of the brain primarily or secondarily affected, the extent of the lesions produced, the intensity and duration of the neuropathological or behavioral alterations induced, the physical mechanism or chemical/biological agent employed to induce the lesion, and the route of administration of such an agent.

The brain regions selectively lesioned for the study of pathogenic mechanisms of AD usually include the basal forebrain region, from which cholinergic nerves diffusely innervate the cortex, or different cortical areas of the cerebral hemispheres and hippocampus. These regions can be lesioned surgically or by the stereotactic injection of different agents. Lesions characteristics depend on the type of agent employed and its capacity to cause selective harm to different types and subtypes of neurons, nerve fibers passing through the affected area, glial cells, and blood vessels. When a neurotoxin-lesion model is designed or selected, the target area must be well known (including its neuronal and glial types, regional neuronal circuits and their connections with other brain areas), as should the mechanism of action of the agent inducing the lesion. Both factors are of capital importance for understanding the primary and secondary induced lesions, i.e., with respect to the anatomical areas involved, the neurons and glia affected, and the markers likely to be affected. A lesion in the basal forebrain could produce changes in different areas of the cortex where basalocortical axons terminate, which of course will depend on the location and size of the lesion induced. The damage will also differ depending on whether a toxin selective for cholinergic nerves (e.g., 192 IgG-saporin) or one that affects both cholinergic and noncholinergic nerves is used (e.g., quisqualic or kainic acids) (6,11,16–19,21). A "local response" will be seen within the damaged basal forebrain region, "proximal responses" in the non-damaged structures neighboring the lesioned basal forebrain area, and "remote responses" in the ipsilateral cortex innervated by the lesioned forebrain area, in other ipsilateral and contralateral cortices, and in the contralateral basal forebrain (21,22).

Other models exist in which a toxic agent induces lesions throughout the brain, leading to mental and behavioral changes reminiscent of AD. Many authors believe that the chronic intake of alcohol or aluminum, or intravenous injections of aluminum, can induce neurochemical and neurophysiological changes in different rodents that are similar to those seen in AD (23–26). These models have been used to clarify a number of pathogenic features of AD and to assess the efficacy of certain anticholinesterase drugs.

Neurotoxins can also induce different lesional models depending on the administration route and protocol. For example, Aβ, colchicine, cholinotoxin AF64, and aluminum provide different models depending on how they are administered (6,12,18,23,24,27–30). When administered intraparenchymally they cause selective lesions, when delivered intracerebroventricularly they induce diffuse lesions in the periventricular regions (although some areas are more affected than others, such as the hippocampus, especially in rodents), and when administered orally, intramuscularly, or intraperitoneally, thereby reaching the brain via the bloodstream, they induce widespread, diffuse lesions in the cerebral hemispheres and/or the basal brain.

The following sections discuss the most widely used models, starting with those that cause selective lesions in defined brain areas. This includes two sections: one on lesions of the anterior basal brain and one on cortical lesions. Discussion is offered on the specific characteristics of these models, which are dependent on the types of neurons damaged and the type of effects the lesion-inducing agent causes, before focusing on models that might be used when administrating toxins via different routes.

2.2. Simple and Combined Models: Models in Aged Animals

Many of these models can be combined to study different combinations of pathogenic mechanisms, e.g., lesion-induced models by double injection of Aβ and oxidative stress enhancers (31), basalocortical cholinergic system lesional models combined with Aβ implantation models (in study in our laboratory) or basalocortical cholinergic system lesional models combined with cortical lesions (32). However, such complex systems have rarely been used, partially because of the greater variability in the results obtained. Many models are modified to increase the neuronal involution, to enhance neuroprotection mechanisms, or to speed recovery from the lesions induced. An example of the first of these is the potentiation of the Aβ neurotoxicity in Aβ implantation models achieved via the reduction of Aβ clearance. This can be brought about by inducing hyperhomocysteinemia through chronic methionine consumption (33). The effects of amyloid or neurotoxin-induced lesions can also be potentiated through ischemia caused by the occlusion of the blood vessels irrigating the part of the brain affected by the lesion (34).

Since no model with all the characteristics of AD is available, it cannot be claimed that any of the models developed is always the first choice in the study of AD pathogeny or in the assessment of potential treatments. However, when pathological changes are being examined, the model that directly or indirectly involves changes most similar to those seen in humans should be chosen. When curative or disease-modifying therapies are to be assessed, the model that reflects changes most similar to those seen in humans should be used. For example, cholinergic lesion models are a priority in the study of the cortical repercussions of cholinergic involution, and for testing the efficacy of anticholinesterase, nicotinic, and nootropic drugs (35,36).

2.3. Factors at the Species and Individual Level

AD models produced by lesions have different characteristics depending on the type of mammal involved, its physiopathological characteristics, and its maintenance (37–41). Lesions induced in the basal forebrain and those induced by toxins in the cerebral cortex in mammals of different orders (rodents, lagomorphs, primates, etc.) differ in their associated neuropathological and behavioral features (42). Even in different rodent species (rats, mice, guinea pigs), nucleus basalis magnocellularis (nbm) lesions induced by Aβ or toxins can produce models with quite different characteristics (38). The rabbit is much more sensitive to aluminum and okadaic acid insult than rodents, and neuropathological alterations more like those of human AD can be obtained in this species. On the contrary, the neuropathological and behavioral changes seen in rats with nbm lesions are more similar to those seen naturally in aged rabbits than in humans with AD.

The characteristics (e.g., age) and maintenance (e.g., housing conditions) of the animals are very important when configuring the intensity of the lesion and its progress (38–46). Older animals are generally more sensitive to lesion induction and experience more intense cortical cholinergic involution than younger animals with the same dose of neurotoxin (21,47–50). Further, many lesion-induction protocols in which short- or long-term recovery occurs in younger animals achieve no such effect in older animals (21,47). Naturally, this is because of the diminished plastic and adaptive capacity of aged neurons (51). It should also be remembered that aged animals show an array of histochemical (e.g., an increase in oxidative stress and apoptosis) and cellular (especially, glial) changes compared with young animals (52–58). When studying aged animals, it should be taken in mind that many of the cellular changes associated with normal aging are also seen in AD.

The term "aged animal" is quite imprecise and needs to be applied correctly in each species. Many studies have considered animals far from the ends of their lives as aged. In many strains of Wistar rat, cortical cholinergic involution and gliosis do not appear until animals are 24–30 months old (their life expectancy is some 36 months). Between the ages of 3 and 20–24 months, no significant age-related differences are seen in the response to excitotoxic lesions produced in the basal brain or the cortex. However, many models involving animals less than 3 months of age have very different characteristics to those involving adult animals. Young animals are capable of far more plastic change than older animals, and, therefore, any extrapolation of observations in young animals should be made with extreme caution.

The model to use should be chosen with the aim of the study in mind, which might be to determine the effects of some variable on a particular anatomical region, on a system, on a neurotransmitter, or on a cellular mechanism. In addition, variables of the model itself should be monitored to confirm that it has behaved correctly; this should allow comparisons with previous studies.

If results are to be valid, the lesioned animals must show homogeneous morphological and functional behaviors. In many models, this is achieved with respect to all the changes induced, while in others it is not. Individual factors that can condition the intensity of the lesion produced and its evolution (including short- and long-term recovery or increasing involution) have been studied scarcely. It would appear logical that they should be related to the individual's capacity for neuroprotection at the time a lesion is induced. In humans too, each individual has characteristics that affect the progress of AD, rendering the pathological process at work in them slightly different. This can be very important in the pathogeny and treatment of the disease. It would therefore be of great interest to study the causes of these individual variations in models; the results could throw light on the pathogeny of AD, and perhaps open up new therapeutic possibilities. Individual variation is more important in some lesion models than others. A clear example of this is seen in the different long-term outcomes in models with excitotoxic lesions in the basal brain.

Kainic acid, when administered subcutaneously or intraperitoneally to produce models of epilepsy or AD, produces a series of functional and neuropathological changes that vary depending on the dose and on interindividual differences in, e.g., APP and tau content of cells, and in the extent of gliosis and apoptosis (59). The reason for these variations, however, is unknown. In our own work, in which we have followed rats with different nmb lesions for up to 3 years (21), we have observed a number of important relationships between the changes induced in different individuals and their progress. An important characteristic in some models is the fact that, after a short time (hours or days), facilitation of morphological (neuronal reinnervation) and functional (the reassignment of functions among the remaining circuits) recovery may occur. These changes are very important when trying to assess the efficacy of neuroprotective medications or neuroplasticity treatments. In such cases, models in which no spontaneous recovery occurs should be used.

The housing provided can also modify the progress of lesioned experimental animals (even aged lesioned animals). Animals maintained in enriched environments show a greater capacity to recover or at least increase their synaptic contact count in different cortical regions (54).

3. Alzheimer Models Produced by Lesions or Dysfunction of the Basal Forebrain: The "Cholinergic" Models

Different types of lesion or dysfunction of the forebrain cholinergic neurons have been provoked to produce a wide variety of lesion-induced vertebrate models of AD. Some involve highly selective and focal cholinergic alterations in restricted areas of the basal forebrain, while others entail more general (cholinergic and noncholinergic) damage to large areas of this anatomical region. Many studies have been performed on the cellular (neuronal and glial) and molecular modifications promoted by lesions in the basal forebrain (local responses), as well as in selected cortical areas (remote responses), the latter alterations mainly caused by cholinergic denervation following damage to the basal forebrain neurons. Lesions in the septum provoke higher cholinergic hypofunctioning in the hippocampus and amygdala. Neuronal damage in the diagonal band of Broca provokes cholinergic alterations in the gyrus cingularis and other limbic structures. Manipulation of the neurons of the nmb leads to cholinergic changes mainly in the neocortex. Cognitive deficits and altered behavior associated with these models have been studied in many settings. As discussed later, the outcome of these models greatly varies depending on the used methodology. The best known and used types are summarized below after a brief commentary on the characteristics of the basal forebrain region.

3.1. The Cholinergic Centers of the Basal Forebrain

The basal forebrain (Fig. 1) is the anatomical region ranging from the septum to the midbrain, which passes under the anterior commissure and groups together both the telencephalic and diencephalic structures. The special characteristics of the "cholinergic centers" in the basal forebrain are largely responsible for the features of the cholinergic models of AD produced by lesion or dysfunction in this region. To understand the cholinergic impairment generated by different basal forebrain lesions, as well as the morphological, biochemical, and behavioral effects induced in brain cortical areas, in-depth knowledge of the basal forebrain of the model species is required.

Shute and Lewis (60,61) and Krnjevic and Silver (62) were the first to describe the cholinergic component of the forebrain in normal and experimental animals. The cholinergic centers are bilateral areas with a high density of cholinergic isodendritic neurons that show no defined cytoarchitecture or well-defined limits. Classically, three cholinergic groups ("centers" or "nuclei") from the anterior-medial to the posterior-lateral area of the basal forebrain are contemplated: the medial septal nucleus, the nuclei of the diagonal band of Broca (dbB; with a vertical and a horizontal set of neurons), and the nbm in rodents, or the nucleus basalis of Meynert (nbM) in primates (including man) (63–65). In the anterior area of the basal forebrain, the first cholinergic group of each brain hemisphere is located in the septum, a medial structure of the brain. The two groups, the left and the right cholinergic septal centers, are topographically very close to one another and septal lesions are always "bilateral" in their effects. In posterior areas of the basal forebrain, the centers are more laterally displaced and lesions at these levels will cause "unilateral" effects. However, the neurons of the posterior areas of the dbB and the nbm/nbM are more intermeshed with the neurons of neighboring brain structures. Laterally, they are in close relationship with the amygdala and the medial area of the temporal lobe dorsally with the globus pallidus and the striatum, and ventrally with the hypothalamus. This general arrangement varies to some degree from species to species. The most outstanding of these neuronal centers are those in which large cholinergic neurons appear. In areas of the basal forebrain of some species, the density of certain kinds of neurons is sufficiently great to contemplate the existence of neuronal centers or gray nuclei, although morphological

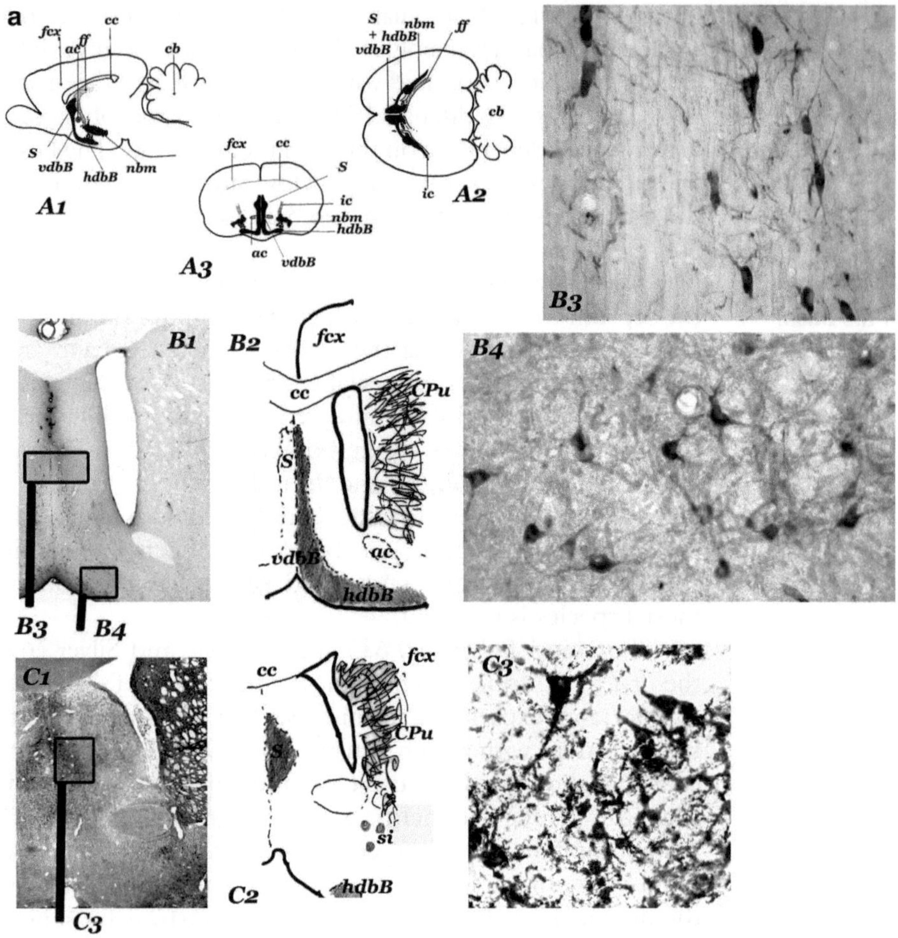

Fig. 1. (Panels 1a and b) Cholinergic "centers" and neurons of the anterior basal forebrain of rat. (**a**) Diagrams showing the sagittal (**a1**), horizontal (**a2**), and frontal (**a3**) projections of the cholinergic centers-septum (s); vertical (vdbB); and horizontal (hdbB) regions of the diagonal band of Broca; nucleus basalis magnocellularis (nbm). In **a2**, the projection of the anterior part of the fimbria/fornix (ff) is represented in the right hemisphere and the projections of the internal capsule (ic) is represented in the left hemisphere. "Septum" (s) – different groups of neurons located in the septal region of the brain – and diagonal band of Broca (dbB) form a continuum of cholinergic cells (observed in **b**, **c**, and **d**). The nucleus basalis magnocellularis (nbm) is difficult to define. Their cholinergic neurons are located near the internal capsule (ic) and close to the globus pallidus (gp), many of them entering these structures. nbm starts near the anterior commissure (ac) and ends near the optic tract (ot), where the globus pallidus disappears (**d**, **e**, **f**). hdbB and nbm are connected by a "network" of isolated cholinergic neurons located in the "substantia innaminata" (a region ventral to the pallidal structures). (**b–f**) Panoramic views (**b1–f1**), diagrams (**b2–f2**), and cholinergic neurons corresponding to frontal sections at different anteroposterior levels (**b** = + 0.5 mm; **c** = - 0.5 mm; **d** = - 0.7 mm; **e** = - 1.3 mm; **f** = - 2.5 mm; the 0.0 mm the plane passes through the bregma–Paxinos and Watson Atlas (229)). At these levels, s and dbB are observed in **b**, **c**, and **d** and nbm (from the most anterior part to the posterior end) in **d**, **e** and **f**. In **b**, the section has been incubated with a ChAT monoclonal antibody and in **f**, with a NGFr monoclonal antibody. **c**, **d**, and **e** are AchE histochemical stained sections. **b3** and **c3** show different types of septal cholinergic neurons. **b4** and **d5** show cholinergic neurons of the vdbB and hdbB, respectively. In **d**, different types of cholinergic neurons in neighbouring areas of the nbm can be observed: cholinergic neurons in the substantia innominata (**d3**), and in the internal capsule (**d4**, *left*) and in the globus pallidus (**d4**, *right*). The magnocellular neurons of the central regions of the nbm can be observed in **e3**; they are located in the "ventral pallidum," entering the globus pallidus (gp) and the internal capsule (ic). Other cholinergic neurons are located in ventral areas (si). The most posterior neurons of the nbm, close to the optic tract can be observed in **f3**. (**a1–3** = × 2; **b1–f1** = × 4,5; **b3–4, c3, d3–5, e3** and **f3** = × 45).

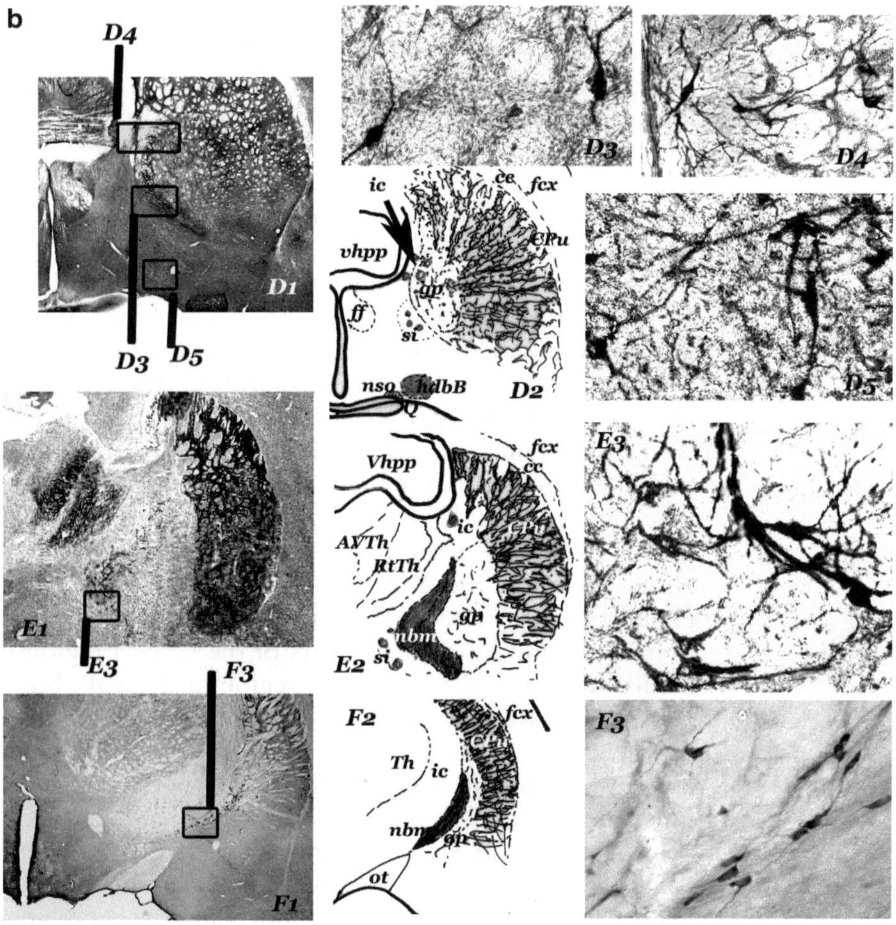

Fig. 1. (continued) *ac, anterior commisure; AVth, anteroventral Thalamus; cb, cerebellum; cc, corpus callosum; CPu, caudate putamen; dbB, diagonal band of Broca; fcx, frontoparietal cortex; ff, fimbria/formix; gp, globus pallidus; hdbB, horizontal region of dbB; ic, internal capsule; nbm, nucleus basalis magnocellularis; ot, optic tract; RtTh, reticular nucleus of the thalamus; s, septum; si, substantia innominata; Th, thalamus; vdbB, vertical region of dbB; Vhpp, ventral hippocampus.*

characteristics, such as the poor definition of the borders and the high number of intermeshed neurons and fibers from neighboring anatomical regions, may hamper interpreting some neuronal groups in this way.

The nucleus basalis is extremely difficult to define in both primates and rodents. Cholinergic neurons presumably belonging to this nucleus are located near the internal capsule and close to the globus pallidus, many of them entering these structures. However, other important clusters are also laid down in areas ventral to the pallidum-striatal complex. These anatomical areas are very poorly defined and referred to by many histologists as the ventral pallidum and substantia innominata. The nbm (or nbM), or their cholinergic neurons, can be understood to start near the anterior commissure and to end near the optical tract where the globus pallidus disappears in serial coronal sections.

Histochemical research on the cholinergic groups of the basal forebrain performed since 1980 have provided a new definition of the cholinergic sectors, the demonstration of different sets of cortical projecting neurons with neurotransmitters other than acetylcholine (ACh) intermeshed among the cholinergic neurons, and a better understanding of the cortical targets, as well as the relationships between cholinergic groups and neighboring extra forebrain structures. Mesulan et al. (66,67) proposed an alternative nomenclature for the cholinergic sectors of the basal forebrain in primates and rodents. These authors describe six sectors (Ch1–Ch6) in the basal forebrain. Sectors Ch1 and Ch2 – the medial septal nucleus and the vertical limb nucleus of the dbB in traditional nomenclature – are the major sources of cholinergic innervation for the hippocampus. Ch3, which mainly innervates the olfactory bulb, corresponds to the lateral part of the old horizontal region of the dbB. Ch4 contains the broadest group of cholinergic neurons, since it includes not only the traditional nbm/nbM and both lateral parts of the vertical region plus the medial section of the horizontal region of the dbB, but also the cholinergic groups or isolated cholinergic neurons observed in the neighboring areas (globus pallidus, substantia innominata, and ansa lenticularis) plus the preoptic nucleus. Ch4 neurons innervate neocortical areas and the amygdala. Sectors Ch1–Ch4 provide an extensive and continuous system of cholinergic neurons that give rise to a topographically organized cholinergic regulatory projection system, which innervates the entire neocortical mantle, as well as many limbic and olfactory structures (i.e., "the ascending cholinergic system") (60,65,66,68). All these projecting neurons show the highest density of p75 NGFr (the low-affinity nerve growth factor receptor) (69,70). In primates, Ch4 can be divided into subsectors depending on the cortical targets, but in rodents this division is less clear. Two other cholinergic sectors, Ch5 (contained mostly within the pedunculopontine nucleus of the pontomesencephalon) and Ch6 (within the periventricular tegmental area), provide cholinergic innervation to the thalamus and cannot be included in the ascending cholinergic system. However, these two neuronal groups are essential components of the ascending reticular activating system and could indirectly regulate cortical activity via a noncholinergic system between the thalamus and cortex. All the cholinergic groups or sectors of the basal forebrain and upper brain stem are reciprocally connected with the thalamus, neocortex, hippocampus, and limbic structures, as well as with one another.

Clusters of noncholinergic neurons are intermingled with cholinergic neurons in certain areas of the basal forebrain (71–78). The best-characterized neurons are those that show immunoreactivity against γ-aminobutyric acid (GABA) (74,78), galanin (alone or in cotransmission with ACh) (72), or peptides such as somatostatin (75,77), neurotensin, or enkephalins (73,75). These noncholinergic neurons have precise projections to different cortical areas such as the GABA set of neurons of the septum, innervating the hippocampus, or the neurotensin basalis group that innervates the frontoparietal cortex. The death or dysfunction of these noncholinergic neurons of the basal forebrain might explain the different symptoms seen in several mental disorders. Some of the cognitive deficits produced in experimental animals with basal forebrain manipulations may also be related to the damage of these noncholinergic cells; this might account for the problematic results returned by some basal forebrain lesion models.

All of the cholinergic and noncholinergic projecting neurons described earlier are considered by Alheid and Heimer (79) as one complex – the "magnocellular corticopetal complex" – of the three morphofunctional complexes into which the basal forebrain as a whole is organized. The other two complexes are the "striatopallidal system" and the "extended amygdala." The three systems have neurons that are closely intermeshed in the basal forebrain. The striatopallidal system is a set of epidermal growth factor-immunoreactive neurons located in the basal forebrain that form a rostroventral extension of both the caudate-putamen and globus pallidus. The extended amygdala groups all the neurons immunoreactive against angiotensin II from the temporal lobe to the stria terminalis; the highest density of these neurons is seen in the centromedial amygdaloid complex and the bed nucleus of the stria terminalis. The magnocellular corticopetal system, which is composed of all the cholinergic and noncholinergic neurons projecting to the different cortical areas, is interwoven with these other two systems, with no clear boundaries apparent. A very large number of connections among these basal forebrain systems exist. All the corticofugal pathways and corticopetal feedback loops of the basal forebrain systems are thought to be of great importance in neuropsychiatric disorders (79). The rough rostrocaudal topographical organization of the cholinergic projections to the cortex that characterized the first descriptions of the basal forebrain (64,80), has been modified to reflect functional models of this brain region, in which a more diffuse cholinergic innervation from the basal forebrain to the cortical and noncortical regions is contemplated (68,81,82). On the basis of the permanent interrelationship between these cortical and basal structures, several theories on learning and memory have been proposed, including the "hippocampal memory indexing theory" (83). In this theory, the cholinergic centers control the processing and encoding of information stored in neocortical modules indexed in the hippocampus. The diffuse topographical organization of the basal forebrain's cholinergic projection into the cortex allows both forebrain neurons to randomly innervate different cortical areas and some cortical areas to receive cholinergic terminals from neurons located in different regions of the basal forebrain. This diffuse cholinergic innervation suggests that cortical innervation is widespread but largely undifferentiated; the functional importance and significance of this is, however, unknown.

Most of the axons of the cholinergic neurons in the septal areas run through the fimbria/fornix to the hippocampal formation. The axons of the neurons of the dbB form a diffuse projection to the limbic and neighboring brain regions. The axons of the nbm/nbM project to the lateral part of the corpus callosum and enter it. Different axon collaterals run from this body and penetrate the different areas of the ipsilateral neocortex; a few cross to the contralateral side. Some axon collaterals of the nbm/nbM run to other neuronal subcortical centers. This special architecture allows the surgical manipulation of the main cholinergic efferents of the septum possible, although it is very difficult. The results obtained can be very variable when lesions are induced with physical or chemical agents in other cholinergic centers. Transection of the fimbria/fornix produces not only cholinergic denervation of the hippocampus, but also interrupts the limbic circuits in the connection between the hippocampus and mammillary bodies. This should be remembered when cognitive and behavioral tests are chosen to evaluate cognitive impairment in AD models produced by fimbia/fornix lesions.

A dense network of cholinergic fibers is seen in the cortex (Figs. 2b and i). The cholinergic fiber terminals show specific patterns within each cortical area and layer. Axon terminals end at neurons of both intrinsic and extrinsic cortical circuits and blood vessels. The nicotinic and muscarinic cholinergic receptors also show a laminar distribution. All cortical neurons respond to ACh. All this suggests the existence of a functional cholinergic organization in the cortex. The fibers belong to neurons of different origin, including local cholinergic neurons. The results of different morphofunctional studies have suggested that the cholinergic system is not a selective activating system, but rather a reticular, diffuse system for globally modulating cortical functions (81,84).

In summary, the basal forebrain is an anatomical region in close relationship with pallidal, striatal, amygdalar, and hypothalamic structures, in which a magnocellular corticopetal system exists. In this corticopetal system, the most important sets of neurons regulating cortical functions are cholinergic. These neurons can be grouped into three functional cholinergic groups, similar to, but more diffusely located, than the classic nucleus of the septum, diagonal band, and nucleus basalis. Each group specifically ends in a part of the cortex with good topographic correlation, although diffuse innervation also exists. Intermeshed with these cholinergic neurons are noncholinergic neurons with GABA,

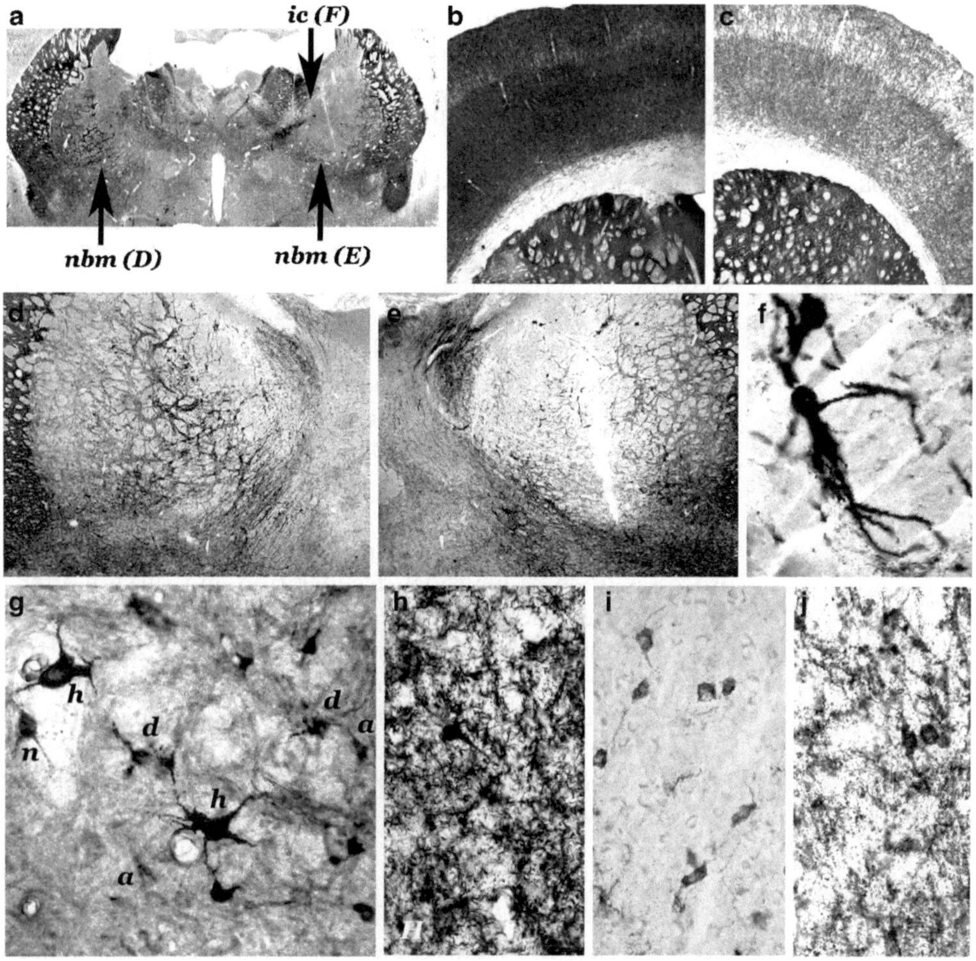

Fig. 2. Morphohistochemical changes in lesion-induced models of AD in rats. (**a–f**) Unilateral lesion induced in the nbm of the right hemisphere by a stereotactic injection of 50 nmols of quisqualic acid; 45 days postlesion induction of the lesion. Coronal sections; AChE enzymohistochemical staining. An intense decrease in AChE-positive neurons is observed in the lesioned side (**a**, **e**) in comparison with the non-lesioned side (**a**, **d**). A very few number of AChE-positive neurons can be observed in the neighbouring areas of the lesion core, such as the two hypertrophic neurons located in the internal capsule (**a**, ic; **f**). Intense AChE-positive network in the motor and somatosensory frontoparietal cortex of the contralateral unlesioned side (**b**) and weak AChE histochemical staining of the ipsilateral cortex (**c**). (**g**) Unilateral lesion induced in the nbm by an injection of 30 nmols of quisqualic acid; 7 days of post-lesion evolution, NGFr immunostaining. Different types of surviving AChE-positive neurons lateral to the "core" of great necrosis: normal (n), hypertrophic (h), atrophic (a), and dystrophic (d) neurons. (**h–j**) Unilateral lesion induced in the nbm by an injection of 40 nmols of ibotenic acid; 90 days postlesion induction. A dense network of AChE-positive fibers in the layer V of the motor frontoparietal cortex of the contralateral non-lesioned side (**h**) of one animal, and a negative (**i**) and a very weak AChE positive (**j**) areas of the cortex ipsilateral to the lesion of two different animals of this model. AChE histochemical positivity strongly increases in cortical non-pyramidal neurons in the ipsilateral side (**f**, **j**). (**a** = × 1.5; **b–c** = × 5.5; **d–e** = × 10; **f–j** = × 45) (methods in (6,14,28)) (Reproduced from Current Alzheimer Research, 2004, with permission from Bentham Science Pub. Ltd (6)).

neurotensin, somatostatin, and galanin as their neurotransmitters. Selective lesions only affect a restricted part of the cortex in a direct fashion, but different collateral axons can provoke modifications in other basal centers. The cortical effect of a cholinergic lesion or dysfunction (especially when evaluated by cognitive tests) is the result of the direct effects on the cortical innervation and on the innervation related to modifications in connected centers, or of damage to neighboring noncholinergic neurons.

3.2. Main Types of Cholinergic Models Produced by Basal Forebrain Lesion or Dysfunction

3.2.1. AD Models Produced by Fimbria/Fornix Transection (Septohippocampal Cholinergic Denervation) (85–87)

Cholinergic neurons of the septal area of the basal forebrain innervate the hippocampus. Their axons reach the hippocampal formation in the fimbria/fornix, a well-defined anatomical structure easily accessible to surgical manipulation. Different AD models can be obtained after unilateral or bilateral, total or partial transection of the fimbria/fornix. Different degrees of cholinergic impairment in the hippocampus and in the basal forebrain have been obtained, as have different alterations in the performance of selected cognitive and behavioral tasks. These models have been very useful in the study of the different roles of cholinergic inputs on hippocampal function in different species. Moreover, this kind of model has been widely used to study various neuronal involutive and adaptive processes, such as neuronal degeneration and survival after axotomy, the free and induced regeneration of axons, and the ability of adult axons to innervate new brain tissue. Many factors (species, age, neurotrophins, and other lesions) regulate the development of the deficits induced by a fimbria/fornix lesion, and recovery varies in the different models. Morphofunctional cholinergic and noncholinergic involution and recovery, and cognitive impairment and recovery, have not always been recorded as parallel processes. Controversy exists regarding the cognitive functions of the hippocampus and the role of the cholinergic system in the regulation of these functions. Morphofunctional recovery is different in different species. Monkeys studied 1.5 years after fornix transection (88) show essentially the same pattern of hippocampal cholinergic fiber loss as do those that have survived less time. This indicates that the residual choline acetyltransferase (ChAT)-immunoreactive fibers, many of which reach the hippocampal formation through a ventral cholinergic pathway, are incapable of reinnervating the denervated portions of the hippocampal formation. This appears to distinguish the monkey from the rat, for which substantial sprouting and reinnervation of cholinergic fibers has been reported after the induction of similar lesions (87,89). The ability of the lesioned axons to innervate grafts surgically implanted in the fimbria has also been studied (89). After performing unilateral transections in adult rats, foetal hippocampus or autologous peripheral nerves have been implanted into the transection site and the regeneration of the fimbria fibers analyzed (87). Prominent innervations by central acetylcholinesterase (AChE)-staining axons were seen after 2–3 weeks in both types of implant; this was sustained until at least 8 months. Reinnervation of the adjacent host hippocampal terminal zone was also apparent, but was sparse compared with the innervation in the implants.

3.2.2. AD Models Produced by Basal Forebrain Electrolytic Lesions (90–93)

These kinds of lesion-induced models, in which the death of neurons over wide areas of the basal forebrain was produced as well as alterations to fibers passing by the target area, were practically abandoned several years ago. However, some modern studies combining these and other types of lesion have provided results useful for differentiating between cholinergic and noncholinergic effects (90–93). This kind of lesion has been used in newborn animals (93).

3.2.3. AD Models Produced by Basal Forebrain Excitotoxic Lesions (6,21,22,94–103)

In these models, nonselective cholinergic lesions/dysfunctions of the forebrain cholinergic centers, including septum (94,95), dbB (94) and nbm/nbM (6,21,22,96–103) (Figs. 2–5), are induced by stereotactic injections of excitotoxins, which leads to the dysfunction and/or the death of the neurons around the injection site. In selected studies, excitotoxins have been administered by i.c.v. injection; this induces lesions in the basal forebrain of less

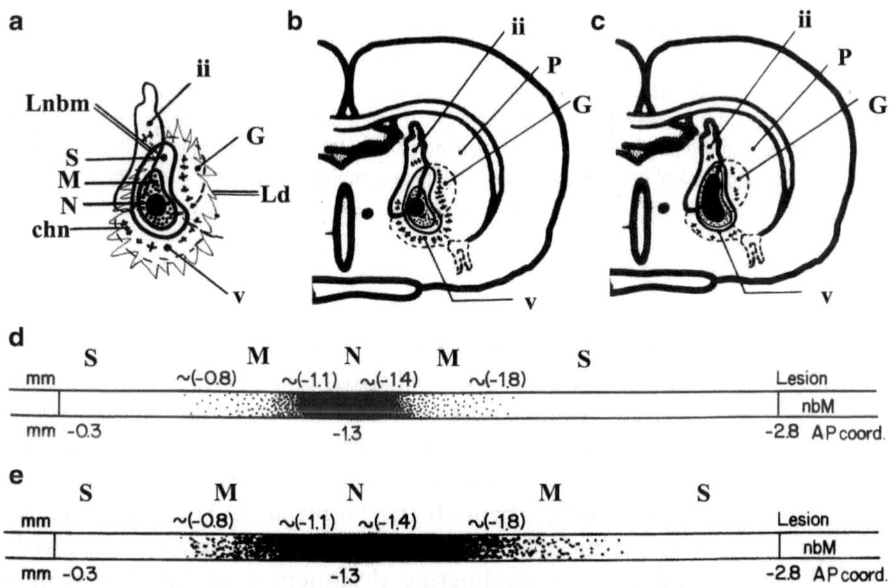

Fig. 3. Diagrams of the damaged nbm areas after excitotoxic injection. (**2-a**) The injection of neurotoxins into the main region of the nbm generated superimposed areas with different degrees of neuronal damage: a "core" of great necrosis, with a very small number of surviving cells (N), a region with mixed necrotic and surviving cells (M), and an area with a high number of surviving cells (S). The dimensions of these areas depend on the type and dose of toxin used and other individual factors (see text). The surviving neurons show different sizes, shape, and reactivities. Surrounding the limits of the nbm (Lnbm), cholinergic neurons are seen in the upper part of the internal capsule, globus pallidus, and infrapallidal areas (chn). In some models, these show hypertrophic/hyperreactive characteristics. Noncholinergic neurons of the basal forebrain and regions other than the nbm are also damaged (Ld = diffuse limit of possible neuronal damage). (**2-b, c**) Diagrams showing the effect of mild (30Q in B) and serious (60Q in C) nbm lesions in the coronal plane, 1.3 mm. (**2-d, e**) Anteroposterior representation of cholinergic damage in the nbm for these two models (areas N, M and S). The cholinergic neurons of the anterior and the most posterior part of the nbm are always preserved. Ap coord = anteroposterior coordinates from the Atlas of Paxinos and Watson (229).

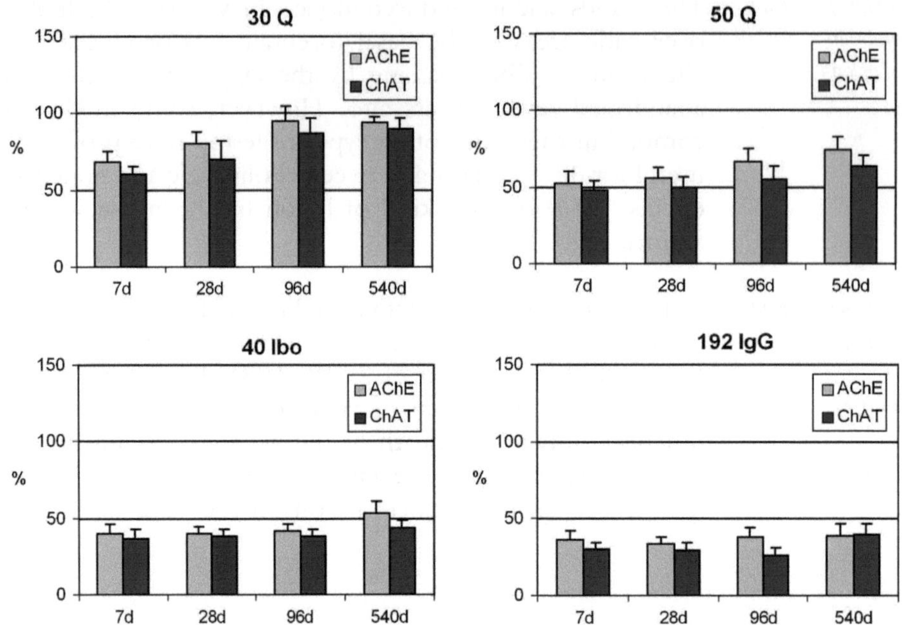

Fig. 4. Changes (up to 540 days postlesion induction) of biochemically determined cortical AChE and ChAT activities in the frontoparietal cortex of rats after different unilateral lesions in the nbm. Activities ipsilateral to the nbm lesion are represented as a percentage of the activity on the contralateral unlesioned side. A rapid and almost total recovery in the model obtained by injection of 30 nmols of quisqualic acid (30Q) and a slow and partial recovery after 50 nmols (50Q) are observed. Non-recovery in the models obtained by injection of 40 nmols of ibotenic acid (40 Ibo) or 5 pg of 192-IgG-saponin is shown. The histochemical AChE-positive neuronal loss in the nbm was 60.2 ± 4.2; 72.2 ± 4.0; 74.6 ± 3.6 and 81.1 ± 2.8 in the 30Q, 50Q, 40Ibo, and 192 IgG, respectively. (n = 12 animals/model/postlesion day. Methods described in (28)).

importance than those seen at cortical level (mainly in the hippocampus). Neuronal damage is produced by the disturbance of ionic homeostasis at the cellular level. Neurotoxic changes induced by abnormal Ca^{2+} influx also occur (104). Theoretically, this type of basal forebrain lesion does not cause alterations to fibers passing through the target area. The excitotoxins are not cholinergic-selective, but these models are associated with important cortical cholinergic dysfunction or hypofunction with no other apparent important deficits. The most widely used excitotoxins are substances that act as agonists on the different glutamatergic receptors, such as *N*-methyl-D-aspartate (NMDA) on NMDA receptors, and ibotenic (IBO), quisqualic (QUIS), α-amino-3hydroxy-5methyl-4isoxazole-propionic (AMPA), and kainic acids on non-NMDA glutamatergic receptors. Not all of these excitotoxins have the same effect; great differences in morphological, histochemical, and biochemical cortical alterations, as well as cognitive and behavioral alterations (22,27,36) have been observed in apparently similar models. Further, in individual experimental animals of certain models, the recovery of functional

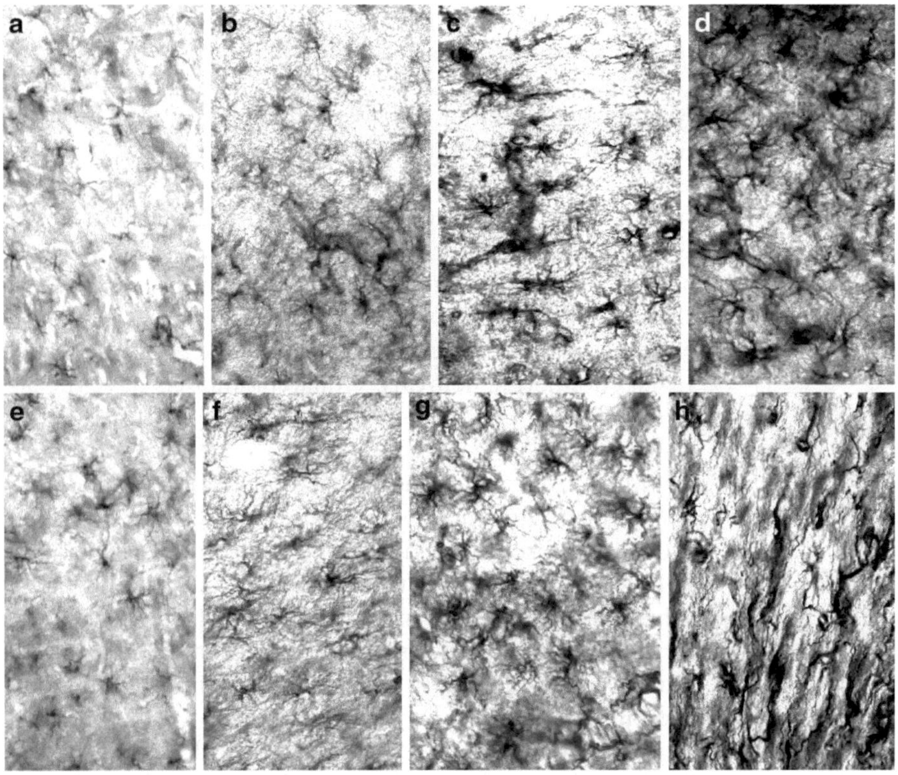

Fig. 5 Photomicrographs of GFAP immunoreactivity in rats of a lesion-induced AD model by unilateral stereotactic injection of 50 nmols of quisqualic acid. All the sections were simultaneously incubated to show the differences in astroglial GFAP immunoreactivity. (**a–c**) Frontoparietal cortex, motor area layer V of 4-month-old rats, 30 days (**a** = unlesioned contralateral side; **b** = lesioned side) and 540 days (**c** = lesioned side) after nbm lesion. (**d**) nbm lesion area, 30 days postlesion. (**e–g**) Frontoparietal cortex, motor area layer V of 20-month-old rats, 30 days (**e** = unlesioned contralateral side; **f** = lesioned side), and 540 days (**g** = lesioned side). (**h**) Cortical scar produced by the needle; 180 postlesion days. × 45. Methods described in (28).

deficits sometimes occurs (6,16,105). The reasons for these differences – most of which have not been well studied – include: (a) the existence of different subtypes of glutamate receptor on cholinergic and noncholinergic neurons in the target basal forebrain area (106), (b) the existence of different subsets of noncholinergic neurons affected by the toxins in the lesion area (or neighboring damaged regions) (73–78), (c) differences in the innervation of brain areas other than the main target zones of the damaged cholinergic neurons (79), (d) poor correspondence between the behavioral task performed and the cognitive function assessed, and (e) the poorly understood role of the cholinergic system in cognitive function (81,107,108). Other neurotoxins (not specifically cholinergic) have occasionally been administered by stereotactic injection into the basal forebrain, such as quinolic acid (109), a natural metabolite of tryptophan (109), colchicine (28,110,111), and other vinca alkaloid substances that bind tubulin and disrupt proximodistal (axoplasmic) intraneuronal transport (112).

Excitotoxins are microinjected (0.2–0.5 μl) using Hamilton syringes manipulated in a stereotactic apparatus. Both unilateral and bilateral cholinergic-impaired models can be produced after unilateral or bilateral stereotactic injection. Series of injections can provoke different degrees of cholinergic deficits in the cortex. Lesions in the dbB or the nbm on one side of the brain provoke high ipsilateral denervation, but only very small contralateral changes (Fig. 2). However, injections into the medial septum provoke bilateral effects since the toxin acts on the left and right septal cholinergic neurons. Unilateral models have the advantage of an internal control – the undamaged contralateral forebrain system.

Most excitotoxic models have been developed in rodents, mainly rats, but other species have also been used. Rabbits could provide an interesting new model since selective damage to the nbm leads to the deposition of Aβ (113), something not seen in other animals. A very small number of studies have been performed in nonhuman primates. In these mammals, the basal forebrain lesion is generally produced by immunotoxins rather than excitotoxins (114). The main reasons for the limited use of these models have been reviewed by Wenk (115). Briefly, the basal forebrain lesions induced are sensitive to cholinergic antagonists but show no response to agonists. Moreover, the memory deficits are small and do not correlate with the degree of cholinergic cell loss, spontaneous recovery occurs in a very short time, and cognitive impairment is more related to attention deficits. Even so, healthy, aged monkeys are often the first choice for testing drugs (116–118) or when studying amyloid production in neurodegeneration (119).

3.2.4. AD Models Produced by Cholinotoxic Lesions

In these models, selective cholinergic lesions/dysfunctions of basal forebrain cholinergic centers are produced by stereotactic injections of chemical (ethylcholine aziridinium – AF64A-) (115,120–125) or immunochemical (192 IgG-Saporin) (11–13,126,127) cholinergic toxins. These models have been deemed among the best representatives of AD, since they theoretically preserve noncholinergic neurons located in, as well as axons passing through, the targeted cholinergic centers.

The cholinotoxin AF64A binds to the high-affinity choline uptake system (HACU) and impairs cholinergic neurons. Theoretically, this neurotoxin is very selective in its production of cholinergic models, but its use is now restricted because of the high variability of the induced deficits. It has been administered intracerebroventricularly and intraparenchymally and very different results have been obtained at different doses in different brain sites (115,120–125). All neurotoxins show a more or less strict correlation between dose and effect in cholinergic cells, but AF64A has unexpected and variable effects. Certainly, high dose i.c.v. injections (>2 nmol) result in widespread histological damage to cholinergic neurons in many cortical and basal forebrain regions (e.g., neocortex, hippocampus, septum, and nbm/nbM). However, when low doses (1–2 nmol) are administered, only subsets of cholinergic cells with long projection axons are affected (mainly in septum, dbB, and nbm/nbM); the cholinergic interneurons of most regions are not affected. Moreover, cholinergic damage varies in these damaged forebrain centers and their corresponding targets to such an extent that, in different forms of this type of model, cholinergic neuronal death, short-term cholinergic dysfunction, and cholinergic recovery can all be observed. The alterations in the cholinergic septohippocampal system are smaller and reversible after only a short time (127), perhaps because of a septal compensatory effect. After i.c.v. infusion, some models show deficits on certain cognitive tasks designed to test memory, learning, or attention. Memory, however, recovers after a short time (128). Probably, the best behavioral tests for measuring cognitive deficits in these models are those related to spatial cognition (129). Intraparenchymal injection of AF64A (123,124) produces gross tissue disruption and perhaps damage to noncholinergic neurons (in the injection site and neighboring areas) (130). To obtain specific cholinergic models with aziridinium compounds, careful study of the dose-related effects (taking into account that there is a very narrow dose range for producing specific effects (125)) at each of the desired specific targets will be needed. The study of the different subsets of cholinergic neurons or of the different characteristics of the cholinergic system of different neurons in each forebrain region could be studied using these models. When the theoretical and practical problems associated with their use are minimized, they might be used to study different aspects of the effects of cholinergic drugs (131) or neurotrophic factors (132).

The basalocortical cholinergic neurons are heavily dependent on a specific neurotrophic system, thereby allowing selective cholinergic lesion-induced models of AD. The maintenance of the morphological and functional characteristics of the basalocortical cholinergic cells (69,70) is a function of nerve growth factor (NGF); the low affinity NGF receptor, or p75 NGFr, is the key point at which NGF exerts its effect. With this knowledge, a specific immunotoxin (192 IgG-saporin) was constructed by conjugating a monoclonal antibody against p75 NGFr (192 IgG) to a neuronal toxin, the ribosome-inactivating protein extracted from the plant *Saponaria officinalis* (saporin) (133). This complex molecule selectively and irreversibly binds to p75 NGF receptors. It is then endocytosed and transported to the neuronal cholinergic bodies where its ribosome-toxic effect is exerted and protein synthesis is impeded

(133,134). Depending on the dose of immunotoxin provided and the route of administration, important dysfunction or even destruction of the cholinergic basal forebrain neurons can be achieved. The theoretical selectivity of the primary damage induced in cholinergic neurons in 192 IgG-saporin models has led to their widespread use. 192 IgG-saporin has been used to study the short- and long-term morphofunctional and cognitive impairments induced by cholinergic involution in normal and diseased adult animals of both sexes (135), in newborn animals (136) and in aged animals (50). As with the AF64A toxin, this immunotoxin can be injected into the forebrain parenchyma or the target cortical areas, or i.c.v.-administered (137). When injected into different areas of the forebrain, primary septal, dbB and/or nbm/nbM lesions, and secondary cortical alterations can be produced. I.c.v.-administered, primary lesions in all areas of the basal forebrain (of different intensity) have been induced together with intrinsic and extrinsic cholinergic cortical alterations and cerebellar lesions, since Purkinje cells also have p75 NGF receptors (126). Different mammal species, including primate species, have been used to obtain AD lesion models using 192 IgG-saporin. The injection of the immunotoxin into different regions of the basal forebrain produces models with selective cortical cholinergic deficits (11,12,126,127) (Fig. 4). In general, noncholinergic neurons are not affected, but on many occasions nonspecific structural damage occurs, mainly long after the lesion is induced. No local basal forebrain damage other than neuronal cholinergic loss is seen at 1–5 months after lesion induction, but at 11 months, extensive tissue loss is observed at the lesion site(s). Cognitive deficits produced by immunotoxin injection into the basal forebrain have been described as being of variable intensity and long development and to depend on the coexistence of other lesions. In a study comparing the long- and short-term effects of single and combined septum/nbm immunotoxic lesions (38), no important behavioral deficits in spatial learning or memory in the radial arm and water maze tasks were appreciated. Only double lesion (septum + nbm) rats showed behavioral impairment; single lesion (septum or nbm) rats remained behaviorally normal. The authors of this study suggest that deficits appear only when damage is intense and when selective cholinergic and nonselective noncholinergic alterations have been produced. Despite the many studies that have used these immunotoxin models and the importance of their results, many discrepancies exist between the cognitive and morphofunctional deficits recorded. The variability of the cognitive responses recorded after theoretically similar lesions have been induced, is related to the different means of evaluating cognitive impairment and the size of the initial immunotoxic lesion (138–141). The different cortical responses of neocortical and limbic structures (123,124,142,143) are also factors that condition the development of deficits after nbm/nbM and septal lesions. Immunotoxins and AF64A have been also used to assess the morphofunctional changes of selected brain areas (e.g., suprachiasmatic nucleus) after cholinergic denervation (144). As seen after partial or total excitotoxic lesions of the forebrain nuclei, partial or total cortical cholinergic impairments, as well as partial or total recoveries of the cholinergic deficits and cognitive alterations, have been described in these cholinotoxic lesion-induced models (127,142,143). "Partial" models have been useful in studying cortical cholinergic involution, as well as natural and induced regeneration or neuronal reorganization (127,145). The effects of neuronal grafts (146) or drugs (e.g., anticholinesterase drugs (147,148) and NGF-like factors (149)) designed to induce cortical cholinergic or cognitive recovery have been tested with these models. The rat has been the animal most used, but others species have been employed in some studies. Many rodents have been used to compare cognitive differences after nbm lesions (150). Rhesus monkeys have been used to assess primate basalocortical cholinergic function (114,115,151). Only the rabbit has been used to evaluate the relationship between two AD hallmarks, namely cholinergic impairment and amyloid deposits (113). Sheep have been used to evaluate the cholinergic mechanisms in offspring recognition (152). Lesions induced by immunotoxins in striatal and brain stem cells (also reported in these models (153)) are now being examined given their possible importance in studies of AD pathogeny or treatment.

4. Morphofunctional Neuronal and Glial Alterations and Cognitive Impairments in Basal Forebrain Lesion AD Models: Changes in Lesion Patterns over Time

The experimental manipulation of the basal forebrain provokes a wide variety of structural and functional changes in both the basal forebrain itself and the cortical areas innervated by its damaged neurons. These cortical lesions can be induced in different areas of the neocortex, paleocortex and/or archicortex. Different morphological and biochemical variables have been used to assess cholinergic and noncholinergic neuron impairment or death, as well as glial morphofunctional changes at both the basal and cortical anatomical levels. These changes provoke parallel cognitive impairments that can be assessed via behavioral tasks. In every cholinergic AD model, a specific pattern of basal lesion/dysfunction, cortical cholinergic and noncholinergic alterations and cognitive impairment is always observed for some time following the controlled induction of a lesion. From this moment, different involutive and/or adaptive processes or changes occur in the internal mechanisms of the directly or indirectly damaged cells

(Figs. 2f, 6.1.3.2G and 3). The repercussions in the neuronal basal forebrain and cortical circuits, as well as in the glial cells of both regions, condition the short- and long-term structural and cognitive outcomes of these patterns. The intensity and duration of changes largely depend on factors related to the degree of basal forebrain damage, the lesion-inducing agent, and how the lesion was produced. On many occasions, the correlation between neuroanatomical alterations and cognitive impairment is very difficult to establish.

4.1. Factors Determining the Lesion Pattern Characterizing Cholinergic AD Models

The most important factors that determine the characteristics of each model are discussed below.

4.1.1. The Method and Protocol Used to Produce the Model

When a protocol is strictly followed to produce a forebrain lesion model, the results obtained by different authors show no great differences. However, those obtained with different (even slightly different) methods or protocols can differ widely. For example, microinjections of different excitotoxins into the same area of the basal forebrain apparently produce the same basal forebrain and cortical lesions yet can provide different models with different degrees of cholinergic dysfunction and associated behavioral impairment depending on the toxin used and its dose (Fig. 4). Ibotenic, quisqualic, or kainic acids are commonly employed to produce similar models, but the animals lesioned with each toxin show strong neuropathological and cognitive differences in in-depth analyses (6,22,105). Such differences are often attributed to the different damage sustained by noncholinergic neurons. When the same model is used but the results obtained differ to those previously published, the reason can lie in the use of different animal strains but also in how the models were actually produced. For example, microinjections using different types of apparatus, at different rates of infusion or pressure, and at different levels of neurotoxin diffusion and local reactivity produce different models. Further, the systems or methods used to evaluate variables (mainly in the performance of tasks) can have a great influence on the actual degree of change detected.

4.1.2. Species and Strain

Different species show specific characteristics with respect to their neural circuits and the cognitive roles played by their basal forebrain and cortical areas. Similar basal forebrain lesions in different species can produce different models with very distinct features. Some species are more sensitive to certain toxins, e.g., the rabbit is very sensitive to aluminum (26). On some occasions, different laboratories have obtained different results when using the same model because different subspecies or strains were used (37,38). The age of the animals is also important (see later). Age-dependent morphofunctional characteristics are also strain-dependent (40). For example, different strains of Wistar and Long Evans rats and of New Zealand white rabbit show different characteristics.

4.1.3. The Location and Extent of Basal Forebrain Lesion

The basal forebrain is a continuum of cholinergic centers with neurons running from the most anterior part of the septum to the most posterior area of the pallidal/subpallidal region. The greatest number of neurons is concentrated at particular levels (the classic cholinergic centers) that can be stereotactically damaged. A single injection may only cause a small lesion and therefore only a small amount of cholinergic damage (Fig. 3). To induce large lesions, two points of toxin injection are often required (99), mainly at the nbm/nbM level. However, a lesion induced in such a way that extensive diffusion of the toxin occurs could cause important cortical changes and behavioral impairments not related to the desired nbm/nbM damage; care should therefore be exercised. The neurons of neighboring structures, such as the striatopallidal and amygdalar complexes, are the main candidates for suffering these side effects. Cortical impairments are topographically related to the forebrain lesion induced; any cognitive impairment would be theoretically closely related to the higher brain functions associated with this area. More intense deficits in cognitive tasks might therefore be expected in all models involving wide basal-forebrain cholinergic damage, and more selective deficits in models with more restricted basal-forebrain lesions. Nonetheless, different relationships among the different directly and indirectly altered basalocortical and cortical circuits, as well as the adaptive neuronal changes that occur, could produce unexpected modifications in some behavioral performances. The behavioral results obtained with different models, including septum and/or nbm/nbM lesions, should be theoretically different, but they can sometimes show unexpected similarities (or even unexpected differences). Clear differences have been reported between general cholinergic forebrain damage and selective septum or nbm/nbM lesion models. However, a small number of studies report similar cognitive impairments after septum or nbm/nbM lesion (154). The reasons for this could lie in adaptive changes and/or in the very different kinds of tasks used to evaluate cognitive impairment. The i.c.v. infusion of cholinotoxins provokes a general cholinergic impairment of all forebrain cholinergic centers in such a way that involution of the neocortex and limbic structures occurs simultaneously. Many authors consider this kind of model to be closely related to AD, but the obtained results have shown very different impairments of the septohippocampal and nbm–cortical systems. A greater ability to adapt or recover from involutive changes in the septohippocampal system might be responsible for these differences.

4.1.4. Type and Concentration of the Neurotoxin Employed (6,16–18,22, 98,99102,105)

Each neurotoxin causes, directly and indirectly, specific neuronal and glial alterations. In general, the lesion pattern and its short-time development are dose-dependent and related to the toxin, but in the mid- or long-term, very different features can arise – generally by this time irrespective of the lesion-inducing agent (Fig. 4). Total recovery, partial recovery or nonrecovery of deficits have been reported, on some occasions with important discrepancies between cortical morphohistochemical features and behavior (6,18,22,25,87–89,96,97,100–102,127,155).

4.1.5. Unilateral or Bilateral Nature of the Lesion (16,22,102,156)

Surgical lesions of and stereotactic microinjections in the basal forebrain can produce models with unilateral or bilateral deficits. Unilateral models have the advantage that each animal has an internal control, i.e., the undamaged contralateral side. Microinjection of excitotoxins is the choice for producing different models with impairments on one or both sides of the brain. However, the manipulation of one side of the septum can lead to some degree of impairment on the contralateral side due to the contralateral diffusion of the toxic agent. Over the long term, some contralateral changes are induced in nbm/nbM lesions (6). Some toxins (e.g., ibotenic acid) are not very useful for producing bilateral nbm/nbM damage since they are associated with high mortality rates.

4.1.6. Animal Age

Age is a major factor not only in the production of morphofunctional and cognitive impairments but in the course of any deficit produced (20,21,47–58,89,116–118,149,157–159). The neurons of aged animals are generally more susceptible to involution due to the impairment of their adaptive/restorative mechanisms. Neuronal plasticity is diminished or abolished in aging (157), and both astrocytosis and microgliosis occur (with a corresponding increase in cytokine production) (21) (Figs. 6 and 7).

4.1.7. Other Phenotypic Factors

Each animal used in each model possesses special characteristics that condition the involutive and adaptive responses to the lesion induced. Statistical analysis of the individual factors that

Fig. 6. (continued) (**d**) *4 m/13 m-post-L animals.* Pattern of astrocyte response is similar to that of the previous group. (**e**) *4 m/20 m-post-L animals.* In the cortex, reactivity was lower than in the 13 m-post-L group, but in the nbm region, reactivity was maintained. The proximal reaction was variable, but lower in most cases. (**f**) *20 m/1.5 m-post-L animals.* Pattern similar to the 4 m/1.5 m-post-L group (diagram **c**), but with a smaller number of H/H astrocytes and less reactivity near the cortical scar. Scale bar =1 mm (Reproduced from Brain Research, 2000, with permission from Elsevier (14)) *cc, corpus callosum; cg, cingulum; cg1, cingulate cortex, area 1; cg2, cingulate cortex, area 2; CPu, caudate putamen (striatum); G1, granular insular cortex; GP, globus pallidus; ic, internal capsule; M1, primary motor cortex; M2, secondary motor cortex; nbm, nucleus basalis magnocellularis or basal nucleus of Meynert; S1BF, primary somatosensory cortex, barrel field; S1DZ, primary somatosensory cortex, dysgranular region; S1FL, primary somatosensory cortex, forelimb region; S1HL, primary somatosensory cortex, hind limb region; VP, ventral pallidum or substantia innominata.*

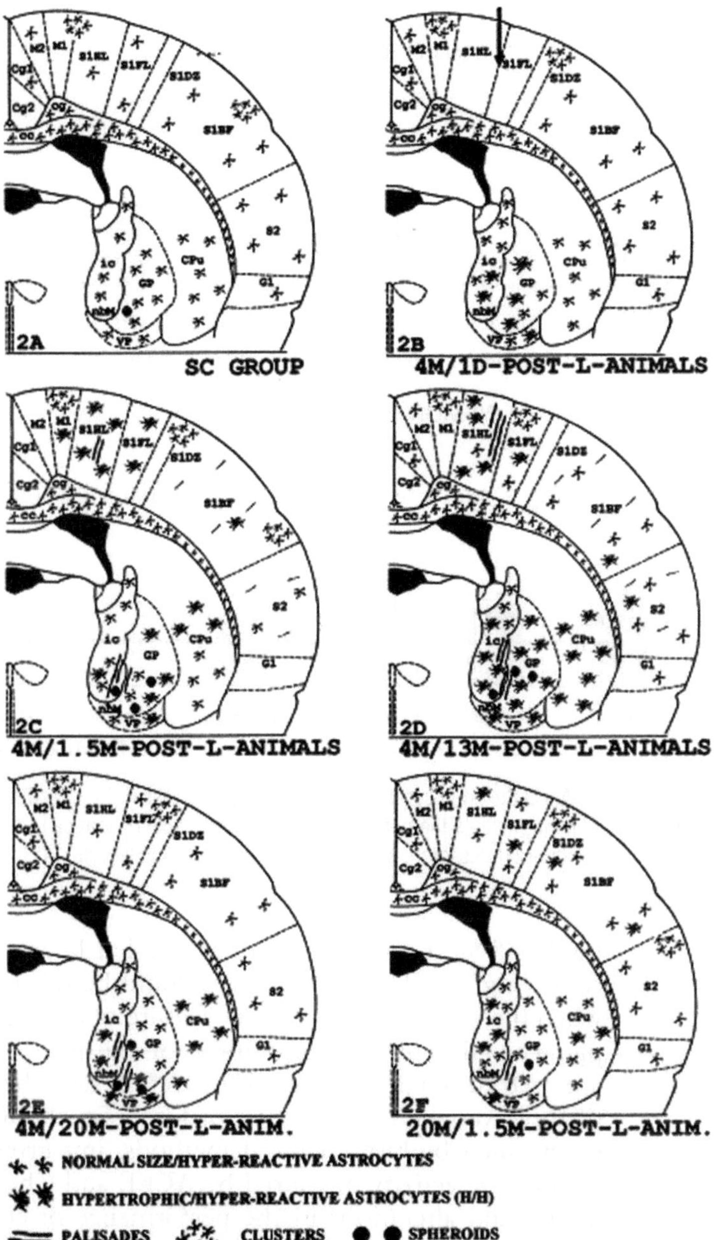

Fig. 6. Diagram of GFAP immunoreactivity after nbm lesioning (cross-sections at −1.4 mm posterior to the bregma, behind the injection trajectory arrow; anatomical regions following the atlas from Paxinos and Watson (229)). (**a**) *SC groups (4–17-month-old rats).* Astrocytes were homogeneous and of normal appearance, and hypertrophic elements were not observed. Hyperreactive astrocytes appeared in the cingulum (cg) and callosal body (cc). Some randomly located clusters were observed in the cortex. Some small spheroid deposits in the nbm were observed in one 17-month-old rat. (**b**) *4 m/1d-post-L animals.* The image of the cortex is quite similar to those of the corresponding SC group; only negative immunoreactivity is observed near the injection trajectory in S1HL and S1FL, and some isolated H/H elements were randomly observed. Heterogeneous astrocytes were observed in the nbm, ic, GP and VP. (**c**) *4 m/1.5 m-post-L animals.* Local, proximal and remote glial responses. Normal and H/H astrocytes (isolated, in clusters and palisades) in the damaged region (nbm, ic, VP, GP), in the proximal area (GP, CPu) and in the cortex. The scar observed in S1HL and S1FL is surrounded by very reactive astrocytes. Some spheroids were observed in the nbm.

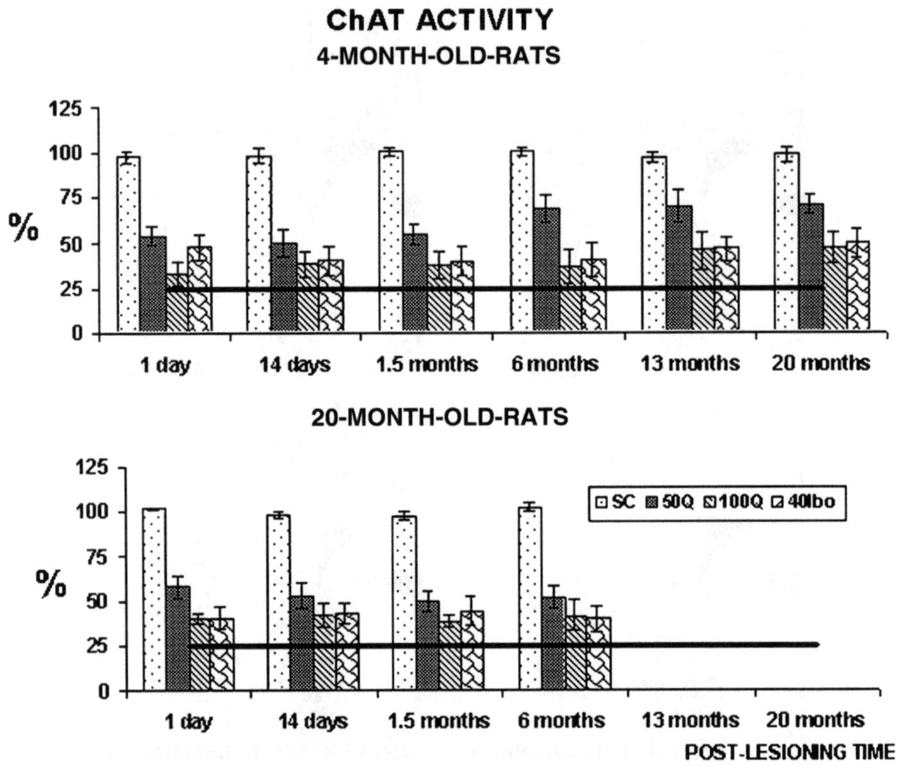

Fig. 7. Choline acetyltransferase (ChAT) in the brain cortex ipsilateral to the nbm lesion (% = ChAT activity in the contralateral cortex in each animal). The line represents the percentage of enzymatic activity that can be attributed to cortical cholinergic neurons (Reproduced from Brain Research, 2000, with permission from Elsevier (14)).

might affect a model has often revealed large standard deviations (Fig. 8). These differences could be of great importance when the variable tested is a key feature of the model. For example, when different concentrations of quisqualic or ibotenic acid are used to obtain nbm lesion models in rats, the quantification of three variables – reduction of forebrain cholinergic neurons determined by histochemical methods, impairment of cortical cholinergic activity measured by AChE and ChAT biochemical assays, and alterations in the performance of cognitive tests – reveals the main features of different models as well as the individual differences among the animals of each model. If models are produced following a strict protocol, and the postmortem analyses of the forebrain lesions of each animal do not suggest the existence of variations caused by technical problems, the results of the basalocortical lesion of each animal must be related to the toxicity of the neurotoxin used and the susceptibility of the animal to it. Postlesional cortical impairment and cognitive deficits are related to the toxin and the lesion induced. However, the short- and long-term outcomes for the surviving basal forebrain cholinergic neurons and the capacity for cholinergic and

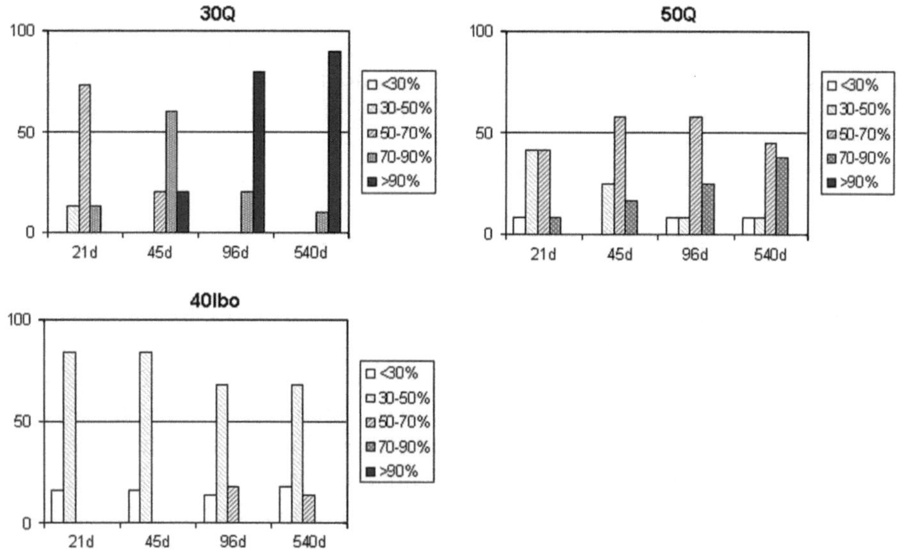

Fig. 8. Analysis of the evolution of the cortical ChAT deficits induced by nbm lesions (30Q and 50Q = nbm injections of 30 and 50 nmols of quisqualic acid; 40 Ibo = nbm injection of 40 nmols of ibotenic acid). Animals have been arbitrarily considered in five groups according their biochemical ChAT activity in the ipsilateral lesioned side (<30%; 30–50%; 50–70%; 70–90% and >90%); being the 100% the value of the ChAT activity in the unlesioned contralateral side of each animal. The percentages of animals belonging to each group have been represented in each model at 21, 45, 96 and 540 post-lesion days. In the 30Q model, most of animals (90% approximately) recovered at 540 days (>90% group), but at 45 days a broad dispersion of results was observed (animals were mostly included in the 70–90% group but some of them were in the 50–70% and >90% groups). In the 50Q model, 18% of animals remained in the <30% and the 30–50% groups of ChAT activity, but a slow and partial recovery is observed in general. In the 40Ibo model, no recovery is observed except in a small number of animals included in the 50–70% group at 96 and 540 postlesion days. In well-controlled models, important differences can be related to individual factors. (n = 26 animals/model/post-lesion day. Methods described in (6,28)).

cognitive recovery are related to the plastic adaptive properties of the animals at the species and individual levels. Indeed, in some of these cholinergic models, some animals recover and others do not (Fig. 8). These models may be useful for studying factors or mechanisms involved in the recovery of neuronal functions. In most models, the homogeneity and repeatability of the results are of prime importance, especially in the study of involutive or restorative neuronal mechanisms, of their consequences in neuronal circuit functioning, and of the effects of putative pharmacological treatments. The study of important involutive or adaptive variations in individual model animals may be of interest when studying the cellular and molecular basis of certain features of the pathogeny and plastic responses of neurons.

Many studies have been performed on aged animals, comparing the results of a subgroup of cognitively unimpaired animals with those of impaired animals maintained under similar conditions (40).

4.2. Neuronal and Glial Alterations in Basal Forebrain Models

4.2.1. Morphohistochemical and Biochemical Changes in the Basal Forebrain

Depending on the type of lesion, a number of cholinergic and noncholinergic types of neuron are altered and either quickly die or begin to show a variable degree of dysfunction (Fig. 3). Some dysfunctional neurons (most of them of dystrophic appearance) may survive but others will involute and die over the next few hours to days (Fig. 2). The volume of damaged tissue with histological signs of neuronal atrophy or dystrophy and gliosis is closely related to the number of each type of neuron showing dysfunction, apoptosis or necrosis, and depends on the method and protocol used. These cellular lesions/dysfunctions are a consequence of the specific susceptibility of each kind of neuron to the toxic agent. The size and neuropathological characteristics of the focal lesions produced by stereotactic injection of toxins depend on the kind of toxin and the dose used (Fig. 3), as well as the administration procedure (e.g., infusion volume). All neurotoxins cause intense necrosis at the injection site, but the damage caused in neighboring areas depends on factors related to the toxin (type, dose, diffusion, and administration protocol) and animal involved (species, age, strain, and individual factors). The exact concentration of the toxin at each point in the surrounding area, which is almost impossible to know, depends on its degree of diffusion. The size of the lesioned zone needs to be controlled by carefully following a very strict protocol (6).

The loss of cholinergic neurons (Figs. 2 and 3) is revealed by biochemical and histochemical assessment of cholinergic markers such as AChE, ChAT, and the p75 NGF receptor. On many occasions, it is hard to make an accurate assessment of cholinergic and noncholinergic neuronal loss. The basal forebrain is an extensive and ill-defined region, its cholinergic neurons are easy to characterize but dispersed, and its noncholinergic neurons are difficult to characterize and localize. In addition, the neuron count is dependent on the enumeration method and the result dependent on many technical factors (e.g., the thickness of the histological sections, the enzymatic or immunohistochemical staining used, and the quantification procedure).

Lesions on one side in the septum lead to an important loss of cholinergic cells in the lesioned area itself, but also to a degree of loss or dysfunction in the contralateral septum (21). In small nbm lesions restricted cholinergic neuronal loss occurs, but if the lesion is large, the surrounding forebrain and striatum pallidum structures can show signs of damage depending on the neurotoxin used (6,22,102). The most neurotoxic chemical compounds (those which produce the greatest lesion volumes and the greatest parvocellular neuronal damage) appear to be kainate, quinolinate, and ibotenate (6,99). NMDA causes a medium-size lesion, while quisqualate causes little damage. Cholinotoxins always cause atrophy and the death of noncholinergic neurons. Injections of immunotoxins result in the almost total loss of cholinergic neurons over a radius of approximately 1 mm around the injection site (as in excitotoxic models), but only cause an average 40–50% cholinergic neuronal loss in other nbm regions (11). This percentage can be higher in excitotoxic models. Such injections also provoke noncholinergic lesions and gliosis (11,30,13).

In all the basal forebrain models produced by focal lesions, and in many of the basal forebrain models produced by general lesions after i.c.v. administration of cholinotoxins, a variable number of cholinergic (and noncholinergic) neurons can survive for long periods (Fig. 2). Starting at the moment of lesioning, involutive and adaptive responses can be observed in many models. The greater or lesser synthesis of necrotic factors by damaged neurons, the production of cytokines and growth factors by astroglial and microglial cells, changes in intracellular/extracellular calcium concentrations, changes in nitric oxide-dependent signaling pathways, and the oxidative stress in the area are, among other factors, all important elements in the production of CNS damage and its clinical course (159–165). Thus, the reduction in the number of cholinergic neurons is intensified in many basal forebrain lesion models (22,28,102,157). Further, in all models, a number of cholinergic cells survive immediate assault and – over the following days and weeks – undergo plastic changes and show hypertrophy and hyperactivity thought to represent an attempt to recover. These changes include signs of local regeneration (axonal sprouting, dendritic growth, and increase in p75 NGRF immunoreactivity), and are seen despite the negative changes in cortical deficits (22,102,157). As an example, in studies in which cholinergic neuronal loss has been quantified after excitotoxic lesions in the nbm of the rat, 20–50% survival of cholinergic neurons has been recorded after the administration of 60–100 nmols of quisqualic or 40 nmols of ibotenic acid (higher doses provoke the death of more than 40% animals) (22,102,157,166,167). In septal lesion models 20–40% neuronal survival has been recorded. After fimbria/fornix transection, 80–100% cholinergic neuronal survival has been seen. These data are, however, only an approximation of the true cholinergic damage since each area of the different centers of the basal forebrain experiences its own level of neuronal loss and dysfunction (22,102).

In most experiments, cholinergic basal forebrain impairment is not directly studied. Rather, an indirect evaluation is made based on the biochemical changes associated with cortical cholinergic impairment. In studies comparing different excitotoxic models, dose response curves have shown a strict correlation between biochemical cortical changes and nbm neuronal loss (21,102), whereas other studies have shown strong discrepancies between these variables (6,16–18). This suggests that in some models axonal sprouting is responsible for cholinergic recovery and cognitive improvement. An analysis of the number and characteristics of the cholinergic nbm neurons suggests that both the number surviving (about 60%) and their adaptive capacity (existence of hypertrophy/cholinergic hyperactivities) are limiting factors in recovery. A slightly greater number of surviving cells in the nbm has been observed under nootropic treatment (35), similar to that seen with NGF or ganglioside GM1 (90,100,112,149,157,167).

Noncholinergic neuronal loss almost certainly occurs (168), though quantitative biochemical and histochemical data for lesion sites have rarely been collected. However, the combined results for the nbm and cortex have led several authors to declare these changes more important than those affecting the cholinergic neurons. For example, Wenk et al. (169) claim that the behavioral impairment seen in the nbm-lesioned models is related to the reduction in neurotensin-transmitting basalocortical neurons rather than to the reduction in cholinergic neurons. Damage to GABAergic (107) and peptidergic projecting neurons (73–77,168) seems to be linked to cognitive impairments in some models. Gliosis is also observed in the lesioned areas of the basal forebrain (21,170). Gliosis usually develops after lesion induction but over a period varying from days to weeks. In toxin-induced models, the severity of astrogliosis and microgliosis is related to the toxin used. Quisqualic acid seems to be the least active against the astroglia (21).

4.2.2. Morphohistochemical and Biochemical Changes in the Cortices: Short- and Long-term Development, Recovery and Non-recovery of Cholinergic Impairment

All basal forebrain lesion models show large deficits in cholinergic markers in the different brain cortices innervated by the damaged centers of the basal forebrain with AChE, ChAT, and HACU activities decreasing by 75% or more (6,16,96,98–102,105) (Fig. 4). All these cholinergic markers are presynaptic, although AChE can be found at postsynaptic neuronal sites or in blood vessel walls (171,172). The number of nicotinic and muscarinic receptors on the cortical neurons and other cells decreases only by about 20% (171–173); the most important reductions occur among the M2 receptor subtype (considered presynaptic to some extent) (173).

Cortical deficits and behavioral impairments do not always correlate well. In toxin-induced models, ibotenic acid promotes fewer cortical deficits and more behavioral impairments than quisqualic acid or AMPA (21,105,169). In rodent models, it is impossible to obtain complete cortical cholinergic deficit, not only because of cortical innervation by cholinergic centers located outside the basal forebrain (68,81,83), but because of the existence of intrinsic cortical cholinergic cells (21,171), which can become hyperactivated after cortical deficits (21).

Long-term changes in cortical deficits and behavioral impairments can be very variable. In many studies, important discrepancies have been seen between the morphological and biochemical changes of the cortex and the extent of cognitive impairment. For example, similar rat excitotoxic nbm lesion models characterized by (i) a lack of recovery of all the evaluated morphofunctional and cognitive parameters (98), (ii) persistent cholinergic impairment with behavioral recovery (96,97,174), or (iii) partial or total recovery of both cholinergic cortical deficits and cognitive impairment (18,22,98,101,102,127,155) have been described. Cholinergic recovery in the cortex has been detected by biochemical (ChAT and AChE activity) and histochemical methods (AChE histochemistry) in some models (6,18,21), whereas other models show biochemical recovery without histochemical recovery of cholinergic innervation or cholinergic axonal sprouting (157,175). Recovery from bilateral lesions is less likely (the capacity to recover from cognitive impairment is also reduced) (18,27,97). Contradictory results similar to those mentioned for the nbm-lesioned models have been reported for the recovery of cholinergic fibers and/or biochemical variables in septal-lesioned models (toxin-induced models and fimbria/fornix transections) (87–89).

It is very likely that spontaneous recovery depends on the capacity of the basal forebrain neurons to enhance their plasticity. The involved processes depend on (natural and synthetic) neurotrophic factors. In rats with nbm lesions, NGF increases ChAT activity and ACh release in the cortex, and improves spatial memory deficits (157,166,167,175). However, the effects are short-lived, and a reversal occurs some weeks after NGF withdrawal (167). Continuous neurotrophic support has been suggested for the amelioration of cholinergic hypofunctioning (167,176). NGF and GM1 ganglioside have a beneficial effect, accelerating spontaneous recovery in unilaterally lesioned rats (100,112,177). Nootropic agents seem to have the same effect. Amiracetam, oxiracetam (178), and pyritinol (35) cause enhanced cortical cholinergic recovery, but special circumstances may have existed in the animals treated. When using pyritinol in quisqualic acid nbm models (35), differences in the action of this compound have been demonstrated depending on the cholinergic deficit induced. In the spontaneous recovery model (30 nmols), animals show total recovery in less time than controls (21 days compared with 45 days), but in the no-recovery model (50 nmol), no changes are observed. Enhancement of recovery has been seen in terms of AChE, ChAT and HACU activities, and also in the histochemical densities of the M2 receptors and the cortical AChE-positive fiber network. Very intensely damaged animals are unable to recover. Cell implants (stem cells, cholinergic neurons, or cells producing neurotrophic factors) (87,176,179,180) have been assayed for the promotion of cell survival, axonal sprouting, axonal growth, and the reinnervation of the cortical cells. Some sustained improvements in cholinergic and cognitive impairments have been obtained in some models, but we are far from being able to use this approach in the treatment of AD.

Noncholinergic alterations of the basal forebrain also provoke neuronal alterations in the cortices and cognitive impairment. However, only in a small number of studies has the importance of the loss of these basalocortical neurons been examined. A reduction in GABAergic function in some models produced by ibotenic injection (107), and reductions in neuropeptides such as neurotensin, somatostatin, and neuropeptide Y in others (73,168,181), show that cognitive deficits are also the consequence of alterations other than those involving cholinergic neurons.

Glucose metabolism is reduced after lesion induction (182), but recovers irrespective of cholinergic markers recovering or not. Changes in amyloid precursor protein (APP) metabolism have been studied to better understand AD. Different modifications have been described, but early and persistent increases in APP synthesis are always observed (183–186). Rabbit models can be very useful for studying AD because amyloid deposits are produced.

Astrogliosis occurs in most models, mainly in the long term (weeks to months). The presence of gliosis must always be studied given its important relationship with both involutive and adaptive neuronal responses (21). In general, microglia do not increase in number either in the short or long term.

4.3. Cognitive Impairments in Basal Forebrain Lesioned AD Models

The cognitive changes observed include deficits in learning, memory, and attention. Cognitive function in lesioned animals has been studied using a wide variety of apparatuses specifically designed to test the ability to solve a problem or perform a task. Varying degrees of deficit in passive and active avoidance (187–189), and in the performance of tests such as the T-maze (190), radial arm maze (191) and Morris water maze test (154,191) have been observed in different models. However, in many models, unexpected results have been obtained, including a lack of cognitive impairment after the induction of a basal forebrain lesion, strong discrepancies between cortical cholinergic and cognitive impairments, or large differences between the recovery of cortical cholinergic deficits and the restoration of cognitive functions. The best-known examples are the different cognitive impairments and their development seen in quisqualic acid, AMPA or ibotenic acid nbm lesion models with similar cortical cholinergic deficits (105). Quisqualic acid or AMPA lesions cause no deficits in spatial learning acquisition and performance in the water maze task, while ibotenic acid nbm injections provoke great impairment in this and other cognitive tests (16,102,105,189,192). Persistent cholinergic impairment with behavioral recovery (96,97) has been observed in some ibotenic acid-lesion models. The use of selective cholinergic toxins (such as AF64A and 192 IgG immunotoxin) has produced cognitive impairments models with similar problems of interpretation. Impairments in the Morris water maze have been seen in some cases but not in others

(139,150). Great controversy exists with respect to the interpretation of such results. The role of ACh and the basal forebrain, the methods for producing basal forebrain lesions, the results obtained with these methods, and the cognitive tests used have all been questioned. Some reasons for this have already been mentioned, but three are particularly important:

1. Cognitive processes are still poorly understood, with each species showing differences; further, the methods to evaluate them are still poorly developed. The theoretically well-defined and differentiated cognitive processes studied in these models are often closely related since, to some extent, they involve the same neuronal circuitry and mechanisms. Further, all animal behavior tests can be questioned since they can never unequivocally discriminate between the cognitive processes of learning, memory, attention, and motivation. Moreover, the way tests are performed can vary from one laboratory to another (11,193).
2. The cognitive functions of cortical ACh are not well understood, and can vary from one species to another (81,114).
3. Many of the functional consequences of lesions in the basal forebrain in some AD models are not related to cortical cholinergic damage but to lesion/dysfunction of other types of cholinergic and noncholinergic neurons, e.g., (a) the noncholinergic neurons of the basal forebrain projecting to the cortices (16,65,68,71–73,105,169), (b) the cholinergic and noncholinergic intermeshed neurons in the basal forebrain that communicate with noncortical and cortical areas other than the main area of cortical projection, and (c) the interneurons controlling the neuronal circuits in the basal forebrain, and (d) the neurons of neighboring areas (such as the hypothalamus and globus pallidus), which can be affected by the large diffusion patterns of some toxins (6).

4.4. Special Characteristics of Aged Animal Basal Forebrain Lesion Models

In these models, cholinergic and noncholinergic impairment is produced in the special background of physiological aging, in which variable deficits in neuronal, glial, and vascular structures and systems exist (Figs. 6 and 7). All these deficits should be assessed and taken into account when interpreting the impairments observed in the model. Different adaptive or involutive neuronal and glial responses have been observed in different individuals belonging to the same experimental and control groups (49,191,194,195).

A variable degree of age-induced neuronal death or dysfunction has been shown in the basal forebrain of rodents. The number of cholinergic neurons in this area is reduced by only a small percentage in rats until 20 months of age (21,157,158). However, they show morphological changes and reductions in p75 NGF receptor immunoreactivity (157) and NGF sensitivity (52,158), suggesting they may have suffered some involution. This dysfunction correlates well with the cognitive impairment observed in some studies (19,158), but in others, cognitively unimpaired and impaired animals of the same age have been found. The morphological, histochemical, and biochemical alterations in the cortex include reductions in the number of neurons (159) and synaptic connections (54,81), gliosis (21,53), and cholinergic impairment, including downregulation of nicotinic and muscarinic receptors (196), reductions in ChAT and AChE levels (196), and a reduction in ACh release (197). The cognitive changes associated with aging correlate well with this cortical involution (19) and involve a wide variety of deficits in the performance of many tasks. Aged animals suffer impairment of learning and memory with respect to both "general" and "specific" information (158,198,199).

Lesions in the basal forebrain of aged rats, especially in the nbm, are associated with greater neuron loss and more profound cortical deficits than in young animals (21,50,144,157,191). Lesions in the basal forebrain made with quisqualic and ibotenic acid seem to provoke similar impairments in the first week, with a similar increase in NGF receptor immunoreactivity despite the low level of this receptor in controls (157). However, the cortical deficits in nicotinic and muscarinic receptors (191), ChAT and AChE are more intense in younger than in older rats (Fig. 7) (21). After total immunolesion, NGF levels in the cortex of aged rats do not increase, although in young rats the concentration is doubled. The development of lesions is also very different in aged and young animals: doses of quisqualic or ibotenic acids that allow for a partial cortical recovery in young rats allow no such recovery in aged animals (21). The existence of gliosis in aged rodents is of great theoretical and practical importance; the course of cortical deficits in these models has been related to the development of gliosis (21).

5. Alzheimer Models Produced by Lesion or Dysfunction of Cortical Areas

Different kinds of lesion have been produced in the cortex of different mammals to study the possible pathogenic mechanisms involved in AD, or to test possible AD treatments. These lesions have been of variable intensity and/or specificity. Controlled cortical contusions have been produced in the brains of some mammals to study neurodegenerative processes (200,201). For example, after craniotomy, lesions in the cortex can be induced by time- and intensity-controlled pressure exertion, but the models obtained are not very specific. Surgical and physical procedures have also been used to partially or totally eliminate specific areas of the neocortex, archicortex, or palaeocortex to study the changes in cognition and behavior, as well as the morphofunctional changes in the unlesioned area of the lesioned cortex or in interconnected unlesioned cortical areas (202,203). However, these models, in which gross tissue lesions are produced, are of little use in AD research. Cell-selective lesions in the neocortex, hippocampus, entorhinal cortex, and subiculum are of much greater interest. Differential neuronal responses and cognitive deficits following aspiration, electrolytic or excitotoxic lesions have been observed in different cortical areas in rats (202–204). Such cortical models have been produced using the same neurotoxins mentioned in the section on cholinergic models. Consequently, different types of specific cholinergic cortical model as well as nonspecific cholinergic/noncholinergic cortical models can be contemplated.

AFG4A and the 192 IgG-saporin immunotoxin lead to cortical cholinergic neuron impairment when administered intracerebroventricularly or intraparenchymally. As mentioned above, i.c.v. administration of cholinotoxins induces lesions of variable intensity in all areas of the basal forebrain and in different cortical regions, although mainly in the hippocampus. Intraparenchymal injections of AFG4A in the neocortex and hippocampus produce results similar to those obtained with basal forebrain injection, but in some cases, more pronounced atrophy of the cholinergic forebrain cholinergic fibers is observed (205,206). Moreover, the intrinsic cholinergic neurons in the cortical areas (a subset of great importance in rodents) suffer great involution. As with the AF64A toxin, the immunotoxins can be i.c.v. administered or injected into the target cortical areas. In both cases, the targets are the p75 NGF receptors of the cholinergic nerve endings in the neocortex, hippocampus, and cingulate cortex. I.c.v. administration also acts at the somatic receptors (207), as well as causes cholinergic neuron impairment or even death in all the basal forebrain centers (although those in the striatum, ventral pallidum and amygdale are spared) (14,207). The main disadvantage is that Purkinje cells strongly express p75 NGF receptors, and the impairment or loss of these cerebellar neurons is observed when high doses of immunotoxin are administered (126). Since the cerebellum is involved in motor control, learning, associative and nonassociative learning processes, and attention, this interferes in the performance on some behavioral tests, and therefore in the interpretation of their results (208,209).

Excitotoxins and noncholinergic selective neurotoxins provoke wide neurotoxic morphological and functional changes when injected intraparenchymally (59). The role of different cortical areas in cognition, as well as their anatomical connections, has been studied after restricted neurotoxin injection. Inducing excitotoxic lesion in the hippocampus leads to neuronal changes in the entorhinal cortex since the key components of processes that may ultimately lead to the expression of AD proteins and neuronal death are induced (203).

6. Alzheimer Models Produced by Amyloid Injections

No animal produces amyloid plaques so similar to those seen in human AD that it is rendered the main model for AD research. In some species, Aβ deposition has been observed in a few aged individuals, or has been described in a few experimental animals. Aβ or Aβ peptides (mainly $A\beta_{25-35}$) have been administered either intracerebroventricularly (7,8,210–212) or intraparenchymally (9,213,214) in different mammals (mainly rodents) to study a wide variety of cellular processes possibly involved in neurodegeneration. Protofibrillar-fibrillar $A\beta_{1-42}$ injected into the entorhinal cortex of the rat (9) or different combinations of Aβ ($A\beta_{1-42}$, $A\beta_{1-43}$, $A\beta_{1-40}$) injected in the hippocampus (10), provoke neuronal lesions, impair synaptic function and plasticity, and cause cognitive deficits, thereby constituting promising animal models of AD. Soluble $A\beta_{25-35}$ injection provokes transient synaptic and spatial learning deficits (215). With a single i.c.v. injection or continuous infusion (using micropump systems) these compounds provoke different degrees of lesion/dysfunction in the basalocortical cholinergic neurons, as well as cortical cholinergic changes such as modifications in ACh release (7,8,27), and a reduction of muscarinic receptors (210). Two weeks after i.c.v. injection of $A\beta_{25-35}$, a degree of cholinergic hyperactivity is observed in the nbm and cortex, suggesting that increased cholinergic function may be responsible for muscarinic receptor downregulation (210). The models obtained by i.c.v. injection of this substance are mainly used to study alterations induced in the hippocampus and their regulation. In these models, however, no special attention has ever been paid to the alterations induced in the ventricular walls (necrosis of the ependymal cells) or the reactive responses in the circumventricular brain structures to either the amyloid peptide or its solvents; this might inhibit the interpretation of results. When injected into the nbm, these peptides seem to induce hypofunction of the cholinergic neurons, which is maintained until the disappearance of the amyloid fibrils. At this time, ChAT immunoreactivity is recovered (27).

Microglial Aβ phagocytosis has been studied in models obtained by intrahippocampal amyloid injection and ventricular infusion of microglial cells (214).

Many neuropathological features of the lesion-induced amyloid models of AD are very different to those characterize amyloid transgenic models (216).

7. Other Lesion-Induced Alzheimer Models

Separate book chapters are dedicated to other types of lesion-induced AD models produced by different kind of chemicals. Some of them present morphohistochemical and neurochemical alterations similar to those described in this chapter together with other neuropathological modifications observed in AD. For this reason, it is very interesting to compare the results of different models to get more informed conclusions on neurodegenerative, neuroprotective, and restorative mechanisms possibly existing in AD. In this sense, we consider of special interest the following models:

(a) Ethanol-induced dementia models. As in humans, the chronic intake of ethanol in rodents (25,26,217–219) provokes a time-dependent partial or total degeneration of the cortical cholinergic basal forebrain systems with important damage to the nucleus basalis and cortical cholinergic denervation. This model has been used to study general methods to compensate for the so-called "syndrome of partial cholinergic differentiation of the cortical mantle" seen in many dementing disorders including AD (217).

(b) AD models produced by aluminum. Controversy has surrounded the possible role of aluminum in the pathogenesis of AD for four decades. Aluminum induces different cortical and subcortical brain lesions, the characteristics of which depend on the species used and age of the animals, on the type of aluminum compound employed, and on the administration route and protocol (23,24,220,221). Chronic exposure over 12 months in mice results in the deposition of Aβ (24) similar to that seen in congophilic amyloid angiopathy in humans. Chronic ingestion of aluminium in aged rats induces oxidative stress, hyperphosphorylation of tau and granovacuolar degeneration without amyloid deposition (220). Rabbits, particularly sensitive to aluminum, develop severe neurological changes including neurofibrillary degeneration which show similarities to neurofibrillary tangles (23). In aged New Zealand white rabbits, intravenous injections of aluminum maltoate produce a lesion model of AD (221) in which cortical neurons show accumulations of tau immunopositive intraneuronal filaments, oxidative stress and apoptosis, and extraneuronal Aβ deposits exist.

(c) AD models produced by okadaic acid. Microinjections of okadaic acid in the cortex of rats activate pro-apoptotic processes observed in AD (222,223) and originate neuritic dystrophia (224). Injections into the nbm reduces ACh levels and spatial memory deficits (225).

(d) AD cholinergic models associated to ischemic events. Based in the hypothesis that ischemic events play a critical role in the pathogenesis of dementia and the acceleration of cognitive impairment, special lesion models of AD have been described. The injection of microspheres (approximately 50 μm in diameter; 5,000 microspheres/100 μl) (226) into the right common carotid artery of rats produces alterations of the microvasculature and the parenchyma of the brain cortex. The ipsilateral medial prefrontal and in frontoparietal cortices show a clear reduction in the density of cholinergic fibers, and in the basal forebrain, neuronal alterations and gliosis is produced. Thus, treated animals show impairments in attention performance. Other selective cognitive impairments (working memory, etc.) can be produced by lesion when cerebral hypoperfusion is originated (227). Partial cortical devascularization induces a loss of cortical cholinergic axon terminals and a retrograde degeneration of basal forebrain neurons. In this type of model, free and drug-induced cholinergic recovery has been successfully studied (228). Cerebral ischemia, instead of hypoxia, combined with continuous i.c.v. administration of Aβ impairs spatial memory in rats (34).

8. Conclusions

Lesion-induced vertebrate models of AD provide a very useful experimental approach for studying the pathogenesis of AD and for testing possible preventive or palliative treatments. The "cholinergic theory" of geriatric dysfunction and AD has allowed the design of a considerable number of AD models obtained by inducing lesions/dysfunctions of the cholinergic neurons of the basal forebrain. These lesions can be focal and selective (induced by surgery or stereotactic injections) or more general (produced by i.c.v. injections of selective cholinotoxins, such as AF64A and 192 IgG-saporin). These models have been of prime importance in obtaining information on cortical cholinergic innervation and on how changes to it can affect neuron functioning and cognitive processes. They have also been very useful for studying possible treatments that seek to preserve cholinergic innervation or its recovery. These studies on cholinergic systems have also been of use in investigating other disorders in which cholinergic systems are impaired. AD models obtained by cortical lesion have helped reveal the role of the cortical areas in cognitive processes and have shown that morphofunctional alterations occur in different areas when changes are induced in other regions in which afferent or efferent neurons are located. A few AD models involving more extensive cortical and subcortical lesions induced by the controlled administration of neurotoxins (aluminum and okaidaic acid) are important in the study of pathogenic mechanisms.

Despite the criticism these models have received, they remain an extremely important experimental tool in the study of the pathogenesis and possible treatment of AD since no type of AD model – not even transgenic models – shows the full range of neuropathological and cognitive disturbances that characterize human AD. The cholinergic models are, at this moment, our most important tool for studying nonamyloid-related aspects of AD.

Improving these lesion-induced AD models could further increase our knowledge of this disease. The development of new models with neuropathological alterations related to processes other than cholinergic impairment, a greater use of aged animals, and a more in-depth study of the results obtained with all models, could help disclose the complex mechanisms associated with it.

The most appropriate model must always be chosen when pathological aspects of a neuronal mechanisms are under study, and researchers must know their chosen model very well, not only with respect to the pattern of lesion development (the complete series of neuropathological and cognitive changes) but also the short- and long-term changes induced by involutive and adaptive neuronal and glial processes. The lesions produced must be carefully studied; indeed, all the general and individual, neuronal and glial factors that intervene in the production and course of the lesion/dysfunction should be studied to obtain full benefit from lesion-induced models of AD.

Acknowledgments

The authors acknowledge the invaluable collaboration of Drs. L Rivas, MA Barca, I. Moradillo, and A. Toledano-Diaz for so many years in studying lesion-induced models of Alzheimer's Dementia.

References

1. Toledano A (1988) Hypotheses concerning the aetiology of Alzheimer's disease. Pharmacopsychiatry 21:17–25
2. Sivaprakasam K (2006) Towards a unifying hypothesis of Alzheimer's disease: Cholinergic system linked to plaques, tangles and neuroinflammation. Curr Med Chem 13:2179–2188
3. Bartus RT, Dean RL, Beer B, et al. (1982) The cholinergic hypothesis of geriatric memory dysfunction. Science 217:408–417
4. Hardy J (2006) Alzheimer's disease: The amyloid cascade hypothesis: An update and reappraisal. J Alzheimers Dis 9:151–153
5. Sorrentino G, Bonavita V (2007) Neurodegeneration and Alzheimer's Disease: The lesson from tauopathies. Neurol Sci 28: 63–71
6. Toledano A, Álvarez MI (2004) Lesions and dysfunctions of the nucleus basalis as Alzheimer's disease models: General and

critical overview and analysis of the long-term changes in several excitotoxic models. Curr Alz Res 1:189–214
7. Giovannelli L, Casamenti F, Scali C, et al. (1995) Differential effects of amyloid peptides beta(1–40) and beta(25–35) injections into the rat nucleus basalis. Neuroscience 66:781–792
8. Giovannelli L, Scali, Faussone-Pellegrini MS, et al. (1998) Long-term changes in the aggregation state and toxic effects of beta-amyloid injected into the rat brain. Neuroscience 87:349–357
9. Sipos E, Kurunczi A, Kasza A, et al. (2007) Beta-amyloid pathology in the entorhinal cortex of rats induces memory deficits: Implications for Alzheimer's disease. Neuroscience 147:28–36
10. Stéphan A, Laroche S, Davis S (2001) Generation of aggregated beta-amyloid in the rat hippocampus impairs synaptic transmission and plasticity and causes memory deficits. J Neurosci 21:5703–5714
11. Perry TA, Hodges H, Gray JA (2001) Behavioural, histological and immunocytochemical consequences following 192-IgG saporin immunolesions of the basal forebrain cholinergic system. Brain Res Bull 54:29–48
12. Pizzo DP, Waite JJ, Thal LJ, et al. (1999) Intraparenchymal infusions of 192 IgG-saporin: Development of a method for selective and discrete lesioning of cholinergic basal forebrain nuclei. J Neurosci Methods 91:9–19
13. Torres EM, Perry TA, Blockland A, et al. (1994) Behavioural, histochemical and biochemical consequences of selective immunolesions in discrete regions of the basal forebrain cholinergic system. Neuroscience 63:95–122
14. Rossner S, Schliebs R, Härting W, et al. (1995) 192 IgG-saporin-induced selective lesion of cholinergic basal forebrain system: Neurochemical effects on cholinergic neurotransmission in rat cerebral cortex and hippocampus. Brain Res Bull 38:371–381
15. Yuede CM, Dong H, Csernansky JG (2007) Anti-dementia drugs and hippocampal-dependent memory in rodents. Behav Pharmacol 18:347–363
16. Dunnett SB, Whishow IQ, Jones GH, et al. (1987) Behavioural, biochemical and histochemical effects of different neurotoxic amino acids injected into nucleus basalis magnocellularis of rats. Neuroscience 20:653–669
17. Smith G (1988) Animal models od Alzheimer's disease: Experimental cholinergic denervation. Brain Res Rev 13:103–118
18. Casamenti F, Di Patre PL, Bartolini L, et al. (1988) Unilateral and bilateral nucleus basalis lesions: Differences in neurochemical and behavioural recovery. Neuroscience 24:209–215
19. Wellman CL, Sengelaub DR (1991) Cortical neuroanatomical correlates of behavioral deficits produced by lesion of the basal forebrain in rats. Behav Neural Biol 56:1–24
20. Solomon PR, Flynn D, Mirak J, et al. (1998) Five-year retention of the classically conditioned eyeblink response in young adult, middle-aged, and older humans. Psychol Aging 13:186–192
21. Monzon-Mayor M, Álvarez MI, Arbelo-Galván JF, et al. (2000) Long-term evolution of local, proximal and remote astrocyte responses after diverse nucleus basalis lesioning (an experimental Alzheimer model). GFAP immunocytochemical study. Brain Res 865:245–258
22. Toledano A (1992) Evolution of cholinergic cortical innervation after nbM lesioning. (An experimental Alzheimer model). In: Knezevic S, Spilich G, Wisniewsky E (eds) Neurodevelopment, aging and cognition. Birkhäuser, Boston, MA/Basel, Switzerland/Berlin, pp 173–198
23. Savory J, Herman MM, Ghribi O (2006) Mechanisms of aluminum-induced neurodegeneration in animals: Implications for Alzheimer's disease. J Alzheimers Dis 10: 135–144
24. Rodella LF, Ricci F, Borsani E, et al. (2008) Aluminium exposure induces Alzheimer's disease-like histopathological alterations in mouse brain. Histol Histopathol 23:433–439
25. Arendt T (1993) The cholinergic deafferentation of the cerebral cortex induced by chronic consumption of alcohol: Revesal by cholinergic drugs and transplantation. In: Hunt WA, Nixon SJ (eds) Alcohol induced brain damage. NIH Publications N° 93–3549, US Departament of Health and Human Services, Rockville, IN, pp 431–460
26. Arendt T (1994) Impairment in memory function and neurodegenerative changes in the cholinergic basal forebrain system induced by chronic intake of ethanol. J Neural Transm 44[Suppl]:173–187
27. Casamenti F, Prosperi C, Scale C, et al. (1998) Morphological, biochemical and bechavioural changes induced by neurotoxic and inflammatory insults to the nucleus basalis. Int J Dev Neuroscience 16:705–714
28. Shaughnessy LW, Barone S, Mundy WR, et al. (1994) Comparison of intracraneal infusions of colchicine and ibotenic acid as models of neurodegeneration in the basal forebrain. Brain Res 637:15–26

29. Gower AJ, Rousseau D, Jamsin P, et al. (1989) Behavioural and histological effects of low concentrations of intraventricular AF64A. Eur J Pharmacol 166:271–281
30. Leanza G, Nilsson OG, Wiley RG, et al. (1995) Selective lesioning of the basal forebrain cholinergic system by intraventricular 192 IgG-saporin: Behavioural, biochemical and stereological studies in the rat. Eur J Neurosci 7:329–343
31. Lecanu L, Greeson J, Papadopoulos V (2006) Beta-amyloid and oxidative stress jointly induce neuronal death, amyloid deposits, gliosis, and memory impairment in the rat brain. Pharmacology 76:19–33
32. Aguilar JS, Jerusalinsky D, Sotckert M, et al. (1982) Localization of hippocampal muscarinic receptors after kainic acid lesion of CA3 and Fimbria-Fornix Transection. Brain Res 247:335–340
33. Leo G, Genedani S, Filaferro M, et al. (2007) Hyper-homocysteinemia alters amyloid peptide-clusterin interactions and neuroglial network morphology and function in the caudate after intrastriatal injection of amyloid peptides. Curr Alzheimer Res 4:305–313
34. Iwasaki K, Egashira N, Hatip-Al-Khatib I, et al. (2006) Cerebral ischemia combined with beta-amyloid impairs spatial memory in the eight-arm radial maze task in rats. Brain Res 1097:216–223
35. Toledano A, Bentura ML (1994) Pyritinol facilitates the recovery of cortical cholinergic deficits caused by nucleus basalis lesions. J Neural Transm [Park Dis Dement Sec] 7: 195–209
36. Pepeu G (2001) Overview and perspective on the therapy of Alzheimer's disease from a preclinical viewpoint. Progs Neuropsychopharmacol Biol Psychiatr 25:193–209
37. Bartus RT (1988) The need for common perspectives in the development and use of animal models for age-related cognitive and neurodegenerative disorders. Neurobiol Aging 9: 445–451
38. Pshennikova MG, Popkova EV, Khomenko IP, et al. (2005) Resistance to neurodegenerative brain damage in August and Wistar rats. Bull Exp Biol Med 139:540–542
39. Pndiki S, Stamatakis A, Frangkouli A, et al. (2006) Effects of neonatal handling on the basal forebrain cholinergic system of adult male and female rats. Neurosci 142:305–314
40. Vaucher E, Aumnot N, Pearson D, et al. (2001) Amyloid Beta peptide levels and its effects on hippocampal acetylcholine release in aged, cognitively-impaired and –unimpared rats. J Chem Neuroanat 21:323–329
41. Works SJ, Wilson RE, Wellman CL (2004) Age-dependent effect of cholinergic lesion on dendritic morphology in rat frontal cortex. Neurobiol Aging 24:963–974
42. Anger WK (1991) Animal test systems to study behaviorual dysfunctions of neurodegenerative disorders. Neurotoxicol 12:403–413
43. Chida Y, Sudo N, Mori J, Kubo C (2006) Social isolation stress impairs passive avoidance learning in senescence-accelerated mouse (SAM). Brain Res 1067:201–208
44. Bigl V, Arendt T, Biesold D (1990) The nucleus basalis of Meynert during ageing and dementing neuropsychiatric disorders. In: Steriade M, Biesold D (eds) Brain cholinergic system. Oxford University Press, Oxford, pp 364–386
45. Michalek H, Fortuna S, Pintor A (1988) Age-related differences in brain cholin acetyltranspherase, cholinesterases and muscarinic receptor sites in two strains of rats. Neurobiol Aging 10:143–148
46. Chen Y, Fenoglio KA, Dubé CM, et al. (2006) Cellular and molecular mechanisms of hippocampal activation by acute stress are age-dependent. Mol Psychiatr 11: 992–1002
47. Zawai N, Arendash GW, Wecker L (1992) Basal forebrain cholinergic neurons in aged rat brain are more susceptible to ibotenate-induced degeneration than neurons in young adult brain. Brain Res 589:333–337
48. Pepeu G, Marconcini PI (1994) Dysfunction of the brain cholinergic system during aging and after lesions of the nucleus basalis of Meynert. J Neural Transm 44[Suppl]:189–194
49. Wellman CL, Sengelaub DR (1995) Alterations in dendritic morphology of frontal cortical neurons after basal forebrain lesions in adult and aged rats. Brain Res 69:48–58
50. Gu Z, Yu J, Perez-Polo JR (1998) Responses in the aged rat brain after total immunolesion. J Neurosci Res 54:7–16
51. Williams B, Granholm AC, Sambamurti K (2007) Age-dependent loss of NGF signaling in the rat basal forebrain is due to disrupted MAPK activation. Neurosci Lett 413: 110–114
52. Fischer W, Gage FH, Bjorklund A (1989) Degenerative changes in forebrain cholinergic nuclei correlate with cognitive impairments in aged rats. Eur J Neurosci 1: 34–45
53. Peinado MA, Martínez M, Pedrosa JA, et al. (1993) Quantitative morphological changes in neurons and glia in the frontal lobe of the aging rat. Anat Rec 237:104–108

54. Saito S, Kobayashi S, Ohashi Y, et al. (1994) Decreased synaptic density in aged brains and its prevention by rearing under enriched environment as revealed by synaptophysin content. J Neurosci Res 39:57–62
55. MacDonald NJ, Decorti F, Pappas TC, et al. (2000) Cytokine/neurotrophin interaction in the aged central nervous system. J Anat 197:543–551
56. Nicolle MM, Gonzalez J, Sugaya K, et al. (2001) Signatures of hippocampal oxidative stress in aged spatial learning-impaired rodents. Neuroscience 107:415–431
57. von Bohlen O, Zacher C, Gass P, Unsicker K (2006) Age-related alterations in hippocampal spines and deficiencies in spatial memory in mice. J Neurosci Res 83:525–531
58. Bertoni-Freddari C, Guiuli C, Pieri C, et al. (1986) Quantitative investigation of the morphological plasticity of synaptic function in rat dentare gyrus during aging. Brain Res 366:187–192
59. Palop JJ, CHin J, Roberson ED, et al. (2007) Aberrant excitatory neuronal activity and compensatory remodelling of inhibitory hippocampal circuits in mouse models of Alzheimer's Disease. Neuron 55:697–711
60. Shute CD, Lewis PR (1963) Cholinergic neurons pathways in the brain. Nature 99: 1160–1164
61. Shute CD, Lewis PR (1967) The ascending cholinergic reticular system: Neocortical, olfactory and subcortical projections. Brain 90:497–520
62. Krnjevic K, Silver A (1965) A histochemical study of cholinergic fibers in the cerebral cortex. J Anat 99:711–759
63. Johnston MV, McKinney M, Coyle JT (1981) Neocortical cholinergic innervation: A description of extrinsic and intrinsic components in the rat. Exp Brain Res 43: 159–172
64. Butcher LL, Woolf NJ (1986) Central cholinergic systems: Synopsis of anatomy and overview of physiology and pathology. In: Scheibel AB, Weschler AF, Brazier MAB (eds) The biological substrates of Alzheimer's disease. Academic, Orlando, FL/San Diego, CA, pp 73–86
65. Schwaber JS, Rogers WT, Satoh K, et al. (1987) Distribution and organization of cholinergic neurons of the rat forebrain demonstrated by computerized data acquisition and three dimensional reconstruction. J Comp Neurol 263:309–325
66. Mesulam MM, Mufson EJ, Levey AI, et al. (1983) Cholinergic innervation of cortex by the basal forebrain: Cytochemistry and cortical connections of the septal area, diagonal band nuclei, nucleus basalis (substancia innominate) and hypothalamus in the rhesus monkey. J Comp Neurol 214:170–197
67. Mesulam MM, Mufson EJ, Weiner EJ, et al. (1983) Central cholinergic pathway in the rat: An overview based on an alternative nomenclature (Ch1-Ch6). Neuroscience 10: 1185–1201
68. Rye DB, Wainer BH, Mesulam MM, et al. (1984) Cortical projections arising from the basal forebrain: A study of cholinergic and noncholinergic components employing combined retrograde tracing and immunohistochemical localization of choline acetyltransferase Neuroscience 13:627–643
69. Batchelor PE, Armstrong DM, Blaker SN, et al. (1989) Nerve growth factor receptor and choline acetyltransferase colocalization in neurons within the rat forebrain: Response to fimbria-fomix transection. J Comp Neurol 284:187–204
70. Woolf NJ, Gould E, Butcher LL (1989) Nerve growth factor receptor is associated with cholinergic neurons of the basal forebrain but not the pontomesencephalon. Neuroscience 30:143–152
71. Bennett-Clarke CA, Joshep SA (1986) Immunocytochemical localization of somatostatin in human brain. Peptides 7:877–884
72. Crowley JN, Wenk GL (1989) Co-existence of galanin involved in memory processes and dementia? Trends Neurosci 12:278–281
73. Unger JW, Schmidt Y (1994) Neuropeptide Y and somatostatin in the neocortex of young and aging rats: Response to nucleus basalis lesions. J Chem Neuroanat 7:25–34
74. Köller C, Chan-Palay V, Wu JY (1984) Septal neurons containing glutamic acid decarboxylase immunoreactivity project to the hippocampal region in the rat brain. Anat Embryol 169:41–44
75. Senut MC, Menetrey D, Lamour Y (1989) Cholinergic and peptidergic projections from the medial septum and the nucleus of the diagonal band of Broca to dorsal hippocampus, cingulate cortex and olfactory bulb: A comibined wheat germagglutinin- apohorseradish peroxidase- gold immunohistochemical study. Neuroscience 30:385–403
76. Bogdanovic N, Nilsson L, Adem A, et al. (1993) Decrease of somatostatin receptor binding in the rat cerebral cortex after ibotenic acid lesion of the nucleus basalis magnocellularis: A quantitative autoradiographic study. Brain Res 628:31–38

77. Walker LC, Koliatsos VE, Kitt CA, et al. (1989) Peptidergic neurons in the basal forebrain magnocellular complex of the rhe¬sus monkey. J Comp Neurol 280:272–282
78. Brashear HR, Zaborsky L, Heimer L (1984) Distribution of gabaergic and cholinergic neurons in the rat diagonal band. Neuroscience 17:439–451
79. Alheid GF, Heimer L (1988) New perspectives in basal forebrain organization of special relevance for neuropsichiatric disorders: The striatopallidal, amyg¬daloid and corticopetal components of substantia innominata. Neuroscience 27:1–39
80. Bigl V, Woolf NJ, Butcher LL (1982) Cholinergic projections from the basal forebrain to frontal, parietal, temporal, occipital and cingulate cortices: A combined fluorescent tracer and acetylcholinesterase analysis. Brain Res Bull 8:727–749
81. Sarter M, Bruno JP (1997) Cognitive functions of cortical acetylcholine: Toward a unifying hypothesis. Brain Res Rev 23:28–46
82. Barskerville KA, Chang HT, Herron P (1993) Topography of cholinergic afferents from the nucleus basalis of Meynert to representational areas of sensorimotor cortices in the rat. J Comp Neurol 335:552–562
83. Teyler TJ, DiScena P (1986) The hippocampal memory indexing theory. Behav Neurosci 100:147–154
84. Mallet PE, Beninger RJ, Flesher SN, et al. (1995) Nucleus basalis lesions: Implications of basoamygdaloid cholinergic pathways in memory. Brain Res Bull 36:51–56
85. Aguilar JS, Jerusalinsky D, Stockert M, et al. (1982) Localization of Hippocampal Muscarinic Receptors after kainic acid lesion of CA3 and fimbria-fornix transection. Brain Res 247:335–340
86. Sabato UC, Aguilar JS, Medina JH, et al. (1981) Changes in rat hippocampal benzodiazepine receptors and lack of changes in muscarinic receptors after fimbria-fornix lesions. Neurosci Lett 27:193–198
87. Wendt JS, Fagg GE, Cotman CW (1983) Regeneration of rat hippocampal fimbria fibers after fimbria transection and peripheral nerve or fetal hippocampal implantation. Exp Neurol 79:452–461
88. Alonso JR, U HS, Amaral DG (1996) Cholinergic innervation of the primate hippocampal formation: II. Effects of fimbria/formix transection. J Comp Neurol 375:527–551
89. He Y, Yao Z, Gu Y, et al. (1992) Nerve growth factor promotes collateral sprouting of cholinergic fibers in the septohippocampal cholinergic system of aged rats with fimbria transection. Brain Res 586:27–35
90. Lescaudron L, Stein DG (1999) Differences in memory impaiment and response to GM1 ganglioside treatment following electrolytic or ibotenic acid lesions of the nucleus basalis magnocellularis. Restor Neurol Neurosci 15:25–37
91. Popovic N, Popovic M, Jovanova-Nesic K, et al. (2002) Effect of neural transplantation on depressive behavior in rats with lesioned nucleus basalis magnocellularis. Int J Neurosci 112:105–115
92. Vale-Martínez A, Guillazo-Blanch G, Marti-Nicolovius M, et al. (2002) Electrolytic and ibotenic acid lesions of the nucleus basalis magnocellularis interrupt long-term retention, but not acquisition of two-way active avoidance, in rats. Exp Brain Res 142:52–66
93. Nishimura A, Homann CF, Johnston MV, et al. (2002) Neonatal electrolytic lesions of the basal forebrain stunt plasticity in mouse barrel field cortex. Int J Dev Neurosci 20:481–489
94. Marston HM, West HL, Wilkinson LS, et al. (1994) Effects of excitotoxic lesions of the septum and vertical limb nucleus of the diagonal band of Broca on conditional visual discrimination: Relationships between performance and choline acetyltransferase activity in the cingulate cortex. J Neurosci 14:2009–2019
95. Mulder J, Harkany T, Czollner K, et al. (2005) Galantamine-induced behavioral recovery after sublethal excitotoxic lesions to the rat medial septum. Behav Brain Res 163:33–41
96. Bartus RT, Flicker C, Dean RL, et al. (1985) Selective memory loss following nucleus basalis lesions: Long term behavioral recovery despite persistent cholinergic deficiencies. Pharmacol Biochem Behav 23:125–135
97. Bartus RT, Pontecorco MJ, Flicker C, et al. (1986) Behavioral recovery following bilateral lesions of the nucleus basalis does not occur spontaneously. Pharmacol Biochem Behav 24:1287–1292
98. EI-Defrawy SR, Coloma E, Jhamandas K, et al. (1986) Lack of recovery of cortical cholinergic function following quinolic or ibotenic acid injections into the nucleus basalis magnocellularis in rats. Exp Neurol 91:628–633
99. EI-Defrawy SR, Coloma F, Jhamandas K, et al. (1985) Functional and neurochemical cortical cholinergic impairment following neurotoxic lesions of the nucleus basalis magnocellularis in the rat. Neurobiol Aging 6: 325–330

100. Casamenti F, Di Patre PL, Milan F, et al. (1989) Effects of nerve growth factor and GM 1 ganglioside on the number and size of cholinergic neurons in rats with unilateral lesion of the nucleus basalis. Neurosci Lett 103:87–91
101. Gardiner IM, De Belleroche J, Prom BK, et al. (1987) Effect of lesion of the nucleus basalis of rat on acetylcholine release in cerebral cortex. Time course of compensatory events. Brain Res 407:263–271
102. Toledano A (1993) Impairment of the cholinergic systems of the nucleus basalis- An animal model for damage of the cholinergic system in Alzheimer's Disease. In: Maurer K (ed) Neurochemistry, neuropathology, neuroimaging, neuropsychology, genetics of Dementias. Vieweg Verlag, Braunschweig, Germany, pp 1–24
103. Haroutunian V, Kanof PD, Davis KL (1986) Partial reversal of lesion induced deficits in cortical cholinergic markers by nerve growth factor. Brain Res 386:397–399
104. Olney JW (1994) New mechamisms of excitatory transmitter neurotoxicity. J Neural Tramsm 43[Supp]:47–51
105. Dunnet SB, Everitt BJ, Robbins TW (1991) The basal forebrain-cortical cholinergic system: Interpreting the functional consequences of excitotoxic lesions. Trend Neurosci 4: 494–501
106. Schwart R, Köhler Ch (1980) Evidence against and exclusive role of glutamate in kainic acid neurotoxicity. Neurosci Lett 9:243–249
107. Rossner S, Schliebs R, Bigl V (1994) Ibotenic acid lesions of nucleus basalis magnocellularis differencitally affects cholinergic, glutamatergic and GABAergic markers in cortical rat brain regions. Brain Res 668:85–99
108. Kilgard M (2003) Cholinergic modulation of skill learning and plasticity. Neuron 38: 678–680
109. Bregman RJ, El-Defraway SR, Jhamandas K, et al. (1985) Quinolic acid neurotoxicity in the nucleus basalis antagonized by kynurenic acid. Neurobiol Aging 6:331–336
110. Petterson GM, McGinty JF (1988) Direct neurotoxic effects of colchicine and cholinergic neurons in medial septum and striatum. Neurosci Lett 94:46–51
111. Tilson HA, Schwartz RD, Ali SF, et al. (1989) Colchicine administered into the area of the nucleus basalis decreases cortical nicotinic cholinergicreceptorslabelledby[3H]-acetylcholine. Neuropharmacology 28:855–861
112. Di Patre PL, Abbamondi A, Bartolini L, et al. (1989) GM1 ganglioside counteracts cholinergic and behavioral deficits induced in the rat by intracerebral injections of vincristine. Eur J Pharm 162:43–50
113. Beach TG, Potter PE, Kuo YM, et al. (2000) Cholinergic deafferentation of the rabbit cortex: A new animal model of Abeta deposition. Neurosci Lett 283:9–12
114. Ridley RM, Barefoot HC, Maclean CJ, et al. (1999) Different effects on learning ability after injection of the cholinergic immunotoxin ME20. 4IgG-saporin into the diagonal band of Broca, basal nucleus of Meynert, or both in monkeys. Behav Neurosci 113:303–315
115. Wenk GL (1993) A primate model of Alzheimer`s disease. Behav Brain Res 7: 117–122
116. Kordower JH, Winn SR, Liu YT, et al. (1994) The aged monkey basal forebrain: Rescue and sprouting of axotomized basal forebrain neurons after grafts of encapsulated cells secreting human nerve growth factor. Proc Natl Acad Sci U S A 91:10898–10902
117. Bontempi B, Whelan KT, Risbrough VB, et al. (2001) SIB-1553A, (+/-)-4-[[2-(1-methyl-2-pyrrolidinyl)ethyl]thio]phenol hydrochloride, a subtype-selective ligand for nicotinic acetylcholine receptors with putative cognitive-enhancing properties: Effects on working and reference memory performances in aged rodents and nonhuman primates. J Pharmacol Exp Ther 299:297–306
118. Buccafusco JJ, Jackson WJ, Stone JD, et al. (2003) Sex dimorphisms in the cognitive-enhancing action of the Alzheimer's drug donepezil in aged rhesus monkeys. Neuropharmacology 44:381–389
119. Sani S, Traul D, Klink A, et al. (2003) Distribution, progression and chemical composition of cortical amyloid-beta deposits in aged rhesus monkeys: Similarities to the human. Acta Neuropathol (Berl) 105:145–156
120. Fisher A, Hanin L (1986) Potential animal model for senile dementia of Alzheimer's type, with emphasis on AF64A-induced cholinotoxicity. Ann Rev Pharmacol Toxicol 26:161–181
121. Hanin I (1996) The AF64A model of cholinergic hypofunction: An update. Life Sci 8:1955–1964
122. Lorens SA, Kindel G, Dong XW, et al. (1991) Septal choline acetyltransferase immunoreactive neurons: Dose-dependent effects of AF64A. Brain Res Bull 26:965–971
123. Dawson VL, Wamsley JK (1990) Hippocampal muscarinic supersensitivity after AF64A medial septal lesion excludes M1 receptors. Brain Res Bull 25:311–317
124. Dawson VL, Hunt ME, Wamsley JK (1992) Alterations in cortical muscarinic receptors

following cholinotoxin (AF64A) lesion of the rat nucleus basalis magnocellularis. Neurobiol Aging 13:25–32

125. Kozlowski MR, Arbogast RE (1986) Specific toxic effects of ethylcholine nitrogen mustard on cholinergic neurons of the nucleus basalis of Meynert. Brain Res 372:45–54

126. Waite JJ, Chen AD, Wardlow ML, et al. (1995) 192 immunoglobulin G-saporin produces graded behavioural and biochemical changes accompanying the loss of cholinergic neurons of the basal forebrains and cerebellar Purkinje cells. Neuroscience 65:463–476

127. Hartonian I, de Lacalle S (2005) Compensatory changes in cortical cholinergic innervation in the rat following an immunotoxic lesion. Restor Neurol Neurosci 23:87–96

128. El Tamer A, Corey J, Wulfert E, et al. (1992) Reversible cholinergic changes induced by AF64A in rat hippocampus and possible septal compensatory effect. Neuropharmacology 31:397–402

129. Lim DK, Oh YH, Kim HS (2001) Impairments of learning and memory following intracerebroventricular administration of AF64A in rats. Arch Pharm Res 24:234–239

130. McGurk SR, Hortgraves SL, Kelly PH, et al. (1987) Is ethylcholine mustard aziridinium ion specific cholinergic neurotoxin? Neuroscience 22:215–224

131. Lermontova N, Lukoyanov N, Serkova T, et al. (1998) Effects of tacrine on deficits in active avoidance performance induced by AF64A in rats. Mol Chem Neuropathol 33:51–61

132. Willson CA, Hanin I (1997) Brain-derived neurotrophic factor induced stimulation of septal choline acetyltransferase activity in ethylcholine mustard aziridinium treated rats. Neurosci Lett 229:149–152

133. Wiley RG (1992) Neural lesioning with ribosome-inactivating proteins: Suicide transport and immunolesioning. Trends Neurosci 15:285–290

134. Book AA, Wiley RG, Schweitzer JB (1992) Specificity of 192 IgG-saporin for NGF receptor-positive cholinergic basal forebrain neurons in the rat. Brain Res 590:350–355

135. Galani R, Jeltsch H, Lehmann O, et al. (2002) Effects of 192 IgG-saporin on acetylcholinesterase histochemistry in male and female rats. Brain Res Bull 58:179–186

136. Ricceri L, Hohmann C, Berger-Sweeney J (2002) Early neonatal 192 IgG saporin induces learning impairments and disrupts cortical morphogenesis in rats. Brain Res 954:160–172

137. Dudkin KN, Chueva V, Makarov FN, et al. (2005) Impairments in working memory and decision-taking processes in monkeys in a model of Alzheimer's disease. Neurosci Behav Physiol 35:281–289

138. Wenk GL, Stoehr JD, Quintana G, et al. (1994) Behavioural, biochemical, histological, and electrophysiological effects of 192 IgG-saporin injections into the basal forebrain of rats. J Neurosci 14:5986–5995

139. Berger-Sweeney J, Heckers S, Mesulam MM, et al. (1994) Differential effects on spatial navigation of immunotoxin-in¬duced cholinergic lesions of the medial septal area and nu¬cleus basalis magnocellularis. J Neurosci 14:4507–4519

140. Lehmann O, Grottick AJ, Cassel JC, et al. (2003) A double dissociation between serial reaction time and radial maze performance in rats subjected to 192 IgG-saporin lesions of the nucleus basalis and/or the septal region. Eur J Neurosci 18:651–666

141. McGaughy J, Dalley JW, Morrison CH, et al. (2002) Selective behavioral and neurochemical effects of cholinergic lesions produced by intrabasalis infusions of 192 IgG-saporin on attentional performance in a five-choice serial reaction time task. J Neurosci 22:1905–1913

142. Helm KA, Han JS, Gallagher M (2002) Effects of cholinergic lesions produced by infusions of 192 IgG-saporin on glucocorticoid receptor mRNA expression in hippocampus and medial prefrontal cortex of the rat. Neuroscience 115:765–774

143. Tomaszewicz M, Rossner S, Schliebs R, et al. (2003) Changes in cortical acetyl-CoA metabolism after selective basal forebrain cholinergic degeneration by 192 IgG-saporin. J Neurochem 87:318–324

144. Beaule C, Amir S (2002) Effect of 192 IgG-saporin on circadian activity rhythms, expression of P75 neurotrophin receptors, calbindin-D28K, and light-induced Fos in the suprachiasmatic nucleus in rats. Exp Neurol 176:377–389

145. Ricceri L (2003) Behavioral patterns under cholinergic control during development: Lessons learned from the selective immunotoxin 192 IgG saporin. Neurosci Biobehav Rev 27:377–384

146. Cassel JC, Gaurivaud M, Lazarus C, et al. (2002) Grafts of fetal septal cells after cholinergic immunotoxic denervation of the hippocampus: A functional dissociation between dorsal and ventral implantation sites. Neuroscience 113:871–882

147. Bassant MH, Poindessous-Jazat F, Schmidt BH (2000) Sustained effect of metrifonate on cerebral glucose metabolism after immunolesion of basal forebrain cholinergic neurons in rats. Eur J Pharmacol 387:151–162

148. Ballmaier M, Casamenti F, Scali C, et al. (2002) Rivastigmine antagonizes deficits in prepulse inhibition induced by selectiveimmunolesioning of cholinergic neurons in nucleus basalis magnocellularis. Neuroscience 114:91–98

149. Wortwein G, Yu J, Toliver-Kinsky T, et al. (1998) Responses of young and aged rat CNS to partial cholinergic immunolesions and NGF treatment. J Neurosci Res 52: 322–333

150. Berger-Sweeney J, Stearns NA, Murg SL, et al. (2001) Selective immunolesions of cholinergic neurons in mice: Effects on neuroanatomy, neurochemistry, and behavior. J Neurosci 21:8164–8173

151. Barefoot HC, Baker HF, Ridley RM (2000) Synergistic effects of unilateral immunolesions of the cholinergic projections from the basal forebrain and contralateral ablations of the inferotemporal cortex and hippocampus in monkeys. Neuroscience 98:243–251

152. Ferreira G, Meurisse M, Gervais R, et al. (2001) Extensive immunolesions of basal forebrain cholinergic system impair offspring recognition in sheep. Neuroscience 106:103–116

153. Calza L, Giuliani A, Fernandez M, et al. (2003) Neural stem cells and cholinergic neurons: Regulation by immunolesion and treatment with mitogens, retinoic acid, and nerve growth factor. Proc Natl Acad Sci U S A 100:7325–7330

154. Hepler DJ, Olton DS, Wenk GL, ct al. (1985) Lesions in nucleus basalis magnocellularis and media] septal area of rats produce qualitatively similar memory impairment. J Neurosci 5:866–873

155. Gage FH, Bojorklund A, Stenevi U (1983) Reinnervation of the partially deaffer¬ented hippocampus by compensatory collateral sprouting from aparent cholin¬ergic and adrenergic afferents. Brain Res 268:27–37

156. Orzi F, Diana G, Casamenti F, et al. (1988) Local cerebral glucose utilization following unilateral and bilateral lesions of the nucleus basalis magnocellularis in the rat. Brain Res 462:99–103

157. Unger JW, Schmidt Y (1992) Quisqualic acid induced lesion of the nucleus basalis of Meynert in young and aging rats: Plasticity of surviving NGF receptor-positive cholinergic neurons. Exp Neurol 117:269–277

158. Riekkimen P Jr, Miettinen R, Sirvio J, et al. (1990) The correlation of passive avoidance deficit in aged rat with the loss of nucleus basalis choline acetyltransferase-positive neurons. Brain Res Bull 25:415–417

159. Wellman CL, Logue SF, Sengelaub DR (1995) Maze learning and morphology of frontal cortex in adult and aged basal forebrain–lesioned rats. Behav Neurosci 109:837–850

160. Ding M, Haglid KG, Hamberger A (2000) Quantitative immunochemistry on neuronal loss, reactive gliosis and BBB damage in cortex/striatum and hippocampus/amygdala after systemic kainic acid administration. Neurochem Int 36:313–318

161. Hailer NP, Wirjatijasa F, Roser N, et al. (2001) Astrocytic factors protect neuronal integrity and reduce microglial activation in an in vitro model of N-methyl-D-aspartate-induced excitotoxic injury in organotypic hippocampal slice cultures. Eur J Neurosci 14:315–326

162. Penkowa M, Molinero A, Carrasco J, et al. (2001) Interleukin-6 deficiency reduces the brain inflammatory response and increases oxidative stress and neurodegeneration after kainic acid-induced seizures. Neuroscience 102:805–818

163. Acarin L, Paris J, Gonzalez B, et al. (2002) Glial expression of small heat shock proteins following an excitotoxic lesion in the immature rat brain. Glia 38:1–14

164. Little AR, Benkovic SA, Miller DB, et al. (2002) Chemically induced neuronal damage and gliosis: Enhanced expression of the proinflammatory chemokine, monocyte chemoattractant protein (MCP)-1, without a corresponding increase in proinflammatory cytokines. Neuroscience 115:307–320

165. Calabrese V, Bates TE, Stella AM (2000) NO synthase and NO-dependent signal pathways in brain aging and neurodegenerative disorders: The role of oxidant/antioxidant balance. Neurochem Res 25:1315–1341

166. Winkler J, Thal LJ (1995) Effects of nerve growth factor treatment on rats with lesions of the nucleus basalis magnocellularis produced by ibotenic acid, quisqualic acid and AMPA. Exp Neurol 136:234–250

167. Winkler J, Power AE, Ramirez GA, et al. (1998) Short-term and complete reversal of NGF effects in rat with lesions of the nucleus basalis magnocellularis. Brain Res 788:1–12

168. Yamamoto M, Chikuma T, Kato T (2003) Changes in the levels of neuropeptides and their metabolizing enzymes in the brain regions of nucleus basalis magnocellularis-lesioned rats. J Pharmacol Sci 92:400–410

169. Wenk GL, Markowska AL, Olton DS (1989) Basal forebrain lesions and mem¬ory: Alterations in neurotensin not acetylcholine, may cause amnesia. Behav Neurosci 103:765–769

170. Oliveira A, Hodges H, Rezaie P (2003) Excitotoxic lesioning of the rat basal forebrain with S-AMPA: Consequent mineralization and associated glial response. Exp Neurol 179:127–138
171. Levey AJ, Wainer BH, Rye DB, et al. (1984) Choline acetyltransferase immunoreactive neurons intrinsic to rodent cortex and distinction from acetylcholinesterase-positive neurons. Neuroscience 13:341–353
172. De Belleroche J, Gardiner IM, Hamilton MH, et al. (1985) Analysis of muscarinic receptor concentration and subtypes following lesion of rat substantia innominata. Brain Res 340:201–209
173. Mash DC, Flynn DD, Potter LT (1985) Loss of M2 muscarinic receptors in the cerebral cortex in Alzheimer's disease and experimental cholinergic denervation. Science 228: 1115–1117
174. Wenk GL, Olton DS (1984) Recovery of neocortical choline acetyltransferase activity following iobotenic acid injection into the nucleus basalis of Meyner in rats. Brain Res 293:184–186
175. Henderson Z (1991) Sprouting of cholinergic axons does not occur in the cerebral cortex after nucleus basalis lesions. Neuroscience 44:149–156
176. Winkler J, Thal LJ, Gage FH, et al. (1998) Cholinergic strategies for Alzheimer's disease. J Mol Med 76:555–567
177. Casamenti F, Scali C, Giovannelli L, et al. (1994) Effect of nerve growth factor and GM 1 ganglioside on the recovery of cholinergic neurons after a lesion of the nucleus basalis in aging rats. J Neural Transm 7:177–193
178. Bartolini L, Casamenti F, Pepeu G (1996) Aniracetam restores object recognition impaired by age, scopolamine, and nucleus basalis lesions. Pharmacol Biochem Behav 53: 277–283
179. Hodges H, Allen Y, Kershow T, et al. (1991) Effects of cholinergic-rich neural grafts on radial maze performance of rats after excitotoxic lesions of the forebrain cholinergic projection system- I Ameloriation of cognitive deficits by transplants into cortex and hippocampus but not into basal forebrain. Neuroscience 45:587–607
180. Hodges H, Allen Y, Sinden J, et al. (1991) The effects of cholinergic drugs and cholinergic-rich foetal neuronal transplants on alcohol-induced deficits in radial maze performance in rats. Behav Brain Res 43:7–28
181. Zhang ZJ, Lappi DA, Wrenn CC, et al. (1998) Selective lesion of the cholinergic basal forebrain causes a loss of cortical neuropeptide Y and somatostatin neurons. Brain Res 800:198–206
182. Katsumi Y, Hayashi T, Oyanagi C, et al. (2000) Glucose metabolism in the rat frontal cortex recovered without the recovery of choline acetyltransferase activity after lesioning of the nucleus basalis magnocellularis. Neurosci Lett 280:9–12
183. Harkany T, Dijkstra IM, Oosterink BJ, et al. (2000) Increased amyloid precursor protein expression and serotonergic sprouting following excitotoxic lesion of the rat magnocellular nucleus basalis: Neuroprotection by Ca(2+) antagonist nimodipine. Neuroscience 101:101–114
184. Wallace W, Ahlers ST, Gotlib J, et al. (1993) Amyloid precursor protein in the cerebral cortex is rapidly and persistently induced by loss of subcortical innervation. Proc Natl Acad Sci U S A 90:8712–8716
185. Haroutunian V, Greig N, Pei XF, et al. (1997) Pharmacological modulation of Alzheimer's beta-amyloid precursor protein levels in the CSF of rats with forebrain cholinergic system lesions. Brain Res Mol 46:161–168
186. Lin L, Georgievska B, Mattsson A, et al. (1999) Cognitive changes and modified processing of amyloid precursor protein in the cortical and hippocampal system after cholinergic synapse loss and muscarinic receptor activation. Proc Natl Acad Sci U S A 96: 12108–12113
187. Friedman E, Lerer B, Kuster J (1983) Loss of cholinergic neurons in the rat neocortex produces deficits in passive avoidance learning. Pharmacol Biochem Behav 19:309–312
188. Miyamoto M, Shintani M, Nagaoka A, et al. (1985) Lesioning of the rat basal forebrain leads to memory impairments in passive and active avoidance. Brain Res 328:97–104
189. Connor DJ, Langlais PJ, Thal LJ (1991) Behavioral impair¬ments after lesions of the nucleus basalis by ibotenic acid and quisqualic acid. Brain Res 555:84–90
190. Salomone ID, Beart PM, Alpert JE, et al. (1984) Impairment in T-maze reinforced alternation performance following nu¬cleus basalis magnocellularis lesions in rats. Behav Brain Res 13:63–70
191. Wellman CL, Pelleymounter MA (1999) Differential effects of nucleus basalis lesions in young adult and aging rats. Neurobiol Aging 20:381–393
192. Markowska AL, Wenk GL, Olton DS (1990) Nucleus basalis magnocellularis and memory: Differential effects of two neu¬rotoxins. Behav Neural Biol 54:13–26

193. Bartus RT (1988) The need for common perspectives in the development and use of animal models for age-related cognitive and neurodegenerative disorders Neurobiol Aging 9:445–451
194. Toledano A, Rivas L, Rodríguez-Arellano JJ (1993) Morphological and histochemical characteristics of mitochondria during senile involution of the cerebellar cortex. Bull Molec Biol Med 18:1–23
195. Toledano A, Barca MA, Moradillo I (1988) Mitochondrial dehydrogenases during sensescence of the cerebellar cortex. A comparative electron microscope study. Cell Molec Biol 34:175–189
196. Araujo DM, Lapchak PA, Meaney MJ, et al. (1990) Effects of aging on nicotinic and muscarinic autoreceptor function in the rat brain: Relationship to presynaptic cholinergic markers and binding sites. J Neurosci 10:3069–3078
197. Wu CF, Bertorelli R, Sacconi M, et al. (1988) Decrease of brain acetylcholine release in aging freely-moving rats detected by microdialysis. Neurobiol Aging 9:357–361
198. Fisher W, Chen KS, Gage FH, et al. (1991) Progressive decline in spacial learning and integrity of forebrain cholinergic neurons in rats during aging. Neurobiol Aging 13:9–23
199. Lee JM, Ross ER, Gower A, et al. (1994) Spatial leaming deficits in the aged rat: Neuroanatomical and neuro¬chemical correlates. Brain Res Bull 33:489–500
200. Weber JT (2007) Experimental models of repetitive brain injuries. Prog Brain Res 161:253–261
201. Inestrosa NC, Reyes AE, Chacón MA, et al. (2005) Human-like rodent amyloid-beta-peptide determines Alzheimer pathology in aged wild-type Octodon degu. Neurobiol Aging 26:1023–1028
202. Glenn MJ, Lehmann H, Mumby DG, Woodside B (2005) Differential fos expression following aspiration, electrolytic, or excitotoxic lesions of the perirhinal cortex in rats. Behav Neurosci 19:806–813
203. Hernández-Ortega K, Ferrera P, Arias C (2007) Sequential expresión of cell-cycle regulators and Alzheimer's diseases-related proteins in entorhinal cortex after hippocampal excitotoxis damage. J Neurosci Res 85:1744–1751
204. Glenn MJ, Nesbitt C, Mumby DG (2003) Perirhinal cortex lesions produce variable patterns of retrograde amnesia in rats. Behav Brain Res 41:183–193
205. Blaker WD, Goodwin SD (1987) Biochemical and behavioral effects of intrahippocampal AF64A in rats. Pharmacol Biochem Behav 28:157–163
206. Mouton PR, Arendash GW (1990) Atrophy of cholinergic neurons within the rat nucleus basalis magnocellularis following intracortical AF64A infusion. Neurosci Lett 111:52–57
207. Heckers S, Ohtake T, Wiley RG, et al. (1994) Complete and selective cholinergic denervation of rat neocortex and hippocampus but not amygdala by an immunotoxin against the p75 NGF-receptor. J Neurosci 14:1271–1289
208. Raymond JL, Lisberger SG, Mauk MI (1996) The cerebel¬lum: A neuronal learning machine. Science 272:1126–1131
209. Gottwald B, Mihazlocic Z, Wilde B, et al. (2003) Does the cerebellum contribute to specificaspectsofattention? Neuropsychologia 41:1452–1460
210. Pavía J, Albrech J, Álvarez MI, et al. (2002) Repeat intracerebroventricular administration of β-amyloid 25–35 to rats decreases muscarinic receptors in cerebral cortex. Neurosci Lett 278:69–72
211. Chen L, Yamada K, Nabeshima T, Sokabe M (2006) Alpha7 Nicotinic acetylcholine receptor as a target to rescue deficit in hippocampal LTP induction in beta-amyloid infused rats. Neuropharmacology 50:254–268
212. Yamada M, Chiba T, Sasabe J, et al. (2005) Implanted cannula-mediated repetitive administration of Abeta25–35 into the mouse cerebral ventricle effectively impairs spatial working memory. Behav Brain Res 164:139–146
213. Li LY, Li JT, Wu QY, et al. (2008) Transplantation of NGF-gene-modified bone marrow stromal cells into a rat model of Alzheimer' Disease. J Mol Neurosci 34:157–163
214. Takata K, Kitamura Y, Yanagisawa D, et al. (2007) Microglial transplantation increases amyloid-beta clearance in Alzheimer model rats. FEBS Lett 581:475–478
215. Holscher C, Gengler S, Gault VA, et al. (2007) Soluble beta-amyloid [25–35] reversibly impairs hippocampal synaptic plasticity and spatial learning. Eur J Pharmacol 561:85–90
216. Coles B, Wilton LA, Good M, et al. (2008) Potassium channels in hippocampal neurones are absent in a transgenic but not in a chemical model of Alzheimer's disease. Brain Res 1190:1–14
217. Arendt TH, Bigl V, Schuyens MM (1989) Ethanol as a neurotoxin- A model of the syndrome of cholinergic deafferentation of the cortical mantle. Brain Dysfunct 2:169–180
218. Floyd EA, Young-Seigler AC, Ford BD, et al. (1997) Chronic ethanol ingestion produces cholinergic hypofunction in rat brain. Alcohol 14:93–98

219. Fernandes PA, Ribeiro AM, Pereira RF, et al. (2002) Chronic ethanol intake and ageing effects on cortical and basal forebrain cholinergic parameters: Morphometric and biochemical studies. Addict Biol 7:29–36
220. Walton JR (2007) An aluminum-based rat model for Alzheimer's disease exhibits oxidative damage, inhibition of PP2A activity, hyperphosphorylated tau, and granulovacuolar degeneration. J Inorg Biochem 101:1275–1284
221. Bharathi I, Shamasundar NM, Sathyanarayana Rao TS, et al. (2006) A new insight on Al-maltolate-treated aged rabbit as Alzheimer's animal model. Brain Res Rev 52:275–292
222. Chen LQ, Wei JS, Lei ZN, et al. (2005) Induction of Bcl-2 and Bax was related to hyperphosphorylation of tau and neuronal death induced by okadaic acid in rat brain. Anat Rec A Discov Mol Cell Evol Biol 287:1236–1245
223. Yoon S, Choi J, Huh JW, et al. (2006) Calpain activation in okadaic-acid-induced neurodegeneration. Neuroreport 17:689–692
224. Lee J, Hong H, Im J, et al. (2000) The formation of PHF-1 and SMI-31 positive dystrophic neurites in rat hippocampus following acute injection of okadaic acid. Neurosci Lett 282:49–52
225. Tian Q, Lin ZQ, Wang XC, et al. (2004) Injection of okadaic acid into the meynert nucleus basalis of rat brain induces decreased acetylcholine level and spatial memory deficit. Neuroscience 126:277–284
226. Craft TK, Mahoney JH, Devries AC, Sarter M (2005) Microsphere embolism-induced cortical cholinergic deafferentation and impairments in attentional performance. Eur J Neurosci 21:3117–3132
227. Shibata M, Yamasaki N, Miyakawa T, et al. (2007) Selective impairment of working memory in a mouse model of chronic cerebral hypoperfusion. Stroke 38:2826–2832
228. Ginestet L, Ferrario JE, Raisman-Vozari R, et al. (2007) Donepezil induces a cholinergic sprouting in basocortical degeneration. J Neurochem 102:434–440
229. Paxinos G, Watson C (1998) The rat brain in stereotaxic coordinates, 4th edn. Academic, San Diego, CA/London

Chapter 17

Aβ Infusion and Related Models of Alzheimer Dementia

Patricia A. Lawlor and Deborah Young

Abstract

Alzheimer's disease (AD) is a progressive neurodegenerative disorder characterized by a decline in cognitive function and the presence of neuropathological hallmarks, including the accumulation of extracellular Aβ. Aspects of AD can be modeled in rodents by direct intracerebral injection of Aβ. This causes learning and memory deficits in treated animals, with the severity of the deficits observed dependent on the species of Aβ infused and the time interval between Aβ administration and behavioral testing. Variation in the reported behavioral and neuropathological consequences of Aβ infusion can also be attributed to the aggregation state and type of Aβ preparation used (synthetic or recombinant), the duration of the infusion (acute or chronic), peptide concentration, and even the solvent used to dilute the peptide. More recently, the use of viral vector gene transfer technology has allowed the development of "somatic transgenic" models, whereby genes putatively involved in AD pathogenesis can be selectively overexpressed in specific brain regions involved in AD. Although this promising strategy has been shown to result in the development of both cognitive deficits and Aβ deposits in treated animals, these genetic models require further characterization to show reproducible development of behavioral deficits and neuropathology prior to their widespread adoption as a reliable and useful model of AD.

Key words: Aβ peptide, Cognitive deficit, Hippocampus, Viral vector, Adeno-associated vector (AAV)

1. Introduction

Alzheimer's disease (AD), the leading cause of dementia in the elderly, is a progressive neurodegenerative disorder identified clinically by a characteristic decline in cognitive function. Symptoms of AD dementia usually begin with impairments in episodic memory whereby patients have difficulty forming new memories of events. As the disease progresses, deficits in semantic (facts and general knowledge) and remote memory (e.g., childhood memories) become apparent, as do deficits in visuo-spatial ability, and attentional and executive function (1–4). Patients can also display other noncognitive behavioral symptoms, such as anxiety, aggression,

Peter Paul De Deyn and Debby Van Dam (eds.), *Animal Models of Dementia*, Neuromethods, vol. 48,
DOI 10.1007/978-1-60761-898-0_17, © Springer Science+Business Media, LLC 2011

depression, agitation, hallucinations, circadian rhythm disturbances, wandering, and personality disturbances (5–10). These cognitive and behavioral changes are accompanied by characteristic neuropathological changes – both familial and sporadic forms of AD display the same hallmarks including accumulation of extracellular amyloid plaques and intracellular neurofibrillary tangles (NFTs), synaptic and neuronal loss, and inflammation (characterized by astrocytic and microglial activation).

An important early discovery in understanding the pathogenesis of AD was that the primary component of the amyloid deposits found in AD brain was an approximately 40-residue long peptide, now known as amyloid-β peptide (Aβ) (11–15). Subsequent studies established that Aβ is the product of proteolytic processing of amyloid precursor protein (APP) following sequential cleavage by beta- and gamma-secretases (16–18). This processing of APP to create Aβ is heterogeneous, with gamma-secretase cleavage generating Aβ peptides ranging in length from 38–43 amino acids, although under normal conditions of APP processing, the two major forms of Aβ that have been observed are Aβ40 and Aβ42. Numerous AD-linked mutations in APP and presenilins 1 and 2 (19–22) alter APP metabolism to increase Aβ production and deposition, resulting in accumulation of Aβ42, the Aβ species essential for the formation of parenchymal and vascular amyloid deposits (23), and proposed to initiate the cascade leading to AD (21).

The Aβ hypothesis of AD posits that regardless of whether the disease is familial or sporadic, cerebral accumulation of Aβ to form amyloid plaques is the primary driver of AD pathogenesis (21, 24), and additional disease processes (NFT formation and inflammation) result from the imbalance between Aβ production and clearance. In light of recent studies, the hypothesis has been modified – it is now thought that synaptic failure can be attributed to the presence of soluble Aβ oligomers (25–27), so recent modifications of the amyloid hypothesis suggest that rather than the amyloid plaques being solely and primarily responsible for AD dementia, soluble and diffusible products of APP processing also play a critical role in disease pathogenesis (28–31).

Importantly, the identification of the major proteins like Aβ involved in AD pathogenesis has allowed creation of rodent models based on elevation of Aβ levels in brain. Initially, this was achieved by direct intracerebral or intraventricular injection of Aβ into rodent brain. Elucidation of gene mutations involved in familial AD allowed development of AD transgenic mice engineered to overexpress these gene mutations. More recently, the use of viral vector gene transfer technology has allowed the development of "somatic transgenic" models, whereby genes putatively involved in AD pathogenesis can be selectively overexpressed in specific brain regions involved in AD. A complete animal model of AD would need to replicate all the cognitive, behavioral, and neuropathological features of the disease, and account for the contribution of risk factors such as aging. Not one model recapitulates all these features of AD – at best, current models are partial only, allowing us to model specific components of AD pathology. The scope of this chapter is to summarize and provide the reader with methodologies for setting up nontransgenic Aβ rodent models of AD based on infusion of Aβ – either direct infusion of Aβ into rodent brain or use of viral vectors to overexpress APP or selected Aβ species.

2. The Effect of Aβ Peptide Infusion on Learning and Memory Behavior in Rodents

The Aβ-related cognitive deficits and some neuropathological features of AD can be induced by intracerebral or intraventricular infusion of Aβ peptides into the rodent brain. Infusion models were originally developed under the assumption that brain Aβ deposition is required to induce behavioral impairments, however, studies demonstrating the disruptive action of soluble species of Aβ on long-term potentiation (LTP) and behavior (27, 32–34) suggest that acute Aβ infusion is sufficient to induce some learning and memory impairments. Overall, infusion of Aβ peptides can lead to both spatial and nonspatial learning and memory deficits (35–39) and behavioral alterations such as decreased activity and exploration (35). However, the severity of the deficits observed varies markedly depending on species of Aβ used, and the time interval between peptide administration and behavioral testing. Another layer of complexity to be considered when interpreting the data is that cognitive deficits can be measured at different stages of the learning and memory process – acquisition, consolidation, or retention. Therefore, Aβ peptides may impact on different aspects of learning and memory depending on the time interval between peptide administration and testing.

Aβ40 and Aβ42 are the two most common lengths of Aβ found in both normal human brain and in the cortical and vascular deposits of AD patients. Hence, many studies have been conducted using synthetic forms of these peptides infused into the ventricles or hippocampus of both rats and mice. Comprising over 90% of the total Aβ load, Aβ40 has been the focus of many *in vivo* studies, with the results overall supporting a role for Aβ40 in development of cognitive deficits, although the effects observed are dependent on the Aβ40 administration regime. For example, Aβ40 administered immediately after acquisition in an active avoidance testing paradigm resulted in consolidation deficits but no effect on short-term spatial memory using the radial arm maze (40) or operant conditioning (41). Multiple, repeated injections, on the other hand, produced performance decrements several weeks later (41). Chronic, repeated dosing also produced deficits in the Morris water maze and passive avoidance acquisition and retention (38, 42–47). Aβ42, the species essential for plaque deposition, is the second most abundant form of Aβ and is generally considered the most toxic form. Following both acute and chronic infusion into rodent brain, this too has been observed to have deleterious effects on both short- and long-term memory and consolidation processes, depending on the Aβ administration regime. For example, deficits in acquisition and retention of spatial memory, passive avoidance, and spontaneous alternation tasks have been observed following infusion of Aβ42 (37, 48–52).

Other groups have demonstrated that intraventricular injection of other Aβ species, including Aβ25-35 and Aβ1-28, into rats and mice impairs memory (35, 36, 50, 53–55). Although species like Aβ25–35 are easy to synthesize, readily form toxic aggregates, and can induce learning and memory deficits in rodents, it is unclear whether these species play a significant role in human disease (15).

In general, acute injections of Aβ cause deleterious effects on learning and memory consolidation rather than short-term memory or retrieval processes. However, effects on both short-term and long-term memory acquisition and retention have been observed when behavioral testing is carried out several days after Aβ administration or following long-term chronic Aβ administration.

3. Neuropathology Observed Following Exogenous Aβ Administration

In addition to measurable deleterious effects on learning and memory tasks in rodents, exogenous administration of Aβ species can lead to neuropathological changes reminiscent of human AD, although the full complexity of the human pathology is not reproduced and these pathologies are not widespread as in the human condition. Accumulation of Aβ deposits in brain parenchyma have been reported – these deposits are localized to the injection site and can be induced by infusion of aggregated Aβ species (52, 56, 57), by injection of the soluble form of putative fibrillogenic Aβ (58–62) or following infusion of Aβ deposits into rodent brain (34, 63, 64). These deposits can be associated with inflammation whereby activated astrocytes and microglial cells surround and infiltrate the deposit (56, 61, 65–67). Local cell loss, proximal to the injection site only (68–70), has also been observed. Other changes, such as local production of oxidative stress mediators (66) and cytokines (71), alterations in *N*-methyl d-aspartate (72), and acetylcholine receptors (52) and neuropeptides (39, 73) have also been observed.

4. Variation Between Studies

Overall, within the published literature there is a wide variation in the reported behavioral and neuropathological effects of Aβ infusion. These inconsistencies in cognitive deficits and neuropathology that have been observed across Aβ infusion models may be due in part to variations in methodologies used by different groups. It is increasingly clear that not only the species of peptide infused (Aβ40, Aβ42, or Aβ25–35), but the aggregation state also has a major impact on its toxicity and actions *in vivo*. Variations in effects of Aβ are also dependent on the type of Aβ preparation (synthetic vs. recombinant peptide; fresh vs. aged Aβ), the duration of the infusion (acute vs. chronic dosing), the site of infusion (e.g., ventricle, hippocampus, entorhinal cortex), the time interval between Aβ administration and behavioral testing, peptide concentration, and even the solvent used to dilute peptides. These methodological differences need to be considered when interpreting data from Aβ infusion studies and prior to embarking on any *in vivo* study. The impact of these points on the model are elaborated later.

4.1. Aggregation State of Peptide

Behavioral data obtained using the Aβ infusion model in rodents demonstrates that interactions between Aβ and synaptic processes are dependent not only on the species of Aβ infused, but also on the assembly state of that peptide, that is, whether a soluble, diffusible form was infused or more fibrillar and aggregated Aβ. Aβ can exist as monomers, oligomers, protofibrils, and fibrils. In general, oligomeric Aβ is more toxic than either monomeric or fibrillar Aβ – the demonstration of the integral role of Aβ oligomers in neuronal dysfunction was achieved by showing that oligomeric Aβ at picomolar concentrations, in the absence of monomers or fibrils, potently inhibits LTP in the mammalian hippocampus (29, 31, 32). Additionally, memory deficits in APP transgenic mice were correlated with the presence of a specific 56 kDa soluble assembly of Aβ (Aβ*56) (74). When isolated from mice, these soluble assemblies disrupted memory in young rats following intraventricular administration.

However, when assessing the activity of different Aβ assemblies, one thing that must be borne in mind is the dynamic nature of Aβ aggregation – it is difficult to definitively attribute activity to a discrete species as Aβ preparations may change over the time course of the experiment. Intermediates have the potential to further associate to higher-order aggregates or dissociate to lower-order species. Aggregation state can be affected by temperature, concentration, pH, and oxidation state (75–77). Additionally, overaging and isoelectric precipitation, phenomena that may inadvertently reduce toxicity of Aβ, are also important. Overaggregation, or overaging, results in the production of large clumps of fibrils, which limit the interaction of fibrils with neurons, thereby reducing its toxicity to neurons. Around its isoelectric point of pH 5.5, Aβ attains a β-sheet conformation resulting in an amorphous, nontoxic precipitate (78).

Importantly, the different forms of Aβ can affect synaptic and neuronal function with a different time course, that is, either very rapidly after the intracerebral infusion of soluble Aβ or after a delay of several days or even months with intracerebral infusion of aggregative forms of Aβ. It has been proposed that injection of soluble (monomeric or oligomeric) forms of Aβ induces an immediate but transient impairment of memory because of direct detrimental effects on synaptic plasticity. The later onset and longer-lasting effects on memory induced by injections of aggregated insoluble forms of Aβ (or by injection of the soluble but highly fibrillogenic Aβ42) are most likely mediated by longer-term alterations in gene expression and receptor levels, as well as inflammatory reactions induced by fibrillar Aβ (34, 61, 79, 80).

Overall, it is advisable that the assembly state of the Aβ used in any *in vivo* study be assessed at both the start and completion of experiments. It is likely that a lack of understanding of Aβ conformation and aggregation state and lack of rigorous characterization of the starting material in early studies of Aβ toxicity contributed to conflicting results between studies (78).

4.2. Synthetic and Recombinant Peptides

Synthetic Aβ peptides are generated using solid-phase peptide synthesis techniques (SPPS) like Fmoc-SPPS (81). Aβ peptides of any specified sequence can be made-to-order by commercial peptide synthesis companies (e.g., GenScript). Additionally, some companies (e.g., Novagen) supply readymade Aβ peptides of defined sequence, along with appropriate antisense or scrambled control sequences. In particular, however, synthetic Aβ42 is difficult to make and batch-to-batch variability is a real issue. Peptide lots, even from the same manufacturer, can vary in both the degree and speed of *in vitro* aggregation. It is conceivable that a lack of aggregation due to a "bad" batch of peptide may lead to reports that Aβ in fact lacks toxicity (78, 82, 83). It is possible that different batches of synthetic peptides contain undetected secondary, tertiary, or quaternary structures that act as a seed and drive the bulk of a peptide solution down a particular aggregation pathway. Conventional quality control methods used by manufacturing companies such as mass spectrometry focus primarily on chemical purity rather than structural heterogeneity, so it is possible that these structural seeds could go undetected during synthesis and quality control but have an effect on the activity and toxicity of the peptide (84). Another important consideration in the use of synthetic Aβ peptides is the artificially high concentrations needed to achieve aggregation. Various studies have suggested that synthetic

Aβ peptides require a minimum concentration of ~10 μM to achieve fibrillization (85), far higher than the threshold required by naturally occurring species and considerably higher than the levels found in brain and cerebrospinal fluid of AD patients. Another disadvantage of using synthetic peptides is that they are relatively insoluble and must often be solubilized in highly nonphysiological solvents (e.g., acetonitrile, trifluoroacetic acid, sodium hydroxide) (29).

An alternative to the manufacture of synthetic peptides is generation of recombinant Aβ peptides; these are commercially available (e.g., through rPeptide). Stably transfected cells secrete the peptides into the culture media, which is then concentrated to produce purified peptides. In contrast to synthetic assemblies, recombinant Aβ species can begin to assemble into oligomers at much lower concentrations (nanomolar levels), thereby overcoming the problem of using synthetic peptides at high, nonphysiological concentrations (29).

4.3. Time Considerations – Acute, Repeated, and Chronic Administration of Aβ

To measure the acute effect of a particular Aβ species, the peptide can be administered by a single stereotactic injection (35, 72); determination of the effect of chronic Aβ presence requires repetitive injections through an implanted cannula (36) or for continuous dosing, attachment of an osmotic mini-pump or micro-infusion pump to an implanted cannula is required (37, 38, 73).

By necessity, direct intracerebral or direct intraventricular injection of Aβ is usually undertaken while the animal is anesthetized. This must be taken into account if the aim of Aβ infusion is to assess the impact on learning and memory behaviors as anesthetics have potent effects on learning and memory, some of which can linger for hours after the animal has seemingly awakened. Therefore, direct injection is not ideal to assess the acute effect of Aβ on learning and memory, but can be used to examine longer-term effects.

The need to examine the acute effects of a single Aβ infusion on behavior can be accommodated by the use of an in-dwelling cannula implanted into the brain – the cannula is implanted while the animal is under anesthesia and the animal is allowed to recover for a few days before proceeding with the Aβ infusion. Long-term implantation of a cannula (and pump) can also be used to assess the impact of repeated doses or chronic Aβ administration on memory acquisition, formation, and retention. Use of an in-dwelling cannula has the advantage that the Aβ can be made up fresh each day, in contrast to an osmotic pump, which is filled with Aβ once prior to implantation, with the possibility that the Aβ changes over time.

Use of an osmotic mini-pump to deliver Aβ has the advantage of providing the continuous release and presence of Aβ in the brain throughout the experiment. Brain infusion kits containing osmotic mini-pumps and required accessories such as cannulae and tubing for delivery of compounds directly into the brain are available from ALZET. Prior to setting up an *in vivo* study, the researcher must consider the time period over which Aβ is to be infused as this affects the capacity of the pump selected for use. Depending on the length of the study, the spent pump may need to be removed after its pumping lifetime has ended to prevent local irritation of tissue caused by leakage of residual concentrated salt solution from the pump. Removal or replacement of the pump requires the animal to be anesthetized, which may have a bearing on behavioral experiments. It must also be borne in mind that each pump can only be filled once and the pump contents may change over time, especially as the Aβ solution is effectively being incubated at 37°C once it has been implanted into the animal. Explanting the pump at the end of its life allows assessment of the activity and stability of the residual Aβ and comparison with the solution that was initially loaded.

5. Advantages and Limitations of the Aβ Infusion Model

Although by no means a complete model of AD, and despite variation in results between labs, the intracerebral Aβ infusion model can contribute to AD research on at least three levels. First, data from this model provides support for the hypothesis of the pivotal pathogenic role of Aβ and allows us to gain insight into the mechanisms and effects of Aβ toxicity *in vivo*. Second, given the plethora of data from infusion models describing the effects of various Aβ species on different parameters of memory (acquisition, consolidation, and retention), this model can be used to study the effect of Aβ on cognition and can be used to test new therapies that potentially alleviate the cognitive symptoms of AD. Third, since infusion of peptide has been shown to induce inflammation and microglial activation, this model could serve not only to provide insights into the role of inflammation in AD, but could be used to test the protective effects of pharmacological modulation of microglial signaling (86).

The Aβ infusion model also offers some advantages over the use of transgenics. Current transgenic models of AD utilize overexpression of mutant human APP and PS1 to increase Aβ production and recapitulate AD cognitive deficits and pathologies (87–90). However, overexpression of APP results not only in increased production of both Aβ40 and Aβ42, but in elevated levels of other APP fragments, which can have neuroprotective (91,92), neurotoxic (93), or signaling functions (94) and influence learning and memory (95–98). Use of direct Aβ infusion allows the dissemination of the contribution of each Aβ peptide to the dementia and pathology associated with AD, allowing researchers to directly test the effect of different Aβ species on the development of symptoms. The infusion model in particular gives the researcher the ability to administer defined amounts of Aβ of known sequence and length or to introduce controlled co-factors related to plaque development. An additional advantage is the ability to accelerate the experimental timeline rather than waiting for transgenic rodents to age (34, 99). It can take 6–12 months to observe deposition of plaques in transgenic mice (89, 100, 101). Following Aβ infusion, plaque deposition, although local only, can be observed in a timeline of weeks (58, 64).

There are, however, limitations to the use of the Aβ infusion model. By necessity, and in contrast to the use of transgenics, the infusion approach involves inevitable injury to the brain associated with the invasive procedure (injection into brain parenchyma or implantation of a cannula) required to deliver Aβ into the brain. Additionally, the vehicle in which Aβ is dissolved (quite often a nonphysiological solvent) can affect Aβ neurotoxicity. Both these factors could contribute to the inflammatory processes induced by the infusion procedure. To a certain extent, however, these limitations can be overcome by adjusting the infusion rate, volume of injection, and time delay between infusion and examination of the animal to minimize these confounding effects (34). It also needs to be ensured that controls, such as infusion of vehicle only or a scrambled peptide sequence, are run concurrently. Additionally, use of the Aβ infusion model does not take into account the contribution of factors other than Aβ known to be critical in AD pathogenesis, such as neurofibrillary tangles and perturbations in vasculature, and largely bypasses the effect of aging on AD progression.

6. Localized Overexpression of AD-Associated Genes

Viral vector-mediated gene delivery has opened up new possibilities for the development of animal models of neurodegenerative disorders. Localized *in vivo* gene transfer of putative pathogenic genes has emerged as a viable alternative strategy to transgenic technology for the generation of genetic models of neurological disease. One vector system particularly useful for this application is adeno-associated virus (AAV), recombinant vectors derived from nonpathogenic, replication-deficient members of the *Parvovirus* family. These vectors are particularly efficient at transducing neurons and following infusion of AAV into the brain stable, long-lasting neuronal transgene expression is observed, with no apparent toxicity (102–105). AAV vectors are well established as versatile tools, allowing upregulation or knockdown of gene expression in specific brain regions, and can be used not only as vehicles for neurological gene therapy (106–109) but for *in vivo* functional genomics studies (110–112) and to create animal models of neurodegenerative disease.

The use of viral vectors to overexpress genes implicated in disease pathogenesis has already been successfully implemented to generate nontransgenic rat models of both Parkinson's (113, 114) and Huntington's disease (115–117), with successful transfer of the method to a primate Parkinson's model (118). Similar strategies have been undertaken to produce rodent models of AD; however, these models have yet to be fully characterized for development of behavioral deficits and associated neuropathological features. The first reported viral vector-mediated models of AD were based on overexpression of a mutant form of APP. In one study, lentiviral-mediated overexpression of $APP751_{SWE}$ in mouse hippocampus led to an increase in Aβ40 expression at 10 days and 4 weeks postinfusion (119). However, these short-term time points were the only ones examined, and Aβ40 expression levels determined via immunohistochemistry were the only parameter measured. Notably, most of the Aβ-specific signals observed with immunohistochemistry appeared to be intracellular, suggested by the authors to be a processed, presecretory form of Aβ located in the endoplasmic reticulum. The effect of lentiviral-mediated APP overexpression on hippocampal-dependent behavioral tasks was not undertaken.

In a similar study, an AAV2 vector was used to overexpress mutant APP ($APP695_{SWE}$) bilaterally in rat hippocampus (120) for prolonged periods of time (up to 15 months). Detectable expression of both APP and Aβ42 was seen at 6 and 12 months postinfusion. Significant behavioral deficits (in both acquisition and retention of the Morris water maze task) were observed at 12 months post-AAV. Notably at this time point, there was no evidence of plaque formation, neuronal loss, or Fluorojade labeling (to indicate presence of dying neurons). There was no measurable decrease in synaptophysin immunostaining, indicating no detectable loss of synaptic arborization following long-term APP overexpression. All these observations confirm data from AD transgenic mice that development of cognitive deficits is not necessarily accompanied by overt neuropathological changes. Additionally, gliosis, indicative of inflammation, was not observed. Notably, in models utilizing mutant APP overexpression, production of multiple fragments of APP processing would occur – any cognitive deficits or pathology observed cannot be attributed to a specific Aβ fragment.

In contrast, studies undertaken in our laboratory have used AAV-mediated gene transfer to create a rat model of AD based on overexpression and secretion of specific Aβ peptides (121). AAV1 vectors encoding BRI-Aβ cDNAs, fusions between human Aβ peptides, and the BRI protein involved in amyloid deposition in British and Danish familial dementia (122, 123) were used to achieve high-level hippocampal expression and secretion of the specific encoded Aβ peptide, either Aβ40 or Aβ42, in the absence of APP overexpression (23, 124). Notably, use of these BRI-Aβ fusions results in enhanced Aβ secretion, distinguishing this approach from overexpression of Aβ minigenes, a strategy that generates high levels of intracellular Aβ but minimal secreted Aβ (125). Animals were infused bilaterally in the hippocampus with the AAV1 vectors and tested for development of behavioral deficits 3 months after AAV infusion. At this time point, a range of behavioral impairments were found, with learning and memory deficits in the Morris water maze, passive avoidance, and novel object recognition tasks. At the conclusion of behavioral testing, brain tissue was removed and analyzed for Aβ levels and evidence of extracellular Aβ deposition. Only AAV-BRI-Aβ42 animals developed extracellular Aβ deposits, located within the hippocampus. These diffuse deposits were immunopositive for Aβ, but did not stain with thioflavin S or Congo Red and were not associated with astrogliosis or proliferation of microglia, indicating that they are "noncored" diffuse structures similar to the diffuse Aβ deposits observed in humans that are primarily composed of Aβ42 and not associated with significant reactive pathology. Expression for 9 months enhanced Aβ42 accumulation but still did not result in formation of cored plaques. This data showed that viral delivery of BRI-Aβ42 can foster considerable Aβ accumulation in a relatively short timeframe, and confirms both transgenic Aβ *Drosophila* (126) and transgenic BRI-Aβ mouse studies (23) where visible Aβ deposits were obtained only with Aβ42, but not Aβ40, overexpression. Overall, the results demonstrated that while virally mediated overexpression of Aβ42 alone is sufficient to initiate plaque deposition, both Aβ40 and Aβ42 levels contribute to the development of cognitive deficits. The lack of correlation between the presence of Aβ plaques and the severity of cognitive dysfunction observed confirmed observations from AD transgenic mice that measurable behavioral deficits are not dependent on the presence of Aβ plaques (89, 100, 101, 127) and may correlate better with other Aβ assemblies such as oligomers (128–131).

7. Advantages of Using Viral Vector-Mediated Gene Transfer

Viral vector-mediated gene transfer of APP or Aβ is a tool that can be used to model two important aspects of AD pathology *in vivo*. First, the model can be used to study the effect of Aβ on various parameters of memory formation and can be used to test new therapies that potentially alleviate the cognitive symptoms of AD. Second, this model could be used to study the effects of particular Aβ species on deposition of amyloid and testing of therapeutics aimed at clearing Aβ. An advantage over the Aβ infusion model is that depending on the Aβ construct encoded, the AAV model can be used to examine the effects of either secreted Aβ (i.e., with use of the BRI fusions) or the effect of intracellular accumulation of Aβ. An additional advantage over Aβ infusion is that the effect of chronic Aβ exposure can be examined without the need for a long-term cannula implanted into the brain, a procedure with inherent technical problems such as maintaining cannula patency and sterility over an extended time period. A single infusion of AAV allowed us to examine the effect of Aβ overexpression for up to 9 months.

The use of viral-vector-mediated gene transfer to create genetic models of neurodegenerative disease also offers some advantages over traditional transgenic technology. First and foremost, use of viral-vector technology affords the researcher temporal and spatial control over the onset of transgene expression. This means that the consequence of transgene expression specifically in (aged) adult brain can be examined, without onset during embryonic development as is the case with many transgenics. The experimenter can also choose where in the brain the transgene is expressed – for example, in AD either the hippocampus and/or cortex could be targeted. Potentially shortened timelines are another advantage of using this technology – it is not necessary to wait for animals to age as the onset of pathology is faster than observed in AD transgenic mice. A single researcher skilled in stereotaxic surgery can easily generate 10–12 vector-injected animals per day. Finally, viral vector-mediated delivery is not limited by species – transgenic techniques are so far largely limited to mice whereas use of viral vector technology means it has been possible to generate rat and even primate models of neurodegenerative disease (118).

Arguably, the main disadvantage of using this method is the difficulty in reproducibility between labs – this is largely due to variability between viral vector stocks. Also, inevitable damage is caused by infusion of the AAV vector into brain parenchyma, similar to what is observed in Aβ infusion models; however, this is not an issue with the use of transgenics. Although viral vector technology can be used to model cognitive deficits reminiscent of AD, it must be borne in mind that the encoded Aβ is present continuously throughout testing (i.e., during acquisition, consolidation, and retention testing) meaning that this model cannot be used to distinguish between effects of different Aβ species on these different parameters of memory formation.

8. Methods

8.1. Preparation of Aβ Peptide

The parameters discussed above must be taken into account when preparing Aβ peptides for *in vivo* use. Once the length and form of peptide has been decided, it is recommended that a single batch sufficient for the entire study be purchased or manufactured. If different batches have to be used, then the best practice is to use a sample of Aβ of a known high quality and establish its level of toxicity in an appropriate assay and qualify subsequent new lots of peptide based on comparative performance in the same assay (15).

The lyophilized Aβ peptide usually requires dissolution in a solvent such as hexafluoroisopropanol (HFIP; 1 mM). The Aβ is then aliquoted into sterile tubes and the HFIP removed under vacuum. The resulting peptide aliquots should be stored at –20°C under dessicating conditions. When required, an aliquot of peptide is resuspended in DMSO to 5 mM. For use in an unaggregated state, further dilutions can be made in saline.

Preparation of aggregated forms of Aβ (oligomers and fibrils) requires further incubation under specific conditions. For example, Dahlgren and co-workers defined conditions for formation of aggregated forms of Aβ42 (84). To prepare oligomers, Ham's F12 media was added to the Aβ42 peptide in DMSO to bring the peptide to a final concentration of 100 μM. This was then incubated at 4°C for 24 h (i.e., incubated at physiological salt concentration and physiological pH) resulting in production of an oligomeric Aβ42 stock. For preparation of fibrils, low salt and low pH conditions were used – the peptide in DMSO was diluted in 10 mM HCl to bring the peptide to a final concentration of 100 μM and incubated for 24 h at 37°C. Once prepared, fibrillar forms of Aβ can be characterized using transmission electron microscopy (57) or atomic force microscopy (84).

8.2. Infusion of Aβ into the Brain

8.2.1. Single Injection Directly into Brain

The anesthetized rat is placed in a stereotaxic such that its head is in a flat position (adjusted by moving the incisor bar up or down). The fur on its scalp is trimmed using scissors, the skin cleaned with 1% iodine/30% ethanol and a rostral-caudal incision made in the scalp using a scalpel and the underlying skull surface exposed. Stereotaxic coordinates relative to bregma are marked out on the skull surface and burr holes drilled using a dental drill (0.7 mm drill bit) to enable infusion of the Aβ into the appropriate structure (hippocampus or lateral ventricle). A Hamilton syringe with a 27G needle attached to a continuous infusion delivery pump is lowered into the site to the required dorso-ventral level. A total of 2 μl of Aβ is infused at a rate of 0.5 μl/min, and the needle left in place for an additional 5 min following the end of the infusion. At the end of the infusion, the needle is slowly withdrawn and the rat's scalp sutured.

8.2.2. Infusion of Aβ via an Intra-Ventricular Cannula

The rat is placed in a stereotaxic and an incision made in its scalp as detailed above. Stereotaxic coordinates to enable implantation of a guide cannula (22G, Plastics One Inc, C313G/SPC) above the lateral ventricle are identified based on coordinates relative to bregma (anterior–posterior 0.8 mm, medial–lateral 1.6 mm, dorsal–ventral 3.5 mm) and a dental drill (0.9 mm drill bit) used to drill a hole in the skull. Three additional holes are drilled for placement of stabilizing screws. The cannula guide is implanted to the appropriate depth and cemented into place, along with three skull screws, using Vertex dental cement. A dummy internal cannula 4 mm in length (Plastics One Inc, C313DC/1/SPC) is placed into the cannula guide to prevent it from becoming blocked. Animals are left for at least 1 week to recover prior to infusion of Aβ.

Peptides are injected into the ventricle via the cannula in awake animals using a Hamilton syringe (23G) connected to an internal cannula 5 mm in length (Plastics One Inc C313I/SPC) with 5 cm of polyethylene tubing. The rat is immobilized by wrapping it in a soft sheet and the dummy cannula removed. The internal cannula is placed into the cannula guide and peptide (2–5 μl) injected slowly into the ventricle. On completion of the infusion, the cannula is left in place for a further minute to ensure that the contents have diffused away, before slowly withdrawing and returning the rat to the home cage. This procedure can be performed by one person, provided the rats are accustomed to being handled. The cannula placement and needle track should be verified as correct at the time of sectioning the brain.

8.2.3. Infusion of Aβ Using Osmotic Mini-Pumps

Brain infusion kits containing osmotic mini-pumps and required accessories such as cannulae and tubing for delivery of compounds directly into the brain are available from ALZET. The reader is directed to their website (www.alzet.com) for detailed instructions on priming, filling, and implantation of these pumps. Briefly, an L-shaped cannula (one arm is inserted into the brain, the other arm receives the tubing from the pump) is implanted into the ventricle and cemented in place along with stabilizing screws as detailed above. The pump, prefilled with the required peptide, is implanted into a subcutaneous pocket created between the shoulder blades and tubing tunneled through from the scalp incision to the pump to connect the pump and cannula.

Depending on the length of the study, the spent pump may also need to be removed after its pumping lifetime has ended to prevent local irritation of tissue caused by leakage of residual concentrated salt solution from the pump. An additional reason for explanting the pump would be to assess the activity and stability of the residual Aβ and compare this to the solution that was initially loaded. Correct cannula placement can be verified during brain sectioning.

8.3. Generation of AAV Vectors

This section will briefly outline the production, purification, and titration methods used for generation of AAV vector stocks. For detailed outlines of the various AAV production, purification, and titration methods, the reader is directed to several comprehensive reviews in this area (99, 132–134). The methods outlined below detail the preparation of 5 × 15 cm plate batch of AAV – transfection of this number of cells is sufficient to generate 200 μl of AAV with a titer of 10^{12}–10^{13} viral genomes/ml.

8.3.1. Expression Constructs and Packaging Plasmids

Appropriate cDNAs for the transgene of interest need to be subcloned into an AAV expression plasmid (e.g., Stratagene) also containing the required promoter (e.g., CAG; hybrid CMV-chicken beta-actin promoter) and containing required regulatory elements (e.g., SAR (scaffold attachment region) element, WPRE (woodchuck hepatitis virus post-transcriptional-regulatory element), and bovine growth hormone polyadenylation signal flanked by AAV2 inverted terminal repeats).

Also required for AAV packaging are both an AAV helper plasmid (encoding AAV *rep* and *cap* genes) and a plasmid encoding adenoviral helper functions (e.g., pFdelta6). High-quality plasmid DNA (i.e., with no contaminating RNA present) should be prepared using Qiagen plasmid kits as per the manufacturer's instructions and diluted to a final concentration of 1–2 μg/μl.

8.3.2. Transfection

Human embryonic kidney (HEK293) cells are commonly used as a producer cell line for AAV preparation. These cells should be maintained in high glucose DMEM supplemented with 10% fetal bovine serum, nonessential amino acids and sodium pyruvate, split 1:5 every 3–4 days, and should be low passage (<p30) for use in vector preparation.

The calcium phosphate transfection method is used to introduce all the necessary plasmids for AAV production into the HEK293 cells; for more detailed instructions on $CaPO_4$ transfection, see (102). Briefly, cells are plated out in complete DMEM at 2.0–2.2 × 10^7 cells per 15 cm plate 24 h before transfection (cells should be 70% confluent by the time of transfection). Three hours before the transfection, the media is replaced with 25 ml of pre-warmed IMDM. The transfection mixture (water, $CaCl_2$, plasmids) is made up and filtered through a 0.2-μm syringe filter. This is added to an equal volume of 2× HeBs buffer while vortexing. A very fine white precipitate forms within a few minutes. The transfection solution is then added to the cells (5 ml to each plate). The media should be removed and replaced with 25 ml of fresh DMEM 16 h after transfection.

8.3.3. Harvesting and Purification

Cells are harvested 60 h after transfection. The cells are washed once in 1× PBS, then detached from the dish by pipetting up and down with 1× PBS. Cells are pelleted at 600 *g* for 30 min, resuspended in lysis buffer (150 mM NaCl, 50 mM Tris pH 8.5), and frozen at –20°C. The following day, the cells are lysed by adding sodium deoxycholate (0.5%) and endonuclease (50 U/ml) and incubating at 37°C for 1 h, before removing cell debris by centrifugation at 3,000 *g* × 15 min, 4°C. The supernatant is transferred to a fresh tube and frozen at –20°C until purification.

AAV vectors are purified from the clarified cell lysate by ultracentrifugation through an iodixanol density gradient. The gradient is subjected to ultracentrifugation at 250,000 *g* for 90 min at 18°C. To retrieve the vector, a needle is stuck through the side of the tube and 3.5 ml of AAV-containing iodixanol removed with a syringe and concentrated using a 15 ml 100,000 MWCO concentrator (Millipore). The concentrated AAV vector is washed with PBS several times to remove iodixanol, removed from the concentrator and sterilized using a 0.2 µm 13 mm syringe filter.

8.3.4. Quality Control – Determination of Genomic Titer, SDS-PAGE

Given that the amount of Aβ produced by transduced cells is governed by the number of genome copies taken up by the cell, it is important to determine the genomic titer of the vector stock prior to any infusion. We employ a real-time PCR method using primers specific to a constitutive regulatory module in the AAV expression cassette, WPRE, to determine the number of genome-containing vector particles in the vector stock. Titers are typically in the range of 10^{12}–10^{13} viral genomes/ml. For a full description of the method, the reader is directed to (102).

In addition to determining the titer of the vector stock, it is important to ensure that the vector is pure and does not contain significant amounts of proteins other than the AAV viral capsid proteins. Protein content of the vector stock can be analyzed by subjecting a sample (10 µl) of the vector to SDS-PAGE. The sample is run on a 12% acrylamide gel and then stained with Coomassie blue – the only bands that should be visible on the stained gel are VP1, VP2, and VP3.

8.4. Infusion of AAV Vectors into Rat Brain

A 250–300 g rat is anesthetized and placed in a Kopf stereotaxic frame with its head in a flat position, as described above. Following an incision into the scalp and clearance of the underlying membranes, stereotaxic coordinates to enable infusions into the appropriate structure are identified based on coordinates relative to bregma and a dental drill (0.7 mm drill bit) used to drill holes in the skull. A Hamilton syringe with a 27G needle attached to a continuous infusion delivery pump is lowered into the site to the required dorso-ventral level. A total of 3 µl of AAV vector is infused at a rate of 70 nl/min. The needle is left in place for an additional 5 min following the end of the infusion and then slowly withdrawn and the rat's scalp sutured.

8.5. Measuring the Effect of Aβ on Behavior

The effects of Aβ on cognition can be investigated at different stages of the learning and memory process – acquisition, consolidation, and retention – using a variety of behavioral tests such as the Morris water maze, passive avoidance, and novel object recognition.

8.5.1. Morris Water Maze

This is the most widely used paradigm for the evaluation of visuo-spatial learning and memory skills in rodents, the highest cognitive level appreciable in rodents (10), and can be used to measure both acquisition (learning) and retention of information. The water maze consists of a circular pool filled with warm (26°C) water and containing a hidden escape platform submerged below the surface of the water (a small amount of nontoxic dye is added to the water to camouflage the platform). Visual cues (e.g., black cardboard shapes on white curtains) are placed around the room and remain constant throughout testing. The rat must learn the location of the hidden platform by referring to the visual cues placed around the room.

The test procedure consists of three parts: acquisition training, a probe (retention) trial, and visible platform training. Acquisition training takes place over several days with each rat trained over multiple trials to learn the location of the platform. A trial consists of allowing the rat a certain length of time (e.g., 60 s) to find the platform. If not located in this time, the animal is guided to the platform by the experimenter before being returned to the home cage. The latency and path length to find the platform and the number of platform crossings are recorded for each trial using tracking software (e.g., Noldus Ethovision) connected to a video camera mounted in the ceiling above the pool. Twenty-four to 48 h after acquisition training, animals undergo a retention (probe) trial during which the platform is removed from the pool but the visual cues are still visible. The time spent by the rat in the quadrant of the pool that previously held the platform (target quadrant) is measured.

Animals also undergo visible platform training – visual cues are removed from the wall and the platform placed in a different quadrant of the pool, raised above the waterline so that it is clearly visible to the rat. The time taken by the rat to swim to the visible platform is recorded. This ensures that any observed deficits in learning are not merely the result of poor eyesight.

8.5.2. Passive Avoidance

This is used as a test of associational learning. Existence of memories is inferred from changes in behavior following an aversive experience. The equipment consists of two chambers (one well-lit, the other dark) connected via a guillotine door. Rats are placed in the less-preferred light chamber and (baseline) time to enter the (preferred) dark chamber recorded. Once the rat fully enters the dark compartment, the door is closed and the animal receives a small electrical shock (e.g., 1.0 mA for 3 s) before being returned to the home cage. Some time later (e.g., 24 h), the animal is placed in the light compartment with free access to the dark compartment and the time taken to enter the dark compartment is measured (step-through latency). No foot shock is given this time. This paradigm produces baseline latencies of 10–15 s and step-through latencies of 100–200 s in normal rats.

8.5.3. Novel Object Recognition

This memory test is based on the rat's natural propensity to explore novel objects and can be used to test both short-term memory and long-term retention, depending on the time delay between trials. Rats are given at least 2 days to habituate to the test environment, that is, placed in the test chamber for 5 min with no objects to explore. During the first trial, the rat is placed in the test chamber now containing two identical objects (O1 and O2; objects never encountered before and of no natural significance). The rat is placed equidistant from the two objects and the time spent exploring each object (nose within 2 cm of object) recorded for several minutes. In the second trial (e.g., 24 h later), the rat is returned to the test chamber and given one familiar (O3) and one novel (N) object to explore. Time spent exploring each object is recorded and a discrimination ratio calculated based on exploration times (N - O3/N + O3). When objects are identical, the amount of time spent exploring each object is generally the same. However, the ability to remember which object was previously seen and which is new is reflected by more time being spent with the novel object than the old one, providing an assessment of the ability to form and recall memories.

8.5.4. Open Field

To ensure that Aβ or potential therapeutic treatments do not affect motor skills or induce anxiety, which may influence performance in learning tasks, each rat is allowed several minutes of free movement in an open field. The frequency, duration, and speed of motor behaviors (e.g., walking, sniffing, grooming, rearing, defecating) is recorded. Thigmotaxia can be assessed by determining the fraction of time the rat spends within 10 cm of the wall of the tank.

8.6 Aβ ELISA

As outlined previously, a major disadvantage of AAV infusion models is the variability between laboratories, largely attributable to differences in titer and purity between AAV preparations. In contrast to the Aβ infusion model where a known amount of Aβ is circulating in the brain, the amount of Aβ produced following AAV-mediated gene transfer will vary depending on the titer and transduction efficiency of the vector used. To validate the model, the amount of Aβ produced following infusion of a known amount of vector needs to be biochemically determined using ELISA techniques. Briefly, on sacrifice of the animal, brain tissue is removed and frozen immediately on dry ice. Tissue is then homogenized in radio-immunoprecipitation assay (RIPA) buffer and ultracentrifuged to separate RIPA-soluble from insoluble fractions. RIPA-insoluble proteins can be extracted using 70% FA, neutralized with Tris base buffer and samples diluted (from 10- to 5,000-fold). Extracted Aβ is then measured using a sandwich ELISA system as described previously (121, 135) – Aβ42 capture with mAb 2.1.3 (mAb40.2,) and detection with HRP-conjugated mAb Ab9 (human Aβ1–16 specific); Aβ40 capture with mAb Ab9 and detection with HRP-conjugated mAb 13.1.1 (mAβ40.1). The reader is also directed to the following review article for further detail on analysis of Aβ peptides (15).

9. Conclusions

Although the use of transgenic mouse models has largely supplanted the use of Aβ infusion as a relevant model of AD, this method still has a place in AD research. Its greatest advantage is that it can be used to test the effect of a known Aβ peptide species at a defined concentration and at a defined time point after infusion. However, since widespread neuropathology is not readily observed following Aβ infusion, this model is not ideal for studies on plaque deposition and clearance. As an alternative to both Aβ infusion and AD transgenic mice, viral-vector-mediated gene delivery can be used to obtain localized overexpression of Aβ. Although this promising strategy has been shown to result in development of both cognitive deficits and Aβ deposits in treated animals, these genetic models require further characterization to show reproducible development of behavioral deficits and neuropathology prior to uptake by the AD research community as a reliable and useful model of AD.

References

1. Dodart JC, May P (2005) Overview on rodent models of Alzheimer's disease. Curr Protoc Neurosci 33:9.22.1–9.22.16
2. Albert MS (1996) Cognitive and neurobiologic markers of early Alzheimer disease. Proc Natl Acad Sci U S A 93: 13547–13551
3. Almkvist O (1996) Neuropsychological features of early Alzheimer's disease: preclinical and clinical stages. Acta Neurol Scand Suppl 165:63–71
4. Fleischman DA, Gabrieli J (1999) Long-term memory in Alzheimer's disease. Curr Opin Neurobiol 9:240–144
5. Ferris SH, Kluger A (1997) Assessing cognition in Alzheimer disease research. Alzheimer Dis Assoc Disord 11(Suppl 6):45–49
6. Perry RJ, Hodges JR (1999) Attention and executive deficits in Alzheimer's disease. A critical review. Brain 122:383–404
7. Collie A, Maruff P (2000) The neuropsychology of preclinical Alzheimer's disease and mild cognitive impairment. Neurosci Biobehav Rev 24:365–374
8. Cummings JL (2000) Cognitive and behavioral heterogeneity in Alzheimer's disease: seeking the neurobiological basis. Neurobiol Aging 21:845–861
9. Chung JA, Cummings JL (2000) Neurobehavioral and neuropsychiatric symptoms in Alzheimer's disease: characteristics and treatment. Neurol Clin 18:829–846

10. Van Dam D, De Deyn PP (2006) Drug discovery in dementia: the role of rodent models. Nat Rev Drug Discov 5:956–970
11. Glenner GG, Wong CW (1984) Alzheimer's disease: initial report of the purification and characterization of a novel cerebrovascular amyloid protein. Biochem Biophys Res Commun 120:885–890
12. Glenner GG, Wong CW (1984) Alzheimer's disease and Down's syndrome: sharing of a unique cerebrovascular amyloid fibril protein. Biochem Biophys Res Commun 122:1131–115
13. Golde TE, Eckman CB, Younkin SG (2000) Biochemical detection of Abeta isoforms: implications for pathogenesis, diagnosis, and treatment of Alzheimer's disease. Biochim Biophys Acta 1502:172–187
14. Masters CL, Simms G, Weinman NA, Multhaup G, McDonald BL, Beyreuther K (1985) Amyloid plaque core protein in Alzheimer disease and Down syndrome. Proc Natl Acad Sci U S A 82:4245–4249
15. Findeis MA (2007) The role of amyloid beta peptide 42 in Alzheimer's disease. Pharmacol Ther 116:266–286
16. Younkin SG (1998) The role of A beta 42 in Alzheimer's disease. J Physiol Paris 92: 289–292
17. Selkoe DJ (2001) Clearing the brain's amyloid cobwebs. Neuron 32:177–180
18. Selkoe DJ (2001) Alzheimer's disease: genes, proteins, and therapy. Physiol Rev 81:741–766
19. Chartier-Harlin MC, Crawford F, Houlden H et al (1991) Early-onset Alzheimer's disease caused by mutations at codon 717 of the beta-amyloid precursor protein gene. Nature 353: 844–846
20. Citron M, Eckman CB, Diehl TS et al (1998) Additive effects of PS1 and APP mutations on secretion of the 42-residue amyloid beta-protein. Neurobiol Dis 5:107–116
21. Hardy J, Selkoe DJ (2002) The amyloid hypothesis of Alzheimer's disease: progress and problems on the road to therapeutics. Science 297:353–356
22. Mullan M, Crawford F, Axelman K et al (1992) A pathogenic mutation for proba le Alzheimer's disease in the APP gene at the N-terminus of beta-amyloid. Nat Genet 1:345–347
23. McGowan E, Pickford F, Kim J et al (2005) Abeta42 is essential for parenchymal and vascular amyloid deposition in mice. Neuron 47:191–199
24. Selkoe DJ (2000) Toward a comprehensive theory for Alzheimer's disease. Hypothesis: Alzheimer's disease is caused by the cerebral accumulation and cytotoxicity of amyloid beta-protein. Ann N Y Acad Sci 924:17–25
25. Hartley DM, Walsh DM, Ye CP et al (1999) Protofibrillar intermediates of amyloid beta-protein induce acute electrophysiological changes and progressive neurotoxicity in cortical neurons. J Neurosci 19:8876–8884
26. Lambert MP, Barlow AK, Chromy BA et al (1998) Diffusible, nonfibrillar ligands derived from Abeta1–42 are potent central nervous system neurotoxins. Proc Natl Acad Sci U S A 95:6448–6453
27. Lue LF, Kuo YM, Roher AE et al (1999) Soluble amyloid beta peptide concentration as a predictor of synaptic change in Alzheimer's disease. Am J Pathol 155:853–862
28. Selkoe DJ (2002) Alzheimer's disease is a synaptic failure. Science 298:789–791
29. Selkoe DJ (2008) Soluble oligomers of the amyloid beta-protein impair synaptic plasticity and behavior. Behav Brain Res 192: 106–113
30. Lacor PN, Buniel MC, Chang L et al (2004) Synaptic targeting by Alzheimer's-related amyloid beta oligomers. J Neurosci 24: 10191–10200
31. Walsh DM, Selkoe DJ (2004) Oligomers on the brain: the emerging role of soluble protein aggregates in neurodegeneration. Protein Pept Lett 11:213–228
32. Walsh DM, Klyubin I, Fadeeva JV et al (2002) Naturally secreted oligomers of amyloid beta protein potently inhibit hippocampal long-term potentiation in vivo. Nature 416:535–539
33. Walsh DM, Selkoe DJ (2004) Deciphering the molecular basis of memory failure in Alzheimer's disease. Neuron 44:181–193
34. Stephan A, Phillips AG (2005) A case for a non-transgenic animal model of Alzheimer's disease. Genes Brain Behav 4:157–172
35. Harkany T, O'Mahony S, Kelly JP et al (1998) Beta-amyloid (Phe(SO3H)24)25–35 in rat nucleus basalis induces behavioral dysfunctions, impairs learning and memory and disrupts cortical cholinergic innervation. Behav Brain Res 90:133–145
36. Yamada M, Chiba T, Sasabe J et al (2005) Implanted cannula-mediated repetitive administration of Abeta 25–35 into the mouse cerebral ventricle effectively impairs spatial working memory. Behav Brain Res 164:139–146
37. Nakamura S, Murayama N, Noshita T, Annoura H, Ohno T (2001) Progressive brain dysfunction following intracerebroventricular infusion of beta(1–42)-amyloid peptide. Brain Res 912:128–136

38. Olariu A, Yamada K, Mamiya T, Hefco V, Nabeshima T (2002) Memory impairment induced by chronic intracerebroventricular infusion of beta-amyloid (1–40) involves downregulation of protein kinase C. Brain Res 957:278–286
39. Nag S, Yee BK, Tang F (1999) Reduction in somatostatin and substance P levels and choline acetyltransferase activity in the cortex and hippocampus of the rat after chronic intracerebroventricular infusion of beta-amyloid (1–40). Brain Res Bull 50:251–262
40. McDonald MP, Dahl EE, Overmier JB, Mantyh P, Cleary J (1994) Effects of an exogenous beta-amyloid peptide on retention for spatial learning. Behav Neural Biol 62:60–67
41. Cleary J, Hittner JM, Semotuk M, Mantyh P, O'Hare E (1995) Beta-amyloid(1–40) effects on behavior and memory. Brain Res 682:69–74
42. Nabeshima T, Nitta A (1994) Memory impairment and neuronal dysfunction induced by beta-amyloid protein in rats. Tohoku J Exp Med 174:241–249
43. Lee CL, Kuo TF, Wang JJ, Pan TM (2007) Red mold rice ameliorates impairment of memory and learning ability in intracerebroventricular amyloid beta-infused rat by repressing amyloid beta accumulation. J Neurosci Res 85:3171–3182
44. Tang J, Xu H, Fan X et al (2008) Embryonic stem cell-derived neural precursor cells improve memory dysfunction in Abeta (1–40) injured rats. Neurosci Res 62:86–96
45. Yamaguchi Y, Miyashita H, Tsunekawa H et al (2006) Effects of a novel cognitive enhancer, spiro [imidazo-[1,2-a]pyridine-3,2-indan]-2(3H)-one (ZSET1446), on learning impairments induced by amyloid-beta1–40 in the rat. J Pharmacol Exp Ther 317:1079–1087
46. Yamada K, Tanaka T, Senzaki K, Kameyama T, Nabeshima T (1998) Propentofylline improves learning and memory deficits in rats induced by beta-amyloid protein-(1–40). Eur J Pharmacol 349:15–22
47. Tanaka T, Yamada K, Senzaki K et al (1998) NC-1900, an active fragment analog of arginine vasopressin, improves learning and memory deficits induced by beta-amyloid protein in rats. Eur J Pharmacol 352:135–142
48. Yamada K, Tanaka T, Mamiya T, Shiotani T, Kameyama T, Nabeshima T (1999) Improvement by nefiracetam of beta-amyloid-(1–42)-induced learning and memory impairments in rats. Br J Pharmacol 126:235–244
49. Yamada K, Tanaka T, Han D, Senzaki K, Kameyama T, Nabeshima T (1999) Protective effects of idebenone and alpha-tocopherol on beta-amyloid-(1–42)-induced learning and memory deficits in rats: implication of oxidative stress in beta-amyloid-induced neurotoxicity in vivo. Eur J Neurosci 11:83–90
50. Mazzola C, Micale V, Drago F (2003) Amnesia induced by beta-amyloid fragments is counteracted by cannabinoid CB1 receptor blockade. Eur J Pharmacol 477:219–225
51. Jhoo JH, Kim HC, Nabeshima T et al (2004) Beta-amyloid (1–42)-induced learning and memory deficits in mice: involvement of oxidative burdens in the hippocampus and cerebral cortex. Behav Brain Res 155:185–196
52. Liu RY, Gu R, Qi XL et al (2008) Decreased nicotinic receptors and cognitive deficit in rats intracerebroventricularly injected with beta-amyloid peptide(1–42) and fed a high-cholesterol diet. J Neurosci Res 86:183–193
53. Flood JF, Morley JE, Roberts E (1991) Amnestic effects in mice of four synthetic peptides homologous to amyloid beta protein from patients with Alzheimer disease. Proc Natl Acad Sci U S A 88:3363–3366
54. Olariu A, Tran MH, Yamada K, Mizuno M, Hefco V, Nabeshima T (2001) Memory deficits and increased emotionality induced by beta-amyloid (25–35) are correlated with the reduced acetylcholine release and altered phorbol dibutyrate binding in the hippocampus. J Neural Transm 108:1065–1079
55. Stepanichev MY, Moiseeva YV, Lazareva NA, Onufriev MV, Gulyaeva NV (2003) Single intracerebroventricular administration of amyloid-beta (25–35) peptide induces impairment in short-term rather than long-term memory in rats. Brain Res Bull 61:197–205
56. Giovannelli L, Scali C, Faussone-Pellegrini MS, Pepeu G, Casamenti F (1998) Long-term changes in the aggregation state and toxic effects of beta-amyloid injected into the rat brain. Neuroscience 87:349–357
57. Sipos E, Kurunczi A, Kasza A et al (2007) Beta-amyloid pathology in the entorhinal cortex of rats induces memory deficits: implications for Alzheimer's disease. Neuroscience 147:28–36
58. Frautschy SA, Yang F, Calderon L, Cole GM (1996) Rodent models of Alzheimer's disease: rat A beta infusion approaches to amyloid deposits. Neurobiol Aging 17:311–321
59. Frautschy SA, Horn DL, Sigel JJ et al (1998) Protease inhibitor coinfusion with amyloid beta-protein results in enhanced deposition and toxicity in rat brain. J Neurosci 18:8311–8321
60. Malm T, Ort M, Tahtivaara L et al (2006) Beta-Amyloid infusion results in delayed and

age-dependent learning deficits without role of inflammation or beta-amyloid deposits. Proc Natl Acad Sci U S A 103:8852–8857

61. Stephan A, Laroche S, Davis S (2001) Generation of aggregated beta-amyloid in the rat hippocampus impairs synaptic transmission and plasticity and causes memory deficits. J Neurosci 21:5703–5714
62. Begum AN, Yang F, Teng E et al (2008) Use of copper and insulin-resistance to accelerate cognitive deficits and synaptic protein loss in a rat Abeta-infusion Alzheimer's disease model. J Alzheimers Dis 15:625–640
63. Frautschy SA, Baird A, Cole GM (1991) Effects of injected Alzheimer beta-amyloid cores in rat brain. Proc Natl Acad Sci U S A 88:8362–8366
64. Frautschy SA, Cole GM, Baird A (1992) Phagocytosis and deposition of vascular beta-amyloid in rat brains injected with Alzheimer beta-amyloid. Am J Pathol 140:1389–1399
65. Weldon DT, Rogers SD, Ghilardi JR et al (1998) Fibrillar beta-amyloid induces microglial phagocytosis, expression of inducible nitric oxide synthase, and loss of a select population of neurons in the rat CNS in vivo. J Neurosci 18:2161–2173
66. Klein AM, Kowall NW, Ferrante RJ (1999) Neurotoxicity and oxidative damage of beta amyloid 1–42 versus beta amyloid 1–40 in the mouse cerebral cortex. Ann N Y Acad Sci 893:314–320
67. Scali C, Prosperi C, Giovannelli L, Bianchi L, Pepeu G, Casamenti F (1999) Beta(1–40) amyloid peptide injection into the nucleus basalis of rats induces microglia reaction and enhances cortical gamma-aminobutyric acid release in vivo. Brain Res 831:319–321
68. Emre M, Geula C, Ransil BJ, Mesulam MM (1992) The acute neurotoxicity and effects upon cholinergic axons of intracerebrally injected beta-amyloid in the rat brain. Neurobiol Aging 13:553–559
69. Kowall NW, Beal MF, Busciglio J, Duffy LK, Yankner BA (1991) An in vivo model for the neurodegenerative effects of beta amyloid and protection by substance P. Proc Natl Acad Sci U S A 88:7247–7251
70. Waite J, Cole GM, Frautschy SA, Connor DJ, Thal LJ (1992) Solvent effects on beta protein toxicity in vivo. Neurobiol Aging 13:595–599
71. Sigurdsson EM, Lee JM, Dong XW, Hejna MJ, Lorens SA (1997) Bilateral injections of amyloid-beta 25–35 into the amygdala of young Fischer rats: behavioral, neurochemical, and time dependent histopathological effects. Neurobiol Aging 18:591–608
72. Harkany T, Abraham I, Timmerman W et al (2000) Beta-amyloid neurotoxicity is mediated by a glutamate-triggered excitotoxic cascade in rat nucleus basalis. Eur J Neurosci 12:2735–2745
73. Nag S, Yee BK, Tang F (1999) Chronic intracerebroventricular infusion of beta-amyloid (1–40) results in a selective loss of neuropeptides in addition to a reduction in choline acetyltransferase activity in the cortical mantle and hippocampus in the rat. Ann N Y Acad Sci 897:420–422
74. Lesne S, Koh MT, Kotilinek L et al (2006) A specific amyloid-beta protein assembly in the brain impairs memory. Nature 440: 352–357
75. Howlett DR, Jennings KH, Lee DC et al (1995) Aggregation state and neurotoxic properties of Alzheimer beta-amyloid peptide. Neurodegeneration 4:23–32
76. Iversen LL, Mortishire-Smith RJ, Pollack SJ, Shearman MS (1995) The toxicity in vitro of beta-amyloid protein. Biochem J 311:1–16
77. Fezoui Y, Hartley DM, Harper JD et al (2000) An improved method of preparing the amyloid beta-protein for fibrillogenesis and neurotoxicity experiments. Amyloid 7:166–178
78. Walsh DM, Hartley DM, Selkoe DJ (2003) The many faces of Aß: structures and activity. Curr Med Chem Immun Endocr Metab Agents 3:277–291
79. Snyder EM, Nong Y, Almeida CG et al (2005) Regulation of NMDA receptor trafficking by amyloid-beta. Nat Neurosci 8:1051–1058
80. Cleary JP, Walsh DM, Hofmeister JJ et al (2005) Natural oligomers of the amyloid-beta protein specifically disrupt cognitive function. Nat Neurosci 8:79–84
81. Tickler AK, Clippingdale AB, Wade JD (2004) Amyloid-beta as a "difficult sequence" in solid phase peptide synthesis. Protein Pept Lett 11:377–384
82. May PC, Gitter BD, Waters DC et al (1992) Beta-Amyloid peptide in vitro toxicity: lot-to-lot variability. Neurobiol Aging 13:605–607
83. Brining SK (1997) Predicting the in vitro toxicity of synthetic beta-amyloid (1–40). Neurobiol Aging 18:581–589
84. Dahlgren KN, Manelli AM, Stine WB Jr, Baker LK, Krafft GA, LaDu MJ (2002) Oligomeric and fibrillar species of amyloid-beta peptides differentially affect neuronal viability. J Biol Chem 277:32046–32053
85. Harper JD, Lansbury PT Jr (1997) Models of amyloid seeding in Alzheimer's disease and scrapie: mechanistic truths and physiological consequences of the time-dependent solubility of amyloid proteins. Annu Rev Biochem 66:385–407
86. McLarnon JG, Ryu JK (2008) Relevance of abeta1–42 intrahippocampal injection as an

animal model of inflamed Alzheimer's disease brain. Curr Alzheimer Res 5:475–480
87. Hsiao K, Chapman P, Nilsen S et al (1996) Correlative memory deficits, Abeta elevation, and amyloid plaques in transgenic mice. Science 274:99–102
88. Borchelt DR, Ratovitski T, van Lare J et al (1997) Accelerated amyloid deposition in the brains of transgenic mice coexpressing mutant presenilin 1 and amyloid precursor proteins. Neuron 19:939–945
89. Holcomb L, Gordon MN, McGowan E et al (1998) Accelerated Alzheimer-type phenotype in transgenic mice carrying both mutant amyloid precursor protein and presenilin 1 transgenes. Nat Med 4:97–100
90. Gordon MN, Holcomb LA, Jantzen PT et al (2002) Time course of the development of Alzheimer-like pathology in the doubly transgenic PS1+APP mouse. Exp Neurol 173:183–195
91. Furukawa K, Sopher BL, Rydel RE et al (1996) Increased activity-regulating and neuroprotective efficacy of alpha-secretase-derived secreted amyloid precursor protein conferred by a C-terminal heparin-binding domain. J Neurochem 67:1882–1896
92. Barger SW, Mattson MP (1996) Induction of neuroprotective kappa B-dependent transcription by secreted forms of the Alzheimer's beta-amyloid precursor. Brain Res Mol Brain Res 40:116–126
93. Lu DC, Rabizadeh S, Chandra S et al (2000) A second cytotoxic proteolytic peptide derived from amyloid beta-protein precursor. Nat Med 6:397–404
94. LaFerla FM (2002) Calcium dyshomeostasis and intracellular signalling in Alzheimer's disease. Nat Rev Neurosci 3:862–872
95. Dodart JC, Mathis C, Ungerer A (2000) The beta-amyloid precursor protein and its derivatives: from biology to learning and memory processes. Rev Neurosci 11:75–93
96. Turner PR, O,Connor K, Tate WP, Abraham WC (2003) Roles of amyloid precursor protein and its fragments in regulating neural activity, plasticity and memory. Prog Neurobiol 70:1–32
97. Bour A, Little S, Dodart JC, Kelche C, Mathis C (2004) A secreted form of the beta-amyloid precursor protein (sAPP695) improves spatial recognition memory in OF1 mice. Neurobiol Learn Mem 81:27–38
98. Galvan V, Gorostiza OF, Banwait S et al (2006) Reversal of Alzheimer's-like pathology and behavior in human APP transgenic mice by mutation of Asp664. Proc Natl Acad Sci U S A 103:7130–7135
99. Ulusoy A, Bjorklund T, Hermening S, Kirik D (2008) In vivo gene delivery for development of mammalian models for Parkinson's disease. Exp Neurol 209:89–100
100. Holcomb LA, Gordon MN, Jantzen P, Hsiao K, Duff K, Morgan D (1999) Behavioral changes in transgenic mice expressing both amyloid precursor protein and presenilin-1 mutations: lack of association with amyloid deposits. Behav Genet 29:177–185
101. Oddo S, Caccamo A, Kitazawa M, Tseng BP, LaFerla FM (2003) Amyloid deposition precedes tangle formation in a triple transgenic model of Alzheimer's disease. Neurobiol Aging 24:1063–1070
102. During MJ, Young D, Baer K, Lawlor P, Klugmann M (2003) Development and optimization of adeno-associated virus vector transfer into the central nervous system. Methods Mol Med 76:221–236
103. Tenenbaum L, Chtarto A, Lehtonen E, Velu T, Brotchi J, Levivier M (2004) Recombinant AAV-mediated gene delivery to the central nervous system. J Gene Med 6:Suppl 1: S212– S222
104. McCown TJ (2005) Adeno-associated virus (AAV) vectors in the CNS. Curr Gene Ther 5:333–338
105. Wang C, Wang CM, Clark KR, Sferra TJ (2003) Recombinant AAV serotype 1 transduction efficiency and tropism in the murine brain. Gene Ther 10:1528–1534
106. Mandel RJ, Burger C (2004) Clinical trials in neurological disorders using AAV vectors: promises and challenges. Curr Opin Mol Ther 6:482–490
107. Bankiewicz KS, Forsayeth J, Eberling JL et al (2006) Long-term clinical improvement in MPTP-lesioned primates after gene therapy with AAV-hAADC. Mol Ther 14:564–570
108. Kaplitt MG, Feigin A, Tang C et al (2007) Safety and tolerability of gene therapy with an adeno-associated virus (AAV) borne GAD gene for Parkinson's disease: an open labl, phase I trial. Lancet 369:2097–2105
109. Giacobini E, Becker RE (2007) One hundred years after the discovery of Alzheimer's disease. A turning point for therapy? J Alzheimers Dis 12:37–52
110. During MJ, Cao L, Zuzga DS et al (2003) Glucagon-like peptide-1 receptor is involved in learning and neuroprotection. Nat Med 9:1173–1179
111. Mouravlev A, Dunning J, Young D, During MJ (2006) Somatic gene transfer of cAMP response element-binding protein attenuates memory impairment in aging rats. Proc Natl Acad Sci U S A 103:4705–4710

112. Szumlinski KK, Lominac KD, Oleson EB et al (2005) Homer2 is necessary for EtOH-induced neuroplasticity. J Neurosci 25:7054–7061

113. Kirik D, Rosenblad C, Burger C et al (2002) Parkinson-like neurodegeneration induced by targeted overexpression of alpha-synuclein in the nigrostriatal system. J Neurosci 22: 2780–2791

114. Lo Bianco C, Ridet JL, Schneider BL, Deglon N, Aebischer P (2002) Alpha -Synucleinopathy and selective dopaminergic neuron loss in a rat lentiviral-based model of Parkinson's disease. Proc Natl Acad Sci U S A 99:10813–10818

115. Senut MC, Suhr ST, Kaspar B, Gage FH (2000) Intraneuronal aggregate formation and cell death after viral expression of expanded polyglutamine tracts in the adult rat brain. J Neurosci 20:219–229

116. de Almeida LP, Ross CA, Zala D, Aebischer P, Deglon N (2002) Lentiviral-mediated delivery of mutant huntingtin in the striatum of rats induces a selective neuropathology modulated by polyglutamine repeat size, huntingtin expression levels, and protein length. J Neurosci 22(9):3473–3483

117. DiFiglia M, Sena-Esteves M, Chase K et al (2007) Therapeutic silencing of mutant huntingtin with siRNA attenuates striatal and cortical neuropathology and behavioral deficits. Proc Natl Acad Sci U S A 104:17204–17209

118. Kirik D, Annett LE, Burger C, Muzyczka N, Mandel RJ, Bjorklund A (2003) Nigrostriatal alpha-synucleinopathy induced by viral vector-mediated overexpression of human alpha-synuclein: a new primate model of Parkinson's disease. Proc Natl Acad Sci U S A 100:2884–2889

119. Hong CS, Goins WF, Goss JR, Burton EA, Glorioso JC (2006) Herpes simplex virus RNAi and neprilysin gene transfer vectors reduce accumulation of Alzheimer's disease-related amyloid-beta peptide in vivo. Gene Ther 13:1068–1079

120. Gong Y, Meyer EM, Meyers CA, Klein RL, King MA, Hughes JA (2006) Memory-related deficits following selective hippocampal expression of Swedish mutation amyloid precursor protein in the rat. Exp Neurol 200: 371–377

121. Lawlor PA, Bland RJ, Das P et al (2007) Novel rat Alzheimer's disease models based on AAV-mediated gene transfer to selectively increase hippocampal Abeta levels. Mol Neurodegener 2:11

122. Vidal R, Frangione B, Rostagno A et al (1999) A stop-codon mutation in the BRI gene associated with familial British dementia. Nature 399:776–781

123. Vidal R, Revesz T, Rostagno A et al (2000) A decamer duplication in the 3′ region of the BRI gene originates an amyloid peptide that is associated with dementia in a Danish kindred. Proc Natl Acad Sci U S A 97:4920–4925

124. Lewis PA, Piper S, Baker M et al (2001) Expression of BRI-amyloid beta peptide fusion proteins: a novel method for specific high-level expression of amyloid beta peptides. Biochim Biophys Acta 1537:58–62

125. LaFerla FM, Tinkle BT, Bieberich CJ, Haudenschild CC, Jay G (1995) The Alzheimer's A beta peptide induces neurodegeneration and apoptotic cell death in transgenic mice. Nat Genet 9:21–30

126. Iijima K, Liu HP, Chiang AS, Hearn SA, Konsolaki M, Zhong Y (2004) Dissecting the pathological effects of human Abeta 40 and Abeta42 in Drosophila: a potential model for Alzheimer's disease. Proc Natl Acad Sci U S A 101:6623–6628

127. Oddo S, Caccamo A, Shepherd JD et al (2003) Triple-transgenic model of Alzheimer's disease with plaques and tangles: intracellular Abeta and synaptic dysfunction. Neuron 39:409–421

128. Chapman PF, White GL, Jones MW et al (1999) Impaired synaptic plasticity and learning in aged amyloid precursor protein transgenic mice. Nat Neurosci 2:271–276

129. Hsia AY, Masliah E, McConlogue L et al (1999) Plaque-independent disruption of neural circuits in Alzheimer's disease mouse models. Proc Natl Acad Sci U S A 96:3228–3233

130. Westerman MA, Cooper-Blacketer D, Mariash A et al (2002) The relationship between Abeta and memory in the Tg2576 mouse model of Alzheimer's disease. J Neurosci 22:1858–1867

131. Jacobsen JS, Wu CC, Redwine JM et al (2006) Early-onset behavioral and synaptic deficits in a mouse model of Alzheimer's disease. Proc Natl Acad Sci U S A 103:5161–5166

132. Lu Y (2004) Recombinant adeno-associated virus as delivery vector for gene therapy – a review. Stem Cells Dev 13:133–145

133. Aucoin MG, Perrier M, Kamen AA (2008) Critical assessment of current adeno-associated viral vector production and quantification methods. Biotechnol Adv 26:73–88

134. Grieger JC, Choi VW, Samulski RJ (2006) Production and characterization of adeno-associated viral vectors. Nat Protoc 1:1412–1428

135. Levites Y, Das P, Price RW et al (2006) Anti-Abeta42- and anti-Abeta 40-specific mAbs attenuate amyloid deposition in an Alzheimer disease mouse model. J Clin Invest 116:193–201

Chapter 18

APP-Based Transgenic Models: The PDAPP Model

Jacob M. Basak and David M. Holtzman

Abstract

The PDAPP mouse has been a very useful model for studying mechanisms underlying amyloid-β (Aβ) metabolism, aggregation, and deposition, and for aiding in the development of new diagnostic and therapeutic approaches for Alzheimer's disease (AD). One of the initial difficulties in producing a mouse model that was useful to study AD was creating an easily manipulated small animal model that demonstrated some of the key *in vivo* neuropathological hallmarks of the disease. The PDAPP model was the first transgenic mouse to overexpress the human amyloid precursor protein (hAPP) that successfully recapitulated several neuropathological features characteristic of AD. These include Aβ deposition in both diffuse and neuritic plaques, cerebral amyloid angiopathy, astrocytosis, microgliosis, hippocampal atrophy, synaptic alterations, and behavioral deficits. Many of the histological, biochemical, and structural alterations present in the PDAPP mouse closely resemble the changes found in the brain of AD patients, especially the temporal and spatial-specific deposition of Aβ in the brain. The animals, however, did not develop tauopathy or significant neuronal cell death. Because of these properties, the PDAPP mouse has proven to be an attractive model to study the disease process underlying aspects of AD that are related to Aβ aggregation and its consequences. The PDAPP mouse has been extensively used to study the effect of genetic factors that modify AD, as well as Aβ-binding proteins and their effect on Aβ deposition. It has also been widely used to characterize the potential use of active and passive immunization in both the diagnosis and treatment of AD.

Key words: PDAPP mouse model, Amyloid beta (Aβ), Platelet-derived growth factor (PDGF)-β, human amyloid precursor protein (hAPP), Amyloid plaque, Cerebral amyloid angiopathy (CAA), Neuritic dystrophy, Apolipoprotein E, Clusterin, Immunization

1. Introduction

Mouse models have proven to be valuable tools for both elucidating the pathophysiology of various disease states and for studying new treatments. The study of Alzheimer's disease (AD) in particular has benefited in recent years from the development of multiple mouse models that exhibit various characteristics of the

Peter Paul De Deyn and Debby Van Dam (eds.), *Animal Models of Dementia*, Neuromethods, vol. 48,
DOI 10.1007/978-1-60761-898-0_18, © Springer Science+Business Media, LLC 2011

disease. An initial challenge in studying AD *in vivo* was creating a mouse model that successfully displayed the neuropathological hallmarks of AD, including extracellular amyloid-β (Aβ) deposition, inflammation, neuritic dystrophy, intracellular neurofibrillary tangles, and neuronal as well as synaptic loss. Investigators began to develop such a model by overexpressing the human amyloid precursor protein (hAPP) with or without mutations that result in early-onset forms of familial, autosomal dominant AD. Early attempts at developing hAPP transgenic mice did not achieve their main objective as the brains of these animals had low levels of Aβ deposition or no Aβ accumulation at all (1–7). However, in 1995 a team of scientists from Athena Neurosciences greatly facilitated the study of AD in mice by developing the PDAPP mouse model (8). This was the first APP-based transgenic mouse to display many of the pathological features of AD, including Aβ deposition in both diffuse and neuritic plaques, astrocytosis, microgliosis, hippocampal atrophy, synaptic alterations, and behavioral deficits. This chapter will summarize the design and characterization of the PDAPP mouse, and also highlight how this model has been used to understand the mechanisms underlying the pathogenesis of AD, as well as to develop better diagnostics and treatments.

2. Design and Biochemical Characterization of the PDAPP Model

Like many other AD mouse models, the generation of the PDAPP mouse relied on previous genetic studies that revealed that multiple mutations in the hAPP gene could cause familial, autosomal dominant forms of AD. The genetic construct used in creating the PDAPP mouse consists of an hAPP minigene with a single-point mutation that contains a phenylalanine substituted for valine at residue 717 (V717F, also known as the Indiana mutation (9)), expressed under the control of the platelet-derived growth factor (PDGF)-β promoter (8). The PDGF-β promoter was chosen because it preferably targets expression of the transgene to neurons (10). The V717F mutation has been shown to increase the relative production of the more amyloidogenic $A\beta_{42}$ in relation to other shorter Aβ species such as $A\beta_{40}$, which most likely plays a significant role in the propensity for Aβ to aggregate in the brain of PDAPP mice (11). To allow for the alternative splicing of exons 7 and 8 that is known to occur with human and mouse APP, the hAPP (V717F) construct also contains introns 6–8 (8, 12). Analysis of protein expression showed that the hAPP protein in PDAPP mice is expressed at a level that is approximately 10-fold higher than both the levels of APP found in the human brain and the amount of endogenous mouse APP in the PDAPP brain (8).

Studies have also shown that the hAPP protein is expressed predominantly in the brain of PDAPP mice, demonstrating that most of the human Aβ produced is of CNS origin (13).

A very useful feature of the PDAPP model lies in the fact that the deposition of Aβ occurs in an age-dependent and region-specific manner that is similar to what is found in the human AD brain. Biochemical analysis of Aβ levels in the PDAPP mouse brain demonstrated a progressive increase in the total Aβ levels found in both the hippocampus and cortex beginning around 6–8 months of age in heterozygous mice, with greater levels occurring in the hippocampus (14). This accumulation correlated with the steady increase in immunoreactive Aβ deposition that occurred as the mice aged (see the next section). The accumulation of $A\beta_{42}$ that occurred dramatically increased with age, likely as a result of the increased levels of $A\beta_{42}$ aggregating in plaques (14, 15). The mechanism underlying the age-dependent nature of cerebral Aβ deposition in these mice is unlikely to be due to increased production of Aβ with age, as experiments proved the processing of hAPP to Aβ to remain relatively stable as the PDAPP mice age (14). Just as age seems to play a prominent role in Aβ build-up, the region of the PDAPP brain is also important in determining where Aβ deposits. Specific regions of the cortex and hippocampus exhibited significantly greater Aβ accumulation than the thalamus and cerebellum (14, 16, 17). Experiments have shown that even though the level of hAPP is greater in both the thalamus and cerebellum of PDAPP homozygous mice in comparison to the level of hAPP in the cortex of heterozygous mice, the heterozygous cortex developed greater levels of Aβ accumulation (14). Therefore, Aβ deposition occurs in specific regions of the brain in a manner that in some cases is not completely predicted by hAPP levels.

The distribution and metabolism of Aβ in the PDAPP mouse has been closely analyzed in specific compartments of the CNS and in the plasma. Both cerebral spinal fluid (CSF) and plasma contained measurable levels of soluble Aβ that exist in a close equilibrium in mice without plaque deposition (18). However, once plaques began to accumulate, CSF Aβ levels, particularly $A\beta_{42}$, decreased and the correlation between plasma and CSF Aβ dissipated, suggesting that Aβ deposition alters the equilibrium between these compartments by increasing the propensity of Aβ to remain in the CNS, with Aβ aggregates acting as a local "sink" for soluble Aβ (15, 18). Subsequent data were obtained using *in vivo* microdialysis that showed that plaque deposition is associated with altered soluble Aβ homeostasis in the brain interstitial fluid (ISF). The half-life of ISF Aβ was about twofold longer in 12-month-old PDAPP mice with plaques than in 3-month-old mice without plaques (15). Finally, Aβ accumulation also occurred in the cerebral vessels as PDAPP mice aged, resulting in cerebral amyloid angiopathy (CAA) (19, 20). Interestingly, the deposition of Aβ in vessels resulted in a significant decrease in the $A\beta_{42}/A\beta_{40}$ ratio in the Aβ aggregates located within the vessel wall, which is similar to what is found in human brains with CAA (20). Therefore, unlike in the brain parenchyma where $A\beta_{42}$ is the most abundant species present in plaques, $A\beta_{40}$ appears to be the primary depositing species of Aβ in vessels.

3. Histopathology of the PDAPP Mouse Brain

Many of the pathologic changes that occur in the PDAPP mouse brain resemble the histological hallmarks that occur in the human AD brain. The progressive development of both neuritic and diffuse Aβ deposits is the primary histological feature in the PDAPP brain, along with secondary changes in parenchymal and synaptic structure. On average, Aβ deposition visible by immunostaining starts to occur around 6–8 months of age and increases progressively as the animals age (14, 16, 21). However, Aβ immunoreactivity has been detected in some homozygous animals at ages as young as 3–4 months old (21). Furthermore, multiple studies have revealed that the age in which visible Aβ deposition begins varies significantly, as heterozygous mice up to 12 months of age have shown no Aβ accumulation (13, 18, 22, 23). Aβ deposits tend to first occur in the cingulate cortex and then soon after or concurrently appear in the molecular layer of the dentate gyrus, the CA1 region of the hippocampus, the entorhinal cortex, and the corpus callosum (8, 16). Very little deposition has been found in the cerebellum and thalamus (8, 17). Multiple studies have shown that the greatest Aβ load occurs in the molecular layer of the dentate gyrus and the lateral entorhinal cortex, suggesting that the lateral perforant pathway in these animals is particularly subject to Aβ deposition (16, 17, 24). Approximately 10% of the Aβ deposits covering the brain stained with thioflavine-S, which is synonymous with Aβ aggregation into amyloid fibrils or fibrillar plaques that have a β-pleated sheet structure (8, 25). The morphology of these plaques varied from irregular immunoreactive deposits that lack a central amyloid core to spherical complex structures with a dense amyloid core (see Fig. 1) (8, 25, 26). Also, fibrillar (thioflavine-S-positive) plaque accumulation associated with CAA occurred in the wall of cerebral arterioles starting around 15 months of age (19, 20, 27, 28). The progression of CAA in the PDAPP brain was associated with the development of microhemorrhages as the mice age (19), presumably due to CAA-associated vessel damage.

Dystrophic neurites and changes in the neuropil structure are closely associated with plaque deposition in the PDAPP brain (8, 26).

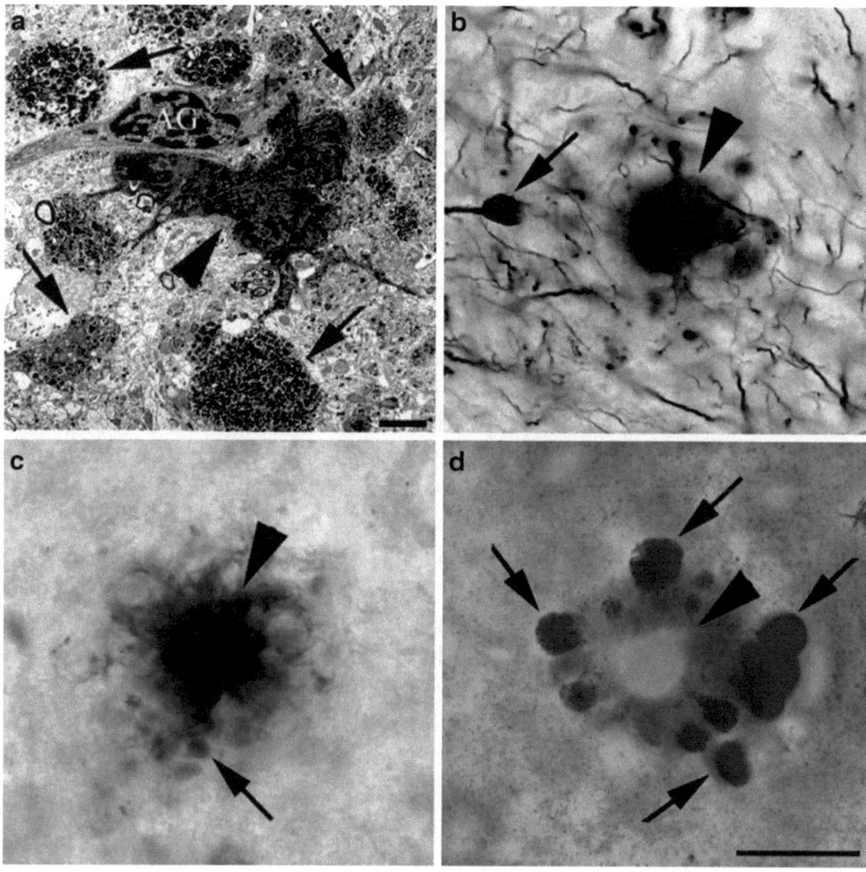

Fig. 1. Neuritic plaque morphology present in PDAPP mice. (**a**) Appearance of an amyloid plaque by transmission electron microscopy highlights the extracellular amyloid deposit (arrowhead), dystrophic neurites (arrows), and astroglial cells (AG) associated with the plaque. Plaques stained by Bielschowsky (**b**), Gallyas (**c**), and de Olmos (**d**) silver staining methods display the dense amyloid core (arrowheads) and dystrophic neurites (arrows). The section in (**d**) is also stained with thioflavine-S to emphasize fibrillar Aβ present in the amyloid core (Reprinted from (60). Copyright 2003 with permission of Wiley-Liss, Inc., a subsidiary of John Wiley & Sons, Inc.).

An extensive comparison of both human AD brains and the PDAPP brain has shown structural similarities between the dystrophic neurites in both cases, including the presence in dystrophic neurites of dense multilaminar bodies, dense-core and clear synaptic vesicles, and neurofilament accumulation (26). Both astrogliosis and elevated levels of activated microglia were prominent in the regions with extensive Aβ immunoreactivity (8, 16, 26, 29). Interestingly, as in the human brain, gliosis was mostly centered around fibrillar plaques, not diffuse plaques (26). Finally, it should be noted that though extracellular Aβ accumulation is extensive in the PDAPP mice, intracellular tau pathology is minimal. Staining for phosphorylated tau and phosphorylated neurofilaments showed a very slight increase in a minority of dystrophic neurites as the PDAPP mice age (26, 30). Most dystrophic neurites in PDAPP mice do not contain abnormal tau immunoreactivity (26, 30). Paired helical filaments characteristic of neurofibrillary tangles have not been observed in the PDAPP brain (8, 26, 30), nor in other APP transgenic mice that develop Aβ deposits unless the human tau gene is also introduced into mice (31).

Pathological analysis of the PDAPP brain also revealed structural alterations in the neuronal parenchyma and synaptic network both before and after Aβ deposition occurs. Multiple studies have reported hippocampal atrophy present in the PDAPP mice as early as 3 months of age, thus before plaque accumulation occurs (21, 32, 33). Whether brain changes such as hippocampal atrophy, which are present in young PDAPP prior to Aβ aggregation, are due to overproduction of APP or APP fragments, soluble forms of Aβ, or another cause has not been clearly demonstrated. Hippocampal atrophy was particularly prominent in the dentate gyrus (33). High-resolution magnetic resonance spectroscopy illustrated that there is a 12% reduction in hippocampal volume in 100-day-old PDAPP mice in comparison to wild-type mice that remained constant as the mice age (33). However, hippocampal atrophy is not likely due to a decrease in neurons, as another study has shown that there is no neuronal loss in the entorhinal cortex, cingulate cortex, or CA1 region of the hippocampus (16). Whether or not synaptic density is altered in the PDAPP brain is still debatable, as a few studies looking at synaptophysin immunoreactivity have described either a decrease in synapses (8, 21), or no change (16). Another report showed that the level of cholinergic nerve terminals in the cerebral cortex was decreased in young PDAPP mice, yet analysis of these same mice did not show a loss of cholinergic cell bodies (34). Therefore, the possibility still remains that Aβ accumulation and deposition in the PDAPP brain either directly or indirectly causes modification of neuronal networks that may lead to some of the behavioral changes and synaptic dysfunction that exist in these mice, particularly as they age (see below).

4. Behavioral Deficits and Synaptic Dysfunction in the PDAPP Model

Behavioral analysis of the PDAPP mouse has revealed the presence of multiple age-dependent and independent deficits that exist in both learning and memory paradigms. Initial observations of these mice demonstrated that both heterozygous and homozygous animals exhibit significant impairment in a radial-arm maze task. A substantial increase in the number of reference and working memory errors occurred in 3, 6, and 9- to 10-month-old PDAPP mice in comparison to wild-type mice (35). In an object recognition task, homozygous 6-month-old mice, and both heterozygous and homozygous 9- to 10-month-old mice showed significant impairments indicative of an age-dependent decline in recognition performance (35). Combining both behavioral data and the pathological characterization of these mice, the authors of this study concluded that the decline in object recognition capability correlates with the progression of Aβ deposition, while spatial reference and working memory deficits are independent of Aβ accumulation and may relate to hippocampal atrophy and synaptic loss due to other reasons such as APP overexpression (21). Another study using a modification of the water maze in which the mice must learn successive platform locations to test working memory demonstrated that PDAPP mice have an age-independent deficit in learning the location of the first platform, and also an age-dependent deficit in learning later platform locations (36). This age-related decline in learning ability correlated with amyloid plaque load and was apparent using both a cross-sectional study of mice of various ages as well as using a longitudinal approach in which the same PDAPP mouse was tested repeatedly with age (36). Using a similar water maze test, another research team used swim distance as the measure of learning performance and showed that both young and old mice were significantly impaired in spatial learning (37). Importantly, there were both age-dependent and independent learning abnormalities. The age-dependent effects are probably related, at least in part, to Aβ accumulation. Other tests show that alterations occur in the PDAPP mice in behavioral patterns not related to cognition, including changes in core body temperature, sleep-wake states, and fear and eye-blink conditioning (32, 38, 39). Therefore, the behavioral phenotype present in PDAPP mice is complex with possibly a developmental defect that leads to learning and other impairments. This defect may be due to overproduction of APP or APP fragments.

A small amount of information has been published demonstrating changes in synaptic function in the PDAPP brain. Electrophysiological studies have shown that different changes in synaptic activity occur in young versus old PDAPP mice (40–42). Young mice (4- to 5 months old) had a greater level of paired-pulse facilitation (PPF) and long-term potentiation (LTP) that decayed at a much more rapid rate than in wild-type mice (40). In older mice (27- to 29 months old), there was also a reduction in PPF, but the difference in LTP in comparison to wild-type control mice was not as robust as in the younger mice. The synaptic response in older mice was also greatly reduced (40). Another study analyzed the extent of cholinergic dysfunction in PDAPP mice and found that basal acetylcholine (ACh) levels were lower in PDAPP mice (from 2 to 6 months of age) compared to nontransgenic mice (43). Also, significant alterations were found in the levels of ACh released following both pharmacological and environmental stimuli (43). Therefore, it appears that in the PDAPP mice both synaptic activity and the extent of neurotransmitter release are affected, though the relationship between these changes and behavior is not yet clear.

5. The PDAPP Model and the Study of Aβ-Binding Proteins

The PDAPP mouse model has played an important role in studying mechanisms underlying AD pathogenesis. In particular, the model has been utilized to examine the *in vivo* relationship between Aβ and its binding proteins, including apolipoprotein E (apoE), clusterin (also known as apolipoprotein J), and α-1-antichymotrypsin. In humans, apoE genotype is the most important genetic risk factor for AD (44, 45). PDAPP mice were crossbred with apoE$^{-/-}$ mice. PDAPP mice deficient in murine apoE had a dramatic dose-dependent reduction in the extent of Aβ immunoreactivity in both the hippocampus and cortex (46–49), with almost no Aβ deposition in the cortex of apoE$^{-/-}$ mice up to 22 months of age (see Fig. 2) (47). Interestingly, the brains of PDAPP/apoE$^{-/-}$ mice

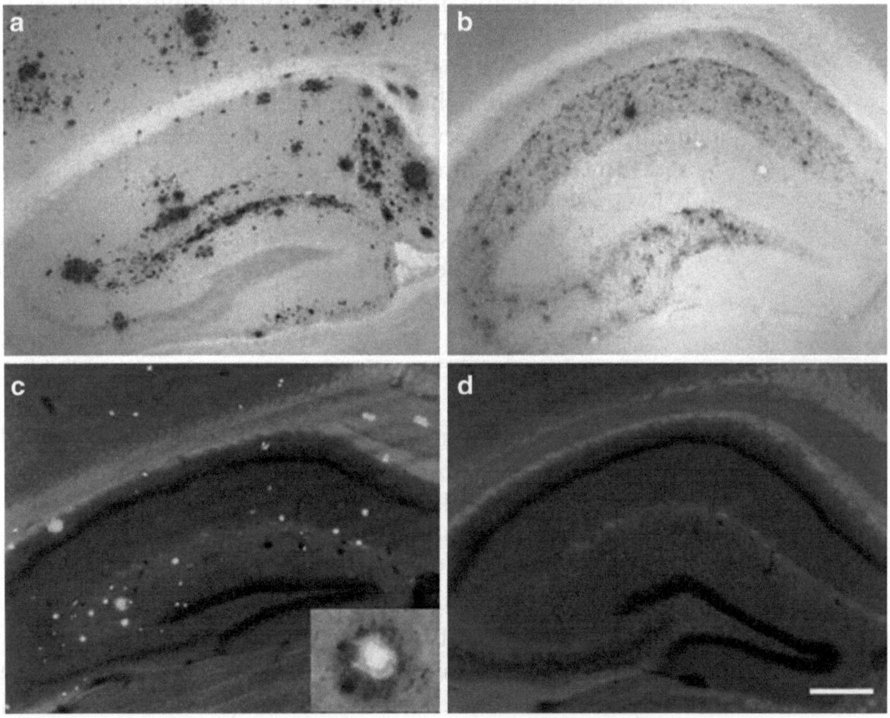

Fig. 2. The role of mouse apoE on Aβ and amyloid deposition. Brains from PDAPP mice that express murine apoE have significantly greater levels of Aβ deposition (**a**) and thioflavine-S stained plaques (**c**) in both the hippocampus and cortex in comparison to mice lacking apoE (**b**,**d**). PDAPP, apoE$^{-/-}$ mice are virtually deficient in thioflavine-S positive deposits (**d**). The inset in (**c**) shows a thioflavine-S plaque surrounded by dystrophic neurites (Reprinted with permission from Macmillan Publishers Ltd., Molecular Psychiatry, (61), 2002).

also showed a virtual complete absence of thioflavine-S-positive plaques up to 22 months of age (46, 47). The anatomical distribution of Aβ that did deposit in the hippocampus of the PDAPP/apoE$^{-/-}$ mice was different from PDAPP/apoE$^{+/+}$ mice, with most deposition occurring in the hilus of the dentate gyrus and very little in the molecular layer (48, 49). Also, very few neuritic plaques were found in the hippocampus of the PDAPP/apoE$^{-/-}$ mice (49), and almost no CAA or microhemorrhages were present in these mice through 24 months of age (19). Therefore, these studies demonstrate that murine apoE is necessary for the deposition of fibrillar Aβ that is present in both neuritic plaques and CAA in PDAPP mice.

The PDAPP mouse has also been crossed to clusterin$^{-/-}$ mice. Brains from PDAPP mice lacking clusterin did not show a change in the levels of Aβ deposition or in the age at which deposits occur (50). However, these brains do have a marked decrease in fibrillar plaque levels and neuritic dystrophy (50, 51). Interestingly, PDAPP mice deficient in both apoE and clusterin (apoE$^{-/-}$/clusterin$^{-/-}$) have earlier Aβ deposition and increased levels of both Aβ and amyloid in comparison to apoE$^{-/-}$ and clusterin$^{-/-}$ mice (51). The elimination half-life of ISF Aβ measured by *in vivo* microdialysis is also significantly decreased in the PDAPP/apoE$^{-/-}$/clusterin$^{-/-}$ mice and in PDAPP/apoE$^{-/-}$ mice (51). These findings suggest that apoE and clusterin are able to cooperatively regulate Aβ deposition, and that apoE plays a role in regulating the metabolism of extracellular Aβ. Finally, PDAPP mice crossed with α-1-antichymotrypsin transgenic mice had a greater plaque load and density than normal PDAPP mice, along with increased Aβ levels in the brain (52). Alpha-1-antichymotrypsin, therefore, may play a role in promoting Aβ deposition and fibrillogenesis in the PDAPP mice.

The effect of human apoE on Aβ metabolism has also been analyzed *in vivo* by crossing PDAPP mice with mice transgenic for the apoE isoforms (apoE2, apoE3, and apoE4) (48, 49, 53). Unlike the effect of murine apoE observed in the PDAPP mice, human apoE appears to delay Aβ deposition by approximately 9 months relative to when murine apoE is present (48, 49, 53). The apoE2, apoE3, and apoE4 PDAPP mice had no immunoreactive Aβ deposits in the hippocampus up to 12 months of age, and the total amount of Aβ measured biochemically was lower in apoE4 mice compared to both murine PDAPP/apoE$^{+/+}$ and PDAPP/apoE$^{-/-}$ mice (48, 49, 53). By age 15 months, however, both apoE4 and apoE3 expressing PDAPP mice did begin to develop both diffuse and thioflavine-S-positive deposits in a similar distribution as in brains with mouse apoE, though, to a lesser extent (49). Interestingly, apoE4 mice had a significantly greater amount of Aβ deposits and a higher density of thioflavine-S-positive plaques compared to apoE3 mice at similar ages (49). Though little data

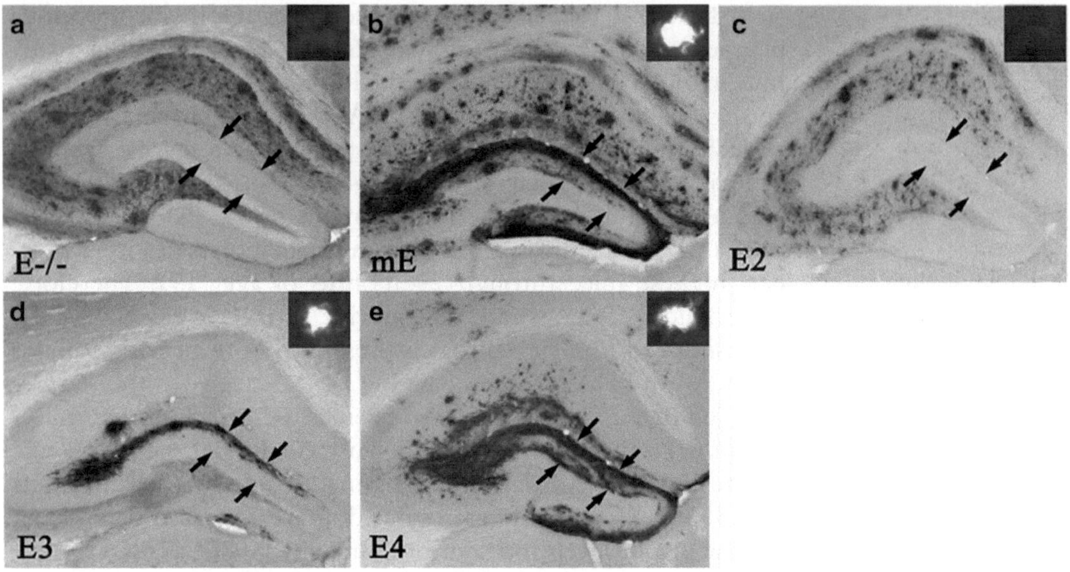

Fig. 3. Effect of human apoE expression on Aβ deposition in PDAPP mice. At 18 months of age, Aβ deposition is greater in the presence of apoE4 (**e**) in comparison to apoE3 (**d**). Both apoE3 and apoE4 mice also have less Aβ deposition than that observed for animals that express murine (m) apoE (**b**). Prominent Aβ deposition is seen in the molecular layer of the dentate gyrus (arrows). Mice that express apoE2 (**c**) have a pattern of deposition that resembles what is found in mice that lack apoE (**a**), with a notable lack of Aβ deposition in the molecular layer of the dentate gyrus. Unlike what is seen in apoE3 and apoE4 mice, Aβ deposits in the apoE2 mice do not stain with thioflavine-S (inset) (Reprinted from (53) with permission from Elsevier).

have been shown on the effect of apoE2, one study did show that in 18-month-old PDAPP transgenic mice expressing apoE2, there were solely thioflavine-S-negative deposits in a hippocampal distribution similar to the apoE$^{-/-}$ mice (53) (see Fig. 3). Therefore, in PDAPP mice, human apoE appears to effect the deposition of Aβ in an isoform-specific manner (E4 > E3 > E2).

6. The PDAPP Model and Alzheimer's Disease Therapeutics

Recently, both passive and active immunization targeting Aβ have gained attention as possible methods of treating AD. The PDAPP mouse was the first AD model used to study the efficacy of active Aβ immunization. In the first active immunization study, PDAPP mice were subcutaneously immunized with synthetic $A\beta_{42}$ peptide starting at 6 weeks of age and monthly thereafter. Remarkably, Aβ deposition was almost completely prevented by age 13 months (54). Also, mice immunized after the onset of Aβ deposition had a significant reduction in Aβ plaque load and progression (54). Passive immunization with anti-Aβ antibodies has also been studied in PDAPP mice. Several studies, in which mice were treated with distinct Aβ-specific antibodies, have shown that some antibodies are able to either prevent or reduce Aβ plaque load (13, 22, 55). Passive immunization has also been proven to improve memory and learning deficits in PDAPP mice in a manner that is not dependent on plaque reduction (37, 56). For example, administration of the anti-Aβ antibody m266 resulted in an improvement on both the object recognition task and the working memory task within 24–72 h after administration. These results suggest that certain forms of soluble Aβ may be responsible for these effects (56). Furthermore, a study using the PDAPP mouse has opened up the possibility of using anti-Aβ antibody administration as a method of determining Aβ deposition and plaque burden in the brain (57). Prior to the administration of the Aβ antibody m266, no correlation was seen between plasma Aβ levels and amyloid burden in the hippocampus of PDAPP mice. However, following parenteral administration of m266, a rapid increase in plasma Aβ was observed that correlated well with the amyloid plaque load in the brain. This study showed that by measuring Aβ plasma levels after m266 administration, it was possible to predict the relative amount of plaque burden in PDAPP mice that was then confirmed using immunohistochemistry and thioflavine-S staining. Based on these observations in the PDAPP mouse, anti-Aβ antibody administration may be worth studying in humans as a possible biomarker to assess the extent of amyloid plaque burden.

The PDAPP mouse has also played an important role in elucidating the various possible mechanisms responsible for antibody-mediated Aβ clearance. One hypothesis is that the antibodies directly cross the blood-brain-barrier and bind to Aβ, resulting in clearance of the peptide via an antibody mediated effect involving microglia. This postulation is supported by the observation that certain antibodies appear to decorate plaques in the PDAPP brain, and an ex vivo assay using microglia cells cultured with PDAPP brain sections showed cells with Aβ-containing phagocytic vesicles (55). Also, multiphoton microscopy analysis has been able to directly show the removal of Aβ plaques when an antibody is applied directly to the surface of the PDAPP mouse brain (27). Another hypothesis suggests that a nonmicroglia-mediated mechanism may play a role in amyloid removal by certain antibodies, as antibody fragments of one antibody were effective in clearing Aβ deposits without the need for microglial involvement (58). A third mechanism of antibody-mediated clearance may rely on the sequestration of soluble Aβ in the periphery and CNS, as the peripheral administration of the m266 antibody to the PDAPP mouse prevents the degradation of plasma Aβ and may increase the transport of Aβ from the CNS to the plasma (13, 56). Though it is still not entirely clear how Aβ immunization is able to result in clearance of Aβ deposits or improve behavior, studies using the PDAPP mouse have been able to highlight various mechanistic possibilities.

Finally, γ-secretase inhibitors have been shown to be a viable means of reducing Aβ levels in the brains of PDAPP mice. In one study, the efficacy of *N*-[*N*-(3,5-difluoropneacetyl)-L-alanyl]-*S*-phenylglycine *t*-butyl ester (DAPT) was analyzed *in vivo* using the PDAPP mouse (59). Subcutaneous injection of DAPT (100 mg/kg) led to peak levels of the compound in the brain after 3 h that were further sustained at a level greater than the IC_{50} for the first 18 h after injection. A reduction in cortical Aβ levels was observed that corresponded to DAPT levels in the brain, including a peak inhibition of Aβ of 40% occurring at 3 h after injection. DAPT treatment also exhibited a dose-dependent effect on the reduction of both total cortical Aβ and cortical $A\beta_{42}$ deposition, along with stabilization of the carboxy-terminal APP fragments. In another study using *in vivo* microdialysis in PDAPP mice, subcutaneous injection with the γ-secretase inhibitor LY411575 (3 mg/kg) led to peak drug concentrations in the brain at around 15 min after injection (15). This treatment led to an approximate 80% decrease in ISF Aβ after 6 h in both 3-month and 12-month-old animals. Both α- and β-C-terminal APP fragments were reduced in the mice treated with LY411575, demonstrating reduced levels of γ-secretase cleavage (15). These studies provide evidence for the rapid metabolism of soluble Aβ, and also highlight the usefulness of the PDAPP model in characterizing the effects of small molecule inhibitors that target APP cleavage events.

7. Summary

One of the initial challenges in studying AD *in vivo* was creating a mouse model that adequately modeled the disease process. The PDAPP mouse was the first hAPP-based transgenic mouse model of AD that successfully reproduced many of the pathological hallmarks of this disease, including Aβ deposition in both diffuse and neuritic plaques, astrocytosis, microgliosis, hippocampal atrophy, synaptic alterations, and behavioral deficits. Because many of these characteristics closely resemble the changes found in the human AD brain, the PDAPP mouse has proven to be an attractive model to study the pathogenesis underlying certain aspects of AD. This model has been extensively used to look at the mechanism underlying Aβ deposition in the brain, especially with regard to the role Aβ-binding proteins play in the process. Also, the PDAPP mouse has played a critical role in furthering the development of new diagnostic and therapeutic approaches towards treating AD, with particular emphasis on both antibody-mediated immunization techniques as well as the use of secretase inhibitors.

Acknowledgments

This work was supported by NIH grants AG13956 and P01-NS32636.

References

1. Quon D, Wang Y, Catalano R, Scardina JM, Murakami K, Cordell B (1991) Formation of beta-amyloid protein deposits in brains of transgenic mice. Nature 352:239–241
2. Sandhu FA, Salim M, Zain SB (1991) Expression of the human beta-amyloid protein of Alzheimer's disease specifically in the brains of transgenic mice. J Biol Chem 266: 21331–21334
3. Kammesheidt A, Boyce FM, Spanoyannis AF et al (1992) Deposition of beta/A4 immunoreactivity and neuronal pathology in transgenic mice expressing the carboxyl-terminal fragment of the Alzheimer amyloid precursor in the brain. Proc Natl Acad Sci U S A 89: 10857–10861
4. Lamb BT, Sisodia SS, Lawler AM et al (1993) Introduction and expression of the 400 kilobase amyloid precursor protein gene in transgenic mice [corrected]. Nat Genet 5:22–30
5. Pearson BE, Choi TK (1993) Expression of the human beta-amyloid precursor protein gene from a yeast artificial chromosome in transgenic mice. Proc Natl Acad Sci U S A 90: 10578–10582
6. Higgins LS, Holtzman DM, Rabin J, Mobley WC, Cordell B (1994) Transgenic mouse brain histopathology resembles early Alzheimer's disease. Ann Neurol 35:598–607
7. Mucke L, Masliah E, Johnson WB et al (1994) Synaptotrophic effects of human amyloid beta protein precursors in the cortex of transgenic mice. Brain Res 666:151–167
8. Games D, Adams D, Alessandrini R et al (1995) Alzheimer-type neuropathology in transgenic mice overexpressing V717F beta-amyloid precursor protein. Nature 373:523–527
9. Murrell J, Farlow M, Ghetti B, Benson MD (1991) A mutation in the amyloid precursor protein associated with hereditary Alzheimer's disease. Science 254:97–99
10. Sasahara M, Fries JW, Raines EW et al (1991) PDGF B-chain in neurons of the central nervous system, posterior pituitary, and in a transgenic model. Cell 64:217–227
11. Suzuki N, Cheung TT, Cai XD et al (1994) An increased percentage of long amyloid beta protein secreted by familial amyloid beta protein precursor (beta APP717) mutants. Science 264:1336–1340
12. Rockenstein EM, McConlogue L, Tan H, Power M, Masliah E, Mucke L (1995) Levels and alternative splicing of amyloid beta protein precursor (APP) transcripts in brains of APP transgenic mice and humans with Alzheimer's disease. J Biol Chem 270:28257–28267
13. DeMattos RB, Bales KR, Cummins DJ, Dodart JC, Paul SM, Holtzman DM (2001) Peripheral anti-A beta antibody alters CNS and plasma A beta clearance and decreases brain A beta burden in a mouse model of Alzheimer's disease. Proc Natl Acad Sci U S A 98:8850–8855
14. Johnson-Wood K, Lee M, Motter R et al (1997) Amyloid precursor protein processing and A beta42 deposition in a transgenic mouse model of Alzheimer disease. Proc Natl Acad Sci U S A 94:1550–1555
15. Cirrito JR, May PC, O'Dell MA et al (2003) In vivo assessment of brain interstitial fluid with microdialysis reveals plaque-associated changes in amyloid-beta metabolism and half-life. J Neurosci 23:8844–8853
16. Irizarry MC, Soriano F, McNamara M et al (1997) Abeta deposition is associated with neuropil changes, but not with overt neuronal loss in the human amyloid precursor protein V717F (PDAPP) transgenic mouse. J Neurosci 17:7053–7059
17. Su Y, Ni B (1998) Selective deposition of amyloid-beta protein in the entorhinal-dentate projection of a transgenic mouse model of Alzheimer's disease. J Neurosci Res 53: 177–186
18. DeMattos RB, Bales KR, Parsadanian M et al (2002) Plaque-associated disruption of CSF and plasma amyloid-beta (Abeta) equilibrium in a mouse model of Alzheimer's disease. J Neurochem 81:229–236
19. Fryer JD, Taylor JW, DeMattos RB et al (2003) Apolipoprotein E markedly facilitates age-dependent cerebral amyloid angiopathy and spontaneous hemorrhage in amyloid precursor protein transgenic mice. J Neurosci 23:7889–7896

20. Racke MM, Boone LI, Hepburn DL et al (2005) Exacerbation of cerebral amyloid angiopathy-associated microhemorrhage in amyloid precursor protein transgenic mice by immunotherapy is dependent on antibody recognition of deposited forms of amyloid beta. J Neurosci 25:629–636
21. Dodart JC, Mathis C, Saura J, Bales KR, Paul SM, Ungerer A (2000) Neuroanatomical abnormalities in behaviorally characterized APP(V717F) transgenic mice. Neurobiol Dis 7:71–85
22. Weiner HL, Lemere CA, Maron R et al (2000) Nasal administration of amyloid-beta peptide decreases cerebral amyloid burden in a mouse model of Alzheimer's disease. Ann Neurol 48:567–579
23. Fishman CE, Cummins DJ, Bales KR et al (2001) Statistical aspects of quantitative image analysis of beta-amyloid in the APP(V717F) transgenic mouse model of Alzheimer's disease. J Neurosci Methods 108:145–152
24. Reilly JF, Games D, Rydel RE et al (2003) Amyloid deposition in the hippocampus and entorhinal cortex: quantitative analysis of a transgenic mouse model. Proc Natl Acad Sci U S A 100:4837–4842
25. Bussiere T, Bard F, Barbour R et al (2004) Morphological characterization of Thioflavin-S-positive amyloid plaques in transgenic Alzheimer mice and effect of passive Abeta immunotherapy on their clearance. Am J Pathol 165:987–995
26. Masliah E, Sisk A, Mallory M, Mucke L, Schenk D, Games D (1996) Comparison of neurodegenerative pathology in transgenic mice overexpressing V717F beta-amyloid precursor protein and Alzheimer's disease. J Neurosci 16:5795–5811
27. Bacskai BJ, Kajdasz ST, Christie RH et al (2001) Imaging of amyloid-beta deposits in brains of living mice permits direct observation of clearance of plaques with immunotherapy. Nat Med 7:369–372
28. Brendza RP, Bacskai BJ, Cirrito JR et al (2005) Anti-Abeta antibody treatment promotes the rapid recovery of amyloid-associated neuritic dystrophy in PDAPP transgenic mice. J Clin Invest 115:428–433
29. Murphy GM Jr., Zhao F, Yang L, Cordell B (2000) Expression of macrophage colony-stimulating factor receptor is increased in the AbetaPP(V717F) transgenic mouse model of Alzheimer's disease. Am J Pathol 157: 895–904
30. Masliah E, Sisk A, Mallory M, Games D (2001) Neurofibrillary pathology in transgenic mice overexpressing V717F beta-amyloid precursor protein. J Neuropathol Exp Neurol 60:357–368
31. Lewis J, Dickson DW, Lin WL et al (2001) Enhanced neurofibrillary degeneration in transgenic mice expressing mutant tau and APP. Science 293:1487–1491
32. Weiss C, Venkatasubramanian PN, Aguado AS et al (2002) Impaired eyeblink conditioning and decreased hippocampal volume in PDAPP V717F mice. Neurobiol Dis 11:425–433
33. Redwine JM, Kosofsky B, Jacobs RE et al (2003) Dentate gyrus volume is reduced before onset of plaque formation in PDAPP mice: a magnetic resonance microscopy and stereologic analysis. Proc Natl Acad Sci U S A 100:1381–1386
34. German DC, Yazdani U, Speciale SG, Pasbakhsh P, Games D, Liang CL (2003) Cholinergic neuropathology in a mouse model of Alzheimer's disease. J Comp Neurol 462: 371–381
35. Dodart JC, Meziane H, Mathis C, Bales KR, Paul SM, Ungerer A (1999) Behavioral disturbances in transgenic mice overexpressing the V717F beta-amyloid precursor protein. Behav Neurosci 113:982–990
36. Chen G, Chen KS, Knox J et al (2000) A learning deficit related to age and beta-amyloid plaques in a mouse model of Alzheimer's disease. Nature 408:975–979
37. Hartman RE, Izumi Y, Bales KR, Paul SM, Wozniak DF, Holtzman DM (2005) Treatment with an amyloid-beta antibody ameliorates plaque load, learning deficits, and hippocampal long-term potentiation in a mouse model of Alzheimer's disease. J Neurosci 25: 6213–6220
38. Gerlai R, Fitch T, Bales KR, Gitter BD (2002) Behavioral impairment of APP(V717F) mice in fear conditioning: is it only cognition? Behav Brain Res 136:503–509
39. Huitron-Resendiz S, Sanchez-Alavez M, Gallegos R et al (2002) Age-independent and age-related deficits in visuospatial learning, sleep-wake states, thermoregulation and motor activity in PDAPP mice. Brain Res 928: 126–137
40. Larson J, Lynch G, Games D, Seubert P (1999) Alterations in synaptic transmission and long-term potentiation in hippocampal slices from young and aged PDAPP mice. Brain Res 840:23–35
41. Giacchino J, Criado JR, Games D, Henriksen S (2000) In vivo synaptic transmission in young and aged amyloid precursor protein transgenic mice. Brain Res 876:185–190
42. Sanchez-Alavez M, Gallegos RA, Kalafut MA, Games D, Henriksen SJ, Criado JR (2002)

Loss of medial septal modulation of dentate gyrus physiology in young mice overexpressing human beta-amyloid precursor protein. Neurosci Lett 330:45–48

43. Bales KR, Tzavara ET, Wu S et al (2006) Cholinergic dysfunction in a mouse model of Alzheimer disease is reversed by an anti-A beta antibody. J Clin Invest 116:825–832
44. Corder EH, Saunders AM, Strittmatter WJ et al (1993) Gene dose of apolipoprotein E type 4 allele and the risk of Alzheimer's disease in late onset families. Science 261:921–923
45. Strittmatter WJ, Saunders AM, Schmechel D et al (1993) Apolipoprotein E: high-avidity binding to beta-amyloid and increased frequency of type 4 allele in late-onset familial Alzheimer disease. Proc Natl Acad Sci U S A 90:1977–1981
46. Bales KR, Verina T, Dodel RC et al (1997) Lack of apolipoprotein E dramatically reduces amyloid beta-peptide deposition. Nat Genet 17:263–264
47. Bales KR, Verina T, Cummins DJ et al (1999) Apolipoprotein E is essential for amyloid deposition in the APP(V717F) transgenic mouse model of Alzheimer's disease. Proc Natl Acad Sci U S A 96:15233–15238
48. Holtzman DM, Bales KR, Wu S et al (1999) Expression of human apolipoprotein E reduces amyloid-beta deposition in a mouse model of Alzheimer's disease. J Clin Invest 103:R15–R21
49. Holtzman DM, Bales KR, Tenkova T et al (2000) Apolipoprotein E isoform-dependent amyloid deposition and neuritic degeneration in a mouse model of Alzheimer's disease. Proc Natl Acad Sci U S A 97:2892–2897
50. DeMattos RB, O'Dell M A, Parsadanian M et al (2002) Clusterin promotes amyloid plaque formation and is critical for neuritic toxicity in a mouse model of Alzheimer's disease. Proc Natl Acad Sci U S A 99:10843–10848
51. DeMattos RB, Cirrito JR, Parsadanian M et al (2004) ApoE and clusterin cooperatively suppress Abeta levels and deposition: evidence that ApoE regulates extracellular Abeta metabolism in vivo. Neuron 41:193–202
52. Nilsson LN, Bales KR, DiCarlo G et al. (2001) Alpha-1-antichymotrypsin promotes beta-sheet amyloid plaque deposition in a transgenic mouse model of Alzheimer's disease. J Neurosci 21:1444–1451
53. Fagan AM, Watson M, Parsadanian M, Bales KR, Paul SM, Holtzman DM (2002) Human and murine ApoE markedly alters A beta metabolism before and after plaque formation in a mouse model of Alzheimer's disease. Neurobiol Dis 9:305–318
54. Schenk D, Barbour R, Dunn W et al (1999) Immunization with amyloid-beta attenuates Alzheimer-disease-like pathology in the PDAPP mouse. Nature 400:173–177
55. Bard F, Cannon C, Barbour R et al (2000) Peripherally administered antibodies against amyloid beta-peptide enter the central nervous system and reduce pathology in a mouse model of Alzheimer disease. Nat Med 6:916–919
56. Dodart JC, Bales KR, Gannon KS et al (2002) Immunization reverses memory deficits without reducing brain Abeta burden in Alzheimer's disease model. Nat Neurosci 5:452–457
57. DeMattos RB, Bales KR, Cummins DJ, Paul SM, Holtzman DM (2002) Brain to plasma amyloid-beta efflux: a measure of brain amyloid burden in a mouse model of Alzheimer's disease. Science 295:2264–2267
58. Bacskai BJ, Kajdasz ST, McLellan ME et al (2002) Non-Fc-mediated mechanisms are involved in clearance of amyloid-beta in vivo by immunotherapy. J Neurosci 22:7873–7878
59. Dovey HF, John V, Anderson JP et al (2001) Functional gamma-secretase inhibitors reduce beta-amyloid peptide levels in brain. J Neurochem 76:173–181
60. Brendza RP, O'Brien C, Simmons K et al (2003) PDAPP; YFP double transgenic mice: a tool to study amyloid-beta associated changes in axonal, dendritic, and synaptic structures. J Comp Neurol 456:375–383
61. Brendza RP, Bales KR, Paul SM, Holtzman DM (2002) Role of apoE/Abeta interactions in Alzheimer's disease: insights from transgenic mouse models. Mol Psychiatry 7:132–135

Chapter 19

APP-Based Transgenic Models: The Tg2576 Model

Robert M.J. Deacon

Abstract

The search for a good animal, preferably mouse, model of Alzheimer's disease (AD) is one of the most imperative in present medical research, given the increasing prevalence of this disorder in an aging population and its enormous social, economic, and personal impact. In 1996, Karen Hsiao and colleagues showed that transgenic mice, which overexpress a human APP cDNA transgene from a Swedish family whose members showed early onset, familial AD (Tg2576 mice) were cognitively impaired at around 10 months of age and also showed pathophysiological characteristics of AD, notably deposits of amyloid. Much work followed this original publication, yet even now there is only limited consensus as to when, and to what extent, cognition is impaired in this model. This work is reviewed here, as well as research on the neuropathophysiological mechanisms by which the mice become cognitively impaired. The fundamental validity of the Tg2576 model and the role of β-amyloid in the disease processes seen in both Tg2576 mice and humans with AD are also considered.

Most research suggests that Tg2576 mice are moderately cognitively impaired by 12 months of age, and this laboratory therefore first started work on a cohort this age. Further work was done on cohorts aged 3, 9, and 21 months. Marked cognitive deficits were found in three paradigms: reference memory in a Y-maze motivated by escape from shallow water (paddling Y-maze); social memory for a juvenile mouse; and habituation to an open field over 24 h. Only the former has so far been shown unambiguously to be age-dependent. The latter two tests are simple to perform, so hold great promise as potential screens for new therapeutic agents. Reference memory in an appetitively motivated T-maze with both response and visual cues may also reveal cognitive impairments of an acceptable magnitude. However, the development of a new Tg2576 line based on a 129 mouse substrain with good behavioral characteristics may hold the greatest promise for future test development.

Key words: Tg2576, dementia, Alzheimer's disease, Mice, Transgenic, β-amyloid, Behavior

1. Introduction

In 1906, a German physician, Alois Alzheimer, presented the results of a postmortem examination of the brain of one of his patients, 55-year-old Auguste Deter, at a meeting in Tubingen. The following year, a publication followed, describing the characteristic

Peter Paul De Deyn and Debby Van Dam (eds.), *Animal Models of Dementia*, Neuromethods, vol. 48,
DOI 10.1007/978-1-60761-898-0_19, © Springer Science+Business Media, LLC 2011

pathology of the disease, which was named after him by Emil Kraepelin (1–3). Ninety years after the initial recognition of this disease, behavioral and pathophysiological results from the Tg2576 mouse model of Alzheimer's disease (AD) were reported (4). Since then, two major themes have developed. First, despite this model being over 12 years old, there is still disagreement as to when and to what extent cognitive impairments arise. Second, many other genetically modified models of AD have been developed, some of which may better model the disease and its neuropathology. This latter area will not be addressed in detail in the present chapter, which will concentrate on the behavioral and neuropathophysiological characteristics of the Tg2576 mouse, and discuss its validity as a model of AD. This mouse is also called APPsw, APP695SWE, or most completely Tg(HuAPP695. K670N/M671L)2576. A similar transgenic construct is known as the APP23 mouse; it shows greater overexpression of soluble β-amyloid (Aβ) than the Tg2576 mouse.

Sporadic AD accounts for the majority of Alzheimer's patients. The incidence increases with age, beginning around 60 (1%). By age 85, up to 30% of the population may be affected. Familial AD, around 10% of all AD cases, has an earlier onset and originates from genetic abnormalities. A germane example is a large Swedish family, which suffers from a double mutation in the gene expressing amyloid precursor protein (APP). This mutant gene was used to create the Tg2576 mouse model of AD.

Both types of AD are characterized by a progressive loss of memory and the ability to perform activities of daily living, such as making a meal, selecting the correct TV channel, telling the date, tying up shoe laces, and shopping. Only much later do motor abnormalities sometimes develop. Psychotic episodes and delusions also can occur, for example Alzheimer's first patient would sometimes scream that someone was trying to kill her. Examples of Capgras syndrome even happen, where the patient believes that a close relative, spouse, or friend is an impostor. Death is generally due to systemic illnesses such as pneumonia. Patients with AD may also be affected by vascular dementia caused by multiple small strokes.

The pathological hallmarks of AD are intercellular amyloid plaques (especially dense cored ones, which are unique to AD) and intracellular neurofibrillary tangles formed by hyperphosphorylation of an intracellular protein called tau, which is associated with microtubules. Neuronal loss and brain atrophy are also characteristic of this disease. Indeed, the progressive thinning of the medial temporal lobe, as measured by imaging techniques, is closely correlated with disease severity (5, 6).

For more information and reviews of AD and its treatment, see (7–10).

2. Validity of the Model

This mouse bears the "Swedish" familial AD mutation, in which there is a double amino acid substitution in APP, Lys 670 to Asn and Met 671 to Leu. This results in overproduction of Aβ (11), and several changes characteristic of AD, such as plaques, increased levels of both soluble and insoluble Aβ, and abnormal synaptic plasticity (4, 12). Microglia are activated (13), which leads to inflammation (14). In addition, oxidative stress is increased (15).

However, many features of AD are not faithfully reproduced. Notably, there is relatively little change in the cholinergic system (16), although it has been reported that cholinergic regulation of hippocampal neuron excitability was impaired in Tg2576 mice (17), and cholinergic neurotransmission was reduced (18). Moreover, Aβ is a powerful activator of α7 nicotinic cholinergic receptors (19). Increased glycosylation of acetylcholinesterase was seen in Tg2576 mice at 4, 8, and 12 months of age (20). Impaired REM sleep was seen in Tg2576 mice at 6 but not 2 months, correlating with a reduction in the numbers of pedunculopontine tegmentum choline acetyltransferase-positive neurons (21). This reduction was selective; however, other cholinergic groups involved in REM sleep were unchanged.

Neurofibrillary tangles are not seen in Tg2576 mice, and little neuronal loss or gross brain atrophy occurs (22). The amyloid peptides found in Tg2576 mice are also chemically and physically different from those found in AD (23).

Moreover, familial AD is a small proportion (10% or less) of the total number of cases of AD, the majority being of the sporadic type. Thus, the Tg2576 mouse, derived from a familial AD model, cannot identically replicate most forms of AD. However, since excess Aβ production and deposition always occurs in AD, the mechanism by which this occurs may not be crucial.

A convincing model of AD, in terms of face validity, should show age-related changes in both cognition and neuropathology, and these should be temporally correlated. There seems to be better agreement as to when changes in neuropathology occur than when behavior changes; most researchers report significant neuropathology by 10–12 months of age (4, 24, 25).

2.1. Is Amyloid Causing the Behavioral Changes, and If So, in What Form?

The Tg2576 mouse was originally created to test the "amyloid hypothesis," namely that the deficits in AD are due to amyloid (rather than, for example, neurofibrillary tangles related to hyperphosphorylation of tau protein) (9). Originally, the amyloid being considered was in the form of plaques, which are easily identifiable at the histological level. Indeed, along with neurofibrillary tangles, these were the classical markers of AD established by Alois Alzheimer in 1907 (1). Plaques can exist as both diffuse and dense-cored forms. However, this hypothesis has fundamental flaws and has been heavily criticized (9); amyloid plaque load does not correlate with cognitive decline in AD (26, 27), nor does plaque count correlate with cognitive deficits in mouse AD models (28, 29), although see (30). Indeed, deficits in contextual fear conditioning have been seen in Tg2576 mice at 5 months of age (19, 31–33), when Aβ levels are elevated but plaques are not yet present (4).

At first, fibrillar forms of Aβ were thought to be responsible for AD pathology (34–36), although McLean et al. suggested that they might be soluble forms (37).

Increasingly, accumulating evidence suggested that soluble rather than aggregated plaque amyloid might be primarily involved in AD (38–40). Westerman and colleagues were the first group to publish *in vivo* evidence that smaller forms of amyloid were involved in cognitive decline in Tg2576 mice (41). Hardy and Selkoe (42) noted that the extent of dementia in AD correlated better with amyloid assayed biochemically (which also detects soluble forms of amyloid) than with histologically determined plaque counts. Palop et al. (43) found that cognitive impairments correlated with the relative abundance of Aβ, but this did not depend on the amount deposited as plaques. At present, oligomeric and soluble forms of Aβ are the prime candidates for causing the cognitive alterations in AD and the mouse models of it. Soluble Aβ is present as early as 4 months in the Tg2576 model (20).

There is increasing evidence that oligomeric (trimers to dodecamers) forms of Aβ are primarily responsible for its toxicity, rather than the insoluble fibrils that are found in amyloid senile plaques (44, 45). Deposits of plaques and insoluble forms of amyloid may be merely end stages in a degenerative process that is initially started by oligomeric forms of amyloid (45–47). Oligomeric rather than monomeric or fibrillar Aβ inhibits the activity of calcineurin, a phosphatase that is elevated and correlated with memory dysfunction in Tg2576 mice, due to inhibition of the transcription factor CREB (44). All soluble amyloid oligomers share a common structure, which probably mediates their toxicity, since an oligomer-specific antibody blocks their action (46, 48). The distribution of soluble amyloid oligomers in human AD brain is distinct to that of fibrillar amyloid. In control, nondemented brains with cored plaques, oligomer-specific antibody staining was almost absent (48). The key element in Tg2576 mouse cognitive deterioration may be a 56-kDa soluble Aβ oligomer, the concentration of which closely temporally mirrors the behavioral impairments (49). These authors noted that (in their work; but see later) the cognitive impairment in a Morris water maze first became evident at 6 months of age, yet remained stable for months afterwards, a time when the concentration of all other species of amyloid was increasing rapidly. However, Dineley et al. (33) noted that the 56-kDa oligomer is present at 6 months in Tg2576 mice but absent 1 month earlier, although deficits in contextual fear conditioning existed at this time point.

Further evidence that amyloid, or at least in most cases the insoluble fibrillar forms, may not be crucial in cognitive impairment is that the acetylcholinesterase inhibitors, physostigmine and donepezil, improved cognition without affecting Aβ deposition. Although donepezil may influence Aβ characteristics and effects under some circumstances, in the latter experiment it appeared not to (50). Treatment with Gingko biloba reduced spatial learning, and memory deficits yet had no effect on the deposition of hippocampal Aβ plaques (51). Dietary supplementation with the antioxidant α-lipoic acid produced improvements in cognition in treated Tg2576 mice without affecting levels of soluble or insoluble amyloid. (52). Moreover, levels of soluble Aβ are substantially elevated in the brains of Tg2576 mice by the age of 6–8 months, when substantial cognitive changes are often not reported, while by 10–11 months, amyloid plaques have started to form (4) and most researchers report behavioral impairments by then. At 16 months, there is a 4–8% amyloid burden in hippocampal regions of the Tg2576 mouse brain (53), which is a similar amyloid burden to that found in human AD brains, and almost all researchers agree that the mice are then cognitively impaired.

However, chronic nicotine treatment reduced amyloid load in Tg2576 mice but only the insoluble forms, including plaques, were affected; soluble amyloid was not reduced (54). Unfortunately, no behavioral work was done to see if nicotine reduced cognitive deficits.

Further evidence that Aβ is not crucial to the memory deficits in mouse models of AD is that vaccinated PDAPP mice improve in both an object recognition task and a holeboard learning and memory task without a change in amyloid load (55). Tg2576 mice lacking the catalytic subunit Nox2 of NADPH oxidase did not show the normal oxidative stress, cerebrovascular dysfunction, or behavioral deficits normally seen at 12–15 months, yet these improvements occurred without reductions in brain Aβ levels or amyloid plaques (56).

However, some reports suggest that amyloid burden is related to cognitive deterioration in Tg2576 mice. Asuni et al. (57) showed that vaccination of Tg2576 mice with an Aβ derivative from age 11 to 24 months reduced amyloid burden by approximately 30%, while cognitive performance on the radial maze and a modified version of the Hebb-Williams maze improved to wild-type levels. Although this overall result suggests a causative role for Aβ in cognitive impairment, this was not supported by the fact that cognitive performance was not correlated with reductions in amyloid levels between individual mice. The authors suggested that oligomeric forms of Aβ might be the moiety responsible for cognitive impairment. Jensen et al. (58), working on the APP/PS1 double transgenic mouse model, also suggested that their amyloid vaccine attenuated the accumulation of small oligomeric forms of amyloid. They obtained ambiguous results, in that, while chronic immunotherapy reduced cognitive impairment at 15–16.5 months of age, there was no decrease in (or correlation to) amyloid deposition. Yet, factor analysis showed that brain amyloid deposition measures loaded heavily with key cognitive measures. Morgan et al. (59) also showed a partial reduction in amyloid burden and improved cognition when Tg2576 mice were vaccinated with Aβ. Although an overall decrease in Aβ was not seen following treatment with an Aβ antibody, the inverse relationship between amyloid load and cognitive function was eliminated, suggesting that the two were related (60). In an experiment comparing immunization of Tg2576 mice with two different Aβ derivatives, cognition on a radial arm maze was improved in both groups but only one showed a reduction in amyloid plaques (61). The phosphodiesterase-4 inhibitor, rolipram, attenuated impaired contextual fear conditioning at 5 months of age in Tg2576 mice, but the authors concluded that the effect of rolipram was independent of amyloid as rolipram-treated controls also improved (31). However, the γ-secretase inhibitor *N*-[*N*-(3,5-difluorophenacetyl)-Lalanyl]-S-phenylglycine-t-butylester also attenuated learning impairment but this was not seen in the controls and was accompanied by reduced brain Aβ load. Exercise (3 weeks on running wheels) produced significant decreases in soluble Aβ40 and soluble fibrillar Aβ in 17- to 19-month-old Tg2576 mice (62). Exercise has been reported to delay the onset of dementia and AD in people over 65 (63).

2.2. At What Age Does Cognitive Impairment Occur in the Tg2576 Mouse?

The Tg2576 mouse shows an impairment in cognition, which is similar to, but (at least initially) less severe than, that caused by complete hippocampal lesions (64). In line with the role of the hippocampus in spatial learning and memory (65), and the extensive Aβ deposition occurring in this brain region in Tg2576 mice (66), it is spatial forms of learning that are primarily affected. Indeed, some forms of memory are preserved (33, 67–69) or even enhanced in Tg2576 mice, probably due to elimination of competing hippocampally driven response tendencies (70). King et al. (69) found no impairment in active or passive avoidance at 3 and 9 months; in our group, no impairment in passive avoidance was seen even at 22 months (71). However, King et al. also failed to detect impairments in water maze acquisition at 9 months, a time when most groups find deficits. In a version of the Barnes maze, they found impairments between wild-type and Tg2576 mice as early as 3 months in females, but neither males nor females were impaired at 9 months in this paradigm. This ambiguity may have arisen because this paradigm does not show very steep learning curves.

The Morris water maze is a standard test of spatial memory. Hsiao et al. (4) found that Tg2576 mice were impaired at 9–10 months, while Westerman et al. (41) and Lesne (49) found impairment at 6 months. King et al. (69) observed deficits in water maze retention at 3 months of age; Arendash et al. (72) showed a comparable water maze deficit in 5-month-old Tg mice. Arendash and King (73), however, failed to find any deficit on water maze acquisition and retention from 3 months to 19 months. King et al. (74) observed deficits on a visible (i.e. nonspatial) water maze in 9-month-old mice, while Horgan et al. found impairment in this paradigm at 6 months. Using a circular platform maze derived from the Barnes maze, where a mouse can escape from an aversive environment through a hole in the floor to a safe refuge, Pompl et al. showed a deficit in Tg2576 mice at 7 months of age, but only during the reversal learning phase of the test (75).

Y-maze alternation is another test sensitive to cognitive changes (despite its drawbacks) (71). As for the water maze, Hsiao et al. (4) found that Tg2576 mice were impaired at 9–10 months; Ognibene et al. (76) at 7–12 months, while Arendash and King (73) even found decreased alternation at 3 months. King et al. (69), however, found no difference between transgenics and controls on Y-maze alternation at 3 and 9 months. Another type of Y-maze paradigm is that devised by Dellu et al. (77); if mice are allowed to explore only two arms of a transparent Y-maze, they prefer to visit the novel arm when retested 30 min later. Park et al. (56) found that whereas there was no difference in novel arm recognition between young (3–4 months) wild-type and Tg2576 mice, a deficit had arisen by 12–15 months. In an operant task, performance impairment was seen at 7–8 months (78). Impaired REM sleep was seen in Tg2576 mice by 6 months (21) and accompanied by selective cholinergic neuronal loss in the tegmental area. Impaired contextual fear conditioning was impaired at 5 months (33). This was unlikely due to the increased activity sometimes seen in Tg2576 mice, as conditioning to an acoustic cue was unaffected. Comery et al. (31) also found impaired learning of this paradigm at 5 months of age. Increased impulsivity, which may model secondary aspects of AD, was seen in 12-month-old Tg2576 mice (79).

2.3. Behavioral Work Done in Oxford on the Tg2576 Mouse

In work done in the author's laboratory at Oxford, both age-related and nonage-related behavioral changes were found in Tg2576 mice (71). Mice were supplied by Boehringer Ingelheim, the offspring of Tg2576 hybrid BL6/SJL male mice crossed with F1B6/SJL females. The Tg2576 mice failed to show a deficit on many of the cognitive tests we performed, even at 12 or 21–23 months of age when they should have been impaired. This is not unusual, as some reports show intact learning and memory even at 15 months (80). In some of our tests, however, even the controls failed to show good learning. These tests have been extensively validated in C57BL/6 mice (64, 81–84). This poor performance might have been due to insufficient backcrossing into C57BL/6 mice. Surprisingly, in spite of the wild-type and Tg2576 performance being very poor on the spontaneous alternation T-maze task, pure SJL mice alternated well.

The three cognitive tests, which showed a notable impairment in mice older than 9 months, were social memory for a juvenile, open field habituation, and a Y-maze escape task motivated by shallow water (the "paddling Y-maze") (71, 85). The latter was age-related, deficits appearing at 10 months; it is not yet known if the former two are. A reference memory test in a T-maze cued by both position and objects (86) showed a potential age-related deficit, but as only two 23-month-old mice were tested, further work needs to be done to confirm impairment at later ages; at 7–8 months, no difference between Tg2576 mice and controls was seen.

Tests of species-typical behaviors showed that Tg2576 mice had a deficit in burrowing behavior at all ages. Burrowing is a newly developed test in which mice spontaneously empty a tube filled with e.g. food pellets (83, 87). Burrowing deficits were present as early as 3 months of age, before extensive Aβ deposition is present in the brain. Burrowing is known to be an extremely sensitive test; it is affected by a dose of lipopolysaccharide three orders of magnitude below that necessary to cause a change in body temperature (88). An age-independent deficit was also seen in nest construction (89), even as early as 3 months, but only when mice were group-housed; most individually housed mice in either group made reasonable nests. This may have been because an individual mouse needs to make a nest more than group housed mice because it has no cagemates to huddle with. Burrowing and nesting both detect the effects of prion (scrapie) infection at a similar early time in the course of the disease (81). It would be extremely interesting to discover whether there is an age at which burrowing and nesting are not affected, to determine whether the effects at 3 months, when no other behavioral deficits exist, is due to task sensitivity or an innate effect of the genotype, probably unrelated to Aβ.

2.4. The Future

The background strain can markedly influence the behavioral expression of a mutation. It has been shown that hybrid 129/Sv × C57BL/6 mice perform better than either of the parental strains in a water maze (90). An interesting development is the new

Tg2576 mouse from Taconic farms (model number 002789) (http://www.taconic.com/wmspage.cfm?parm1=941), which is backcrossed at least 16 times on to one of the behaviorally superior 129 strains (129S6/SvEv). This 129 substrain performs well in the water maze, especially reversal learning (91), although it has a more anxious phenotype than the C57BL/6 mouse (92). It is also wholly pigmented, whereas the mixed B6/SJL genetic background of model 001349 (traditional Tg2576 mice) can produce pink-eyed animals or homozygosity for the Pde6brd1 retinal degeneration mutation.

In conclusion, if I were a clinician, would I prescribe a drug that had been shown to attenuate behavioral deficits in year-old Tg2576 mice? Certainly, not least because there would be few alternatives, but I would consider the chances of a successful outcome to be no more than 50% at best. This mouse has been highly productive in AD research, but there may well be better alternatives.

Acknowledgment

R. Deacon is supported by grant GR065438MA from the Wellcome Trust to the Oxford OXION group.

References

1. Alzheimer A (1907) Uber eine eigenartige Erkrankung der Hirnrinde. Allgemeine Z Psychiatrie Psychisch-Gerichtliche Med 64: 146–148
2. Small DH, Cappai R (2006) Alois Alzheimer and Alzheimer's disease: a centennial perspective. J Neurochem 99:708–710
3. Alzheimer A, Stelzmann RA, Schnitzlein HN, Murtagh FR (1995) An English translation of Alzheimer's 1907 paper, 'Uber eine eigenartige Erkankung (sic) der Hirnrinde'. Clin Anat 8:429–431
4. Hsiao K, Chapman P, Nilsen S et al (1996) Correlative memory deficits, Abeta elevation, and amyloid plaques in transgenic mice. Science 274:99–102
5. Smith D (2002) Imaging the progression of Alzheimer pathology through the brain. Proc Natl Acad Sci 99 4135–4137
6. Rusinek H, Endo Y, De Santi S et al (2004) Atrophy rate in medial temporal lobe during progression of Alzheimer disease. Neurology 63:2354–2359
7. Melnikova I (2007) Therapies for Alzheimer's disease. Nature Rev Drug Discov 6:341–2
8. Cummings JL (2004) Alzheimer's Disease. N Engl J Med 351:56–67
9. Marx J (2007) A new take on tau. Science 316:1416–1417
10. Emilien G, Durlach C, Minaker K et al (eds) (2004) Alzheimer disease: neuropsychology and pharmacology. Birkhäuser, Basel
11. Duff K, Suleman F (2004) Transgenic mouse models of Alzheimer's disease: How useful have they been for therapeutic development? Brief Funct Genomic Proteomic 3:47–59
12. Chapman PF, White GL, Jones MJ et al (1999) Impaired synaptic plasticity and learning in aged amyloid precursor protein transgenic mice. Nat Neurosci 2:271–276
13. Frautschy SA, Yang F, Irrizzary, M et al (1998) Microglial response to amyloid plaques in APPsw transgenic mice. Am J Pathol 152: 307–317
14. Benzing WC, Wujek JR, Ward EK et al (1999) Evidence for glial-mediated inflammation in APPsw transgenic mice. Neurobiol Aging 20: 581–590
15. Smith, M.A, Hirai, K, Hsiao K et al (1998) Amyloid-ß deposition in Alzheimer transgenic

mice is associated with oxidative stress. J Neurochem 70:2212–2215

16. Gau J-T, Steinhilb ML, Kao T-C et al (2002) Stable β-secretase activity and presynaptic cholinergic markers during progressive central nervous system amyloidogenesis in Tg2576 mice. Am J Pathol 160:731–738
17. Ohno M, Sametsky EA, Younin LH et al (2004) BACE1 deficiency rescues memory deficits and cholinergic dysfunction in a mouse model of Alzheimer's disease. Neuron 41:27–33
18. Apelt J, Kumar A, Schliebs R (2002) Impairment of cholinergic neurotransmission in adult and aged transgenic Tg2576 mouse brain expressing the Swedish mutation of human β-amyloid precursor protein. Brain Res 953:17–30
19. Dineley KT, Bell KA, Bui D, Sweatt JD (2002) Beta-amyloid peptide activates alpha 7 nicotinic acetylcholine receptors expressed in Xenopus oocytes. J Biol Chem 277:25056–6
20. Fodero LR, Sáez-Valero J, McLean CA et al (2002) Altered glycosylation of acetylcholinesterase in APP (SW) Tg2576 transgenic mice occurs prior to amyloid plaque deposition. J Neurochem 81:441–448
21. Zhang B, Veasey SC, Wood MA et al (2005) Impaired rapid eye movement sleep in the Tg2576 APP murine model of Alzheimer's disease with injury to pedunculopontine cholinergic neurons. Am J Pathol 167:1361–1369
22. Irizarry MC, McNamara M, Fedorchak K et al (1997) APPSW transgenic mice develop age-related Aβ deposits and neuropil abnormalities, but no neuronal loss in CA1. J Neuropathol Exp Neurol 56:965–973
23. Kalback W, Watson MD, Kokjohn, TA et al (2002) APP transgenic mice Tg2576 accumulate Aβ peptides that are distinct from the chemically modified and insoluble peptides deposited in Alzheimer's disease senile plaques. Biochemistry 41:922–928
24. Kawarabayashi T, Younkin LH, Saido TC et al (2001) Age-dependent changes in brain, CSF, and plasma amyloid (beta) protein in the Tg2576 transgenic mouse model of Alzheimer's disease. J Neurosci 21:372–381
25. Praticò D, Uryu, K, Leight, S, Trojanowski JQ, Lee, VM-Y (2001) Increased lipid peroxidation precedes amyloid plaque formation in an animal model of Alzheimer amyloidosis. J Neurosci 21:4183–4187
26. Nagy Z, Esiri MM, Jobst KA et al (1995) Relative roles of plaques and tangles in the dementia of Alzheimer's disease: correlations using three sets of neuropathological criteria. Dementia 6:21–31
27. Terry RD, Masliah E, Hansen LA (1999) The neuropathology of Alzheimer disease and the structural basis of its cognitive alterations. In: Terry RD, Katzman R, Bick KC, Sisodia SS (eds.) Alzheimer disease, 2nd edn. Lippincott Williams Wilkins, Philadelphia, pp. 187–206
28. Moechars D, Dewachter I, Lorent K et al (1999) Early phenotypic changes in transgenic mice that over-express different mutants of amyloid precursor protein. J Biol Chem 274:6483–6492
29. Hsia AY (1999) Plaque-independent disruption of neural circuits in Alzheimer's disease mouse modes. Proc Natl Acad Sci U S A 96: 3228–3233
30. Stern EA, Backsai BJ, Hickey GA et al (2004) Cortical synaptic integration in vivo is disrupted by amyloid-β plaques. J Neurosci 24: 4535–4540
31. Comery TA, Martone RL, Aschmies S et al (2005) Acute gamma-secretase inhibition improves contextual fear conditioning in the Tg2576 mouse model of Alzheimer's disease. J Neurosci 25:8898–8902
32. Corcoran KA, Lu Y, Turner RS, Maren S (2002) Overexpression of hAPPswe impairs rewarded alternation and contextual fear conditioning in a transgenic mouse model of Alzheimer's disease. Learn Mem 9:243–252
33. Dineley KT, Hogan D, Zhang WR et al (2007) Acute inhibition of calcineurin restores associative learning and memory in Tg2576 APP transgenic mice. Neurobiol Learn Mem 88 (2):217–224
34. Lorenzo A, Yanker BA (1994) β-amyloid neurotoxicity requires fibril formation and is inhibited by congo red. Proc Natl Acad Sci U S A 91:12243–12247
35. Hartley DM (1999) Protofibrillar intermediates of amyloid β-protein induce acute electrophysiological changes and progressive neurotoxicity in cortical neurons. J Neurosci 19:8876–8884
36. Walsh DM, Hartley DM, Kusumoto Y et al (1999) Amyloid β-protein fibrillogenesis. Structure and biological activity of protofibrillar intermediates. J Biol Chem 274:25945–25952
37. McLean CA, Cherny RA, Fraser FW et al (1999) Soluble pool of Aβ amyloid as a determinant of severity of neurodegeneration in Alzheimer's disease. Ann Neurol 46: 860–866
38. Ashe KH (2001) Learning and memory in transgenic mice modeling Alzheimer's disease. Learn Mem 8:301–308
39. Morris R, Mucke L (2006) Alzheimer's disease: a needle from the haystack. Nature 440:284–285
40. Klein WL, Krafft GA, Finch CE (2001) Targeting small Aβoligomers: the solution to

an Alzheimer's disease conundrum? Trends Neurosci 24:219–224
41. Westerman MA, Cooper-Blacketer D, Mariash A et al (2002) The relationship between Abeta and memory in the Tg2576 mouse model of Alzheimer's disease. J Neurosci 22:1858–1867
42. Hardy J, Selkoe DJ (2002) The amyloid hypothesis of Alzheimer's disease: progress and problems on the road to therapeutics. Science 297:353–356
43. Palop JJ, Jones B, Kekonius L et al (2003) Neuronal depletion of calcium-dependent proteins in the dentate gyrus is tightly linked to Alzheimer's disease-related cognitive deficits. Proc Natl Acad Sci U S A 100:9572–9577
44. Reese LC, Zhang W, Dineley KT et al (2008) Selective induction of calcineurin activity and signaling by oligomeric amyloid beta. Aging cell 7:824–835
45. Lambert MP, Barlow AK, Chromy BA et al (1998) Diffusible, nonfibrillar ligands derived from Abeta1–42 are potent central nervous system neurotoxins. Proc Natl Acad Sci U S A 95:6448–6453
46. Carrell RW, Mushunjea A, Zhoua A (2008) Serpins show structural basis for oligomer toxicity and amyloid ubiquity. FEBS Lett 582:2537–2541
47. Glabe CG (2006) Common mechanisms of amyloid oligomer pathogenesis in degenerative disease. Neurobiol Aging 27:570–575
48. Kayed R, Head E, Thompson JL et al (2003) Common structure of soluble amyloid oligomers implies common mechanism of pathogenesis. Science 300:486–489
49. Lesne S, Koh MT, Kotilinek L et al (2006) A specific amyloid-beta protein assembly in the brain impairs memory. Nature 440:352–357
50. Dong H, Csernansky CA, Martin MV et al (2005) Acetylcholinesterase inhibitors ameliorate behavioral deficits in the Tg2576 mouse model of Alzheimer's disease. Psychopharmacology 181:145–152
51. Stackman RW, Eckenstein F, Frei B et al (2003) Prevention of age-related spatial memory deficits in a transgenic mouse model of Alzheimer's disease by chronic Ginkgo biloba treatment. Exp Neurol 184:510–520
52. Quinn JF, Bussiere JR, Hammond RS et al (2007) Chronic dietary α-lipoic acid reduces deficits in hippocampal memory of aged Tg2576 mice. Neurobiol Aging 28:213–225
53. Irizarry MC, Locascio JJ, Hyman BT (2001) β-Site APP cleaving enzyme mRNA expression in APP transgenic mice. Am J Pathol 158:173–177
54. Nordberg A, Hellstrom-Lindahl E, Lee M et al (2002) Chronic nicotine treatment reduces b-amyloidosis in the brain of a mouse model of Alzheimer's disease (APPsw). J Neurochem 81:655–658
55. Dodart JC, Bales KR, Gannon KS et al (2002) Immunization reverses memory deficits without reducing brain Abeta burden in Alzheimer's disease model. Nat Neurosci 5: 452–457
56. Park L, Zhou P, Pitstick R et al (2008) Nox2-derived radicals contribute to neurovascular and behavioral dysfunction in mice overexpressing the amyloid precursor protein. Proc Natl Acad Sci U S A 105:1347–1352
57. Asuni AA, Boutajangout A, Scholtzova H (2006) Vaccination of Alzheimer's model mice with Aβ derivative in alum adjuvant reduces Aβ burden without microhemorrhages. Eur J Neurosci 24:2530–2542
58. Jensen MT, Mottin MD, Cracchiolo JR et al (2005) Lifelong immunization with human β-amyloid (1–42) protects Alzheimer's transgenic mice against cognitive impairment throughout aging. Neuroscience 130:667–684
59. Morgan D, Diamond DM, Gottschall PE et al (2000) A beta peptide vaccination prevents memory loss in an animal model of Alzheimer's disease. Nature 408:982–985
60. Kotilinek LA, Bacskai B, Westerman M et al (2002) Reversible memory loss in a mouse transgenic model of Alzheimer's disease. J Neurosci 22:6331–6335.
61. Sigurdsson EM, Knudsen E, Asuni, A et al (2004) An attenuated immune response is sufficient to enhance cognition in an Alzheimer's disease mouse model immunized with amyloid-β derivatives. J Neurosci 24:6277–6282
62. Nichol KE, Poon WW, Parachikova A et al (2008) Exercise alters the immune profile in Tg2576 Alzheimer mice toward a response coincident with improved cognitive performance and decreased amyloid. J. Neuroinflamm 5:13 doi:10.1186/1742-2094-5-13
63. Larson EB, Wang L, Bowen JD et al (2006) Exercise is associated with reduced risk for incident dementia among persons 65 years of age and older. Ann Intern Med 144:73–81
64. Deacon RMJ, Bannerman DM, Kirby BP et al (2002) Effects of cytotoxic hippocampal lesions in mice on a cognitive test battery. Brain Res 133:57–68
65. O'Keefe J, Nadel L (1978) The hippocampus as a cognitive map. Clarendon Press, Oxford
66. Lehman EJ, Kulnane LS, Lamb BT (2003) Alterations in β-amyloid production and deposition in brain regions of two transgenic models. Neurobiol Aging 24:645–653
67. Hale G, Good M (2005) Impaired visuospatial recognition memory but normal object

novelty detection and relative familiarity judgments in adult mice expressing the APPswe Alzheimer's disease mutation. Behav Neurosci 119:884–891

68. McNee M, McArthur R, Goodwin A (2000) Y-maze odor discrimination in Tg_{2576} APP-overexpressing transgenic mice. Poster presented at measuring behavior 2000, 3rd international conference on methods and techniques in behavioral research, Nijmegen
69. King DL, Arendash GW, Crawford F et al (1999) Progressive and gender-dependent cognitive impairment in the APP_{SW} transgenic mouse model for Alzheimer's disease. Behav Brain Res 103:145–162
70. Middei S, Geracitano R, Caprioli A et al (2004) Preserved fronto-striatal plasticity and enhanced procedural learning in a transgenic mouse model of Alzheimer's disease overexpressing mutant hAPPswe. Learn Mem 11:447–452
71. Deacon RMJ, Cholerton LL, Talbot K et al (2008) Age-dependent and -independent behavioral deficits in Tg2576 mice. Behav Brain Res 189:126–138
72. Arendash GW, Lewis J, Leighty RE et al (2004) Multi-metric behavioural comparison of APP_{SW} and P301L models for Alzheimer's Disease: linkage of poorer cognitive performance to tau pathology in forebrain. Brain Res 1012:9–41
73. Arendash GW, King DL (2002) Behavioural characterization of the Tg2676 transgenic model of Alzheimer's disease through 19 months. Physiol Behav 75:627–642
74. Horgan J, Miguel-Hidalgo JJ, Thrasher M, Bissette G (2007) Longitudinal brain corticotropin releasing factor and somatostatin in a transgenic mouse (TG2576) model of Alzheimer's disease. J Alzheimers Dis 12:115–127
75. Pompl PN, Mullan MN, Bjugstad K, Arendash GW (1999) Adaptation of the circular platform spatial memory task for mice: use in detecting cognitive impairment in the APPsw transgenic mouse model for Alzheimer's disease. J Neurosci Methods 87:87–95
76. Ognibene E, Middei S, Daniele S et al (2005) Aspects of spatial memory and behavioral disinhibition in Tg2576 transgenic mice as a model of Alzheimer's disease. Behav Brain Res 156:225–232
77. Dellu F, Contarino A, Simon H et al (2000) Genetic differences in response to novelty and spatial memory using a two-trial recognition task in mice. Neurobiol Learn Mem 73:31–48
78. Cleary JP, Ghilardi JR, Maggio JE et al (2000). Immunization against Aβ in APP-overexpressing transgenic mice. Soc Neurosci Abs 26:497
79. Adriani W, Ognibene E, Heuland E et al (2006) Motor impulsivity in APP-SWE mice: a model of Alzheimer's disease. Behav Pharmacol 17:525–533
80. Bizon J, Prescott S, Michelle M, Nicolle MM (2007) Intact spatial learning in adult Tg2576 mice. Neurobiol Aging 28: 440–446
81. Cunningham C, Deacon RMJ, Chan K et al (2005) Neuropathologically distinct prion strains give rise to similar temporal profiles of behavioural deficits. Neurobiol Dis 18: 258–269
82. Cunningham C, Deacon R, Wells H et al (2003) Synaptic changes characterize early behavioural signs in the ME7 model of murine prion disease. Eur J Neurosci 17:2147–2155
83. Deacon RMJ, Raley JM, Perry VH, Rawlins JNP (2001) Burrowing into prion disease. Neuroreport 12:2053–2057
84. Guenther K, Deacon RMJ, Perry VH, Rawlins JNP (2001) Early behavioural changes in scrapie-affected mice and the influence of dapsone. Eur J Neurosci 14:401–409
85. Deacon RMJ, Koros E, Bornemann KD, Rawlins JNP (2008) Aged Tg2576 mice are impaired on social memory and open field habituation tests. Behav Brain Res 197:466–468
86. Deacon RMJ (2006) Appetitive position discrimination in the T-maze. Nat Protoc 1:13–15
87. Deacon RMJ (2006) Burrowing in rodents: a sensitive method for detecting behavioral dysfunction. Nat Protoc 1:118–121
88. Teeling JL, Felton LM, Deacon RMJ et al (2007) Sub-pyrogenic systemic inflammation impacts on brain and behavior, independent of cytokines. Brain Behav Immun 21:836–850
89. Deacon RMJ (2006) Assessing nest building in mice. Nat Protoc 1:1117–1119
90. Wolfer DP, Müller U, Stagliar M, Lipp HP (1997) Assessing the effects of the 129/Sv genetic background on swimming navigation learning in transgenic mutants: a study using mice with a modified beta-amyloid precursor protein gene. Brain Res 771:1–13
91. Clapcote SJ, Roder JC (2004) Survey of embryonic stem cell line source strains in the water maze reveals superior reversal learning of 129S6/SvEvTac mice. Behav Brain Res 152:35–48
92. Holmes A, Lit Q, Murphy DL et al (2003) Abnormal anxiety-related behavior in serotonin transporter null mutant mice: the influence of genetic background. Genes Brain Behav 2:365–380

Chapter 20

APP-Based Transgenic Models: The APP23 Model

Debby Van Dam and Peter Paul De Deyn

Abstract

Animal models are considered essential in research for elucidation of human disease processes and subsequent testing of potential therapeutic strategies. This is especially true for neurodegenerative disorders, in which the first steps in pathogenesis are often not accessible in human patients. Alzheimer's disease (AD) is vastly becoming a major medical and social-economical problem in our aging society. Valid animal models for this uniquely human condition should exhibit histopathological, biochemical, cognitive, and behavioral alterations observed in AD patients. Major progress has been made since the understanding of the genetic basis of AD and the development and improvement of transgenic mouse models.

All AD models developed up to date are partial, but nevertheless, essential in further unraveling the nature, and spatial and temporal development of the complex molecular pathology underlying this condition. One of the more recent transgenic attempts to model AD is the APP23 transgenic mouse. This paper describes the development and assessment of this human amyloid precursor protein overexpression model. We summarize histopathological and biochemical, cognitive and behavioral observations made in heterozygous APP23 mice, thereby emphasizing the model's contribution to clarification of neurodegenerative disease mechanisms. In addition, the first therapeutic interventions assessing predictive validity in the APP23 model are included.

Key words: APP23, Amyloid pathology, Cognitive decline, Circadian rhythm disturbances, Aggression, Cholinergic deficit, Face validity, Predictive validity

1. Introduction

Laboratory animal research contributes in a major way to the understanding of human pathophysiological processes and will – undoubtedly – continue to have a tremendous impact on the progress in biomedical research. Merit and utility of a model to unravel the etiology and pathogenesis of a human condition and to examine potential therapeutic interventions should be assessed through rigorous validation. A valid animal model should resemble the human condition it intents to mimic, in etiology, pathophysiology,

Peter Paul De Deyn and Debby Van Dam (eds.), *Animal Models of Dementia*, Neuromethods, vol. 48,
DOI 10.1007/978-1-60761-898-0_20, © Springer Science+Business Media, LLC 2011

and symptomatology, and it should exhibit concordant effects of therapeutic interventions as human patients (1). Based upon the various aspects of cognitive impairment and neuropathological hallmarks of dementia syndromes, different strategies to model these conditions have been employed. Although a range of diseases leads to dementia, laboratory animal research in this field has mainly focused on the development of models for Alzheimer's disease (AD). This is probably attributable to the high – and still escalating – prevalence of this neurodegenerative condition due to demographic changes and our still increasing life expectancy. In addition, AD is characterized by a number of well-defined neuropathological modifications that form an ideal base of operations for the development of animal models. Up to date, however, no AD model has succeeded in mimicking all of the histopathological, biochemical and neurochemical, cognitive and behavioral hallmarks characterizing the human condition. Hence, all AD models are partial, but nonetheless, indispensable to the further unraveling of the character, and spatial and temporal development of the complex molecular pathology underlying this disease state.

Capital progress has emanated from the use of pronuclear microinjection, and the development of embryonic stem cell and gene targeting techniques in the late 1970s–early 1980s, resulting in the development of transgenic mice. Transgenic mouse technology indisputably impacted our knowledge of normal and altered gene expression, regulation, and function. Numerous transgenic models are designed with the intention to understand how genes contribute to the pathogenesis of a certain human disease and to mimic human symptomatology thereby providing a reliable tool to screen new therapeutic interventions. Particularly for brain-related disorders, like AD, valid animal models are crucial to assess early biochemical and patho(physio)logical processes that are not easily accessible in human patients.

Transgenic AD mouse models often typically overexpress wild-type or mutated alleles of the early onset AD-linked genes: amyloid precursor protein (APP) or one of the presenilin genes (PSEN1 or PSEN2). The first transgenic models were based on the APP gene, mainly because of APP's central double role as a precursor of amyloid β peptide (Aβ), the main component of AD-typical amyloid plaques, and as the target gene of the first mutations described in early-onset AD families. The first models failed to develop typical AD deposits, which could, at least partially, be attributed to insufficient transgene expression levels. Models with at least a 2- to 3-fold overexpression of the transgene, like the APP23 model, succeeded in reproducing pathological features characteristic of AD (2).

This chapter will focus on the different steps in the validation process of the APP23 model and its utility to assess patho(physio) logical mechanisms and therapeutic hypotheses.

2. Construction of the APP23 Model

The APP23 transgenic mouse model was first described by Sturchler-Pierrat and colleagues in 1997 (3). The neuron-specific murine Thy-1 promoter drives the expression of human APP_{751} cDNA carrying the Swedish double mutation (K670N/M671L), known to cause early-onset AD. The model was originally established in a C57BL/6xDBA2 background, but has been continuously backcrossed to C57BL/6 to create an isogenic line. The transgene is seven times overexpressed at mRNA level compared to endogenous mouse APP levels. The spatial expression pattern shows the highest expression levels in neocortex and hippocampus, regions that typically exhibit AD pathology.

3. Validation at the Histopathological Level

Upon histopathological examination, AD brain discloses striking ultrastructural lesions in most cortical and subcortical areas, being neurofibrillary tangles (NFT) and senile amyloid plaques. Plaques often are surrounded by distorted and dystrophic neurites, and activated microglia, indicative of inflammation. Eventually, these lesions lead to neurodegeneration and substantial neuronal loss (4).

In APP23 brain, the first scarce Aβ deposits appear in heterozygous animals in the frontal cortex and subiculum around the age of 6 months. With aging, plaques further increase in size and number until eventually, an extensive area of neocortex and hippocampus is occupied by Aβ deposits in 24-month-old APP23 mice. At this age, deposits can also be perceived in thalamus, olfactory nucleus, and caudate putamen. In very old mice (2–3 years), most, if not all, brain regions contain some plaques. As in AD, the obvious exception is the cerebellum. Plaques are also present in white matter such as the corpus callosum and the fimbria fornix. Deposits display immunoreactivity with different Aβ-specific antibodies, and most plaques are intensely stained with methenamine silver and show Congo red birefrigerence, indicative of Aβ fibrils and dense core plaques. Diffuse deposits appear only after the age of 20 months. APP23 amyloid plaques resemble human plaque morphology, although they often contain a larger compact core surrounded by a smaller halo of more disperse amyloid. In addition, some biochemical and biophysical differences are evident. Human amyloid cores are extremely resistant to detergents and denaturating agents, while APP23 amyloid fibrils are more soluble, most likely based on the lack of covalent crosslinking. Because of these differences in solubility, vigilance is required when assessing therapeutic interventions by their effect on amyloid burden (3, 5, 6).

Deposition of Aβ in cerebral vasculature (cerebral amyloid angiopathy, CAA) can be detected in about 90% of AD patients (7). In APP23 mice aged 1 year or older, Aβ deposits can be observed in the cerebral vasculature of, primarily, leptomeninx, thalamus, cortex, and hippocampus. Interestingly, given the neuron-specific Thy1 promoter, all human APP and Aβ must be of neuronal origin (8). CAA in APP23 brain is associated with vessel wall weakening, aneurismal vasodilatation, and hemorrhage (9).

The amount of both $A\beta_{1-40}$ and $A\beta_{1-42}$ insoluble peptides in brain extracts exponentially increases with aging in parallel to the growing plaque burden (10, 11), whereas the brain levels of soluble Aβ were continuously elevated when compared to the control levels (10).

Analogous to the human condition, plaque deposition and CAA in APP23 brain is accompanied by inflammatory reactions. Hypertrophic astrocytes and microglial activation are closely associated with compact plaques, but not with diffuse plaques (3, 5, 6, 12, 13). Microglial processes persistently invade the amyloid plaques, but their cell bodies are positioned at the plaque periphery, and no evidence of phagocytotic internalization and degradation of Aβ fibrils was found (14).

Plaque deposition results in substantial disruption of normal cytoarchitecture and/or neuronal loss dependent on the brain region considered. Neurodegeneration is prominent around compact plaques with Aβ deposition prompting aberrant axonal sprouting, dystrophic neurites, ectopic terminal formation, and disruption of neuronal connectivity. The distorted neurites that surround the dense core plaques stain positive for hyperphosphorylated microtubule-associated protein tau. In addition, APP23 brain shows immunoreactivity with Alz50 antibody, which recognizes a conformational epitope of tau indicating early structural changes in the assembly into filaments. Nevertheless, like all other single transgenic APP models, the APP23 model fails to develop paired helical filaments and NFT (3, 5, 6).

Like in human AD brain, a significant neuronal loss (14–25%) inversely correlating with plaque load was observed in the hippocampal CA1 region of 14- to 18-month-old APP23 mice (15). No quantitative evidence of neuronal loss in neocortex as a whole was observed (15), although in-depth analysis showed a selective loss of entorhinal neurons (5). While plaque formation causes severe disruption of cytoarchitecture in neocortex, no significant neuronal loss was observed (15). Upon reexamination of neocortical neuron numbers in young (2–3 months), adult (8 months), and aged (27 months) APP23, the neocortex of young and adult APP23 mice was shown to retain 10–15% more neurons than age-matched controls. Despite robust Aβ deposition, the neocortical synaptic bouton number is maintained indicating either that cerebral amyloidosis in APP23 mice is insufficient to induce synaptic loss or that compensatory synaptotrophic effects may be present (16). These observations complied with the growth and survival promoting, and neuroprotective character of (soluble) APP. Aged APP23 brain was characterized by significant gliogenesis, predominantly around Aβ plaques (17).

4. Validation at the Neurochemical Level

The nucleus basalis of Meynert located in the basal forebrain and providing the major cholinergic input into the neocortex experiences extensive neuronal loss in AD brain (18). Consequently, the neocortex suffers a loss of cholinergic fibers and receptors, as well as a reduction of cholinergic system-related enzymes (19). Cholinergic changes in APP23 brain consist of a moderate decrease in cortical cholinergic enzyme activity, and a severe disruption of the cholinergic fiber network and significant reduction of total cholinergic fiber length in the vicinity of plaques in aged APP23 mice (20).

In 7- to 8-month-old APP23 mice, specific activity levels of choline acetyltransferase (ChAT) and acetylcholinesterase (AChE) were determined in brain regions with known importance in cholinergic transmission and clinical relevance to AD patients. The significantly reduced levels of ChAT (approx. 30% lower than wild-type brain) and AChE (approx. 40% lower than wild-type brain) in the basal forebrain confirmed the presence of an early cholinergic deficit in the APP23 model (21), as previously suggested by shrinkage of cholinergic neurons in basal forebrain nuclei Ch1 and Ch2 in 8-month-old APP23 mice (20). Besides a cholinergic deficit, other neurotransmitter systems seem to be affected in APP23 brain as well (21). Significant alterations in compounds of the noradrenergic and serotonergic system were noted, in addition to changed levels of the inhibitory neurotransmitter glycine and several nonneurotransmitter amino acids.

The presence and distribution of several neuropeptides (cholecystokinin, dynorphin, enkephalin galanin, neuropeptide Y, and substance P) was immunohistochemically assessed in several hippocampal and neocortical regions of 21- to 27-month-old APP23 and WT mice. Marked neuropeptide changes were observed in hippocampus and ventral cortex. Dystrophic neurites in the close vicinity of Aβ plaques showed enrichment of neuropeptide levels, which may reflect attempts to counteract degenerative processes (22).

5. Validation at the Cognitive and Behavioral Levels

Progressive cognitive impairment is the cardinal clinical feature of AD. In humans, different modalities of memory are classified in a hierarchical structure from higher to lower memory components as suggested by Tulving: episodic, semantic, and procedural memory (23). Higher memory functions are often impaired in the early stage of dementia, whereas procedural memory is conserved, even in advanced stages of the disease. Higher hierarchical memory requires the more complex, and hence, more vulnerable, brain systems, and is first affected in AD (24).

As a valid phenocopy of the human condition, Alzheimer models should therefore develop age-dependent memory disturbances reminiscent of the dementia process in AD. Learning and memory in rodents can be studied using very distinct experimental paradigms, which assess different forms or aspects of learning and memory.

The Morris water maze (MWM) is a widely used tool in behavioral neuroscience to assess hippocampus-dependent spatial learning and memory in rodents. The experimental setup generally comprises a circular pool filled with opaque water and surrounded by a number of invariable visual cues. A small escape platform is hidden under the water surface at a fixed position. Animals have to learn to locate the platform using extra-maze cues during a series of training trials. Memory acquisition is reflected in learning curves depicting escape latency or path length. This acquisition phase is usually followed by a probe trial in order to assess storage and retrieval of information and spatial accuracy in the trained animal (25).

The MWM paradigm was employed by three research teams to study cognitive alterations in the APP23 model; however, each with slightly different training protocols and maze characteristics, as well as experimental groups (11, 26–28). While at the early age of 6–8 weeks, APP23 mice were still able to reach the same level of performance as control littermates (11), from the age of 3 months onwards, APP23 mice displayed a major decline in visual spatial learning capacities as tested in the MWM paradigm (11, 27). Visible-platform MWM trials excluded loss of motivation, and motor or sensory disturbances (26–28). Since Aβ plaque formation is first observed in 6-month-old heterozygous mice, the cognitive decline disclosing at the early age of 3 months is most likely independent of plaque formation. Interestingly, a negative correlation existed between probe trial performance (percentage of time in target quadrant) and whole brain level of soluble $A\beta_{1-42}$ peptides in 3-month-old APP23 mice (11).

Since AD patients typically exhibit short-term memory deficits in the early stages of the clinical disease process, a plus-shaped water maze or cross-maze was used to assess both reference and a correlate of working memory. Six-month-old APP23 mice exhibit slower learning curves during the acquisition phase, but nevertheless, consolidation of the learned spatial information appeared intact during the probe trial. The subsequent delayed matching-to-place task indicated impairment of short-term memory components, which may be based on reduced response-flexibility (29).

Analogous observations were reported employing the dry-land Barnes maze. APP23 mice aged 12 months were slower in finding the escape tunnel across the initial acquisition phase, whereas they performed equally to their wild-type littermates during the probe trials and reversal trials. When dissecting these acquisition deficits, it became evident that APP23 mice displayed a delay in switching from a random to a serial or spatial search strategy to locate the escape route (30).

Spatial learning was also assessed in two food-baited dry-land mazes, namely the classical 8-arm radial arm maze and an intricate complex labyrinth-like maze. While 10-month-old APP23 performed equally well as their control littermates in the simple radial maze, they showed no sign of improved performance in the complex paradigm (31).

Although eventually not suited for screening of hippocampus-dependent learning and memory in mouse models with a rapidly progressive phenotype, an odor-paired associate task was employed to assess nonspatial hippocampus-dependent learning in the APP23 model. APP23 mice were impaired in remembering the association between the different odors during the training phase. Due to the old age of the mice by the end of this time-consuming paradigm, both wild-type and APP23 mice showed poor performance during the subsequent transitivity and symmetry tasks (32).

Two research teams investigated passive avoidance (PA) learning in a step-through box (11, 27). In contrast to the MWM task, PA learning is mainly based on nonspatial abilities. While PA learning remained unaffected in young and adult mice (11), only at the advanced age of 25 months, APP23 mice were impaired in acquiring the PA response (27).

In conclusion, it is clear that not all types of learning and memory are affected simultaneously in the APP23 mouse model. The model appears to mimic the dementia process based on this differential impairment of memory components in AD, as higher-order cognitive functions, as tested e.g., in the MWM, are affected prior to cognitively less demanding tasks.

Performance on several cognitive and behavioral paradigms can be affected by neuromotor deficits. Motor performance and equilibrium were evaluated using an accelerating rotarod apparatus (11, 26), the stationary beam task, and a coat hanger test (26). Besides some minor differences between 3- and 6-month-old

APP23 mice and control littermates evident in one study using the rotarod, which did not affect MWM performance in these same animals (11), the employed neuromotor tests could not discriminate between APP23 mice and wild-type animals (26).

For many decades, the detrimental progressive cognitive disability was the main focus of attention and efforts in AD-related clinical and preclinical research. Approximately in the last decade, noncognitive symptoms of AD, generally known as behavioral and psychological signs and symptoms of dementia (BPSD), have progressively entered the spotlight of clinical AD research. BPSD form an integral part of the disease process and significantly impact patients, their families and caregivers, and society at large, since BPSD-related problems increase patients' disability, while they are not adequately treatable at the moment, and therefore, most frequently motivate placement in long-term care facilities. Hence, much effort is put into the assessment of BPSD-related phenomena in AD models. Based in the Behavioral Pathology in Alzheimer's Disease Rating Scale (i.e., Behave-AD), BPSD can be categorized into seven clusters: delusions, hallucinations, aggression, activity disturbances, diurnal rhythm disturbances, affective disturbances, and anxiety (33). Besides these BPSD, AD patients also present with other behavioral problems, including altered ingestive behavior, changes in sexual behavior, and stereotyped behavior (34). Unmistakably, several of these symptoms are relatively exclusive to human beings and, therefore, difficult or impossible to assess in animals, let alone in rodents.

Circadian activity patterns were assessed by placing a home cage with a single-housed APP23 or control mouse (aged 3, 6, or 12 months) between infrared sensors continuously recording horizontal locomotor activity. While activity profiles of 3-month-old APP23 mice still largely resembled wild-type patterns, 6-month-old APP23 mice showed the first indication of increased activity during the second half of the dark, i.e., active phase. This bimodal overnight activity pattern, reminiscent of sundowning behavior in AD patients, became even more pronounced at the age of 12 months. The APP23 model was therefore shown to develop age-dependent diurnal rhythm disturbances and sundowning-like phenomena (35).

Physical aggression occurs in 20–65% of AD patients (36, 37). Aggressive behavior is distressing to patients and caregivers, thereby often motivating institutionalization, overmedication, and physical restraint. The use of antipsychotics requires a delicate balance between benefit and risk of substantial adverse effects (38). Male aggressive behavior was studied in 6- and 12-month-old APP23 mice and control littermates using an isolation-induced/resident-intruder protocol. At both ages, APP23 mice showed increased aggression as exemplified by a higher mean number of attacks and shorter latency to the first attack of the intruder mouse (39).

Anxiety and depression seem to cluster in one neuropsychiatric subsyndrome of affective behaviors. Depression, a common neuropsychiatric syndrome associated with poorer quality of life, greater disability in activities of daily living, a faster cognitive decline, a high rate of nursing home placement, relatively higher mortality, and a higher frequency of depression and burden in caregivers is estimated to occur in 20–50% of AD patients (40). Anxiety symptoms, which are also associated with psychiatric morbidity, disability, increase healthcare utilization, and mortality, have been reported in 50–70% of subjects (41–43). Although the evaluation of mood-related behaviors in animal models is not straightforward and should be critically interpreted, several behavioral paradigms were developed to assess depression-related symptoms and anxiety in rodents. Depression-related symptoms were studied in the APP23 model using three different behavioral paradigms: the Porsolt forced swim test and the tail suspension test, which are both based on behavioral despair when confronted with inescapable stress, and the sucrose preference test, which is based on anhedonia, defined as the decreased capacity to experience pleasure of any kind. Neither protocol indicated the development of depression-related symptoms in APP23 mice aged 3, 6, and 12 months (44). Concerning anxiety-related behavior, some contradictory observations have been reported. Anxiety was evaluated using an elevated plus maze (EPM), an emergence-type test, and by dark-light exploration in a two-compartment dark-light transition box (DLTB). All these tests are based on transitions from/between a secure environment (enclosed arms in EPM, dark compartment in DLTB, enclosed object in the emergence test) to/and a potentially threatening and frightening environment (open arms in EPM, light chamber of DLTB, arena surrounding object in emergence-type test). While female APP23 aged 16 months did not perform different from wild-type littermates on the EPM and emergence test (26), 24-month-old APP23 mice spent significantly more time in the open arms of the EPM compared to control littermates. Latter observations were potentially linked to increased general arousal. Again, no differences were noted in the emergence test (28). Also the dark-light transition box failed to distinguish APP23 mice and control littermates (11). On the contrary, our group described an increased latency to the first open arm entry of the EMP in 12-month-old male APP23 mice, indicative of increased anxiety levels (44). Analogously, 6-month-old males displayed augmented conditioned freezing responses during both context- and cue-dependent fear conditioning (44). Latter paradigm also confirmed that these mice were still able to learn the association between both stimuli equally well as controls, which is not unanticipated since associative learning is one of the simplest learning processes (23). Exploratory and novelty seeking may relate to anxiety levels. Using an open field (OF) paradigm, our group showed decreased exploration (significantly lower total path length and fewer entries in the center circle) in male 6-month-old APP23 mice during a 10-min recording in the active phase of the animals day-night cycle, potentially related to increased anxiety levels (11). Contrastingly, another research group that assessed OF exploration during 5-min-recording periods over 3 days observed no differences in 16-month-old females (26). When tested at the age of 24 months, however, APP23 females showed continuously high levels of exploration in the OF on the three recording days, whereas exploration levels decreased on the second and third day in control mice (28). These observations might be related to the continuous arousal of APP23 mice in a cage activity paradigm, while WT mice showed adaptation to the new surroundings over a 3-day-recording period (35). Three 5-min recordings in a photocell actimeter (45 × 45 cm arena, surrounded by several infrared photocell detectors) showed significant reduction of ambulatory and stereotyped movements, hence hypoactivity, in female APP23 mice at 16 months of age (26), whereas hyperactivity was noted at the age of 24 months (28). In young male APP23 mice (age 3 months), locomotor activity in a new environment was assessed during 3 h (recording started 3 h after light onset) using circular corridors equipped with photoelectric cells and food and water suspended in the middle of the cage. Authors observed augmented locomotor activity related to novelty seeking (45). Clearly, the choice of paradigm or protocol, the gender and age of the mice studied, in addition to the fact whether or not animals are scored during the active phase of their day-night cycle may well influence the outcome of the behavioral alterations, thereby motivating prudent interpretation of these behavioral observations. Nevertheless, the circadian rhythm disturbances and sundowning-like phenomena, as well as the increased aggressive behavior, appear clear-cut phenotypes that greatly add to the face validity of the APP23 mouse as a model of BPSD-related disturbances.

Other behavioral alterations that are not directly covered by the definition and clusters of BPSD according to the Behave-AD (33, 34) have been studied in the APP23 mouse as well.

Dementing patients frequently develop appetite or eating disturbances (prevalence rate around 65%) (46, 47), which may range from increased food consumption, decreased food consumption, attempts to eat inedible substances, increased preference for sweet things, becoming more picky about their food choices, to becoming less fussy (46). Our research groups used conditioning boxes equipped with a pellet feeder to provide 20-mg dustless food pellets and a water bottle with optical lickometer. Photocell sensors detected pellet removal and the number of licks at the drinking tube. Eating and drinking patterns were recorded over a period of 1 week under a 12h/12h light-dark cycle in 3, 6, and 12-month-old

APP23 mice and control littermates. Both genotypes showed a clear circadian rhythm in food and water intake. Overall, APP23 mice took more pellets and performed more lick responses than wild-type mice at the three ages tested. An extended meal and drink pattern analysis indicated a higher number of meals and drink sessions, and quicker initiation of the subsequent meal or drinking session (48). Despite these augmented levels of ingestive behavior, APP23 mice weighed significantly less than age- and gender-matched control animals (26, 27, 48). From the age of weaning onward, male APP23 mice weighed on average 14% less than wild-type littermates (48). The existence of a hypermetabolic state in AD has been hypothesized by a limited number of studies (49–51). The observations in the APP23 model appear to support this hypothesis and require further examination of this presumed hypermetabolic state.

Stereotypic behaviors have been described in AD patients (52, 53). Analogously, stereotyped behavior, in particular popping behavior, defined as rearing and jumping in a continuous fashion against the wall of the home cage, has been described in the APP23 model (26, 54).

Prevalence rates of inappropriate sexual behavior range from 2 to 15% depending on the study considered (55, 56). Male sexual behavior was observed by confronting the APP23 or control male with an ovariectomized and hormonally primed stimulus female. No differences in social and genital investigation, or mounting behavior were noted in a 3-month-old cohort (44).

6. Assessing Predictive Validity

As a subsequent step in the validation process of APP23 mice, predictive validity of the model was assessed at both the cognitive and BPSD-related level. Symptomatic treatment effects of clinically relevant doses of donepezil, rivastigmine, galantamine, and memantine, four internationally approved drugs for symptomatic treatment of AD patients, were evaluated in 4-month-old APP23 mice and control littermates. The previously described deficit in MWM learning was used as an outcome parameter (11). All four drugs were able to improve learning and memory performance in the APP23 mice compared to sham-treated counterparts, thereby substantiating the predictive validity of the APP23 model (57). Based on the isolation-induced resident-intruder paradigm showing increased aggressive behavior in male APP23 mice (39), the predictive validity with regard to BPSD-related behavioral alterations was confirmed using risperidone, which is currently the first choice antipsychotic agent for treatment of BPSD in AD patients (58). The sensitivity of both cognitive deficits and aggressive behavior to pharmacological modulation by drugs with proven efficacy in the clinical setting underpins the predictive validity of the APP23 model as a preclinical tool for evaluation of novel pharmaceuticals targeting cognitive or BPSD-related alterations.

7. Conclusions

Over the past decade, transgenic mouse models, like the APP23 model, have become indispensable tools in neurodegenerative disease research. They provide easy accessible subjects to discern the effects of genetic and environmental factors on pathogenesis, and to understand the role of complex interactions of molecular and cellular processes and mechanisms on the progression of the disease.

The development of transgenic models has significantly increased the need for thoroughly scrutinized, validated, and refined test batteries, evaluating multiple levels of cognitive function and behavior in mice. We should, however, bear in mind that exact replication of all aspects of AD cognition and behavior in these animal models might not be attainable due to fundamental dissimilarities in phylogenesis and behavior between man and mouse. The extrapolation of findings in animal models to the human condition is based on morphological and physiological homologies between species. As a consequence, differences should be taken into account when generalizing animal findings to the human condition.

All things considered, we can conclude that the APP23 model is a highly valuable model for AD research, and neurodegenerative disease in general. Further neurochemical, cognitive, behavioral, biochemical and pathophysiological assessments, and appraisal of innovating pharmacological interventions will most probably further illustrate the validity of this transgenic model.

Acknowledgments

This work was financed by the Fund for Scientific Research – Flanders (FWO, G.0164.09), Agreement between the University of Antwerp and the Institute Born-Bunge, Interuniversity Poles of Attraction (IUAP Network P6/43), Methusalem excellence grant of the Flemish Government, Neurosearch Antwerp, the Antwerp Medical Research Foundation, and the Thomas Riellaerts Research fund. DVD is a postdoctoral fellow of the FWO.

References

1. Van Dam D, De Deyn PP (2006) Drug discovery in dementia: the role of rodent models. Nat Rev Drug Discov 5:956–970.
2. Guénette SY, Tanzi RE (1999) Progress toward valid transgenic mouse models for Alzheimer's disease. Neurobiol Aging 20:201–211.
3. Sturchler-Pierrat C, Abramowski D, Duke M et al (1997) Two amyloid precursor protein transgenic mouse models with Alzheimer disease-like pathology. Proc Natl Acad Sci U S A 94:13287–13292.
4. Selkoe DJ (2001) Alzheimer's disease: genes, proteins, and therapy. Physiol Rev 81:741–766.
5. Bornemann KD, Staufenbiel M (2000) Transgenic mouse models of Alzheimer's disease. Ann N Y Acad Sci 908:260–266.
6. Sturchler-Pierrat C, Staufenbiel M (2000) Pathogenic mechanisms of Alzheimer's disease analyzed in the APP23 transgenic mouse model. Ann N Y Acad Sci 920:134–139.
7. Jellinger KA (2002) Alzheimer disease and cerebrovascular pathology: an update. J Neural Transm 109:813–8136.
8. Calhoun ME, Burgermeister P, Phinney AL et al (1999) Neuronal overexpression of mutant amyloid precursor protein results in prominent deposition of cerebrovascular amyloid. Proc Natl Acad Sci U S A 96:14088–14093.
9. Winkler DT, Bondolfi L, Herzig MC et al (2001) Spontaneous hemorrhagic stroke in a mouse model of cerebral amyloid angiopathy. J Neurosci 21:1619–1627.
10. Kuo YM, Beach TG, Sue LI et al (2001) The evolution of A beta peptide burden in the APP23 transgenic mice: implications for A beta deposition in Alzheimer disease. Mol Med 7:609–618.
11. Van Dam D, D'Hooge R, Staufenbiel M, Van Ginneken C, Van Meir F, De Deyn PP (2003) Age-dependent cognitive decline in the APP23 model precedes amyloid deposition. Eur J Neurosci 17:388–396.
12. Stalder M, Phinney A, Probst A, Sommer B, Staufenbiel M, Jucker M (1999) Association of microglia with amyloid plaques in brains of APP23 transgenic mice. Am J Pathol 154:1673–1684.
13. Bornemann KD, Wiederhold KH, Pauli C et al (2001) Abeta-induced inflammatory processes in microglia cells of APP23 transgenic mice. Am J Pathol 158:63–73.
14. Stalder M, Deller T, Staufenbiel M (2001) Jucker M. 3D-reconstruction of microglia and amyloid in APP23 transgenic mice: no evidence of intracellular amyloid. Neurobiol Aging 22:427–434.
15. Calhoun ME, Wiederhold KH, Abramowski D et al (1998) Neuron loss in APP transgenic mice. Nature 395:755–756.
16. Boncristiano S, Calhoun ME, Howard V et al (2005) Neocortical synaptic bouton number is maintained despite robust amyloid deposition in APP23 transgenic mice. Neurobiol Aging 26:607–613.
17. Bondolfi L, Calhoun M, Ermini F et al (2002) Amyloid-associated neuron loss and gliogenesis in the neocortex of amyloid precursor protein transgenic mice. J Neurosci 22:515–522.
18. Whitehouse PJ, Price DL, Clark AW, Coyle JT, DeLong MR (1981) Alzheimer disease: evidence for selective loss of cholinergic neurons in the nucleus basalis. Ann Neurol 10:122–126.
19. Coyle JT, Price DL, DeLong MR (1983) Alzheimer's disease: a disorder of cortical cholinergic innervation. Science 219:1184–1190.
20. Boncristiano S, Calhoun ME, Kelly PH et al (2002) Cholinergic changes in the APP23 transgenic mouse model of cerebral amyloidosis. J Neurosci 22:3234–3243.
21. Van Dam D, Marescau B, Engelborghs S et al (2005) Analysis of cholinergic markers, biogenic amines, and amino acids in the CNS of two APP overexpression mouse models. Neurochem Int 46:409–422.
22. Diez M, Danner S, Frey P et al (2003) Neuropeptide alterations in the hippocampal formation and cortex of transgenic mice overexpressing beta-amyloid precursor protein (APP) with the Swedish double mutation (APP23). Neurobiol Dis 14:579–594.
23. Tulving E (1987) Multiple memory systems and consciousness. Hum Neurobiol 6:67–80.
24. Shinosaki K, Nishikawa T, Takeda M (2000) Neurobiological basis of behavioral and psychological symptoms in dementia of the Alzheimer type. Psychiatry Clin Neurosci 54:611–620.
25. D'Hooge R, De Deyn PP (2001) Applications of the Morris water maze in the study of learning and memory. Brain Res Brain Res Rev 36:60–90.
26. Lalonde R, Dumont M, Staufenbiel M, Sturchler-Pierrat C, Strazielle C (2002) Spatial learning, exploration, anxiety, and motor coordination in female APP23 transgenic mice with the Swedish mutation. Brain Res 956:36–44.

27. Kelly PH, Bondolfi L, Hunziker D et al (2003) Progressive age-related impairment of cognitive behavior in APP23 transgenic mice. Neurobiol Aging 24:365–378.
28. Dumont M, Strazielle C, Staufenbiel M, Lalonde R (2004) Spatial learning and exploration of environmental stimuli in 24-month-old female APP23 transgenic mice with the Swedish mutation. Brain Res 1024:113–121.
29. Vloeberghs E, Van Dam D, D'Hooge R, Staufenbiel M, De Deyn PP (2006) APP23 mice display working memory impairment in the plus-shaped water maze. Neurosci Lett 407:6–10.
30. Prut L, Abramowski D, Krucker T et al (2007) Aged APP23 mice show a delay in switching to the use of a strategy in the Barnes maze. Behav Brain Res 179:107–110.
31. Hellweg R, Lohmann P, Huber R, Kühl A, Riepe MW (2006) Spatial navigation in complex and radial mazes in APP23 animals and neurotrophin signaling as a biological marker of early impairment. Learn Mem 13:63–71.
32. Van Dijck A, Vloeberghs E, Van Dam D, Staufenbiel M, De Deyn PP (2008) Evaluation of the APP23-model for Alzheimer's disease in the odour paired-associate test for hippocampus-dependent memory. Behav Brain Res 190:147–151.
33. Reisberg B, Borenstein J, Salob SP, Ferris SH, Franssen E, Georgotas A (1987) Behavioral symptoms in Alzheimer's disease: phenomenology and treatment. J Clin Psychiatry 48 Suppl:9–15.
34. De Deyn PP, Engelborghs S, Saerens J et al (2005) The Middelheim Frontality Score: a behavioural assessment scale that discriminates frontotemporal dementia from Alzheimer's disease. Int J Geriatr Psychiatry 20:70–79.
35. Vloeberghs E, Van Dam D, Engelborghs S, Nagels G, Staufenbiel M, De Deyn PP (2004) Altered circadian locomotor activity in APP23 mice: a model for BPSD disturbances. Eur J Neurosci 20:2757–2766.
36. Ryden MB (1988) Aggressive behavior in persons with dementia who live in the community. Alzheimer Dis Assoc Disord 2:342–355.
37. Burns A, Jacoby R, Levy R (1990) Psychiatric phenomena in Alzheimer's disease. IV: disorders of behaviour. Br J Psychiatry 157:86–94.
38. Ballard C, Waite J (2006) The effectiveness of atypical antipsychotics for the treatment of aggression and psychosis in Alzheimer's disease. Cochrane Database Syst Rev 25(1):CD003476.
39. Vloeberghs E, Van Dam D, Coen K, Staufenbiel M, De Deyn PP (2006) Aggressive male APP23 mice modeling behavioral alterations in dementia. Behav Neurosci 120:1380–1383.
40. Starkstein SE, Mizrahi R (2006) Depression in Alzheimer's disease. Expert Rev Neurother 6:887–895.
41. Teri L, Ferretti LE, Gibbons LE et al (1999) Anxiety of Alzheimer's disease: prevalence, and comorbidity. J Gerontol A Biol Sci Med Sci 54:M348–M352.
42. Ferretti L, McCurry SM, Logsdon R, Gibbons L, Teri L (2001) Anxiety and Alzheimer's disease. J Geriatr Psychiatry Neurol 14:52–58.
43. Mega MS, Cummings JL, Fiorello T, Gornbein J (1996) The spectrum of behavioral changes in Alzheimer's disease. Neurology 46:130–135.
44. Vloeberghs E, Van Dam D, Franck F, Staufenbiel M, De Deyn PP (2007) Mood and male sexual behaviour in the APP23 model of Alzheimer's disease. Behav Brain Res 180:146–1451.
45. Senechal Y, Prut L, Kelly PH et al (2008) Increased exploratory activity of APP23 mice in a novel environment is reversed by siRNA. Brain Res 1243:124–133.
46. Cullen P, Abid F, Patel A, Coope B, Ballard CG (1997) Eating disorders in dementia. Int J Geriatr Psychiatry 12:559–562.
47. Mirakhur A, Craig D, Hart DJ, McLlroy SP, Passmore AP (2004) Behavioural and psychological syndromes in Alzheimer's disease. Int J Geriatr Psychiatry 19:1035–1039.
48. Vloeberghs E, Van Dam D, Franck F et al (2008) Altered ingestive behavior, weight changes, and intact olfactory sense in an APP overexpression model. Behav Neurosci 122:491–497.
49. Wolf-Klein GP, Silverstone FA, Levy AP (1992) Nutritional patterns and weight change in Alzheimer patients. Int Psychogeriatr 4:103–118.
50. Wolf-Klein GP, Silverstone FA, Lansey SC et al (1995) Energy requirements in Alzheimer's disease patients. Nutrition 11:264–268.
51. Wang PN, Yang CL, Lin KN, Chen WT, Chwang LC, Liu HC (2004) Weight loss, nutritional status and physical activity in patients with Alzheimer's disease. A controlled study. J Neurol 251:314–320.
52. Snowden JS, Bathgate D, Varma A, Blackshaw A, Gibbons ZC, Neary D (2001) Distinct behavioural profiles in frontotemporal dementia and semantic dementia. J Neurol Neurosurg Psychiatry 70:323–332.
53. Nyatsanza S, Shetty T, Gregory C, Lough S, Dawson K, Hodges JR (2003) A study of stereotypic behaviours in Alzheimer's disease and frontal and temporal variant frontotemporal dementia. J Neurol Neurosurg Psychiatry 74:1398–1402.
54. Lalonde R, Dumont M, Staufenbiel M, Strazielle C (2005) Neurobehavioral characterization of

APP23 transgenic mice with the SHIRPA primary screen. Behav Brain Res 157:91–98.
55. Alagiakrishnan K, Lim D, Brahim A et al (2005) Sexually inappropriate behaviour in demented elderly people. Postgrad Med J 81:463–466.
56. Tsai SJ, Hwang JP, Yang CH, Liu KM, Lirng JF (1999) Inappropriate sexual behaviors in dementia: a preliminary report. Alzheimer Dis Assoc Disord 13:60–62.
57. Van Dam D, Abramowski D, Staufenbiel M, De Deyn PP (2005) Symptomatic effect of donepezil, rivastigmine, galantamine and memantine on cognitive deficits in the APP23 model. Psychopharmacology (Berl) 180:177–190.
58. Vloeberghs E, Coen K, Van Dam D, De Deyn PP (2008) Validation of the APP23 transgenic mouse model of Alzheimer's disease through evaluation of risperidone treatment on aggressive behaviour. Arzneimittelforschung 58:265–268.

Chapter 21

Presenilin-Based Transgenic Models of Alzheimer's Dementia

Yuji Yoshiike and Akihiko Takashima

Abstract

Since the identification of the mutations in presenilin 1 and presenilin 2 genes more than a decade ago, a great deal of research has filled the gap in our knowledge of mutations underlying various phenotypes of Alzheimer's disease (AD) that appear relatively early in the life of presenilin (PS) mutation carriers. Various molecular and cell biological techniques that showed functional differences between wild-type and mutant PS emerged during this time. In this chapter, we review this research by roughly categorizing findings that are similar or support a certain hypothesis. Sect. 1 discusses the function of PS as the component of γ-secretase, which generates amyloid β (Aβ). We also present a short history of how PS mutations were first considered to produce more Aβ42 (gain of function) and later found to produce less Aβ40, resulting in a higher Aβ42/Aβ40 ratio (loss as gain of function). This produces a condition in which Aβ is prone to aggregate, supporting the amyloid hypothesis of AD. Sect. 2 summarizes PS function vis-à-vis Notch signaling, which was identified by the gene-knockout approach. A hypothesis is discussed that suggests a partial loss of function, mainly based on various AD-like phenotypes observed in conditional double PS knockout mice. The first half of Sect. 3 is devoted to a review of various abnormalities related to the intracellular calcium regulation in cell and animal transgenic models of PS. The remainder of Sect. 3 discusses other potential mechanisms of PS dysfunction caused by mutations. These include abnormalities in protein trafficking, β-catenin/cadherin-related activities, and posttranslational modifications, the latter of which include endoproteolytic cleavage of PS itself, GSK-3β-dependent phosphorylation of tau, autophagy-based protein degradation, neprilysin-mediated Aβ metabolism, and alteration of unfolded protein response signaling. Any one or more of these PS dysfunctions may underlie the pathogenesis of familial AD and perhaps sporadic AD, if linked to mode of actions caused by nongenetic risk factors such as aging. Finally, we suggest the importance of bridging nonlinear dynamics of memory with molecular neuroscience of AD from multidimensional perspectives.

Key words: Presenilin, Aβ, APP, Tau, Gain of function, Loss as gain of function, Loss of function, Familial Alzheimer's disease, Notch signaling, Trafficking, GSK-3β, Autophagy, β-catenin, N-cadherin, Neprilysin, Unfolded protein response, Transgenic, Knockout, Knock-in, Aging, γ-secretase, Calcium, Excitotoxicity, Amyloid hypothesis, Presenilin hypothesis, Calcium hypothesis, Amyloid oligomer

Peter Paul De Deyn and Debby Van Dam (eds.), *Animal Models of Dementia*, Neuromethods, vol. 48,
DOI 10.1007/978-1-60761-898-0_21, © Springer Science+Business Media, LLC 2011

1. The Amyloid Hypothesis: Role of Familial Alzheimer's Disease-Linked Presenilin Mutations

1.1. Formulation of the Amyloid Cascade Hypothesis

The amyloid hypothesis is the most popular hypothesis concerning the pathogenesis of Alzheimer's disease (AD). This hypothesis states that the accumulation of longer form(s) of amyloid β (Aβ) in the brain is the primary influence driving AD pathogenesis (1, 2). One of the known physiological functions of presenilin (PS) is to cleave Aβ from its precursor protein (i.e., amyloid precursor protein, APP). We summarize first how the amyloid hypothesis was formulated within this context.

The first missense mutation associated with familial cases of AD (FAD) was found in the APP gene (3). Located on the long arm of chromosome 21, this mutation causes an amino-acid substitution from Val to Ile close to the carboxy terminus of Aβ. A few years later, Aβ was isolated from conditioned medium of human neuroblastoma cells transfected with constructs expressing wild-type APP or APP717 mutants linked to FAD (4). Analysis of Aβ by ELISA demonstrated that the APP717 mutations consistently caused a 1.5- to 1.9-fold increase in the percentage of longer Aβ species generated. The hydrophobic nature of a few residues at the carboxy terminus of Aβ is thought to make longer Aβ species, such as Aβ42, to be more prone to aggregate than shorter species like Aβ40, which constitutes the major species of the heterogeneous species generated (5, 6).

Increased levels of longer Aβ species generated from mutant APP indicated that FAD-linked APP mutants may cause AD by fostering amyloid deposition and the formation of senile plaques, a pathological hallmark believed to be directly correlated with disease symptoms (4). Immunostaining with carboxy terminus-specific antibodies against both Aβ40 and Aβ42(43) revealed that diffuse plaques, representing the earliest stage of Aβ deposition in the cortex of AD patients, exclusively stained positive for Aβ42(43) but negative for Aβ40 (7). This shows that one of the first pathological Aβ species deposited is Aβ42(43), the longer form (7). To test the absolute requirement of Aβ42 for amyloid deposition, transgenic models were generated that express Aβ40 or Aβ42 in the absence of human APP overexpression (8). While mice expressing high levels of Aβ40 did not develop overt amyloid pathology, mice that express a 5- to 10-fold lower Aβ42 transgene compared to mice expressing Aβ40 accumulated insoluble Aβ42 and developed both diffuse and compact amyloid plaques, confirming that Aβ42 is an essential component of amyloid deposition in mice. According to the modified version of the amyloid hypothesis, soluble amyloid oligomers, not amyloid deposits, are toxic and may contribute to AD pathogenesis (9, 10).

1.2. From "Gain of Function" to "Loss as Gain of Function"

A year after the identification of a FAD-linked mutation in the APP gene, genetic linkage analysis pointed to a FAD locus on chromosome 14 (11). Three years later, a PS1 gene bearing five different missense mutations was found to cosegregate with early-onset FAD (12). Several point mutations in a similar gene (i.e., PS2) were soon identified on chromosome 1 (13–15). On the basis of these results, it was established that mutations in the genes encoding APP, PS1, and PS2 cause early-onset, autosomal dominant AD. To establish that PS mutations act as gain-of-function dominants, researchers constructed mice expressing wild-type and mutant PS genes (16). Overexpression of mutant, but not wild-type PS1, selectively increases brain Aβ42(43), suggesting that PS mutations probably cause FAD through a gain of function that increases the amount of Aβ42(43) in the brain (16). This assumption was supported by a report of elevated Aβ42(43) in plasma from subjects with FAD-linked PS1, PS2, and APP mutations (17).

Another group established transfected cell and transgenic mouse models that coexpress human PS and APP genes and analyzed quantitatively the effects of PS expression on APP processing (18). In the transfected cells, PS1 and PS2 mutations caused a highly significant increase in Aβ42 secretion in all mutant clones. Likewise, mutant but not wild-type PS1 transgenic mice showed significant overproduction of Aβ42 in the brain. It was also demonstrated that transgenic animals that express both mutant PS1 (A246E) and Swedish mutant APP (APPswe) develop numerous amyloid deposits much earlier than age-matched control mice expressing APPswe and wild-type human PS1 or APPswe alone (19). These results provided evidence for the view that FAD-linked mutant PS1 causes AD through a pathogenic mechanism that accelerates the rate of Aβ deposition in brain. Consistent with this idea is the finding of accelerated plaque accumulation and associative learning deficits in mice carrying both the PS1 A246E mutation and APP K670N/M671L (20).

In another series of transgenic mice that coexpress APPswe with two FAD-PS1 variants that differentially accelerate amyloid pathology in the brain, a direct correlation was found between the concentration of Aβ42 and the rate of amyloid deposition (21). They also showed that the shift in Aβ42/Aβ40 ratio associated with the expression of FAD-PS1 variants is due to a specific elevation in steady-state levels of Aβ42, while maintaining a constant level of Aβ40 (21). Other evidence supporting the idea that clinically manifested PS1 mutations cause a gain of function is found in studies investigating whether PS1 is involved in normal APP processing. In neuronal cultures derived from PS1-deficient mouse embryos, cleavage of the extracellular domain of APP by α- and β-secretase was not affected by the absence of PS1, whereas cleavage of the transmembrane domain of APP by γ-secretase was partially prevented (22). This incomplete cleavage caused carboxy-terminal fragments (CTF) of APP to accumulate and a fivefold drop in the production of Aβ peptides. One caveat was the important residual γ-secretase activity observed, which led to the question: Do cells that are completely devoid of PS1 and PS2 maintain this γ-secretase activity? This question was subsequently addressed using stem cells.

Because PS-null mice die early in embryogenesis, pluripotent lines of embryonic stem cells from PS-null blastocysts were generated by mating of $PS1^{+/-} \times PS2^{-/-}$ mice (23, 24). After confirming the absence of PS1 and PS2 by Western blotting, Herreman et al. expressed APP695 harboring the K595N/M596L (Swedish) mutation in embryonic stem cells using recombinant Semliki Forest virus. In contrast to results obtained with single-knockout cells, no Aβ was detected in the conditioned media of $PS1^{-/-}PS2^{-/-}$ cells. Similar observations were reported by comparing Aβ production from $PS1^{+/+}PS2^{+/+}$, $PS1^{-/-}PS2^{+/+}$, $PS1^{+/-}PS2^{-/-}$, and $PS1^{-/-}PS2^{-/-}$ blastocyst cultures (25).

The reports described are just part of the evidence that more or less supports the view that increased production of Aβ42 due to mutations in PS genes contributes to AD pathogenesis. This view, which is often referred to as "gain of function" as a result of PS mutations, is consistent with the consequences of APP mutations and therefore strongly supports the amyloid hypothesis.

Contemporaneously, other reports appeared that were less well recognized but nonetheless described the phenotypes of FAD mutations from another perspective, one that focused on the ratio of Aβ species. One study using cell models showed that the FAD-linked APP mutation V715M significantly reduces Aβ40 secretion without strongly affecting Aβ42 production (26). The significance of an enhanced Aβ42/total Aβ ratio in cells expressing PS1 mutants compared to that in wild-type cells was also recognized (27). *In vitro* aggregation kinetics of a mixture of Aβ40 and Aβ42 also lends support to this view. With a specific ratio in which Aβ42 is increased and Aβ40 is decreased, not only did we observe accelerated nucleation, but we also observed induced elongation and cytotoxicity (28). Using PS1/PS2 mouse embryonic fibroblast cells and transient transfection with PS constructs, Schroeter et al. suggested that some FAD-causing PS mutations reside in proteins possessing low catalytic activity (29). Similarly, most PS2 mutations cause a significant decrease in Aβ40 and increase in the Aβ42/Aβ40 ratio (30). Another comprehensive analysis found that mutations such as Dexon9 and L166P in PS1 cause a reduction in Aβ40 production, whereas G384 mutant significantly increases Aβ42, suggesting that the different mutations in PS affect γ-secretase structure or function in different ways (31). In HEK293 cells stably expressing PS with or without various mutations, all mutations significantly increased the Aβ42/Aβ40 ratio by decreasing Aβ40 and causing the accumulation of APP CTF fragments, a sign of decreased PS activity (32). In this system, a significant increase in the absolute levels of Aβ42 was observed for only half of the mutations tested. Moreover, this study also showed that the age of onset of PS1-linked FAD correlated inversely with Aβ42/Aβ40 and absolute levels of Aβ42, but directly with Aβ40 levels. Cell lines from PS1/PS2-deficient fibroblasts stably transfected with either wild-type PS1 or different mutants also showed that the total amount of Aβ secreted by FAD mutant PS1-expressing cells is significantly reduced, whereas the Aβ42/Aβ40 ratio is increased relative to wild-type PS1-expressing cells (33).

These consistent *in vitro* observations on the consequences of PS1 mutations on Aβ42/Aβ40 ratios were further corroborated with mouse models. PS1 knock-in mice, in which exon 10, which encodes most of the hydrophilic loop sequence, was artificially deleted from the endogenous PS1 gene, were reported to be viable but exhibited drastically reduced γ-secretase cleavage at the Aβ40 site but not at the Aβ42 site (34). Surprisingly, this reduction of Aβ40 was associated with exacerbated plaque pathology when expressed on an APP transgenic background, suggesting that decreased Aβ40, not increased Aβ42, is likely to be the cause of accelerated plaque deposition in these animals. The same group also expressed the PS1 FAD-linked mutant M146V knock-in allele, either on wild-type PS1 or PS1 null backgrounds, and crossed these mice with Tg2576 APP transgenic mice (35). Introducing the PS1 M146V mutation on a Tg2576 background caused an earlier onset of plaque pathology, whereas removing wild-type PS1 in the presence of the PS1 M146V mutation greatly exacerbated the amyloid burden. This was attributed to a reduction of γ-secretase activity rather than an increase in Aβ42. In another study, BRI-Aβ40 mice that selectively express a fusion protein between the BRI protein, involved in amyloid deposition in Familial British (FBD) and Danish Dementia (FDD) and Aβ40 were crossed with both Tg2576 mice and BRI-Aβ42 mice expressing a fusion protein between the BRI protein and Aβ42 (8, 36). In the bitransgenic mice, the increased steady-state levels of Aβ40 decreased Aβ deposition by 60–90%, demonstrating that Aβ42 and Aβ40 have opposing effects on amyloid deposition: Aβ42 promotes amyloid deposition, whereas Aβ40 inhibits it. Furthermore, because increasing Aβ40 levels protected BRI-Aβ40/Tg2576 mice from the premature-death phenotype observed in Tg2576 mice, the authors suggested that strategies that preferentially target Aβ40 might actually worsen the disease course and that a selective increase in Aβ40 levels may actually reduce the risk for development of AD (36). However, since increasing Aβ40 levels (BRI-Aβ40/BRI-Aβ42 mice) robustly exacerbated the premature-death phenotype in BRI-Aβ42 mice, their assumption may not simply apply in terms of mortality.

Regardless, these studies importantly observed that PS mutation-linked FAD is not always caused by increased Aβ42 (i.e., gain of function), but instead is more often caused by reduced Aβ40, resulting in an increased Aβ42/Aβ40 ratio (i.e., loss as gain of function). Although consensus has not been reached at this point, there is certainly a trend toward considering PS mutations in AD pathogenesis as more than a simple gain-of-function assumption to a loss-as-a-gain-of-function (37–39).

2. The Presenilin Hypothesis of Alzheimer's Disease

2.1. Understanding the Physiological Function of Presenilin

A possible counterpart of the original gain-of-function hypothesis is the "loss-of-function" hypothesis, which is based on the idea that FAD-linked mutations cause PS dysfunction (38). There is a short history of investigations that led to the formation of this hypothesis, which occurred independent of the amyloid hypothesis. We imagine that the loss-of-function hypothesis was primarily formulated on the basis of the plausible logic that understanding the physiological function of PS should reveal the underlying mechanisms of AD. Indeed, FAD is caused by the dysfunction of PS as a result of gene mutation. One way to understand the normal biological function of a protein is by knocking out the gene responsible for producing the protein in organisms (40). A targeted null mutation in the murine homolog of PS1 ($PS1^{-/-}$ mice) causes mice to die shortly after birth (41). Homozygous $PS1^{-/-}$ mice have deformed skeletons, extensive CNS hemorrhages, thin ventricular zone, and bilateral cerebral cavitations, suggesting that PS1 is required for proper formation of the axial skeleton, normal neurogenesis, and neuronal survival (41).

PS1 is known to be a functional homologue of *Caenorhabditis elegans sel-12*, which facilitates signaling mediated by the Notch/LIN-12 family of receptors (42). Another group generated $PS1^{-/-}$ mice with the goal of examining potential roles of PS1 in facilitating Notch-mediated signaling during mammalian embryogenesis (42). $PS1^{-/-}$ embryos exhibited abnormal patterning of the axial skeleton and spinal ganglia as well as reduced expression of mRNA encoding Notch1 and Delta-like 1 (Dll1), a vertebrate Notch ligand in the presomitic mesoderm (42). This suggested that PS1 is required for the spatiotemporal expression of Notch1 and Dll1, which are essential for somite segmentation and maintenance of somite borders (42). Another group reported that $PS1^{-/-}$ mice develop a cortical dysplasia resembling human type 2 lissencephaly, with leptomeningeal fibrosis and migration of cortical-plate neurons beyond their normal position into the marginal zone and subarachnoid space (43). This disorder of neuronal migration is associated with the disappearance of the Cajal-Retzius pioneer neurons accompanied by disorganization of Notch1 immunoreactivity on the neuronal cell membranes (43). Because PS1 is normally expressed in the leptomeninges, the authors suggested that the loss of Cajal-Retzius cells was caused by a defective trophic interaction with leptomeningeal cells, possibly involving disruption of Notch signaling. Another study also showed that PS1-mediated Notch-signaling is required to induce Dll1 expression in the caudal half of the somite, indicating that PS1-dependent activation of the Notch-signaling pathway is necessary to regulate Dll1 expression and to establish the rostrocaudal polarity of somites (44). On the other hand, in the ventricular zone of the $PS1^{-/-}$ mice generated by Shen et al. (41), expression of the Notch1 downstream effector gene Hes5 is reduced and expression of the Notch1 ligand Dll1 is elevated, whereas expression of Notch1 is unchanged (45). The level of Dll1 transcripts is also increased in the presomitic mesoderm of $PS1^{-/-}$ embryos, while the level of Notch1 transcripts is unchanged (45). Since their mice also showed premature differentiation of neural progenitor cells, the authors suggested that PS1 controls neuronal differentiation in association with the downregulation of Notch signaling during neurogenesis (45). Abnormal blood vessel development in $PS1^{-/-}$ mouse embryos has also been reported, suggesting the essential roles of PS1 in angiogenesis as well (46).

Mice homozygous for a targeted null mutation in PS2 ($PS2^{-/-}$ mice) exhibit no obvious defects, whereas loss of PS2 on a PS1 null background ($PS1^{-/-}PS2^{-/-}$ mice) leads to early embryonic lethality and multiple early patterning defects, including lack of somite segmentation, disorganization of the trunk ventral neural tube, midbrain mesenchyme cell loss, anterior neuropore closure delays, and abnormal heart and second brachial arch development (47). Additionally, in $PS1^{-/-}PS2^{-/-}$ mice, Hes5 expression is undetectable and Dll1 is expressed ectopically in the neural tube and brain of embryos, supporting the premise that PS plays essential roles in Notch signaling (47). Another report showed that $PS2^{-/-}$ mice develop only mild pulmonary fibrosis and hemorrhage with age (24). Although $PS1^{+/-}PS2^{-/-}$ mice survive with relatively good health, $PS1^{-/-}PS2^{-/-}$ mice display a phenotype closely resembling full Notch1 deficiency, suggesting that PS1 is essential and PS2 is redundant for normal Notch signaling during mammalian embryological development (24). $PS1^{+/-}PS2^{-/-}$ mice also develop splenomegaly with severe granulocyte infiltration, suggesting a critical role of PS in myelopoiesis (48). Six months after birth, the majority of $PS1^{+/-}PS2^{-/-}$ mice develop an autoimmune disease characterized by dermatitis, glomerulonephritis, keratitis, vasculitis, hypergammaglobulinemia, and benign skin hyperplasia, indicating that skin and immune systems are especially sensitive targets of impaired PS function (49).

There are other mouse models in which the Thy-1 promoter was used to drive PS1 expression in order to rescue the embryonic lethality of PS1 knockout mice, resulting in organ-specific knockouts (50, 51). In a skin-specific deficient model, loss of PS1 caused epidermal dysplasia and skin cancer, which was suggested to result from β-catenin accumulation (50). In a kidney-specific deficient model, lack of both PS1 and PS2 resulted in altered nephrogenesis, most likely because of deficient Notch signaling (51). Overall, it is clear that genetic deficiencies in PS affect a variety of physiological functions, mainly via the Notch signaling pathway, but other pathways might contribute as well (40).

2.2. Partial Loss of Presenilin Function as a Mechanistic Model of Alzheimer's Disease

Having seen a great deal of evidence suggesting that genetic deficiencies in PS cause a variety of developmental abnormalities, mainly via Notch signaling, one might question how these abnormalities may or may not relate to the pathogenesis of AD. An attempt to answer this question was made by overcoming the embryonic lethal characteristics of PS knockout mice again genetically. Selective inactivation of PS1 in the forebrain has been achieved by crossing floxed PS1 gene mice (PS1cKO) with mice expressing Cre recombinase under the control of the αCaMKII promoter (52, 53). PS1cKO mice show no gross abnormalities but do accumulate the CTF of APP and display reduced generation of Aβ (52). The expression of Notch downstream effecter genes, Hes1, Hes5, and Dll1, is unaffected in the cerebral cortex of these mice (52). Although basal synaptic transmission and synaptic plasticity, a hypothetical cellular coordinate of memory, such as long-term potentiation (LTP) and long-term depression (LTD) at CA1 synapses are normal, PS1cKO mice exhibit subtle but significant deficits in long-term spatial memory (53). Another report showed that PS1cKO mice have a pronounced deficiency in enrichment-induced neurogenesis in the dentate gyrus, even though this reduction in neurogenesis does not result in LTP or learning deficits, indicating that the addition of new neurons is not required for memory formation (53). However, the higher extent of memory retention in PS1cKO mice than in control mice after postlearning enrichment suggests that adult dentate neurogenesis may play a role in the periodic clearance of outdated hippocampal memory traces after cortical memory consolidation, thereby ensuring that the hippocampus is continuously available to process new memories (53). Because PS1cKO mice still possess γ-secretase activity of PS2, the possible contribution of Aβ to their memory deficits cannot be excluded. Thus, forebrain-specific PS1cKO mice were generated on a $PS2^{-/-}$ background (PS1cKO $PS2^{-/-}$ or PScDKO) (54). PScDKO mice lacking both PS1 and PS2 in the postnatal forebrain exhibit impaired hippocampal memory and synaptic plasticity linked to specific reductions in NMDA receptor-mediated responses (54). They also show reduced synaptic levels of NMDA receptors, αCAM-KII, cAMP response element binding protein (CREB) binding protein (CBP), as well as CREB/CBP target gene products such as c-fos and brain-derived neurotrophic factor (BDNF) (54). With increasing age, these mutant mice develop striking neurodegeneration of the cerebral cortex and worsening impairments of memory and synaptic function (54). Neurodegeneration is accompanied by increased levels of p25 that activates cyclin-dependent kinase 5 (Cdk5), one of the kinases phosphorylating tau, and hyperphosphorylated tau (54). The authors speculated that PS is needed for the correct transport and insertion of the affected NMDA receptors into the postsynaptic membrane (54).

Decreased NMDA receptor-signaling could explain the decreased αCaMKII activity and the changes in LTP and synaptic transmission observed in PScDKO mice. Abnormal Notch signaling could also contribute to the overall phenotype of these mice. CBP contains a consensus sequence site for the CBF-1 transcription factor, which is regulated by Notch (54). CBP is an essential cofactor for the transcription factor CREB, and deficiencies of these factors on their own lead to neurodegeneration (54). Although these speculations (54) still have to be confirmed, especially in the context of age-dependency, the various phenotypes of PScDKO mice that are also seen in AD patients (with the exception of Aβ deposition), as well as the observation that PS mutations do not increase Aβ42 but decrease Aβ40, have provided the main rationale for proposing that partial loss of PS function underlies memory impairment and neurodegeneration in the pathogenesis of AD (55).

The molecular events that may underlie the observed phenotypes were further investigated (56). ELISA analysis showed that both PS1cKO and PScDKO mice had reduced levels of Aβ40 and Aβ42, although any change in the ratio of the two Aβ species was not mentioned. Microarray, quantitative real-time reverse transcription PCR, and Western blot analyses confirmed the elevated levels of glial fibrillary acidic protein, complement component C1q, and cathepsin S (56), the upregulation of which has been associated with inflammatory responses in various neurodegenerative processes (56). Strong microglial activation was also observed in the hippocampus and deep cortical layers of PScDKO mice (56). These results support the premise that the memory impairment and neurodegeneration in PScDKO mice are not caused by Aβ accumulation and that loss of PS function leads to differential upregulation of inflammatory markers in the cerebral cortex.

In *C. elegans* lacking *sel-12*, six types of FAD-linked mutant PS1 had a reduced ability to rescue the phenotype, suggesting that mutant PS1 has lower than normal PS activity (57). A similar study also showed that human wild-type PS1, not mutant PS1, fully rescued the *sel-12*-deficient phenotype, namely an egg-laying defect, supporting the hypothesis that PS1 mutations cause a partial loss of PS1 function (58). Mutations of *sel-12* in *C. elegans* result in a defect in temperature memory that is caused by the loss of PS function in two cholinergic interneurons that display neurite morphology defects (59). The morphology of the affected neurons in *sel-12* mutant animals can be rescued by expressing only wild-type human PS1, not FAD mutant PS1 A246E, in these cells (59). In contrast, in PS1$^{-/-}$ mice, embryonic lethality and axial skeletal defects were efficiently rescued by both human wild-type PS1 and A246E PS1 to similar degrees, findings consistent with the view that FAD-linked PS1 mutants retain sufficient normal function during mammalian embryonic development (60, 61). Furthermore, a 50% reduction of PS1 activity in PS1$^{+/-}$ mice does not lead to Aβ42(43) increase, whereas expression of human mutant PS1 on a murine PS1 null background is sufficient to elevate Aβ42(43), supporting the idea that gain-of-function activity results from the PS1 mutation (61).

Although these pioneering reports in mice described much about the developmental abnormalities of these mice and about Aβ deposition, they did not rigorously examine the cognitive functions of these mice. In fact, the lack of progressive spatial learning impairment in mice expressing the mutated human PS1 (M146L and L286V) transgene in the Morris water maze test indicates that other AD-associated genes like APP may be required for full phenotypic expression of mutant PS1 alleles (62). At 3 months of age, PS1 M146V knock-in mice exhibited impaired hippocampus-dependent associative learning as measured with a contextual fear conditioning paradigm; this was correlated with reduced adult neurogenesis in the dentate gyrus, indicating that impaired adult neurogenesis may contribute to the memory deficit associated with FAD (63). Mice with postnatal neuron-specific PS1 deficiency (PS1 n$^{-/-}$) were generated by means of loxP/Cre recombinase-mediated deletion (64). In adult PS1 n$^{-/-}$mice, levels of endogenous brain Aβ are greatly reduced, concomitant with accumulation of APP CTF (64). Crossing APP V717I transgenic mice with PS1 n$^{-/-}$ mice effectively prevented the amyloid pathology and LTP deficit in APP V717I transgenic mice (64). However, the cognitive defect, assessed by the object recognition test, in APP single transgenic mice could not be rescued by crossing these mice with PS1 n$^{-/-}$ mice. Similarly, conditional inactivation of PS1 in APP transgenic mice (PS1cKO;APP Tg) also effectively prevents the accumulation of Aβ and inflammatory responses, although it also causes an age-dependent accumulation of APP CTF (65). Short-term PS1 inactivation in young PS1cKO; APP Tg mice rescued deficits in contextual fear conditioning and serial spatial reversal learning in a water maze (65). However, longer-term PS1 inactivation in older PS1cKO;APP Tg mice failed to rescue contextual memory and hippocampal LTP, and had a decreasing ameliorative effect on spatial memory impairment. Taken together, these observations highlight the complex functional relationship of mutant PS and APP to synaptic plasticity and memory in the adult and aging brain (64). Because of the complex nature of memory in the course of AD progression, it makes it difficult to determine which genetically manipulated animal is a better model for AD or FAD.

3. Various Forms of Presenilin Malfunction in Alzheimer's Disease Pathogenesis

3.1. Calcium Dysregulation Hypothesis of Presenilin Mutations

Most reports that support the "gain-of-function" or "loss-as-a-gain-of-function" hypotheses are based on the idea that amyloid deposition results from an increased Aβ42/Aβ40 ratio. The recent discovery of a soluble oligomer as the most toxic species of Aβ requires these hypotheses to be adjusted so that they address whether PS mutations indeed promote the generation of Aβ oligomers and in what ways these mutations cause memory deficits. In addition, it still needs to be established how the lack of, or reduction in, normal physiological PS function (i.e., loss-of-function) mediated by Notch signaling and by other potential molecules cause AD-like phenotypes in aging animals. We will discuss these issues more in detail later in this section.

One shared view of both hypotheses is the relatively simplistic view that the mechanisms underlying PS mutation-related FAD may also apply to sporadic AD. If this were the case, then the increase in Aβ42/Aβ40 ratio and disruption of Notch signaling or other PS function in sporadic AD patients should be demonstrated before expanding the argument beyond genetics. The highest risk factor for AD is advanced age, not genetic dysfunction.

An AD pathogenesis hypothesis that takes into account both genetic and aging risk factors is the calcium dysregulation hypothesis (66). During aging, and particularly in neurodegenerative disorders, cellular calcium-regulating systems are compromised, resulting in synaptic dysfunction, impaired plasticity, and neurodegeneration. From the perspective of cell and animal models, we briefly summarize here only the genetic aspect, specifically those relating to PS mutations, of perturbed calcium homeostasis.

To explore a possible link between a genetic defect that causes AD and excitotoxic neuronal degeneration, Guo et al. generated PS1 mutant knock-in mice (PS1 M146V KI) (67). Although these mice displayed no overt mutant phenotype, their hippocampus was hypersensitive to seizure-induced synaptic degeneration and necrotic neuronal death. Moreover, cultured hippocampal neurons from PS1 M146V KI mice had increased vulnerability to glutamate-induced death, which was correlated with perturbed calcium homeostatis, increased oxidative stress, and mitochondrial dysfunction. Agents that suppress calcium influx/release and antioxidants protected neurons against excitotoxic action of the PS mutation. The same group showed that synaptosomes prepared from transgenic PS1 mutant mice exhibited elevated cytoplasmic calcium levels following exposure to depolarizing agents and Aβ (68). Mitochondrial dysfunction and caspase activation following exposures to Aβ were exacerbated in synaptosomes from PS1 mutant mice. Agents that buffer cytoplasmic calcium or prevent calcium release from the endoplasmic reticulum (ER) protected synaptosomes against the adverse effect of PS1 mutations on mitochondrial function.

In an electrophysiological study using hippocampal slices, medium and late afterhyperpolarizations in CA3 pyramidal cells were larger in mice overexpressing PS1 M146L or M146V than in wild-type and nontransgenic control mice (69). Calcium responses to depolarization and synaptic potentiation of the CA3-to-CA1 projection were also stronger in M146L mice than in nontransgenic littermates, demonstrating disruption of the control of intracellular calcium and electrophysiological dysfunction in PS1 mutant mice. Primary neurons from another line of PS1 mutant transgenic mice (L286V) also showed increased vulnerability to both excitotoxic and hypoxic-hypoglycemic damage compared to neurons from wild-type PS1 transgenic mice or nontransgenic mice (70). In addition, less excitotoxic damage occurred in neurons from PS1 knockout mice and wild-type mice in which PS1 gene expression was knocked down by antisense treatment (70).

Accelerated neuronal death was demonstrated in the hippocampus of mutant PS1 mice after peripheral administration of kainic acid in comparison to nontransgenic mice (71). Moreover, another line of human mutant PS1 transgenic mice (A246E), not human wild-type PS1 mice, showed (1) lower excitotoxic threshold for kainic acid *in vivo*, (2) greater hippocampal LTP in brain slices, and (3) increased glutamate-induced intracellular calcium levels in isolated neurons (71). Prominent higher calcium responses were triggered by thapsigargin and bradykinin, indicating that mutant PS modulates the dynamic release and storage of calcium ions in the ER. In reaction to glutamate, overfilled calcium stores resulted in higher than normal cytosolic calcium levels, explaining the facilitated LTP and enhanced excitotoxicity. Lower excitotoxic threshold for kainic acid was also observed in transgenic mutant human PS2 (N141I) mice; this lower threshold was prevented by dantrolene, an inhibitor of calcium release from the ER (71).

Fibroblasts obtained from mutant PS1 (M146V) knock-in mice showed a marked potentiation in the amplitude of calcium transients evoked by bradykinin, a cell surface receptor agonist (72). These cells also showed significant impairments in capacitative calcium entry (CCE), an important cellular signaling pathway wherein depletion of intracellular calcium stores triggers an influx of extracellular calcium into the cytosol. Notably, deficits in CCE were evident after bradykinin stimulation, but not if intracellular calcium stores were completely depleted with thapsigargin (72). Treatment with ionomycin and thapsigargin revealed that calcium levels within the ER were significantly increased in mutant PS knock-in cells (72). Collectively, these observations suggest that the overfilling of calcium stores represents the fundamental cellular defect underlying alterations in calcium signaling conferred by PS mutations (72). Another group followed up this study using stable PS1 SY5Y cell lines (73). PS1 knockout and inactive mutant PS1 D257A potentiated CCE, whereas mutant PS1 M146L and mutant PS2 N141I (primary embryonic cortical neurons derived from transgenic mice) significantly reduced CCE compared to wild-type PS. While inhibition of CCE by its antagonist SFK96365 selectively increased Aβ42, increased accumulation of Aβ had no effect on CCE. Thus, reduced CCE is likely to be an early cellular event leading to increased Aβ42 generation associated with FAD mutant PS (73).

Electrophysiological recordings with two-photon imaging revealed that mutant PS knock-in (PS KI M146V) mice displayed exaggerated ER calcium signals relative to nontransgenic mice (74). In nontransgenic mice, ryanodine receptors (RYRs) contributed modestly to inositol triphosphate (IP3)-evoked calcium release, whereas the exaggerated signals in PS1 KI (M146V) mice resulted primarily from enhanced RYR-mediated calcium release and were associated with increased RYR expression across all ages. The consequences of this abnormality in PS1 KI (M146V) mice include increased vulnerability of neurons to excitotoxic, metabolic, and oxidative insults (66).

In another study, recombinant wild-type PS protein but not recombinant PS1 M146V and PS2 N141I mutant proteins were shown to form low-conductance divalent-cation-permeable ion channels in lipid bilayers (75). By using embryonic fibroblasts derived from PS1/2 double knockout (DKO) mice, Tu et al. found that PS accounts for about 80% of passive calcium leak from the ER (75). Moreover, deficient calcium signaling in DKO fibroblast cells was rescued by the expression of either wild-type PS1 or PS2 but not by the expression of either PS1 M146V or PS2 N141I mutants. Interestingly, other PS1 mutants such as D257A and Δexon9 behave more like wild-type PS1 (75); thus, not all FAD-linked mutations in PS depend on this functional difference.

Unlike CCE, the ER calcium leak function of PS is independent of γ-secretase activity. Thus, PS might have another function other than those related to proteolytic activities, since PS2, which is considered to be a functional redundant of PS1, does not perfectly rescue PS1-deficient phenotypes linked with proteolytic functions (25). Determining differences, if any, between PS1 and PS2 in ER calcium leak function, may help us understand further the function of PS. The same group examined a series of PS1 mutants and discovered that L116P, A246E, E273A, G384A, and P436Q FAD mutations in PS1 abolished ER calcium leak function (76). In contrast, A79V FAD mutation or frontal temporal dementia-associated mutations (L113P, G183V, and Rins352) did not appear to affect ER calcium leak function, indicating that the observed effects are disease-specific. All these studies point to a calcium dysregulation hypothesis.

It is noteworthy that high vulnerability to kainate-induced seizures and hyperexcitable neural networks was also reported in APP transgenic mice carrying Swedish and Indiana FAD mutations (77). Another view shared by both the amyloid hypothesis and the presenilin hypothesis is that surface NMDA receptors and NMDA receptor-mediated signaling at synapses is reduced (54, 55, 78). However, this view appears to contradict the well-known efficacy of an NMDA receptor antagonist, memantine, on AD patients (79). Memantine's mode of action can more easily be described by the calcium dysregulation hypothesis. The homeostatic effects of calcium dysregulation, however, still need to be established. In the context of this hypothesis, we are curious of how the calcium dysregulation that occurs as a result of gene mutations or aging leads to memory deficits.

3.2. Potential Mechanisms of PS Dysfunction Caused by FAD Mutations

The genetic linkage of PS mutations to FAD has led many to propose various functions for PS, such as in γ-secretase, Notch signaling, and calcium homeostasis. These processes represent just a few of the many processes that PS may be involved in, processes that can be affected by PS mutations. In this section, we summarize other proposed dysfunctions of PS.

Posttranslational modification is a way to modulate and more specifically to determine the function of a protein. PS1, an eight transmembrane protein, was reported to be endoproteolytically processed to amino-terminal and a carboxy-terminal derivatives that accumulate to saturable levels in the brains of PS1 transgenic mice, independent of the expression of PS1 holoprotein (80). The absolute amounts of accumulated derivatives generated from the FAD-linked PS1 variants A246E or M146L accumulated to a significantly higher degree than the fragments derived from wild-type PS1 (81). Moreover, the FAD-linked Δexon9 PS1 variant, a polypeptide that is not subject to endoproteolytic cleavage *in vivo*, also accumulated in greater amounts than the fragments generated from wild-type human PS1 (81). These reports suggest that differences in the metabolism of PS1 variants may underlie the mechanism of FAD.

The roles of PS1 in protein trafficking and metabolism have been reported. PS1-deficient neurons fail to secrete Aβ, and the rate of appearance of soluble APP derivatives in the conditioned medium is increased (82). APP CTF accumulates in these neurons, indicating that PS1 promotes intramembrane cleavage and/or degradation of membrane-bound CTFs. Moreover, the maturation of TrkB and BDNF-inducible TrkB autophosphorylation is severely compromised in PS1-deficient neurons (82). In a separate study using neuroblastoma cells, PS1 was shown to regulate the biogenesis of APP-containing vesicles from the trans-Golgi network and the ER (83). PS1 deficiency or the expression of loss-of-function variants (D385A) led to robust vesicle formation and cell surface accumulation of APP, whereas FAD-linked PS1 mutants caused reduced vesicle release and APP accumulation at cell surface, especially at axonal terminals (83). Another study revealed a unique function of PS in the pigmentation of retinal pigment epithelium and epidermal melanocytes (84). According to this study, PS deficiency led to aberrant accumulation of tyrosinase-containing post-Golgi vesicles and of tyrosinase CTF. Furthermore, the PS1 M146V FAD mutant exhibited a partial loss of function in pigment synthesis. These reports suggest that PS1 plays an essential role in modulating trafficking and metabolism of a selected set of membrane and secretory proteins in neurons.

PS1 has also been known to form a complex with β-catenin *in vivo*, which increases β-catenin stability (85). β-catenin is a subunit of the cadherin protein complex. Interestingly, FAD mutations (H163R or I143T) in PS1 reduce the ability of PS1 to stabilize β-catenin, leading to increased degradation of β-catenin in the brains of PS1 transgenic mice (85). β-catenin levels are also markedly reduced in the brains of FAD patients harboring mutant PS1 (G209V, C410Y, M139I, H163R) (85). Moreover, loss of β-catenin in hippocampal neurons increases their vulnerability to apoptosis induced by Aβ (85). Indeed, PS1 knockout mice that were rescued through neuronal expression of a human PS1 transgene (either wild-type or A246E) developed spontaneous skin cancers (50). Further examination revealed that PS1 null keratinocytes exhibited higher cytosolic β-catenin and β-catenin/lymphoid enhancer factor-1/T cell factor (β-catenin/LEF)-mediated signaling (50) and that tumors contained nuclear β-catenin (50). This effect was reversed by reintroducing wild-type PS1, but not by introducing a PS1 mutant active in Notch processing but defective in β-catenin binding (50). This report suggested that deregulation of the β-catenin pathway contributes to the skin tumor phenotype; its relevance to FAD patients, however, is unknown.

Another report showed that PS functions as a scaffold that rapidly couples β-catenin phosphorylation through two sequential kinase activities independent of the Wnt-regulated Axin/CK1α complex (86). They also showed that PS deficiency increased β-catenin stability *in vitro* and *in vivo* by disconnecting the stepwise phosphorylation of β-catenin. These findings highlight an aspect of β-catenin regulation outside of the canonical Wnt-regulated pathway and a function of PS separate from intramembrane proteolysis. In another study, PS1-dependent γ-secretase protease activity was shown to promote an ε-like cleavage of N-cadherin to produce its intracellular domain, N-Cad/CTF2 (87). N-Cad/CTF2 binds CBP and promotes its proteasomal degradation, inhibiting CRE-dependent transactivation (87). Thus, the PS1-dependent ε-cleavage product N-Cad/CTF2 functions as a potent repressor of CBP/CREB-mediated transcription. In addition, PS1 mutations (Y115H, M146L, A246E, E280A, E280G, Δexon9, G384A, and D257A) inhibit N-Cad/CTF2 production and upregulate CREB-mediated transcription (87). These results appear to contradict observations in PScDKO mice (54), or simply indicate that deficiency does not phenotypically correspond to mutations. Interestingly, NMDA receptor agonists stimulate N-Cad/CTF2 production, presumably leading to the suppression of CBP/CREB-mediated transcription, which is again inconsistent with simultaneous reductions in NMDA receptor subunits and CBP observed in PScDKO mice (54, 87). If the consequences of NMDA receptor inhibition resemble the phenotypes resulting from PS mutations, PS deficiency, or AD, then it seems necessary to explain the symptomatic efficacy of memantine, an NMDA receptor antagonist, in AD patients (79).

One proposed function of PS1 is the regulation of tau phosphorylation through glycogen synthase kinase-3β (GSK-3β) (88). In human brain, levels of PS1, tau, and GSK-3β are intercorrelated (88). Direct binding of PS1 with tau and GSK-3β was also confirmed at the cellular level (88). Furthermore, FAD-linked mutant forms (C263R and P264L) of PS1 show an increased ability to bind GSK-3β and to phosphorylate tau more than wild-type PS1 (88), suggesting that an enhanced association of mutant PS1 with GSK-3β might result in a higher probability that its other binding partner, tau, will get phosphorylated. PS1 also forms complexes with the p85 subunit of PI3-kinase (PI3K) and promotes cadherin/PI3K association (89). Preincubation of mouse embryonic fibroblasts with a metal ion chelating agent, ethylene glycol-bis(beta-aminoethyl ether)-*N*-tetraacetic acid (EGTA), which inhibits this association, prevents PS1-induced PI3K/Akt activation, indicating that PS1 stimulates PI3K/Akt signaling by promoting cadherin/PI3K association (89). By activating PI3K/Akt signaling, PS1 promotes phosphorylation/inactivation of GSK-3, suppresses GSK-3-dependent phosphorylation of tau at residues that are overphosphorylated in AD, and prevents apoptosis of confluent cells (89). In contrast, PS1 FAD mutations inhibit PS1-dependent PI3K/Akt activation, thus promoting GSK-3 activity and tau overphosphorylation at AD-related residues, indicating that a possible function of PS1 is to prevent development of AD pathology by activating the PI3K/Akt signaling pathway (89). This study was followed by *in vivo* experiments showing that PS1 mutant (I213T) knock-in mice formed hyperphosphorylated tau inclusions that react with various phospho-dependent tau antibodies and with Alz50, which recognizes the conformational change of tau into paired helical filament (PHF) that is often observed in neurofibrillary tangles (NFT), a pathological hallmark of AD and tauopathies (90, 91). Some neurons exhibited Congo red birefringence and Thioflavin T reactivity, both of which are histological criteria for NFT. Biochemical analysis of the samples revealed SDS-insoluble tau, which under electron microscopy examination, resembled tau fibrils. On the other hand, accumulation of murine Aβ42(43) but not Aβ40 was reported in this PS1 mutant (I213T) knock-in mouse (92). These results indicate that PS1 mutant (I213T) knock-in mice exhibit NFT-like tau pathology in the absence of Aβ deposition, suggesting that PS1 mutations contribute to the onset of AD not only by enhancing Aβ42 production but also by accelerating the formation and accumulation of phosphorylated tau.

Hyperphosphorylation of tau caused by PS mutations is phenotypically consistent with observations in PScDKO mice. Along this line, it is not too difficult to imagine that PS mutations trigger pathogenesis via multiple modes. Needless to say, accumulations of both Aβ and hyperphosphorylated tau are two major pathological hallmarks of AD. Accumulation of misfolded proteins such as Aβ and tau is a characteristic of not only AD but also of other neurodegenerative diseases. Since degradation is another type of protein modification, impaired degradation of misfolded proteins has been put forward as a mechanism for accumulation of such proteins.

The neuron-specific cell adhesion molecule telencephalin (TLN) has also been identified as a binding partner of PS in the brain (93). The first transmembrane domain and carboxy-terminus of PS1, different from the γ-secretase catalytic site, form a binding pocket with the transmembrane domain of TLN. In $PS1^{-/-}$ hippocampal neurons, TLN accumulates in intracellular structures bearing characteristics of autophagic vacuoles (94). This accumulation of TLN is suppressed by adenoviral expression of wild-type, FAD-linked mutants, and γ-secretase-inactive D257A mutant PS1, indicating that this phenotype is independent of γ-secretase activity (94). Absence of endosomal/lysosomal proteins in the vacuoles suggests that the TLN-positive vacuoles fail to fuse with endosomes/lysosomes, preventing their acidification and further degradation. Collectively, PS1 deficiency affects the turnover of TLN through autophagic vacuoles in a γ-secretase-independent fashion, most likely by an impaired capability to fuse with lysosomes (94), although such a partial loss of function of PS1 is unlikely to be the process underlying the cause of FAD by mutations.

Neprilysin was identified as a proteolytic enzyme that degrades Aβ *in vivo* (95). One report suggests that PS is also involved in the regulation of Aβ degradation via neprilysin (96). PS-deficient cells and chronic treatment of wild-type $PS^{+/+}$ fibroblasts with DAPT, a γ-secretase inhibitor, failed to degrade Aβ and caused drastic reductions in the transcription, expression, and activity of neprilysin (96). Reduction in neprilysin activity was also observed in brain homogenates from PScDKO mice (96). Neprilysin activity was restored by transient expression of wild-type PS1 or PS2 and by expression of amyloid intracellular domain (AICD), which transactivated neprilysin gene promoters (96). These results suggested that the PS-dependent regulation of neprilysin, mediated by AICDs, provides a physiological means for modulating Aβ levels with varying levels of γ-secretase activity. The reproducibility of this result, however, has been controversial (97, 98). To propose that this mechanism underlies the cause of FAD, it will be necessary at least to confirm that the variation in γ-secretase activity caused by FAD-linked mutations in PS directly or indirectly influences neprilysin activity.

Unfolded-protein response (UPR) is a subcellular function that is invoked to respond to an increased amount of unfolded proteins in the ER under conditions that cause ER stress (99). Involvement of PS in UPR was also proposed (99). PS1 mutations lead to decreased expression of GRP78/Bip, a molecular chaperone, present in the ER that can promote protein folding (99). GRP78 levels are reduced in the brains of AD patients (99). The downregulation of UPR signaling by PS1 mutations is caused by disturbed function of IRE1, which is the proximal sensor of conditions in the ER lumen (99), where PS also localizes. Overexpression of GRP78 in neuroblastoma cells bearing PS1 mutations almost completely restores resistance to ER stress to the level of cells expressing wild-type PS1 (99). These results show that mutations in PS1 may increase vulnerability to ER stress by altering the UPR signaling pathway. There is still more to be done, one task of which would be to identify stress *in vivo* that promotes ER stress leading to accumulation of proteins, which may or may not lead to dementia.

4. Conclusions

Unlike studies of Aβ and tau that were initiated when histopathology examination identified Aβ and tau aggregates in the brains of demented patients, studies of PS were initiated when PS gene-linkage to FAD was discovered. Therefore, the primary focus of PS studies has been on the pathogenic mechanism of FAD instead of on that of sporadic AD, which at present represents more than 95% of all AD cases. The function of a protein is determined by its structure, the folding of which depends on the amino acid sequence encoded by a gene. The central dogma promises the determination of specific polypeptide sequences but not all the functions of proteins in the context of an individual whose behavior is inarguably spatiotemporal. For example, although certain mutations in PS are dominantly linked with FAD, one cannot predict exactly when AD will begin, just because an individual carries one of these mutations. Aging is a major risk factor for AD. Age-dependent accumulation of Aβ and phosphorylated tau, synapse loss, and memory impairment occur in several independent animal models characterized by differentially manipulated PS gene(s). In discussions about which is a better model of AD and what mechanism arises from the analysis of each model, one should keep in mind that most AD patients lack mutations at the genes identified. Thus, all the AD-like phenotypes of gene-manipulated models might be caused by mechanisms distinct from those of sporadic AD, even though they could still be considered as an acceleration of aging-associated phenotypes by mutant transgenes. Pathophysiological investigations of brain aging-associated memory dynamics that used nontransgenic animals represented one of the main foci of AD research before the identification of FAD-linked genes. Based on genetic discoveries, molecular and cell biological approaches certainly facilitated our understanding of the pathogenesis of early-onset FAD. We propose that it is now the time to pose questions about possible links between nonlinear dynamics of memory and molecular neuroscience and, specifically, to begin identifying the common and uncommon characteristics of nontransgenic aged and gene-manipulated brains at multiple levels: molecular, synaptic, network, systems, and organismal.

References

1. Hardy J, Allsop D (1991) Amyloid deposition as the central event in the aetiology of Alzheimer's disease. Trends Pharmacol Sci 12:383–388
2. Hardy J, Selkoe DJ (2002) The amyloid hypothesis of Alzheimer's disease: progress and problems on the road to therapeutics. Science 297:353–356

3. Goate A, Chartier-Harlin MC, Mullan M et al (1991) Segregation of a missense mutation in the amyloid precursor protein gene with familial Alzheimer's disease. Nature 349:704–706
4. Suzuki N, Cheung TT, Cai XD et al (1994) An increased percentage of long amyloid β protein secreted by familial amyloid β protein precursor (β APP717) mutants. Science 264:1336–1340
5. Jarrett JT, Berger EP, Lansbury PT Jr (1993) The carboxy terminus of the β amyloid protein is critical for the seeding of amyloid formation: implications for the pathogenesis of Alzheimer's disease. Biochemistry 32:4693–4697
6. Jarrett JT, Lansbury PT Jr (1993) Seeding "one-dimensional crystallization" of amyloid: a pathogenic mechanism in Alzheimer's disease and scrapie? Cell 73:1055–1058
7. Iwatsubo T, Odaka A, Suzuki N et al (1994) Visualization of Aβ42(43) and Aβ40 in senile plaques with end-specific Aβ monoclonals: evidence that an initially deposited species is Aβ42(43). Neuron 13:45–53
8. McGowan E, Pickford F, Kim J et al (2005) Aβ42 is essential for parenchymal and vascular amyloid deposition in mice. Neuron 47: 191–199
9. Walsh DM, Selkoe DJ (2004) Deciphering the molecular basis of memory failure in Alzheimer's disease. Neuron 44:181–193
10. Kayed R, Head E, Thompson JL et al (2003) Common structure of soluble amyloid oligomers implies common mechanism of pathogenesis. Science 300:486–489
11. Schellenberg GD, Bird TD, Wijsman EM et al (1992) Genetic linkage evidence for a familial Alzheimer's disease locus on chromosome 14. Science 258:668–671
12. Sherrington R, Rogaev EI, Liang Y et al (1995) Cloning of a gene bearing missense mutations in early-onset familial Alzheimer's disease. Nature 375:754–760
13. Levy-Lahad E, Wijsman EM, Nemens E et al (1995) A familial Alzheimer's disease locus on chromosome 1. Science 269:970–973
14. Levy-Lahad E, Wasco W, Poorkaj P et al (1995) Candidate gene for the chromosome 1 familial Alzheimer's disease locus. Science 269:973–977
15. Rogaev EI, Sherrington R, Rogaeva EA et al (1995) Familial Alzheimer's disease in kindreds with missense mutations in a gene on chromosome 1 related to the Alzheimer's disease type 3 gene. Nature 376:775–778
16. Duff K, Eckman C, Zehr C et al (1996) Increased Aβ42(43) in brains of mice expressing mutant presenilin 1. Nature 383:710–713
17. Scheuner D, Eckman C, Jensen M et al (1996) Secreted amyloid β-protein similar to that in the senile plaques of Alzheimer's disease is increased in vivo by the presenilin 1 and 2 and APP mutations linked to familial Alzheimer's disease. Nat Med 2:864–870
18. Citron M, Westaway D, Xia W et al (1997) Mutant presenilins of Alzheimer's disease increase production of 42-residue amyloid β-protein in both transfected cells and transgenic mice. Nat Med 3:67–72
19. Borchelt DR, Ratovitski T, van Lare J et al (1997) Accelerated amyloid deposition in the brains of transgenic mice coexpressing mutant presenilin 1 and amyloid precursor proteins. Neuron 19:939–945
20. Dineley KT, Xia X, Bui D et al (2002) Accelerated plaque accumulation, associative learning deficits, and up-regulation of α 7 nicotinic receptor protein in transgenic mice co-expressing mutant human presenilin 1 and amyloid precursor proteins. J Biol Chem 277:22768–22780
21. Jankowsky JL, Fadale DJ, Anderson J et al (2004) Mutant presenilins specifically elevate the levels of the 42 residue β-amyloid peptide in vivo: evidence for augmentation of a 42-specific γ secretase. Hum Mol Genet 13:159–170
22. De Strooper B, Saftig P, Craessaerts K et al (1998) Deficiency of presenilin-1 inhibits the normal cleavage of amyloid precursor protein. Nature 391:387–390
23. Herreman A, Serneels L, Annaert W et al (2000) Total inactivation of γ-secretase activity in presenilin-deficient embryonic stem cells. Nat Cell Biol 2:461–462
24. Herreman A, Hartmann D, Annaert W et al (1999) Presenilin 2 deficiency causes a mild pulmonary phenotype and no changes in amyloid precursor protein processing but enhances the embryonic lethal phenotype of presenilin 1 deficiency. Proc Natl Acad Sci U S A 96:11872–11877
25. Zhang Z, Nadeau P, Song W et al (2000) Presenilins are required for γ-secretase cleavage of β-APP and transmembrane cleavage of Notch-1. Nat Cell Biol 2:463–465
26. Ancolio K, Dumanchin C, Barelli H et al (1999) Unusual phenotypic alteration of beta amyloid precursor protein (βAPP) maturation by a new Val-715 à Met βAPP-770 mutation responsible for probable early-onset Alzheimer's disease. Proc Natl Acad Sci U S A 96:4119–4124
27. Murayama O, Tomita T, Nihonmatsu N et al (1999) Enhancement of amyloid β 42 secretion by 28 different presenilin 1 mutations of familial Alzheimer's disease. Neurosci Lett 265:61–63

28. Yoshiike Y, Chui DH, Akagi T et al (2003) Specific compositions of amyloid-β peptides as the determinant of toxic β-aggregation. J Biol Chem 278:23648–23655
29. Schroeter EH, Ilagan MX, Brunkan AL et al (2003) A presenilin dimer at the core of the γ-secretase enzyme: insights from parallel analysis of Notch 1 and APP proteolysis. Proc Natl Acad Sci U S A 100:13075–13080
30. Walker ES, Martinez M, Brunkan AL et al (2005) Presenilin 2 familial Alzheimer's disease mutations result in partial loss of function and dramatic changes in Aβ 42/40 ratios. J Neurochem 92:294–301
31. Bentahir M, Nyabi O, Verhamme J et al (2006) Presenilin clinical mutations can affect γ-secretase activity by different mechanisms. J Neurochem 96:732–742
32. Kumar-Singh S, Theuns J, Van Broeck B et al (2006) Mean age-of-onset of familial alzheimer disease caused by presenilin mutations correlates with both increased Aβ42 and decreased Aβ40. Hum Mutat 27:686–695
33. Shimojo M, Sahara N, Murayama M et al (2007) Decreased Aβ secretion by cells expressing familial Alzheimer's disease-linked mutant presenilin 1. Neurosci Res 57:446–453
34. Deng Y, Tarassishin L, Kallhoff V et al (2006) Deletion of presenilin 1 hydrophilic loop sequence leads to impaired γ-secretase activity and exacerbated amyloid pathology. J Neurosci 26:3845–3854
35. Wang R, Wang B, He W et al (2006) Wild-type presenilin 1 protects against Alzheimer disease mutation-induced amyloid pathology. J Biol Chem 281:15330–15336
36. Kim J, Onstead L, Randle S et al (2007) Aβ40 inhibits amyloid deposition in vivo. J Neurosci 27:627–633
37. Wolfe MS (2007) When loss is gain: reduced presenilin proteolytic function leads to increased Aβ42/Aβ40. Talking Point on the role of presenilin mutations in Alzheimer disease. EMBO Rep 8:136–140
38. De Strooper B (2007) Loss-of-function presenilin mutations in Alzheimer disease. Talking Point on the role of presenilin mutations in Alzheimer disease. EMBO Rep 8:141–146
39. Selkoe DJ, Wolfe MS (2007) Presenilin: running with scissors in the membrane. Cell 131:215–221
40. Marjaux E, Hartmann D, De Strooper B (2004) Presenilins in memory, Alzheimer's disease, and therapy. Neuron 42:189–192
41. Shen J, Bronson RT, Chen DF et al (1997) Skeletal and CNS defects in presenilin-1-deficient mice. Cell 89:629–639
42. Wong PC, Zheng H, Chen H et al (1997) Presenilin 1 is required for Notch1 and Dll1 expression in the paraxial mesoderm. Nature 387:288–292
43. Hartmann D, De Strooper B, Saftig P (1999) Presenilin-1 deficiency leads to loss of Cajal-Retzius neurons and cortical dysplasia similar to human type 2 lissencephaly. Curr Biol 9:719–727
44. Takahashi Y, Koizumi K, Takagi A et al (2000) Mesp2 initiates somite segmentation through the Notch signalling pathway. Nat Genet 25:390–396
45. Handler M, Yang X, Shen J (2000) Presenilin-1 regulates neuronal differentiation during neurogenesis. Development 127:2593–2606
46. Nakajima M, Yuasa S, Ueno M et al (2003) Abnormal blood vessel development in mice lacking presenilin-1. Mech Dev 120:657–667
47. Donoviel DB, Hadjantonakis AK, Ikeda M et al (1999) Mice lacking both presenilin genes exhibit early embryonic patterning defects. Genes Dev 13:2801–2810
48. Qyang Y, Chambers SM, Wang P et al (2004) Myeloproliferative disease in mice with reduced presenilin gene dosage: effect of γ-secretase blockage. Biochemistry 43:5352–5359
49. Tournoy J, Bossuyt X, Snellinx A et al (2004) Partial loss of presenilins causes seborrheic keratosis and autoimmune disease in mice. Hum Mol Genet 13:1321–1331
50. Xia X, Qian S, Soriano S et al (2001) Loss of presenilin 1 is associated with enhanced β-catenin signaling and skin tumorigenesis. Proc Natl Acad Sci U S A 98:10863–10868
51. Wang P, Pereira FA, Beasley D et al (2003) Presenilins are required for the formation of comma- and S-shaped bodies during nephrogenesis. Development 130:5019–5029
52. Yu H, Saura CA, Choi SY et al (2001) APP processing and synaptic plasticity in presenilin-1 conditional knockout mice. Neuron 31:713–726
53. Feng R, Rampon C, Tang YP et al (2001) Deficient neurogenesis in forebrain-specific presenilin-1 knockout mice is associated with reduced clearance of hippocampal memory traces. Neuron 32:911–926
54. Saura CA, Choi SY, Beglopoulos V et al (2004) Loss of presenilin function causes impairments of memory and synaptic plasticity followed by age-dependent neurodegeneration. Neuron 42:23–36
55. Shen J, Kelleher RJ III (2007) The presenilin hypothesis of Alzheimer's disease: evidence for a loss-of-function pathogenic mechanism. Proc Natl Acad Sci U S A 104:403–409

56. Beglopoulos V, Sun X, Saura CA et al (2004) Reduced β-amyloid production and increased inflammatory responses in presenilin conditional knock-out mice. J Biol Chem 279: 46907–46914
57. Levitan D, Doyle TG, Brousseau D et al (1996) Assessment of normal and mutant human presenilin function in Caenorhabditis elegans. Proc Natl Acad Sci U S A 93:14940–14944
58. Baumeister R, Leimer U, Zweckbronner I et al (1997) Human presenilin-1, but not familial Alzheimer's disease (FAD) mutants, facilitate Caenorhabditis elegans Notch signalling independently of proteolytic processing. Genes Funct 1:149–159
59. Wittenburg N, Eimer S, Lakowski B et al (2000) Presenilin is required for proper morphology and function of neurons in C. elegans. Nature 406:306–309
60. Davis JA, Naruse S, Chen H et al (1998) An Alzheimer's disease-linked PS1 variant rescues the developmental abnormalities of PS1-deficient embryos. Neuron 20:603–609
61. Qian S, Jiang P, Guan XM et al (1998) Mutant human presenilin 1 protects presenilin 1 null mouse against embryonic lethality and elevates Aβ1–42/43 expression. Neuron 20:611–617
62. Janus C, D'Amelio S, Amitay O et al (2000) Spatial learning in transgenic mice expressing human presenilin 1 (PS1) transgenes. Neurobiol Aging 21:541–549
63. Wang R, Dineley KT, Sweatt JD et al (2004) Presenilin 1 familial Alzheimer's disease mutation leads to defective associative learning and impaired adult neurogenesis. Neuroscience 126:305–312
64. Dewachter I, Reversé D, Caluwaerts N et al (2002) Neuronal deficiency of presenilin 1 inhibits amyloid plaque formation and corrects hippocampal long-term potentiation but not a cognitive defect of amyloid precursor protein [V717I] transgenic mice. J Neurosci 22:3445–3453
65. Saura CA, Chen G, Malkani S et al (2005) Conditional inactivation of presenilin 1 prevents amyloid accumulation and temporarily rescues contextual and spatial working memory impairments in amyloid precursor protein transgenic mice. J Neurosci 25:6755–6764
66. Mattson MP (2007) Calcium and neurodegeneration. Aging Cell 6:337–350
67. Guo Q, Fu W, Sopher BL et al (1999) Increased vulnerability of hippocampal neurons to excitotoxic necrosis in presenilin-1 mutant knock-in mice. Nat Med 5:101–106
68. Begley JG, Duan W, Chan S et al (1999) Altered calcium homeostasis and mitochondrial dysfunction in cortical synaptic compartments of presenilin-1 mutant mice. J Neurochem 72:1030–1039
69. Barrow PA, Empson RM, Gladwell SJ et al (2000) Functional phenotype in transgenic mice expressing mutant human presenilin-1. Neurobiol Dis 7:119–126
70. Grilli M, Diodato E, Lozza G et al (2000) Presenilin-1 regulates the neuronal threshold to excitotoxicity both physiologically and pathologically. Proc Natl Acad Sci U S A 97: 12822–12827
71. Schneider I, Reverse D, Dewachter I et al (2001) Mutant presenilins disturb neuronal calcium homeostasis in the brain of transgenic mice, decreasing the threshold for excitotoxicity and facilitating long-term potentiation. J Biol Chem 276:11539–11544
72. Leissring MA, Akbari Y, Fanger CM et al (2000) Capacitative calcium entry deficits and elevated luminal calcium content in mutant presenilin-1 knockin mice. J Cell Biol 149: 793–798
73. Yoo AS, Cheng I, Chung S et al (2000) Presenilin-mediated modulation of capacitative calcium entry. Neuron 27:561–572
74. Stutzmann GE, Smith I, Caccamo A et al (2006) Enhanced ryanodine receptor recruitment contributes to Ca^{2+} disruptions in young, adult, and aged Alzheimer's disease mice. J Neurosci 26:5180–5189
75. Tu H, Nelson O, Bezprozvanny A et al (2006) Presenilins form ER Ca^{2+} leak channels, a function disrupted by familial Alzheimer's disease-linked mutations. Cell 126:981–993
76. Nelson O, Tu H, Lei T et al (2007) Familial Alzheimer disease-linked mutations specifically disrupt Ca^{2+} leak function of presenilin 1. J Clin Invest 117:1230–1239
77. Palop JJ, Chin J, Roberson ED et al (2007) Aberrant excitatory neuronal activity and compensatory remodeling of inhibitory hippocampal circuits in mouse models of Alzheimer's disease. Neuron 55:697–711
78. Snyder EM, Nong Y, Almeida CG et al (2005) Regulation of NMDA receptor trafficking by amyloid-β. Nat Neurosci 8:1051–1058
79. Parsons CG, Stöffler A, Danysz W (2007) Memantine: a NMDA receptor antagonist that improves memory by restoration of homeostasis in the glutamatergic system-too little activation is bad, too much is even worse. Neuropharmacology 53:699–723

80. Thinakaran G, Borchelt DR, Lee MK et al (1996) Endoproteolysis of presenilin 1 and accumulation of processed derivatives in vivo. Neuron 17:181–190
81. Lee MK, Borchelt DR, Kim G et al (1997) Hyperaccumulation of FAD-linked presenilin 1 variants in vivo. Nat Med 3:756–760
82. Naruse S, Thinakaran G, Luo JJ et al (1998) Effects of PS1 deficiency on membrane protein trafficking in neurons. Neuron 21:1213–1221
83. Cai D, Leem JY, Greenfield JP et al (2003) Presenilin-1 regulates intracellular trafficking and cell surface delivery of β-amyloid precursor protein. J Biol Chem 278:3446–3454
84. Wang R, Tang P, Wang P et al (2006) Regulation of tyrosinase trafficking and processing by presenilins: partial loss of function by familial Alzheimer's disease mutation. Proc Natl Acad Sci U S A 103:353–358
85. Zhang Z, Hartmann H, Do VM et al (1998) Destabilization of β-catenin by mutations in presenilin-1 potentiates neuronal apoptosis. Nature 395:698–702
86. Kang DE, Soriano S, Xia X et al (2002) Presenilin couples the paired phosphorylation of β-catenin independent of axin: implications for β-catenin activation in tumorigenesis. Cell 110:751–762
87. Marambaud P, Wen PH, Dutt A et al (2003) A CBP binding transcriptional repressor produced by the PS1/epsilon-cleavage of N-cadherin is inhibited by PS1 FAD mutations. Cell 114:635–645
88. Takashima A, Murayama M, Murayama O et al (1998) Presenilin 1 associates with glycogen synthase kinase-3β and its substrate tau. Proc Natl Acad Sci U S A 95:9637–9641
89. Baki L, Shioi J, Wen P et al (2004) PS1 activates PI3K thus inhibiting GSK-3 activity and tau overphosphorylation: effects of FAD mutations. EMBO J 23:2586–2596
90. Tanemura K, Chui DH, Fukuda T et al (2006) Formation of tau inclusions in knock-in mice with familial Alzheimer disease (FAD) mutation of presenilin 1 (PS1). J Biol Chem 281: 5037–5041
91. Braak H, Braak E, Bohl J (1993) Staging of Alzheimer-related cortical destruction. Eur Neurol 33:403–408
92. Nakano Y, Kondoh G, Kudo T et al (1999) Accumulation of murine amyloidβ42 in a gene-dosage-dependent manner in PS1 'knock-in' mice. Eur J Neurosci 11:2577–2581
93. Annaert WG, Esselens C, Baert V et al (2001) Interaction with telencephalin and the amyloid precursor protein predicts a ring structure for presenilins. Neuron 32:579–589
94. Esselens C, Oorschot V, Baert V et al (2004) Presenilin 1 mediates the turnover of telencephalin in hippocampal neurons via an autophagic degradative pathway. J Cell Biol 166:1041–1054
95. Iwata N, Tsubuki S, Takaki Y et al (2001) Metabolic regulation of brain Aβ by neprilysin. Science 292:1550–1552
96. Pardossi-Piquard R, Dunys J, Kawarai T et al (2005) Presenilin-dependent transcriptional control of the Aβ-degrading enzyme neprilysin by intracellular domains of βAPP and APLP. Neuron 46:541–554
97. Chen AC, Selkoe DJ (2007) Response to: Pardossi-Piquard et al., "Presenilin-dependent transcriptional control of the Aβ-degrading enzyme neprilysin by intracellular domains of βAPP and APLP." Neuron 46:541–554. Neuron 53:479–483
98. Pardossi-Piquard R, Dunys J, Kawarai T et al (2007) Response to correspondence: Pardossi-Piquard et al., "Presenilin-dependent transcriptional control of the Aβ-degrading enzyme neprilysin by intracellular domains of βAPP and APLP." Neuron 46, 541–554. Neuron 53:483–486
99. Katayama T, Imaizumi K, Sato N et al (1999) Presenilin-1 mutations downregulate the signalling pathway of the unfolded-protein response. Nat Cell Biol 1:479–485

Chapter 22

APOE-Based Models of "Pre-Dementia"

Patrick M. Sullivan

Abstract

Producing a valid animal model of apolipoprotein E (*APOE*)-based dementia is critical to understanding the etiology and progression of late-onset Alzheimer's disease (AD). Unfortunately, no such model exists. Herein, I review all past and present attempts to create an *APOE*-based model with suggestions on how to reproduce AD in a mouse. Our ability to succeed in recreating late-onset AD will depend on the identification of new genetic markers in addition to the validation of nongenetic factors thought to increase the risk of dementia. These new factors can then be incorporated into an *APOE*-expressing animal for mechanistic studies and identification of new drug targets.

Key words: Apolipoprotein E, Predementia, *APOE*-deficient mice, *APOE* transgenic mice, *APOE* knock-in mice

1. Introduction

The term "dementia" refers to multiple etiologies and is therefore difficult to define. The simplest definition for dementia therefore can be a progressive loss of brain function, characterized by memory, behavior, and communication problems. Loss of brain function translates into loss of neuronal function; therefore a true animal model of dementia must exhibit signs of progressive neuronal dysfunction. There are no published reports of apolipoprotein E (*APOE*)-based models that exhibit progressive neuronal dysfunction. There are, however, several reports of *APOE* models that portray indirect measures of neuronal dysfunction or ones that might be characterized as exhibiting pre-dementia or mild cognitive impairment (MCI). The following review summarizes the attempts to classify various *APOE* animal models of "pre-dementia."

Peter Paul De Deyn and Debby Van Dam (eds.), *Animal Models of Dementia*, Neuromethods, vol. 48,
DOI 10.1007/978-1-60761-898-0_22, © Springer Science+Business Media, LLC 2011

2. APOE Link to Dementia

The first evidence to demonstrate an association between the *APOE* gene and dementia was genetic linkage analysis performed at Duke University in 1993, where the *APOE* locus on human chromosome 19 repeatedly showed a high correlation with both disease onset and frequency of Alzheimer's disease (AD). AD is the most common form of dementia. It was later shown that the common *APOE* genetic polymorphisms were linked to 90% of cases <65 years of age and 60% of AD cases >65 years (1). The human *APOE* gene encodes one of three alleles designated as *APOE*2*, *E*3*, or *E*4*, which in the Caucasian population occurs at frequencies of 7.3%, 78.3%, and 14.3%, respectively (2). The presence of an *APOE*4* allele decreases the average age of onset and increases the risk of AD (3, 4). The *APOE*2* allele lowers the risk of AD and increases the average age of onset compared to the *APOE*3* allele (5). Each *APOE* isoform differs by a single amino acid at positions 112 and 158, which results in significant structural and functional differences (6). Since the early 1970s, *APOE* had been studied extensively because of its relationship to heart disease, and also because its function as a lipid metabolism protein became well established. Today, multiple theories abound to explain the function of *APOE* in the brain. The *APOE* made in the brain (primarily of glial origin) is separate from *APOE* made in the periphery (primarily of hepatocellular origin). ApoE is secreted as a 34 KDa glycoprotein, where it is thought to exist as the major apolipoprotein and lipid carrier in the brain (7). Brain *APOE* is secreted primarily by astrocytes, and it resides on high-density lipoprotein (HDL)-like particles. Due to its amphipathic nature it is drawn to lipids while still retaining the ability to bind receptors at the cell surface (e.g., neuronal *APOE* receptors). *In vitro* studies show that each isoform exhibits significant differences in affinity for the low-density lipoprotein (LDL) receptor (8, 9), however, these studies were performed with nonneuronal cell types. The difficulty of performing *in vivo APOE*–receptor studies in the brain limits our understanding of this interaction. Furthermore, each isoform has been shown to exhibit significant difference in protein stability, redox status, immune activation, and other purported functions, underscoring the complexity of *APOE* biology. ApoE-based animal models at minimum provide us with the opportunity to test cell-based theories in a system-wide approach where all the necessary parts are able to interact in a human-like fashion.

3. APOE-Deficient Models

The *APOE*-deficient (–/–) mouse was the first animal used to study *APOE's* role in dementia (or more specifically AD). The first studies in *APOE* (–/–) mice showed age-dependent deficits in specific neuronal markers such as synaptophysin and microtubule-associated-2 protein (10). Other studies showed reduced choline acetyltransferase activity (11), and long-term potentiation (LTP) deficits in *APOE* (–/–) mice (12) compared to wild-type mice. The *APOE* (–/–) mice also exhibited behavioral deficits in spatial memory that could be rescued by infusion of recombinant *APOE*3 or E4 (10). Multiple studies were performed with these animals (for review see (10) and (13)) demonstrating cognitive deficits similar to human "pre-dementia." However, other groups were unable to replicate these findings, casting doubt over the utility of these animals for future dementia studies (14, 15). It has been suggested that differences in genetic background or environment might explain the discrepancy observed in varying studies of *APOE* (–/–) mice. It became apparent that refining our understanding of *APOE's* relationship with dementia would require transgenic mice expressing all three isoforms.

4. Transgenic APOE Models

The majority of disease-linked genes involve a mutated (disease causation) and a normal gene. The *APOE* polymorphism is unique in that one allele (*E*4*) is linked to disease, another (*E*3*) is neutral, and the least common allele (*E*2*) is considered protective. Therefore, a good animal model of *APOE*-based dementia should include all three alleles as it is equally important to determine why *E*4* is harmful as why *E*2* is protective (with *E*3* serving as the control). This led to the creation of *APOE* transgenic mice expressing all or a subset of the *APOE* alleles via standard pronuclear injection of human *APOE* DNA. The first attempt was performed by Bowman et al. wherein human *APOE* cDNAs driven by the transferrin (TF) promoter were injected into wild-type mice (not crossed to *APOE* (–/–) mice) and analyzed for brain *APOE* expression. Very little characterization of these lines was performed, other than to show the mice expressed brain *APOE* RNA and *APOE*3 protein (16). A more extensive study was performed using human *APOE* transgenic mice created by Xu et al. These mice were created via insertion of human *APOE* genomic fragments driven by 3–15 kb of upstream regulatory sequences for AD modeling studies (17). Like most transgenic animals, *APOE* expression was variable in both the amount of *APOE* protein produced and the spatial distribution pattern as dictated by each founder line. Human *APOE* could be detected in a subset of neurons by immunocytochemistry, which varied by *APOE* genotype (founder-specific) and was detected sparingly in astrocytes (18). Today we know that brain *APOE* is synthesized predominantly in glial cells (i.e., astrocytes) and can only be detected in neurons post-injury. Even in the post-injury state, neuronal *APOE* expression is difficult to detect, and its importance in AD pathogenesis remains to be determined. In an attempt to answer this question, transgenic mice containing the neuron-specific enolase (NSE), an upstream promoter of a human *APOE* cDNA, was created (19). These mice expressed *APOE* at extraordinarily high levels in neurons (only), and exhibited behavioral deficits (19) and an inability to protect against kainic acid-induced injury (20). To avoid confounding effects from mouse *APOE* expression, these mice were crossed to an *APOE* (−/−) mouse, which adds a different layer of complexity to the overall phenotype. The NSE–*APOE/APOE* (−/−) doubly transgenic mice do not express *APOE* in any glial cells or peripheral cells (e.g., hepatocytes and macrophages). Therefore, these mice have the same phenotype of *APOE* (−/−) mice, which is hypercholesterolemia and chronic inflammation. This along with an extremely high level of neuronal *APOE* expression makes interpretation of the resultant phenotype difficult. Other attempts to address the *APOE* neuronal expression theory used promoters such as platelet-derived growth factor (PDGF), thy1, and phosphoglycerate kinase (PGK) to derive expression of human *APOE4* in a wild-type mouse *APOE* background (21). Mice that had the highest level of neuronal *APOE4* expression coupled with advanced age showed evidence for increases in hyperphosphorylated tau (22). No comparison to similar mice expressing *APOE3* or E2 was published. A different group did look at *APOE2* expression (driven by PDGF or TF) in mice with an *APOE* (−/−) background showing that *APOE2* expression restored levels of synaptophysin that were reduced in *APOE* (−/−) mice (23).

In an attempt to create an animal that more closely matches the "normal" pattern of brain *APOE* expression, glial fibrillary acidic protein (GFAP) transgenic mice were made. These mice express *APOE* only in astrocytes, and because they are crossed with *APOE* (−/−) mice they lack peripheral expression of *APOE* (24). There are no overt differences between GFAP-E3 or E4 mice and no brain pathology suggestive of dementia in these mice. However, the GFAP-E4 mice do show reduced spine density compared to the GFAP-E3 mice (25). The GFAP mice suffer the same limitations as other transgenic mice that must be crossed on to an *APOE* (−/−) background (see above).

At the time, use of transgenic mice for studying human *APOE* expression provided us with valuable clues to the function of brain

APOE. We also learned that transgenic technology produced several caveats. This method typically produces mice with varying levels of *APOE* expression due to differences in chromosomal location and copy number of the transgene. The use of artificial promoters also results in imprecise spatial and temporal expression of the transgene in both brain and other tissues where *APOE* is normally expressed (e.g., liver, kidney, gonads, adrenals). Since all transgenic animals must be crossed to an *APOE* (−/−) animal to remove the influence of mouse *APOE*, the peripheral cholesterol levels (~600 mg/dl) and the subsequent systemic inflammation can lead to unknown isoform-specific consequences in the central nervous system (CNS). Phenotypic differences due to transgene insertional mutagenesis cannot be ruled out, and expression of unwanted sequences within the transgenic construct can complicate the interpretation of the data. Therefore, it is important when designing experiments to investigate *APOE* isoform-specific effects, that a model with fewer confounding factors should be used.

5. APOE "Knock-In" Mice

One way to avoid confounding effects from an *APOE* (−/−) background is to create *APOE* "knock-in" mice. Hamanaka et al. (26) created *APOE4* "knock-in" mice by targeting the human *APOE4* cDNA to the mouse *APOE* locus. Unfortunately, this strategy resulted in *APOE4* mice exhibiting a hypercholesterolemic phenotype (i.e., plasma total cholesterol was 2.5-fold higher than wild-type mouse controls). This may have been due to the lack of intronic enhancers and/or genomic placement of the neomycin-resistant gene. To address this concern, a separate group bred these mice to ZP3-cre mice to remove the neomycin gene, which effectively eliminated the hypercholesterolemia phenotype from the *APOE4* mice. No overt differences between *APOE3* and E4 mice in this group have been identified (27).

Possibly the simplest and least complicated *APOE* animal model available today are the human *APOE* targeted replacement (TR) mice. In contrast to the "Hamanaka et al." strategy, these mice were created by replacing the mouse *APOE* gene with a human *APOE* genomic fragment (28–30). These animals express human *APOE* at physiological levels in both a temporal and spatial pattern similar to wild-type mice, nonhuman primates, and humans (31). This similarity in expression pattern across species is due to the fact that the human *APOE* gene is driven by endogenous mouse regulatory sequences, thus keeping the biofeedback loop intact for studying modulation by extrinsic factors (e.g., diet, oxidative stress, and injury). Although this model has been criticized (32) for not including human regulatory sequences, there are no animal models that contain sufficiently large segments (e.g., 100–200 kb) of the human *APOE* gene for gene expression studies. The TR mice are currently our best alternative for modeling human *APOE* expression *in vivo*. Furthermore, the *APOE3* and E4 TR mice are normolipidemic when maintained on a normal rodent chow diet and exhibit no overt pathology in any tissue examined. Thus, like human *APOE4* carriers, other genetic and nongenetic factors are required to precipitate disease. Most importantly, all three *APOE* lines (E2, E3, and E4) were made in identical fashion and thus can be used for *APOE* isoform-specific comparison studies.

We have shown that the *APOE4* TR mice, in the absence of injury, model early susceptibilities to cognitive impairment analogous to humans (33). The *APOE4* mice display reduced excitatory postsynaptic currents, stunted neuronal morphology, a twofold reduction in spines (synapses) (33), reduced LTP (34), and memory deficits (35) when compared to *APOE3* mice at a young age. The *APOE4* mice are also more susceptible to amyloid β peptide (Aβ)-induced repression of LTP (36), and exhibit increased amyloid burden compared to *APOE3* mice (37), suggesting a susceptibility to injury (Aβ)-induced dementia. Thus, the TR mice are ideal for studying *APOE* isoform-specific effects in dementia as they model early characteristics of dementia, and the *APOE* expression profile is both spatially and temporally similar to humans. Like humans, TR mice synthesize the majority of their *APOE* in the liver as a 34 kDa lipophilic glycoprotein that binds to lipoprotein particles and is required for maintaining lipid homeostasis. The *APOE* synthesized in the brain is derived from astrocytes, and is incorporated into HDL-like particles that are very similar to the HDL-like particles analyzed from nondemented human brain samples (38). Although the function of brain-derived *APOE* remains unknown, it is reasonable to suggest that brain *APOE* shares overlapping functions with liver-derived *APOE* (i.e., maintain lipid and cholesterol homeostasis).

Collaborative studies with these mice suggest that *APOE4* is less able to repair damaged neuronal tissue, irrespective of whether the damage was via traumatic brain injury (39) or entorhinal cortex lesion (40).

Other *APOE*-based models have been created to better understand the relationship between isoform-specific structure and function (41), as well as the implication of truncated forms of human *APOE* (42). Further insight will likely come from crossbreeding *APOE* mice to other AD gene candidates (e.g., APP, amyloid precursor protein, low-density lipoprotein receptor (LDLR), lipoprotein receptor-related protein (LRP) and CYP24S, cytochrome P hydroxylase), as *APOE4* alone is not sufficient to cause dementia. Likewise, designing experiments to study nongenetic risk factors (e.g., diet, stress, and aging) of AD in the human *APOE* animal model may result in conversion of a "pre-dementia" phenotype to one more closely resembling early- to mid-stage AD.

6. Conclusions

Although there are several *APOE*-based animal models of "pre-dementia," none of them reproduces the full spectrum of neuropathology observed in humans or meets the definition of "progressive neuronal dysfunction." The few animals that do exhibit histopathological hallmarks of AD (i.e., plaques, tangles, and neuronal cell loss) are artificially produced typically by overexpression of mutant genes that do not reflect the genetics of the common forms of dementia (e.g., late-onset AD). These animal models are valuable for studying the rare forms of familial dementia but should be used with caution when studying the etiology and progression of the more common forms of dementia. Since it is thought that the only way to treat dementia is through prevention (or in the early stages of disease), this underscores the need for an animal model that mimics the common disease phenotype. The value of a good animal model cannot be understated. Multiple clinical trials have failed, and the blame generally falls on the use of a poor model. Poor models result in incorrect targets for drug design and erroneous biomarkers of the disease. We do not always have to understand the mechanisms responsible for disease initiation to produce a drug that works, however, our successes would be greatly enhanced if we were able to create a valid model of disease.

References

1. Rubinsztein DC and Easton DF (1999) Apolipoprotein E genetic variation and Alzheimer's disease: a meta-analysis. Dement Geriatr Cogn Dis 10:199–209
2. Hallman DM, Boerwinkle E, Saha N, et al. (1991) The apolipoprotein E polymorphism: a comparison of allele frequencies and effects in nine populations. Am J Hum Genet 49: 338–349
3. Saunders AM, Strittmatter WJ, Schmechel D, et al. (1993) Association of apolipoprotein E allele epsilon 4 with late-onset familial and sporadic Alzheimer's disease. Neurology 43: 1467–1472
4. Poirier J, Davignon J, Bouthillier D, Kogan S, Bertrand P, Gauthier S (1993) Apolipoprotein E polymorphism and Alzheimer's disease. Lancet 342:697–699
5. Corder EH, Saunders AM, Risch NJ, et al. (1994) Protective effect of apolipoprotein E type 2 allele for late onset Alzheimer disease. Nat Genet 7:180–184
6. Weisgraber KH (1994) Apolipoprotein E: structure-function relationships. Adv Protein Chem 45:249–302
7. Mahley RW (1988) Apolipoprotein E: cholesterol transport protein with expanding role in cell biology. Science 240:622–630
8. Weisgraber KH, Innerarity TL, Mahley RW (1982) Abnormal lipoprotein receptor-binding activity of the human E apoprotein due to cysteine-arginine interchange at a single site. J Biol Chem 257:2518–2521
9. Bohnet K, Pillot T, Visvikis S, Sabolovic N, Siest G (1996) Apolipoprotein (apo) E genotype and *APOE* concentration determine

binding of normal very low density lipoproteins to HepG2 cell surface receptors. J Lipid Res 37:1316–1324

10. Masliah E, Mallory M, Veinbergs I, Miller A, Samuel W (1996) Alterations in apolipoprotein E expression during aging and neurodegeneration. Prog Neurobiol 50:493–503
11. Gordon I, Grauer E, Genis I, Sehayek E, Michaelson DM (1995) Memory deficits and cholinergic impairments in apolipoprotein E-deficient mice. Neurosci Lett 199:1–4
12. Veinbergs I, Mante M, Jung MW, Van Uden E, Masliah E (1999) Synaptotagmin and synaptic transmission alterations in apolipoprotein E-deficient mice. Prog Neuropsychopharmacol Biol Psychiatr 23:519–531
13. Teter B (2004) ApoE-dependent plasticity in Alzheimer's disease. J Mol Neurosci 23: 167–179
14. Fagan AM, Murphy BA, Patel SN, et al. (1998) Evidence for normal aging of the septo-hippocampal cholinergic system in *APOE* (–/–) mice but impaired clearance of axonal degeneration products following injury. Exp Neurol 151:314–325
15. Anderson R, Barnes JC, Bliss TV, et al. (1998) Behavioural, physiological and morphological analysis of a line of apolipoprotein E knockout mouse. Neuroscience 85:93–110
16. Bowman BH, Yang F, Buchanan JM, et al. (1996) Human APOE protein localized in brains of transgenic mice. Neurosci Lett 219: 57–59
17. Xu PT, Schmechel D, Rothrock-Christian T, et al. (1996) Human apolipoprotein E2, E3, and E4 isoform-specific transgenic mice: human-like pattern of glial and neuronal immunoreactivity in central nervous system not observed in wild-type mice. Neurobiol Dis 3:229–245
18. Xu PT, Gilbert JR, Qiu HL, et al. (1998) Regionally specific neuronal expression of human APOE gene in transgenic mice. Neurosci Lett 246:65–68
19. Raber J, Wong D, Buttini M, et al. (1998) Isoform-specific effects of human apolipoprotein E on brain function revealed in ApoE knockout mice: increased susceptibility of females. Proc Natl Acad Sci USA 95: 10914–10919
20. Buttini M, Orth M, Bellosta S, et al. (1999) Expression of human apolipoprotein E3 or E4 in the brains of Apoe–/– mice: isoform-specific effects on neurodegeneration. J Neurosci 19:4867–4880
21. Tesseur I, Van Dorpe J, Bruynseels K, et al. (2000) Prominent axonopathy and disruption of axonal transport in transgenic mice expressing human apolipoprotein E4 in neurons of brain and spinal cord. Am J Pathol 157:1495–1510
22. Tesseur I, Van Dorpe J, Spittaels K, Van den Haute C, Moechars D, Van Leuven F (2000) Expression of human apolipoprotein E4 in neurons causes hyperphosphorylation of protein tau in the brains of transgenic mice. Am J Pathol 156: 951–964
23. Georgopoulos S, McKee A, Kan HY, Zannis VI (2002) Generation and characterization of two transgenic mouse lines expressing human ApoE2 in neurons and glial cells. Biochemistry 41: 9293–9301
24. Sun Y, Wu S, Bu G, et al. (1998) Glial fibrillary acidic protein-apolipoprotein E (*APOE*) transgenic mice: astrocyte-specific expression and differing biological effects of astrocyte-secreted *APOE3* and *APOE4* lipoproteins. J Neurosci 18:3261–3272
25. Ji Y, Gong Y, Gan W, Beach T, Holtzman DM, Wisniewski T (2003) Apolipoprotein E isoform-specific regulation of dendritic spine morphology in apolipoprotein E transgenic mice and Alzheimer's disease patients. Neuroscience 122:305–315
26. Hamanaka H, Katoh-Fukui Y, Suzuki K, et al. (2000) Altered cholesterol metabolism in human apolipoprotein E4 knock-in mice. Hum Mol Genet 9:353–361
27. Mann KM, Thorngate FE, Katoh-Fukui Y, et al. (2004) Independent effects of APOE on cholesterol metabolism and brain beta levels in an Alzheimer disease mouse model. Hum Mol Genet 13:1959–1968
28. Sullivan PM, Mezdour H, Aratani Y, et al. (1997) Targeted replacement of the mouse apolipoprotein E gene with the common human APOE3 allele enhances diet-induced hypercholesterolemia and atherosclerosis. J Biol Chem 272:17972–17980
29. Sullivan PM, Mezdour H, Quarfordt SH, Maeda N (1998) Type III hyperlipoproteinemia and spontaneous atherosclerosis in mice resulting from gene replacement of mouse Apoe with human Apoe*2. J Clin Invest 102:130–135
30. Knouff C, Hinsdale ME, Mezdour H, et al. (1999) Apo E structure determines VLDL clearance and atherosclerosis risk in mice. J Clin Invest 103:1579–1586
31. Sullivan PM, Mace BE, Maeda N, Schmechel D (2004) Marked regional differences of brain human apolipoprotein E expression in targeted replacement mice. Neuroscience 124:725–733

32. Maloney B, Ge YW, Alley GM, Lahiri DK (2007) Important differences between human and mouse APOE gene promoters: limitation of mouse APOE model in studying Alzheimer's disease. J Neurochem 103:1237–1257
33. Wang C, Wilson WA, Moore SD, et al. (2005) Human *APOE4* targeted replacement mice display reduced excitatory synaptic activity and dendritic arborization. Neurobiol Dis 18: 390–398
34. Trommer BL, Shah C, Yun SH, et al. (2004) ApoE isoform affects LTP in human targeted replacement mice. Neuroreport 15:2655–2658
35. Grootendorst JBA, Vogel E, Kelche C, et al. (2005) Human *APOE* targeted replacement mouse lines: h-*APOE4* and h-*APOE3* mice differ on spatial memory performance and avoidance behavior. Behav Brain Res 159:1–14
36. Trommer BL, Shah C, Yun SH, et al. (2005) ApoE isoform-specific effects on LTP: blockade by oligomeric amyloid-beta1–42. Neurobiol Dis 18:75–82
37. Fryer JD, Simmons K, Parsadanian M, et al. (2005) Human apolipoprotein E4 alters the amyloid-beta 40:42 ratio and promotes the formation of cerebral amyloid angiopathy in an amyloid precursor protein transgenic model. J Neurosci 25:2803–2810
38. Morikawa M, Fryer JD, Sullivan PM, et al. (2005) Production and characterization of astrocyte-derived human apolipoprotein E isoforms from immortalized astrocytes and their interactions with amyloid-beta. Neurobiol Dis 19:66–76
39. Wang H, Durham L, Dawson H, et al. (2007) An apolipoprotein E-based therapeutic improves outcome and reduces Alzheimer's disease pathology following closed head injury: evidence of pharmacogenomic interaction. Neuroscience 144:1324–1333
40. Blain JF, Sullivan PM, Poirier J (2006) A deficit in astroglial organization causes the impaired reactive sprouting in human apolipoprotein E4 targeted replacement mice. Neurobiol Dis 21:505–514
41. Raffai RL, Weisgraber K (2002) Hypomorphic apolipoprotein E mice: a new model of conditional gene repair to examine apolipoprotein E-mediated metabolism. J Biol Chem 277: 11064–11068
42. Harris FM, Brecht WJ, Xu Q, et al. (2003) Carboxyl-terminal-truncated apolipoprotein E4 causes Alzheimer's disease-like neurodegeneration and behavioral deficits in transgenic mice. Proc Natl Acad Sci USA 100: 10966–10971

Chapter 23

TAU Models

Nicolas Sergeant and Luc Buée

Abstract

Tau pathology refers to molecular mechanisms leading to the intracellular aggregation of abnormally modified Tau protein isoforms and to the propagation of this degenerating process along neuronal circuitry. Tau proteins belong to the family of microtubule-associated proteins. Tau is strongly expressed in neurons, localized in the axon and is essential for neuronal architecture, plasticity and network. In animal models, overexpression of mutated Tau proteins is often used to induce a Tau pathology, leading to motor and/or cognitive dysfunction. These animal models are essential to understand how neuronal degeneration and Tau aggregation are related. Indeed, Tau pathology is certainly a good therapeutic target, but untangling Tau remains a major therapeutic challenge.

Key words: Alzheimer's disease, Neurofibrillary degeneration, Microtubule-associated protein Tau, Tauopathies, Tau pathology

1. Introduction

In Alzheimer's disease (AD), microtubule-associated protein Tau is the major component of paired helical filaments inside degenerating neurons referred to as neurofibrillary tangles (NFT). However, aggregation of Tau inside neurons or glial cells is observed in more than 20 neurological disorders, which together form the so-called Tauopathies. The *Tau/MAPT* gene locus is associated with the development of frontotemporal dementia, and microdeletion in the locus region is associated with mental retardation. Besides its most well-known function as being a microtubule-polymerizing and -stabilizing factor, Tau is implicated in the architecture and organization of the paralleled order of microtubules and it also important for the regulation axonal transport through the regulation of motor proteins and vesicular transport. When aggregated, one of the major questions that

Peter Paul De Deyn and Debby Van Dam (eds.), *Animal Models of Dementia*, Neuromethods, vol. 48,
DOI 10.1007/978-1-60761-898-0_23, © Springer Science+Business Media, LLC 2011

remain to be answered is whether there is a loss or gain of toxic functions. Recent development of transgenic animal models may be useful to address this question.

2. Microtubule-Associated Tau

2.1. Tau/MAPT Gene

The human *Tau/MAPT* gene is unique, and contains 16 exons located over more than 150 kb on the long arm of chromosome 17 at band position 17q21 (1, 2) (Fig. 1). Although the mechanism underlying the neuronal expression of Tau is not known, the expression of Cre recombinase under the control of the murine Tau promoter results in an almost completely restricted expression of the recombinase in neurons (3). In humans, Tau expression is also found in muscle and scarcely in other tissues or organs (4). Several polymorphisms are identified and found to be in complete linkage disequilibrium with one another, defining an extended haplotype that covers the entire *MAPT* gene (5) and even spans to a region covering ~1.8 Mb (6) (Fig. 1a). The H2 haplotype is rarer than the H1 haplotype (73–79%) in healthy individuals, and results from H1 by the inversion of a ~970 kb

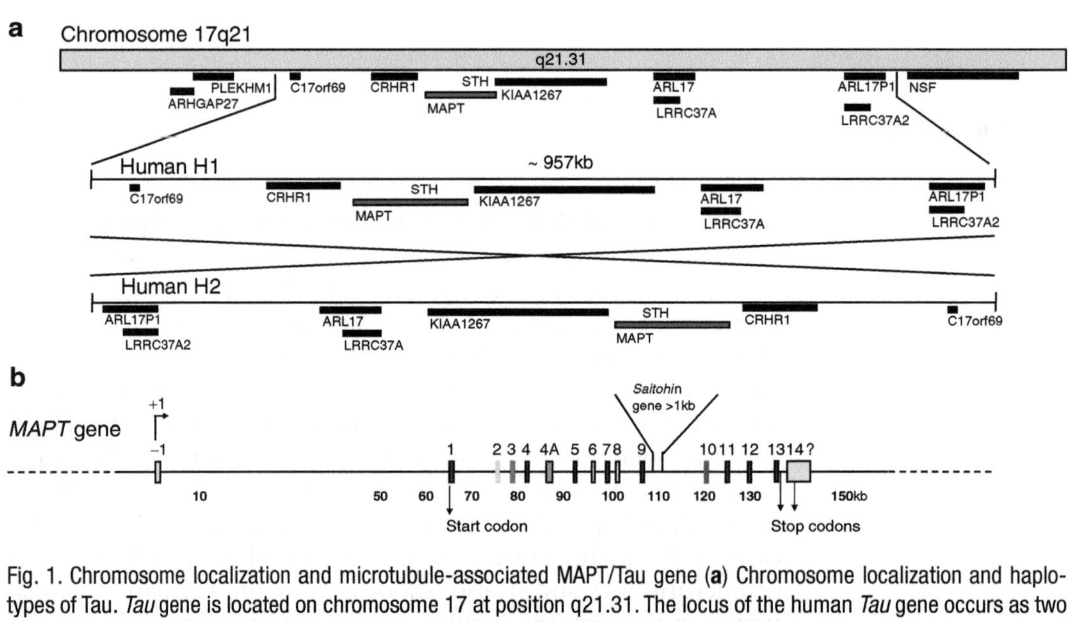

Fig. 1. Chromosome localization and microtubule-associated MAPT/Tau gene (**a**) Chromosome localization and haplotypes of Tau. *Tau* gene is located on chromosome 17 at position q21.31. The locus of the human *Tau* gene occurs as two haplotypes H1 and the inverted version of H1, H2. No recombination is observed between H1 and H2 over a region of approximately 1.5 Mb. Tau gene locus and the surrounding genes are represented. H1 and its ~970 kb inverted H2 haplotype are represented. (**b**) Tau gene. Tau gene spans more than 130 kb and is composed of 16 exons. Transcript initiation starts at exon -1 and the start codon is located in exon 1. They are two alternate stop codons located following exon 13 or inside exon 14. The 3′ ending of exon 14 is not completely characterized in humans. The initiation of transcription is indicated by +1. The promoter region of Tau encompasses the sequence upstream exon -1, the exon -1, and the intron 0. The promoter is GC-rich and contains SP1 and AP2 binding-motifs proximal to exon -1.

segment (7). The H2 haplotype is associated with an increased risk of de novo microdeletion and developmental delay and learning disability (8, 9). Specific H1 sub-haplotypes are associated with the risk of developing AD, Parkinsonism syndromes such as corticobasal degeneration and progressive supranuclear palsy and likely influence the profile of Tau isoform expression (10–12).

2.2. Posttranscriptional Modifications

The Tau primary transcript contains 16 exons (Fig. 2a). In the central nervous system (CNS), two transcripts of 2 kb and 6 kb arise from the use of two alternative polyadenylation sites, the 2 kb mRNA targets Tau to the nucleus and the 6 kb encodes the major form in axons (13, 14). In the human brain, Tau exons 4A and 8 are skipped. Exon 4A is found in bovine, human, and

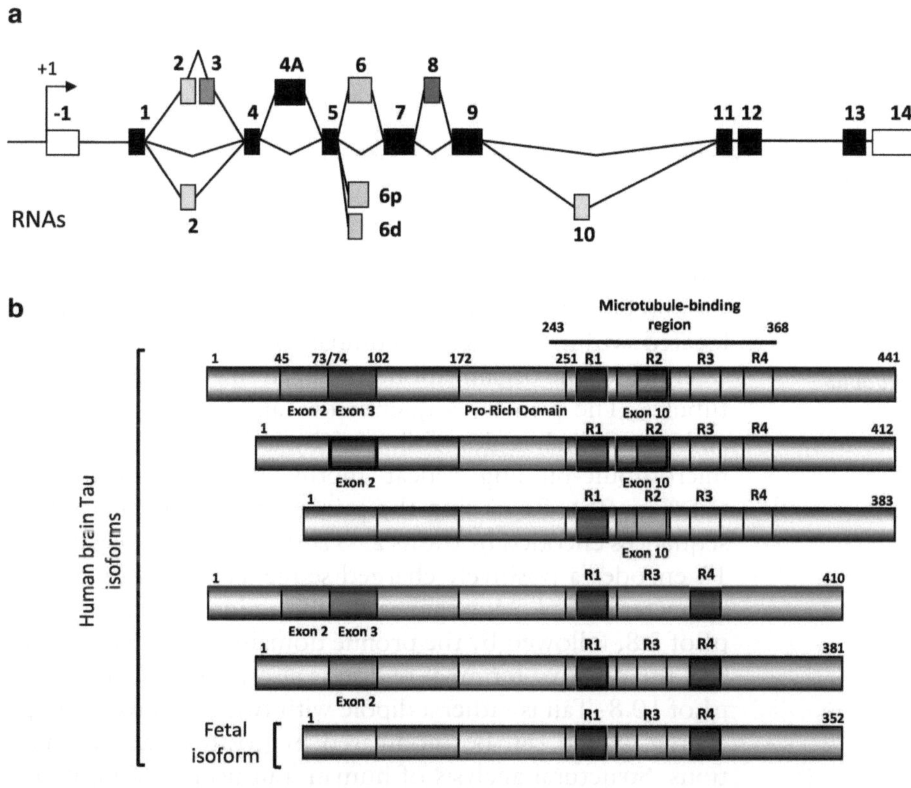

Fig. 2. Alternative splicing and Tau proteins in the central nervous system (**a**) Alternatively spliced cassettes. Several mRNAs are generated by alternative splicing of exons 2, 3, 4A and 10. The 6p and 6d are cryptic splicing sites rarely used. The selection of those sites is supposed to generate carboxy-truncated Tau proteins (for review see (13)). Exon 4A is excluded in the brain and included in the spinal cord and peripheral nervous system. Exon 8 is found included only in muscle and found also in other species such as bovine Tau. (**b**) Tau protein isoforms in the central nervous system. In the human brain, six major Tau isoforms are generated from the alterative splicing of exons 2, 3 and 10. Exon 3 is always included with exon 2. The exon 10 encodes an additional microtubule-binding motif numbered R1 to R4 that together defines the microtubule-binding region. Half of Tau proteins contain three microtubule-binding motifs (light grey) and the other half (dark grey) has four microtubule-binding motifs. Upstream the microtubule-binding is the proline-rich domain containing several phosphorylation sites that regulate Tau binding to microtubules.

rodent peripheral tissues with a high degree of homology. Exon 8 is alternative and found in striated muscle, spinal cord and pituitary (4). Noteworthy, Tau expression is tenfold higher in brain and spinal cord than in skeletal muscle. Cryptic splicing sites are described in exon 6 that generate Tau mRNA lacking the remaining 3′ exon cassettes (15). Those are found in muscle and the spinal cord, but the presence of the protein remains to be determined. Exon −1 and adjacent intron are part of the promoter. Exon −1 is transcribed but not translated. Exons 1, 4, 5, 7, 9, 11, 12, and 13 are constitutive exons. Exon 14 is part of the 3′ untranslated region of Tau mRNA (2, 16, 17). Exons 2, 3, and 10 are alternatively spliced and are adult brain-specific (18). Exon 3 is never included independently of exon 2, although exon 2 can be included independently of exon 3 (2). Thus, alternative splicing of these three major exons allows for six mRNAs (2–3–10-; 2+3–10–; 2+3+10–; 2–3–10+; 2+3–10+; 2+3+10+) and in the human brain, the Tau primary transcript comprises six mRNAs translated into six protein isoforms (16, 17, 19) (Fig. 2b).

3. Tau Proteins, Posttranslational Modification, and Functions

3.1. Tau Functions

Tau was first discovered in the mid-1970s as a factor promoting tubulin polymerization (20). Tau is a neuronal protein essentially located within the axonal compartment. Its structure makes it essential for the organization, stabilization and dynamics of microtubules. The primary sequence of Tau can be subdivided in an amino-terminal region followed by a proline-rich domain, the microtubule-binding repeat motifs and the carboxy-terminal tail (Fig. 2b). Regarding the primary structure, the polypeptide sequences encoded by exons 2/3 add to Tau acidity, whereas exon 10 encodes a positively charged sequence that adds to the basic character of Tau. More generally, the amino-terminal region has a pI of 3.8, followed by the proline domain, which has a pI of 11.4, and the carboxy-terminal region is also positively charged with a pI of 10.8. Tau is rather a dipole with two domains with opposite charge, which can be modulated by posttranslational modifications. Structural analysis of human Tau using several biophysical methods showed that in solution Tau behaves as an unfolded protein (21). However, functions of Tau are distributed both in the amino- and the repeat-domains.

The amino-terminal region together with the proline-rich domain is referred to as the "projection domain." This unstructured and negatively charged region detaches from the surface of microtubules (22) and can interact with the plasma membrane or cytoskeletal proteins (23). Tau may also contribute to the spacing in-between microtubule lattice and also to the parallel ordered organization of microtubules in axons (24). The spacing may be dependent upon the presence of additional amino-terminal sequences such as exons 2, 3 or 4A. The latter is included only in the spinal cord in peripheral nerve tissue (25). More recently, using a biophysical assay, Rosenberg and colleagues suggest that the amino-terminal regions of two Tau molecules, each one individually binding to a microtubule, form an electrostatic "zipper" (26). The amino-terminal region of Tau also interacts with a growing panel of proteins including motor proteins such as kinesin-1 (27) and dynactin–dynein complex (28), SH3-containing tyrosine kinases such as the phospholipase C-gamma 1 or the p85α subunit of PI-3K (29). Recently, it has been established that Tau regulates the motility of dynein and kinesin motor proteins by an isoform-dependent mechanism. Indeed, the shortest Tau isoform lacking exons 2, 3 and 10 impedes the motility of both kinesin and dynein, whereas the longest Tau isoforms with all exons less affect motor protein motility (30). Thus, the axonal transport of vesicles may be finely tuned by the ratio of Tau isoforms expressed. In neurodegenerative disorders, a modified pattern of Tau isoform expression/ratio, due to Tau aggregation for instance, may profoundly affect the axonal transport and could possibly lead to neurodegeneration (31).

The carboxy-terminal region, which is the basic region of Tau protein, is characterized by the presence of three or four repeat motifs, depending on the inclusion or not of the exon 10 encoding sequence. The repeat motifs are corresponding to the microtubule-binding domains. Apart from binding to microtubules, the repeat-domains of Tau interact with the histone deacetylase 6 (HDAC6), and HDAC6 is suggested to regulate Tau phosphorylation. The inhibition of the proteasome activity by MG132 enhances the interaction of HDAC6 to Tau. Interestingly, in this condition, HDAC6/Tau complexes are observed in perinuclear aggresomes (32). The proteasomal degradation of Tau is also sensitive to chaperone and co-chaperones Hsc70 and BAG-1 (Bcl2-associated athanogene-1) respectively that are interactors of Tau proteins (33, 34). The repeat-domains also interact more strongly with apolipoprotein E3 (ApoE3) than the E4 (ApoE4) isovariant (35). The functionality of Tau–ApoE interaction remains unknown. However, a triple transgenic mouse model of AD crossed with a knock-in ApoE4 mouse showed a strong influence of ApoE4 upon the topographical distribution of amyloid deposits, but surprisingly, the Tau pathology was strongly reduced (36). Together, Tau interacting partners may profoundly regulate its function but also modulate the Tau pathology. The physiologic and pathologic functions of Tau are also regulated by posttranslational modifications such as phosphorylation.

3.2. Tau Phosphorylation

Microtubule-associated proteins Tau are phosphoproteins (37). There are 85 potential phosphorylation sites on the longest Tau isoform. Phosphorylation sites were characterized using phospho-dependant Tau antibodies, phospho-peptide mapping, mass spectrometry, or nuclear magnetic resonance (NMR). According to the latest extensive analysis of Tau phosphorylation (38) and that of previously published reviews (39, 40), 71 among the 85 putative phosphorylation sites can be phosphorylated in physiological or pathological conditions (for review see (41)). Most of the phosphorylation sites are surrounding the microtubule-binding domains, in the proline-rich region and carboxy-terminal region of Tau. A total of more than 20 protein kinases can phosphorylate Tau proteins. This includes four groups of protein kinases: (1) the proline-directed protein kinases (PDPK), which phosphorylate Tau on Serines or Threonines that are followed by a Proline residue. This group includes cyclin-dependant kinase 2 and 5 (42–44), MAPK (45) and several SAPKs (46, 47). (2) The non-PDPK group includes Tau-tubulin kinases 1 and 2 (48–50), Casein kinases 1, 2 and 1δ (38, 51), DYRK1A (52–55), the phosphorylase kinase (56), Rho kinase (57), PKA, PKB/AKT, PKC, PKN (for review see (40)). (3) The third group includes protein kinases that phosphorylate Tau on Serine or Threonine residues followed or not by a Proline. Glycogen protein kinases (GSK3α and GSK3β) belong to this group and have a recognition motif (SXXXS or SXXXD/E) (58). This group also includes mitogen and stress-activated protein kinase MSK1, which belongs to the AGC kinases group. AGC kinases preferentially phosphorylate serine and threonine residues that lie in RXRXXS/T motifs (59). The S6 kinases p70 and p90 (RSK1 and RSK2) also phosphorylate Tau as well as serum- and glucocorticoid-induced kinase-1 (SGK1) (59). (4) The fourth group corresponds to tyrosine protein kinases such as Src kinases, C-abl and c-met (60) (see Diane Hanger Web site http://cnr.iop.kcl.ac.uk/hangerlab/tautable).

4. Tau Pathology

4.1. Tau in Neurofibrillary Degeneration

Neurofibrillary degeneration, as first described by Aloïs Alzheimer, consists of the intraneuronal accumulation of proteinaceous fibrils forming flame-shaped neurons. Fibrils were later demonstrated to correspond to paired helical filaments (PHF) indicating the well-structured organization of the molecular constituent (61). This material is insoluble and this biochemical property was used to isolate the material and to generate antibodies. The major antigen of PHF was shown to correspond to Tau protein, namely PHF-Tau (for review see (62)). Similarly, antibodies raised against isolated PHF-stained Tau proteins (63). Proteomic isolation and characterization of the most aggregated components of PHF showed that the core of PHF was mostly composed of the microtubule-binding motif lacking the exon 10 encoding sequences (64–66).

The most striking difference between postmortem Tau and PHF-Tau proteins is the molecular weight, as Tau protein-derived from postmortem control individual are resolved as six main bands (45–67 kDa), whereas PHF-Tau comprise four main bands between 60 and 74 kDa. Two-dimensional gel electrophoresis followed by Western-blotting is a useful method to demonstrate that PHF-Tau is more acidic than normal Tau (67, 68). Increased Tau acidity is due to hyper- and abnormal phosphorylation in degenerating neurons. A number of studies, using phosphatase, antibodies against PHF-Tau – thereafter showed to be phospho-dependant Tau antibodies – and mass-spectrometry analyses, together demonstrated that PHF-Tau is hyperphosphorylated on "native" phospho-sites and abnormally phosphorylated on pathologically related sites. The former observation was demonstrated using phospho-dependant antibodies and two-dimensional gel electrophoresis on human brain tissue obtained from tumor resections and, therefore, without postmortem/surgery delay (69, 70). Abnormal phospho-sites were demonstrated using antibodies that recognize PHF-Tau and not normal or "native" Tau, such as AT100 or AP422 that recognize the phosphorylated Threonines 212–214 and Serine 217, and the Serine 422, respectively. Nonetheless, to generate the pathological phospho-sites at AT100 epitope a sequential phosphorylation by two kinases (71) is necessary suggesting that several kinases may be deregulated in AD.

In AD disease, Tau pathology is found in brain areas affected with aging and several additional brains areas are also affected. However, the Tau pathology is following cortico-cortical connections with a stereotyped, sequential and hierarchical pathway affecting large pyramidal neurons. Braak and Braak proposed a six-stage spatiotemporal progression of Tau pathology in AD (72). A similar scheme of spatiotemporal spreading into ten stages is also described according to the brain regions sequentially affected: transentorhinal cortex (S1), entorhinal (S2), hippocampus (S3), anterior temporal cortex (S4), inferior temporal cortex (S5), mid-temporal cortex (S6), polymodal association areas (prefrontal, parietal inferior, temporal superior) (S7), unimodal areas (S8), primary motor (S9a) or sensory (S9b, S9c) areas and all neocortical areas (S10). Up to stage 6, the disease could be asymptomatic. In all the cases of our study, stage 7 individuals with two polymodal association areas affected by Tau pathology were cognitively impaired.

Yet, the cortical factors that trigger the cortical-spreading of Tau pathology in AD remain ill-defined (73–75). However, recent data suggest that neurofibrillary degeneration cortical spreading could follow a transmissible prion-like process. In fact, purified aggregates of PHF-Tau were purified from transgenic mice. Intracranial injection of this preparation was done in a mouse model, which overexpresses human Tau protein but does not display Tau pathology. Following the injection, development of a Tau pathology was observed. This pathology progressed from the injection site to neighboring brain structures (76).

5. Inherited Frontotemporal Dementia Linked to Chromosome 17 (FTDP) and Tau Gene Mutations

Historically, frontotemporal dementia (FTD) was often classified as a form of Pick's disease, even when Pick cells or Pick bodies were not found (77). However, this denomination may involve different subgroups of pathologies, and the Lund and Manchester groups published in 1994 a consensus on Clinical and Neuropathological Criteria for Frontotemporal Dementia (1994). This publication clarified the position of Pick's disease within FTD, and several of the reported cases of familial Pick's disease were probably cases of familial FTD. Indeed it is difficult to ascertain families that have the classic pathological features of Pick's disease from the literature (78), because they often have unusual clinical features.

In 1994, Wilhelmsen and colleagues described an autosomal dominantly inherited disease related to familial FTD, characterized by adult-onset behavioural disturbances, frontal lobe dementia, parkinsonism, and amyotrophy (79). They demonstrated a genetic linkage between this pathology, designated disinhibition–dementia–parkinsonism–amyotrophy complex (DDPAC) and chromosome 17q21–22 (79). Since then, several families sharing strong clinical and pathological features and for which there is a linkage with chromosome 17q22–22 have been described (80–83). They have been included in a group of pathologies referred to as frontotemporal dementia with parkinsonism linked to chromosome 17 (FTDP-17) (84).

Although clinical heterogeneity exists between and within the families with FTDP-17, usual symptoms include behavioural changes, loss of frontal executive functions, language deficit, and hyperorality. Parkinsonism and amyotrophy are described in some families, but are not consistent features. Neuropathologically, brains of FTD patients exhibit an atrophy of frontal and temporal lobes, severe neuronal cell loss, gray and white matter gliosis, and a superficial laminar spongiosis. One of the main characteristics is the filamentous pathology affecting the neuronal cells, or both neuronal and glial cells in some cases. The absence of amyloid aggregates is usually established (84, 85).

FTDP-17 has been related to mutations on the Tau gene (86–89). Tau mutations always segregate with the pathology and are not found in the control subjects, suggesting their pathogenic role. A total of 44 mutations are indexed. Thirty one missense mutations in coding regions, three silent mutations L284L, N296N and S305S, two single amino acid deletions ΔK280 and ΔN296 and nine intronic mutations in the splicing region following exon 10 at position +3, +11, +12, +13, +14, +16, +19, +29 and +33 have been reported (for review see http://www.molgen.ua.ac.be/FTDmutations). More recently, a Tau gene deletion encompassing exons 6 to 9 has been described in a French family and associated with the development of FTD (90).

Depending on their functional effects, Tau protein mutations may be divided into three groups: mutations affecting the alternative splicing of exon 10, and leading to changes in the proportion of 4R- and 3R-Tau isoforms, and mutations modifying Tau interactions with microtubules and those facilitating aggregation. Mutations of the Tau gene and their involvement in FTDP-17 emphasize the fact that abnormal Tau proteins may play a central role in the etiopathogenesis of neurodegenerative disorders, without any implication of the amyloid cascade. The functional effects of the mutations suggest that a reduced ability of Tau to interact with microtubules may be upstream of hyperphosphorylation and aggregation. These mutations may also lead to an increase in free cytoplasmic Tau (especially the 4R-Tau isoforms), and therefore facilitating their aggregation into filaments (91). Finally, some mutations may have a direct effect on Tau fibrillogenesis (92).

6. Tau Transgenic Mice

6.1. Tau-Knockout Mice

Tau expression was suppressed in different mouse models by MAPT deletion or invalidation (93–95). All of them appear physically normal and are able to reproduce. They do not display any change in central or peripheral nervous systems (94, 95). In one model, slight changes were seen in axonal diameter (93) and motor and cognitive deficits were also reported (96). Moreover, delay in axonal maturation was also reported in primary neuronal cell cultures from Tau-deficient mice (94). Tau deficiency is likely to be compensated by other microtubule-associated proteins such as MAP1A.

6.2. Wild-Type Tau Transgenic Mice

When 3R Tau isoforms are overexpressed in transgenic mice, hyperphosphorylated Tau mainly accumulates in spinal cord neurons. Main pathological observations were axonal degeneration, diminished number of microtubules, and reduced axonal transport

(97, 98). Similar data were observed in transgenic mice expressing the longest human brain Tau isoform under the control of the human Thy-1 promoter. Hyperphosphorylated human Tau protein was present in nerve cell bodies, axons and dendrites (99–101).

Interestingly, transgenic mice bearing the human MAPT mini-gene were also developed (102). This model overexpresses human Tau proteins two- to threefold higher than murine Tau, but it does not exhibit any Tau pathology. However, surprisingly, when this model was crossbred with the murine MAPT gene deletion background model (95), the offspring displayed Tau pathology in a time-dependent manner, suggesting that murine Tau proteins may act as inhibitors of Tau aggregation (103). Independently, crossbreeding between other Tau-deficient mice (94) and transgenic mice with MAPT mini-gene was also used for understanding Tau splicing in early developmental stages (from 3R to 4R isoforms) (104). There is no data yet available on aged animals in these strains.

6.3. Mutated Tau Transgenic Mice

With the discovery of FTDP-17 mutations on MAPT, numerous transgenic models using these mutations were developed. They all allow for the development of Tau pathology characterized by Tau aggregation and neurofibrillary degeneration. Tau transgene is under various promoters (2′,3′-cyclic nucleotide 3′-phosphodiesterase, CaMKII, PDGF, Prion or Thy1.2) under an inducible system or not.

They display various phenotypes, with the most prominent one being motor deficits. In Table 1, we summarize the general phenotype of these models. In the next section, we describe four representative mutated Tau transgenic mice: rTg4510, $Tau_{RD}/\Delta K280$, K3 and Thy-Tau22[105–108], which allow for better understanding of Tau pathology.

In rTg4510 Tau transgenic mice, P301L mutant Tau protein is expressed in an inducible way. Following P301L Tau expression, these mice develop NFTs, neuron loss and behavioural impairment in a time-dependent manner (105). However, it is possible to switch off the expression of the transgene at a given time. Suppression of P301L Tau expression in rTg4510 Tau transgenic mice, which normally express the mutant protein at a high level, reverses behavioural impairments in these mice, although NFT formation continues. It suggested that NFT formation could be dissociated from neuronal dysfunction. In fact, soluble Tau rather than NFTs may be neurotoxic (105). Thus, NFT are unlikely to be the major toxic Tau species, at least in the early stages of Tauopathy. Two forms of Tau multimers (140 and 170 kDa), whose molecular weight suggests an oligomeric aggregate, were recently described in rTg4510 Tau transgenic mice. They accumulate early in the pathogenic cascade and their levels

Table 1
Main Tau transgenic mouse models

Strain name	Promoter	Mutation Tau construct	Phenotype	References
P301S Tau	Thy1.2	P301S 2-3-10+	Neuron, motor	Allen et al., 2002 (116)
]T-279	MAPT	N279K 2+3+10+	Neuron, glial, motor	Dawson et al., 2007 (117)
]pR5	Thy1.2	P301L 2+3+10+	Neuron, cogn	Götz et al., 2001 (118)
](line 12) Pl Tg	2′,3′-cyclic nucleotide 3′-phosphodiesterase	P301L 2+3-10+	Glial, motor	Higuchi et al., 2005 (119)
]K3	Thy1.2	K369I 2+3-10+	Neuron, motor, park	Ittner et al., 2008 (107)
]TauV337M	Thy1	V337M 2+3+10-	Neuron, Cogn	Lambourne et al., 2007 (120)
]Thy-Tau30	Thy1.2	G272V/P301S 2+3-10+	Neur, motor	Leroy et al., 2007 (111)
]JNPL3	PrP	P301L 2-3-10+	Neuron, Motor	Lewis et al., 2000 (121)
]VLW	Thy1	G272V P301L R406W	Neuron	Lim et al., 2001 (122)
]TauRD/ΔK280 Proaggregation	CaMKII-driven rTA+tetOp (binary system)	ΔK280, 4R construct	Neuron, motor	Mocanu et al., 2008 (108)
]TauRD/ΔK280/2P Antiaggregation	CaMKII-driven rTA+tetOp (binary system)	ΔK280/I277P/I308P, 4R construct	4R construct expression	Mocanu et al., 2008 (108)
]Tg-TauP301L	PrP	P301L 2+3+10+	Neuron, glial, cogn	Murakami et al., 2006 (123)
]rTg4510	(inducible) CaMKII-driven rTA+tetOp (binary system)	P301L 2-3-10+	Neuron, cogn, motor	Santacruz et al., 2005 (105)
]Thy-Tau22	Thy1.2	G272V/P301S 2+3-10+	Neuron, cogn	Schindowski et al., 2006 (106)
]Tg214	PDGF	V337M 2+3+10+ (myc/FlAG-tagged)	Neuron, cogn	Tanemura et al., 2001 (124)
]R406W Tg	CaMKII	R406W 2+3+10+ (myc/FlAG-tagged)	Neuron, cogn	Tatebayashi et al., 2002 (125)

]Abbreviations: **4R**, four microtubule-binding domains; **CamKII**, Calcium calmodulin protein kinase II; **Cogn**, cognitive impairment; **Glial**, gliofibrillary tangles; **MAPT**, MAPT/Tau gene; **Motor**, motor deficits; **Neuron**, neurofibrillary or neuronal-like tangles; **Park**, parkinsonism; **PDGF**, platelet-derived growth factor; **PrP**, Prion protein correlated consistently with memory loss at various ages in the rTg4510 mouse model (109). These Tau oligomers were also found in the brains of JNPL3 mice and patients presenting with FTDP-17 (109).

Tau_{RD}/ΔK280 is also an inducible mouse model of Tauopathy, which is based on the expression of only the 4R Tau microtubule-binding domains carrying the ΔK280 mutation (deletion of both amino- and carboxy terminal parts of Tau protein). The ΔK280 mutation accelerates Tau aggregation, and is referred to as a pro-aggregation mutant. Low expression of ΔK280 Tau repeats leads to Tau aggregation followed by astrogliosis and loss of synapses and neurons (108). Thus, Tau aggregates are required for neuronal death. It is interesting to note that expression of ΔK280 Tau repeats in Tau_{RD}/ΔK280 is very low (about 70% of endogenous murine Tau). When the transgene is switched off, ΔK280 Tau repeats disappear and only aggregated murine Tau proteins are found. Thus, this construct acts as a nucleation factor for Tau aggregation (108).

The K3 transgenic mouse strain expresses human Tau carrying the K369I mutation under the Thy1 promoter (107). The K369I Tau mutation was found in a family of patients presenting with Pick's disease without parkinsonism and amyotrophy (110). K3 mice show early-onset memory impairment, Pick body-like inclusions, amyotrophy and parkinsonism in the absence of overt neurodegeneration. Amyotrophy is likely to be related to Tau expression in the sciatic nerve as described in Leroy et al. (111). Moreover, Tau transgene is mainly expressed in the substantia nigra, and such expression leads to an early-onset parkinsonism phenotype. Interestingly, motor performance of young, but not old K3 mice improves upon l-dopa treatment (107). This model emphasizes that Tau mutations are responsible for some phenotypes (Pick-body like inclusions and memory impairment) but not others, which are likely to be related to the insertion site of the transgene in the murine genome.

Thy-Tau22 mouse transgenic line exhibits progressive neuron-specific AD-like Tau pathology devoid of any motor deficits (Fig. 3). In this model, a progressive development of NFT is observed in the hippocampus and amygdala, which parallels behavioural impairments as well as electrophysiological alterations (106) (Van der Jeugd et al., submitted). These latter changes are observed despite any striking loss of neuronal/synaptic markers until 12 months of age in the hippocampus (Van der Jeugd et al., submitted). In addition, in the hippocampus, hyper- and abnormally phosphorylated Tau species accumulate within the somatodendritic area (106, 112), supporting a possible influence on hippocampal-dependent plasticity always confirmed by behavioural and electrophysiological evaluations. Indeed, a careful evaluation of learning and memory at 12 months of age indicates

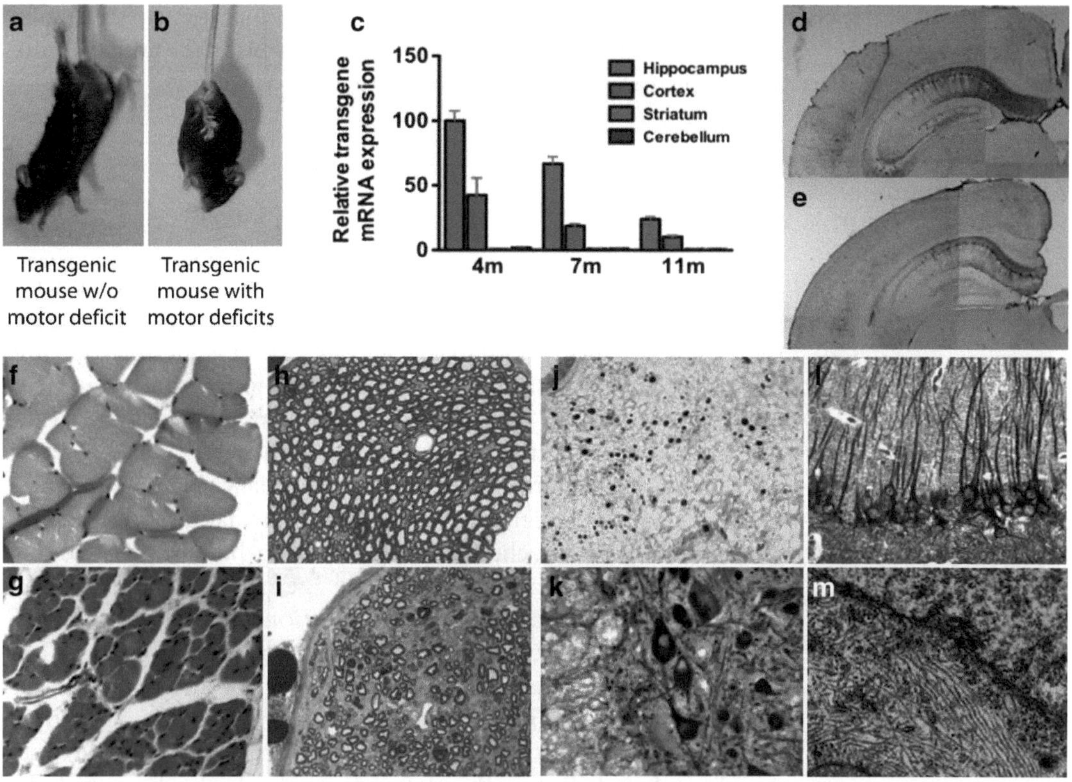

Fig. 3. Examples of typical features in Tau transgenic mice (**a**) Tail test in a 12-month-old transgenic mouse showing no motor deficit. (**b**) Tail test showing the abnormal limb-clasping reflex in a 12-month-old transgenic mouse exhibiting motor deficits. (**c**) Tau immunoblotting in brain regions of Tau transgenic mice. (**d**) AT8-immunoreactivity in brain section of Tau transgenic mouse. (**e**) AT100-immunoreactivity in brain section of Tau transgenic mouse. (**f–g**) Tissue section showing the cross-sectional aspect of muscle fibres in the quadriceps femoris in wild-type (**f**) and transgenic mice with motor deficits (**g**). (**h–i**) Semithin transverse sections of the sciatic nerve of 12-month-old wild-type (**h**) and transgenic mice with motor deficits (**i**) mice. High magnifications of tau immunohistochemistry showing the strong expression of human tau in the sciatic nerve (**j**), the spinal cord (**k**) and the hippocampus (**l**). Ultrastructural aspects of fibrillary inclusions in the hippocampus (**m**).

alterations in various spatial and nonspatial tests. These alterations were sustained by long-term depression and depotentiation defects within the hippocampus (Van der Jeugd et al., submitted).

Interestingly, at that time point, Thy-Tau22 mice exhibit septo-hippocampal Tau pathology accompanied by altered retrograde transport from hippocampus to medial septum and the loss of cholinergic neurons within this latter area (113).

In summary, NFD with accumulation of abnormal Tau species in the Thy-Tau22 transgenic model leads to synaptic plasticity defects underlying learning and memory alterations without striking neuronal loss and is thus a unique model to study molecular dysfunctions resulting from NFD.

Altogether, this section indicates that a large number of animal models are available. They allow for both understanding

Tau pathology and developing innovative therapeutic strategies. All models have strengths and weaknesses but they are still valuable tools for studying Tau biology and pathology.

7. Conclusions

In the present review, we indicate that Tau is clearly first a microtubule-associated protein and then a key factor in dementia. Tau models have shown the role of Tau proteins in microtubule dynamics and its dysfunctions. However, the role of Tau in neurodegeneration may be trickier than a simple defect in axonal transport due to microtubule depolymerization.

Finally, propagation way of neurofibrillary degeneration is a real question. Is it related to specific neuronal subpopulations, genetics or other mechanisms? Recent data have suggested that Tau aggregates may act as a prion-like agent. In their experiment, the authors performed intracranial injections of PHF from P301S transgenic brain extracts in hTau40 transgenic mice. They observed Tau aggregation at the injection site and then cortical spreading of neurofibrillary degeneration in a time-dependent manner (76). This shows that, similar to prion disease, exogenously induced Tauopathy depends on both the host and the source of the agent, suggesting the existence of polymorphic Tau-PHF strains reminiscent of prion strains. It is known that Tau may be secreted and then captured by neighbouring neurons (114, 115).

Thus, in conclusion, Tau models are really needed to explore the role in neurodegeneration. It is clear that more models are needed to fully understand all of Tau facets.

Acknowledgments

This work was supported by Inserm, CNRS, IMPRT, University Lille 2, Lille County Hospital (CHR-Lille), Région Nord/Pas-de-Calais, FEDER, ADERMA, PRIM, DN2M and grants from Association Française contre les Myopathies (AFM2006-1579 and AFM2007-1043), ANR-05-BLANC-0320-01, ANR-08-MNPS-002 AMYTOXTAU, France Alzheimer and Fédération pour la Recherche sur le Cerveau and from the European Community: APOPIS (contract LSHM-CT-2003-503330), MEMOSAD (contract 200611) & cNEUPRO (contract LSHM-CT-2007-037950). We also thank patients and their families and all members of the laboratory.

References

1. Neve RL, Harris P, Kosik KS, Kurnit DM, Donlon TA (1986) Identification of cDNA clones for the human microtubule-associated protein tau and chromosomal localization of the genes for tau and microtubule-associated protein 2. Brain Res 387:271–280
2. Andreadis A, Brown WM, Kosik KS (1992) Structure and novel exons of the human tau gene. Biochemistry 31:10626–10633
3. Muramatsu K, Hashimoto Y, Uemura T, et al. (2008) Neuron-specific recombination by Cre recombinase inserted into the murine tau locus. Biochem Biophys Res Commun 370:419–423
4. Castle JC, Zhang C, Shah JK, et al. (2008) Expression of 24,426 human alternative splicing events and predicted cis regulation in 48 tissues and cell lines. Nat Genet 40:1416–1425
5. Schraen-Maschke S, Dhaenens CM, Delacourte A, Sablonniere B (2004) Microtubule-associated protein tau gene: a risk factor in human neurodegenerative diseases. Neurobiol Dis 15: 449–460
6. Pittman AM, Myers AJ, Duckworth J, et al. (2004) The structure of the tau haplotype in controls and in progressive supranuclear palsy. Hum Mol Genet 13:1267–1274
7. Stefansson H, Helgason A, Thorleifsson G, et al. (2005) A common inversion under selection in Europeans. Nat Genet 37:129–137
8. Shaw-Smith C. (2006) Oesophageal atresia, tracheo-oesophageal fistula, and the VACTERL association: review of genetics and epidemiology. J Med Genet 43:545–554
9. Zody MC, Garber M, Adams DJ, et al. (2006) DNA sequence of human chromosome 17 and analysis of rearrangement in the human lineage. Nature 440:1045–1049
10. Rademakers R, Melquist S, Cruts M, et al. (2005) High-density SNP haplotyping suggests altered regulation of tau gene expression in progressive supranuclear palsy. Hum Mol Genet 14:3281–3292
11. Myers AJ, Pittman AM, Zhao AS, et al. (2007) The MAPT H1c risk haplotype is associated with increased expression of tau and especially of 4 repeat containing transcripts. Neurobiol Dis 25:561–570
12. Caffrey TM, Joachim C, Wade-Martins R (2008) Haplotype-specific expression of the N-terminal exons 2 and 3 at the human MAPT locus. Neurobiol Aging 29:1923–1929
13. Andreadis A. (2005) Tau gene alternative splicing: expression patterns, regulation and modulation of function in normal brain and neurodegenerative diseases. Biochim Biophys Acta 1739:91–103
14. Gallo JM, Noble W, Martin TR (2007) RNA and protein-dependent mechanisms in tauopathies: consequences for therapeutic strategies. Cell Mol Life Sci 64:1701–1714
15. Wei ML, Andreadis A (1998) Splicing of a regulated exon reveals additional complexity in the axonal microtubule-associated protein tau. J Neurochem 70:1346–1356
16. Goedert M, Spillantini MG, Jakes R, Rutherford D, Crowther RA (1989) Multiple isoforms of human microtubule-associated protein tau: sequences and localization in neurofibrillary tangles of Alzheimer's disease. Neuron 3:519–526
17. Goedert M, Spillantini MG, Potier MC, Ulrich J, Crowther RA (1989) Cloning and sequencing of the cDNA encoding an isoform of microtubule-associated protein tau containing four tandem repeats: differential expression of tau protein mRNAs in human brain. EMBO J 8:393–399
18. Andreadis A, Broderick JA, Kosik KS (1995) Relative exon affinities and suboptimal splice site signals lead to non-equivalence of two cassette exons. Nucleic Acids Res 23:3585–3593
19. Himmler A, Drechsel D, Kirschner MW, Martin DW, Jr (1989) Tau consists of a set of proteins with repeated C-terminal microtubule-binding domains and variable N-terminal domains. Mol Cell Biol 9:1381–1388
20. Weingarten MD, Lockwood AH, Hwo SY, Kirschner MW (1975) A protein factor essential for microtubule assembly. Proc Natl Acad Sci U S A 72:1858–1862
21. Schweers O, Schonbrunn-Hanebeck E, Marx A, Mandelkow E (1994) Structural studies of tau protein and Alzheimer paired helical filaments show no evidence for beta-structure. J Biol Chem 269:24290–24297
22. Hirokawa N, Shiomura Y, Okabe S (1988) Tau proteins: the molecular structure and mode of binding on microtubules. J Cell Biol 107:1449–1459
23. Brandt R, Leger J, Lee G (1995) Interaction of tau with the neural plasma membrane mediated by tau's amino-terminal projection domain. J Cell Biol 131:1327–1340
24. Chen J, Kanai Y, Cowan NJ, Hirokawa N (1992) Projection domains of MAP2 and tau determine spacings between microtubules in dendrites and axons. Nature 360:674–677
25. Georgieff IS, Liem RK, Couchie D, Mavilia C, Nunez J, Shelanski ML (1993) Expression of

high molecular weight tau in the central and peripheral nervous systems. J Cell Sci 105 (Pt 3):729–737
26. Rosenberg KJ, Ross JL, Feinstein HE, Feinstein SC, Israelachvili J (2008) Complementary dimerization of microtubule-associated tau protein: implications for microtubule bundling and tau-mediated pathogenesis. Proc Natl Acad Sci U S A 105:7445–7450.
27. Utton MA, Noble WJ, Hill JE, Anderton BH, Hanger DP (2005) Molecular motors implicated in the axonal transport of tau and alpha-synuclein. J Cell Sci 118:4645–4654
28. Magnani E, Fan J, Gasparini L, et al. (2007) Interaction of tau protein with the dynactin complex. EMBO J 26:4546–4554
29. Reynolds CH, Garwood CJ, Wray S, et al. (2008) Phosphorylation regulates tau interactions with Src homology 3 domains of phosphatidylinositol 3-kinase, phospholipase Cgamma1, Grb2, and Src family kinases. J Biol Chem 283:18177–18186
30. Dixit R, Ross JL, Goldman YE, Holzbaur EL (2008) Differential regulation of dynein and kinesin motor proteins by tau. Science 319:1086–1089
31. Crosby AH (2003) Disruption of cellular transport: a common cause of neurodegeneration? Lancet Neurol 2:311–316
32. Ding H, Dolan PJ, Johnson GV (2008) Histone deacetylase 6 interacts with the microtubule-associated protein tau. J Neurochem 106:2119–2130
33. Elliott E, Laufer O, Ginzburg I (2009) BAG-1M is upregulated in hippocampus of Alzheimer's disease patients and associates with tau and APP proteins. J Neurochem 109(4):1168–1178
34. Elliott E, Tsvetkov P, Ginzburg I (2007) BAG-1 associates with Hsc70.Tau complex and regulates the proteasomal degradation of Tau protein. J Biol Chem 282:37276–37284
35. Huang DY, Weisgraber KH, Goedert M, Saunders AM, Roses AD, Strittmatter WJ (1995) ApoE3 binding to tau tandem repeat I is abolished by tau serine262 phosphorylation. Neurosci Lett 192:209–212
36. Oddo S (2008) The ubiquitin-proteasome system in Alzheimer's disease. J Cell Mol Med 12:363–373
37. Butler M, Shelanski ML (1986) Microheterogeneity of microtubule-associated tau proteins is due to differences in phosphorylation. J Neurochem 47:1517–1522
38. Hanger DP, Byers HL, Wray S, et al. (2007) Novel phosphorylation sites in tau from Alzheimer brain support a role for casein kinase 1 in disease pathogenesis. J Biol Chem 282:23645–23654
39. Buee L, Bussiere T, Buee-Scherrer V, Delacourte A, Hof PR (2000) Tau protein isoforms, phosphorylation and role in neurodegenerative disorders. Brain Res Brain Res Rev 33:95–130
40. Sergeant N, Bretteville A, Hamdane M, et al. (2008) Biochemistry of Tau in Alzheimer's disease and related neurological disorders. Expert Rev Proteomics 5:207–224
41. Hanger DP, Anderton BH, Noble W (2009) Tau phosphorylation: the therapeutic challenge for neurodegenerative disease. Trends Mol Med 15:112–119
42. Baumann K, Mandelkow EM, Biernat J, Piwnica-Worms H, Mandelkow E (1993) Abnormal Alzheimer-like phosphorylation of tau-protein by cyclin-dependent kinases cdk2 and cdk5. FEBS Lett 336:417–424
43. Vincent I, Jicha G, Rosado M, Dickson DW (1997) Aberrant expression of mitotic cdc2/cyclin B1 kinase in degenerating neurons of Alzheimer's disease brain. J Neurosci 17:3588–3598
44. Hamdane M, Sambo AV, Delobel P, et al. (2003) Mitotic-like tau phosphorylation by p25-Cdk5 kinase complex. J Biol Chem 278:34026–34034
45. Holzer M, Gartner U, Klinz FJ, Narz F, Heumann R, Arendt T (2001) Activation of mitogen-activated protein kinase cascade and phosphorylation of cytoskeletal proteins after neurone-specific activation of p21ras. I. Mitogen-activated protein kinase cascade. Neuroscience 105:1031–1040
46. Goedert M, Hasegawa M, Jakes R, Lawler S, Cuenda A, Cohen P (1997) Phosphorylation of microtubule-associated protein tau by stress-activated protein kinases. FEBS Lett 409:57–62
47. Buee-Scherrer V, Goedert M (2002) Phosphorylation of microtubule-associated protein tau by stress-activated protein kinases in intact cells. FEBS Lett 515:151–154
48. Tomizawa K, Omori A, Ohtake A, Sato K, Takahashi M (2001) Tau-tubulin kinase phosphorylates tau at Ser-208 and Ser-210, sites found in paired helical filament-tau. FEBS Lett 492:221–227
49. Sato S, Cerny RL, Buescher JL, Ikezu T (2006) Tau-tubulin kinase 1 (TTBK1), a neuron-specific tau kinase candidate, is involved in tau phosphorylation and aggregation. J Neurochem 98:1573–1584
50. Kitano-Takahashi M, Morita H, Kondo S, et al. (2007) Expression, purification and crystallization of a human tau-tubulin kinase 2 that phosphorylates tau protein. Acta Crystallogr Sect F Struct Biol Cryst Commun 63:602–604

51. Greenwood JA, Scott CW, Spreen RC, Caputo CB, Johnson GV. (1994) Casein kinase II preferentially phosphorylates human tau isoforms containing an amino-terminal insert: Identification of threonine 39 as the primary phosphate acceptor. J Biol Chem 269: 4373–4380
52. Drewes G, Ebneth A, Preuss U, Mandelkow EM, Mandelkow E (1997) MARK, a novel family of protein kinases that phosphorylate microtubule-associated proteins and trigger microtubule disruption. Cell 89:297–308
53. Drewes G (2004) MARKing tau for tangles and toxicity. Trends Biochem Sci 29:548–555
54. Woods YL, Cohen P, Becker W, et al. (2001) The kinase DYRK phosphorylates protein-synthesis initiation factor eIF2Bepsilon at Ser539 and the microtubule-associated protein tau at Thr212: potential role for DYRK as a glycogen synthase kinase 3-priming kinase. Biochem J 355:609–615
55. Liu F, Liang Z, Wegiel J, et al. (2008) Overexpression of Dyrk1A contributes to neurofibrillary degeneration in Down syndrome. FASEB J 22:3224–3233
56. Paudel HK (1997) The regulatory Ser262 of microtubule-associated protein tau is phosphorylated by phosphorylase kinase. J Biol Chem 272:1777–1785
57. Amano M, Kaneko T, Maeda A, et al. (2003) Identification of Tau and MAP2 as novel substrates of Rho-kinase and myosin phosphatase. J Neurochem 87:780–790
58. Hanger DP, Hughes K, Woodgett JR, Brion JP, Anderton BH (1992) Glycogen synthase kinase-3 induces Alzheimer's disease-like phosphorylation of tau: generation of paired helical filament epitopes and neuronal localisation of the kinase. Neurosci Lett 147:58–62
59. Virdee K, Yoshida H, Peak-Chew S, Goedert M (2007) Phosphorylation of human microtubule-associated protein tau by protein kinases of the AGC subfamily. FEBS Lett 581:2657–2662
60. Kinoshita E, Kinoshita-Kikuta E, Matsubara M, et al. (2008) Separation of phosphoprotein isotypes having the same number of phosphate groups using phosphate-affinity SDS-PAGE. Proteomics 8:2994–3003
61. Kidd M (1954) Alzheimer's Disease-an Electron Microscopical Study. Brain 87:307–320
62. Brion JP (2006) Immunological demonstration of tau protein in neurofibrillary tangles of Alzheimer's disease. J Alzheimers Dis 9:177–185
63. Delacourte A, Defossez A (1986) Biochemical characterization of an immune serum which specifically marks neurons in neurofibrillary degeneration in Alzheimer's disease. C R Acad Sci III 303:439–444
64. Kondo J, Honda T, Mori H, et al. (1988) The carboxyl third of tau is tightly bound to paired helical filaments. Neuron 1:827–834
65. Wischik CM, Novak M, Thogersen HC, et al. (1988) Isolation of a fragment of tau derived from the core of the paired helical filament of Alzheimer disease. Proc Natl Acad Sci U S A 85:4506–4510
66. Jakes R, Novak M, Davison M, Wischik CM (1991) Identification of 3- and 4-repeat tau isoforms within the PHF in Alzheimer's disease. EMBO J 10:2725–2729.
67. Ksiezak-Reding H, Binder LI, Yen SH (1988) Immunochemical and biochemical characterization of tau proteins in normal and Alzheimer's disease brains with Alz 50 and Tau-1. J Biol Chem 263:7948–7953
68. Bretteville A, Ando K, Ghestem A, et al. (2009) Two-dimensional electrophoresis of tau mutants reveals specific phosphorylation pattern likely linked to early tau conformational changes. PLoS ONE 4:e4843
69. Matsuo ES, Shin RW, Billingsley ML, et al. (1994) Biopsy-derived adult human brain tau is phosphorylated at many of the same sites as Alzheimer's disease paired helical filament tau. Neuron 13:989–1002
70. Sergeant N, Bussiere T, Vermersch P, Lejeune JP, Delacourte A (1995) Isoelectric point differentiates PHF-tau from biopsy-derived human brain tau proteins. Neuroreport 6: 2217–2220
71. Yoshida H, Goedert M (2006) Sequential phosphorylation of tau protein by cAMP-dependent protein kinase and SAPK4/p38delta or JNK2 in the presence of heparin generates the AT100 epitope. J Neurochem 99:154–164
72. Braak H, Braak E (1991) Neuropathological stageing of Alzheimer-related changes. Acta Neuropathol 82:239–259
73. Delacourte A, David JP, Sergeant N, et al. (1999) The biochemical pathway of neurofibrillary degeneration in aging and Alzheimer's disease. Neurology 52:1158–1165
74. Delacourte A, Sergeant N, Champain D, et al. (2002) Nonoverlapping but synergetic tau and APP pathologies in sporadic Alzheimer's disease. Neurology 59:398–407
75. Duyckaerts C, Colle MA, Delatour B, Hauw JJ (1999) Alzheimer's disease: lesions and their progression. Rev Neurol (Paris) 155(Suppl 4): S17–S27
76. Clavaguera F, Bolmont T, Crowther RA, et al. (2009) Transmission and spreading of tauopathy in transgenic mouse brain. Nat Cell Biol 11:909–913

77. Constantinidis J, Richard J, Tissot R (1974) Pick's disease: histological and clinical correlations. Eur Neurol 11:208–217
78. Brown J (1992) Pick's disease. Baillieres Clin Neurol 1:535–557
79. Wilhelmsen KC, Lynch T, Pavlou E, Higgins M, Nygaard TG (1994) Localization of disinhibition-dementia-parkinsonism-amyotrophy complex to 17q21–22. Am J Hum Genet 55:1159–1165
80. Wijker M, Wszolek ZK, Wolters EC, et al. (1996) Localization of the gene for rapidly progressive autosomal dominant parkinsonism and dementia with pallido-ponto-nigral degeneration to chromosome 17q21. Hum Mol Genet 5:151–154
81. Bird TD, Wijsman EM, Nochlin D, et al. (1997) Chromosome 17 and hereditary dementia: linkage studies in three non-Alzheimer families and kindreds with late-onset FAD. Neurology 48:949–954
82. Heutink P, Stevens M, Rizzu P, et al. (1997) Hereditary frontotemporal dementia is linked to chromosome 17q21–q22: a genetic and clinicopathological study of three Dutch families. Ann Neurol 41:150–159
83. Murrell JR, Spillantini MG, Zolo P, et al. (1999) Tau gene mutation G389R causes a tauopathy with abundant pick body-like inclusions and axonal deposits. J Neuropathol Exp Neurol 58:1207–1226
84. Foster NL, Wilhelmsen K, Sima AA, Jones MZ, D'Amato CJ, Gilman S (1997) Frontotemporal dementia and parkinsonism linked to chromosome 17: a consensus conference. Conference Participants. Ann Neurol 41:706–715
85. Spillantini MG, Bird TD, Ghetti B (1998) Frontotemporal dementia and Parkinsonism linked to chromosome 17: a new group of tauopathies. Brain Pathol 8:387–402
86. Dumanchin C, Camuzat A, Campion D, et al. (1998) Segregation of a missense mutation in the microtubule-associated protein tau gene with familial frontotemporal dementia and parkinsonism. Hum Mol Genet 7:1825–1829
87. Hutton M, Lendon CL, Rizzu P, et al. (1998) Association of missense and 5'-splice-site mutations in tau with the inherited dementia FTDP-17. Nature 393:702–705
88. Poorkaj P, Bird TD, Wijsman E, et al. (1998) Tau is a candidate gene for chromosome 17 frontotemporal dementia. Ann Neurol 43:815–825
89. Spillantini MG, Goedert M (2001) Tau gene mutations and tau pathology in frontotemporal dementia and parkinsonism linked to chromosome 17. Adv Exp Med Biol 487: 21–37
90. Rovelet-Lecrux A, Lecourtois M, Thomas-Anterion C, et al. (2009) Partial deletion of the MAPT gene: a novel mechanism of FTDP-17. Hum Mutat 30:E591–E602
91. Yen SH, Hutton M, DeTure M, Ko LW, Nacharaju P (1999) Fibrillogenesis of tau: insights from tau missense mutations in FTDP-17. Brain Pathol 9:695–705
92. Chiti F, Stefani M, Taddei N, Ramponi G, Dobson CM (2003) Rationalization of the effects of mutations on peptide and protein aggregation rates. Nature 424:805–808
93. Harada A, Oguchi K, Okabe S, et al. (1994) Altered microtubule organization in small-calibre axons of mice lacking tau protein. Nature 369:488–491
94. Dawson HN, Ferreira A, Eyster MV, Ghoshal N, Binder LI, Vitek MP (2001) Inhibition of neuronal maturation in primary hippocampal neurons from tau deficient mice. J Cell Sci 114:1179–1187
95. Tucker KL, Meyer M, Barde YA (2001) Neurotrophins are required for nerve growth during development. Nat Neurosci 4:29–37
96. Ikegami S, Harada A, Hirokawa N (2000) Muscle weakness, hyperactivity, and impairment in fear conditioning in tau-deficient mice. Neurosci Lett 279:129–132
97. Brion JP (1999) Neurofibrillary tangles and early modification of the neuronal cytoskeleton in Alzheimer's disease and in experimental models. Bull Mem Acad R Med Belg 154:287–294
98. Ishihara T, Hong M, Zhang B, et al. (1999) Age-dependent emergence and progression of a tauopathy in transgenic mice overexpressing the shortest human tau isoform. Neuron 24:751–762
99. Gotz J, Probst A, Spillantini MG, et al. (1995) Somatodendritic localization and hyperphosphorylation of tau protein in transgenic mice expressing the longest human brain tau isoform. EMBO J 14:1304–1313
100. Spittaels K, Van den Haute C, Van Dorpe J, et al. (1999) Prominent axonopathy in the brain and spinal cord of transgenic mice overexpressing four-repeat human tau protein. Am J Pathol 155:2153–2165
101. Probst A, Gotz J, Wiederhold KH, et al. (2000) Axonopathy and amyotrophy in mice transgenic for human four-repeat tau protein. Acta Neuropathol 99:469–481
102. Duff K, Knight H, Refolo LM, et al. (2000) Characterization of pathology in transgenic

mice over-expressing human genomic and cDNA tau transgenes. Neurobiol Dis 7:87–98
103. Andorfer C, Acker CM, Kress Y, Hof PR, Duff K, Davies P (2005) Cell-cycle reentry and cell death in transgenic mice expressing nonmutant human tau isoforms. J Neurosci 25:5446–5454
104. McMillan P, Korvatska E, Poorkaj P, et al. (2008) Tau isoform regulation is region- and cell-specific in mouse brain. J Comp Neurol 511:788–803
105. Santacruz K, Lewis J, Spires T, et al. (2005) Tau suppression in a neurodegenerative mouse model improves memory function. Science 309:476–481
106. Schindowski K, Bretteville A, Leroy K, et al. (2006) Alzheimer's disease-like tau neuropathology leads to memory deficits and loss of functional synapses in a novel mutated tau transgenic mouse without any motor deficits. Am J Pathol 169:599–616
107. Ittner LM, Fath T, Ke YD, et al. (2008) Parkinsonism and impaired axonal transport in a mouse model of frontotemporal dementia. Proc Natl Acad Sci U S A 105: 15997–16002
108. Mocanu MM, Nissen A, Eckermann K, et al. (2008) The potential for beta-structure in the repeat domain of tau protein determines aggregation, synaptic decay, neuronal loss, and coassembly with endogenous Tau in inducible mouse models of tauopathy. J Neurosci 28:737–748
109. Berger Z, Roder H, Hanna A, et al. (2007) Accumulation of pathological tau species and memory loss in a conditional model of tauopathy. J Neurosci 27:3650–3662
110. Neumann M, Schulz-Schaeffer W, Crowther RA, et al. (2001) Pick's disease associated with the novel Tau gene mutation K369I. Ann Neurol 50:503–513
111. Leroy K, Bretteville A, Schindowski K, et al. (2007) Early axonopathy preceding neurofibrillary tangles in mutant tau transgenic mice. Am J Pathol 171:976–992
112. Schindowski K, Belarbi K, Bretteville A, Ando K, Buee L (2008) Neurogenesis and cell cycle-reactivated neuronal death during pathogenic tau aggregation. Genes Brain Behav 7(Suppl 1):92–100
113. Belarbi K, Schindowski K, Burnouf S, et al. (2009) Early Tau pathology involving the septo-hippocampal pathway in a Tau transgenic model: relevance to Alzheimer's disease. Curr Alzheimer Res 6:152–157
114. Frost B, Jacks RL, Diamond MI (2009) Propagation of tau misfolding from the outside to the inside of a cell. J Biol Chem 284:12845–12852
115. Frost B, Ollesch J, Wille H, Diamond MI (2009) Conformational diversity of wild-type Tau fibrils specified by templated conformation change. J Biol Chem 284:3546–3551
116. Allen B, Ingram E, Takao M, Smith MJ, Jakes R, Virdee K, Yoshida H, Holzer M, Craxton M, Emson PC, Atzori C, Migheli A, Crowther RA, Ghetti B, Spillantini MG, Goedert M (2002) Abundant tau filaments and nonapoptotic neurodegeneration in transgenic mice expressing human P301S tau protein. J Neurosci 22:9340–9351
117. Dawson HN, Cantillana V, Chen L, Vitek MP (2007) The tau N279K exon 10 splicing mutation recapitulates frontotemporal dementia and parkinsonism linked to chromosome 17 tauopathy in a mouse model. J Neurosci 27:9155–9168
118. Götz J, Chen F, Barmettler R, Nitsch RM (2001) Tau filament formation in transgenic mice expressing P301L tau. J Biol Chem 276:529–534
119. Higuchi M, Zhang B, Forman MS, Yoshiyama Y, Trojanowski JQ, Lee VM (2005) Axonal degeneration induced by targeted expression of mutant human tau in oligodendrocytes of transgenic mice that model glial tauopathies. J Neurosci 25:9434–9443
120. Lambourne SL, Humby T, Isles AR, Emson PC, Spillantini MG, Wilkinson LS. (2007) Impairments in impulse control in mice transgenic for the human FTDP-17 tauV337M mutation are exacerbated by age. Hum Mol Genet 16: 1708–1719
121. Lewis J, McGowan E, Rockwood J, Melrose H, Nacharaju P, Van Slegtenhorst M, Gwinn Hardy K, Paul Murphy M, Baker M, Yu X, Duff K, Hardy J, Corral A, Lin WL, Yen SH, Dickson DW, Davies P, Hutton M (2000) Neurofibrillary tangles, amyotrophy and progressive motor disturbance in mice expressing mutant (P301L) tau protein. Nat Genet 25:402–405
122. Lim F, Hernandez F, Lucas JJ, Gomez-Ramos P, Moran MA, Avila J (2001) FTDP-17 mutations in tau transgenic mice provoke lysosomal abnormalities and Tau filaments in forebrain. Mol Cell Neurosci 18:702–714
123. Murakami T, Paitel E, Kawarabayashi T, Ikeda M, Chishti MA, Janus C, Matsubara E, Sasaki A, Kawarai T, Phinney AL, Harigaya Y, Horne P, Egashira N, Mishima K, Hanna A, Yang J, Iwasaki K, Takahashi M, Fujiwara M, Ishiguro K, Bergeron C, Carlson GA, Abe K, Westaway D, St George-Hyslop P, Shoji M (2006) Cortical neuronal and glial pathology in TgTauP301L transgenic mice: neuronal degeneration,

memory disturbance, and phenotypic variation. Am J Pathol 169:1365–1375

124. Tanemura K, Akagi T, Murayama M, Kikuchi N, Murayama O, Hashikawa T, Yoshiike Y, Park JM, Matsuda K, Nakao S, Sun X, Sato S, Yamaguchi H, Takashima A (2001) Formation of filamentous tau aggregations in transgenic mice expressing V337M human tau. Neurobiol Dis 8:1036–1045

125. Tatebayashi Y, Miyasaka T, Chui DH, Akagi T, Mishima K, Iwasaki K, Fujiwara M, Tanemura K, Murayama M, Ishiguro K, Planel E, Sato S, Hashikawa T, Takashima A (2002) Tau filament formation and associative memory deficit in aged mice expressing mutant (R406W) human tau. Proc Natl Acad Sci U S A 99: 13896–13901

Chapter 24

The 3xTg-AD Mouse Model: Reproducing and Modulating Plaque and Tangle Pathology

Michael Sy, Masashi Kitazawa, and Frank LaFerla

Abstract

Alzheimer's disease (AD) is a devastating disease, and the most common form of dementia to afflict the elderly population. The disease causes a slow but progressive neurodegeneration, leading to memory impairments and dysfunction in other cognitive domains. The molecular mechanism of disease development and progression has not yet been fully established, nor have any cures or effective, long-lasting treatments been developed. Various transgenic mouse models of AD have proven to be invaluable tools for elucidating disease mechanisms and for providing a platform to evaluate therapeutic strategies. In this chapter, we discuss findings from the 3xTg-AD mouse model, which develops both plaque and tangle pathologies, the two major pathological hallmarks of AD. Studies using the 3xTg-AD mice have revealed a strong interaction between amyloid-beta (Aβ) and tau, which synergistically drive the pathogenesis in the brain.

Key words: Alzheimer disease, Amyloid-beta, Tau, 3xTg-AD mice, Animal models

1. Introduction

Alzheimer's disease (AD) is the leading cause of dementia in the elderly population, afflicting nearly 24 million people worldwide (1). AD patients suffer through a progressive cognitive decline, characterized by the devastating loss of semantic and episodic memory, impairments in judgment, spatial orientation, and language. The disease process is believed to be triggered by the accumulation of the amyloid-beta (Aβ) peptide, which lies upstream of the other pathological changes. AD brains are marked by two pathological hallmarks, amyloid plaques (comprised of Aβ) and neurofibrillary tangles (NFTs), which are made mostly of hyperphosphorylated tau protein (2, 3). Other pathological changes occur as well, including the occurrence of dystrophic neurites,

Peter Paul De Deyn and Debby Van Dam (eds.), *Animal Models of Dementia*, Neuromethods, vol. 48,
DOI 10.1007/978-1-60761-898-0_24,

synapse loss, and neuron loss. Currently, no effective treatment exists that either halts or slows the progression of the disease. Thus, there is a great need to more fully understand the disease and to develop and test potential therapies for clearing the pathology and restoring cognition. In the past 2 decades, research has advanced significantly due to the identification of genetic components of AD, and the subsequent development of transgenic mouse models that mimic AD pathology and cognitive decline. These transgenic mice have helped expand our understanding of AD pathogenesis and have provided insights into possible therapies.

2. Transgenic Mice and AD

Genetic analysis of rare, autosomal dominant, early-onset AD identified Aβ as the main culprit behind the pathogenesis of the disease (4). Mutations in the presenilins (PS-1, PS-2) or the amyloid precursor protein (APP) lead to either increased production of total Aβ, the longer more amyloidogenic Aβ1-42, or increase the aggregation of Aβ (5). These findings heralded the amyloid cascade hypothesis, which stipulates that the accumulation of Aβ is the main driving force behind the development of AD (6). Additional genetic evidence provides further support for the amyloid cascade hypothesis with the discovery that patients with APP gene duplications and Down syndrome develop Aβ plaque and AD-like pathologies (7, 8).

The mutations associated with early-onset AD quickly found their way into transgenic mouse models, and the findings from these models have largely supported the amyloid cascade hypothesis (9). Transgenic mice overexpressing various mutant APP constructs develop Aβ pathology with some also displaying cognitive decline reminiscent of human AD (10–12). Aβ deposits in progressive fashion and eventually develops into extracellular plaques. Similarly, transgenic APP mice exhibit impairments in behavioral tasks. In contrast, mutations in PS1 that cause FAD in humans do not lead to amyloid deposition and impaired behavioral phenotype in mice despite disruptions in calcium homeostasis and increases in $Aβ_{1-42}$, a form of Aβ more prone to aggregation (13, 14). However, crossing APP and PS1 mice exacerbates the Aβ pathology and accelerates the development of cognitive impairments (15, 16). Thus, the introduction of multiple transgenes into mice has expanded our understanding of the role that mutations in PS1 and the subsequent increase in $Aβ_{1-42}$ might play in the development of AD. Even though multiple mutations are not expressed in humans, the introduction of multiple transgenes in mice does provide relevant insight into the pathogenesis of AD.

Despite the introduction of the APP and PS1 transgenes in an attempt to better recapitulate AD pathology, these mice still lack tau pathology. Hence, it became obvious that more aggressive approaches would be necessary, including the introduction of additional transgenes to mimic both Aβ and tau pathology in a mouse. One means of achieving this goal is to cross independent transgenic lines (17). Alternatively, Aβ and tau pathology can be induced by injections of pathological protein (18). Although both of these approaches create models that mimic the Aβ and tau pathology found in AD, they pose certain practical difficulties such as poor breeding, the appearance of motor deficits unrelated to AD, and short life expectancy. These unfortunate constraints preclude the use of these models in various applications including learning and memory studies.

3. The 3xTg-AD Mouse Model

To further study the interaction between Aβ and tau and their impact on AD pathogenesis, we sought to develop a transgenic mouse model that exhibits both Aβ and tau pathology. To achieve this goal, APP and tau transgenes were co-microinjected into single-cell embryos from mice with a knock-in of PS1 with a M146V mutation ($PS1_{M146V}$-KI) (19). The Swedish mutation of APP (APPswe) containing the double mutation KM670/671NL was used to induce the development of Aβ pathology. To ensure the development of pathological tau lesions, instead of using wild-type tau, a mutant form of tau associated with frontotemporal dementia with parkinsonism-17 (FTDP-17) was used. In FTDP-17 patients, tau deposits into NFTs similar to AD, but the deposits develop in different brain regions (frontal-temporally). Likely as a result of the spatial positioning of pathology, unlike AD where memory deficits are first observed, in FTD, executive functioning is impaired and behavioral symptoms (personality changes) are observed (20). Nevertheless, mutations in tau generally cause tau to aggregate more easily. Hence, the human tauP301L transgene was included in the 3xTg-AD mice.

Both the APPswe and tau_{P301L} transgenes were under the control of the Thy1.2 promoter, which limits expression of transgenes largely to the CNS (21). These transgenes were co-microinjected into single-cell embryos from $PS1_{M146V}$-KI mice. As noted before, the inclusion of the PS1 mutation in mice increases the production of $A\beta_{1-42}$ and accelerates the development of pathology. Thus, with the 3xTg-AD mouse model, Aβ and tau pathology accumulates rapidly, allowing study of both pathologies in the mouse lifespan. Yet, pathology still appears progressively and in relevant brain regions, making the study of different stages of the disease possible.

In addition to the development of relevant AD-like pathology, the 3xTg-AD mice boast excellent colony-maintenance characteristics.

Both the APPswe and tauP301L genes integrated into a single locus and sort as if a single gene. In addition, $PS1_{M146V}$ is "knocked-in" into the endogenous locus so essentially the mice breed as if they contained only a single transgene, despite harboring three transgenes.

3.1. Aβ Pathology in the 3xTg-AD Mice

As noted before, Aβ and tau pathology develop progressively and in a brain-region-specific manner in the 3xTg-AD mouse model. Thus, tracking the development of pathology added support to the amyloid cascade hypothesis. Despite equal transgene expression of both APP and human tau, accumulation of Aβ precedes tau accumulation in the 3xTg-AD mice, and it is likely that Aβ accumulation drives the development of tau pathology (19).

Aβ pathology in the 3xTg-AD mice first appears intraneuronally at 3–4 months of age in the neocortex. At this point in time, these accumulations are detectable by various anti-Aβ antibodies including $Aβ_{1-42}$-specific antibodies, and most of the Aβ is still detergent-soluble. That soluble intraneuronal Aβ accumulation precedes plaque formation is consistent with other transgenic mouse models (22). Similarly, brain affected by Down syndrome shows evidence of increased Aβ production likely due to the increase in APP copy number and also forms intraneuronal Aβ deposits before extracellular accumulations develop (23). As the mice age, Aβ continues to accumulate as levels of both Aβ1-40 and $Aβ_{1-42}$ increase. By 6 months, intraneuronal Aβ can be found in the hippocampus in the CA1 region and also in the amygdala. By 6 months, some of the intraneuronal Aβ is oligomeric in nature as the deposits are detectable by the antibody A11 (24).

Around 9 months of age, extracellular deposits of Aβ first begin to appear in the cortex. As extracellular plaques of Aβ develop and enlarge in size and number, intraneuronal Aβ begins to recede, but exactly what triggers the shift of Aβ from intracellular stores to extracellular space is not clear. Nevertheless, Aβ continues to accumulate extracellularly such that by 12 months of age, extracellular deposits can be seen in the hippocampus and some plaques become thioflavin-S positive, indicating the development of β-pleated sheet characteristics (25). By 15 months of age, extracellular plaques are widespread throughout the hippocampus and cortex. At this stage, large amounts of Aβ are found in detergent-insoluble fractions and likely fibrillar forms of Aβ dominate. However, even with the development of plaques, oligomeric forms of Aβ and other soluble Aβ species persist (24).

3.2. Tau Pathology in the 3xTg-AD Mice

Similar to the deposition of Aβ, tau pathology also follows a temporal and brain region-specific manner. Immunoreactivity for the antibody HT7, which detects human tau, is evident only in the somatic-dendritic compartment in pyramidal neurons of the CA1 region as early as 4 months of age (26). Mimicking the pattern observed in human brains, tau later spreads to cortical regions (19). As the mice age, reactivity with the conformation-specific MC-1 antibody appears first, followed by phosphorylation of tau. The MC-1 antibody detects a pathological conformation state of tau protein observed in the early stages of AD brains (27). The phosphorylation of tau at certain epitopes can be detected using monoclonal antibodies against specific phosphorylation sites. For example, the AT180 antibody detects tau phosphorylated at Thr231, AT8 detects tau phosphorylated at Ser202 and Thr205, and PHF-1 detects tau phosphorylated at Ser396 and Ser404. In the 3xTg-AD mice, phosphorylation at these sites appears in a sequential manner. Site-specific phosphorylation of tau at the AT180 and AT8 occurs before immunoreactivity with PHF-1 is observed (19). Although AT180 and AT8 immunoreactivity is apparent at 10–15 months of age, PHF-1 is only apparent at 15+ months of age. Tau accumulation is also apparent by Gallyas silver staining and Thioflavin-S at 12 months of age indicating that tau is depositing into aggregates (19). The presence of NFTs was also confirmed by electron microscopy analysis (Fig. 1).

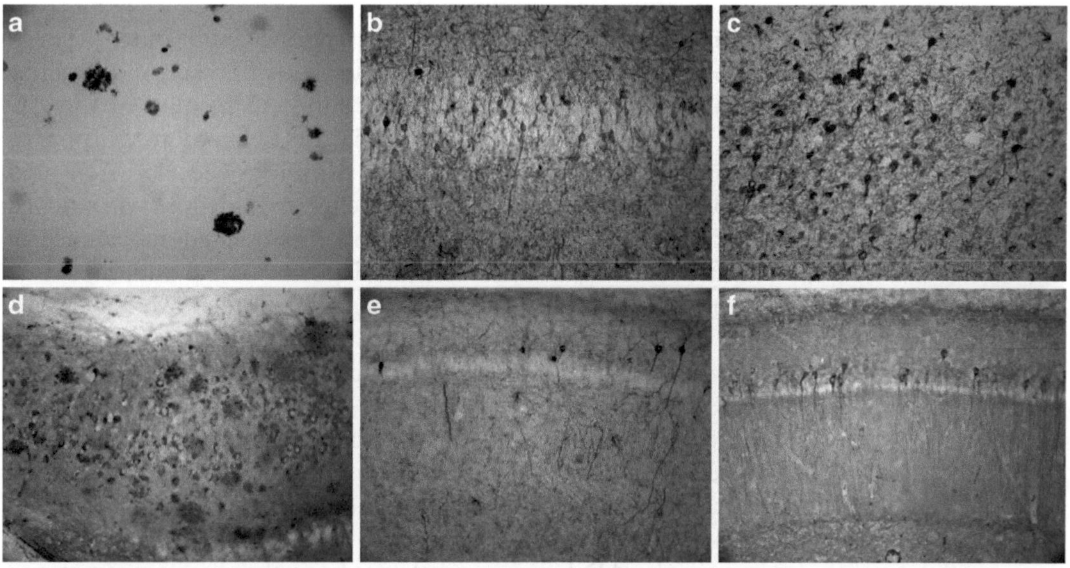

Fig. 1. Representative AD pathology in the 3xTg-AD mice. (**a**) 6E10 immunoreactivity in 6-month-old 3xTg-AD mice reveals intraneuronal Aβ pathology in the subiculum; (**b**) 6E10 immunoreactivity in 15-month-old 3xTg-AD mouse shows the development of extracellular plaques; (**c**) 1560 immunoreactivity in human AD brain of extracellular Aβ plaques; (**d**) HT7 immunoreactivity in 6-month-old 3xTg-AD mouse brain reveals intraneuronal tau accumulation in CA1 region of the hippocampus; (**e**) HT7 immunoreactivity in 12-month-old 3xTg-AD shows increased tau accumulation with age; (**f**) AT8 immunoreactivity in human AD brain showing tau phosphorylation; (**g**) AT8 immunoreactivity first appears in 3xTg-AD mice at 10–12 months of age in the hippocampus; (**h**) PHF-1 immunoreactivity follows at 15+ months of age; (**j**) PHF-1 immunoreactivity is apparent in human AD.

3.3. Synaptic and Cognitive Deficits First Appear Alongside Intraneuronal Aβ Deposition

Synaptic deficits are the best correlate of the memory deficits in AD (28). Although plaque number does not correlate well with cognitive impairments, various Aβ species induce synaptic dysfunction. For example, Aβ oligomers induce LTP deficits (29). Synaptic dysfunction and impaired LTP is observed as early as 6 months in 3xTg-AD mice at a time when oligomeric Aβ is readily apparent (19). Furthermore, at this time point, Aβ oligomers appear intracellularly, indicating a possibility that these intraneuronal accumulations of Aβ contribute to the synaptic dysfunction.

As expected from the early synaptic dysfunction observed, behavioral deficits are apparent in 3xTg-AD mice at an early age. By the time synaptic dysfunction is apparent at 6 months of age, behavioral deficits manifest as retention impairments in the Morris water maze task (30). Fear conditioning is also impaired at this age. These impairments correlate with the appearance of intraneuronal Aβ in the hippocampus and amygdala, brain areas involved in memory and fear conditioning respectively. Thus, it is likely that intraneuronal Aβ plays a role in early behavioral impairments. This finding, however, does not preclude extracellular Aβ also contributing to the development of cognitive dysfunction. Notably, even as mice age and intraneuronal Aβ deposition wanes and gives way to extracellular deposits, behavioral deficits increase, and expand to include impairment in learning and memory (31).

3.4. Inflammation in the 3xTg-AD Mice

Inflammation is an invariant feature of the AD brain and is believed to be in response to the accumulation of fibrillar Aβ. Reactive microglia and astrocytes localize around Aβ plaques in the AD brain (32, 33). These activated glia are believed to be phagocytosing Aβ in an attempt to reduce pathology (34, 35). Activated microglia first appear around extracellular Aβ deposits around 12 months of age (26). The spatial and time-related appearance of activated microglia with fibrillar deposits of Aβ indicates that microglia activation may be induced by fibrillar Aβ. However, there are indications that inflammatory processes begin earlier. Elevated levels of tumor necrosis factor-α (TNFα) and monocyte chemoattractant protein-1 (MCP-1) were found in the entorhinal cortex at 6 months of age (36). Furthermore, primary microglia can be activated by monomeric and oligomeric forms of Aβ, which are present well before the appearance of extracellular Aβ deposits at 12 months (37, 38). Thus, it is likely that inflammatory processes in the AD brain occur throughout the disease process and may even contribute to the progression of the disease.

Even while activated glial cells are clearing Aβ, they are releasing various inflammatory mediators such as cytokines and chemokines that may actually be promoting pathology. Indeed, numerous cytokines have been shown to be elevated in the AD brain (39–42). These cytokines may play various roles in driving AD pathology including inducing tau phosphorylation. Cytokines have been shown to be able to induce tau phosphorylation in mice (43).

Similarly, lipopolysaccharide (LPS)-induced increases in inflammation in the 3xTg-AD mice did not alter Aβ pathology, but tau phosphorylation was increased via a Cdk5-dependent mechanism (26). In addition to increases in tau phosphorylation, increased microglial activation was observed along with increases in Interleukin 1β (IL-1β). It is likely that the increased cytokines in the LPS-induced inflammatory state are responsible for the increase in tau phosphorylation. In the AD brain, Aβ accumulation induces inflammation and the subsequent increases in cytokines observed in AD brains may contribute to the development of tau phosphorylation. Thus, inflammation represents a pathway by which Aβ might induce tau pathology.

3.5. Modulation of AD Pathology in the 3xTg-AD Mice

In addition to modeling various aspects of AD, mouse models represent an important platform for the ultimate goal of developing or identifying treatments for AD. Preclinical trials of various treatments in the 3xTg-AD mouse model allow evaluation of the efficacy of reducing both Aβ and tau pathology. In this section, we will discuss several therapeutic treatments that modulate Aβ and/or tau pathology in the 3xTg-AD mice, and their limitations and potentials for treating AD in humans.

As described above, the genetic and pathological evidence suggests a central role for Aβ in the pathogenesis of AD (44). The work supporting the amyloid cascade hypothesis is compelling enough such that attempts to either reduce production of or increase degradation of Aβ have become prominent goals in the treatment of AD. One proposed method of reducing Aβ load is the direct removal of Aβ by vaccination or immunotherapy. The first Aβ immunotherapy was described in 1999, and the treatment was effective in reducing Aβ plaques in APP transgenic mice (45). Subsequent Aβ immunotherapy in other mouse models has shown similar effectiveness in reducing Aβ load and restoring behavioral impairments (46, 47). To study the efficacy of Aβ immunotherapy as a treatment for clearing both Aβ and tau pathology, antibodies against Aβ were administered to 3xTg-AD mice. Immunotherapy against Aβ was effective in clearing Aβ plaques, early intraneuronal Aβ deposits, and early tau aggregates from the brains of 3xTg-AD mice, as well as rescuing cognitive deficits (30, 48). Interestingly, Aβ immunotherapy does not clear hyperphosphorylated tau aggregates indicating that later tau aggregations are more permanent (48). In addition to providing evidence that Aβ immunotherapy may be useful for treating AD, immunotherapy has provided further evidence for the amyloid cascade hypothesis because tau is cleared even though immunotherapy is directed only against Aβ. Furthermore, after clearance of Aβ and tau using Aβ immunotherapy, pathology reemerges in a hierarchical fashion with Aβ deposition occurring first, followed by the reemergence of tau accumulation (48). Thus, Aβ accumulation appears to be able to induce tau accumulation in the 3xTg-AD mice.

One pathway by which Aβ may influence tau accumulation may be by inducing proteasome dysfunction. Proteasome activity can be a major clearance mechanism for tau protein (49). As noted above, Aβ immunotherapy is able to also clear tau pathology. Further studies showed that clearance of tau pathology with immunotherapy is mediated by the proteasome (48). In addition, hyperphosphorylated tau is not cleared by Aβ immunotherapy, and this is consistent with studies showing clearance of unphosphorylated tau by the proteasome system (50). Moreover, proteasome function has been shown to be directly impaired by Aβ oligomers in the 3xTg-AD mice (51). These results show that Aβ likely drives tau pathology by impairing proteasome function, leading to tau accumulation. Clearance of Aβ using immunotherapy can reverse this proteasome dysfunction leading to a restoration of tau aggregation. Clinical trials using Aβ immunotherapy have shown promise with indications of removal of Aβ in patients (52). However, clinical trials were stopped because 6% of patients developed subacute meningoencephalitis (53). Nevertheless, work continues on developing safer immunotherapy approaches against Aβ that will not induce dangerous side effects (54). For example, a novel gene therapy using a herpes simplex virus delivering $A\beta_{1-42}$ and Interleukin 4 (IL-4) induces a T-helper 2 (Th2) antibody response that leads to clearance of Aβ pathology and decreases in tau phosphorylation in the 3xTg-AD mice (55). With this new form of immunotherapy, improvements in the Barnes maze task were observed, and new techniques such as this novel gene therapy may help clear pathology and treat AD.

In addition to immunotherapy, several pharmacological approaches have been used as treatments for AD. In particular, the acetylcholine receptor has been an important target for the treatment of AD. Acetylcholine receptors are divided into two major subgroups: nicotinic acetylcholine receptors (nAChR) and muscarinic acetylcholine receptors (mAChR). Selective loss of the nAChR along with a loss of cholinergic neurons is a notable and invariant feature in the AD brain (56, 57). Treatments for AD have focused on restoring acetylcholine levels in the AD brain and, until recently, the only major treatment for AD was the administration of acetylcholinesterase inhibitors. Because nAChR are selectively lost in the AD brain, the chronic use of nicotine in cigarette smokers has been studied as a risk factor for AD. Although epidemiological studies surrounding the use of nicotine and the risk of AD have provided nebulous results, some evidence exists that suggests that nicotine may be able to modulate Aβ and tau pathology differentially (58, 59).

Studies in transgenic mouse models of AD have given more insight into the role nicotine might play in AD pathogenesis. In APP transgenic mice, chronic nicotine administration led to decreases in Aβ plaque load, although soluble Aβ levels were unchanged (60).

Unfortunately, effects of nicotine on tau pathology could not be studied because of lack of tau pathology in these mice. In contrast, the 3xTg-AD mouse model provides an opportunity to study nicotine and its effects on tau pathology. In addition, the 3xTg-AD model mimics the loss of α7nAChR that is observed in human AD and provides a good model for observing nicotinic effects on AD pathology (61). Similar to the studies in APP mice, administration of nicotine to 3xTg-AD mice did not alter soluble Aβ pathology (61). In contrast, increases in tau phosphorylation via activation of p38-mitogen-activated- protein kinase pathway was observed (61). Thus, nicotine may play a role in driving tau pathology in AD.

In contrast to the nicotinic receptors, mAChR are preserved in the AD brain, and they present an enticing target for restoring cholinergic hypofunction in AD (62). The M1 mAChR subtype, the predominant mAChR in the CNS, is involved in short-term memory processes, and represents a promising target (63). In addition, M1 agonists have shown ability to modulate both Aβ and tau pathology (64). However, their clinical use has been limited because of poor selectivity for the M1 receptor, poor bioavailability, poor activity, and narrow safety margin (63). The development of newer, highly selective M1 agonists presents an opportunity to study the effects of M1 agonists on AD pathology (65). The new M1 agonist AF267b was administered to the 3xTg-AD mice and was effective in reducing Aβ pathology and reducing tau phosphorylation in the hippocampus and cortex, but not in the amygdala (66). Accordingly, AF267b restored cognitive impairments on cortical and hippocampal-dependent tasks but impairments in contextual fear conditioning persisted (66). The M1 agonist was able to exert these effects by activating disintegrin and metalloproteinase 17 (ADAM17), shifting APP processing toward a nonamyloidogenic pathway, thereby reducing Aβ. The reduction in tau phosphorylation was mediated by a reduction in glycogen synthase kinase-3β (GSK-3β) activity. Because reducing GSK-3β activity was effective in reducing tau phosphorylation, lithium, a known GSK-3β inhibitor, was administered to the 3xTg-AD mice. Lithium is commonly used to treat bipolar disease, but because of its inhibitory activity against GSK-3β, it may have other uses. In 3xTg-AD mice, lithium was able to reduce tau phosphorylation, but Aβ load was not altered and cognitive function was not restored (67). Thus, it appears that reducing tau pathology with lithium is likely not sufficient to restore cognitive deficits.

Because direct application of Aβ induces synaptic dysfunction, clearance of Aβ is likely needed to restore cognitive function (68, 69). However, as noted before, reducing Aβ pathology often leads to reduced tau pathology. Immunotherapy leads to clearance of early tau pathology, whereas M1 agonists reduce tau accumulation (48, 66). Thus, although various direct applications of Aβ in vitro induce synaptic dysfunction, reductions in tau may also be responsible for rescuing cognitive deficits. Indeed, reductions of soluble tau and Aβ are necessary to restore behavioral function in 3xTg-AD mice (70). In contrast, reducing soluble Aβ alone is insufficient to rescue cognitive function. Furthermore, the presence of tau may facilitate Aβ-induced cognitive dysfunction. Reductions in endogenous tau levels can prevent cognitive impairments typically seen in transgenic APP mice (71). Thus, reduction of tau pathology may be in an important therapeutic target for treating AD.

4. Lifestyle and AD

Although Aβ accumulation may be the major insult in the pathogenesis of AD, patients with sporadic AD do not have elevated Aβ production. Thus, it is likely that many environmental and lifestyle factors contribute to the risk of developing AD. For example, mice exposed to environment enrichment or prior learning have improved cognition compared to mice who live in standard living units (72, 73). In the 3xTg-AD mouse model, mice involved in repeated learning and training tasks in the Morris water maze show reduced Aβ pathology including reduced plaque load compared to naive mice that were never exposed to behavioral tasks (31). In addition, Aβ*56, an oligomer of Aβ that correlates with development of cognitive loss, and tau phosphorylation via reduced GSK-3β activity was decreased in mice given repeated training. Accompanying these reductions in pathology, mice performed better on cognitive tasks when exposed to early and repeated learning. These studies provide evidence that constant mental activity can have pathology modifying effects.

In contrast to the beneficial effects of mental activity, a stressful lifestyle may increase risk for developing AD. During stressful conditions, glucocorticoids (usually cortisol in primates) are released by the body and exert far-ranging effects. In addition to altering fat distribution and immune suppressive effects, glucocorticoids are able to cross the blood-brain barrier and exert important influence on learning and memory (74). Furthermore, cortisol levels appear to be increased in the AD brain and may play a role in disease pathogenesis (75–77). To study the effects that glucocorticoids may have on Aβ and tau pathology, glucocorticoids were administered to 3xTg-AD mice. After 1 week of glucocorticoid treatment, increases in Aβ accumulation through increased β-APP cleaving enzyme (BACE) activity was observed (78). In addition, tau accumulation was increased. Because stress hormones appear to exacerbate pathology in 3xTg-AD mice, reducing stress or the pharmacological reduction of glucorticoids may be a viable treatment for AD.

Another important component of one's lifestyle is diet, and certain diets may increase or reduce the risk of AD. Docosahexaenoic acid (DHA) is an important fatty acid obtained largely from dietary intake, which composes around 60% of the fatty acids in neuronal membranes. As DHA is particularly concentrated in synaptic membranes (79), it has been implicated to be important in healthy brain functioning. Furthermore, diets rich in DHA can have disease-modifying effects as APP transgenic mice fed DHA-rich diets exhibit reduced amyloid and dendritic pathology (80, 81). In the 3xTg-AD mice, DHA supplementation of the diet decreased Aβ accumulation due to decreases in steady-state PS1 (82). In addition, tau accumulation was decreased in mice fed the DHA diet. When docosapentaenoic acid (DPAn-6) was combined with DHA in the diet, tau phosphorylation was decreased via reduced c-Jun N-terminal kinase (JNK) activity. These results provide evidence that diet can alter AD pathology in mice and may be important in both treatment and prevention.

5. Concluding Remarks

The 3xTg-AD mouse model has been instrumental in evaluating the amyloid cascade hypothesis *in vivo*. Because the model reproduces Aβ and tau pathology, studies using the 3xTg-AD model have shown that Aβ pathology likely both precedes and can promote tau pathology. Looking forward, the 3xTg-AD mouse model provides an excellent platform for evaluating potential disease-modifying treatments because studies on both Aβ and tau can be performed.

References

1. Ferri CP, Prince M, Brayne C, et al. (2005) Global prevalence of dementia: a Delphi consensus study. Lancet 366:2112–2117.
2. Lee VM, Goedert M, Trojanowski JQ (2001) Neurodegenerative tauopathies. Annu Rev Neurosci 24:1121–1159.
3. Selkoe DJ (2003) Toward a remembrance of things past: deciphering Alzheimer disease. Harvey Lect 99:23–45.
4. Hardy J, Allsop D (1991) Amyloid deposition as the central event in the aetiology of Alzheimer's disease. Trends Pharmacol Sci 12:383–388.
5. Scheuner D, Eckman C, Jensen M, et al. (1996) Secreted amyloid beta-protein similar to that in the senile plaques of Alzheimer's disease is increased in vivo by the presenilin 1 and 2 and APP mutations linked to familial Alzheimer's disease. Nat Med 2:864–870.
6. Hardy JA, Higgins GA (1992) Alzheimer's disease: the amyloid cascade hypothesis. Science 256:184–185.
7. Rovelet-Lecrux A, Hannequin D, Raux G, et al. (2006) APP locus duplication causes autosomal dominant early-onset Alzheimer disease with cerebral amyloid angiopathy. Nat Genet 38:24–26.
8. Glenner GG, Wong CW (1984) Alzheimer's disease and Down's syndrome: sharing of a unique cerebrovascular amyloid fibril protein. Biochem Biophys Res Commun 122:1131–1135.
9. Games D, Buttini M, Kobayashi D, Schenk D, Seubert P (2006) Mice as models: transgenic

approaches and Alzheimer's disease. J Alzheimers Dis 9:133–149.
10. Games D, Adams D, Alessandrini R, et al. (1995) Alzheimer-type neuropathology in transgenic mice overexpressing V717F beta-amyloid precursor protein. Nature 373:523–527.
11. Hsiao K, Chapman P, Nilsen S, et al. (1996) Correlative memory deficits, Abeta elevation, and amyloid plaques in transgenic mice. Science 274:99–102.
12. Sturchler-Pierrat C, Abramowski D, Duke M, et al. (1997) Two amyloid precursor protein transgenic mouse models with Alzheimer disease-like pathology. Proc Natl Acad Sci U S A 94:13287–13292.
13. Suzuki N, Cheung TT, Cai XD, et al. (1994) An increased percentage of long amyloid beta protein secreted by familial amyloid beta protein precursor (beta APP717) mutants. Science 264:1336–1340.
14. Duff K, Eckman C, Zehr C, et al. (1996) Increased amyloid-beta42(43) in brains of mice expressing mutant presenilin 1. Nature 383:710–713.
15. Borchelt DR, Ratovitski T, van Lare J, et al. (1997) Accelerated amyloid deposition in the brains of transgenic mice coexpressing mutant presenilin 1 and amyloid precursor proteins. Neuron 19:939–945.
16. Holcomb L, Gordon MN, McGowan E, et al. (1998) Accelerated Alzheimer-type phenotype in transgenic mice carrying both mutant amyloid precursor protein and presenilin 1 transgenes. Nat Med 4:97–100.
17. Lewis J, Dickson DW, Lin WL, et al. (2001) Enhanced neurofibrillary degeneration in transgenic mice expressing mutant tau and APP. Science 293:1487–1491.
18. Gotz J, Chen F, van Dorpe J, Nitsch RM (2001) Formation of neurofibrillary tangles in P301l tau transgenic mice induced by Abeta 42 fibrils. Science 293:1491–1495.
19. Oddo S, Caccamo A, Shepherd JD, et al. (2003) Triple-transgenic model of Alzheimer's disease with plaques and tangles: intracellular Abeta and synaptic dysfunction. Neuron 39:409–421.
20. Goedert M (2005) Tau gene mutations and their effects. Mov Disord 20(Suppl 12): S45–S52.
21. Caroni P (1997) Overexpression of growth-associated proteins in the neurons of adult transgenic mice. J Neurosci Methods 71:3–9.
22. Shie FS, LeBoeuf RC, Jin LW (2003) Early intraneuronal Abeta deposition in the hippocampus of APP transgenic mice. Neuroreport 14:123–129.
23. Gyure KA, Durham R, Stewart WF, Smialek JE, Troncoso JC (2001) Intraneuronal abeta-amyloid precedes development of amyloid plaques in Down syndrome. Arch Pathol Lab Med 125:489–492.
24. Oddo S, Caccamo A, Tran L, et al. (2006) Temporal profile of amyloid-beta (Abeta) oligomerization in an in vivo model of Alzheimer disease: a link between Abeta and tau pathology. J Biol Chem 281:1599–1604.
25. Oddo S, Caccamo A, Smith IF, Green KN, LaFerla FM (2006) A dynamic relationship between intracellular and extracellular pools of Abeta. Am J Pathol 168:184–194.
26. Kitazawa M, Oddo S, Yamasaki TR, Green KN, LaFerla FM (2005) Lipopolysaccharide-induced inflammation exacerbates tau pathology by a cyclin-dependent kinase 5-mediated pathway in a transgenic model of Alzheimer's disease. J Neurosci 25:8843–8853.
27. Jicha GA, Bowser R, Kazam IG, Davies P (1997) Alz-50 and MC-1, a new monoclonal antibody raised to paired helical filaments, recognize conformational epitopes on recombinant tau. J Neurosci Res 48:128–132.
28. Scheff SW, Scott SA, DeKosky ST (1991) Quantitation of synaptic density in the septal nuclei of young and aged Fischer 344 rats. Neurobiol Aging 12:3–12.
29. Walsh DM, Klyubin I, Fadeeva JV, et al. (2002) Naturally secreted oligomers of amyloid beta protein potently inhibit hippocampal long-term potentiation in vivo. Nature 416:535–539.
30. Billings LM, Oddo S, Green KN, McGaugh JL, LaFerla FM (2005) Intraneuronal Abeta causes the onset of early Alzheimer's disease-related cognitive deficits in transgenic mice. Neuron 45:675–688.
31. Billings LM, Green KN, McGaugh JL, LaFerla FM (2007) Learning decreases A beta*56 and tau pathology and ameliorates behavioral decline in 3xTg-AD mice. J Neurosci 27: 751–761.
32. Rogers J, Lue LF (2001) Microglial chemotaxis, activation, and phagocytosis of amyloid beta-peptide as linked phenomena in Alzheimer's disease. Neurochem Int 39: 333–340.
33. Wyss-Coray T, Mucke L (2002) Inflammation in neurodegenerative disease-a double-edged sword. Neuron 35:419–432.
34. Akiyama H, Schwab C, Kondo H, et al. (1996) Granules in glial cells of patients with Alzheimer's disease are immunopositive for C-terminal sequences of beta-amyloid protein. Neurosci Lett 206:169–172.

35. Funato H, Yoshimura M, Yamazaki T, et al. (1998) Astrocytes containing amyloid beta-protein (Abeta)-positive granules are associated with Abeta40-positive diffuse plaques in the aged human brain. Am J Pathol 152: 983–992.
36. Janelsins MC, Mastrangelo MA, Oddo S, LaFerla FM, Federoff HJ, Bowers WJ (2005) Early correlation of microglial activation with enhanced tumor necrosis factor-alpha and monocyte chemoattractant protein-1 expression specifically within the entorhinal cortex of triple transgenic Alzheimer's disease mice. J Neuroinflammation 2:23.
37. Parvathy S, Rajadas J, Ryan H, Vaziri S, Anderson L, Murphy GM, Jr (2009) Abeta peptide conformation determines uptake and interleukin-1alpha expression by primary microglial cells. Neurobiol Aging 30(11): 1792–1804.
38. Lindberg C, Selenica ML, Westlind-Danielsson A, Schultzberg M (2005) Beta-amyloid protein structure determines the nature of cytokine release from rat microglia. J Mol Neurosci 27:1–12.
39. Griffin WS, Stanley LC, Ling C, et al. (1989) Brain interleukin 1 and S-100 immunoreactivity are elevated in Down syndrome and Alzheimer disease. Proc Natl Acad Sci U S A 86:7611–7615.
40. Patel AJ, Jen A, Wickenden C, Jen LS, Gentleman SM, de Silva HA (1997) Glia-derived cytokines and the biogenesis of amyloid plaques in Alzheimer's disease. Mol Psychiatry 2:130–132.
41. Hesselgesser J, Horuk R (1999) Chemokine and chemokine receptor expression in the central nervous system. J Neurovirol 5:13–26.
42. Lukiw WJ, Bazan NG (2000) Neuroinflammatory signaling upregulation in Alzheimer's disease. Neurochem Res 25:1173–1184.
43. Sheng JG, Zhu SG, Jones RA, Griffin WS, Mrak RE (2000) Interleukin-1 promotes expression and phosphorylation of neurofilament and tau proteins in vivo. Exp Neurol 163:388–391.
44. Hardy J (2006) Has the amyloid cascade hypothesis for Alzheimer's disease been proved? Curr Alzheimer Res 3:71–73.
45. Schenk D, Barbour R, Dunn W, et al. (1999) Immunization with amyloid-beta attenuates Alzheimer-disease-like pathology in the PDAPP mouse. Nature 400:173–177.
46. Morgan D, Diamond DM, Gottschall PE, et al. (2000) A beta peptide vaccination prevents memory loss in an animal model of Alzheimer's disease. Nature 408:982–985.
47. Janus C, Pearson J, McLaurin J, et al. (2000) A beta peptide immunization reduces behavioural impairment and plaques in a model of Alzheimer's disease. Nature 408:979–982.
48. Oddo S, Billings L, Kesslak JP, Cribbs DH, LaFerla FM (2004) Abeta immunotherapy leads to clearance of early, but not late, hyperphosphorylated tau aggregates via the proteasome. Neuron 43:321–332.
49. David DC, Layfield R, Serpell L, Narain Y, Goedert M, Spillantini MG (2002) Proteasomal degradation of tau protein. J Neurochem 83: 176–185.
50. Goldbaum O, Oppermann M, Handschuh M, et al. (2003) Proteasome inhibition stabilizes tau inclusions in oligodendroglial cells that occur after treatment with okadaic acid. J Neurosci 23:8872–8880.
51. Tseng BP, Green KN, Chan JL, Blurton-Jones M, Laferla FM (2007) Abeta inhibits the proteasome and enhances amyloid and tau accumulation. Neurobiol Aging 29(11):1607–1618.
52. Nicoll JA, Barton E, Boche D, et al. (2006) Abeta species removal after abeta42 immunization. J Neuropathol Exp Neurol 65:1040–1048.
53. Orgogozo JM, Gilman S, Dartigues JF, et al. (2003) Subacute meningoencephalitis in a subset of patients with AD after Abeta42 immunization. Neurology 61:46–54.
54. Lemere CA, Maier M, Peng Y, Jiang L, Seabrook TJ (2007) Novel Abeta immunogens: is shorter better? Curr Alzheimer Res 4:427–436.
55. Frazer ME, Hughes JE, Mastrangelo MA, Tibbens JL, Federoff HJ, Bowers WJ (2008) Reduced pathology and improved behavioral performance in Alzheimer's disease mice vaccinated with HSV amplicons expressing amyloid-beta and interleukin-4. Mol Ther 16:845–853.
56. Davies P, Maloney AJ (1976) Selective loss of central cholinergic neurons in Alzheimer's disease. Lancet 2:1403.
57. McGeer PL, McGeer EG, Suzuki J, Dolman CE, Nagai T (1984) Aging, Alzheimer's disease, and the cholinergic system of the basal forebrain. Neurology 34:741–745.
58. Ulrich J, Johannson-Locher G, Seiler WO, Stahelin HB (1997) Does smoking protect from Alzheimer's disease? Alzheimer-type changes in 301 unselected brains from patients with known smoking history. Acta Neuropathol 94:450–454.
59. Court JA, Johnson M, Religa D, et al. (2005) Attenuation of Abeta deposition in the entorhinal cortex of normal elderly individuals

associated with tobacco smoking. Neuropathol Appl Neurobiol 31:522–535.

60. Nordberg A, Hellstrom-Lindahl E, Lee M, et al. (2002) Chronic nicotine treatment reduces beta-amyloidosis in the brain of a mouse model of Alzheimer's disease (APPsw). J Neurochem 81:655–658.
61. Oddo S, Caccamo A, Green KN, et al. (2005) Chronic nicotine administration exacerbates tau pathology in a transgenic model of Alzheimer's disease. Proc Natl Acad Sci U S A 102:3046–3051.
62. Svensson AL, Alafuzoff I, Nordberg A (1992) Characterization of muscarinic receptor subtypes in Alzheimer and control brain cortices by selective muscarinic antagonists. Brain Res 596:142–148.
63. Fisher A (2000) Therapeutic strategies in Alzheimer's disease: M1 muscarinic agonists. Jpn J Pharmacol 84:101–112.
64. Fisher A, Pittel Z, Haring R, et al. (2003) M1 muscarinic agonists can modulate some of the hallmarks in Alzheimer's disease: implications in future therapy. J Mol Neurosci 20:349–356.
65. Fisher A, Brandeis R, Bar-Ner RH, et al. (2002) AF150(S) and AF267B: M1 muscarinic agonists as innovative therapies for Alzheimer's disease. J Mol Neurosci 19:145–153.
66. Caccamo A, Oddo S, Billings LM, et al. (2006) M1 receptors play a central role in modulating AD-like pathology in transgenic mice. Neuron 49:671–682.
67. Caccamo A, Oddo S, Tran LX, LaFerla FM (2007) Lithium reduces tau phosphorylation but not A beta or working memory deficits in a transgenic model with both plaques and tangles. Am J Pathol 170:1669–1675.
68. Lambert MP, Barlow AK, Chromy BA, et al. (1998) Diffusible, nonfibrillar ligands derived from Abeta1–42 are potent central nervous system neurotoxins. Proc Natl Acad Sci U S A 95:6448–6453.
69. Snyder EM, Nong Y, Almeida CG, et al. (2005) Regulation of NMDA receptor trafficking by amyloid-beta. Nat Neurosci 8:1051–1058.
70. Oddo S, Vasilevko V, Caccamo A, Kitazawa M, Cribbs DH, LaFerla FM (2006) Reduction of soluble Abeta and tau, but not soluble Abeta alone, ameliorates cognitive decline in transgenic mice with plaques and tangles. J Biol Chem 281:39413–39423.
71. Roberson ED, Scearce-Levie K, Palop JJ, et al. (2007) Reducing endogenous tau ameliorates amyloid beta-induced deficits in an Alzheimer's disease mouse model. Science 316:750–754.
72. Churchill JD, Galvez R, Colcombe S, Swain RA, Kramer AF, Greenough WT (2002) Exercise, experience and the aging brain. Neurobiol Aging 23:941–955.
73. Fillit HM, Butler RN, O'Connell AW, et al. (2002) Achieving and maintaining cognitive vitality with aging. Mayo Clin Proc 77:681–696
74. Roozendaal B, Phillips RG, Power AE, Brooke SM, Sapolsky RM, McGaugh JL (2001) Memory retrieval impairment induced by hippocampal CA3 lesions is blocked by adrenocortical suppression. Nat Neurosci 4:1169–1171.
75. Davis KL, Davis BM, Greenwald BS, et al. (1986) Cortisol and Alzheimer's disease, I: basal studies. Am J Psychiatry 143:300–305.
76. Masugi F, Ogihara T, Sakaguchi K, et al. (1989) High plasma levels of cortisol in patients with senile dementia of the Alzheimer's type. Methods Find Exp Clin Pharmacol 11:707–710.
77. Swanwick GR, Kirby M, Bruce I, et al. (1998) Hypothalamic-pituitary-adrenal axis dysfunction in Alzheimer's disease: lack of association between longitudinal and cross-sectional findings. Am J Psychiatry 155:286–159.
78. Green KN, Billings LM, Roozendaal B, McGaugh JL, LaFerla FM (2006) Glucocorticoids increase amyloid-beta and tau pathology in a mouse model of Alzheimer's disease. J Neurosci 26:9047–9056.
79. Bazan NG, Scott BL (1990) Dietary omega-3 fatty acids and accumulation of docosahexaenoic acid in rod photoreceptor cells of the retina and at synapses. Ups J Med Sci Suppl 48:97–107.
80. Calon F, Lim GP, Yang F, et al. (2004) Docosahexaenoic acid protects from dendritic pathology in an Alzheimer's disease mouse model. Neuron 43:633–645.
81. Lim GP, Calon F, Morihara T, et al. (2005) A diet enriched with the omega-3 fatty acid docosahexaenoic acid reduces amyloid burden in an aged Alzheimer mouse model. J Neurosci 25:3032–3040.
82. Green KN, Martinez-Coria H, Khashwji H, et al. (2007) Dietary docosahexaenoic acid and docosapentaenoic acid ameliorate amyloid-beta and tau pathology via a mechanism involving presenilin 1 levels. J Neurosci 27:4385–4395.

Part V

Animal Models of Non-Alzheimer Neurodegenerative Disease

Chapter 25

Cognitive Dysfunction in Genetic Mouse Models of Parkinsonism

Sheila M. Fleming, J. David Jentsch, and Marie-FranÇoise Chesselet

Abstract

Parkinson's disease (PD) is primarily recognized as a motor disorder; however, patients also present with a wide range of nonmotor manifestations. Cognitive dysfunctions in nondemented PD patients can occur early in the disease and primarily consist of deficits in executive function. Because it can be assessed with noninvasive measurement tools, cognitive dysfunction could be evaluated to determine the effects of potential disease-modifying agents in patients. A challenge is to reproduce these deficits in animals for preclinical drug testing. Genetic mouse models of PD have been generated based on mutations causing rare familial forms of PD. Although only a few models show extensive nigrostriatal dopamine cell loss, several present extensive anomalies in functions that are also altered in premanifest phases of PD. Here we review the few studies that have so far investigated cognitive function in these new models.

Key words: Alpha-synuclein, Reversal learning, Parkin, Prefrontal cortex, Locus coeruleus

1. Introduction

Parkinson's disease (PD) is primarily characterized by motor impairments including tremor, rigidity, bradykinesia, and postural instability that are associated with the progressive loss of dopamine neurons in the substantia nigra pars compacta. However, in addition to the motor symptoms, patients also develop a wide variety of nonmotor symptoms such as cognitive dysfunction, autonomic dysfunction, sleep disorders, and olfactory impairments indicating PD is a systemic disorder. Both environmental and genetic factors have been linked to PD and suggest that sporadic PD may arise from an interaction between environmental toxins combined with a genetic susceptibility.

Peter Paul De Deyn and Debby Van Dam (eds.), *Animal Models of Dementia*, Neuromethods, vol. 48,
DOI 10.1007/978-1-60761-898-0_25, © Springer Science+Business Media, LLC 2011

2. Cognitive Dysfunction in Early PD

The nonmotor symptoms associated with PD are eliciting growing interest because many of these symptoms may appear early in the disease process and greatly contribute to the overall quality of life of patients with the disorder. Among these nonmotor symptoms, cognitive dysfunction includes subtle impairments in implicit memory, attention, visuospatial skills, and executive function (1, 2). The estimated prevalence of cognitive dysfunction in PD ranges from 20% to 40% (3–5). Several studies show that PD patients, as a group, exhibit impairments in executive function similar to patients with frontal lobe lesions (6), including increased perseverative errors in the Wisconsin Card Sorting Task (7), attention deficits such as ignoring irrelevant stimuli properties (8), and attentional set shifting (9). Patients also have shown impaired habit learning (1). Deficits in executive function and habit learning have been associated with alterations in the striatum (1, 10), and dysfunction within the prefrontal cortex (11, 12). Dopamine agonists can have both beneficial and detrimental effects on cognitive function in patients and in healthy volunteers (13–16), suggesting that extranigral pathology may contribute, at least in part, to the early cognitive dysfunction in PD. Indeed, Braak and colleagues proposed the concept of progressive pathology in PD through analysis of α-synuclein pathology in the brains of PD patients at varying stages of the disease and of individuals without overt neurological disorders (17). These authors have shown that early α-synuclein pathology begins in subcortical nuclei including the dorsal motor nucleus X, olfactory regions, and in the locus coeruleus prior to the development of α-synuclein pathology in the substantia nigra that leads to manifest PD (17). The potential dysfunction of the locus coeruleus in the early stages of PD has important implications because the locus coeruleus has diffuse projections to cortical areas, including the prefrontal cortex. At later stages, progressive pathology to the cerebral cortex itself may underlie the worsening of cognitive dysfunction that is observed in some patients with advanced PD.

3. Cognitive Dysfunction in Toxin Models of Parkinsonism

Traditional models of PD are based on the use of neurotoxins, including 1-methyl-4-phenyl-1,2,3,6-tetrahydropyridine (MPTP) and 6-hydroxydopamine, to kill nigrostriatal dopaminergic neurons. Cognitive dysfunction has been detected in some of these toxin-induced models of Parkinsonism. Depending on the dosing regimen, MPTP-treated nonhuman primates show deficits in a range of tasks that require working memory and/or response control/inhibition (18–21). Similarly, MPTP-treated and 6-hydroxydopamine-treated mice show cognitive impairments in alternation, habituation, and spatial memory (22–24). Although these models have been important for our understanding of cognitive deficits associated with nigrostriatal dopamine cell loss, they lack the broad extranigral pathology observed in PD.

4. Cognitive Dysfunction in Genetic Models of Parkinsonism

Several types of genetic mouse models of PD have been generated. Some models are based on the deletion or inactivation of factors that are important for the differentiation and/or maintenance of nigrostriatal dopaminergic neurons such as Nurrl, Pitx3, and engrailed (25–27). These models exhibit progressive loss of nigrostriatal dopaminergic neurons and show motor and/or affective deficits (25, 28, 29), but none has yet been examined for cognitive dysfunction. More relevant to PD are models based on mutations known to cause rare familial forms of PD. Few lead to overt, progressive nigrostriatal dopaminergic cell loss in the lifetime of the mouse, but they have strong construct validity because they are based on mechanisms known to cause PD in humans. Furthermore, they provide the opportunity to study the early nonmotor symptoms associated with PD (30).

A variety of PD-causing mutations has now been expressed in several different lines of mice (30, 31). Of particular interest are mice that overexpress the presynaptic protein α-synuclein. Mutations and multiplication of the α-synuclein gene are sufficient to induce PD (32–35). In sporadic PD, α-synuclein is a major component of Lewy bodies, the pathological hallmark of PD (36). Alpha-synuclein accumulates in selective neuronal populations throughout the body (17), indicating that it is involved in sporadic as well as familial forms of PD.

4.1. Cognitive Deficits in Mice Overexpressing Alpha-Synuclein

Several laboratories have begun to study cognitive function in different lines of α-synuclein transgenic mice (Table 1). One line of mice that expresses the human A30P α-synuclein mutation and presents with α-synuclein inclusions predominantly in the brain stem and spinal cord develops motor impairments after the first 12 months of life primarily due to motor neuron pathology (37). However, these mice also have α-synuclein inclusions in the amygdala that may be associated with neuropsychiatric impairments such as cognitive dysfunction and anxiety. Indeed, when transgenic and wild-type mice were tested for visuospatial learning and memory in the Morris water maze, for fear conditioning, and for active avoidance at 4 and 12 months of age prior to the onset

Table 1
Cognitive impairments in genetic mouse models of Parkinsonism

α-synuclein transgenic mice	Cognitive impairment
Thy1-aSyn (50)	↓ Cumulative accuracy in reversal learning task
Thy1-A30P (38)	↓ Preference for platform quadrant in Morris water maze
CaM_α-syn (39)	↓ Retention in Morris water maze
Parkin knockout mice	**Cognitive impairment**
Exon 3 deleted (54)	↓ Spontaneous alternation in T-maze
Exon 3 deleted (53)	Impaired spatial learning in Morris water maze

Thy1-aSyn mice overexpressing human wild-type alpha-synuclein under the Thy1 promoter; Thy1-A30P mice that express the human A30P α-synuclein mutation; CaM_α-syn express high levels of human wild-type α-synuclein in midbrain and forebrain regions under the prion promoter

of motor impairments, the transgenic mice displayed age-dependent alterations in all three tasks suggesting cognitive dysfunction and increased anxiety (38) (Table 1).

In a novel conditional mouse model generated to express high levels of human wild-type α-synuclein in midbrain and forebrain regions under the prion promoter, α-synuclein pathology developed in both the substantia nigra and hippocampus leading to motor and cognitive impairments (39). In the Morris water maze, transgenics did not differ from controls in swim speed, acquisition, or in the 24 h probe trial. However, they showed reduced retention following a 7-day probe trial. Although the cognitive deficit appears to be associated with hippocampal dysfunction, which is not prominent in PD patients, it does highlight the relationship between α-synuclein pathology and cognitive deficits (Table 1).

We have extensively characterized a line of mice with broad overexpression of human wild-type α-synuclein under the Thy1 promoter (Thy1-aSyn), exhibiting proteinase K-resistant α-synuclein aggregates in areas similar to those affected in PD, including the locus coeruleus and substantia nigra (40, 41, Hutson et al. in preparation). These mice do not show overt cell loss up to 8 months of age, but exhibit progressive loss of tyrosine hydroxylase-containing terminals in the striatum (Mortazavi et al., in preparation). Thy1-aSyn mice have alterations in striatal dopamine synapses including a chronic increase in basal levels of extracellular dopamine in the striatum (42), abnormal behavioral responses to dopaminergic agonists (43), and abnormal electrophysiological responses to dopaminergic agonists and antagonists in striatal slices (44). In addition, Thy1-aSyn mice show olfactory, cardiovascular, and gastrointestinal dysfunction, all nonmotor symptoms associated with early PD (45–47). Recent data indicates that noradrenaline in the prefrontal cortex is decreased in Thy1-aSyn mice (Hean et al. 2010, unpublished observations), and given the critical role of noradrenaline in modulating aspects of cognition (48), these observations suggest that Thy1-aSyn mice may have impairments in cognitive function.

Because patients show difficulty in strategy switching, we tested Thy1-aSyn and wild-type mice in a reversal learning task that assesses cognitive flexibility (49). Male Thy1-aSyn and wild-type littermates at 4–5 months of age were trained on an operant serial reversal task. Mice were trained to respond into one of two illuminated apertures to obtain food reward and the contingencies were reversed after performance criteria were achieved. In this test we found that Thy1-aSyn mice can learn a simple operant strategy as well as controls, but show greater difficulty in their ability to switch their response at reversal, compared to wild-type mice, although they were able to eventually reach criteria and learn the reversed contingency (50).

Since dopamine agonists can have both beneficial and detrimental effects on cognition (13–15), we assessed the effect of two different doses of l-dopa (15 and 25 mg/kg, i.p.) in Thy1-aSyn and wild-type mice in the reversal learning paradigm (50). In wild-type mice the low dose of l-dopa had no effect on accuracy, while the higher dose reduced accuracy (Table 2). In contrast, both doses had a positive effect on reversal accuracy in the Thy1-aSyn mice. Although l-dopa is known for its affects on the dopamine system, it also increases noradrenaline and it is possible that improvement in the l-dopa studies may be due, in part, to stimulation of noradrenaline receptors, since Thy1-aSyn mice have decreased noradrenaline in the prefrontal cortex (Hean et al. 2010, unpublished observation). To test this hypothesis, we assessed the effect of two different doses of the α2-noradrenergic agonist guanfacine (0.1 and 1.0 mg/kg, i.p.) on the reversal task. Similar to l-dopa, both doses of guanfacine had a positive effect

Table 2
Effect of l-dopa and guanfacine in Thy1-aSyn and wild-type mice in a reversal task

	Cumulative accuracy	
Drug (compared to saline)	**Wild-type**	**Thy1-aSyn**
L-DOPA (15 mg/kg, i.p.)	No effect	↑
L-DOPA (25 mg/kg, i.p.)	↓	↑
Guanfacine (0.1 mg/kg, i.p.)	No effect	↑
Guanfacine (1.0 mg/kg, i.p.)	↓	↑

Thy1-aSyn mice overexpressing human wild-type alpha-synuclein under the Thy1 promoter on reversal accuracy in the Thy1-aSyn mice, while the low dose had no effect and the higher dose reduced accuracy in wild-type mice. This indicates that Thy1-aSyn mice have cognitive impairments that can be improved pharmacologically (Table 2).

4.2. Cognitive Deficits in Mice Lacking Parkin

Mutations in the gene encoding the E3 ubiquitin ligase parkin cause early onset, recessive, familial PD. While, similar to patients with parkin mutations, parkin knockout mice do not have α-synuclein pathology, some have alterations within the nigrostriatal dopamine and noradrenergic systems, which may contribute to cognitive dysfunction (51, 52). Indeed, some lines of parkin knockout mice exhibit impairments in spatial memory (53) and decreased spontaneous alternation (54).

5. Conclusion

Genetic mouse models of Parkinsonism are still in the beginning stages of characterization compared to toxin models like MPTP; however, recent studies are demonstrating their utility as a model for understanding the early phase of PD. Nonmotor symptoms that are unmet medical needs, such as cognitive dysfunction, are the important targets for novel treatments and thus, make these models useful for preclinical studies of novel treatments for PD. These models also present a unique opportunity to examine pathological mechanisms leading to PD-related cognitive dysfunction, which remains poorly understood in humans.

Acknowledgments

We gratefully acknowledge the valuable assistance of Eddie C. Garcia. Funded by Morris K. Udall Parkinson's Disease Research Center of Excellence at UCLA (P50NS38367), the American Parkinson Disease Association, and the American Parkinson Disease Association UCLA Center of Excellence.

References

1. Knowlton BJ, Mangels JA, Squire LR (1996) A neostriatal habit learning system in humans. Science 273:1399–1402.
2. Levin BE, Katzen HL (2005) Early cognitive changes and nondementing behavioral abnormalities in Parkinson's disease. Adv Neurol 96:84–94.
3. Aarsland D, Tandberg E, Larsen JP, Cummings JL (1996) Frequency of dementia in Parkinson disease. Arch Neurol 53:538–542.
4. Aarsland D, Zaccai J, Brayne C (2005) A systematic review of prevalence studies of dementia in Parkinson's disease. Mov Disord 20:1255–1263.

5. Javin CC, Aarsland D, Larsen JP (2005) Cognitive predictors of dementia in Parkinson's disease: a community-based, 4 longitudinal study. J Geriat Psychiatry Neurol 18: 149–154.
6. Taylor AE, Saint-Cyr JA, Lang AE (1986) Frontal lobe dysfunction in Parkinson's disease. The cortical focus of neostriatal outflow. Brain 109:845–883.
7. Canavan AG, Passingham RE, Marsden CD, Quinn N, Wyke M, Polkey CE (1989) The performance on learning tasks of patients in the early stages of Parkinson's disease. Neuropsychologia 27:141–156.
8. Downes JJ, Roberts AC, Sahakian BJ, Evenden JL, Morris RG, Robbins TW (1989) Impaired extra-dimensional shift performance in medicated and unmedicated Parkinson's disease: evidence for a specific attentional dysfunction. Neuropsychologia 27:1329–1343.
9. Owen AM, James M, Leigh PN, et al. (1992) Fronto-striatal cognitive deficits at different stages of Parkinson's disease. Brain 115:1727–1751.
10. Packard MG, McGaugh JL (1996) Inactivation of hippocampus or caudate nucleus with lidocaine differentially affects expression of place and response learning. Neurobiol Learn Mem 65:65–72.
11. Ragozzino ME, Wilcox C, Raso M, Kesner RP (1999) Involvement of rodent prefrontal cortex subregions in strategy switching. Behav Neurosci 113:32–41.
12. Cole BJ, Robbins TW (1992) Forebrain norepinephrine: role in controlled information processing in the rat. Neuropsychopharmacology 7:129–142.
13. Cools R, Barker RA, Sahakian BJ, Robbins TW (2001) Enhanced or impaired cognitive function in Parkinson's disease as a function of dopaminergic medication and task demands. Cereb Cortex 11:1136–1143.
14. Swainson R, Rogers RD, Sahakian BJ, Summers BA, Polkey CE, Robbins TW (2000) Probabilistic learning and reversal deficits in patients with Parkinson's disease or frontal or temporal lobe lesions: possible adverse effects of dopaminergic medication. Neuropsychologia 38:596–612.
15. Gotham AM, Brown RG, Marsden CD (1988) 'Frontal' cognitive function in patients with Parkinson's disease 'on' and 'off' levodopa. Brain 111:299–321.
16. Mehta MA, Manes FF, Magnolfi G, Sahakian BJ, Robbins TW (2004) Impaired set-shifting and dissociable effects on tests of spatial working memory following the dopamine D2 receptor antagonist sulpiride in human volunteers. Psychopharmacology (Berl) 176:331–342.
17. Braak H, Del Tredici K, Rub U, de Vos RA, Jansen Steur EN, Braak E (2003) Staging of brain pathology related to sporadic Parkinson's disease. Neurobiol Aging 24:197–211.
18. Taylor JR, Elsworth JD, Roth RH, Sladek JR Jr, Redmond DE Jr (1990) Cognitive and motor deficits in the acquisition of an object retrieval/detour task in MPTP-treated monkeys. Brain 113:617–637.
19. Schneider JS, Kovelowski CJ 2nd (1990) Chronic exposure to low doses of MPTP. I. Cognitive deficits in motor asymptomatic monkeys. Brain Res 519:122–128.
20. Schneider JS, Pope-Coleman A (1995) Cognitive deficits precede motor deficits in a slowly progressing model of parkinsonism in the monkey. Neurodegeneration 4:245–255.
21. Slovin H, Abeles M, Vaadia E, Haalman I, Prut Y, Bergman H (1999) Frontal cognitive impairments and saccadic deficits in low-dose MPTP-treated monkeys. J Neurophysiol 81:858–874.
22. Tanila H, Björklund M, Riekkinen P Jr (1998) Cognitive changes in mice following moderate MPTP exposure. Brain Res Bull 45:577–582.
23. Dluzen DE, Kreutzberg JD (1993) 1-Methyl-4-phenyl-1,2,3,6-tetrahydropyridine (MPTP) disrupts social memory/recognition processes in the male mouse. Brain Res 609:98–102.
24. De Leonibus E, Pascucci T, Lopez S, Oliverio A, Amalric M, Mele A (2007) Spatial deficits in a mouse model of Parkinson disease. Psychopharmacology (Berl) 194:517–525.
25. Sonnier L, Le Pen G, Hartmann A, et al. (2007) Progressive loss of dopaminergic neurons in the ventral midbrain of adult mice heterozygote for Engrailed1. J Neurosci 27: 1063–1071.
26. Nunes I, Tovmasian LT, Silva RM, Burke RE, Goff SP (2003) Pitx3 is required for development of substantia nigra dopaminergic neurons. Proc Natl Acad Sci U S A 100:4245–4250.
27. Zetterström RH, Solomin L, Jansson L, Hoffer BJ, Olson L, Perlmann T (1997) Dopamine neuron agenesis in Nurr1-deficient mice. Science 276:248–250.
28. Hwang DY, Fleming SM, Ardayfio P, et al. (2005) 3,4-dihydroxyphenylalanine reverses the motor deficits in Pitx3-deficient aphakia mice: behavioral characterization of a novel genetic model of Parkinson's disease. J Neurosci 25:2132–2137.
29. Eells JB, Lipska BK, Yeung SK, Misler JA, Nikodem VM (2002) Nurr1-null heterozygous mice have reduced mesolimbic and mesocortical dopamine levels and increased stress-induced locomotor activity. Behav Brain Res 136:267–275.

30. Fleming SM, Fernagut PO, Chesselet MF (2005) Genetic mouse models of parkinsonism: strengths and limitations. NeuroRx 2:495–503.
31. Manning-Bog AB, Langston JW (2007) Model fusion, the next phase in developing animal models for Parkinson's disease. Neurotox Res 11:219–240.
32. Polymeropoulos MH, Lavedan C, Leroy E, et al. (1997) Mutation in the alpha-synuclein gene identified in families with Parkinson's disease. Science 276:2045–2047.
33. Kruger R, Kuhn W, Muller T, et al. (1998) Ala30Pro mutation in the gene encoding alpha-synuclein in Parkinson's disease. Nat Genet 18:106–108.
34. Singleton AB, Farrer M, Johnson J, et al. (2003) Alpha-Synuclein locus triplication causes Parkinson's disease. Science 302:841.
35. Chartier-Harlin MC, Kachergus J, Roumier C, et al. (2004) Alpha-synuclein locus duplication as a cause of familial Parkinson's disease. Lancet 364:1167–1169
36. Spillantini MG, Schmidt ML, Lee VM, et al. (1997) Alpha-synuclein in Lewy bodies. Nature 388: 839–840.
37. Neumann M, Kahle PJ, Giasson BI, et al. (2002) Misfolded proteinase K-resistant hyperphosphorylated alpha-synuclein in aged transgenic mice with locomotor deterioration and in human alpha-synucleinopathies. J Clin Invest 110:1429–1439.
38. Freichel C, Neumann M, Ballard T, et al. (2007) Age-dependent cognitive decline and amygdala pathology in alpha-synuclein transgenic mice. Neurobiol Aging 28:1421–1435.
39. Nuber S, Petrasch-Parwez E, Winner B, et al. (2008) Neurodegeneration and motor dysfunction in a conditional model of Parkinson's disease. J Neurosci 28:2471–2484.
40. Rockenstein E, Mallory M, Hashimoto M, et al. (2002) Differential neuropathological alterations in transgenic mice expressing alpha-synuclein from the platelet-derived growth factor and Thy-1 promoters. J Neurosci Res 68:568–578.
41. Fernagut PO, Hutson CB, Fleming SM, et al. (2007) Behavioral and histopathological consequences of paraquat intoxication in mice: effects of alpha-synuclein overexpression. Synapse 61, 991–1001.
42. Hean, S, Richter, F, Torres, ES, et al. (2010) Mice overexpressing human alpha synuclein (thy1-aSyn) show dopamine loss, catalepsy and sever motor deficits partially rescued by L-DOPA at 14 months of age. Neurosci Abst 36.
43. Fleming SM, Salcedo J, Hutson CB, et al. (2006) Behavioral effects of dopaminergic agonists in transgenic mice overexpressing human wildtype alpha-synuclein. Neuroscience 142:1245–1253.
44. Wu N, Cepeda C, Masliah E, et al. (2005) Abnormal glutamate and dopamine receptor function in the striatum of α-synuclein-overexpressing mice. Program No. 85.12. AbstractViewer/Itinerary Planner. Wasington, DC: Society for Neuroscience
45. Wang L, Fleming SM, Chesselet M-F, Taché Y (2008) Abnormal colonic motility in mice overexpressing human wild-type α-synuclein. Neuroreport 19:873–876.
46. Fleming SM, Jordan, MC, Masliah E, et al. (2007) Alterations in baroreceptor function in transgenic mice overexpressing human wildtype alpha synuclein. Program No. 50.9. Neuroscience Meeting Planner. San Diego, CA: Society for Neuroscience.
47. Fleming SM, Tetreault NA, Mulligan CM, et al. (2008) Alterations in olfactory function in transgenic mice overexpressing human wild-type alpha synuclein. Eur J Neurosci 28:247–256.
48. Aston-Jones G, Rajkowski J, Cohen J (1999) Role of locus coeruleus in attention and behavioral flexibility. Biol Psychiatry 46:1309–1320.
49. Seu E, Lang A, Rivera RJ, Jentsch JD (2009) Inhibition of the norepinephrine transporter improves behavioral flexibility in rats and monkeys. Psychopharmacology 202:505–519.
50. Fleming SM, Garcia EC, Masliah E, et al. (2008) Impaired reversal learning in transgenic mice overexpressing human wildtype alpha-synuclein. Neuroscience Meeting Planner. Washington DC: Society for Neuroscience.
51. Goldberg MS, Fleming SM, Palacino JJ, et al. (2003) Parkin-deficient mice exhibit nigrostriatal deficits but not loss of dopaminergic neurons. Journal of Biological Chemistry 278: 43628–43635.
52. Von Coelln R, Thomas B, Savitt JM, et al. (2004) Loss of locus coeruleus neurons and reduced startle in parkin null mice. Proc Natl Acad Sci U S A 101:10744–10949.
53. Zhu XR, Maskri L, Herold C, et al. (2007) Non-motor behavioural impairments in parkin-deficient mice. Eur J Neurosci 26:1902–1911.
54. Itier JM, Ibanez P, Mena MA, et al. (2003) Parkin gene inactivation alters behaviour and dopamine neurotransmission in the mouse. Hum Mol Genet 12:2277–2291.

Chapter 26

Mouse Models of Metachromatic Leukodystrophy and Adrenoleukodystrophy

Patrick Aubourg, Caroline Sevin, and Nathalie Cartier

Abstract

Metachromatic leukodystrophy (MLD) and adrenoleukodystrophy (ALD) are two inherited leukodystrophies that result in most cases in rapid destruction of the myelin within the central nervous system. There are no spontaneous animal models of these two leukodystrophies and knockout MLD and ALD mice have been generated by homologous recombination into murine embryonic stem cells. The initial analyses of these two mouse models were disappointing as they do not develop overt cerebral demyelination. Further and in-depth analysis has however revealed new and very important insights on the physiopathogenesis of MLD and ALD at an early stage of the disease. These data have now paved the way to new therapeutic approaches.

Key words: Metachromatic leukodystrophy, Adrenoleukodystrophy, Knockout mouse models, Early-stage pathology, Rotarod

1. Introduction

The term "leukodystrophy" (*leuko* – white, *dystroph* – defective nutrition) was introduced by Bielschowsky and Henneberg in 1928 to describe a heritable and progressive disorder of cerebral white matter. Based on gene identification, more than 15 different leukodystrophies have now been delineated. With the advancement of neuroimaging (brain magnetic resonance imaging, MRI), two to three new leukodystrophies are identified each year. Leukodystrophies are divided into two groups: (1) "dysmyelinating" or "hypomyelinating" leukodystrophies, the prototype of which is Pelizaeus–Merzbacher disease (PMD) due to mutations in the proteolipid gene; (2) "demyelinating" leukodystrophies, to which metachromatic leukodystrophy (MLD) and adrenoleukodystrophy (ALD)

Peter Paul De Deyn and Debby Van Dam (eds.), *Animal Models of Dementia*, Neuromethods, vol. 48,
DOI 10.1007/978-1-60761-898-0_26,

belong. While spontaneous animal models of leukodystrophy have been identified for PMD and globoid cell leukodystrophy (Krabbe disease), a demyelinating leukodystrophy due to the deficiency of a lysosomal enzyme, this is not the case for MLD and ALD. As for many other neurodegenerative diseases, modern molecular genetics came to the rescue to induce conditions that do not spontaneously develop in nonhuman mammals.

2. Metachromatic Leukodystrophy

2.1. MLD in Human

MLD is a lysosomal lipid storage disorder caused by the deficiency of lysosomal arylsulfatase A (ARSA; EC 3.1.6.8) enzyme (1), or, more rarely, of its activator protein saposin B (SAP-B) (2). The crude birth incidence of the disease ranges from 1/43,000 to 1/70,000. The resulting deficiency of ARSA leads to an accumulation of the sphingolipid cerebroside 3-sulfate (termed sulfatide). This lipid is particularly abundant in the myelin of the nervous system, where it constitutes about 4% of all myelin lipids. Myelin is synthesized by oligodendrocytes in the central nervous system (CNS) and by Schwann cells in the peripheral nervous system (PNS). Functionally, the accumulation of sulfatides affects mostly the nervous system, in particular oligodendrocytes, microglia, and Schwann cells, resulting in severe demyelination of the CNS and PNS. However, sulfatides accumulate also in CNS neurons, contributing to additional neuronal dysfunction and degeneration (3, 4). Storage of sulfatides in gall bladder epithelia and renal tubules results in little or no functional impairment. The term "metachromatic" is based on the observation of a change of the absorbance spectrum (i.e., metachromasia) that is observed when sulfatides bind some types of dyes (such as acid cresyl violet). In the CNS, pathological lesions of MLD include severe myelin loss, astrocytosis, microglia activation, and the presence of macrophages filled with Periodic acid-Schiff (PAS)-positive material. Axonal degeneration is often severe, presumably secondary to myelin loss. Significant abnormalities of motor neurons in the spinal cord and neurons in pallidum are present. Peripheral nerves also display severe demyelination with secondary axonal degeneration. Ultrastructural analysis of peripheral nerves show inclusions in Schwann cells related to myelin breakdown. Other inclusions display a periodicity of 5.8 nm and consist of zebra bodies, vacuoles containing irregularly orientated lamellar material and stacks of flattened discs. These inclusions represent the metachromatic sulfatide deposits.

Clinically, signs and symptoms caused by the involvement of CNS and PNS characterize the various forms of MLD. In the majority of the cases, the prognosis is severe, leading to vegetative stage or death within few years after the diagnosis (1). The disease is usually classified in four main phenotypes according to the age of onset. The late infantile (LI), which is the most frequent form (approximately 50%) usually manifests in the second year of life. The juvenile variant, with an onset age between 4 and 12 years, is further subdivided into early juvenile (EJ) and late juvenile (LJ), depending on whether the onset is before or after 6 years of age. The term adult MLD refers to patients with onset of neurological symptoms after the age of 12 years. Neurologic symptoms include gait disturbance and clumsiness due to a combination of cerebellar ataxia, pyramidal signs, and/or peripheral nerve involvement of the lower limbs. Inability to walk and stand up follows rapidly with the onset of truncal hypotonia, tetraparesis, dystonia, and athetosis. Cognitive decline occurs together with gait disturbances and involves initially mostly visuospatial and executive functions that are often underestimated in young patients. Primary visual loss due to optic nerve atrophy or involvement of occipital white matter occurs at a relative advanced stage. As in many "demyelinating" leukodystrophies, MLD patients frequently develop seizures when the disease progresses. While all patients with the infantile form of MLD develop severe peripheral demyelinating neuropathy that can be the first manifestation of the disease for a couple of months, as a rule, the older the patient at onset of symptoms, the less frequent is the peripheral demyelinating neuropathy. Adult patients heterozygous for the I179S mutation of the ARSA gene, frequently present with schizophrenia-like behavioral abnormalities, social dysfunction, and mental decline, while motor abnormalities may be scarce (5).

Brain MRI using a 1.5 tesla magnet shows hyposignal in T1 sequence and hypersignal in FLAIR and T2 sequence that affect the white matter in the centrum ovale, the corpus callosum, and the internal capsules (6). Radial stripes having a tigroid aspect are frequently seen, reflecting likely the accumulation of sulfatides in the Virchow spaces (7). Diffusion tensor imaging is a very sensitive biomarker (fractional anisotropy, measurement of apparent diffusion coefficient or ADC) of demyelination and enables the visualization, (tractography with 3D reconstruction) and characterization (measurement of ADC) of demyelinated fasciculi in two and three dimensions (8).

Motor and sensorimotor nerve conductions of the peripheral nerves (lower and upper limbs) are markedly decreased (<20 m/s) in patients who develop demyelinating peripheral neuropathy. As a consequence of peripheral neuropathy but also of spinal cord and brain involvement, somatosensory-evoked potentials (SSEP) from the lower and upper limbs are markedly delayed or abolished. Similarly, as a consequence of pyramidal tract involvement from motor cortex to the lower part of the spinal cord, motor evoked responses are also delayed or abolished. Brainstem auditory-evoked potentials (BAEPs) show delayed latencies of I–V interwaves that reflect demyelination in the brain stem. Latency of wave I is normal. In contrast, visually evoked responses (P100) can remain normal or nearly normal for years.

ARSA gene mutation allows prediction of the phenotype and the severity of MLD to a relatively good extent, particularly in the late infantile forms of MLD. The genotype–phenotype correlation is mostly based upon the distinction of two functionally different types of mutations of the ARSA gene (9): (1) mutations encoding inactive ARSA (0-allele, *c*.459+1 G>A) and (2) mutations encoding ARSA with residual enzymatic activity (R-alleles). Genotypes comprising two 0-alleles cause the severe late infantile type of MLD. Coincidence of a 0-allele and an R-allele induces predominantly the intermediate juvenile type, whereas two R-alleles usually cause the milder adult type of MLD. More than 120 ARSA mutations have been described to date according to the Human Gene Mutation Database (HGMD, http://www.hgmd.cf.ac.uk/ac/index.php). The null allele *c*.459+1 G>A and the Pro426Leu alleles (usually observed in adult MLD patients presenting with motor symptoms) account respectively for 25% and 18.6% of MLD alleles (3, 9). Most other mutant alleles have been found in single families (10), but there is now an international effort to systematically search for disease-causing mutation in MLD and correlate these ARSA gene mutations, as well as polymorphisms in the normal ARSA gene, with the phenotype.

2.2. MLD in Mice

ARSA-deficient mice lack ARSA activity and develop a disease that resembles MLD, but which is much less severe and, in particular, does not include cerebral demyelination (11). To establish ARSA-deficient mice, murine embryonic stem cells with a null mutation of the ARSA gene were injected into C57BL/6 blastocysts and chimeric male mice were bred with 129/OlaHsd female mice to generate ARSA+/– mice, and subsequently ARSA–/– mice with a pure 129/OlaHsd background.

ARSA-deficient mice show an age-dependent increase in brain sulfatides from the age of 3 months that can be assessed using thin-layer chromatography (11, 12, 13). Measurement of sulfatide isoforms by tandem mass spectrometry allows to identify specific neuronal (C18:0 isoform) and oligodendrocyte (C24:0) species of this lipid. Oligodendrocyte sulfatide species accumulate at higher levels than neuronal sulfatide species (Aubourg et al., unpublished). There is an age-related increase in C24:0 species and a corresponding decrease in C16:0–C20:0 species. Histologically, sulfatide storage can be assessed using Alcian blue (11, 12, 13) staining and starts to be obvious at 9 months of age, mostly in the white matter (corpus callosum, hippocampal fimbria, internal capsule, and optic nerve). On light microscopical level, sulfatide storage in the white matter results in two different morphologies (11, 14): (1) in the form of numerous fine granules arranged immediately adjacent to myelinated nerve fibers that correspond likely to accumulation of sulfatides in oligodendrocytes, and (2) as clusters of larger granules within swollen cells interspersed within the white matter that correspond to astocytes and microglia.

Despite significant accumulation of sulfatides in oligodendrocytes and obvious sulfatide storage in the white matter, ARSA-deficient mice do not develop any sign of cerebral or cerebellar demyelination, even at 18 months of age, when assessed by immunostaining with anti-proteolipid protein (PLP), anti-myelin basic protein (MBP), anti-2,3-cyclic nucleotide 3-phosphodiesterase (CNPAse) antibodies or Luxol fast blue staining. There is however some delay in myelination (15). At 2 weeks of age, ARSA-deficient mice show a substantial reduction in MBP mRNA and protein. This is confirmed by immunohistochemical analysis. MBP mRNA and protein, however, reach normal levels at 3 weeks of age. PLP and myelin and lymphocyte protein (MAL) mRNA are also reduced in ARSA-deficient mice at 2 weeks of age, whereas the level of PLP mRNA is normal at 26 weeks of age. *In situ* hybridization reveals no significant changes in the number of myelinating oligodendrocytes or oligodendrocyte precursor cells. This suggests that oligodendrocyte differentiation is normal. The excess of sulfatide does not seem to affect the survival of normal neural stem cells, at least when these cells are transplanted in the brain of ARSA-deficient mice (16). However, it might affect the differentiation of normal neural stem cells in mature neurons or oligodendrocytes (16).

Neuronal storage of sulfatides is also observed in several layers of cerebral cortex, nuclei of brain stem, diencephalon, spinal cord, and cerebellum (12, 13). Fluoro-Jade B specifically stains degenerating neural cells. Brain sections from 18-month-old untreated MLD mice show few Fluoro-Jade B-positive cells scattered throughout the cerebral and cerebellar cortex, and the pons.

Immunostaining with lectin or Iba1 antibody reveals an increased number of swollen amoeboid cells that characterize activated microglia from the age of 12 months. Similarly, marked astrogliosis is observed progressively from the age of 12 months, mostly in the white matter (corpus callosum, fimbria, cerebellar white matter), hippocampus, and pons of ARSA-deficient mice.

Periodic acid-Schiff (PAS)-reactive material reflecting lipid storage is detected from the age of 12 months, mainly in the cytoplasm of large swollen macrophages within the white matter.

ARSA-deficient mice display progressive loss of their Purkinje cells starting after the age of 14–16 months (12, 16, 17). Examination of cerebellar histology shows that 2-year-old ARSA-deficient mice have lost most of the calbindin immunoreactivity from their Purkinje cell dendrites and show simplified dendritic architecture. Recordings of unitary potentials and stimulation of climbing fibers on cerebellar slices from 2-year-old mice indicate that although the main cerebellar synapses seem to be present and functioning physiologically, the climbing fibers of ARSA-deficient mice may have enhanced effects on Purkinje cell activity (17).

Neuronal damage is dramatic in the inner ear of ARSA-deficient mice (11, 18). Already at 8 months of age, the number of acoustic ganglion cells, as well as their corresponding myelinated nerve fibers, are greatly reduced. Remaining ganglion cells are surrounded by Schwann cells containing large amounts of storage material. In younger mice (6 months), the number of acoustic ganglion cells appears normal, but the neurons and the surrounding Schwann cells show marked sulfatide storage. Sulfatide storage is also observed in the vestibular ganglion, but without reduction of neurons and nerve fibers. A decreased number of neurons is also observed in the ventral cochlear (after 8 months) and trapezoid nucleus (after 20 months).

As a consequence of the inner ear lesions, the amplitude of waves recorded during the study of BAEPs is severely decreased in 6-month-old ARSA-deficient mice (12, 19). The latency of the I–V interpeak is no longer reliably measurable at 6 months, and BAEPs are completely abolished at 9 months of age with a complete disappearance of wave I.

In the peripheral nerves, sulfatide storage is also detected in Schwann cells, but there are no signs of demyelination or axonal damage. On a different but not well-defined background, one group reported that ARSA-deficient mice exhibit reduced nerve conduction velocity, without however significant lesions of myelinated axons (20). These electrophysiological abnormalities have not been observed on the pure 129/OlaHsd background (De Deyn P and Aubourg P, unpublished).

The gall bladder, intrahepatic bile ducts, exocrine pancreatic ducts, respiratory epithelium, and, to a lesser extent, testicular Sertoli cells show sulfatide storage (21). Hepatocytes, pancreatic islets, adrenal glands, and gastric epithelium are unaffected. Apart from some differences, the topographic distribution of the sulfatide storage resembles that of human MLD.

Most distinct behavioral abnormalities only appear beyond 1 year of age, and include impaired motor coordination on the rotarod, abnormal gait, and deafness (11–13, 17, 19, 22). Older mice are generally more active and display a stronger grip. In the open-field, ARSA-deficient mice make less corner entries, and older animals display an increased path length. During the training phase of the passive avoidance task, ARSA-deficient mice show longer step-through latencies, but with better results during the testing phase than the training, indicating successful learning, clearly in contrast with the cognitive decline observed in patients.

In respect to cerebellar ataxia that is observed early in nearly all MLD patients, overt signs of ataxia are not observed in 6-month-old ARSA-deficient mice. Velocity, time and distance moved, and number of ambulatory episodes are similar to normal mice with the same genetic background.

Quantitative gait analysis using video-tracking during open-field exploration reveals however that ARSA-deficient mice display increased hind base width and increased stride length for all paws at 6 months (23). Their covert motor incoordination is evident in a correlation analysis which unveiled decreased harmonization of concurrent gait parameters. Furthermore, various behavioral observations indicate emotional alterations, including reduced proactive anxiety and anhedonia. ARSA-deficient mice spend more time in the open arms of the elevated plus maze, make less corner entries in the open-field test, and display longer step-through latencies during the training phase of the passive avoidance test. Additionally, 6-month-old ARSA-deficient mice show lower response rates in scheduled appetitive conditioning. Six-month-old mice also show decreased response rates in scheduled operant responding that is likely more the consequence of emotional dullness or inattentiveness rather than due to neuromotor defects (23). In conclusion, several data sets confirm the lack of pronounced cognitive deficits in ARSA-deficient mice below the age of 1 year.

One reason for this discrepancy between the human and mouse phenotype could be that sulfatides accumulate to a much smaller extent (sixfold less) in the brain of ARSA-deficient mice compared to the brain of MLD patients (24). Despite the fact that both at the biochemical and histological level, sulfatides accumulate mostly in oligodendrocytes and white matter, it is generally considered that the motor and behavioral abnormalities observed in MLD mice reflect neuronal dysfunction. In fact, it is likely that the motor and behavioral abnormalities observed in MLD mice are the consequences of axonal dysfunction due to primary oligodendrocyte impairment. It is indeed clearly established that once myelination has been achieved, oligodendrocytes serve two functions: (1) to preserve axons, and (2) to maintain myelin integrity. Oligodendrocyte dysfunction can result in axonal dysfunction without demyelination (25, 26). There is also evidence that sulfatides play an important role at the nodes of Ranvier, i.e., at the axonal–glial junction (see below).

ARSA-deficient mice with increased synthesis of sulfatides in neurons or oligodendrocytes and Schwann cells have been generated. The ARSA-deficient mice overexpressing the sulfatide-synthesizing enzymes UDP-galactose:ceramide galactosyltransferase (CGT) and cerebroside sulfotransferase (CST) in neurons show a more marked neuronal sulfatide storage than the ARSA-knockout mouse (27). Transgenic CGT/ARSA(−/−) mice develop more severe neuromotor coordination deficits and weakness of hind- and forelimbs. Light and electron microscopical analyses demonstrate nerve fiber degeneration in their spinal cord, which is well correlated with higher amounts of Alcian blue-positive material. The difference between transgenic and nontransgenic ARSA mice is even more pronounced in the brain. The motor neurons of the facial and hypoglossal nuclei display relatively intense sulfatide storage, which is not a feature of ARSA-deficient mice. As in the forebrain, sulfatide storage is also more prominent in the isocortical lamina 5, CA1 and CA3 regions of the hippocampus, and in the thalamus and amygdaloid nucleus. The area postrema, dorsal vagal nucleus, nucleus of solitary tract, the inferior olive, cerebellar Purkinje cells, and the caudate putamen show however no sulfatide storage. Surprisingly, transgenic CGT/ARSA(−/−) mice do not develop degeneration and loss of Purkinje cells, at least at 6 months of age. Despite the increased accumulation of sulfatides in neurons, neuronal apoptosis cannot be detected using the TUNEL assay. Although already present in ARSA-deficient mice, more significant cortical hyperexcitability, with recurrent spontaneous cortical EEG discharges lasting 5–15 s is observed in transgenic CGT/ASA(−/−) mice (27).

A different strategy was used to generate ARSA-deficient mice with increased synthesis of sulfatides in oligodendrocytes and Schwann cells. Sulfatides are synthesized in the Golgi apparatus by galactose-3-*O*-sulfotransferase-1 (Gal3st1) transferring sulfate from 3′-phosphoadenosine-5′-phosphosulfate (PAPS) to galactosylceramide (GalC). The overexpression of the *Gal3st1* gene under control of the PLP promoter in oligodendrocytes and Schwann cells of ARSA-deficient mice leads also to a significant, more marked increase of sulfatide storage in the nervous system than in simple ARSA-knockout mice (28). These transgenic mice develop myelin pathology in the CNS and particularly in PNS. Until 1 year of age, appearance and behavior of transgenic Gal3st1 ARSA-deficient mice are indistinguishable from nontransgenic ARSA-deficient mice. Older transgenic mice, however, develop behavioral abnormalities not seen in ARSA-deficient mice. When suspended by their tails, transgenic Gal3st1 ARSA-deficient mice often grasp their hind limbs to their body. Later in life, they develop a progressive hind-limb paralysis absent in simple ARSA-deficient mice. Sulfatide storage is clearly more marked in spinal cord and peripheral nerves, as revealed by Alcian blue staining. In the peripheral nerves, Schwann cells are filled with large amount of sulfatides and myelin thickness is reduced. The number of Schwann cells is increased, but there is no absolute evidence of a demyelination/remyelination process. In contrast to ARSA-deficient mice, nerve conduction velocities and compound muscle action potentials (CMAPs) are clearly reduced, while F-wave latency is significantly increased. While the number of lipid-filled macrophages is increased in the optic nerves and corpus callosum, there is however no clear evidence of marked demyelination in the cerebral white matter. One has to rely upon electronic microscopy to demonstrate that the thickness of the myelin sheaths is inhomogeneous in the corpus callosum and that some axons are hypo- or unmyelinated. Thus, the major effect of an increased synthesis of sulfatides in myelin-forming cells is observed in the peripheral nerves and spinal cord, not in the brain. Brain MRI with a 5 or 7 Tesla magnet has not yet been performed to determine if myelin abnormalities can be detected in the brain of transgenic Gal3stl ARSA-deficient mice. Whether the lack of obvious cerebral demyelination results from the relative weakness of the PLP promoter or from other factors is not known.

In the perspective of evaluating enzyme-replacement therapy (ERT), the ARSA-deficient mice has been "humanized." A cysteine-to-serine substitution was introduced into the active site of the human ARSA gene and the resulting inactive Cystein69Serine variant was constitutively expressed in ARSA-deficient mice. This mouse model of MLD is therefore tolerant to the repeated injection of recombinant human ARSA enzyme (29). The phenotype of this new ARSA-deficient mouse has not been described in detail, but seems similar to the knockout ARSA mice (30). In the same line, for evaluation of ERT, a PLP-CST/hASA-c69s/ASA(−/−) mouse has recently been generated. This double transgenic ARSA-deficient mouse expresses the CST enzyme in oligodendrocytes under the control of the PLP promoter and the pathogenic Cystein69Serine mutation of the human ARSA gene.

2.3. Pathogenesis

The physiopathology of MLD is still not fully understood, but significant progress has been made in this field (4, 31). Together with their precursor galactosylceramide (GalCer), sulfatides account for almost one third of myelin lipids and are exclusively found on the extracellular leaflet of the membranes. As all sphingolipids, sulfatides exhibit variation of their structure due to different acyl chain lengths, which can also be 2-hydroxylated. Because neurons synthesize especially ceramides containing C18:0-fatty acids, neuronal sulfatides are enriched in C18:0-species. In contrast, oligodendrocytes are enriched in C22:0/C26:0 sulfatide species. Sulfatides do not seem to have the same topology on oligodendrocytes and neurons. In the myelin sheath (an expansion of oligodendrocyte membrane) the head group of sulfatides faces outwards from the cell. In neurons (and astrocytes), they are localized in intracellular compartments. It is therefore likely that accumulation of sulfatides does not have the same pathogenic effects in oligodendrocytes and neurons.

Although sulfatides are mainly found in oligodendrocytes and Schwann cells, they are also present in neurons and astrocytes. However, it is not clear whether sulfatides are synthesized by neurons or astrocytes themselves or imported, i.e., via lipoprotein endocytosis, particularly apolipoprotein E-containing lipoproteins secreted by astrocytes.

During oligodendrocyte differentiation, sulfatides are first detected at the stage of immature oligodendrocytes, and their synthesis is upregulated before oligodendrocytes wrap myelin around axons. This suggests that sulfatides may not only fulfill a role as a structural component of myelin. The initiation of myelination appears to be stimulated by sulfatides, at least in cultured Schwann cells. Sulfatides bind to components of the extracellular matrix, like tenascin-R or laminin, and can generate signaling via the c-src/fyn kinase pathway. However, decreased synthesis or degradation of sulfatides does not seem to significantly affect myelination and oligodendrocyte survival both in mouse and human, but rather myelin maintenance.

Sulfatides produced by oligodendrocytes likely have an important role in axonal maintenance. As discussed above, most of behavioral and motor abnormalities observed in ARSA-deficient mice are likely the consequence of axonal dysfunction. The role of sulfatides in axonal function is exemplified by a reduced axon caliber and extended axonal protrusions at the nodes of Ranvier in adult CST-deficient mice. In addition, these nodes contain abnormal, enlarged vesicles and an unusual absence of contactin-associated protein (Caspr) and NF155 clusters. As a component of detergent-resistant myelin membranes (lipid rafts), sulfatides could be involved in recruiting proteins to the myelin but also to the axonal membrane. At 18 months of age, ARSA-deficient mice display a 35% decrease in the GalCer content of the brain (12, 32). This decrease could reflect instability of the myelin sheaths and/or result from an increase in the degradation of GalCer (15). The sulfatide/GalCer ratio increases however progressively as ARSA-deficient mice become older (12, 13). The conservation of this ratio might be required to form and to recruit NF155 into stable lipid rafts at axon–glial junctions. NF155 clusters are essential to concentrate Caspr and contactin in the axonal membrane, thereby forming stable axon–glial junctions.

MAL is mistargeted in ARSA-deficient mice (24). In view of its binding to sulfatides and the sulfatide-dependent missorting of MAL in sulfatide-storing kidney cells, MAL-deficiency might also impair transport of sulfatides to "paranodal lipid rafts," causing destabilization of the axon–glial junction.

Sulfatides accumulate in neurons of ARSA-deficient mice. However, loss of neurons does not always correlate with sulfatide storage, as for example cerebellar Purkinje cells degenerate in old ARSA -deficient mice without any indication for lipid accumulation. This raises the important question: To what extent is intracellular lysosomal storage or elevated sulfatide levels in the plasma membrane involved in the pathogenesis of the disease? Secondary changes in lipids that have been observed in ARSA-deficient mice, as in other lysosomal storage disorders, include increase in gangliosides GM2 and GD3 (13) and reduced cholesterol levels in old ARSA-deficient mice (33). Although GM2 accumulation in ARSA-deficient mice is only moderate, it might also affect neuronal function. The slight (15%) reduction in cholesterol level observed in ARSA-deficient mice might however be relevant for the pathogenesis of the disease, given the role of this lipid in signal transduction, particularly in lipid rafts.

Neuronal hyperexcitability in ARSA-deficient mice suggests modulation of electrophysiological properties by sulfatides. This might be relevant for the pathogenesis of MLD. Sulfatides potentially could affect functional properties of ion channels, ion pumps, receptors, or transporters. Sulfatides are thought to modulate fish gill Na^+/K^+-ATPase activity, though it is not known if this applies also to mammalian Na^+/K^+-ATPase.

In vitro, excess of sulfatides changes the morphology of primary microglia to their activated form, and it induces the production of various inflammatory mediators in primary microglia and astrocytes (34). Moreover, sulfatides rapidly trigger the phosphorylation of p38, ERK, and JNK, and they markedly enhance the NF-kappaB and AP1-binding elements. Sulfatide-triggered inflammatory events appear to occur at least in part through an L-selectin-dependent mechanism. L-selectin is dramatically downregulated upon exposure to sulfatide, and inhibition of L-selectin results in suppression of sulfatide-triggered responses. Thus, sulfatide excess may induce and exacerbate the inflammatory response that probably plays a role in the death of neurons and oligodendrocytes.

Finally, sulfatides can trigger the release of intracellular calcium stocks, leading to an increase in intracytoplasmic calcium and, hence, oxidative stress. Increased intracytoplasmic Ca^{2+} could partially account for the selective vulnerability of Purkinje cells in ARSA-deficient mice, since these cells are more sensitive to Ca^{2+} modification than other neuronal cells.

3. X-Linked Adrenoleukodystrophy (ALD)

3.1. ALD in Humans

X-linked adrenoleukodystrophy (X-ALD), a genetic demyelinating disorder secondary to loss of function in the *ABCD1* (ATP-binding cassette, subfamily D, member 1, also called ALD protein) gene, results in defective peroxisomal β-oxidation of very-long-chain fatty acids (VLCFAs) in all tissues (35–37). The disorder primarily affects the adrenal cortex and the nervous system. The pathology in the nervous system is variable, and can include either demyelination, often with an inflammatory component (cerebral ALD) or a slowly progressive axonopathy affecting the ascending and descending tracts in the spinal cord (adrenomyeloneuropathy, AMN). The frequency of X-ALD is approximately 1:17,000 in all geographic and ethnic groups. The cerebral form of ALD affects 65% males, and is more common in childhood between 5 and 10 years of age. Demyelinating lesions within the brain result in rapid neurological deterioration. Cerebral demyelinating lesions usually start in the splenium of the corpus callosum – less frequently in the genu of corpus callosum, internal capsules, or brain stem – and initially show a slow progression over 1 to 3 years. MRI of the brain reveals an abnormal signal (T2, T1, FLAIR) of the affected white matter. In the earliest symptomatic phase of the disease, patients present with subtle neurocognitive deficits (visuospatial deficits in the occipital forms, executive and attention deficits in the frontal forms) without neurologic deficits. After this initial period, demyelination progresses rapidly with a devastating course. Patients lose their ability to read, understand language, and walk within a few weeks. They develop tetraparesis, cerebellar signs, decreased visual acuity with hemianopsia, central deafness, and often seizures. This stage corresponds to the onset of inflammatory lesions with infiltration and accumulation of macrophages and mononuclear cells behind the active edge of demyelinating lesions (38). At this stage, MRI shows marked progression of demyelination and focal disruption of the blood-brain barrier. Most patients enter a vegetative state within 2–4 years of the first symptoms, followed by death in 90% of cases by age 10 years. Of note, some patients develop cerebral demyelination, but never progress to the active neuroinflammatory stage. This indicates that the neuroinflammatory process is in fact an end stage of the disease and not the primary cause of cerebral demyelination.

In contrast, adult males with AMN develop progressive spastic paraparesis, sensory disturbances in the lower limbs, and bladder dysfunction between ages 20 and 30 years. The progression of AMN is usually slow (over a period of 5–15 years), without a relapsing–remitting course as in multiple sclerosis. Brain MRI is normal except for hyperintensity of pyramidal tracts in the brain stem, pons, and internal capsules, and often a grayish aspect of white matter of the centrum ovale on FLAIR sequences and a decreased content in *N*-acetyl-aspartate (NAA) at brain MR spectroscopy is noted. NAA is a marker of axonal integrity. Two thirds of AMN patients also develop a peripheral neuropathy that can be of the demyelinating or axonal type, but that usually results in minimal clinical signs. Of note, 35% of males with AMN develop cerebral ALD in a second stage, usually before the age of 40 years, which ultimately has the same poor prognosis as in boys. In contrast to cerebral ALD, the primary pathology in spinal cord is a noninflammatory axonal degeneration with or without secondary moderate demyelination (39). Seventy percent of all males with ALD have primary adrenocortical insufficiency (Addison's disease). In most instances, this is also associated with cerebral ALD or

AMN. However, Addison's disease can be the first symptom of ALD and precedes neurologic involvement by years or decades. In addition, AMN patients frequently have biological signs of testicular dysfunction (involving both Leydig and Sertoli cells) that infrequently result in clinical signs.

It is estimated that approximately 65% of women who are heterozygous for ALD develop AMN, usually later (40 years) than males. In contrast to AMN male, they never develop cerebral ALD. Overt adrenal gland insufficiency is present in less than 1% of heterozygous ALD women.

All males and 90–95% of heterozygous women for ALD exhibit accumulation of VLCFAs in plasma or fibroblasts, but the level of VLCFA accumulation does not correlate with the phenotype. Similarly, there is no correlation between the *ABCD1* gene mutation and the clinical phenotype that ALD patients present. In fact, affected males within the same family frequently present different clinical forms of ALD, ranging from devastating cerebral ALD with onset at age 4–5 years to mild AMN with moderate motor disability at 65 years of age. The AMN phenotype in ALD males is completely penetrant, i.e., all male newborns with a pathogenic mutation of the *ABCD1* gene, even if they do not develop cerebral ALD, will develop AMN symptoms before the age of 50 years. This indicates that other factors than the loss of ALD protein (ALDP) is necessary for the onset of cerebral demyelinating lesions and neuroinflammation. In the CNS, ALDP is expressed in oligodendrocytes, astrocytes, endothelial cells, and microglia, but not in neurons (40). ALDP is also expressed in adrenocorticotropin-producing cells in the pituitary gland (41). The ALDP structurally represents a half-ABC transporter, with only one hydrophobic transmembrane domain and one hydrophilic nucleotide-binding domain, and presumably has to dimerize in order to become a functional unit. Three other mammalian half-ABC transporters, structurally similar to ALDP, have been identified: (1) the ALD-related protein (ALDR), (2) the 70 kDa peroxisomal membrane protein (PMP70), and (3) the PMP70-related protein (P70R) with, respectively, 63%, 33%, and 25% amino acid identity compared to ALDP and encoded by the *ABCD2*, *ABCD3*, and *ABCD4* genes (42). While there is *in vitro* evidence that ALDP can form homo- as well as hetero-dimers with ALDP, ALDRP, and PMP70 (43), there is no evidence that this occurs *in vivo* (44). Recent studies indicate that ALDP imports not only VLCFA-CoA derivatives in peroxisomes, but in fact a wide range of fatty acyl-CoA (45).

3.2. ALD in Mice

Two mouse models of ALD have been developed (46, 47). They differ essentially in the constructs that were used to inactivate the *ABCD1* gene in murine embryonic stem cell by homologous recombination. For the first ALD mouse (46), the *ABCD1* gene was disrupted in the first exon that is deleted. Selected R1 embryonic stem cells were microinjected in C57BL/6J blastocysts, and embryos transferred to foster mothers by standard procedures. Highly chimeric males were bred to C57BL/6J females and then, heterozygous (ald –/) F1 females were crossed with C57Bl/6J males to obtain hemizygous (ald-/Y) F2 male mutants. These were also bred to heterozygous females to generate homozygous (ald –/–) females in the F3 generation. For the second ALD mouse (47), the *ABCD1* gene was disrupted by a neomycin cassette in the second exon. Selected J1 cells were injected into blastocysts derived from C57BL/6, and the injected blastocysts were transferred into pseudo-pregnant CD-1 female mice to generate chimeras and then male ald –/Y and female ald –/– mouse mutants. Both mouse models of ALD were backcrossed to either a 129/sv or C57BL/6J genetic background for 11 generations. Because of the possibility that ALDP heterodimers with other peroxisomal transporters, a knock-in mouse with pathogenic missense mutation of the *ABCD1* gene has been generated (K. Nave, unpublished results). This mouse exhibits exactly the same phenotype as the knockout mouse.

Despite accumulation of VLCFAs in all tissues, including the CNS, the ALD mouse does not develop any behavioral or neuropathological phenotype until 15–16 months of age, either on a 129/sv or C57BL/6J genetic background. In 6-month-old mutants, adrenal cortex cells display a ballooned morphology and needlelike lipid inclusions, also found in testes and ovaries. The histologic abnormalities of the mutant adrenals are similar to some features of human ALD, in which cells with a ballooned appearance and trilammelar inclusions can be found in the adrenal cortex. Unlike ALD patients, the ALD mouse does not show elevated VLCFAs plasma (only an increase of C26:0 in the lyso-PC (1-hexacosanoyl-2-lyso-sn-3-glycero-phosphorylcholine (26:0-lyso-PC) fraction can be detected. This likely reflects that, in contrast to what is observed in human hepatocytes, ALDP is not expressed in murine hepatocytes, raising the possibility that another peroxisomal ABC transporter might fulfill the role of ALDP in mouse liver. At around 15–16 months of age, the ALD mouse develops moderate motor deficits and neuropathologic lesions in the spinal cord that resemble AMN (48, 49). In the spinal cord, electron microscopy analysis reveals small-caliber axons with unusual electrondense cytoplasm and the disappearance of organelles. Immunoreactivity against amyloid precursor protein and synaptophysin is often found along axonal swellings. Some astrocytosis and recruitment of ramified microglia is observed in the spinal cord, but myelin abnormalities are minimal and limited to the presence of sparse myelin debris, as revealed by Sudan black staining. Axonal abnormalities are seen in the sciatic nerves, but in addition, there are axons with abnormally thick and disorganized myelin sheaths. There is no onion-bulb formation and Schwann cells contain myelin-like figures and osmiophilic inclusions.

At 12 months, ALD mice perform normally at the rotarod, but at 15 months they start to show impairment in their performance. By age 20 months, they fall off the rotarod in less than 20 s on average. In the open-field mobility paradigm, ALD mice exhibit abnormalities on the first interval of testing at 15 months, while at 20 months a significant reduction of all exploratory activities is observed. When compared to control littermates, ALD mice show increased compound muscle action potentials (CMAPs), which are more marked at 20 months. CMAP is representative of fact conducting, mostly in motor fibers. Moderate slowing of sensitive nerve conduction velocity of peripheral nerve is observed. The motor and neuropathologic abnormalities are more marked in the ALD mouse in which the *ABCD1* gene has been disrupted in exon 2 (47), than in the ALD mouse in which the *ABCD1* gene has been disrupted in exon 1 (48). Whether these differences are due to epigenetic factors or environment (animal facility) is unknown.

Very importantly, malonaldehyde–lysine, a consequence of lipoxidative damage to proteins, accumulates in the spinal cord of ALD mice as early as 3.5 months of age when neuropathological abnormalities are undetectable (50). At 12 months, ALD mice accumulate additional proteins modified by oxidative damage arising from metal-catalyzed oxidation and glycoxidation/lipoxidation. There is also evidence of altered expression of enzymatic antixoxidant defenses, as demonstrated by a 1.8-fold increase in the expression glutathione peroxidase 1, whereas the expression of cytosolic and mitochondrial superoxide dismutase genes is decreased. Studies performed in organic spinal cord slice cultures of ALD mice have confirmed these findings. Altogether, these results indicate that oxidative stress damage is an early event in the pathogenic cascade that leads to spinal cord degeneration in ALD mouse.

The ALD mouse, however, does not develop cerebral demyelination, even when backcrossed to the SJL (that is commonly used to induce experimental autoimmune encephalitis, EAE) or Swiss genetic background (Kemp S and Aubourg P, unpublished results). Transgenic ALD mice expressing tumor necrosis factor-alpha in oligodendrocytes also do not develop cerebral demyelination (Aubourg P, unpublished results). Double myelin-associated glycoprotein (MAG) and ALD knockout mice show essentially additive phenotypes (51). The ALDR (*ABCD2*) knockout mouse develops a late-onset cerebellar and sensory ataxia, with loss of cerebellar Purkinje cells and dorsal root ganglia cell degeneration (52). Axonal degeneration is also present in dorsal and ventral columns in the spinal cord. VLCFAs accumulate only in the dorsal root ganglia. Because overexpression of *ABCD2* gene corrects VLCFA accumulation in human ALD fibroblasts and in ALD mouse (49, 53), it has been speculated that the *ABCD2* gene might have an overlapping function with the *ABCD1* (ALD gene).

Double mutants for *ABCD1* and *ABCD2* genes exhibit an earlier onset and more severe disease with, interestingly, the presence of inflammatory infiltrates in the spinal cord composed mainly of T lymphocytes (49). Overexpression of *ABCD2* gene in the ALD mouse corrects most of the neuropathological and motor abnormalities. One must however remain cautious about the interpretation of these data, given the mild phenotype of the ALD mouse, the mild and different phenotype of the ALDR (*ABCD2*) mouse, and that recent data suggests that ALDRP rather plays a role in the degradation of long-chain saturated and omega 9-mono-unsaturated fatty acids and in the synthesis of docosahexanoic acid (DHA) (54).

Altogether, all these studies indicate that the primary phenotype of ALD is AMN and that additional factors (missing in mice) are necessary to trigger cerebral demyelination.

3.3. Pathogenesis

As mentioned before, oligodendrocyte dysfunction can lead to isolated axonal dysfunction without demyelination. Axonopathy is the hallmark of AMN patients and ALD mice. Loss of ALDP function in oligodendrocytes results in the accumulation of VLCFAs that might trigger production of reactive oxygen species or hamper natural defenses against oxidative stress homeostasis in those glial cells. Compared to other brain cell types, oligodendrocytes are more susceptible to oxidative stress possibly because of their high content of iron and reduced levels of glutathione (55). However, ALDP imports fatty acid-CoA other than VLCFA-CoA into peroxisomes. The consequences of a more generalized impairment in fatty acid-CoA metabolism in ALD are unknown. Another factor that could contribute to oxidative stress in oligodendrocytes is the reduced level of plasmalogens that is observed in the white matter of patients with X-ALD (56). Plasmalogens are ether-linked phospholipids, characterized by the presence of a double-bond between the *first* and *second* carbon atom of the long-chain fatty alcohol attached to the sn-1 position of the glycerol backbone. This vinyl-ether bond endows plasmalogens with a unique intramolecular antioxidant property.

VLCFAs accumulate at a lower level in the brain of ALD mice than in the brain of ALD patients. While VLCFA degradation is impaired in X-ALD due to the deficient import of VLCFA-CoA into peroxisomes, synthesis of VLCFA is also enhanced (57). Interestingly, the search for polymorphisms that could contribute to the phenotypic variability of X-ALD led to the identification of a variant in the promoter of the *ELOVL1* gene that is responsible for the synthesis of $C_{26:0}$ starting from $C_{24:0}$ (Kemp S and Aubourg P, unpublished observations). Transgenic ALD mice overexpressing *ELOVL1* are being generated, and it will be interesting to see if these mice develop cerebral demyelination.

Interestingly, the *CNPase-Pex5* knockout mice develop pathology that resumes the different phenotypes of ALD. The CNPase enzyme, the exact function of which remains unknown, is expressed before myelination in oligodendrocyte progenitors as well as in adult oligodendrocytes. The *PEX5* gene encodes a cytosolic transporter (peroxin 5) that brings peroxisomal enzymes harboring the peroxisomal targeting (PTS) signal 1 to the peroxisomal membrane (58). Loss of Pex5 function in human and mice results in generalized impairment of peroxisomal functions and severe phenotype (the Zellweger spectrum) (59, 60). While loss of Pex5 function in oligodendrocytes does not impair their capacity to myelinate, the *CNPase-Pex5* mice that present a selective inactivation of the *PEX5* gene in oligodendrocytes develop axonal damage in the brain around 6 months of age, followed by subcortical demyelination and then neuroinflammation that resembles to that observed in ALD patients with cerebral ALD (61). Plasmalogens are synthesized in peroxisomes (62) and the peroxin 7 (encoded by *PEX7* gene) imports dihydroxyacetone-phosphate-acyltransferase (DHAP-AT) into peroxisomes (63). DHAP-AT is a key enzyme involved in plasmalogen biosynthesis (61). Double *PEX7* and ALD knockout mice develop demyelination in the cerebellar white matter (64). Ongoing studies aim to determine if other secondary peroxisomal dysfunctions ("peroxisomal aging") can trigger cerebral demyelination in ALD.

It was also proposed that lipid antigen presentation might be the trigger for inflammatory demyelination (38), possibly through the binding of "antigenic" lipids to CD1 molecules present at the surface of CNS antigen-presenting cells. Interestingly, mice express only CD1d molecules involved in the presentation of lipid antigens to natural killer T (NKT) cells that express an invariant form of the T cell receptor (TCR)-alpha chain. These NKT cells play an important role in the innate immunity against bacterial, viral, fungal, and parasite infections. A significant amount of evidence indicates also that these NKT cells have a suppressive effect on autoimmunity. By stimulating Toll-like receptors on antigen-presenting cells, CD1-restricted T cells alter cytokine secretion, lipid antigen synthesis, and CD1 protein translation (65–67). In contrast to mice, humans express CD1 a, b, c and e molecules that also present lipid antigens to T lymphocytes. CD1b is markedly expressed in the brain of ALD patients, and interestingly, the cleft of this HLA-like molecule is large enough to bind to mycolic acid, a 80-carbon length fatty acid found in mycobacteria (66, 68). It is therefore possible that the lack of CD1b in ALD mice is another factor precluding the onset of neuroinflammation and severe demyelination.

4. Conclusions

Mouse models of MLD and ALD have highlighted that the absence of a phenotype resembling the most severe form of human disease has its own advantage. It allows to decipher more precisely the early stage of pathogenesis of human disease, and, therefore, to design new potential therapeutic approaches that may have more chance to be of clinical benefits in patients. The studies of MLD and ALD mice have also highlighted two important things. One is the central role of oligodendrocytes in the maintenance of axons. The second is that additional factors than the gene defect itself are necessary to the development of cerebral demyelination, in human as well as in mice.

References

1. Von Figura K, Gieselmann V, Jaeken J (2001) Metachromatic leukodystrophy. In: Scriver CR, Beaudet AL, Sly WS, Valle D (eds) Childs B, Kinzler KW, Vogelstein B (assoc eds) The metabolic and molecular bases of inherited disease, 8th edn. New York: McGraw-Hill, pp 3695–3724
2. Holtschmidt H, Sandhoff K, Kwon HY, et al. (1991) Sulfatide activator protein: alternative splicing that generates three mRNAs and a newly found mutation responsible for a clinical disease. J Biol Chem 266:7556–7560
3. Gieselmann V, Matzner U, Hess B, et al. (1998) Metachromatic leukodystrophy: molecular genetics and an animal model. J Inherit Metab Dis 21:564–574
4. Sevin C, Aubourg P, Cartier N (2007) Enzyme, cell and gene-based therapies for metachromatic leukodystrophy. J Inherit Metab Dis 30:175–183
5. Rauschka H, Colsch B, Baumann N, et al. (2006) Late-onset metachromatic leukodystrophy: genotype strongly influences phenotype. Neurology 67:859–863
6. Sung Kim T, Kim I-E, Sun Kim W, et al. (1997) MR of childhood metachromatic leukodystrophy. AJNR Am J Neuroradiol 18:733–738
7. Van der Voorn JP, Pouwels PJ, Kamphorst W, et al. (2005) Histopathologic correlates of radial stripes on MR images in lysosomal storage disorders. AJNR Am J Neuroradiol 26:442–446
8. Oguz KK, Anlar B, Senbil N, Cila A (2004) Diffusion-weighted imaging findings in juvenile metachromatic leukodystrophy. Neuropediatrics 35:279–282
9. Gieselmann V, Zlotogora J, Harris A, Wenger DA, Morris CP (1994) Molecular genetics of metachromatic leukodystrophy. Hum Mutat 1994:233–242
10. Biffi A, Cesani M, Fumagalli F, et al. (2008) Metachromatic leukodystrophy – mutation analysis provides further evidence of genotype-phenotype correlation. Clin Genet 74:349–357
11. Hess B, Saftig P, Hartmann D, et al. (1996) Phenotype of arylsulfatase A-deficient mice: relationship to human metachromatic leukodystrophy. Proc Natl Acad Sci U S A 93: 14821–14826
12. Sevin C, Benraiss A, Van Dam D, et al. (2006) Intracerebral adeno-associated virus-mediated gene transfer in rapidly progressive forms of metachromatic leukodystrophy. Hum Mol Genet 15:53–64
13. Sevin C, Verot L, Benraiss A, et al. (2007) Partial cure of established disease in an animal model of metachromatic leukodystrophy after intracerebral adeno-associated virus-mediated gene transfer. Gene Ther 14:405–414
14. Wittke D, Hartmann D, Gieselmann V, Lüllmann-Rauch R (2004) Lysosomal sulfatide storage in the brain of arylsulfatase A-deficient mice: cellular alterations and topographic distribution. Acta Neuropathol 108: 261–271
15. Yaghootfam A, Gieselmann V, Eckhardt M (2005) Delay of myelin formation in arylsulphatase A-deficient mice. Eur J Neurosci 21: 711–720
16. Givogri MI, Bottai D, Zhu HL, et al. (2008) Multipotential neural precursors transplanted into the metachromatic leukodystrophy

brain fail to generate oligodendrocytes but contribute to limit brain dysfunction. Dev Neurosci 30:340–357

17. D'Hooge R, Hartmann D, Manil J, et al. (1999) Neuromotor alterations and cerebellar deficits in aged arylsulfatase A-deficient transgenic mice. Neurosci Lett 273:93–96
18. D'Hooge R, Coenen R, Gieselmann V, Lüllmann-Rauch R, De Deyn PP (1999) Decline in brainstem auditory-evoked potentials coincides with loss of spiral ganglion cells in arylsulfatase A-deficient mice. Brain Res 847:352–356
19. Coenen R, Gieselmann V, Lüllmann-Rauch R (2001) Morphological alterations in the inner ear of the arylsulfatase A-deficient mouse. Acta Neuropathol 101:491–498
20. Biffi A, De Palma M, Quattrini A, et al. (2004) Correction of metachromatic leukodystrophy in the mouse model by transplantation of genetically modified hematopoietic stem cells. J Clin Invest 113:1118–1129
21. Schott I, Hartmann D, Gieselmann V, Lüllmann-Rauch R (2001) Sulfatide storage in visceral organs of arylsulfatase A-deficient mice. Virchows Arch 439:90–96
22. D'Hooge R, Van Dam D, Franck F, Gieselmann V, De Deyn PP (2001) Hyperactivity, neuromotor defects, and impaired learning and memory in a mouse model for metachromatic leukodystrophy. Brain Res 907:35–43
23. Stroobants S, Leroy T, Eckhardt M, et al. (2008) Early signs of neurolipidosis-related behavioural alterations in a murine model of metachromatic leukodystrophy. Behav Brain Res 189:306–316
24. Saravanan K, Schaeren-Wiemers N, Klein D, et al. (2004) Specific downregulation and mistargeting of the lipid raft-associated protein MAL in a glycolipid storage disorder. Neurobiol Dis 16:396–406
25. Griffiths I, Klugmann M, Anderson T, et al. (1998) Axonal swellings and degeneration in mice lacking the major proteolipid of myelin. Science 280:1610–1613
26. Lappe-Siefke C, Goebbels S, Gravel M, et al. (2003) Disruption of Cnp1 uncouples oligodendroglial functions in axonal support and myelination. Nat Genet 33:366–374
27. Eckhardt M, Hedayati KK, Pitsch J, et al. (2007) Sulfatide storage in neurons causes hyperexcitability and axonal degeneration in a mouse model of metachromatic leukodystrophy. J Neurosci 27:9009–9021
28. Ramakrishnan H, Hedayati KK, Lüllmann-Rauch R, et al. (2007) Increasing sulfatide synthesis in myelin-forming cells of arylsulfatase A-deficient mice causes demyelination and neurological symptoms reminiscent of human metachromatic leukodystrophy. J Neurosci 27:9482–9490
29. Matzner U, Matthes F, Herbst E, et al. (2007) Induction of tolerance to human arylsulfatase A in a mouse model of metachromatic leukodystrophy. Mol Med 13:471–479
30. Matzner U, Lüllmann-Rauch R, Stroobants S, et al. (2009) Enzyme replacement improves ataxic gait and central nervous system histopathology in a mouse model of metachromatic leukodystrophy. Mol Ther 17(4):600–606
31. Eckhardt M (2008) The role and metabolism of sulfatide in the nervous system. Mol Neurobiol 37:93–103
32. Molander-Melin M, Pernber Z, Franken S, et al.(2004) Accumulation of sulfatide in neuronal and glial cells of arylsulfatase A deficient mice. J Neurocytol 33:417–427
33. Lutjohann D, Harzer K, Gieselmann V, Eckhardt M (2006) Reduced brain cholesterol content in arylsulfatase A-deficient mice. Biochem Biophys Res Commun 34:647–650
34. Jeon SB, Yoon HJ, Park SH, Kim IH, Park EJ (2008) Sulfatide, a major lipid component of myelin sheath, activates inflammatory responses as an endogenous stimulator in brain-resident immune cells. J Immunol 181:8077–8087
35. Mosser J, Douar AM, Sarde CO, et al. (1993) Putative X-linked adrenoleukodystrophy gene shares unexpected homology with ABC transporters. Nature 361:726–730
36. Dubois-Dalcq M, Feigenbaum V, Aubourg P (1999) The neurobiology of X-linked adrenoleukodystrophy, a demyelinating peroxisomal disorder. Trends Neurosci 22:4–12
37. Moser HW, Mahmood A, Raymond GV X-linked adrenoleukodystrophy. Nat Clin Pract Neurol 3:140–151
38. Ito M, Blumberg BM, Mock DJ, et al. (2001) Potential environmental and host participants in the early white matter lesion of adrenoleukodystrophy: morphologic evidence for CD8 cytotoxic T cells, cytolysis of oligodendrocytes, and CD1-mediated lipid antigen presentation. J Neuropathol Exp Neurol 60:1004–1019
39. Powers JM, DeCiero DP, Ito M, Moser AB, Moser HW (2000) Adrenomyeloneuropathy: a neuropathologic review featuring its noninflammatory myelopathy. J Neuropathol Exp Neurol 5989–6102
40. Fouquet F, Zhou JM, Ralston E, et al. (1997) Expression of the adrenoleukodystrophy protein in the human and mouse central nervous system. Neurobiol Dis 3:271–285

41. Höftberger R, Kunze M, Weinhofer I, et al. (2007) Distribution and cellular localization of adrenoleukodystrophy protein in human tissues: implications for X-linked adrenoleukodystrophy. Neurobiol Dis 28:165–174
42. Kemp S, Wanders RJ (2007) X-linked adrenoleukodystrophy: very long-chain fatty acid metabolism, ABC half-transporters and the complicated route to treatment. Mol Genet Metab 9:268–276
43. Liu LX, Janvier K, Berteaux-Lecellier V, et al. (1999) Homo- and heterodimerization of peroxisomal ATP-binding cassette half-transporters. J Biol Chem 274:32738–32743
44. Guimarães CP, Domingues P, Aubourg P, et al. (2004) Mouse liver PMP70 and ALDP: homomeric interactions prevail in vivo. Biochim Biophys Acta 1689:235–243
45. Van Roermund CW, Visser WF, Ijlst L, et al. (2008) The human peroxisomal ABC half transporter ALDP functions as a homodimer and accepts acyl-CoA esters. FASEB J 22: 4201–4208
46. Forss-Petter S, Werner H, Berger J, et al. (1997) Targeted inactivation of the X-linked adrenoleukodystrophy gene in mice. J Neurosci Res 50:829–843
47. Lu JF, Lawler AM, Watkins PA, et al. (1997) A mouse model for X-linked adrenoleukodystrophy. Proc Natl Acad Sci U S A 94:9366–9371
48. Pujol A, Hindelang C, Callizot N, et al. (2002) Late onset neurological phenotype of the X-ALD gene inactivation in mice: a mouse model for adrenomyeloneuropathy. Hum Mol Genet 11:499–505
49. Pujol A, Ferrer I, Camps C, et al. (2004) Functional overlap between ABCD1 (ALD) and ABCD2 (ALDR) transporters: a therapeutic target for X-adrenoleukodystrophy. Hum Mol Genet 13:2997–3006
50. Fourcade S, López-Erauskin J, Galino J, et al. (2008) Early oxidative damage underlying neurodegeneration in X-adrenoleukodystrophy. Hum Mol Genet 17:1762–1773
51. Dumser M, Bauer J, Lassmann H, Berger J, Forss-Petter S (2007) Lack of adrenoleukodystrophy protein enhances oligodendrocyte disturbance and microglia activation in mice with combined Abcd1/Mag deficiency. Acta Neuropathol 114:573–586
52. Ferrer I, Kapfhammer JP, Hindelang C, et al. (2005) Inactivation of the peroxisomal ABCD2 transporter in the mouse leads to late-onset ataxia involving mitochondria, Golgi and endoplasmic reticulum damage. Hum Mol Genet 14:3565–3577
53. Flavigny E, Sanhaj A, Aubourg P, Cartier N (1999) Retroviral-mediated adrenoleukodystrophy-related gene transfer corrects very long chain fatty acid metabolism in adrenoleukodystrophy fibroblasts: implications for therapy. FEBS Lett 448:261–264
54. Fourcade S, Ruiz M, Camps C, et al. (2009) A key role for the peroxisomal ABCD2 transporter in fatty acid homeostasis. Am J Physiol Endocrinol Metab 296:E211–E221
55. Dringen R, Gutterer JM, Hirrlinger J (2000) Glutathione metabolism in brain metabolic interaction between astrocytes and neurons in the defense against reactive oxygen species. Eur J Biochem 267:4912–4916
56. Khan M, Singh J, Singh I (2008) Plasmalogen deficiency in cerebral adrenoleukodystrophy and its modulation by lovastatin. J Neurochem 106:1766–1779
57. Kemp S, Valianpour F, Denis S, et al. (2005) Elongation of very long-chain fatty acids is enhanced in X-linked adrenoleukodystrophy. Mol Genet Metab 84:144–151
58. Platta HW, El Magraoui F, Schlee D, et al. (2007) Ubiquitination of the peroxisomal import receptor Pex5p is required for its recycling. J Cell Biol 177:197–204
59. Baes M, Gressens P, Baumgart E, et al. (1997) A mouse model for Zellweger syndrome. Nat Genet 17:49–57
60. Gould SJ, Valle D (2000) Peroxisome biogenesis disorders: genetics and cell biology. Trends Genet 16:340–345
61. Kassmann CM, Lappe-Siefke C, Baes M, et al. (2007) Axonal loss and neuroinflammation caused by peroxisome-deficient oligodendrocytes. Nat Genet 39:969–976
62. Wanders RJ, Waterham HR (2006) Peroxisomal disorders: the single peroxisomal enzyme deficiencies. Biochim Biophys Acta 1763:1707–1720
63. Purdue PE, Zhang JW, Skoneczny M, Lazarow PB (1997) Rhizomelic chondrodysplasia punctata is caused by deficiency of human PEX7, a homologue of the yeast PTS2 receptor. Nat Genet 15:381–384
64. Brites P, Mooyer PA, El Mrabet L, Waterham HR, Wanders RJ (2009) Plasmalogens participate in very-long-chain fatty acid-induced pathology. Brain 132:482–492

65. Moody DB, Zajonc DM, Wilson IA (2005) Anatomy of CD1-lipid antigen complexes. Nat Rev Immunol 5:387–399
66. Young DC, Moody DB (2006) T-cell recognition of glycolipids presented by CD1 proteins. Glycobiology 16:103R–112R
67. Moody DB (2006) TLR gateways to CD1 function. Nat Immunol 7:811–817
68. Cheng TY, Relloso M, Van Rhijn I, et al. (2006) Role of lipid trimming and CD1 groove size in cellular antigen presentation. EMBO J 25:2989–2999

Chapter 27

Animal Models of Amyotrophic Lateral Sclerosis

Ludo Van Den Bosch

Abstract

Amyotrophic lateral sclerosis (ALS) is a neurodegenerative disorder characterized by the selective death of motor neurons. Although ALS is predominantly a disease of motor system degeneration, cognitive impairments have also been reported. In order to determine the pathogenic mechanisms leading to this neurodegeneration, a large number of different animal models have helped to get insights. These animal models contain mutated genes (spontaneous or induced) or (over)express (mutant) genes, and recapitulate important aspect of this neurodegenerative disease. Although the exact mechanisms are not yet elucidated, animal models have been of great value to delineate potential pathogenic mechanisms that are involved in, and/or are important in causing selective motor neuron degeneration. Moreover, these animal models can be used as first screening assays to select potential therapeutics to treat this fatal disease.

Key words: ALS, Cognitive impairments, Superoxide dismutase 1, Neurodegeneration

1. Introduction

Selective death of motor neurons in the motor cortex, brain stem, and spinal cord is the cause of amyotrophic lateral sclerosis (ALS). This adult-onset, progressive neurodegenerative disorder is sporadic in 90% of cases, while 10% of patients have a familial variant of the disease. Sporadic and familial ALS are clinically indistinguishable and patients suffer from muscle weakness, atrophy, and spasticity, due to the loss of both upper and lower motor neurons. Ultimately, patients become paralyzed and denervation of respiratory muscles leads to the death of the patient, on average 3–5 years after the onset of the first symptoms. As it is very difficult to study a sporadic disease, most research focused on the familial form of the disease and on the genetic causes of familial ALS (for reviews see (1, 2)).

Peter Paul De Deyn and Debby Van Dam (eds.), *Animal Models of Dementia*, Neuromethods, vol. 48, DOI 10.1007/978-1-60761-898-0_27, © Springer Science+Business Media, LLC 2011

By far the most important cause of inherited ALS (ALS1) responsible for about 20% of familial ALS are mutations in the gene on chromosome 21q encoding superoxide dismutase 1 (SOD1) (3). The mutations are present in all five exons encoding this protein of 153 amino acids. Other rare genetic causes of the disease are mutations in the genes encoding alsin (ALS2; (4)), senataxin (ALS4; (5)), vesicle-associated membrane protein B (VAPB; ALS8; (6)), angiogenin (ALS9; (7)), and dynactin (8). More recently, several groups identified mutations in the gene encoding the TAR DNA-binding protein (TDP-43; (9–12)) as a cause of ALS. In addition, (rare) sequence variants associated with ALS have been reported in a number of other genes (e.g., neurofilaments, peripherin, vascular endothelial growth factor (VEGF)); although it is not always clear whether these are involved in the pathogenesis of ALS.

Traditionally, it is thought that cognitive functions are spared during ALS. However, evidence exists that ALS patients show a wide range of cognitive impairments (for a review see (13)). Most patients with ALS have mild cognitive impairment with subtle executive deficits, while ~5% show frontotemporal dementia (FTD) (14, 15). Although the idea that ALS and FTD form a clinical spectrum remains controversial, there is not only clinical overlap between both diseases, but also on the pathological and genetic level. Deposits of TDP-43 are found in both ALS and FTD (16), while genetic variations in the progranulin gene could be the cause of both diseases (17–19).

A large number of animal models have been created in an attempt to unravel the pathogenic mechanisms leading to ALS (Table 1). Moreover, a number of rodent models exist that are characterized by selective motor neuron death (e.g. Pmn, Nmd, Loa, Cra1). These models are also interesting as the (induced or spontaneous) mutations responsible for the phenotype are elucidated. Unfortunately, almost no information is available on cognitive alterations in these animal models. It is not always clear whether there is indeed no effect on other brain regions or whether this aspect was not thoroughly investigated.

In this chapter, we will give an overview of the most important animal models of motor neuron degeneration and we will discuss the pathogenic information, also in relation to cognitive impairments, deduced from these genetic models.

2. ALS1 and Superoxide Dismutase 1

Shortly after the discovery of mutations in the SOD1 gene, a transgenic mouse overexpressing mutant SOD1^{G93A} was created by insertion of multiple copies of human genomic SOD1 into the mouse genome (20). These transgenic mice showed progressive

Table 1
Overview of animal models for motor neuron degeneration

Disease	Gene product	Inheritance	Animal	Genetic modification	Reference
ALS1	Superoxide dismutase 1	Dominant	Mouse	genomic hSOD1 G37R	(22)
				genomic hSOD1 G85R	(21)
				genomic mSOD1 G86R	(23)
				genomic hSOD1 G93A	(20)
				genomic hSOD1 L126Z(stop)	(25)
				genomic hSOD1 L126delTT	(26)
				genomic hSOD1 Quad	(30)
				PrP; cDNA SOD1 G37R	(33)
				Thy-1; cDNA hSOD1 G93A	(39)
				genomic hSOD1 H46R	(27)
			Rat	genomic hSOD1 G93A	(27)
				genomic hSOD1 G93A	(28)
		Dominant/recessive	Mouse	genomic hSOD1 D90A	(24)
ALS2	Alsin	Recessive	Mouse	KO (exon 3)	(41)
				KO (stop codon in exon 3)	(42)
				KO (exon 3 and 4)	(43)
				KO (exon 4)	(44)
ALS8	Vesicle-associated membrane protein	Dominant	Fruit fly	mutant males	(52)
Dynactin	Dynactin	Dominant	Mouse	G59S $p150^{Glued}$ knock-in	(54)
				Thy-1; G59S $p150^{Glued}$	(55)
				Thy-1; G59S $p150^{Glued}$ (X Chr)	(56)
Charcot–Marie–Tooth-2E/1F	Neurofilament-L	Dominant	Mouse	NF-L L394 P	(69)

(continued)

Table 1 (continued)

Disease	Gene product	Inheritance	Animal	Genetic modification	Reference
NA	Peripherin	NA	Mouse	Overexpression	(72)
NA	Vascular endothelial growth factor	NA	Mouse	$VEGF^{\delta/\delta}$	(87)
SMARD	Immunoglobulin μ-binding protein 2	Recessive	Mouse	Nmd mouse	(49, 50)
HRD/SSS	Tubulin-specific chaperone E	Recessive	Mouse	Pmn mouse	(76)
FTDP-tau	Tau	Dominant	Mouse	4R human tau	(78)
				R406W human tau	(79, 80)
				P301L human tau	(81)
				G272V, P301S human tau	(82)
				V337M human tau	(83)
				P301S human tau	(84)
				K257T, P301S human tau	(85)

hSOD1: human superoxide dismutase 1, mSOD1: mouse superoxide dismutase 1, KO: knockout; SMARD: spinal muscular atrophy with respiratory distress, HRD/SSS: hypoparathyroidism–retardation dysmorphism/ Sanjad–Sakati syndrome, FTD: frontotemporal dementia with parkinsonism, NA: not available

hind-limb weakness leading to paralysis and death. The moment of disease onset and the life span of these mice were related to the level of overexpression of mutant SOD1, while overexpression of non-mutated SOD1 gave no overt phenotype. Transgenic mice that (over)expressed human SOD1 containing other mutations (G37R, G85R, or D90A) or mutant (G86R) mouse SOD1 showed a similar phenotype as the mutant $SOD1^{G93A}$ mouse line (21–24). Moreover, transgenic mice overexpressing a C-terminal truncated SOD1 also showed motor neuron degeneration (25, 26). In addition, transgenic rats overexpressing mutant SOD1 (H46R and G93A) were created that also developed an age-dependent degeneration of motor neurons leading to paralysis and death (27, 28). The advantage of this larger rat model is that it allows studies involving complex manipulations of the spinal fluid and spinal cord (e.g., implantation of intrathecal catheters for chronic therapeutic studies, sampling of cerebrospinal fluid).

Since its creation, the mutant SOD1 mouse models have been extensively studied. These transgenic mice proved that mutant SOD1 caused selective motor neuron death by a 'gain of function'. Moreover, these mice were used to study both the pathogenic changes that occur during the disease process, as well as to test possible drugs. Mutant SOD1 transgenic mice were also crossbred with other transgenic mice to learn about the pathogenic mechanisms involved. Using these strategies, it was shown that SOD1-mediated oxidative abnormalities were not the primary cause of mutant SOD1 toxicity. SOD1 is an enzyme that requires copper to catalyze the conversion of toxic superoxide radicals into hydrogen peroxide and oxygen. Copper plays a crucial role in the normal and/or aberrant enzymatic activity of the enzyme and copper loading of SOD1 is performed by a specific copper chaperone (CCS). Crossbreeding of transgenic mutant $SOD1^{G93A}$ mice with knockout mice lacking the CCS did not influence the life span of mutant $SOD1^{G93A}$ mice (29). Another argument that oxidative stress is not the initiating factor was presented by a transgenic mouse overexpressing mutant SOD1 in which the four essential histidines that bind copper were mutated (SOD1-Quad). Two of these mutations are known human mutations causing ALS and the SOD1-Quad mice developed age-dependent motor neuron loss despite the lack of copper binding by SOD1 (30).

SOD1 aggregates were found in ALS patients with SOD1 mutations, as well as in the different mutant SOD1 mouse models. They were present both in neurons and in surrounding glial cells (31). The formation of SOD1 aggregates was one of the first pathological signs and the abundance increased as a function of the disease process (32). Furthermore, detergent-insoluble forms of SOD1 could be detected in the brain stem and spinal cord of the different transgenic mouse models (30) and there was a clear correlation between the formation of SOD1 aggregates and disease (33). Despite these results, it is still an open question whether the formation of SOD1 aggregates is the cause of ALS. Alternatively, the formation of aggregates could be a harmless side effect of the presence of mutant SOD1, or it could even be protective by sequestering aberrant SOD1 protein.

Selective expression of mutant SOD1 in motor neurons (34) or in glial cells (35) was not sufficient to induce pathology, indicating that an interplay between different cell types is necessary to get motor neuron death. However, it was difficult to determine whether mutant SOD1 was always expressed to a sufficiently high level in these experiments. As a consequence, the most important argument for the non-cell autonomous character of motor neuron death was provided through the generation of chimeric animals. In these animals, neighboring cells that did not express mutant SOD1 delayed degeneration and significantly extended survival of mutant SOD1 expressing motor neurons (36). The opposite was found as well: normal motor neurons (not expressing mutant SOD1) surrounded by cells that did express mutant SOD1 were damaged and contained ubiquitin-positive inclusions (36). The identification of the non-neuronal cell type contributing to mutant SOD1-induced motor neuron death was done using a floxed mutant SOD1 gene that can be excised by the action of Cre recombinase. Selective reduction of the amount of mutant SOD1 in microglia and peripheral macrophages had a dramatic effect and delayed significantly the progression of the disease (37). Moreover, reduction of mutant SOD1 expression in astrocytes also affected disease progression. This selective silencing of mutant SOD1 expression in astrocytes was obtained by crossbreeding transgenic mice containing floxed mutant SOD1 with mice expressing Cre driven by the glial fibrillary acidic protein (GFAP) promoter (38). The disease progression in these double transgenic mice was significantly slowed and the effect on survival was comparable to that of the selective removal of mutant SOD1 from postnatal motor neurons (37, 38). Although these data showed that the surrounding cells clearly contribute, it was recently shown by Jaarsma et al. (39) that exclusive neuronal expression of mutant SOD1 can be sufficient to cause selective motor neuron degeneration and paralysis. The motor neurons in these transgenic mice showed clear cytosolic SOD1 aggregates as the dominant pathological feature (39).

There is no clear indication that mutant SOD1 mouse models develop age-dependent cognitive impairments. However, it was reported that overexpression of mutant SOD1 induces changes in prefrontal cortex connectivity and function before the onset of motor disturbances (40). Mutant $SOD1^{G93A}$ mice exhibited a paradoxical selective enhancement of reactivity to spatial changes compared to mice overexpressing wild-type SOD1. It was hypothesized that the high levels of glutamate present in the brain of presymptomatic mice could mediate site-specific and molecular and synaptic changes providing favorable conditions to spatial information processing (40).

3. ALS2 and Alsin

ALS2 is a rare autosomal, recessive form of ALS that is characterized by a juvenile onset of progressive spasticity in the limbs, facial, and pharyngeal muscles. In families of Arabic origin, mutations in the ALS2 gene on chromosome 2 were discovered (4). At present, at least 11 different mutations in the ALS2 gene have been reported leading either to ALS, to primary lateral sclerosis (PLS), or to infantile onset ascending hereditary spastic paraplegia (IAHSP) with only upper motor neurons involved (for a review (1)). All patients are homozygous for the mutant gene and all reported mutations lead to premature termination of the transcript and as a consequence to a truncated protein.

Several groups have generated an alsin knockout mouse (41–44), but these mice had no clear motor phenotype. Primary motor neuron cultures from the first line of alsin knockout mice were more susceptible to oxidative stress (41). The second transgenic mouse model showed an age-dependent, slowly progressive loss of cerebellar Purkinje cells and a disturbance of spinal motor neurons associated with astrocytosis and microglial cell activation, indicating a subclinical dysfunction of the motor system (42). The third mouse model showed mild age-independent hypoactivity and smaller cortical motor neurons. In addition, disturbances in endosomal transport were reported (43). In the fourth transgenic model, slowed movement without muscle weakness and progressive axonal degeneration in the lateral spinal cord was observed (44). In general, it can be concluded from these observations that knocking out alsin in mouse is insufficient to cause major motor deficits, while subtle behavioral changes and a clear loss in cerebellar Purkinje cells were reported.

4. ALS4 and Senataxin

ALS4 is a rare, childhood- or adolescent-onset, autosomal dominant form of ALS, characterized by slow disease progression, limb weakness, severe muscle wasting, and pyramidal signs associated with degeneration of motor neurons in the brain and spinal cord. Typical for ALS4 are the long duration of the disease and the sparing of bulbar and respiratory muscles. Different missense mutations in the gene coding for senataxin (SETX) were found in three unrelated families in Austria, Belgium, and the United States (5). Interestingly, the distal hereditary motor neuropathy (dHMN) phenotype in the Austrian and Belgian families is strikingly similar to the original ALS4 family (45), suggesting that ALS4 and dHMN with pyramidal tract signs may be one and the same disorder (46).

The discovery of missense mutations in the senataxin gene of patients suffering from ALS4 suggests that defects in RNA metabolism are involved in motor neuron death as the protein contains a C-terminal motif typical for DNA/RNA helicases. The autosomal dominant pattern of inheritance suggests that mutations in senataxin result in a gain of function. Moreover, loss-of-function mutations were identified in patients with ataxia-ocular apraxia 2 (AOA2), an autosomal recessive ataxia associated with impairment of eye movement (47).

A common denominator of a number of motor neuron diseases is indeed that proteins playing a role in RNA metabolism are involved. As a consequence, it was suggested that lower motor neurons are selectively vulnerable to defects in RNA metabolism (48). This is illustrated by the neuromuscular degeneration (Nmd) mouse. This mouse line contained a spontaneous autosomal recessive mutation giving rise to neuromuscular degeneration (49) and developed a progressive motor neuron disease, in which muscle atrophy was secondary to motor neuron loss. Homozygous Nmd mice became progressively paralyzed and rarely survived longer than 4 weeks after birth, while no behavioral or cognitive impairments were found.

A mutation that creates a cryptic splice donor site was found in intron 4 of gene encoding the immunoglobulin μ-binding protein 2 (Ighmbp2) (50) that co-localizes with the RNA-processing machinery. The consequence of this mutation is that the majority of the Ighmbp2 transcripts are aberrantly spliced, and that a truncated Ighmbp2 protein is formed. This mouse model became highly relevant after the discovery of Ighmbp2 mutations in patients suffering from spinal muscular atrophy with respiratory distress (SMARD) type 1 (51). SMARD is an autosomal recessive motor neuron disease that affects infants. Patients present with respiratory distress due to diaphragmatic paralysis and progressive muscle weakness with predominantly distal lower limb muscle involvement.

5. ALS8 and VAPB

Originally, a missense mutation in the *VAPB* gene was found in a Brazilian family (6). This mutation results in different clinical manifestations including late-onset spinal muscular atrophy and late-onset atypical ALS with slow progression. To date, there are eight families of which seven are of Portuguese-Brazilian and one is of African-Brazilian origin. There are no mouse models for VAPB, although the protein is highly conserved in the mouse. In Drosophila, hemizygous mutant males exhibit severe motor deficits and a compromised microtubule assembly (52).

6. Dynactin

A G59S mutation located in the microtubule-binding domain of dynactin p150^{Glued} was described as the cause of an autosomal dominant, late-onset motor neuron disease in a large family of European descent (8). Initial *in vitro* microtubule binding studies indicated that mutant p150^{Glued} exhibited a reduced binding efficiency to microtubules (8), which is consistent with a loss of function. In addition, cell culture experiments demonstrated that mutant G59S p150^{Glued} perturbs dynactin function and causes protein aggregation (53). Recently, two mouse models were created, indicating that motor neuron loss is caused by a dominant negative mechanism (54, 55). Lai et al. (54) generated a mutant G59S p150^{Glued} knock-in mouse. Early embryonic lethality was observed in the mutant G59S p150^{Glued} homozygous knock-in mice, while the heterozygous mutant G59S p150^{Glued} mice started to display a motor neuron disease-like phenotype at 10 months of age. This was accompanied by excessive accumulation of cytoskeletal and synaptic vesicle proteins at the neuromuscular junctions, loss of spinal motor neurons, increase of astrogliosis, and shortening of gait (54). Laird et al. (55) and Chevalier-Larsen et al. (56) overexpressed mutant G59S p150^{Glued} under the control of the neuron-specific Thy-1 promoter. These mice also displayed a motor neuron disease phenotype resulting in muscle weakness, paralysis, and/or death, while no cognitive or behavioral abnormalities were reported.

Further evidence that disturbances in axonal transport can cause motor neuron loss is provided by the characterization of two other mouse lines. These mutant mouse lines, Legs at odd angles (Loa) and Cramping 1 (Cra1), arose in two independent mutagenesis experiments in the offspring of *N*-ethyl-*N*-nitrosourea (ENU)–treated mice. These mice manifest progressive motor neuron disorders and show remarkable similarities to specific features of human pathology, including Lewy body–like inclusions containing SOD1, neurofilaments, and ubiquitin. Two different point mutations in dynein were found in the Loa and Cra1 mice (57). Dynein is a motor protein complex that is involved in retrograde transport and moves in the minus-end direction along microtubules.

Similarly, disruption in postnatal motor neurons of the dynactin complex, an activator of cytoplasmic dynein that makes it more processive (58), produced a late-onset, progressive motor neuron disease in mice. This disruption was obtained by overexpression of dynamitin, the p50 subunit of dynactin, leading to neurofilamentous swellings in motor axons and inhibition of retrograde axonal transport (59).

Impairment of retrograde axonal transport could also be involved in ALS as indicated by the observation that this process is disturbed in mutant SOD1 transgenic mice (60, 61). Unexpectedly, crossbreeding of the Loa mice with mutant (G93A) SOD1 mice significantly extended the life span of the mutant $SOD1^{G93A}$ mice by 28% (60). Defects in retrograde transport were no longer found in motor neurons cultured from these double transgenic mice (60). These experiments indicate that axonal transport is restored by combining the dynein mutation with an SOD1 mutation, although the exact explanation for these intriguing findings is not yet clear.

7. Neurofilaments and Peripherin

Neurofilaments (NFs) are the most abundant intermediate filaments in neurons and consist of three subunits: NF-L, NF-M, and NF-H. Accumulation of NFs and concomitant slowing of slow anterograde axonal transport were reported in the mutant $SOD1^{G93A}$ mouse model (62). Moreover, NF accumulations were found in both familial and sporadic ALS cases (63–65). In ALS patients, mutations in the KSP phosphorylation domain of NF-H were found in a limited number of ALS patients (66). Over the years, a large number of transgenic mice with modifications related to NFs were made, in order to determine the role of these intermediate filaments in the pathology of motor neurons (for reviews (67, 68)).

Knockout mice for these different subunits alone or double transgenic mice deficient in two NF subunits did not show a clear phenotype, although in some of these mice a loss of motor axons was detected. Also overexpression of the different NF subunits did not induce motor neuron death. In some of these transgenic mice, NF accumulations in neuronal cell bodies were found, but this did not induce motor neuron death.

However, NF abnormalities can induce selective motor neuron death *in vivo* as indicated by a transgenic mouse expressing a mutant NF-L. In these mice, leucine at position 394 was mutated into a proline and this caused a dominant motor neuron disease (69). These transgenic mice showed massive, selective degeneration of motor neurons accompanied by accumulations of NFs, although no effect on the life span of these mice was reported. This mouse model became highly relevant after the discovery of

NF-L mutations as the cause of a dominantly inherited motor neuropathy: Charcot–Marie–Tooth disease, type 2E (70, 71).

Another intermediate filament protein that could be involved in selective motor neuron death is peripherin. It is mostly expressed in the peripheral nervous system and is upregulated both in the peripheral and central nervous system after injury and by inflammatory cytokines. Transgenic mice that overexpress wild-type peripherin developed motor dysfunctions very late in their life (after 2 years). This phenotype was accompanied by the loss of motor axons and by the appearance of peripherin inclusion bodies in the motor neurons. The onset of motor dysfunction and axonal loss was dramatically accelerated by the absence of NF-L, as revealed by cross-breeding of peripherin overexpressing mice with NF-L knockout mice. These double transgenic mice also showed a dramatic loss of motor neurons (72). The exact mechanism underlying this peripherin-mediated neurodegeneration is not yet clear. Upregulation or suppression of peripherin expression had no effect on disease onset, mortality, and loss of motor neurons in mutant $SOD1^{G37R}$ mice, indicating that peripherin is not a contributing factor to motor neuron disease in this mouse model (73). However, this does not necessarily exclude a role for peripherin in ALS as illustrated by the discovery in ALS patients of a frameshift deletion and a mutation in the peripherin gene (74, 75).

8. Tubulin-Specific Chaperone E Gene and Tau

The tubulin-specific chaperone E gene (Tbce) has an effect on microtubule stability and/or on the polymerization dynamics of microtubules. In the Pmn mouse, point mutations were identified in the Tbce gene on chromosome 13 (76). However, in humans, a deletion in Tbce is responsible for hypoparathyroidism-retardation dysmorphism syndrome (HRD or Sanjad–Sakati syndrome, SSS) or autosomal recessive Kenny–Caffey syndrome (AR-KCS). Although neurological symptoms are part of the HRD/SSS phenotype, this syndrome is clinically variable, involves multiple tissues, and is quite different from the disease observed in the Pmn mouse.

Other mouse models that show a clear motor neuron disease phenotype were generated by changing the microtubule-associated protein tau. This axonal phosphoprotein establishes short cross-bridges between axonal microtubules and, thereby, supports intracellular trafficking, including axonal transport. In neurons affected by a tauopathy, tau is hyperphosphorylated and is located not only in axons but also in cell bodies and dendrites. Tau is the major component of the intracellular filamentous deposits found in a number of neurodegenerative diseases including Alzheimer disease, while mutations in tau are associated with frontotemporal dementia with parkinsonism (FTDP) (77). Several tau transgenic mice showed a progressive motor phenotype with muscle atrophy and paresis. The motor axons had dilations that contained accumulation of NFs, mitochondria, and vesicles (78). Overexpression of mutant tau also induced age-dependent accumulation of insoluble filamentous tau aggregates in neuronal perikarya of spinal cord, a motor phenotype, and a reduced life span (79). These examples indicate that mouse models also have their limitations in that the phenotypes observed in mice do not always exactly replicate the disease process in humans.

Other tau-related animal models developed associative (80) and spatial memory deficits (81, 82). Moreover, reduced synaptic excitability was observed (82, 83), with a limited perturbation in long-term potentiation (84). Recently, a novel transgenic mouse model expressing double mutant tau driven by its natural promoter exhibited clear cognitive deficits reminiscent of the situation in FTDP patients (85).

9. Vascular Endothelial Growth Factor (VEGF)

Survival of motor neurons is critically dependent on the presence of different growth factors. Absence of one of these neurotrophic factors can cause motor neuron death, as is illustrated by the phenotype of the transgenic mice in which the gene for ciliary neurotrophic factor (CNTF) was deleted. CNTF is a cytosolic protein, expressed at high levels in myelinating Schwann cells, promoting survival of motor neurons *in vitro*. These knockout mice had no phenotype during the postnatal weeks, but developed atrophy and loss of motor neurons with increasing age (86).

Another indication that selective motor neuron death can be induced by insufficient growth factor supply was the unexpected phenotype of the $VEGF^{\delta/\delta}$ transgenic mouse. This mouse model was created by deleting the hypoxia-response element (HRE) in the promoter region of the gene encoding VEGF. This modification induced an adult-onset, slowly progressive motor neuron loss leading to muscle atrophy and a motor phenotype, while no cognitive impairments were present (87). Crossbreeding of $VEGF^{\delta/\delta}$ mice with mutant $SOD1^{G93A}$ transgenic mice resulted in earlier motor neuron loss and a reduction of the life span of these double transgenic mice by 14% (88), indicating that low VEGF levels accelerated motor neuron degeneration in the mutant $SOD1^{G93A}$ mouse model. The effect of VEGF is at least partially due to its direct neuroprotective effect on motor neurons as was both shown *in vitro* (89) and *in vivo* by crossbreeding mutant $SOD1^{G93A}$ mice with mice overexpressing the VEGF receptor 2 (also termed Flk-1). Overexpression of this receptor did not only delay the onset of motor impairment by 21%, but it also prolonged survival of the mutant SOD1^{G93A} mice by 8% (90). These studies show that a low VEGF level can induce selective motor neuron death, most likely due to the lack of neuroprotection, although indirect, vascular effects of VEGF cannot be excluded.

In contrast to CNTF and VEGF, knockout mice that lack other growth factors like insulin-like growth factor I (IGF-I), brain-derived neurotrophic factor (BDNF), or glial-derived neurotrophic factor (GDNF) did not show age-related, selective motor neuron loss and/or motor dysfunctions (91–93). This indicates that the selective loss of motor neurons is not a general consequence of neurotrophic factors shortage, although compensatory effects in these transgenic mice cannot be excluded.

10. Conclusions

In conclusion, animal models that show selective motor neuron death and a progressive motor phenotype indicate that a number of pathogenic pathways can lead to this phenotype. Aggregation of mutated proteins, defects in RNA processing, problems with axonal transport, and shortage of CNTF/VEGF can at a certain point in life compromise the survival of motor neurons. Moreover, it is very plausible that in at least some of these motor neuron disorders, a complicated interplay of these and other mechanisms ultimately leads to motor neuron death. As a consequence, it remains a challenge to find out the exact contribution of and interaction between all these different mechanisms, as it will otherwise remain difficult to interfere with this multifactorial disease that is life-threatening. In relation to cognitive impairments, it is striking that in these animal models for motor neuron degeneration only very subtle differences were observed, with the exception of tau-related animal models showing both cognitive and motor phenotypes.

References

1. Gros-Louis F, Gaspar C, Rouleau GA (2006) Genetics of familial and sporadic amyotrophic lateral sclerosis. Biochim Biophys Acta 1762: 956–972
2. Van Den Bosch L, Timmerman V (2006) Genetics of motor neuron disease. Curr Neurol Neurosci Rep 6:423–431
3. Rosen DR, Siddique T, Patterson D, et al. (1993) Mutations in Cu/Zn superoxide dismutase gene are associated with familial amyotrophic lateral sclerosis. Nature 362:59–62
4. Hadano S, Hand CK, Osuga H, et al. (2001) A gene encoding a putative GTPase regulator is mutated in familial amyotrophic lateral sclerosis 2. Nat Genet 29:166–173
5. Chen YZ, Bennett CL, Huynh HM, et al. (2004) DNA/RNA helicase gene mutations in a form of juvenile amyotrophic lateral sclerosis (ALS4). Am J Hum Genet 74:1128–1135
6. Nishimura AL, Mitne-Neto M, Silva HC, et al. (2004) A mutation in the vesicle-trafficking protein VAPB causes late-onset

spinal muscular atrophy and amyotrophic lateral sclerosis. Am J Hum Genet 75:822–831
7. Greenway MJ, Andersen PM, Russ C, et al. (2006) ANG mutations segregate with familial and 'sporadic' amyotrophic lateral sclerosis. Nat Genet 38:411–413
8. Puls I, Jonnakuty C, LaMonte BH, et al. (2003) Mutant dynactin in motor neuron disease. Nat Genet 33:455–456
9. Sreedharan J, Blair IP, Tripathi VB, et al. (2008) TDP-43 mutations in familial and sporadic amyotrophic lateral sclerosis. Science 319:1668–1672
10. Kabashi E, Valdmanis PN, Dion P, et al. (2008) TARDBP mutations in individuals with sporadic and familial amyotrophic lateral sclerosis. Nat Genet 40:472–4
11. Van Deerlin VM, Leverenz JB, Bekris LM, et al. (2008) TARDBP mutations in amyotrophic lateral sclerosis with TDP-43 neuropathology: a genetic and histopathological analysis. Lancet Neurol 7:409–416
12. Gitcho MA, Baloh RH, Chakraverty S, et al. (2008) TDP-43 A315T mutation in familial motor neuron disease. Ann Neurol 63:535–538
13. Phukan J, Pender NP, Hardiman O (2007) Cognitive impairment in amyotrophic lateral sclerosis. Lancet Neurol 6:994–1003
14. Lomen-Hoerth C, Murphy J, Langmore S, Kramer JH, Olney RK, Miller B (2003) Are amyotrophic lateral sclerosis patients cognitively normal? Neurology 60:1094–1097.
15. Barson FP, Kinsella GJ, Ong B, Mathers SE (2000) A neuropsychological investigation of dementia in motor neurone disease (MND). J Neurol Sci 80:107–113
16. Neumann M, Sampathu DM, Kwong LK, et al. (2006) Ubiquitinated TDP-43 in frontotemporal lobar degeneration and amyotrophic lateral sclerosis. Science 314:130–133
17. Cruts M, Gijselinck I, van der Zee J, et al. (2006) Null mutations in progranulin cause ubiquitin-positive frontotemporal dementia linked to chromosome 17q21. Nature 442:920–924
18. Baker M, Mackenzie IR, Pickering-Brown SM, et al. (2006) Mutations in progranulin cause tau-negative frontotemporal dementia linked to chromosome 17. Nature 442:916–919
19. Sleegers K, Brouwers N, Maurer-Stroh S, et al. (2008) Progranulin genetic variability contributes to amyotrophic lateral sclerosis. Neurology 71(4):253–259
20. Gurney ME, Pu H, Chiu AY, et al. (1994) Motor neuron degeneration in mice that express a human Cu,Zn superoxide dismutase mutation. Science 264:1772–1775
21. Bruijn LI, Becher MW, Lee MK, et al. (1997) ALS-linked SOD1 mutant G85R mediates damage to astrocytes and promotes rapidly progressive disease with SOD1-containing inclusions. Neuron 18:327–338
22. Wong PC, Pardo CA, Borchelt DR, et al. (1995) An adverse property of a familial ALS-linked SOD1 mutation causes motor neuron disease characterized by vacuolar degeneration of mitochondria. Neuron 14:1105–1116
23. Ripps ME, Huntley GW, Hof PR, Morrison JH, Gordon JW (1995) Transgenic mice expressing an altered murine superoxide dismutase gene provide an animal model of amyotrophic lateral sclerosis. Proc Natl Acad Sci U S A 92:689–693
24. Jonsson PA, Graffmo KS, Brannstrom T, Nilsson P, Andersen PM, Marklund SL (2006) Motor neuron disease in mice expressing the wild type-like D90A mutant superoxide dismutase-1. J Neuropathol Exp Neurol 65:1126–1136
25. Wang J, Xu G, Li H, et al. (2005) Somatodendritic accumulation of misfolded SOD1-L126Z in motor neurons mediates degeneration: alphaB-crystallin modulates aggregation. Hum Mol Genet 14:2335–2347
26. Watanabe Y, Yasui K, Nakano T, et al. (2005) Mouse motor neuron disease caused by truncated SOD1 with or without C-terminal modification. Brain Res Mol Brain Res 135:12–20
27. Nagai M, Aoki M, Miyoshi I, et al. (2001) Rats expressing human cytosolic copper-zinc superoxide dismutase transgenes with amyotrophic lateral sclerosis: associated mutations develop motor neuron disease. J Neurosci 21: 9246–9254
28. Howland DS, Liu J, She Y, et al. (2002) Focal loss of the glutamate transporter EAAT2 in a transgenic rat model of SOD1 mutant-mediated amyotrophic lateral sclerosis (ALS). Proc Natl Acad Sci U S A 99:1604–1609
29. Subramaniam JR, Lyons WE, Liu J, et al. (2002) Mutant SOD1 causes motor neuron disease independent of copper chaperone- mediated copper loading. Nat Neurosci 5:301–307
30. Wang J, Slunt H, Gonzales V, et al. (2003) Copper-binding-site-null SOD1 causes ALS in transgenic mice: aggregates of non-native SOD1 delineate a common feature. Hum Mol Genet 12:2753–2764
31. Bruijn LI, Houseweart MK, Kato S, et al. (1998) Aggregation and motor neuron toxicity of an ALS-linked SOD1 mutant independent from wild-type SOD1. Science 281:1851–1854
32. Johnston JA, Dalton MJ, Gurney ME, Kopito RR (2000) Formation of high molecular

weight complexes of mutant Cu,Zn-superoxide dismutase in a mouse model for familial amyotrophic lateral sclerosis. Proc Natl Acad Sci U S A 97:12571–12576
33. Wang J, Xu G, Slunt HH, et al. (2005) Coincident thresholds of mutant protein for paralytic disease and protein aggregation caused by restrictively expressed superoxide dismutase cDNA. Neurobiol Dis 20:943–952
34. Pramatarova A, Laganiere J, Roussel J, Brisebois K, Rouleau GA (2001) Neuron-specific expression of mutant superoxide dismutase 1 in transgenic mice does not lead to motor impairment. J Neurosci 21:3369–3374
35. Gong YH, Parsadanian AS, Andreeva A, Snider WD, Elliott JL (2000) Restricted expression of G86R Cu/Zn superoxide dismutase in astrocytes results in astrocytosis but does not cause motoneuron degeneration. J Neurosci 20:660–665
36. Clement AM, Nguyen MD, Roberts EA, et al. (2003) Wild-type nonneuronal cells extend survival of SOD1 mutant motor neurons in ALS mice. Science 302:113–117
37. Boillee S, Yamanaka K, Lobsiger CS, et al. (2006) Onset and progression in inherited ALS determined by motor neurons and microglia. Science 312:1389–1392
38. Yamanaka K, Chun SJ, Boillee S, et al. (2008) Astrocytes as determinants of disease progression in inherited amyotrophic lateral sclerosis. Nat Neurosci 11:251–253
39. Jaarsma D, Teuling E, Haasdijk ED, De Zeeuw CI, Hoogenraad CC (2008) Neuron-specific expression of mutant superoxide dismutase is sufficient to induce amyotrophic lateral sclerosis in transgenic mice. J Neurosci 28:2075–2088
40. Spalloni A, Geracitano R, Berretta N, et al. (2006) Molecular and synaptic changes in the hippocampus underlying superior spatial abilities in pre-symptomatic G93A+/+ mice overexpressing the human Cu/Zn superoxide dismutase (Gly93 ALA) mutation. Exp Neurol 197:505–514
41. Cai H, Lin X, Xie C, et al. (2005) Loss of ALS2 function is insufficient to trigger motor neuron degeneration in knock-out mice but predisposes neurons to oxidative stress. J Neurosci 25:7567–7574
42. Hadano S, Benn SC, Kakuta S, et al. (2006) Mice deficient in the Rab5 guanine nucleotide exchange factor ALS2/alsin exhibit age-dependent neurological deficits and altered endosome trafficking. Hum Mol Genet 15:233–250
43. Devon RS, Orban PC, Gerrow K, et al. (2006) Als2-deficient mice exhibit disturbances in endosome trafficking associated with motor behavioral abnormalities. Proc Natl Acad Sci U S A 103:9595–9600
44. Yamanaka K, Miller TM, McAlonis-Downes M, Chun SJ, Cleveland DW (2006) Progressive spinal axonal degeneration and slowness in ALS2-deficient mice. Ann Neurol 60:95–104
45. Rabin BA, Griffin JW, Crain BJ, Scavina M, Chance PF, Cornblath DR (1999) Autosomal dominant juvenile amyotrophic lateral sclerosis. Brain 122:1539–1550
46. De Jonghe P, Auer-Grumbach M, Irobi J, et al. (2002) Autosomal dominant juvenile ALS and distal hereditary motor neuronopathy with pyramidal tract signs: Synonyms for the same disorder? Brain 125:1320–1325
47. Moreira MC, Klur S, Watanabe M, et al. (2004) Senataxin, the ortholog of a yeast RNA helicase, is mutant in ataxia-ocular apraxia 2. Nat Genet 36:225–227
48. Anderson K, Talbot K (2003) Spinal muscular atrophies reveal motor neuron vulnerability to defects in ribonucleoprotein handling. Curr Opin Neurol 16:595–599
49. Cook SA, Johnson KR, Bronson RT, Davisson MT (1995) Neuromuscular degeneration (nmd): a mutation on mouse chromosome 19 that causes motor neuron degeneration. Mamm Genome 6:187–191
50. Cox GA, Mahaffey CL, Frankel WN (1998) Identification of the mouse neuromuscular degeneration gene and mapping of a second site suppressor allele. Neuron 21:1327–1337
51. Grohmann K, Schuelke M, Diers A, et al. (2001) Mutations in the gene encoding immunoglobulin mu-binding protein 2 cause spinal muscular atrophy with respiratory distress type 1. Nat Genet 29:75–77
52. Pennetta G, Hiesinger PR, Fabian-Fine R, Meinertzhagen IA, Bellen HJ (2002) Drosophila VAP-33A directs bouton formation at neuromuscular junctions in a dosage-dependent manner. Neuron 35:291–306
53. Levy JR, Sumner CJ, Caviston JP, et al. (2006) A motor neuron disease-associated mutation in p150Glued perturbs dynactin function and induces protein aggregation. J Cell Biol 172:733–745
54. Lai C, Lin X, Chandran J, Shim H, Yang WJ, Cai H (2007) The G59S mutation in p150(glued) causes dysfunction of dynactin in mice. J Neurosci 27:13982–13990
55. Laird FM, Farah MH, Ackerley S, et al. (2008) Motor neuron disease occurring in a mutant dynactin mouse model is characterized by defects in vesicular trafficking. J Neurosci 28:1997–2005

56. Chevalier-Larsen ES, Wallace KE, Pennise CR, Holzbaur EL (2008) Lysosomal proliferation and distal degeneration in motor neurons expressing the G59S mutation in the p150Glued subunit of dynactin. Hum Mol Genet 17:1946–1955
57. Hafezparast M, Klocke R, Ruhrberg C, et al. (2003) Mutations in dynein link motor neuron degeneration to defects in retrograde transport. Science 300:808–812
58. King SJ, Schroer TA (2000) Dynactin increases the processivity of the cytoplasmic dynein motor. Nat Cell Biol 2:20–24
59. LaMonte BH, Wallace KE, Holloway BA, et al. (2002) Disruption of dynein/dynactin inhibits axonal transport in motor neurons causing late-onset progressive degeneration. Neuron 34:715–727
60. Kieran D, Hafezparast M, Bohnert S, et al. (2005) A mutation in dynein rescues axonal transport defects and extends the life span of ALS mice. J Cell Biol 169:561–567
61. Murakami T, Nagano I, Hayashi T, et al. (2001) Impaired retrograde axonal transport of adenovirus-mediated E. coli LacZ gene in the mice carrying mutant SOD1 gene. Neurosci Lett 308:149–152
62. Borchelt DR, Wong PC, Becher MW, et al. (1998) Axonal transport of mutant superoxide dismutase 1 and focal axonal abnormalities in the proximal axons of transgenic mice. Neurobiol Dis 5:27–35
63. Hirano A, Donnenfeld H, Sasaki S, Nakano I (1984) Fine structural observations of neurofilamentous changes in amyotrophic lateral sclerosis. J Neuropathol Exp Neurol 43:461–470
64. Rouleau GA, Clark AW, Rooke K, et al. (1996) SOD1 mutation is associated with accumulation of neurofilaments in amyotrophic lateral sclerosis. Ann Neurol 39:128–131
65. Hirano A, Nakano I, Kurland LT, Mulder DW, Holley PW, Saccomanno G (1984) Fine structural study of neurofibrillary changes in a family with amyotrophic lateral sclerosis. J Neuropathol Exp Neurol 43:471–480
66. Al-Chalabi A, Andersen PM, Nilsson P, et al. (1999) Deletions of the heavy neurofilament subunit tail in amyotrophic lateral sclerosis. Hum Mol Genet 8:157–164
67. Julien JP (2001) Amyotrophic lateral sclerosis: unfolding the toxicity of the misfolded. Cell 104:581–591
68. Robertson J, Kriz J, Nguyen MD, Julien JP (2002) Pathways to motor neuron degeneration in transgenic mouse models. Biochimie 84:1151–1160
69. Lee MK, Marszalek JR, Cleveland DW (1994) A mutant neurofilament subunit causes massive, selective motor neuron death: implications for the pathogenesis of human motor neuron disease. Neuron 13:975–988
70. Mersiyanova IV, Perepelov AV, Polyakov AV, et al. (2000) A new variant of Charcot-Marie-Tooth disease type 2 is probably the result of a mutation in the neurofilament-light gene. Am J Hum Genet 67:37–46
71. De Jonghe P, Mersivanova I, Nelis E, et al. (2001) Further evidence that neurofilament light chain gene mutations can cause Charcot-Marie-Tooth disease type 2E. Ann Neurol 49:245–249
72. Beaulieu JM, Nguyen MD, Julien JP (1999) Late onset of motor neurons in mice overexpressing wild-type peripherin. J Cell Biol 147: 531–544
73. Lariviere RC, Beaulieu JM, Nguyen MD, Julien JP (2003) Peripherin is not a contributing factor to motor neuron disease in a mouse model of amyotrophic lateral sclerosis caused by mutant superoxide dismutase. Neurobiol Dis 13:158–166
74. Leung CL, He CZ, Kaufmann P, et al. (2004) A pathogenic peripherin gene mutation in a patient with amyotrophic lateral sclerosis. Brain Pathol 14:290–296
75. Gros-Louis F, Lariviere R, Gowing G, et al. (2004) A frameshift deletion in peripherin gene associated with amyotrophic lateral sclerosis. J Biol Chem 279:45951–45956
76. Martin N, Jaubert J, Gounon P, et al. (2002) A missense mutation in Tbce causes progressive motor neuronopathy in mice. Nat Genet 32:443–447
77. Lee VM, Goedert M, Trojanowski JQ (2001) Neurodegenerative tauopathies. Annu Rev Neurosci 24:1121–1159
78. Spittaels K, Van den Haute C, Van Dorpe J, et al. (1999) Prominent axonopathy in the brain and spinal cord of transgenic mice overexpressing four-repeat human tau protein. Am J Pathol 155:2153–2165
79. Zhang B, Higuchi M, Yoshiyama Y, et al. (2004) Retarded axonal transport of R406W mutant tau in transgenic mice with a neurodegenerative tauopathy. J Neurosci 24: 4657–4667
80. Tatebayashi Y, Miyasaka T, Chui DH, et al. (2002) Tau filament formation and associative memory deficit in aged mice expressing mutant (R406W) human tau. Proc Natl Acad Sci U S A 99:13896–13901
81. Ramsden M, Kotilinek L, Forster C, et al. (2005) Age-dependent neurofibrillary tangle formation, neuron loss, and memory impairment in a mouse model of human tauopathy (P301L). J Neurosci 25:10637–10647

82. Schindowski K, Bretteville A, Leroy K, et al. (2006) Alzheimer's disease-like tau neuropathology leads to memory deficits and loss of functional synapses in a novel mutated tau transgenic mouse without any motor deficits. Am J Pathol 169:599–616
83. Tanemura K, Murayama M, Akagi T, et al. (2002) Neurodegeneration with tau accumulation in a transgenic mouse expressing V337M human tau. J Neurosci 22:133–141
84. Yoshiyama Y, Higuchi M, Zhang B, et al. (2007) Synapse loss and microglial activation precede tangles in a P301S tauopathy mouse model. Neuron 53:337–351
85. Rosenmann H, Grigoriadis N, Eldar-Levy H, et al. (2008) A novel transgenic mouse expressing double mutant tau driven by its natural promoter exhibits tauopathy characteristics. Exp Neurol 212:71–84
86. Masu Y, Wolf E, Holtmann B, Sendtner M, Brem G, Thoenen H (1993) Disruption of the CNTF gene results in motor neuron degeneration. Nature 365:27–32
87. Oosthuyse B, Moons L, Storkebaum E, et al. (2001) Deletion of the hypoxia-response element in the vascular endothelial growth factor promoter causes motor neuron degeneration. Nat Genet 28:131–138
88. Lambrechts D, Storkebaum E, Morimoto M, et al. (2003) VEGF is a modifier of amyotrophic lateral sclerosis in mice and humans and protects motoneurons against ischemic death. Nat Genet 34:383–394
89. Van Den Bosch L, Storkebaum E, Vleminckx V, et al. (2004) Effects of vascular endothelial growth factor (VEGF) on motor neuron degeneration. Neurobiol Dis 17: 21–28
90. Storkebaum E, Lambrechts D, Dewerchin M, et al. (2005) Treatment of motoneuron degeneration by intracerebroventricular delivery of VEGF in a rat model of ALS. Nat Neurosci 8:85–92
91. Gao WQ, Shinsky N, Ingle G, Beck K, Elias KA, Powell-Braxton L (1999) IGF-I deficient mice show reduced peripheral nerve conduction velocities and decreased axonal diameters and respond to exogenous IGF-I treatment. J Neurobiol 39:142–152
92. Ernfors P, Lee KF, Jaenisch R (1994) Mice lacking brain-derived neurotrophic factor develop with sensory deficits. Nature 368:147–150
93. Moore MW, Klein RD, Farinas I, et al. (1996) Renal and neuronal abnormalities in mice lacking GDNF. Nature 382:76–79

Chapter 28

Animal Models of Frontotemporal Dementia

Hana N. Dawson and Daniel T. Laskowitz

Abstract

Frontotemporal dementia (FTD) is a multifaceted syndrome with a high degree of clinical and neuropathological variability, an extensive genetic contribution, and involvement of multiple proteins. FTD accounts for up to 50% of dementias with the onset prior to age 60. The heterogeneous genetic, clinical, and pathological manifestations of FTD have created challenges in generating clinically relevant animal models with which to test new therapeutic approaches. Nevertheless, tau transgenic models have been developed in mice, *Drosophila melanogaster*, and *Caenorhabditis elegans* in the past decade. These models have played an important role in elucidating a number of mechanisms associated with tau FTD-related neurodegeneration, and it is likely that these preclinical models will help to facilitate new therapeutic strategies. It is clear that both wild-type and mutated tau protein are sufficient to elicit tauopathy, although mutated tau increases the severity of the pathology. Furthermore, the aberrant expression and/or incorrect temporal/developmental expression of tau may also cause tau pathology. Importantly, these tau models have clarified some long-held theories pertaining to tau and neurodegeneration. For example, it has been shown that oxidative stress plays a crucial role in FTD and that tau pathology reactivates the cell cycle machinery. Conversely, tau aggregates are not necessary for tau neurotoxicity. However, new models representing other forms of FTD need to be developed and much work still remains before the disease is clearly understood and disease-modifying therapies become available.

Key words: FTD, Animal models, Transgenic, Tau, Progranulin, TDP-43, Neurodegeneration

1. Introduction

Frontotemporal dementia (FTD), previously known as Pick's complex, is a multifaceted syndrome with a high degree of clinical and neuropathological variability, an extensive genetic contribution, and involvement of multiple proteins. FTD accounts for up to 50% of dementias with the onset prior to age 60 (1). The heterogeneous clinical and pathological manifestations of FTD have created challenges in generating clinically relevant animal models with which to test new therapeutic approaches. Nevertheless,

Peter Paul De Deyn and Debby Van Dam (eds.), *Animal Models of Dementia*, Neuromethods, vol. 48,
DOI 10.1007/978-1-60761-898-0_28, © Springer Science+Business Media, LLC 2011

much has been learned from current transgenic models of mice, *Drosophila melanogaster* (Drosophila), and *Caenorhabditis elegans* (C. elegans) in the past decade. Some molecules and mechanisms that play crucial roles in FTD-related neurodegeneration have been delineated; however, much work still remains before the disease is clearly understood and palative therapies are available.

1.1. Background of FTD

The original name for FTD was Pick's disease (PiD), which has led to some confusion in nomenclature. In 1882, Arnold Pick first described PiD as a clinical picture of frontotemporal atrophy (2). Subsequently, Aloysius Alzheimer described this case of PiD as distinct from Alzheimer's disease based on histology characterized by the presence of large, dark-staining aggregates of proteins (Pick bodies) that were present in the absence of plaques or tangles (3). However, we now know that all cases with frontotemporal atrophy do not present with Pick bodies. Eventually, the general clinical syndrome was defined as FTD. Diagnoses of PiD are now reserved for only a subset of FTD that presents postmortem with Pick bodies.

Pathologically, all FTD include frontotemporal lobar atrophy, neuronal loss, gliosis, and superficial spongiosis. Although rarely present at the onset of the disease, cognitive impairment is a constant feature of FTD (4–7). The onset of FTD, however, can present with a variety of clinical manifestations (please see Table 1) with the most common variant presenting first with progressive deterioration of behavior and personality (FTD-bv) (8). FTD often initially presents with language dysfunction, Primary Progressive Aphasia (PPA), where a loss or impairment of the power to use or comprehend words occurs, and semantic dementia (SD), which presents with a comprehension deficit of the meaning of nouns and objects. Finally, the onset of FTD may first present with extrapyramidal symptoms such as progressive supranuclear palsy (PSP; characterized by bulbar symptoms) and corticobasal degeneration (CBD, associated with apraxia). However, after onset, patients may present with one or more of the other initial clinical syndromes causing the distinction between these clinical phenotypes to be unclear. Additionally, several different FTD symptoms may appear in one family, and even vary within a family where a single mutation has been identified, thus emphasizing the fact that other genetic and environmental factors play significant roles in the disease.

Given the clinical heterogeneity of FTD, a more complete understanding of its neuropathology has played a central role in the characterization of the disease process, and ultimately, in the creation of relevant animal models. Neurofibrillary tangles (NFTs) containing hyperphosphorylated aggregates of the microtubule-associated protein tau in cell bodies are abundant in FTD (Table 2).

Table 1
Clinical manifestations of frontotemporal dementia

Syndrome at onset	Description	Pathology[a]
FTD-bv Frontotemporal dementia – behavioral variant	Patients present with disinhibition, apathy, loss of judgment and insight, negligence of personal hygiene, verbal and physical aggressiveness	MNDI, DLH, Pick bodies, CBD
PPA Primary progressive aphasia	Patients become increasingly nonfluent with increasing aphasia (impairment in the use and comprehension of words) and logopenia (word-finding difficulty)	MNDI, CBD, PICK'S BODIES
SD Semantic dementia	Patients remain fluent while developing progressive deterioration of understanding and recognizing words (meaningless speech)	
CBD corticobasal degeneration	Patients develop unilateral rigidity, apraxia, alien hand	CBD, PSP
PSP Progressive supranuclear palsy	Patients develop vertical gaze palsy, falling, axial rigidity, pseudobulbar palsy	

[a] Please see Table 2 for descriptions of postmortem histological classification or pathology

However, some patients with FTD have tau-negative, ubiquitin-positive lesions and some patients have no identifiable lesions at all (Table 2). Therefore, FTD is now further subclassified according to histochemical and immunocytochemical properties of the material accumulated in neuronal cells: (1) Tau-positive variant of FTD (FTD-T), (2) Ubiquitin-positive, tau-negative variant of FTD termed frontotemporal lobar degeneration with ubiquitin inclusions (FTLD-U), and (3) a variant with no identifiable accumulation termed dementia lacking distinctive histology (DLDH).

1.2. FTD Genetics

Great progress has been made over the past decade in defining the genetic basis of FTD. To date, mutations associated with FTD have been identified in the following three genes: *MAPT*, which encodes for tau (9, 10); *PRGN*, which encodes for progranulin (11, 12); and *VCP*, which encodes for the valosin-containing protein (13). In 1997, a consensus conference designated a subset of FTD as FTDP-17 (frontotemporal dementia and parkinsonism linked to chromosome 17 (14), which was soon followed by seminal publications identifying both exonic and intronic mutations in the *MAPT* gene (9, 10, 15). Of the 40-plus known *MAPT* mutations (16), several have been expressed in transgenic mice, Drosophila, and C. elegans, confirming that tau dysfunction is causal in neurodegeneration and dementia.

Table 2
Postmortem pathology in frontotemporal dementia

Histological sub-types of FTD pathology	Syndrome-specific pathology	Histo and immuno reactivity
PD Pick's Disease	Pick bodies (PB) – round or oval compact intracytoplasmic neuronal inclusions, microvacuolation	Bielschowsky but not Gallyas, tau immunoreactivity
CBD Corticobasal degeneration	Ballooned achromatic neurons, intracytoplasmic neuronal inclusions of variable morphology, pervasive gray and white matter threads, astrocytic plaques	Tau positive – intracytoplasmic inclusions, threads, oligodendrocytes and glial plaques Bielschowsky and Galyas positive intracytoplasmic inclusions and threads
PSP Progressive	Predominantly subcortical distribution of intracytoplasmic neuronal inclusions of variable morphology, tufted astrocytes	Tau-positive tangles and tufted astrocytes
MNDI Motor neuron type inclusions	Neuronal intracytoplasmic and nuclear inclusions, cortical microvacuolation, in fascia dentate, and cerebral cortex	Ubiquitin-positive neurons and dystrophic neurites, gliosis, TDP-43-positive neuronal inclusions
DLDH Dementia lacking distinctive histology	Cortical microvacuolation, neuronal loss, and gliosis	Tau and ubiquitin negative, gliosis

[a] All FTD cases present with lobar atrophy, neuronal loss, and gliosis

Biochemically, tau was one of the first proteins identified within the intracellular inclusions in FTD. Tau accumulation can be somal, in dystrophic neurites, and/or in glia and it is the hallmark of several other diseases besides FTD, collectively called tauopathies. Tau protein in these cellular lesions can be hyperphosphorylated, ubiquitinated, and carboxy-terminal truncated.

Although most cases of FTLD-U are sporadic, the efforts of several groups led to the identification of familial disease resulting from the loss-of-function mutations in *PRGN* implicating a haploinsufficiency mechanism (11, 12, 17). While it is poorly understood why a partial loss of PGRN leads to FTLD-U, the marked reduction of PGRN protein in this disease points toward a function of PGRN in neuronal survival. However, as with most FTDs, the age of onset in families segregating with the same *PRGN* mutation varies widely as do the symptoms suggesting the existence of further modifying factors (18).

Interestingly, unlike tau protein, mutated PGRN does not accumulate in the neuronal cells of FTLD-U patients. Instead, the

TAR DNA sequence-binding protein (TDP-43) accumulates in ubiquitin-positive inclusions in FTLD-U patients with *PGRN* mutations (19). Similar to accumulated tau protein in cells, TDP-43, in cell lesions, is hyperphosphorylated, ubiquitinated, and carboxy-terminal truncated (19, 20). TDP-43 is encoded by the *TARDBP* gene located on chromosome 1 and is a highly conserved ubiquitously expressed nuclear protein implicated in repression of gene transcription, inhibition of exon splicing, and interactions with splicing factors and nuclear bodies. Although accumulated TDP-43 is found in approximately half of all FTD cases, no mutations have been associated with the disease. However, mutations in the *TARDBP* gene have been identified in both familial and sporadic amyotrophic lateral sclerosis (ALS) (21–24).

It is important to note that TDP-43 inclusions were also discovered in the autosomal dominant inclusion body myopathy associated with Paget disease of the bone with frontotemporal dementia (PDB-FTD) (25, 26). Although not classified as typical FTD, patients with this disease develop dementia that is typical of FTD and is characterized by language and/or behavioral dysfunction with relative preservation of memory (27, 28). Although no mutations were associated with the *TARDBP* gene in PDB-FTD, mutations in the valosin-containing protein (VCP), which is an essential component of the endoplasmic reticulum-associated degradation process, were associated with this disease (13). No VCP transgenic animal models have been reported to date.

It is likely that the disruption of synaptic function leading to neuronal death is the common neuropathological feature leading to the diverse clinical manifestations associated with FTD. However, there are many pathways that can lead to the final common denominator of synaptic dysfunction, perhaps explaining the wide variety of genes involved. The distinctive patterns that characterize the cognitive and behavioral deficits may reflect the neural circuitry that is affected.

An increased understanding of the neuropathology and genetic influences associated with FTD has aided the development of relevant animal models, which can be used both to enhance our fundamental understanding of disease pathogenesis and to develop screens that aid in the development of new therapeutic strategies. Since it is not possible to perform screens in humans, animal models have become the foundation for studies designed to advance our understanding of disease progression. In the case of FTD, many animal models have been created ranging from the simple *C.* elegans, zebrafish, and *Drosophila* to the more complex mammalian mice and rats. Using these models, abnormal accumulation and misfolding of synaptic and cytoskeletal proteins are being extensively explored as a key pathogenic event leading to neurodegeneration in FTD. Additional models have also been generated using FTD familial mutations.

2. Murine Tau Models

Biochemically, tau was one of the first proteins identified within the intracellular inclusions in FTD. Hence, many of the animal models of FTD to date have been engineered to overexpress either normal human tau or FTDP-17 mutated human tau. Tau protein is a neuron-specific protein that binds and stabilizes microtubules that are involved in axonal transport (29–34). The alternative splicing of exons 2, 3, and 10 of tau is developmentally, physiologically, and spatially regulated. In the fetal brain, only the shortest isoform lacking exons 2, 3, and 10 is expressed. As the brain matures to adulthood, differential splicing of exons 2, 3, and 10 results in the generation of six tau isoforms (mRNA and protein) (35–37). These tau isoforms are typically distinguished by the presence of either three (3R) or four (4R) tandem repeats of the microtubule binding domain. The 4R tau isoforms contain additional amino acids encoded by exon 10 that are absent in the 3R isoforms (38, 39).

NFTs are the intracellular lesions found in the tau-positive subset of FTD and are composed of filaments of hyperphosphorylated tau. NFTs correlate well with cognitive deficits and neuron loss, and have been implicated in mediating neurodegeneration and dementia in FTD and other tauopathies. The association between NFTs, neuron loss, and brain dysfunction in humans has led some to posit that NFTs invariably cause brain dysfunction and neurodegeneration. As a result, the gold standard of tauopathy models has been the generation of NFTs.

2.1. Murine Models Expressing Normal Human Tau

Initially, starting as early as 1995, transgenic mouse models of FTD were created by expressing human tau protein from cDNA constructs of various tau isoforms under the regulation of heterologous promoters. The original theory was that the overexpression of normal tau protein would model clinical neuropathology by generating tau NFTs. Thus, mice expressing the longest (including exons 2, 3, and 10) or shortest (without exons 2, 3, and 10) human tau isoforms were engineered. Human tau protein was expressed under the regulation of the Thy-1 promoter by different groups. The model of Gotz et al. (1995) (40) displayed tau protein expression 10% of endogenous mouse tau levels, whereas the models of Spittaels et al. (1999) (41) and Probst et al. (2000) (42) exhibited tau protein expression 150–400% and 130% of endogenous mouse tau, respectively. Other models were created that expressed human tau under the regulation of (1) the hydroxy–methyl–glutaryl CoA reductase (HMG-CR housekeeping gene) promoter, with protein expression levels of approximately 200% of endogenous mouse tau developed by Brion et al. (1999) (43), and

(2) the murine prion protein (PrP) promoter, protein expression 500–1,500% of endogenous mouse tau developed by Ishihara et al. (1999) (44).

All the models developed somatodendritic relocalization of tau protein, which was believed to be the pretangle stage of tau pathology. Immunopositivity with phosphorylation and conformation-specific antibodies in neurons and astrocytes were also seen in all these models; however, it is important to note that most of these antibodies are also immunoreactive to some extent to normal tau. As with all tau overexpressing models, caution should be taken that observed changes are a result of tau pathology and not a fortuitous artifact of overexpression. Of these models (Spittaels (41), Probst (42) and Ishihara (44)), the ones expressing high levels of human tau also presented with axonal dilations associated with abnormal tau and neurofilament spheroids in the spinal cord. These lesions were functionally associated with motor deficits.

Spittaels et al. further extended these results and developed several transgenic lines with increasing amounts of tau protein overexpression ranging from 150 to 400% of endogenous mouse tau (41). In these lines, a clear correlation was observed between the degree of tau overexpression and increased pathology. Axonopathy in brain and spinal cord were transgene dosage-related but no other characteristics of FTD such as tangles, astrogliosis, ubiquitination, or motor deficits correlated with the level of tau expression. Similarly, the Ishihara et al. model expressing the highest levels of human tau protein not only developed axonopathy, but also motor dysfunction, weight loss, and exhibited a decreased rate of survival (44). Although this model recapitulated many of the clinical features observed in patients with ALS, it failed to model many of the cardinal cognitive and extrapyramidal features of FTD. However, these models clearly demonstrated that increasing levels of tau protein were detrimental to the nervous system.

Interestingly, mice expressing human tau protein from human tau promoters, from the human tau gene at approximately 300% endogenous mouse tau levels (45, 46), and from a cDNA of the longest tau isoform at approximately 100% endogenous mouse tau levels (47) resulted in minimal tau pathology. It would be expected that the expression of tau protein under the regulation of the human tau promoter would be in a more physiological manner temporally and developmentally than when tau protein is expressed from heterologous promoters. It is therefore possible that the tau pathology in mouse models with heterologous promoters may in large part be attributed to nonphysiological patterns of expression. A conclusion that can be drawn from these early models is that the overexpression and/or incorrect temporal/developmental expression of normal tau is by itself causal in tauopathy.

2.2. Murine Models Expressing Mutated Human Tau

Mutations in the tau gene, *MAPT*, were identified in patients with FTDP-17 in 1998 (9, 10, 15). These mutations and their consecutive expression in transgenic animals confirmed that tau dysfunction can cause neurodegeneration and dementia. The FTDP-17 associated tau mutations are located predominantly in the microtubule-binding region, indicating that this region is highly sensitive to disease-causing mutations. Mutations in this region would disrupt the microtubular network with downstream effects on a variety of intracellular processes, such as intracellular transport, metabolism, mRNA trafficking, protein sorting, and targeting. These, in turn, could have potential consequences for the maintenance of neuritic structures and synaptic plasticity. The mutations in the TAU gene fall under two categories: exonic mutations that directly alter the structure and function of tau protein by amino acid alterations, and intronic or exonic mutations that indirectly alter the structure and function of tau protein by altering the splicing of exon 10. Mutations that alter the splicing of exon 10 consequently affect the relative ratio of 4R to 3R microtubule-binding repeats and hence the tau/microtubule interactions.

2.3. Biochemical Mutations

Although NFTs are one of the main pathological hallmarks of tau FTD (48), they were not reproduced until the expression of human tau protein carrying a biochemical FTDP-17 mutation was introduced into animal models. The first model that reproduced NFTs and extensive hyperphosphorylation was the JNPL3 mouse that expressed the P301L mutation under the regulation of the prion promoter (49). The JNPL3 mice express human tau protein at approximately 300% the level of endogenous mouse tau and develop NFTs, neuropil threads, motor deficits, and neuronal cell loss and invariably develop severe motor deficits that lead to hind limb paralysis and death. Furthermore, the mice also develop decreased grooming habits, a behavior that may be indicative of FTD. However, by the time the grooming deficits occur, the mice are in poor physical shape and it is therefore difficult to discern whether this lack of grooming is due to cognitive deterioration or physical deficits. These mice have been widely utilized for understanding some of the mechanisms of FTD tauopathy. Several other models using the same or different FTDP-17 mutations were also engineered, all reproducing similar results. Based on the success of these models, a second generation of models were generated with varying degrees of tau expression by a number of investigators, namely Gotz et al. 2001 (50) (P301L, 70% of endogenous tau), Gotz et al. 2001 (51) (G272V, 1% of endogenous tau), Allen et al. 2002 (52) (P301S, 200% of endogenous tau), Tanemura et al. 2002 (53) (V337M, 100% of endogenous tau), and Tatebayashi et al. 2002 (54) (R406W, 14% of endogenous tau). All the mice developed varying amounts of tau pathology generally in the form of abnormal, tau-reactive nerve cell bodies and dendrites and large numbers of pathologically enlarged axons containing tau-reactive spheroids. A high percentage of the mice also develop glial tau pathology similar to that seen in PSP and CBD. The majority of the tau mutant mice developed motor deficits, ranging from hind limb clasping, to balance beam, rotarod, and in some cases cognitive deficits. A few of the later models also developed severe motor deficits similar to the JNPL3 mice.

Some of the tau pathology and behavior deficits are similar to what is seen in human wild-type tau transgenic mice; however, the mutations accelerate these phenotypes. As in the wild-type transgenic models, the levels of mutated tau overexpression increase the tau pathology and related behavior deficits as homozygous transgenic mice generally show an earlier onset of tau accumulation and motor deficits. All the above-mentioned models express tau protein under the regulation of nonhomologous promoters. It is therefore important to note that our laboratory has generated individual cDNA mutant transgenic mice with all of the above mutations (P301L, G272V, V337M, R406W) expressing human tau at 20–80% of endogenous mouse tau under the regulation of the human tau promoter, without reproducing any of the tauopathy (unpublished data). Thus, as suggested in the discussion of the wild-type transgenic models, it is very likely that tau protein is more physiologically expressed from the human tau promoter than from heterologous nontau promoters and it may be crucial for normal tau function to be expressed in the correct temporospatial pattern.

Although conventional wisdom would suggest that NFTs are primarily responsible for neuronal injury, a recent study in a conditional mouse model of tauopathy (rTg4510 mice) with the P301L mutation has suggested that NFT formation can be dissociated from memory loss and neurodegeneration (55). The rTg4510 mice normally express mutated human tau at approximately 15 times the endogenous mouse tau levels. However, when the expression of human tau protein is downregulated, behavioral impairments in these mice are reversed but the NFTs still remain. This observation suggests that less aggregated tau oligomers may be primarily responsible for neurotoxicity, rather than NFT formation. However, it is important to note that the Tg4510 mice with downregulated expression of P301L human tau still express approximately three times the level of endogenous mouse tau, which is a level of tau expression associated with NFT formation in the original JNPL3 mice.

2.4. Tau Intronic Mutation Models

Tau intronic mutations account for approximately half the mutations observed in FTDP-17 (9, 14, 56, 57). However, modeling FTD by intronic mutations that alter the inclusion of exon 10 and hence alter the 3R to 4R tau ratio is a more difficult undertaking than creating mice from cDNAs. The human tau gene is 140 plus kilobases, making direct mutagenesis difficult. For this reason, our laboratory generated a minigene where the alternatively spliced exons 2, 3, and 10 are surrounded by approximately 400–500 base pairs of endogenous intronic sequences (47). We introduced the N279K splicing mutation into the wild-type minigene and generated transgenic mice. Mice with the N279K mutation express predominantly 4R tau and develop tau accumulation, while control wild-type minigene mice develop no tau pathology. Neuronal death and swollen, beaded axons are observed by 1 year of age in these T-279 mice, and the mice develop progressive cognitive and motor deficits that lead to death. These mice express human tau at approximately 1% of endogenous mouse tau under the regulation of the human tau promoter, confirming that not only is aberrant splicing of exon 10 sufficient for neurodegeneration but that high levels of tau expression are not necessary for neurotoxicity. This mouse model may be particularly useful for discerning the mechanisms of other tauopathies. The tau variant of the *MAPT* gene, the H1 haplotype, has been implicated in such tauopathies as PSP, CBD, and Pick's disease (56–64). Since the majority of the polymorphisms associated with this haplotype are in noncoding regions, it is believed that the H1 haplotype alters the regulation of tau expression, either the total protein levels or exon splicing. This theory is supported by the fact that insoluble tau in several of these tauopathies almost exclusively includes exon 10.

3. Invertebrate Tau Models

Excellent models in invertebrates have contributed significantly to our knowledge of FTD tauopathy. To begin with, tau transgenic *Drosophila* (65) and *C.* elegans (65–67) have confirmed findings in the mouse models that both wild-type and mutated tau protein are sufficient to elicit tauopathy, although the mutated tau increases the severity of the pathology. Furthermore, work in *Drosophila* and *C.* elegans defined that tangle formation is not necessary for tau neurotoxicity. The generation of synaptic abnormalities without tangle formation in these models has led to new theories implicating low molecular weight tau oligomers, appearing prior to tangle formation, as the possible entities that elicit tau toxicity.

Furthermore, owing to the relative ease of genetic manipulation and the cost-effectiveness of *Drosophila*, several important pathways relevant to FTD neurodegeneration have been elucidated in this model. It has long been theorized that oxidative stress plays a role in tauopathies and neurodegeneration. Mel Feanny's (68) group was able to confirm that oxidative stress does play a crucial role in FTD. Tau toxicity and neuronal death was enhanced in the

Drosophila model by genetically downregulating antioxidant defense pathways in the R406W FTDP-17 mutant. Conversely, antioxidant treatment decreased tau-induced neurotoxicity in this model. Moreover, these experiments were able to activate the c-JunN-terminal kinase pathway as well as the cell cycle. This work linked oxidative stress to cell cycle activation and provided evidence for another long-held hypothesis which poses that tau pathology reactivates the cell cycle machinery and mitosis. Furthermore, these results implicated that this activation is most likely downstream of tau phosphorylation (69). Lastly, tau hyperphosphorylation has long been considered necessary for tau-induced neurotoxicity and multiple phosphorylation sites exist within the tau protein. Seminally, work in Drosophila has shown that no one site is sufficient to confirm neurotoxicity but that multiple phosphorylation sites work in concert to promote neurotoxicity (70).

Data obtained from tau transgenic mice, *Drosophila*, and *C.* elegans, have shed doubt on some of the entrenched tauopathy hypotheses. Tau has been identified to play an important role in intracellular trafficking, although its precise role in neurodegeneration remains an area of controversy. Going back as far as 2 decades, two major theories on the functionality of tau protein in neurodegeneration have been proposed. The first theory predicts that misfolded and aggregated tau protein causes a toxic gain of function by hindering normal axonal processes. This contention is supported by *in vitro* experiments demonstrating that mutations in the tau gene either increase the binding of tau to microtubules or increase aggregation of the mutant tau proteins. The second theory predicts that neurodegeneration is due to the loss of tau function caused either by a decrease in tau microtubule binding capabilities or by a decrease of the available pool of tau proteins as a result of aggregation and/or phosphorylation. It is becoming clear from FTD animal models that a combination of the two theories of tauopathy is most likely true and the mechanism of action may depend on the mode of tau dysregulation and/or mutation.

4. Progranulin Animal Models

To date, no transgenic animal models with progranulin mutations have been reported; however, overexpression of PGRN is associated with tumor genesis. In spite of theory that *PGRN* haplodeficiency is causal in neurodegeneration, clinical or histological pathology related to FTD was not noted in the initial publication describing the PGRN knockout mice (71). Further analysis of the PGRN knockout mouse model found no disease characterizing TDP-43 aggregates or disease characterizing carboxy-terminal fragments of TDP-43 (72).

5. TDP-43 Animal Models

Recently, several publications have begun to shed light on TDP-43 and neurodegeneration. It has been shown *in vitro* that human and *Drosophila* TDP-43 proteins that lack the C-terminal domain are unable to affect splicing (73). Furthermore, restricting nuclear-cytoplasmic trafficking of TDP-43 results in accumulation of TDP-43 in insoluble aggregates in cell culture (20, 74). Finally, loss of TDP-43 human cultured cell lines results in dysmorphic nuclear shape, misregulation of the cell cycle, and apoptosis by upregulation of cyclin-dependent kinase 6 (75). To date, TDP-43 models have been reported in zebrafish only. However, both the knockdown of endogenous TDP-43 and the overexpression of wild-type and mutated human TDP-43 did not induce cell death, aggregation, or premature mortality in the zebrafish models (72, 76).

6. Summary

The ultimate goal of developing animal models is to find cures for the FTD diseases. Also of importance is to note that the mechanism of pathological action most likely varies with mutations. For example, it has been shown that peptidylpropyl cis/trans isomerase NIMA-interacting 1 protein (PIN1) catalyzed isomerization can restore the biological function of phosphorylated tau by promoting tau dephosphorylation and PIN1-knockout mice develop tau-related pathologies (77). However, a recent paper by Lim et al. 2008 (78) showed that PIN1 has opposite effects on wild-type than on P301L mutated tau. PIN1 knock-down or knock-out increases tau pathology in wild-type mice, while decreasing tau pathology in P301L mice, and conversely, PIN1 overexpression drastically exacerbated tau-related phenotypes in P301L mice. It is therefore critical to develop relevant models for investigating treatments for various mutations of tauopathy and we cannot assume that treatments that work for one form of tauopathy will automatically translate to other diseases.

References

1. Ballatore C, Lee VM, Trojanowski JQ (2007) Tau-mediated neurodegeneration in Alzheimer's disease and related disorders. Nat Rev Neurosci 8(9):663–672
2. Pick A (1892) Über die Beziehungen der senilen Hirnatrophie zur Aphasie. Prager medicinische Wochenschrift 17:165–167
3. Alzheimer A (1911) Uber eigenartige Krankheitsfalle des spateren Alters. Z Gesamte Neurol Psychia 4:356–385
4. Kertesz A, McMonagle P, Blair M, Davidson W, Munoz DG (2005) The evolution and pathology of frontotemporal dementia. Brain 128(Pt 9):1996–2005

5. Neary D, Snowden JS, Gustafson L et al (1998) Frontotemporal lobar degeneration: a consensus on clinical diagnostic criteria. Neurology 51(6):1546–1554
6. Ratnavalli E, Brayne C, Dawson K, Hodges JR (2002) The prevalence of frontotemporal dementia. Neurology 58(11):1615–1621
7. Rosso SM, Donker KL, Baks T et al (2003) Frontotemporal dementia in The Netherlands: patient characteristics and prevalence estimates from a population-based study. Brain 126(Pt 9):2016–2022
8. Brun A (1987) Frontal lobe degeneration of non-Alzheimer type. I. Neuropathology. Arch Gerontol Geriatr 6(3):193–208
9. Hutton M, Lendon CL, Rizzu P et al (1998) Association of missense and 5'-splice-site mutations in tau with the inherited dementia FTDP-17. Nature 393(6686):702–705
10. Poorkaj P, Bird TD, Wijsman E et al (1998) Tau is a candidate gene for chromosome 17 frontotemporal dementia. Ann Neurol 43(6):815–825
11. Baker M, Mackenzie IR, Pickering-Brown SM et al (2006) Mutations in progranulin cause tau-negative frontotemporal dementia linked to chromosome 17. Nature 442(7105):916–919
12. Cruts M, Gijselinck I, van der ZJ et al (2006) Null mutations in progranulin cause ubiquitin-positive frontotemporal dementia linked to chromosome 17q21. Nature 442(7105): 920–924
13. Watts GD, Wymer J, Kovach MJ et al (2004) Inclusion body myopathy associated with Paget disease of bone and frontotemporal dementia is caused by mutant valosin-containing protein. Nat Genet 36(4):377–381
14. Foster NL, Wilhelmsen K, Sima AA, Jones MZ, D'Amato CJ, Gilman S (1997) Frontotemporal dementia and parkinsonism linked to chromosome 17: a consensus conference. Conference Participants. Ann Neurol 41(6):706–715
15. Spillantini MG, Murrell JR, Goedert M, Farlow MR, Klug A, Ghetti B (1998) Mutation in the tau gene in familial multiple system tauopathy with presenile dementia. Proc Natl Acad Sci USA 95(13):7737–7741
16. Alzheimer Disease & Frontotemporal Dementia Mutation Database (2009). 2-23-2009. Ref Type: Internet Communication
17. Mukherjee O, Pastor P, Cairns NJ et al (2006) HDDD2 is a familial frontotemporal lobar degeneration with ubiquitin-positive, tau-negative inclusions caused by a missense mutation in the signal peptide of progranulin. Ann Neurol 60(3):314–322
18. Gijselinck I, Van BC, Cruts M (2008) Granulin mutations associated with frontotemporal lobar degeneration and related disorders: an update. Hum Mutat 29(12):1373–1386
19. Neumann M, Sampathu DM, Kwong LK et al (2006) Ubiquitinated TDP-43 in frontotemporal lobar degeneration and amyotrophic lateral sclerosis. Science 314(5796):130–133
20. Winton MJ, Igaz LM, Wong MM, Kwong LK, Trojanowski JQ, Lee VM (2008) Disturbance of nuclear and cytoplasmic TAR DNA-binding protein (TDP-43) induces disease-like redistribution, sequestration, and aggregate formation. J Biol Chem 283(19): 13302–13309
21. Arai T, Hasegawa M, Akiyama H et al (2006) TDP-43 is a component of ubiquitin-positive tau-negative inclusions in frontotemporal lobar degeneration and amyotrophic lateral sclerosis. Biochem Biophys Res Commun 351(3):602–611
22. Gitcho MA, Baloh RH, Chakraverty S et al (2008) TDP-43 A315T mutation in familial motor neuron disease. Ann Neurol 63(4): 535–538
23. Kabashi E, Valdmanis PN, Dion P et al (2008) TARDBP mutations in individuals with sporadic and familial amyotrophic lateral sclerosis. Nat Genet 40(5):572–574
24. Yokoseki A, Shiga A, Tan CF et al (2008) TDP-43 mutation in familial amyotrophic lateral sclerosis. Ann Neurol 63(4):538–542
25. Forman MS, Mackenzie IR, Cairns NJ et al (2006) Novel ubiquitin neuropathology in frontotemporal dementia with valosin-containing protein gene mutations. J Neuropathol Exp Neurol 65(6):571–581
26. Neumann M, Mackenzie IR, Cairns NJ et al (2007) TDP-43 in the ubiquitin pathology of frontotemporal dementia with VCP gene mutations. J Neuropathol Exp Neurol 66(2): 152–157
27. Kimonis VE, Kovach MJ, Waggoner B et al (2000) Clinical and molecular studies in a unique family with autosomal dominant limb-girdle muscular dystrophy and Paget disease of bone. Genet Med 2(4):232–241
28. Kovach MJ, Waggoner B, Leal SM et al (2001) Clinical delineation and localization to chromosome 9p13.3-p12 of a unique dominant disorder in four families: hereditary inclusion body myopathy, Paget disease of bone, and frontotemporal dementia. Mol Genet Metab 74(4):458–475
29. Binder LI, Frankfurter A, Rebhun LI (1985) The distribution of tau in the mammalian central nervous system. J Cell Biol 101(4): 1371–1378

30. Caceres A, Kosik KS (1990) Inhibition of neurite polarity by tau antisense oligonucleotides in primary cerebellar neurons. Nature 343(6257):461–463
31. Drubin DG, Caput D, Kirschner MW (1984) Studies on the expression of the microtubule-associated protein, tau, during mouse brain development, with newly isolated complementary DNA probes. J Cell Biol 98(3): 1090–1097
32. Ebneth A, Godemann R, Stamer K, Illenberger S, Trinczek B, Mandelkow E (1998) Overexpression of tau protein inhibits kinesin-dependent trafficking of vesicles, mitochondria, and endoplasmic reticulum: implications for Alzheimer's disease. J Cell Biol 143(3): 777–794
33. Stamer K, Vogel R, Thies E, Mandelkow E, Mandelkow EM (2002) Tau blocks traffic of organelles, neurofilaments, and APP vesicles in neurons and enhances oxidative stress. J Cell Biol 156(6):1051–1063
34. Weingarten MD, Lockwood AH, Hwo SY, Kirschner MW (1975) A protein factor essential for microtubule assembly. Proc Natl Acad Sci U S A 72(5):1858–1862
35. Goedert M, Spillantini MG, Potier MC, Ulrich J, Crowther RA (1989) Cloning and sequencing of the cDNA encoding an isoform of microtubule-associated protein tau containing four tandem repeats: differential expression of tau protein mRNAs in human brain. EMBO J 8(2):393–399
36. Himmler A, Drechsel D, Kirschner MW, Martin DW Jr (1989) Tau consists of a set of proteins with repeated C-terminal microtubule-binding domains and variable N-terminal domains. Mol Cell Biol 9(4):1381–1388
37. Himmler A (1989) Structure of the bovine tau gene: alternatively spliced transcripts generate a protein family. Mol Cell Biol 9(4): 1389–1396
38. Goedert M, Jakes R (1990) Expression of separate isoforms of human tau protein: correlation with the tau pattern in brain and effects on tubulin polymerization. EMBO J 9(13):4225–4230
39. Shahani N, Brandt R (2002) Functions and malfunctions of the tau proteins. Cell Mol Life Sci 59(10):1668–1680
40. Gotz J, Probst A, Spillantini MG et al (1995) Somatodendritic localization and hyperphosphorylation of tau protein in transgenic mice expressing the longest human brain tau isoform. EMBO J 14(7):1304–1313.
41. Spittaels K, Van den HC, Van DJ et al (1999) Prominent axonopathy in the brain and spinal cord of transgenic mice overexpressing four-repeat human tau protein. Am J Pathol 155(6): 2153–2165
42. Probst A, Gotz J, Wiederhold KH et al (2000) Axonopathy and amyotrophy in mice transgenic for human four-repeat tau protein. Acta Neuropathol (Berl) 99(5):469–481
43. Brion JP, Tremp G, Octave JN (1999) Transgenic expression of the shortest human tau affects its compartmentalization and its phosphorylation as in the pretangle stage of Alzheimer's disease. Am J Pathol 154(1): 255–270
44. Ishihara T, Hong M, Zhang B et al (1999) Age-dependent emergence and progression of a tauopathy in transgenic mice overexpressing the shortest human tau isoform. Neuron 24(3):751–762
45. Duff K, Knight H, Refolo LM et al (2000) Characterization of pathology in transgenic mice over-expressing human genomic and cDNA tau transgenes. Neurobiol Dis 7(2):87–98
46. Dawson HN, Ferreira A, Eyster MV, Ghoshal N, Binder LI, Vitek MP (2001) Inhibition of neuronal maturation in primary hippocampal neurons from tau deficient mice. J Cell Sci 114(Pt 6):1179–1187
47. Dawson HN, Cantillana V, Chen L, Vitek MP (2007) The tau N279K exon 10 splicing mutation recapitulates frontotemporal dementia and parkinsonism linked to chromosome 17 tauopathy in a mouse model. J Neurosci 27(34):9155–9168
48. Yoshiyama Y, Lee VM, Trojanowski JQ (2001) Frontotemporal dementia and tauopathy. Curr Neurol Neurosci Rep 1(5):413–421
49. Lewis J, McGowan E, Rockwood J et al (2000) Neurofibrillary tangles, amyotrophy and progressive motor disturbance in mice expressing mutant (P301L) tau protein. Nat Genet 25(4):402–405
50. Gotz J, Chen F, Barmettler R, Nitsch RM (2001) Tau filament formation in transgenic mice expressing P301L tau. J Biol Chem 276(1):529–534
51. Gotz J, Tolnay M, Barmettler R, Chen F, Probst A, Nitsch RM (2001) Oligodendroglial tau filament formation in transgenic mice expressing G272V tau. Eur J Neurosci 13(11):2131–2140
52. Allen B, Ingram E, Takao M et al (2002) Abundant tau filaments and nonapoptotic neurodegeneration in transgenic mice expressing human P301S tau protein. J Neurosci 22(21):9340–9351
53. Tanemura K, Murayama M, Akagi T et al (2002) Neurodegeneration with tau accumulation in a transgenic mouse expressing

V337M human tau. J Neurosci 22(1): 133–141
54. Tatebayashi Y, Miyasaka T, Chui DH et al (2002) Tau filament formation and associative memory deficit in aged mice expressing mutant (R406W) human tau. Proc Natl Acad Sci U S A 99(21):13896–13901
55. SantaCruz K, Lewis J, Spires T et al (2005) Tau suppression in a neurodegenerative mouse model improves memory function. Science 309(5733):476–481
56. Rademakers R, Cruts M, Van BC (2004) The role of tau (MAPT) in frontotemporal dementia and related tauopathies. Hum Mutat 24(4):277–295
57. Reed LA, Wszolek ZK, Hutton M (2001) Phenotypic correlations in FTDP-17. Neurobiol Aging 22(1):89–107
58. Buee L, Bussiere T, Buee-Scherrer V, Delacourte A, Hof PR (2000) Tau protein isoforms, phosphorylation and role in neurodegenerative disorders. Brain Res Brain Res Rev 33(1):95–130
59. Forman MS, Trojanowski JQ, Lee VM (2004) Neurodegenerative diseases: a decade of discoveries paves the way for therapeutic breakthroughs. Nat Med 10(10):1055–1063
60. Goedert M, Hasegawa M (1999) The tauopathies: toward an experimental animal model. Am J Pathol 154(1):1–6
61. Katsuse O, Iseki E, Arai T et al (2003) 4-repeat tauopathy sharing pathological and biochemical features of corticobasal degeneration and progressive supranuclear palsy. Acta Neuropathol (Berl) 106(3):251–260
62. Morris HR, Osaki Y, Holton J et al (2003) Tau exon 10 +16 mutation FTDP-17 presenting clinically as sporadic young onset PSP. Neurology 61(1):102–104
63. Sergeant N, Wattez A, Delacourte A (1999) Neurofibrillary degeneration in progressive supranuclear palsy and corticobasal degeneration: tau pathologies with exclusively "exon 10" isoforms. J Neurochem 72(3):1243–1249
64. Togo T, Sahara N, Yen SH et al (2002) Argyrophilic grain disease is a sporadic 4-repeat tauopathy. J Neuropathol Exp Neurol 61(6):547–556
65. Wittmann CW, Wszolek MF, Shulman JM et al (2001) Tauopathy in Drosophila: neurodegeneration without neurofibrillary tangles. Science 293(5530):711–714
66. Kraemer BC, Zhang B, Leverenz JB, Thomas JH, Trojanowski JQ, Schellenberg GD (2003) Neurodegeneration and defective neurotransmission in a Caenorhabditis elegans model of tauopathy. Proc Natl Acad Sci USA 100(17):9980–9985
67. Miyasaka T, Ding Z, Gengyo-Ando K et al (2005) Progressive neurodegeneration in C. elegans model of tauopathy. Neurobiol Dis 20(2):372–383
68. Dias-Santagata D, Fulga TA, Duttaroy A, Feany MB (2007) Oxidative stress mediates tau-induced neurodegeneration in Drosophila. J Clin Invest 117(1):236–245
69. Illenberger S, Zheng-Fischhofer Q, Preuss U et al (1998) The endogenous and cell cycle-dependent phosphorylation of tau protein in living cells: implications for Alzheimer's disease. Mol Biol Cell 9(6):1495–1512
70. Steinhilb ML, as-Santagata D, Fulga TA, Felch DL, Feany MB (2007) Tau phosphorylation sites work in concert to promote neurotoxicity in vivo. Mol Biol Cell 18(12): 5060–5068
71. Kayasuga Y, Chiba S, Suzuki M et al (2007) Alteration of behavioural phenotype in mice by targeted disruption of the progranulin gene. Behav Brain Res 185(2):110–118
72. Haass C (2009) Proteolytic Processing and Aggregatopm of TAR DNA Binding Protein-43 After Caspase Activation. Alzheimer's & Parkinson's Diseases: 9th International Conference 2009, Prague Czech Republic, 2009
73. Ayala YM, Pantano S, D'Ambrogio A et al (2005) Human, Drosophila, and C.elegans TDP43: nucleic acid binding properties and splicing regulatory function. J Mol Biol 348(3):575–588
74. Winton MJ, Van DV, Kwong LK et al (2008) A90V TDP-43 variant results in the aberrant localization of TDP-43 in vitro. FEBS Lett 582(15):2252–2256
75. Ayala YM, Misteli T, Baralle FE (2008) TDP-43 regulates retinoblastoma protein phosphorylation through the repression of cyclin-dependent kinase 6 expression. Proc Natl Acad Sci USA 105(10):3785–3789
76. Modeling Aspects of Frontotemporal Dementia in Zebrafish. Alzheimer's & Parkinson's Diseases: 9th International Conference 2009, Prague, Czech Republic; 2009.
77. Lu KP, Zhou XZ (2007) The prolyl isomerase PIN1: a pivotal new twist in phosphorylation signalling and disease. Nat Rev Mol Cell Biol 8(11):904–916
78. Lim J, Balastik M, Lee TH et al (2008) Pin1 has opposite effects on wild-type and P301L tau stability and tauopathy. J Clin Invest 118(5):1877–1889

Part VI

Animal Models of Vascular Dementia

Chapter 29

CADASIL: Molecular Mechanisms and Animal Models

Karl J. Fryxell

Abstract

Cerebral autosomal dominant arteriopathy with subcortical infarcts and leukoencephalopathy (CADASIL) is a human genetic syndrome that causes multiple small strokes, due to a single-gene, autosomal dominant mutation with 100% penetrance. CADASIL mutations encode amino acid substitutions that increase or decrease the number of cysteines within the extracellular epidermal growth factor (EGF) repeat domain of the NOTCH3 receptor protein. Histological studies of CADASIL patients have shown a characteristic accumulation of granular osmiophilic material surrounding the arterial smooth muscle cells, and the degeneration and death of some of these smooth muscle cells. The adjacent endothelial cells appear normal. The result is occasional occlusion of small arteries, producing small infarcts throughout the body, particularly in subcortical white matter in the brain in persons over 50 years of age. Several experimental models related to CADASIL have been investigated, including transgenic mice, knockout mice, cell line expression systems, and *Drosophila* mutants. None of these models reproduce all the characteristics of human CADASIL, but taken together they have provided an increasingly detailed picture. It is clear that continuing NOTCH3 signaling is required in adult mammalian arterial smooth muscle cells to maintain their survival, differentiation, and normal responses to injury and mechanical stress, particularly in the smaller arteries. In these respects, the characteristics of CADASIL are consistent with a partial loss of NOTCH3 signaling. However, attempts to confirm a loss of NOTCH3 signaling by CADASIL alleles in cell culture have led to mixed results. CADASIL patients do consistently exhibit extracellular accumulation of granular osmiophilic material that contains the extracellular (but not the intracellular) domain of the NOTCH3 receptor. Moreover, the accumulation of the extracellular domain of NOTCH3 has been reproduced in some transgenic mice, as well as in a cell culture model. This gradual accumulation of extracellular aggregates of the NOTCH3 extracellular domain may interfere with NOTCH signaling and may help to explain the characteristically late onset of symptoms of CADASIL. In any case, the well-defined cellular and molecular defects in CADASIL provide a promising area for further research, and the location of the affected cell type could facilitate future drug treatments for this disease.

Key words: CADASIL, NOTCH3, Stroke

1. Introduction

Stroke is the third most common cause of death and the leading cause of long-term neurological disability in the world. Preventable (lifestyle-related) vascular risk factors for stroke (such as high blood

Peter Paul De Deyn and Debby Van Dam (eds.), *Animal Models of Dementia*, Neuromethods, vol. 48,
DOI 10.1007/978-1-60761-898-0_29, © Springer Science+Business Media, LLC 2011

pressure, atherosclerosis, obesity, and cigarette smoking) account for roughly 50% of the total risk for stroke (1). Genetic factors are believed to account for a significant proportion of the remaining stroke risk, in most cases through complex interactions of multiple genes, age, and environmental factors. Monogenic (single-gene) disorders account for approximately 1% of all ischemic strokes (1). Although these monogenic conditions are relatively rare, they provide valuable models for elucidating the biology of stroke and cerebrovascular disease, as well as the accompanying symptoms such as vascular dementia. The genes responsible for a number of inherited vascular diseases have been identified in recent years (2).

Cerebral autosomal dominant arteriopathy with subcortical infarcts and leukoencephalopathy (CADASIL) is the best known and probably the most common of the monogenic disorders that cause stroke and vascular dementia (3). CADASIL alleles are genetically dominant and have a penetrance of 100% (i.e., all individuals carrying one or more copies of the mutant allele will eventually exhibit this syndrome). Moreover, the molecular nature of the lesions within the NOTCH3 gene (4) are equally well defined – virtually all the mutations that cause CADASIL in humans result in an odd number of cysteine amino acids within the epidermal growth factor (EGF) repeats in the extracellular domain of the NOTCH3 receptor protein, possibly causing mispairing of disulfide bonds (5). An odd number of cysteines in the extracellular EGF repeats also cause a unique phenotype (not caused by other amino acid substitutions) in *Drosophila Notch* (6).

The high penetrance of the mutations that cause CADASIL, and their well-defined molecular basis, suggest that the study of this disorder has advanced to the point where animal models are appropriate (6). However, recent results from these animal models have been mixed at best. To put these results into perspective, it will be necessary for us to first briefly review what is known about the clinical characteristics of CADASIL, the molecular mechanisms of Notch signaling, and the role of this Notch signaling in the development and maintenance of blood vessels.

1.1. Clinical Characteristics of CADASIL: Pathology and Prognosis

CADASIL is a degenerative vascular disease that typically first causes noticeable symptoms in middle-aged individuals. Affected individuals exhibit a variety of symptoms, including migraines, mood disorders, recurrent subcortical ischemic strokes, progressive cognitive decline, dementia, and premature death. The histological characteristics of CADASIL include the progressive degeneration of smooth muscle cells (SMCs) in small arteries, the narrowing of small arteries, the accumulation of granular osmiophilic material (GOM) within the SMC basement membrane, and small, localized lesions throughout the body, particularly in the white matter of the brain, that are apparently due to multiple, small lacunar infarcts (3, 7). GOM accumulation in arterial SMC is one of the most distinguishing features of CADASIL, in small arteries throughout the body (7, 8). GOM accumulation in CADASIL patients does occur in larger arteries, although to a lesser extent (9). GOM accumulation in CADASIL patients has never been reported in veins. Another distinguishing feature of CADASIL is lacunar infarcts in the anterior temporal white matter and external capsule. Strokes involving the territory of a large artery have occasionally been reported in CADASIL patients, but are probably coincidental (10).

CADASIL is a relatively rare condition, but the number of published reports has been increasing in recent decades. The prevalence of CADASIL mutation carriers (including those who are asymptomatic at present) has been estimated to be 4 per 100,000 adults in Western Scotland (11). In most areas, CADASIL is underreported, due to the time, effort, expense, and multiple areas of expertise required to establish a definitive diagnosis. For example, a false-negative report of no family history of relevant symptoms is often incorrectly obtained on the initial patient interview (12). In addition, an initial observation of widespread lesions of white matter is often misdiagnosed as multiple sclerosis (13, 14). Clinical DNA sequencing, if undertaken at all, is often limited to exons 3 and 4 of the NOTCH3 gene, where the first mutations in CADASIL were found (5, 15). However, mutations that cause CADASIL have been found throughout the EGF repeat domain, that is in exons 2–24, corresponding to EGF repeats 1–34 of the NOTCH3 gene (8, 16–25). Another reason that CADASIL is underreported is that patients with CADASIL present with a wide variety of initial symptoms and diagnoses, including acute encephalopathy (26, 27), adjustment disorder (28), affective disorders (29), angiitis (30), bipolar disorder (31), cognitive impairment (32), coma (27), depression (32), dizziness and vertigo (33, 34), facial dystonia (35), impaired visual physiology (36, 37), intermittent neurological symptoms (38), migraine headaches with aura (28, 29, 32), multiple sclerosis (39), Parkinson's disorder (40), postpartum psychiatric disturbance (41), presenile dementia (42), schizophrenia (43), seizures (32), subcortical strokes (32, 42), and sudden hearing loss (44).

The age at onset of the first CADASIL symptoms has varied from 26 to 70 years of age (28, 34). In most cases, the age at first diagnosed stroke is in the 40s (45) or 50s (46). Relatively few individuals with CADASIL experience their first stroke in the 20s or 30s (47). After the first observable stroke, affected individuals show a progressive decline in cognitive (48) and motor functions (28, 46), including periods of chronic progression, acute worsening, stabilization, and improvement (49). Most (78%) CADASIL patients eventually become completely unable to care for themselves, but they can live into their 60s or 70s if they are given sufficient care (45, 46). The life expectancy of CADASIL patients varies between families and nationalities, but is consistently below the average for their national region.

CADASIL is genetically dominant and has a penetrance of 100%; in other words, all individuals carrying the mutant gene eventually exhibit the disease and its progression. Nevertheless, there is variation in symptoms, and some of this variation is correlated within families. For example, the size and progression of CADASIL brain lesions are correlated within families, leading to an estimated heritability of 0.6–0.7 (50). Linear regression analysis confirmed that these variations in T2 lesion volume were age-dependent, but did not appear to be accounted for by the specific identity of the NOTCH3 mutant allele, suggesting that other genes modify the severity of the CADASIL phenotype (50). Consistent with this possibility, anecdotal reports suggest that vertigo may be experienced more often by CADASIL patients in Asia, while migraine may be experienced more often by CADASIL patients in Europe (34). A few particular CADASIL alleles may increase the likelihood of lesions of the corpus callosum (51), or increase the rate of disease progression (46). Nevertheless, large CADASIL families often include both affected individuals with frequent migraine headaches, as well as other CADASIL patients without migraine headaches. The psychiatric manifestations of CADASIL families also frequently vary within families, rather than between families (28).

CADASIL patients are generally heterozygotes, but a single exception to this rule is known – a 54-year-old homozygous CADASIL patient (47). Comparison to age-matched CADASIL patients (all carrying the same R133C allele as the homozygous patient) showed that the homozygous patient had the earliest age at first stroke (compared to nine heterozygotes), was the most severely affected on most neuropsychological tests (compared to eight heterozygotes), had the most severe findings by MRI (compared to seven heterozygotes), and had the greatest accumulation of granular osmiophilic material (GOM) (compared to one heterozygous patient). The homozygous patient also had two heterozygous sons whose age at first stroke was close to their father's (47). Moreover, one of the heterozygous patients showed more rapid progression than the homozygous patient, and was so severely affected that he could not be included in the neuropsychological or MRI tests. In other words, the symptoms showed by the homozygous CADASIL patient were within the range seen in CADASIL heterozygotes, although near the severe end of the scale (47).

CADASIL patients have significantly reduced middle cerebral artery basal blood flow velocity, as well as significantly reduced cerebrovascular CO_2 reactivity. Both these arterial defects become more severe with age and/or disease progression (52).

1.2. Genetic Nomenclature

As mentioned above, "CADASIL" is not a gene name; rather it is an acronym for the name of a genetic disease caused by certain mutations in the human NOTCH3 gene. This review includes similar mutations in human, rat, mouse, fruit fly, and cell culture studies, each of which has its own system of genetic nomenclature (53). Accordingly, specific references to human genes and proteins will be shown in all capital (Roman) letters, but gene names in other species will be italicized, and the corresponding names of specific proteins in those species will be shown in all capital (Roman) letters. General references to genetic pathways and signaling mechanisms will be shown in conventional text (neither italicized nor all capital letters).

2. Molecular Mechanisms of Notch Signaling

The Notch signaling pathway is a short-range channel of communication between adjacent cells that is involved in many fundamental aspects of multicellular life, including proliferation, stem cells and stem cell niche maintenance, cell fate acquisition, differentiation, cell death, and even the consolidation of long-term memory (54–60). Notch signaling is highly conserved throughout the animal kingdom. Both the Notch receptor and its ligands (known as Jagged in mammals) are transmembrane proteins with large extracellular domains, consisting primarily of EGF-like repeats. In mammals, the Notch receptor is processed during protein maturation by protease cleavage (also known as S1 cleavage), and is normally located at the cell surface as a heterodimer (one subunit contains a transmembrane segment). In other words, the extracellular domain associates noncovalently with the membrane-tethered intracellular domain (the latter includes a small extracellular stub). The extracellular domain of the Notch receptor usually contains 36 EGF-repeats, as well as three cysteine-rich Notch/LIN-12 repeats (LNG domain). EGF repeats are a rather common sequence motif in proteins and often bind to calcium ions, which may be necessary for the proper folding of the adjacent EGF repeat (61).

The EGF repeats in the Notch extracellular domain are post-translationally modified by an O-fucosyltransferase (OFUT1) that attaches fucose specifically to certain serine and threonine amino acids that are required for Notch signaling (62). These fucose side chains are then extended by the "Fringe" family of enzymes, which have a fucose-specific β1,3 N-acetylglucosaminyl transferase activity and initiate elongation of the O-linked fucose during Notch processing in the Golgi apparatus (63–65). Both OFUT1 and Fringe are capable of producing both positive and negative effects on Notch signaling (62, 64). Fringe binds to Notch through the EGF repeats 22–36, as well as the LIN-12 repeats (65). Extension of fucose side chains by Fringe modifications modulates the preference of the Notch receptor for its ligands Delta and Jagged/Serrate (66–69).

In mammals, there are four Notch receptor genes (*Notch1*, *Notch2*, *Notch3*, and *Notch4*), all of which encode single-pass transmembrane proteins. These interact with a few ligands, which are also single-pass transmembrane proteins: JAGGED1, JAGGED2, DELTA-LIKE1, DELTA-LIKE3, and DELTA-LIKE4. Notch receptor proteins are located at the cell surface as a heterodimer due to proteolytic cleavage at the S1 site, as mentioned previously. Ligand binding leads to a second proteolytic cleavage of the Notch receptor at the S2 site (on the extracellular surface of the transmembrane segment). S2 cleavage is followed by endocytosis of the Notch extracellular domain, still bound to its protein ligand (70). This endocytosis of the Notch receptor and its ligands is regulated at multiple levels and is essential for both receptor/ligand activation and subsequent degradation (71–74). The membrane-tethered intracellular domain then becomes susceptible to S3 cleavage within the transmembrane segment, releasing the intracellular domain. The freed intracellular domain binds to the CSL (CBF1/Su(H)/LAG1) family of transcription factors (called CBF1/RBP-Jκ in mammals), along with its coactivator Mastermind, translocates to the nucleus, and regulates the transcription of target genes (54, 74–77). In most (perhaps all) cases, all four Notch receptors appear to activate this same intracellular transcription factor complex.

2.1. The Role of Notch Signaling in the Development of Blood Vessels

Genetic studies in vertebrate model systems have revealed a signaling cascade that is responsible for determining arterial and venous cell fates in embryonic blood vessels. This cascade is initiated by vascular endothelial growth factor A (*Vegfa*) signaling (78), which in turn induces signaling through the *Notch1* pathway (79, 80). The Notch and Vegf pathways also regulate blood vessel sprouting (81). Within the developing dorsal aorta, the loss of Notch signaling leads to loss of artery-specific molecular markers, as well as abnormal expression of venous markers (82). Conversely, the ectopic activation of Notch signaling leads to repression of venous cell fate (82). *Notch1* is normally expressed in arteries but not in veins (83). Moreover, *Notch3* and its ligands *Jagged1* and *Jagged2* are expressed in arterial SMCs (but not in veins), while *Notch4* and *Delta4* are expressed in arterial endothelial cells (but not in veins) (83–86). Adult vascular phenotypes can also be produced by gain-of-function *Notch4* alleles (87).

Notch2 was not detectable in the circulatory system. Only *Notch4* and *Delta4* were detectable in capillaries (83).

After the initial development of the blood vessels, the SMCs within those blood vessels remain able to undergo significant changes in phenotype in response to cell signaling events and changes in local environmental conditions (88). Several different *Notch* genes are involved in these cell signaling events, and these *Notch* genes appear to have somewhat different roles. For example, *Notch1* (87, 89–91) and *Notch4* (90, 92, 93) play essential roles in embryonic angiogenesis and vascular remodeling, while *Notch3* functions primarily in the postnatal development and plasticity of SMCs. Taken together, these results indicate that Notch signaling in the circulatory system is required for the normal development and maintenance of arteries and capillaries, but not veins.

In normal mice, *Notch3* is expressed postnatally, in arteries but not in veins, and in SMCs, but not endothelial cells (87, 94, 95). In mouse *Notch3* −/− knockouts, the embryonic stages of vascular development appear to be normal, and these mice develop into viable and fertile adults (96, 97). Mouse *Notch3* −/− knockouts do have defects in the postnatal maturation of arterial SMCs (87, 95), and possibly reduced numbers of thymocytes in the immune system (98). Several lines of evidence (morphological, developmental, gene expression studies) indicate that arterial SMCs in mouse *Notch3* −/− knockouts adopt a venous SMC identity (87, 95). For example, adolescent and adult *Notch3* −/− mutant mice exhibited pronounced defects in arterial maturation that were not evident at birth, including enlarged arteries with a relatively thin and disorganized arterial SMC layer. The abnormal SMC layer was not due to alterations in cell proliferation or cell death, as shown by assays of the percentage of presumptive arterial SMCs that were entering mitosis or were labeled for apoptotic markers. Rather, presumptive arterial SMCs appear to have adopted the identity of normal vein SMCs, based on their morphology as well as their failure to express artery-specific molecular markers. In contrast, the endothelial cells in *Notch3* −/− arteries express normal levels of several artery-specific endothelial cell markers, indicating that the arterial identity of the endothelial cells remains normal in *Notch3* −/− knockouts. In spite of the apparently complete transformation of the identity of their arterial SMCs, *Notch3* −/− mice did not show any obvious brain pathology on histological examination by 12 months of age. Later ages were not examined (95).

Additional observations suggest that the effects of *Notch3* differ between small arteries, *vs.* medium-sized or large arteries (95, 99). The *SM22α* gene is normally expressed in all SMCs, but arterial-specific regulatory elements have been identified within the *SM22α* promoter, which were used to construct an arterial-specific promoter driving the expression of a *LacZ* reporter gene (100). In the large and medium-sized arteries of normal mice, the expression of this *SM22α-LacZ* reporter gene reached a peak at postnatal day 6 (P6), and remained at a high level thereafter. In the small arteries of normal mice, expression of the *SM22α-LacZ* reporter gene was barely detectable at P6, but increased later to reach its peak value at P28, in adolescent mice (99, 101). In the large and medium-sized arteries of *Notch3* −/− knockout mice, the expression of *SM22α-LacZ* was normal until P6, after which expression declined. In the small arteries of *Notch3* −/− knockout mice, *SM22α-LacZ* was not expressed at any age (99). In other words, the maturation of arterial SMCs in larger arteries after P6, and in smaller arteries at all ages, required NOTCH3 signaling. Taken together, these results from mouse genetics indicate that NOTCH3 signaling is: (i) required in arteries but not in veins, (ii) required in SMCs but not in endothelial cells, (iii) required for postnatal maturation of SMCs but not for prenatal development of SMCs, and (iv) has a greater effect on small arteries than on larger arteries. These characteristics are remarkably similar to the symptoms of CADASIL, which suggests (but does not necessarily prove) that CADASIL may be caused by (a) defect(s) in NOTCH3 signaling.

2.2. The Role of Notch Signaling in the Response to Vascular Injury

The neointima is a thickened layer of vascular SMCs that forms after vascular injury, presumably by proliferation of SMCs. This formation of the neointima was significantly decreased in *Hey2* −/− knockout mice, suggesting that Notch signaling was required for normal formation of the neointima (102). Consistent with this view, primary aortic SMCs from *Hey2* −/− knockout mice were found to proliferate at a reduced rate in cell culture (102), and expression of activated forms of *Notch1* or *Notch3* enhanced the proliferation of adult rat vascular SMCs in culture (103). Activated forms of *Notch1* or *Notch3* also inhibited apoptosis and cell migration (103). The effects of activated *Notch3* on SMC proliferation depended on the details of cell culture conditions (84).

The arterial expression of several genes in the Notch pathway, including *Notch1* and *Notch3*, as well as Notch ligands (*Jag1* and *Jag2*) and downstream transcription factors (*Hey1* and *Hey2*) are downregulated within the first 2 days after experimental vascular injury, but are later upregulated 7–14 days after injury (84, 104–106). During the downregulation of NOTCH1 and NOTCH3 receptor expression by mechanical strain *in vitro*, SMC proliferation was reduced and apoptosis increased (107, 108). Conversely, genetically forcing overexpression of the NOTCH1 or NOTCH3 intracellular domain (during mechanical strain in vascular SMCs *in vitro*) increased proliferation and decreased apoptosis (107, 108). Taken together with the developmental studies (see above), these observations suggest that

Notch signaling is required to maintain the normal ability of arterial SMCs to withstand the mechanical strains produced by high and variable arterial pressures.

3. The Molecular Details: CADASIL Mutations and Granular Osmiophilic Material

Almost all the mutations that cause CADASIL in humans result in the gain or loss of a cysteine residue in one of the 34 EGF-like repeats in the extracellular domain of the NOTCH3 receptor (5, 109–116). In a carefully controlled study, 125 patients whose CADASIL diagnosis was confirmed by ultrastructural analysis of skin biopsy specimens were subjected to sequencing of NOTCH3 exons 3–4, and then (if no mutation was found) to further sequencing of exons 2–24. The results showed that patients with a positive skin biopsy had mutations that caused an odd number of cysteines in the EGF repeat domain in 120/125 cases; in the other five cases, no mutation was found. Of the mutations in the EGF repeat domain, 69% were located in exons 3–4 and 86% were located in exons 2–6 (8). Two of the mutations in exon 4 were associated with significantly faster disease progression, but four other mutations in exon 4 were not (46).

Very rarely, patients with some of the characteristics of CADASIL have been found to have amino acid substitutions in the NOTCH3 gene that do not change the number of cysteines in the EGF repeat domain (113, 117, 118). However, these may be polymorphisms unrelated to the disease condition. There is no published case in which these patients have been subjected to a detailed pedigree analysis to show cosegregation of a disease condition with the amino acid substitution. Therefore, the pathology in these cases may be caused by other genes, or even nongenetic factors. It is known that mutations in other genes can cause small vessel angiopathies (119). Moreover, it is also clear that amino acid sequence polymorphisms in the NOTCH3 gene are present in the human population that do not contribute to cerebrovascular disease (120).

The strikingly nonrandom nature of the mutations that cause CADASIL, and particularly the absence in CADASIL patients of frameshift mutations or deletions of the NOTCH3 gene, strongly suggests that the mutations that cause CADASIL are not simply NOTCH3-null alleles (5, 6, 121). A similar argument has been advanced with respect to cysteine mutations in the EGF repeat domain of the *Notch* gene in *Drosophila*, where more powerful genetic tools are available (see next section).

In electron microscope studies, the small arteries of CADASIL patients show extracellular aggregates of GOM in the vicinity of arterial SMCs (122). GOM are typically rounded profiles about 0.1–0.5 μm in diameter (94, 123), are often located in proximity to the basement membrane, and are sometimes located adjacent to the extracellular surface of the SMC plasma membrane, or even in isolation (i.e., within the artery, but not close to any particular structure) (94). It has been shown by immunohistochemical methods that the extracellular domain of the NOTCH3 protein is present in substantial amounts in these GOM aggregates, but the intracellular domain of the NOTCH3 protein is not present (94, 123). In experiments based on conventional embedding methods, the NOTCH3 extracellular domain appeared to be confined to surfaces of the GOM adjacent to SMC plasma membranes (94). However, postembedding immunogold electron microscopy with cryofixed and freeze-substituted biopsies (which allows better antibody access to the interior of molecular aggregates) has shown convincingly that the NOTCH3 extracellular domain is actually uniformly distributed throughout the interior of the GOM (123).

When biopsies of CADASIL arteries were analyzed on SDS polyacrylamide gels under reducing conditions, the results showed selective (and abnormal) accumulation of the extracellular domain of NOTCH3 (210 kDa), which was present in high levels. The 280 kDa NOTCH3 precursor and the 97 kDa intracellular domain showed no evidence of accumulation in these CADASIL biopsies. Under nonreducing conditions, the accumulation of NOTCH3 extracellular domains in CADASIL biopsies migrate in higher molecular weight form(s), suggesting the formation of abnormal disulfide bridges (94). Moreover, similar high molecular weight forms were also observed in a cell culture model of CADASIL, in which amino acid substitutions causing the gain or loss of a cysteine in a peptide portion of the extracellular domain of NOTCH3 were shown to be both necessary and sufficient to form abnormal, cross-linked homodimers of the NOTCH3 extracellular domain, as well as heterodimers of the NOTCH3 extracellular domain with an enzyme named "Lunatic Fringe"(25). These results suggest that Lunatic Fringe may normally have an unpaired cysteine, or alternatively that the unpaired cysteine in CADASIL may disrupt disulfide bond formation in adjacent proteins (25, 94). In any case, all the cysteine mutations tested in cell culture showed small but consistent decreases in the elongation of fucose side chains by Fringe (25). These Fringe modifications are known to modulate the preference of the Notch receptor for its ligands Delta and Jagged/Serrate (66–69).

These results raise several possible (although not mutually exclusive) scenarios. One possible scenario is that aberrant dimerization of the NOTCH3 extracellular domain may interfere with its normal reuptake, leading to the gradual accumulation of extracellular aggregates (GOM) that are characteristic of CADASIL and may interfere with binding to potential Notch ligands (25, 94, 123). Another possible scenario is that cross-linking of the NOTCH3 extracellular domain to enzymes in the Fringe family may inhibit the latter enzymes, which in turn are known to modulate the binding specificity of Notch for its various ligands (74). A third possible scenario is suggested by the observation that Notch signaling and Lunatic Fringe gene expression are coupled together, in an oscillatory negative feedback loop that plays key roles in mammalian development (124). Feedback between NOTCH3 signaling and Lunatic Fringe enzyme activity could gradually magnify the abnormalities in older CADASIL patients, perhaps leading to age-dependent pathology that is unrelated to arteriosclerosis.

4. *Drosophila* Models: Cysteine Mutations in the Extracellular EGF Repeat Domain of *Notch*

Drosophila has a single *Notch* gene, with essentially the same structure and function(s) as mammalian *Notch* genes (74). Hundreds of *Drosophila Notch* mutations are known, many of which have been classified by genetic (fine-structure) map location and phenotype into categories such as *Abruptex*, *Confluens*, *notchoid*, *facet*, and *split* (125). The *Abruptex* (*Ax*) class of *Notch* alleles have a characteristic, tissue-specific dominant visible phenotype (shortened, incomplete wing veins), and have been further subdivided into three subcategories, of which the recessive lethal subcategory (such as N^{Ax59b}, N^{Ax59d}, and N^{AxM1}) have cysteine substitutions in the extracellular EGF repeats and thus may provide a model for CADASIL (6). These "lethal *Abruptex*" alleles have been characterized genetically as "loss-of-function" alleles, but in addition they also show particular phenotypes (for example in the initiation of proneural cluster development) that are characteristic of dominant negative *Notch* alleles (126). These results are not contradictory, because antimorphs (dominant negative alleles) are expected to have loss-of-function phenotypes, as well as an abnormal gain-of-function that allows them to interfere with the wild-type allele (in heterozygotes) (127).

This was clearly demonstrated for one of the lethal *Abruptex* alleles (N^{Ax59b}, which corresponds to the cysteine substitution C972G in the EGF repeat 24 of the *Notch* gene), and was characterized as a loss-of-function allele in some respects (126), but appeared to be an antimorph (dominant negative) in other assays (128). Heterozygotes with one copy of the cysteine substitution C972G have the typical dominant visible phenotype that defines *Abruptex* alleles (shortened, incomplete wing veins). The addition of a downstream frameshift produced a double mutant allele that encoded a truncated protein and completely lacked *Notch* function (128). Heterozygotes carrying one copy of the double mutant allele lost the characteristic dominant visible *Abruptex* phenotype (see above), but gained an alternative dominant visible phenotype that is seen in heterozygotes of null *Notch* alleles or deletions of the *Notch* gene (notches in the wing margins) (125, 128). Therefore, the dominant visible *Abruptex* phenotype is caused in this case by an abnormal function, probably an antimorph (dominant negative) (126, 128).

Another possible *Drosophila* model for CADASIL is provided by the *notchoid-3* mutation (N^{nd3}, previously known as *nd3* or fa^{no}), which has a cysteine substitution in EGF repeat 2 of the *Drosophila Notch* gene (76). In other words, this cysteine substitution is located within the amino terminal region (EGF repeats 1–6) of *Drosophila Notch*, corresponding to the region of human NOTCH where the majority of CADASIL mutations are found (5, 8). Thus, analysis of N^{nd3} may help to elucidate why human CADASIL mutations are clustered in this area (76). On the other hand, CADASIL mutations have been found essentially throughout the EGF repeat domain of the human NOTCH3 gene (8, 16–25), including the EGF repeats 24–29 where most *Abruptex* mutations are located (125). Moreover, *notchoid* alleles have a negligible dominant phenotype (125) and other *notchoid* alleles map to the intracellular domain of Notch (129); in other words, the overall *notchoid* subgroup of alleles cannot be comparable to CADASIL. The lethal *Abruptex* subgroup of alleles does appear to be comparable to CADASIL, at least in the sense that the alleles in this group correspond to cysteine substitutions in the extracellular EGF repeats, and all share a unique, tissue-specific, dominant visible phenotype. One difference is that the lethal *Abruptex* alleles do appear (so far) to map to a more limited domain (EGF repeats 24–29) than CADASIL alleles (EGF repeats 1–34). However, EGF repeats 22–36 are involved in binding to Fringe (65), which as noted above, accumulates as a heterodimer with NOTCH3 in cell culture models of CADASIL (25).

In any case, Wesley and colleagues found that $\text{NOTCH}^{\text{nd3}}$ receptors in *Drosophila* tissue culture cells showed a cold-sensitive reduction in Notch signaling and showed a reduced rate of internalization (76). Interestingly, both $\text{NOTCH}^{\text{nd3}}$ and $\text{NOTCH}^{\text{Ax59d}}$ receptors also accumulated to abnormally high protein levels and showed apparently similar abnormal distributions (low fraction available for cell-surface biotinylation) (76). The detailed molecular consequences of cysteine substitutions on Notch protein processing may be somewhat different in *Drosophila* than in vertebrates, because in *Drosophila*, the colinear full-length form (not the heterodimeric form as in vertebrates) is the predominant form of Notch receptor (130). Nevertheless, the finding that

Drosophila Notch EGF cysteine substitutions cause a reduction in Notch signaling, as well as increased Notch protein accumulation and reduced Notch protein internalization, are all rather similar to the characteristics of CADASIL, particularly given that the accumulation of NOTCH3 in GOM is now recognized as being sufficiently specific and reproducible that it is diagnostic for CADASIL (7, 8).

5. Cell Culture Models: Testing the Molecular Functions of CADASIL Alleles

Joutel and colleagues modified a cloned human NOTCH3 gene, to create the equivalent of five known CADASIL mutations, which were assayed in cell culture (transient transfection) assays: R90C and C212S (located in EGF repeats 2–5), C428S (located in the ligand-binding domain, EGF repeats 10–11), C542Y (EGF repeat 13), and R1006C (EGF repeat 26). Their results showed that all five mutant receptors reached the cell surface in the normally cleaved form (131). Only the C542Y mutant had a significantly reduced number of receptors at the cell surface. Only the C428S mutant lost its Jagged1-binding ability. Both C542Y and C428S exhibited a net significant reduction in ligand-induced reporter gene expression (Notch signaling), but the other three mutant receptors (R90C, C212S, R1006C) appeared to retain normal levels of Notch signaling ability (131). One difficulty with these experiments was that transient transfection experiments create a cell population in which individual cells have varying gene copy numbers, which tends to reduce the ability to detect small changes in gene activity. However, Joutel et al. were able to simulate "heterozygotes" by transfecting with equal amounts of wild-type and mutant NOTCH3 genes, and these experiments confirmed dose-dependent (intermediate) levels of Notch signaling in C428S/+ and C542Y/+ "heterozygotes," providing convincing evidence that Notch signaling by these alleles was significantly different from the R90C, C212S, and R1006C alleles (131). Human patients who were heterozygous for three of these mutations (R90C, C212S, C428S) were compared to each other histologically, and found to have similar amounts of arterial aggregates of the NOTCH3 extracellular domain, indicating that the accumulation of GOM did not require ligand binding by the NOTCH3 receptor (131).

A similar study was conducted by Kalaria and colleagues (132), who genetically engineered the cloned human NOTCH3 gene to create four different mutations (R90C, R133C, C185R, and R449C). These were overexpressed under control of the CMV promoter, by stable transfection in two different cell lines (HEK293 and SH-SY5Y). The rates of processing (assayed by pulse-chase analysis), transport to the cell surface (estimated by immunofluorescence), and Notch signaling (assayed by reporter gene expression) were not significantly different from control experiments, in which the normal NOTCH3 gene was similarly overexpressed (132).

Another such study was conducted by Schanen and colleagues, who genetically engineered the cloned rat *Notch3* gene to create the C187R amino acid substitution (which corresponds to the C185R mutation found in some human CADASIL patients) (133). Recombinant rat *Notch3* genes were driven by an expression vector, in transient transfection experiments with cultured 293T human embryonic kidney cells and NIH 3T3 mouse embryonic fibroblasts. Only a minority of the protein was processed to the mature form in both normal and mutant receptors, presumably due to their overexpression in this system. Nevertheless, there was a significant (three- to fourfold) decrease in the relative abundance of the processed (mature, proteolytically cleaved) form of the mutant receptor, compared to the wild-type recombinant rat *Notch3* controls, indicating a decrease in the efficiency of the furin processing step (133). The other characteristics of the mutant NOTCH3 receptors (such as ligand-binding and Notch-signaling ability) were not significantly different from wild-type controls.

Lundkvist and colleagues stably transfected cultured cells with a mutant mouse *Notch3* gene (R142C, which corresponds to R141C in humans), or the wild-type mouse *Notch3* control gene, driven by a strong constitutive promoter (CMV viral promoter in the pcDNA3.1 expression vector) (134). Under these conditions, the wild-type *Notch3* gene accumulates nearly as much of the uncleaved NOTCH3 precursor protein as of the cleaved mature protein product. The R142C mutant showed a small but reproducible decrease in the proportion of the NOTCH3 protein in the mature (cleaved) form (134).

Kosik and colleagues expressed a peptide portion of the extracellular domain of NOTCH3 in a cell culture model, and were able to show that an odd number of cysteines was both necessary and sufficient to cause the accumulation of abnormal, cross-linked homodimers of the NOTCH3 extracellular domain, as well as heterodimers of the NOTCH3 extracellular domain with an enzyme named "Lunatic Fringe" (25). This experiment differed from the other cell culture studies cited here in two important respects: (i) they expressed a smaller portion of the NOTCH3 protein, which facilitated the use of nonreducing protein electrophoresis, and (ii) they used a variety of electrophoresis systems to determine the identity of proteins crosslinked by disulfide bonds. They found that all three of the cysteine mutations that they tested showed small but consistent decreases in the elongation of fucose side chains by Fringe (25). This result is significant, both because it was consistently obtained with several different

CADASIL alleles, and because Fringe enzyme activity plays a key role in the regulation of Notch signaling (66–69, 124).

6. Transgenic Mouse Models: Testing CADASIL Mutations *In Vivo*

Several transgenic mice that express mutations found in CADASIL patients have been developed, with rather different results. In the first case, a 2.2-kb fragment containing the promoter region and noncoding first exon from the arterial smooth muscle cell-specific mouse SM22α gene was ligated to a 0.5-kb fragment containing an intron from the bovine β-globin gene, plus a 7.0-kb fragment containing the full-length human NOTCH3 cDNA (genetically engineered to have an arginine-to-cysteine missense mutation at amino acid position 90), and a small fragment containing the polyadenylation signal from the SV40 virus (86). The resulting linear DNA fragment was microinjected into fertilized mouse eggs, yielding three independently transformed transgenic (*TghNotch3R90C*) mouse strains. Although the transgene was specifically expressed in arterial SMCs, this transformation method did not control the transgene copy number, amount of expression, or genetic location of insertion in the chromosome(s). The authors reported that the copy number of the transgenes varied from approximately 5–15 across the three *TghNotch3R90C* lines. Nevertheless, the total amount of the mutant protein was <25% of the normal NOTCH3 protein levels in two of the *TghNotch3R90C* lines, and was even lower (undetectable) in the third *TghNotch3R90C* line. Expression of both copies of the endogenous (normal) *Notch3* gene was not affected by the presence of the transgenes (86). The multiple copies of transgenes were likely to be present in tandem arrays (rather than multiple insertion sites, based on the reported Mendelian ratios), in which case recombination may cause their copy number to vary in subsequent mouse generations (135).

The two *TghNotch3R90C* lines with detectable levels of mutant protein expression showed accumulation of GOM-containing mutant NOTCH3 protein in cerebral and peripheral arteries of old (17–20-month-old) *TghNotch3R90C* mice, but not in middle-aged (10-month-old) *TghNotch3R90C* mice, nor in genetically normal control mice of any age (86). Morphological evidence of arterial SMC degeneration was also seen in tail arteries of old *TghNotch3R90C* mice, to a lesser extent in middle-aged *TghNotch3R90C* mice, but not in young (5-month-old) *TghNotch3R90C* mice. Brain lesions were rare – for example, in the “old” age group (17–20-month-old), one infarct and some glial scars were observed in 1/6 *TghNotch3R90C* mice and 0/3 control mice (86).

Old (18 month) and middle-aged (10 month) *TghNotch3R90C* mice had significantly reduced increases in cerebral blood flow in response to two potent vasodilator stimuli (hypercapnia and acetazolamide); the former response was approximately 50% of normal (136). More detailed tests confirmed that the "plateau" (roughly constant cerebral blood flow caused by homeostatic responses of SMCs) was essentially missing in *TghNotch3R90C* mice, confirming that these transgenic mice had severe vascular dysfunction (136). In normal mice, stepwise artificial elevation of the mean arterial blood pressure up to 150 mm-Hg (by intravenous infusion of phenylephrine), or decrease down to 30 mm-Hg (by withdrawal of arterial blood) was used to demonstrate a region of homeostasis (plateau of roughly constant cerebral blood flow) from 50 to 100 mmHg, with steep declines in cerebral blood flow below that range and more gradually increases above that range. Further studies on isolated caudal arteries from the tails of middle-aged (10–11-month-old) *TghNotch3R90C* mice showed significantly decreased flow-induced dilation and significantly increased pressure-induced myogenic tone, compared to age-matched controls (137). Overall, *TghNotch3R90C* mice do appear to recapitulate the early, preclinical phases of CADASIL, including vascular dysfunction, GOM accumulation, and arterial SMC degeneration, but did not progress to the point of having frequent brain infarcts.

The ability of *TghNotch3R90C* to rescue the morphological defects in *Notch3−/−* mice was tested in the middle cerebral arteries of 1-month-old mice (99). *TghNotch3R90C* proved to be almost as effective in rescuing the morphology as a normal human NOTCH3 gene expressed at similar levels (70–80% of the normal mouse Notch gene expression in these experiments). *TghNotch3R90C* was also effective in rescuing the expression of an arterial SMC-specific reporter gene, and a putative Notch signaling reporter construct (99). The sensitivity of this assay was established by showing that the rescued phenotypes were distinct from (closer to normal than) the semidominant phenotypes in *Notch3 +/−* heterozygotes (99). This does prove that *TghNotch3R90C* has Notch signaling activity *in vivo*, but it is unclear whether this activity is quantitatively normal, particularly given previous studies showing that *TghNotch3R90C* produced <25% of the normal NOTCH3 protein levels (86). Smaller arteries and older mice were not investigated, and may produce different results.

An alternative mouse transgenic model of CADASIL was constructed with the "knock-in" method, which does control the transgene copy number, amount of expression, and location in the genome (because it replaces the normal gene with the mutant gene) (138). In this case, a mouse genomic DNA fragment was genetically engineered to have an arginine-to-cysteine missense mutation at amino acid position 142, which corresponds to the position 141 that is mutated in some human CADASIL patients (138). Transgenic mice (which we will refer to here as *TgmNotch3R142C*) showed normal levels of *Notch3* mRNA and protein, as well as morphologically normal arterial SMCs in both heterozygous and homozygous mice from 5 to 20 months of age (138). *TgmNotch3R142C* mice did show a suggestive trend of small, transgene dose-dependent increases in extracellular granular debris. However, this trend was not statistically significant, and some similar debris was also seen in normal mice. *TgmNotch3R142C* mice showed a significant increase in open field rearing (exploratory) behavior compared to controls at 10 months of age, but not at 8 months of age, and were normal in several other behavioral tests (138). The differing results from *TghNotch3R90C* mice versus *TgmNotch3R142C* mice may have been due to differences between the mutation sites (although both sites can cause CADASIL in humans), and/or differences between the human and mouse NOTCH3 proteins (*TghNotch3R90C* mice express human NOTCH3), and/or differences between the genetic backgrounds of the mouse strains used. Another consideration is that human CADASIL patients are typically diagnosed at about 40–60 years of age and show heritable rates of disease progression that appear to be due to additional genes (50) i.e., the genetic background. In contrast, transgenic mice are studied at about 5–20 months of age and come from inbred laboratory strains, in which genetic background effects may be more severe, and most published reports include only one inbred mouse genetic background.

7. Other Syndromes that Involve Cysteine Mutations Within EGF Repeats

Marfan syndrome is caused by cysteine substitutions in the extracellular EGF repeats of the FBN1 gene, which encodes FIBRILLIN-1, a secreted glycoprotein that contains 47 EGF repeats and is the primary constituent of the extracellular microfibrils in the connective tissue matrix of many organs (139). Marfan syndrome often manifests in middle age, shows an autosomal dominant pattern of inheritance, and is preferentially (but not always) caused by amino acid substitutions that alter the cysteines within the EGF repeats of FIBRILLIN-1 (139, 140). Mutant fibroblasts have been cultured from 25 individuals with Marfan syndrome, representing 22 different cysteine substitutions spanning the entire length of the fibrillin-1 protein (139). All but two of these fibroblast cultures showed reduced deposition of newly synthesized extracellular matrix. In 11/23 cultures with reduced matrix deposition, the defect was attributable to delayed or decreased intracellular processing and/or secretion of the mutant FIBRILLIN-1 protein. In an additional 5/23 cultures, the reduced matrix deposition was attributable to instability (degradation) of the mutant protein (139). The other cultures appeared to be normal in these two respects. The authors concluded that cysteine substitutions within the EGF repeat domains disrupt the folding of FIBRILLIN-1, and hence disrupt matrix deposition. However, the detailed effects of misfolding (in terms of protein processing, proteolytic degradation, multimer formation, etc.) vary between cysteine substitutions (139). Similar results were obtained in a second study, in which two of the three cysteine substitutions tested (that caused Marfan syndrome) led to the accumulation of the recombinant protein within the endoplasmic reticulum, but the third did not (141). It was further shown that cysteine substitutions can interfere with the ability of an EGF repeat to bind calcium; in some cases (but not others) the binding of calcium is then a prerequisite for the proper folding of the adjacent EGF repeat (61).

8. Conclusions and Future Directions

CADASIL has emerged as a rather well-defined, autosomal dominant inherited disease that is caused by an odd number of cysteines within the extracellular EGF repeat domain of the NOTCH3 receptor protein. CADASIL is associated with the age-dependent accumulation of GOM surrounding SMCs in small arteries, with minimal involvement of large arteries and no involvement of veins. Death typically results from the cumulative effects of many small strokes, which typically begin in middle age, particularly in subcortical nuclei and white matter. The detailed mechanism of these small arterial infarcts remains to be established. Evidence to date suggests that multiple factors may play a role in the causation of small infarcts in CADASIL, including narrowing of the lumen (evidently due to proliferation of SMCs) (3, 7), reduced cerebral artery blood flow velocity (52), the accumulation of extracellular debris in small arteries (7, 8), and abnormal hemodynamics (failure of CADASIL SMCs to properly regulate cerebral blood pressure) (136).

The prospects for further progress in understanding the molecular basis of CADASIL is suggested by Marfan syndrome, another late-onset, autosomal dominant inherited disease caused by cysteine substitutions within a long tandem array of EGF repeats (139, 140). The causation and symptoms of Marfan syndrome initially suggested the possibility of a single, straightforward molecular mechanism, but cellular and molecular studies have shown that various cysteine substitutions in the target protein (FIBRILLIN-1) interfere with extracellular matrix formation by distinctly different mechanisms, including defects in protein processing, proteolytic degradation, multimer formation, and so on (61, 139, 141). Likewise, studies of cysteine substitutions in the EGF repeat domain of *Drosophila Notch* also have distinct phenotypes, reinforcing the view that subregions in the Notch EGF repeat domain have distinct functions (65, 76). Thus, it should not be surprising that various CADASIL mutations have different molecular consequences in cell culture assays (25, 131–134).

Histological studies have shown that the accumulation of GOM in small arteries occurs in essentially 100% of CADASIL patients, and apparently not in other genetically determined small vessel angiopathies (7). In fact, skin biopsies are now regarded as having diagnostic value equal to DNA sequencing for the diagnosis of CADASIL (7, 8). The development of a cell culture model for the formation of GOM by Kosik and colleagues is thus one of the most promising developments in this field (25). Kosik and colleagues have shown that, when expressing a peptide portion of the extracellular domain of NOTCH3 in cell culture, an odd number of cysteines is both necessary and sufficient to cause the accumulation of abnormal, crosslinked homodimers of the NOTCH3 extracellular domain, as well as heterodimers of the NOTCH3 extracellular domain with an enzyme named "Lunatic Fringe" (25). Hopefully, we will soon learn whether crosslinked Lunatic Fringe is present in GOM in CADASIL patients, or for that matter in transgenic *TghNotch3R90C* mice, which also show age-dependent accumulation of GOM (86).

A variety of other experimental models relevant to CADASIL have been investigated, including *Notch3* knockout mice, Notch signaling reporter cell lines with normal and mutant NOTCH3 receptors, transgenic mice with cysteine substitutions in the NOTCH3 receptor, and *Drosophila* mutants with cysteine substitutions in their Notch receptor. Each of these models has certain advantages and disadvantages (see below).

Much of the debate in this field has focused on whether CADASIL causes a genetic gain or loss of function of NOTCH3, and whether the biochemical mechanism is mediated by a loss of Notch signaling versus side-effects of protein accumulation (GOM). It should be noted that these alternatives are not mutually exclusive. For example, antimorphs (dominant negatives), by definition, have both a gain and a loss of function (i.e., have gained an abnormal function that reduces the normal functioning of the wild-type allele), and that for example is evidently the case with the *Drosophila* lethal *Abruptex* alleles (126, 128). Likewise, one potential effect of the accumulation of the NOTCH3 extracellular domain in GOM would be binding to Notch ligands, which would presumably interfere with Notch signaling (from which the GOM is disconnected).

Studies of *Notch3* knockout mice have clearly shown that NOTCH3 signaling is specifically required by SMCs in arteries but not veins (87, 94, 95), primarily in small (rather than large) arteries

(95, 99), primarily for postnatal maintenance (rather than prenatal development) (87, 95), and that failure of NOTCH3 signaling results in a disorganized SMC layer in small arteries (95). Given this background, it is not surprising that NOTCH3 signaling is also implicated as a prerequisite for a normal SMC response to changes in blood pressure (hemodynamics) (136, 137) and mechanical stress (wound healing) (107, 108). These detailed point-by-point correlations between the effects of NOTCH3 signaling and the characteristics of CADASIL are unlikely to be a coincidence. Moreover, physiological dysfunction in arterial SMCs are the earliest phenotypes observed in both mouse *Notch3−/−* knockouts and in *TgNotch3R90C* mice. This is particularly striking given that the normal NOTCH3 protein is at least fourfold more abundant than the mutant NOTCH3 protein in *TgNotch3R90C* mice (86). These and other results suggest that the mutation of the cysteine in position 90 interferes in some way with NOTCH3 signaling, which in turn reduces the differentiation of adult arterial SMCs and reduces their ability to cope physiologically with the relatively high and variable mechanical pressures encountered in the arterial circulation. This interpretation does not rule out additional (and not mutually exclusive) possibilities, such as a NOTCH3 gain of function of the "dominant negative" variety, or a biochemical mechanism in which the gradual extracellular accumulation of the NOTCH3 extracellular domain is a key initial step.

In human CADASIL patients, the accumulation of GOM surrounding the vascular SMCs, the degeneration and death of these vascular SMCs, and the occlusion of subcortical small arteries, all take a considerable amount of time to develop. Symptomatic strokes are typically first observed in CADASIL patients in their 40s and 50s. These conditions may not have sufficient time to fully develop in an "elderly" (18–22 month-old) mouse. In any case, further analysis of the *Notch3* knockout and *TgNotch3R90C* mouse models should lead to valuable insights into the onset and progression of CADASIL during the early stages, which are also the stages during which therapeutic interventions are most likely to be beneficial. At the biochemical level, the development of a cell culture model for the formation of GOM (25) is an important advance, and further research with this cell culture model should complement the *in vivo* studies.

Acknowledgments

Scientific interest in CADASIL has grown in recent years to the point where it is no longer possible to cite every scientific paper in this area. I apologize in advance to anyone whose contributions may have been inadvertently omitted. I also thank the Editors for their patience and support.

References

1. Razvi SS, Bone I (2006) Single gene disorders causing ischaemic stroke. J Neurol 253: 685–700
2. Wang QK (2005) Update on the molecular genetics of vascular anomalies. Lymphat Res Biol 3:226–233
3. Kalaria RN, Viitanen M, Kalimo H, Dichgans M, Tabira T (2004) The pathogenesis of CADASIL: an update. J Neurol Sci 226:35–39
4. Joutel A, Corpechot C, Ducros A, et al (1996) Notch3 mutations in CADASIL, a hereditary adult-onset condition causing stroke and dementia. Nature 383:707–710
5. Joutel A, Vahedi K, Corpechot C, et al (1997) Strong clustering and stereotyped nature of Notch3 mutations in CADASIL patients. Lancet 350:1511–1515
6. Fryxell KJ, Soderlund M, Jordan TV (2001) An animal model for the molecular genetics of CADASIL. (Cerebral autosomal dominant arteriopathy with subcortical infarcts and leukoencephalopathy). Stroke 32:6–11
7. Ruchoux MM, Brulin P, Leteurtre E, Maurage CA (2000) Skin biopsy value and leukoaraiosis. Ann NY Acad Sci 903:285–292
8. Peters N, Opherk C, Bergmann T, Castro M, Herzog J, Dichgans M (2005) Spectrum of mutations in biopsy-proven CADASIL: implications for diagnostic strategies. Arch Neurol 62:1091–1094
9. Choi EJ, Choi CG, Kim JS (2005) Large cerebral artery involvement in CADASIL. Neurology 65:1322–1324
10. Dichgans M (2007) Genetics of ischaemic stroke. Lancet Neurol 6:149–161
11. Razvi SS, Davidson R, Bone I, Muir KW (2005) The prevalence of cerebral autosomal dominant arteriopathy with subcortical infarcts and leucoencephalopathy (CADASIL) in the west of Scotland. J Neurol Neurosurg Psychiatry 76:739–741
12. Razvi SS, Davidson R, Bone I, Muir KW (2005) Is inadequate family history a barrier to diagnosis in CADASIL? Acta Neurol Scand 112:323–326
13. Pandey T, Abubacker S (2006) Cerebral autosomal dominant arteriopathy with subcortical infarcts and leukoencephalopathy: an imaging mimic of multiple sclerosis. A report of two cases. Med Princ Pract 15:391–395
14. O'Riordan S, Nor AM, Hutchinson M (2002) CADASIL imitating multiple sclerosis: the importance of MRI markers. Mult Scler 8:430–432
15. Oberstein SA, Ferrari MD, Bakker E, et al (1999) Diagnostic Notch3 sequence analysis in CADASIL: three new mutations in Dutch patients. Dutch CADASIL Research Group. Neurology 52:1913–1915
16. Dotti MT, Federico A, Mazzei R, et al (2005) The spectrum of Notch3 mutations in 28 Italian CADASIL families. J Neurol Neurosurg Psychiatry 76:736–738
17. Markus HS, Martin RJ, Simpson MA, et al (2002) Diagnostic strategies in CADASIL. Neurology 59:1134–1138
18. Lee YC, Yang AH, Liu HC, et al (2006) Cerebral autosomal dominant arteriopathy with subcortical infarcts and leukoencephalopathy: two novel mutations in the NOTCH3 gene in Chinese. J Neurol Sci 246:111–115
19. Pescini F, Bianchi S, Salvadori E, et al (2008) A pathogenic mutation on exon 21 of the NOTCH3 gene causing CADASIL in an octogenarian paucisymptomatic patient. J Neurol Sci 267:170–173
20. Vikelis M, Papatriantafyllou J, Karageorgiou CE (2007) A novel CADASIL-causing mutation in a stroke patient. Swiss Med Wkly 137:323–325
21. Ragno M, Cacchio G, Fabrizi GM, et al (2007) Clinical presentation of CADASIL in an Italian patient with a rare Gly528Cys exon 10 NOTCH3 gene mutation. Neurol Sci 28:181–184
22. Oki K, Nagata E, Ishiko A, et al (2007) Novel mutation of the Notch3 gene in a Japanese patient with CADASIL. Eur J Neurol 14:464–466
23. Oliveri RL, Muglia M, De Stefano N, et al (2001) A novel mutation in the Notch3 gene in an Italian family with cerebral autosomal dominant arteriopathy with subcortical infarcts and leukoencephalopathy: genetic and magnetic resonance spectroscopic findings. Arch Neurol 58:1418–1422
24. Ragno M, Fabrizi GM, Cacchio G, et al (2006) Two novel Italian CADASIL families from Central Italy with mutation CGC-TGC at codon 1006 in the exon 19 Notch3 gene. Neurol Sci 27:252–256
25. Arboleda-Velasquez JF, Rampal R, Fung E, et al (2005) CADASIL mutations impair Notch3 glycosylation by Fringe. Hum Mol Genet 14:1631–1639
26. Kleinig TJ, Kimber T, Thompson PD (2007) Acute encephalopathy as the initial symptom of CADASIL. Intern Med J 37:786–787
27. Schon F, Martin RJ, Prevett M, Clough C, Enevoldson TP, Markus HS (2003) "CADASIL coma": an underdiagnosed acute encephalopathy. J Neurol Neurosurg Psychiatry 74: 249–252

28. Dichgans M, Mayer M, Uttner I, et al (1998) The phenotypic spectrum of CADASIL: clinical findings in 102 cases. Ann Neurol 44: 731–739
29. Nishio T, Arima K, Eto K, Ogawa M, Sunohara N (1997) Cerebral autosomal dominant arteriopathy with subcortical infarcts and leukoencephalopathy – report of an autopsied Japanese case. Rinsho Shinkeigaku 37:910–916
30. Williamson EE, Chukwudelunzu FE, Meschia JF, Witte RJ, Dickson DW, Cohen MD (1999) Distinguishing primary angiitis of the central nervous system from cerebral autosomal dominant arteriopathy with subcortical infarcts and leukoencephalopathy: the importance of family history. Arthritis Rheum 42:2243–2248
31. Kumar SK, Mahr G (1997) CADASIL presenting as bipolar disorder. Psychosomatics 38:397–398
32. Desmond DW, Moroney JT, Lynch T, Chan S, Chin SS, Mohr JP (1999) The natural history of CADASIL: a pooled analysis of previously published cases. Stroke 30:1230–1233
33. Matsumoto H, Tsumoto M, Yamamoto T, et al (2005) A case of early stage CADASIL showing only dizziness and vertigo with a novel mutation of Notch 3 gene. Rinsho Shinkeigaku 45:27–31
34. Lv H, Yao S, Zhang W, et al (2004) Clinical features in 4 Chinese families with cerebral autosomal dominant arteriopathy with subcortical infarcts and leukoencephalopathy (CADASIL). Beijing Da Xue Xue Bao 36:496–500
35. Miranda M, Dichgans M, Slachevsky A, et al (2006) CADASIL presenting with a movement disorder: a clinical study of a Chilean kindred. Mov Disord 21:1008–1012
36. Parisi V, Pierelli F, Fattapposta F, et al (2003) Early visual function impairment in CADASIL. Neurology 60:2008–2010
37. Parisi V, Pierelli F, Malandrini A, et al (2000) Visual electrophysiological responses in subjects with cerebral autosomal arteriopathy with subcortical infarcts and leukoencephalopathy (CADASIL). Clin Neurophysiol 111: 1582–1588
38. Mourad A, Levasseur M, Bousser MG, Chabriat H (2006) CADASIL with minimal symptoms after 60 years. Rev Neurol (Paris) 162:827–831
39. Trojano M, Paolicelli D (2001) The differential diagnosis of multiple sclerosis: classification and clinical features of relapsing and progressive neurological syndromes. Neurol Sci 22(Suppl 2):S98–S102
40. Van Gerpen JA, Ahlskog JE, Petty GW (2003) Progressive supranuclear palsy phenotype secondary to CADASIL. Parkinsonism Relat Disord 9:367–369
41. Pantoni L, Pescini F, Inzitari D, Dotti MT (2005) Postpartum psychiatric disturbances as an unrecognized onset of CADASIL. Acta Psychiatr Scand 112:241–242
42. Otto V, Kaps M, Burgmann T, Kompf D (1997) CADASIL: 2 case reports of hereditary multi-infarct dementia. Fortschr Neurol Psychiatr 65:90–95
43. Lagas PA, Juvonen V (2001) Schizophrenia in a patient with cerebral autosomally dominant arteriopathy with subcortical infarcts and leucoencephalopathy (CADASIL disease). Nord J Psychiatry 55:41–42
44. Phillips JS, King JA, Chandran S, Prinsley PR, Dick D (2005) Cerebral autosomal dominant arteriopathy with subcortical infarcts and leukoencephalopathy (CADASIL) presenting with sudden sensorineural hearing loss. J Laryngol Otol 119:148–151
45. Wielaard R, Bornebroek M, Ophoff RA, et al (1995) A four-generation Dutch family with cerebral autosomal dominant arteriopathy with subcortical infarcts and leukoencephalopathy (CADASIL), linked to chromosome 19p13. Clin Neurol Neurosurg 97:307–313
46. Opherk C, Peters N, Herzog J, Luedtke R, Dichgans M (2004) Long-term prognosis and causes of death in CADASIL: a retrospective study in 411 patients. Brain 127:2533–2539
47. Tuominen S, Juvonen V, Amberla K, et al (2001) Phenotype of a homozygous CADASIL patient in comparison to 9 age-matched heterozygous patients with the same R133C Notch3 mutation. Stroke 32:1767–1774
48. Amberla K, Waljas M, Tuominen S, et al (2004) Insidious cognitive decline in CADASIL. Stroke 35:1598–1602
49. Peters N, Herzog J, Opherk C, Dichgans M (2004) A two-year clinical follow-up study in 80 CADASIL subjects: progression patterns and implications for clinical trials. Stroke 35:1603–1608
50. Opherk C, Peters N, Holtmannspotter M, Gschwendtner A, Muller-Myhsok B, Dichgans M (2006) Heritability of MRI lesion volume in CADASIL: evidence for genetic modifiers. Stroke 37:2684–2689
51. Murakami T, Iwatsuki K, Hayashi T, et al (2001) Two Japanese CADASIL families with a R141C mutation in the Notch3 gene. Intern Med 40:1144–1148

52. Pfefferkorn T, von Stuckrad-Barre S, Herzog J, Gasser T, Hamann GF, Dichgans M (2001) Reduced cerebrovascular CO(2) reactivity in CADASIL: A transcranial Doppler sonography study. Stroke 32:17–21
53. Council of Biology (eds) (1994) Scientific style and format, 6th edn. Cambridge University Press, Cambridge, UK
54. Artavanis-Tsakonas S, Rand MD, Lake RJ (1999) Notch signaling: cell fate control and signal integration in development. Science 284:770–776
55. Bishop SA, Klein T, Arias AM, Couso JP (1999) Composite signalling from Serrate and Delta establishes leg segments in Drosophila through Notch. Development 126:2993–3003
56. Kurata S, Go MJ, Artavanis-Tsakonas S, Gehring WJ (2000) Notch signaling and the determination of appendage identity. Proc Natl Acad Sci USA 97:2117–2122
57. Justice NJ, Jan YN (2002) Variations on the Notch pathway in neural development. Curr Opin Neurobiol 12:64–70
58. Costa RM, Drew C, Silva AJ (2005) Notch to remember. Trends Neurosci 28:429–435
59. Lubman OY, Korolev SV, Kopan R (2004) Anchoring Notch genetics and biochemistry; structural analysis of the ankyrin domain sheds light on existing data. Mol Cell 13:619–626
60. Barrick D, Kopan R (2006) The Notch transcription activation complex makes its move. Cell 124:883–885
61. Whiteman P, Willis AC, Warner A, Brown J, Redfield C, Handford PA (2007) Cellular and molecular studies of Marfan syndrome mutations identify co-operative protein folding in the cbEGF12–13 region of fibrillin-1. Hum Mol Genet 16:907–918
62. Okajima T, Irvine KD (2002) Regulation of Notch signaling by O-linked fucose. Cell 111:893–904
63. Bruckner K, Perez L, Clausen H, Cohen S (2000) Glycosyltransferase activity of Fringe modulates Notch–Delta interactions. Nature 406:411–415
64. Moloney DJ, Panin VM, Johnston SH, et al (2000) Fringe is a glycosyltransferase that modifies Notch. Nature 406:369–375
65. Ju B-G, Jeong S, Bae E, et al (2000) Fringe forms a complex with Notch. Nature 405:191–195
66. Irvine KD (1999) Fringe, Notch, and making developmental boundaries. Curr Opin Genet Dev 9:434–441
67. Hicks C, Johnston SH, diSibio G, Collazo A, Vogt TF, Weinmaster G (2000) Fringe differentially modulates Jagged1 and Delta1 signalling through Notch1 and Notch2. Nat Cell Biol 2:515–520
68. Haines N, Irvine KD (2003) Glycosylation regulates Notch signalling. Nat Rev Mol Cell Biol 4:786–797
69. Haltiwanger RS, Lowe JB (2004) Role of glycosylation in development. Annu Rev Biochem 73:491–537
70. Chitnis A (2006) Why Is Delta endocytosis required for effective activation of Notch? Dev Dyn 235:886–894
71. Le Borgne R, Bardin A, Schweisguth F (2005) The roles of receptor and ligand endocytosis in regulating Notch signaling. Development 132:1751–1762
72. Lai EC, Roegiers F, Qin X, Jan YN, Rubin GM (2005) The ubiquitin ligase Drosophila Mind bomb promotes Notch signaling by regulating the localization and activity of Serrate and Delta. Development 132: 2319–2332
73. Parks AL, Klueg KM, Stout JR, Muskavitch MAT (2000) Ligand endocytosis drives receptor dissociation and activation in the Notch pathway. Development 127:1373–1385
74. Bray SJ (2006) Notch signalling: a simple pathway becomes complex. Nat Rev Mol Cell Biol 7:678–689
75. Hsieh JJ-D, Henkel T, Salmon P, Robey E, Peterson MG, Hayward SD (1996) Truncated mammalian Notch1 activates CBF1/RBPJk-repressed genes by a mechanism resembling that of Epstein-Barr virus EBNA2. Mol Cell Biol 16:952–959
76. Bardot B, Mok LP, Thayer T, Ahimou F, Wesley C (2005) The Notch amino terminus regulates protein levels and Delta-induced clustering of Drosophila Notch receptors. Exp Cell Res 304:202–223
77. Schweisguth F (2004) Notch signaling activity. Curr Biol 14:R129–R138
78. Siekmann AF, Covassin L, Lawson ND (2008) Modulation of VEGF signalling output by the Notch pathway. Bioessays 30:303–313
79. Lawson ND, Vogel AM, Weinstein BM (2002) sonic hedgehog and vascular endothelial growth factor act upstream of the Notch pathway during arterial endothelial differentiation. Dev Cell 3:127–136
80. Weinstein BM, Lawson ND (2002) Arteries, veins, Notch, and VEGF. Cold Spring Harb Symp Quant Biol 67:155–162
81. Siekmann AF, Lawson ND (2007) Notch signalling limits angiogenic cell behaviour in developing zebrafish arteries. Nature 445:781–784

82. Lawson ND, Scheer N, Pham VN, et al (2001) Notch signaling is required for arterial-venous differentiation during embryonic vascular development. Development 128:3675–3683
83. Villa N, Walker L, Lindsell CE, Gasson J, Iruela-Arispe ML, Weinmaster G (2001) Vascular expression of Notch pathway receptors and ligands is restricted to arterial vessels. Mech Dev 108:161–164
84. Campos AH, Wang W, Pollman MJ, Gibbons GH (2002) Determinants of Notch-3 receptor expression and signaling in vascular smooth muscle cells: implications in cell-cycle regulation. Circ Res 91:999–1006
85. Iso T, Hamamori Y, Kedes L (2003) Notch signaling in vascular development. Arterioscler Thromb Vasc Biol 23:543–553
86. Ruchoux MM, Domenga V, Brulin P, et al (2003) Transgenic mice expressing mutant Notch3 develop vascular alterations characteristic of cerebral autosomal dominant arteriopathy with subcortical infarcts and leukoencephalopathy. Am J Pathol 162: 329–342
87. Gridley T (2007) Notch signaling in vascular development and physiology. Development 134:2709–2718
88. Owens GK, Kumar MS, Wamhoff BR (2004) Molecular regulation of vascular smooth muscle cell differentiation in development and disease. Physiol Rev 84:767–801
89. Huppert SS, Le A, Schroeter EH, et al (2000) Embryonic lethality in mice homozygous for a processing-deficient allele of Notch1. Nature 405:966–970
90. Krebs LT, Xue Y, Norton CR, et al (2000) Notch signaling is essential for vascular morphogenesis in mice. Genes Dev 14: 1343–1352
91. Limbourg FP, Takeshita K, Radtke F, Bronson RT, Chin MT, Liao JK (2005) Essential role of endothelial Notch1 in angiogenesis. Circulation 111:1826–1832
92. Carlson TR, Yan Y, Wu X, et al (2005) Endothelial expression of constitutively active Notch4 elicits reversible arteriovenous malformations in adult mice. Proc Natl Acad Sci USA 102:9884–9889
93. Uyttendaele H, Ho J, Rossant J, Kitajewski J (2001) Vascular patterning defects associated with expression of activated Notch4 in embryonic endothelium. Proc Natl Acad Sci USA 98:5643–5648
94. Joutel A, Andreux F, Gaulis S, et al (2000) The ectodomain of the Notch3 receptor accumulates within the cerebrovasculature of CADASIL patients. J Clin Invest 105: 597–605
95. Domenga V, Fardoux P, Lacombe P, et al (2004) Notch3 is required for arterial identity and maturation of vascular smooth muscle cells. Genes Dev 18:2730–2735
96. Krebs LT, Xue Y, Norton CR, et al (2003) Characterization of Notch3-deficient mice: normal embryonic development and absence of genetic interactions with a Notch1 mutation. Genesis 37:139–143
97. Mitchell KJ, Pinson KI, Kelly OG, et al (2001) Functional analysis of secreted and transmembrane proteins critical to mouse development. Nat Genet 28:241–249
98. Kitamoto T, Takahashi K, Takimoto H, et al (2005) Functional redundancy of the Notch gene family during mouse embryogenesis: analysis of Notch gene expression in Notch3-deficient mice. Biochem Biophys Res Commun 331:1154–1162
99. Monet M, Domenga V, Lemaire B, et al (2007) The archetypal R90C CADASIL-NOTCH3 mutation retains NOTCH3 function in vivo. Hum Mol Genet 16:982–992
100. Moessler H, Mericskay M, Li Z, Nagl S, Paulin D, Small JV (1996) The SM 22 promoter directs tissue-specific expression in arterial but not in venous or visceral smooth muscle cells in transgenic mice. Development 122:2415–2425
101. Adriani W, Macrì S, Pacifici R, Laviola G (2002) Peculiar vulnerability to nicotine oral self-administration in mice during early adolescence. Neuropsychopharmacology 27: 212–224
102. Sakata Y, Xiang F, Chen Z, et al (2004) Transcription factor CHF1/Hey2 regulates neointimal formation in vivo and vascular smooth muscle proliferation and migration in vitro. Arterioscler Thromb Vasc Biol 24:2069–2074
103. Sweeney C, Morrow D, Birney YA, et al (2004) Notch 1 and 3 receptor signaling modulates vascular smooth muscle cell growth, apoptosis, and migration via a CBF-1/RBP-Jk dependent pathway. FASEB J 18:1421–1423
104. Doi H, Iso T, Yamazaki M, et al (2005) HERP1 inhibits myocardin-induced vascular smooth muscle cell differentiation by interfering with SRF binding to CArG box. Arterioscler Thromb Vasc Biol 25:2328–2334
105. Wang W, Campos AH, Prince CZ, Mou Y, Pollman MJ (2002) Coordinate Notch3-hairy-related transcription factor pathway regulation in response to arterial injury. Mediator role of platelet-derived growth factor and ERK. J Biol Chem 277: 23165–23171

106. Lindner V, Booth C, Prudovsky I, Small D, Maciag T, Liaw L (2001) Members of the Jagged/Notch gene families are expressed in injured arteries and regulate cell phenotype via alterations in cell matrix and cell-cell interaction. Am J Pathol 159:875–883
107. Morrow D, Sweeney C, Birney YA, et al (2005) Cyclic strain inhibits Notch receptor signaling in vascular smooth muscle cells in vitro. Circ Res 96:567–575
108. Morrow D, Scheller A, Birney YA, et al (2005) Notch-mediated CBF-1/RBP-J {kappa} -dependent regulation of human vascular smooth muscle cell phenotype in vitro. Am J Physiol Cell Physiol 289:C1188–C1196
109. Nakamura T, Watanabe H, Hirayama M, et al (2005) CADASIL with NOTCH3 S180C presenting anticipation of onset age and hallucinations. J Neurol Sci 238:87–91
110. Dichgans M, Ludwig H, Muller-Hocker J, Messerschmidt A, Gasser T (2000) Small in-frame deletions and missense mutations in CADASIL: 3D models predict misfolding of Notch3 EGF-like repeat domains. Eur J Hum Genet 8:280–285
111. Ishida C, Sakajiri K, Yoshita M, Joutel A, Cave-Riant F, Yamada M (2006) CADASIL with a novel mutation in exon 7 of NOTCH3 (C388Y). Intern Med 45:981–985
112. Joutel A, Dodick DD, Parisi JE, Cecillon M, Tournier-Lasserve E, Bousser MG (2000) De novo mutation in the Notch3 gene causing CADASIL. Ann Neurol 47:388–391
113. Kim Y, Choi EJ, Choi CG, et al (2006) Characteristics of CADASIL in Korea: a novel cysteine-sparing Notch3 mutation. Neurology 66:1511–1516
114. Kotani N, Hara H, Fujimura H, Miyashita T, Miyaguchi K, Tabira T (2004) A case of CADASIL (cerebral autosomal dominant arteriopathy with subcortical infarcts and lekoencephalopathy) with Notch 3 (Arg169Cys) mutation and typical granular osmiophilic materials in peripheral small arteries. Rinsho Shinkeigaku 44:274–279
115. Kotorii S, Takahashi K, Kamimura K, et al (2001) Mutations of the notch3 gene in non-caucasian patients with suspected CADASIL syndrome. Dement Geriatr Cogn Disord 12:185–193
116. Posada IJ, Garcia-Morales I, Martinez MA, Hoenicka J, Bermejo F (2003) CADASIL: a case with clinical, radiological, histological and genetic diagnoses. Neurologia 18:229–233
117. Fouillade C, Chabriat H, Riant F, et al (2008) Activating NOTCH3 mutation in a patient with small-vessel-disease of the brain. Hum Mutat 29:452
118. Mazzei R, Conforti FL, Lanza PL, et al (2004) A novel Notch3 gene mutation not involving a cysteine residue in an Italian family with CADASIL. Neurology 63:561–564
119. Hagel C, Groden C, Niemeyer R, Stavrou D, Colmant HJ (2004) Subcortical angiopathic encephalopathy in a German kindred suggests an autosomal dominant disorder distinct from CADASIL. Acta Neuropathol 108:231–240
120. Ito D, Tanahashi N, Murata M, et al (2002) Notch3 gene polymorphism and ischaemic cerebrovascular disease. J Neurol Neurosurg Psychiatry 72:382–384
121. Donahue CP, Kosik KS (2004) Distribution pattern of Notch3 mutations suggests a gain-of-function mechanism for CADASIL. Genomics 83:59–65
122. Ruchoux MM, Maurage CA (1997) CADASIL: Cerebral autosomal dominant arteriopathy with subcortical infarcts and leukoencephalopathy. J Neuropathol Exp Neurol 56:947–964
123. Ishiko A, Shimizu A, Nagata E, Takahashi K, Tabira T, Suzuki N (2006) Notch3 ectodomain is a major component of granular osmiophilic material (GOM) in CADASIL. Acta Neuropathol 112:333–339
124. Shifley ET, Vanhorn KM, Perez-Balaguer A, Franklin JD, Weinstein M, Cole SE (2008) Oscillatory lunatic fringe activity is crucial for segmentation of the anterior but not posterior skeleton. Development 135: 899–908
125. Lindsley DL, Zimm GG (1992) The genome of Drosophila melanogaster. Academic Press, New York
126. Brennan K, Tateson R, Lieber T, Couso JP, Zecchini V, Arias AM (1999) The Abruptex mutations of Notch disrupt the establishment of proneural clusters in Drosophila. Dev Biol 216:230–242
127. Muller HJ (1962) Studies in genetics: the selected papers of H. J. Muller. Indiana University Press, Bloomington, IN
128. Kelley MR, Kidd S, Deutsch WA, Young MW (1987) Mutations altering the structure of epidermal growth factor-like coding sequences at the Drosophila Notch locus. Cell 51:539–548
129. Xu T, Rebay I, Fleming RJ, Scottgale NT, Artavanis-Tsakonas S (1990) The Notch locus and the genetic circuitry involved in early Drosophila neurogenesis. Genes Dev 4:464–475
130. Kidd S, Lieber T (2002) Furin cleavage is not a requirement for Drosophila Notch function. Mech Dev 115:41–51

131. Joutel A, Monet M, Domenga V, Riant F, Tournier-Lasserve E (2004) Pathogenic mutations associated with cerebral autosomal dominant arteriopathy with subcortical infarcts and leukoencephalopathy differently affect Jagged1 binding and Notch3 activity via the RBP/JK signaling Pathway. Am J Hum Genet 74:338–347
132. Low WC, Santa Y, Takahashi K, Tabira T, Kalaria RN (2006) CADASIL-causing mutations do not alter Notch3 receptor processing and activation. Neuroreport 17:945–949
133. Haritunians T, Chow T, De Lange RP, et al (2005) Functional analysis of a recurrent missense mutation in Notch3 in CADASIL. J Neurol Neurosurg Psychiatry 76: 1242–1248
134. Karlstrom H, Beatus P, Dannaeus K, Chapman G, Lendahl U, Lundkvist J (2002) A CADASIL-mutated Notch 3 receptor exhibits impaired intracellular trafficking and maturation but normal ligand-induced signaling. Proc Natl Acad Sci USA 99:17119–17124
135. Fryxell KJ (1996) The coevolution of gene family trees. Trends Genet 12:364–369
136. Lacombe P, Oligo C, Domenga V, Tournier-Lasserve E, Joutel A (2005) Impaired cerebral vasoreactivity in a transgenic mouse model of cerebral autosomal dominant arteriopathy with subcortical infarcts and leukoencephalopathy arteriopathy. Stroke 36: 1053–1058
137. Dubroca C, Lacombe P, Domenga V, et al (2005) Impaired vascular mechanotransduction in a transgenic mouse model of CADASIL arteriopathy. Stroke 36:113–117
138. Lundkvist J, Zhu S, Hansson EM, et al (2005) Mice carrying a R142C Notch 3 knock-in mutation do not develop a CADASIL-like phenotype. Genesis 41:13–22
139. Schrijver I, Liu W, Brenn T, Furthmayr H, Francke U (1999) Cysteine substitutions in epidermal growth factor–like domains of Fibrillin-1: distinct effects on biochemical and clinical phenotypes. Am J Hum Genet 65:1007–1020
140. Dietz HC, Saraiva JM, Pyeritz RE, Cutting GR, Francomano CA (1992) Clustering of fibrillin (FBN1) missense mutations in Marfan syndrome patients at cysteine residues in EGF-like domains. Hum Mutat 1:366–374
141. Whiteman P, Handford PA (2003) Defective secretion of recombinant fragments of fibrillin-1: implications of protein misfolding for the pathogenesis of Marfan syndrome and related disorders. Hum Mol Genet 12:727–737

Chapter 30

Spontaneously Hypertensive Rat (SHR): An Animal Model of Vascular Brain Disorder

Francesco Amenta and Daniele Tomassoni

Abstract

Arterial hypertension is a well-known risk factor for stroke and cognitive deterioration of vascular origin, including vascular cognitive impairment (VCI) and vascular dementia (VaD). Patients with ischemic VaD have a significantly greater incidence of hypertension than patients with dementia of the Alzheimer's type. Areas of white matter (WM) hyperintensity and ventricular enlargement are more common in hypertensive subjects compared with that in normotensive individuals. An association between hypertension and reduced cerebral blood flow (CBF) and VCI is documented. Hypertension in midlife is associated with a higher probability of cognitive impairment than hypertension developed later in life. These findings collectively indicate that arterial hypertension is a main cause of vascular brain disorder (VBD).

In this chapter, the main characteristics of cerebrovascular and brain damage in spontaneously hypertensive rats (SHR) and stroke-prone (SP) SHR are reviewed. SHR and SHR-SP are two widely used animal models of hypertension and hypertension plus stroke, respectively. They are normotensive at birth and gradually develop severe hypertension in the first 2–4 months of life. At 6 months, they have developed a sustained hypertension. SHR were the model most extensively investigated for assessing hypertensive brain damage and treatment of it. The time-dependent rise of arterial blood pressure, the occurrence of brain atrophy, loss of nerve cells, and glial reaction are phenomena shared to some extent with hypertensive brain damage in humans. SHR, therefore, can represent a reasonable model of morphological damage induced by hypertension and possibly reversed by appropriate antihypertensive treatment.

From a neurotransmitter point of view, SHR present changes of some specific neurotransmitter systems that may have functional and behavioral relevance. An impaired cholinergic neurotransmission characterizes both SHR and SHR-SP, similarly as reported in patients affected by VaD. The possibility that cholinergic neurotransmission mechanisms contribute to central nervous system plasticity or may improve brain circulation in this animal model is suggested by some studies. SHR are also characterized by a dopaminergic hypofunction and noradrenergic hyperactivity. This causes an imbalance between noradrenergic hyperfunction and dopaminergic hypofunction similarly as occurs in attention-deficit with hyperactivity disorder (ADHD).

The majority of microanatomical, behavioral, and neurochemical data on SHR are in favor of the hypothesis that this strain is a suitable model of VBD. Changes in catecholaminergic transmission put forward SHR as a possible model of ADHD as well. Hence, SHR could represent a multifaced model of two important groups of pathologies, VBD and ADHD. As for most models, researchers should always consider that SHR offer some similarities with corresponding human pathologies, but they do not suffer from the same disease.

Key words: Hypertension, Hypertension plus stroke, Rat model, Vascular dementia

Peter Paul De Deyn and Debby Van Dam (eds.), *Animal Models of Dementia*, Neuromethods, vol. 48, DOI 10.1007/978-1-60761-898-0_30,

1. Introduction

For most of the last century, adult-onset dementia disorders were generally attributed to arteriosclerosis and consequently chronic cerebral ischemia. The increasing capabilities of diagnosing Alzheimer's disease and the demonstration that infarcts more likely than chronic ischemia were the basis of dementia of vascular origin brought about the introduction of the concept of multiinfarct dementia. The definition of multiinfarct dementia was replaced by the term "vascular dementia" (VaD) as a consequence of the recognition that different etiologies apart from multiple infarcts, such as single infarct, episodes of hypotension, leukoaraiosis, incomplete infarction, and hemorrhage in specific brain areas, elaborating cognitive functions can be involved (1, 2). However, in the currently accepted definition of dementia, the presence of memory disturbances is considered as the *sine qua non* condition for a diagnosis of dementia, whereas at least in the early phases of cognitive impairment associated with cerebrovascular disease, executive control dysfunction occurs earlier than the development of other symptoms. This observation led to the development of the concept of Vascular cognitive impairment (VCI), as a part of the broader concept of vascular cognitive disorder (VCD) (1, 2).

Arterial hypertension represents a well-known risk factor not only for stroke but also for VCI (3–6). Several studies have suggested a positive correlation between hypertension and cognitive deterioration (7, 8). Patients with ischemic VaD have a significantly greater incidence of hypertension than patients with dementia of the Alzheimer's type (9). Neuropathological and neuropsychological investigations have shown that areas of white matter (WM) hyperintensity and ventricular enlargement are more common in hypertensive subjects compared with that in normotensive individuals. An association between hypertension and reduced Cerebral blood flow (CBF) and VCD is documented (3–5). Longitudinal studies have also suggested that hypertension in midlife is most associated with a higher probability of cognitive impairment than hypertension developed later in life (10).

Arterial hypertension is a multifactorial and polygenic disease involving complex interactions between genetically determined homeostatic control mechanism and environmental factors (11, 12). Exploration of such a complex syndrome requires the availability of animal models. Essential hypertension is the most common form of arterial hypertension in human beings. Multiple genes contribute to the individual phenotype, each expressing diverse alleles and degree of penetration.

The ideal animal model for hypertension research should have human-like cardiovascular anatomy, hemodynamics, and physiology and should develop a disease course and complications in a timely or even accelerated fashion (13). The phenotype-driven experimental approach takes advantage of a natural variation among inbred strains and crosses to find quantitative traits and determine which genes are responsible (14). The phenotype-driven rat experimental model is the most abundant for hypertension research and has particular potentials for exploration of polygenic hypertension. In general, development of homozygous hypertensive rat strains is achieved by selective breeding of animals displaying the desired phenotype over several generations. Once a trait is fixed, sib mating is maintained for about 20 generations to achieve genetic homogeneity. This approach has been useful to develop fundamental models such as spontaneously hypertensive rat (SHR) (15) or stroke-prone SHR (SHR-SP) derived from the Wistar strain (16, 17).

Much progress in experimental research was made with the development of rat strains with characteristics of human disease to evaluate the pathophysiology and the pharmacological treatment of a given disease. Although an animal model for VBD does not exist, SHR currently represents the best model of essential hypertension, allowing investigation of causes, mechanisms, and pathology of hypertension (18).

In this chapter, the main characteristics of cerebrovascular and brain damage in SHR are reviewed, and the validity of this rat strain as an animal model of VBD is discussed.

2. Development and Main Characteristics of Spontaneously Hypertensive Rats

SHR and SHR-SP are two widely used animal models of hypertension and hypertension plus stroke, respectively (19–22). They are normotensive at birth and gradually develop severe hypertension in the first 2–4 months of life. At 6 months, they show a sustained hypertension (23, 24). The development of the SHR strain began in 1950s by Okamoto and coworkers. The selection of hypertensive rats started by mating male Wistar Kyoto rats (WKY) with mild hypertension (systolic pressure 145–175 mmHg) and females with relatively high blood pressure (systolic pressure 130–140 mmHg). Through their efforts to select hypertensive offspring by repeatedly checking the blood pressure, they finally were able to establish a colony of rats developing hypertension without exception and, in 1963, reported them as SHR (15–17) (Fig. 1).

To develop a suitable model of hypertension-related human cardiovascular pathology, a new substrain of SHR, namely the SHR-SP strain, was established through selective inbreeding of the SHR offspring of animals that had died of stroke. In SHR-SP, more than 80% of the population develops stroke (25). SHR and SHR-SP spontaneously develop moderate-to-severe hypertension and are

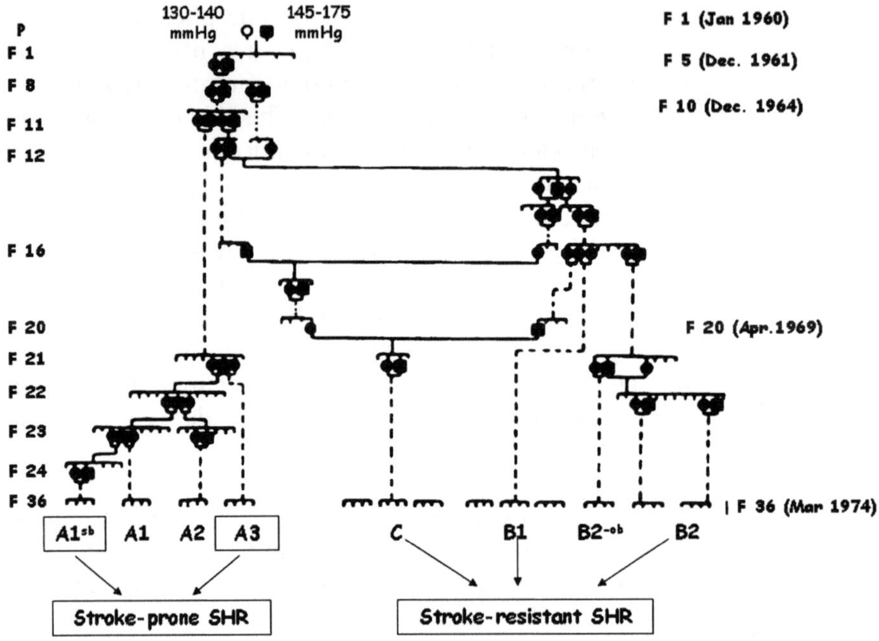

Fig. 1. Genealogy of the spontaneously hypertensive rat (SHR). Modified by (16).

different from the Dahl salt-sensitive or the Sabra-hypertensive strain, which require salt loading or deoxycorticosterone acetate (DOCA) salt administration for the development of a severe hypertension (26, 27). SHR generally develop hypertension with a systolic blood pressure plateau of about 200 mmHg. SHR-SP show a more rapid increase in blood pressure from young age and reach blood pressure values of approximately 240 mmHg. SHR, and to a greater extent SHR-SP, have lower body weight compared with WKY rats. SHR-SP, when compared with SHR, show various symptoms of cerebrovascular involvement such as irritability, hyperkinesis, hypokinesis, and paralysis (17).

The genetic mechanisms of hypertension in SHR have been attributed to both neural and vascular alterations. Individual variations in the genetic background of SHR may influence evolution of hypertension and end-organ damage resulting from elevated arterial pressure (28). Three major genes are involved in early development of hypertension, whereas an additional gene identified on chromosome 10 contributes to development and maintenance of hypertension during aging in SHR (29). Androgens probably modulate hypertension and age-related renal injury in this model (30) by altering natriuretic mechanisms and tubuloglomerular feedback (31) in a manner potentially resembling human hypertension. This prominent role of androgens in SHR suggests that the model could imitate postmenopausal hypertension as well (32).

The possibility to develop a 100% hypertensive strain of rats by selective inbreeding demonstrates that the heredity itself is an extremely important factor. Fingerprint analyses in SHR-SP and WKY disclosed various DNA polymorphisms. Linkage of these polymorphisms has allowed the identification of a Bp1 gene on rat chromosome 10 with a LOD score of 5.10. This gene is linked to the gene encoding angiotensin-converting enzyme (ACE), an enzyme playing a major role in blood pressure homeostasis and which is an important target of antihypertensive drugs (33–35). Analysis using molecular biology techniques and cDNA microarray has identified genes differentially expressed between SHR and WKY rats (36). Twenty-four genes were found to be upregulated and 20 were downregulated in SHR, even though the identification of differentially expressed gene may not be an efficient method for selecting candidate genes for hypertension in the SHR-WKY system (36).

In general, in most phenotype-driven models, hypertension is associated with cardiac hypertrophy, endothelial dysfunction, and renal functional impairment (proteinuria and decreased creatinine clearance). Other complications of hypertension such as cardiac insufficiency and end-stage renal disease are not constantly observed (37). The outcomes seem to depend on the underlying origin, genetic background, and possibly species differences, as well as on the degree of hypertension.

In SHR, cardiac function is preserved at 1 year of age but tends to fail between 18 and 24 months. At this age, 57% of SHR have a myocardial decompensation that can be identified by clinical signs. On echocardiography, left ventricular diameters are increased together with a reduced left ventricular fractional shortening and left ventricular filling pressure (38). SHR with heart failure are characterized by increased myocardial apoptosis and neurohumoral changes in comparison with SHR without heart failure (39, 40).

SHR with heart failure appear to be the ideal model to mimic hypertension-induced heart failure in humans due to the expression both in the animal model and in human beings of relevant cardiac, hemodynamic, and neurohumoral changes during the transition from compensated hypertrophy to failure. Another advantage of this model is that it does not require technical interventions and is not complicated by the mortality rate caused by surgical or pharmacological manipulations. A main limitation of this model is the long time required until the development of heart failure over a period of approximately 2 years.

3. Cerebrovascular Anatomy and Physiology in SHR

SHR were the model most extensively investigated for assessing hypertensive brain damage and treatment of it. The time-dependent rise of arterial blood pressure, the occurrence of brain atrophy, loss of nerve cells, and glial reaction are phenomena shared to some extent with hypertensive brain damage in humans. SHR, therefore, can represent a reasonable model of morphological damage induced by hypertension and possibly reversed by appropriate antihypertensive treatment (23).

In hypertensive humans, CBF is comparable to normotensive individuals. The constancy in CBF is achieved through a rise of cerebrovascular resistance, because the median blood pressure is increased. In chronic hypertension, cerebral blood vessels show hypertrophy and hyperplasia, limiting their capacity to dilate. Hence, chronic hypertension increases resistance of large cerebral arteries, leading to protection of microcirculation from elevated blood pressure (41). While these changes protect the brain from high perfusion pressures, they impair dilation of the vessels at low blood pressures and therefore impair adaptation mechanisms of brain circulation (42, 43).

Cerebrovascular alterations typical of hypertension are closely related to the degree of blood pressure elevation and consist in hypertrophy or remodeling. Hypertrophy consists in thickening of arterial wall with decrease of internal diameter, increase of external diameter and consequently reduction of lumen size. Remodeling consists of smooth muscle rearrangement with decrease of both external and internal diameters and lumen size reduction (44–47). These changes collectively increase the wall-to-lumen ratio and consequently arterial resistance.

In terms of possible anatomical selectivity of cerebrovascular involvement in SHR, frontal cortex of 24-week-old SHR displayed wall hypertrophy and luminal narrowing (Fig. 2) (48, 49). In the striatum, the increase of wall area was not accompanied by luminal narrowing, indicating the occurrence of arterial hypertrophy without vasoconstriction (49). In hippocampal arteries of SHR, luminal narrowing was not accompanied by changes of wall area (49). In line with the hypothesis of a prevalent smooth muscle hypertrophy, cerebral arteries of SHR aged 24 weeks display prominent lumen diameter reduction compared with normotensive WKY rats (Fig. 2) (49). An important parameter in the analysis of vascular changes is the values of wall-to-lumen ratio, which directly relates to vascular resistance (50). It has been emphasized that the increase of the wall-to-lumen ratio is an important factor in the pathogenesis and maintenance of hypertension (51). Assessment of wall-to-lumen ratio, plotted versus lumen diameter of vessels in 24-week-old SHR (Fig. 2), revealed an increased sensitivity of intracerebral vessels to hypertension parallel with the decrease of luminal diameter (49). In general, smallest arterioles were the vessels most sensible to hypertension and may represent candidates to degenerative phenomena leading to brain hypertensive damage. These findings are compatible with data of a significant increase of smooth muscle actin-immunoreactive areas in pial arteries of SHR, without

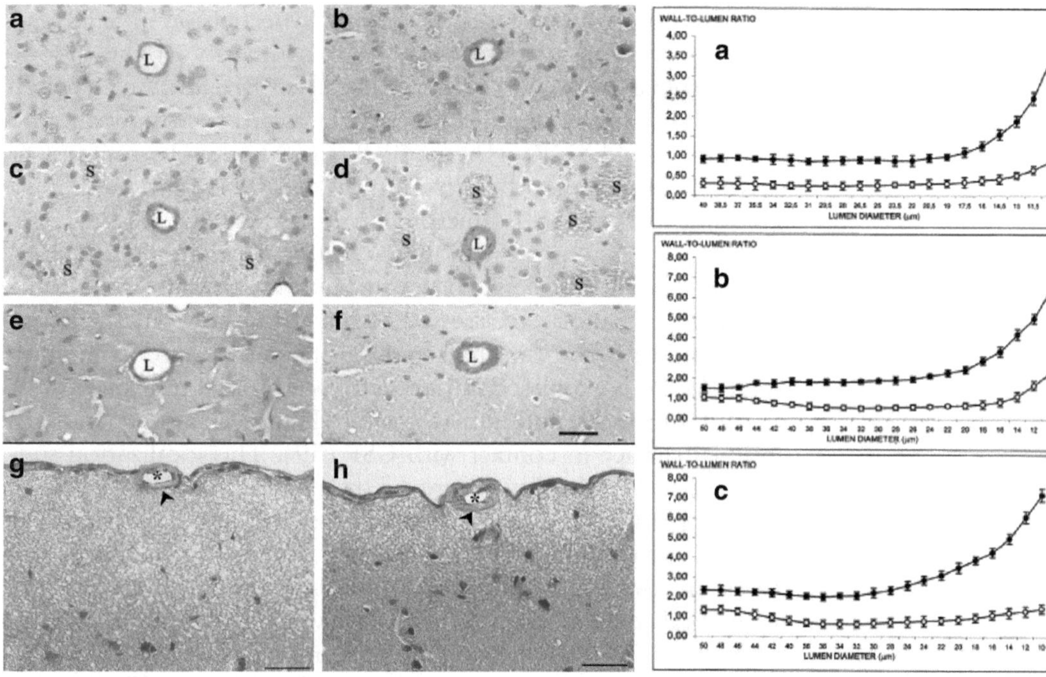

Fig. 2. *Left-hand panel* – Micrographs of arterioles of frontal cortex (pictures **a** and **b**), striatum (pictures **c** and **d**), and hippocampus (pictures **e** and **f**) in Wistar-Kyoto (WKY) rats (**a**, **c**, and **e**) and spontaneously hypertensive rats (SHR) (**b**, **d**, and **f**) aged 6 months. Micrographs of pial artery of 6-month-old WKY rats (**g**) and age-matched SHR rats (**h**). Hematoxylin and eosin. In intracerebral arterioles of frontal cortex, striatum, and hippocampus of SHR, an increased thickness of the wall and luminal narrowing were found in comparison with WKY rats. L, arterial lumen; s, striosomes. Calibration bar: 30 µm. *Right-hand panel* – Plotting of wall area values versus lumen diameter in arteries (lumen diameter <46 µm) and arterioles (lumen diameter between 46 and 10 µm) of frontal cortex (panel **a**), striatum (panel **b**), and hippocampus (panel **c**) in Wistar-Kyoto (WKY) rats (○) and SHR (●). Data are the mean ± S.E. In frontal cortex and striatal vessels, differences were statistically significant ($P < 0.01$ versus WKY rats) in all different-sized vessels examined. In hippocampus, differences were statistically significant ($P < 0.01$ versus WKY rats) since vessels with a lumen diameter of <40 µm (Adapted from (49)).

increase in the collagen content (52). Parenchymal small arteries of SHR-SP undergo fibrinoid necrosis with a decrease of smooth muscle actin, whereas small arteries perfusing WM do not show these changes (53).

A moderate increase in arterial blood pressure induces cerebral blood vessel autoregulation and consequently constriction (54). With severe increase in pressure, the autoregulatory capacity of cerebral vessels is exaggerated and large cerebral arteries dilate (55). This dilatation may lead to increased stress of the vessel wall, with consequent probable blood–brain barrier (BBB) damage. In chronic hypertension, vascular changes might influence the susceptibility of BBB. Adult SHR cerebral vessels are less susceptible to disruption of the BBB induced by rise of blood pressure than WKY rats (56). This does not occur in young SHR, which are as susceptible as WKY rats to cerebral complications of abrupt elevation of blood pressure (57).

It has been suggested that chronic hypertension in SHR may cause more pronounced defects in the integrity of the blood–cerebrospinal fluid (CSF) barrier than in the BBB (58). An increased permeability of the blood–CSF barrier toward sucrose was found in 4-month-old SHR, whereas BBB was found to be resistant to passage of sucrose and lanthanum in both SHR and WKY rats. This indicates that BBB functionality is maintained (58).

The expression of aquaporin 4 (AQP4), a membrane channel protein serving as selective pores through which water crosses the plasma membranes, was assessed by one group to verify the integrity of the BBB and the characteristic of cerebral water content in SHR. AQP4 is expressed in astrocyte foot process near blood vessels in rat (59, 60) and human brain (61), as well as in ependymal and pial surface in contact with CSF (59). This localization suggests that AQP4 plays a critical role in brain water balance. Different studies indicate that the expression of brain aquaporins is sensitive to brain injury, swelling, and other experimental procedures. In rodents, AQP4 is upregulated in astrocytes in response to cerebral edema caused by brain injury (62, 63), focal brain ischemia (64), and hyponatremia (65). In rat models of brain injury, AQP4 is overexpressed in astrocytes at the site where BBB is disrupted (62). Figure 3 shows AQP4 immunohistochemistry in frontal cortex of 6-month-old SHR and age-matched WKY rats. An intense immunoreaction is noticeable in cerebral vessels (Fig. 3), with a remarkable increase of immune reaction in cerebral vessels of SHR (Fig. 3b). These findings suggest that an increased expression of AQP4 in the brain of SHR may be implicated in impaired autoregulation of CBF and water content in this animal model. Based on these data, the occurrence of a cytotoxic edema in SHR cannot be excluded. This may contribute to the development of brain lesions documented in SHR.

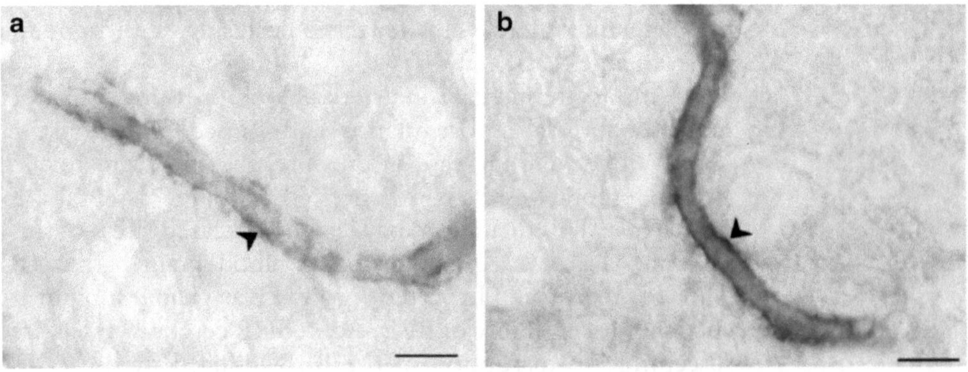

Fig. 3. Micrographs of arterioles of cerebrocortical white matter (WM) in a Wistar-Kyoto (WKY) rat (**a**) and spontaneously hypertensive rat (SHR) (**b**) of 6 months. Aquaporin 4 (AQP4) immunohistochemistry. Note in SHR the increased expression of AQP4 immunoreactivity indicating enhanced blood–brain barrier permeability. Calibration bar: 15 μm.

4. Gross and Microanatomy of SHR Brain

Ventricular enlargement accompanied by loss of brain tissue, and decrease of brain weight and volume of grey matter were reported in 8-week-old SHR (66–68). Altered anatomical organization of hypothalamic regions (69) and gross morphological brain differences between SHR and age-matched normotensive WKY rats were reported (70). At 14 weeks of age, SHR display little differences in brain morphology compared with age-matched WKY rats (71). Around the age of 26 weeks, brain microanatomical changes become obvious in SHR (68, 71, 72). In 6–7-month-old SHR, ventricular volume is twofold greater in SHR than in WKY rats; the volume of the entire brain and of different gray matter areas as well as the thickness of cerebral cortex were 11–25% less in SHR (Fig. 4). Some investigators reported the same neuronal density in SHR and WKY rats at 6–7 months of age (67). However, the presence of less neurons in different brain areas of SHR is obvious if density/number of nerve cell profiles is corrected by the volume, which is decreased in SHR compared with their normotensive cohorts (68, 71). A more recent study employing magnetic resonance imaging (MRI) has reported significantly smaller caudate-putamen and cerebellar vermis in 4–6-week-old SHR compared with age-matched normotensive WKY rats (Fig. 5) (73).

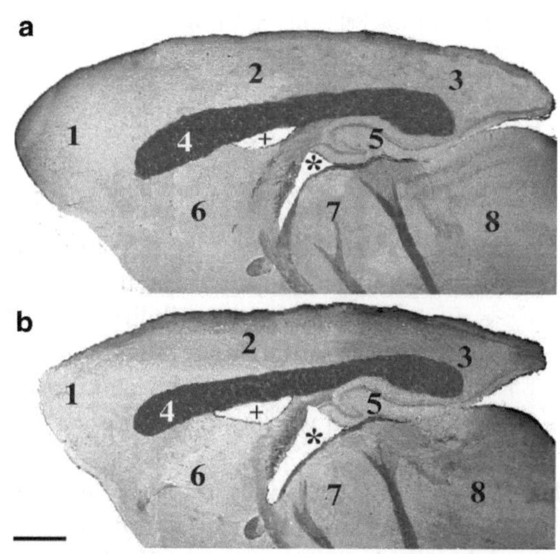

Fig. 4. Parasagittal section of brain of a 6-month-old normotensive male Wistar-Kyoto (WKY) rats (**a**) and age-matched spontaneously hypertensive rats (SHR) stained with Luxol fast blue. Note in SHR the enlargement of lateral ventricle (+) and of the third ventricle (asterisk), a decreased thickness of cerebral cortex (2) and a reduced size of corpus callosum (4). 1: frontal cortex; 3: occipital cortex; 5: hippocampus; 6: striatum; 7: dienchephalon; 8: midbrain. Calibration bar: 1.5 mm (Adapted from (72) with permission from Wiley-Blackwell).

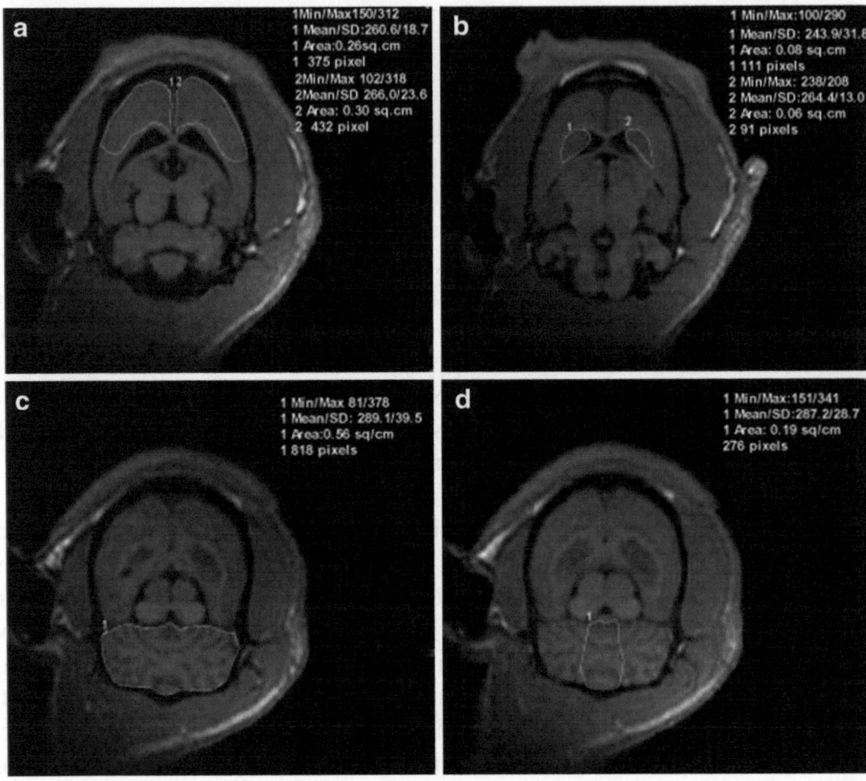

Fig. 5. Magnetic resonance imaging (MRI) data from coronal sections of the brains of spontaneously hypertensive (SHR) and Wistar-Kyoto (WKY) rats. Image **a** shows the prefrontal area; **b** shows caudate nucleus; **c** shows the cerebellum; **d** shows the vermis cerebelli (Adapted from (71) with permission of publication from Elsevier).

The hippocampus, a key area for learning and memory, is smaller in 6-month-old SHR compared with that in normotensive cohorts (66, 74–76), as well as the density of nuclei in the CA1 subfield in SHR aged 6–7 months (67) (Fig. 6). Data of local cerebral glucose utilization in most gray matter areas of SHR (77, 78) suggest depressed neuronal activity in SHR around age 6 months. A greater susceptibility of SHR to brain ischemia is documented. Brain infarct size and behavioral abnormalities caused by middle cerebral artery (MCA) occlusion were greater in SHR than in WKY rats 3 weeks after occlusion of the right MCA (79).

4.1. White Matter Changes

WM changes are often associated with risk factors for stroke such as age and arterial hypertension (80). In aging and hypertension, the occurrence of WM lesions increases as documented by neuroimaging techniques, such as computerized tomography or magnetic resonances imaging (81, 82). WM lesions are closely associated with cognitive impairment and gait disorders in the elderly, and, therefore, may represent the anatomical correlate of complications of cerebrovascular disease and a predictive factor for the development of cognitive impairment and stroke (83).

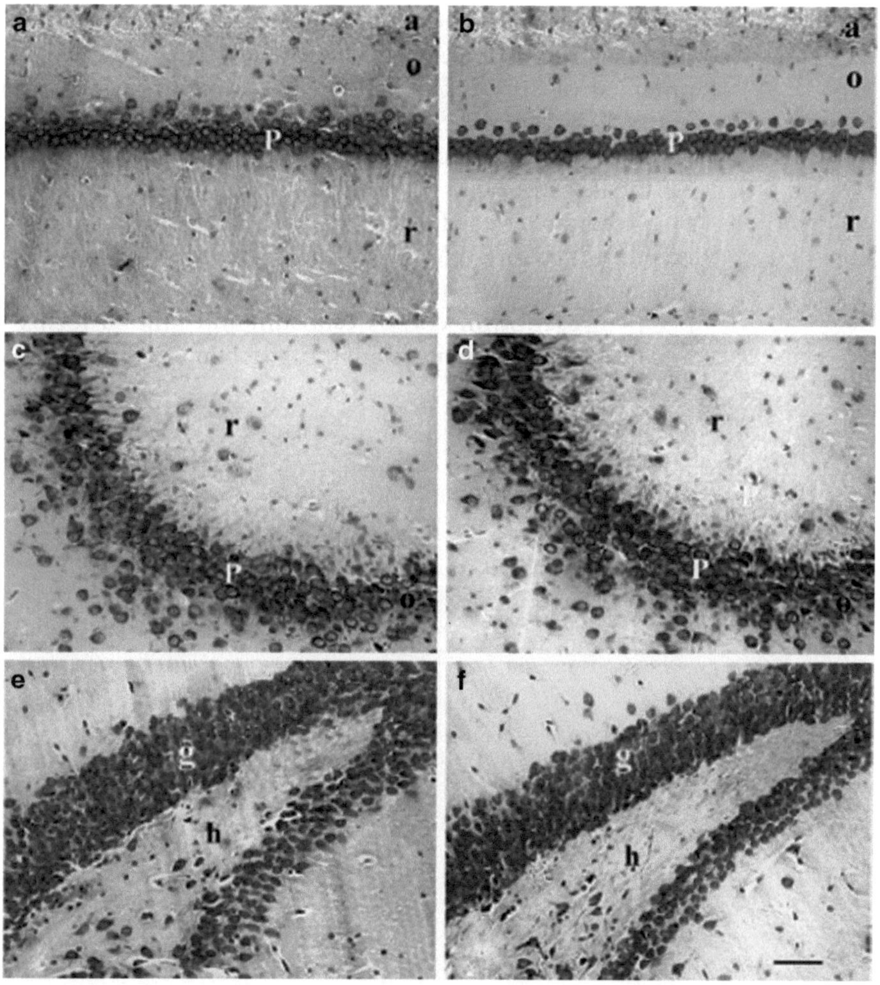

Fig. 6. Sections of the CA1 subfield (**a**, **b**), CA3 subfield (**c**, **d**) and dentate gyrus (DG) (**e**, **f**) of the hippocampus of 6-month-old normotensive Wistar-Kyoto (WKY) rats (**a**, **c** and **e**) and age-matched spontaneously hypertensive rats (SHR) (**b**, **d** and **f**). Cresyl Violet staining. Note in SHR the decreased density of nerve cell profiles in the CA1 subfield and to a lesser extent in the DG. No changes in the number of nerve cell profiles can be seen in the CA3 subfield between WKY rats and SHR. P: pyramidal neurons; a: alveus; o: stratum oriens; r: stratum radiatum; g: granule neurons; h: hilum. Scale bar: 50 μm (Adapted from (75)).

Data on WM in SHR are sparse. Smaller volumes of anterior commissure and genu of the corpus callosum were reported in 6–7-month-old SHR in comparison with WKY (67). A more systematic microanatomical study revealed a remarkable decrease in volume of internal capsule and striosomes, a moderate reduction in volume of the corpus callosum, and no changes in volume of the external capsule and of WM of hippocampus (68). The loss of forebrain WM in SHR is more pronounced in areas bearing fibers originating or ending in the cerebral cortex. These changes probably reflect the reduction of afferent or efferent cerebrocortical fibers to cerebral cortex (68). WM lesions characterized by disarrangement of nerve fibers, occurrence of vacuoles and/or disappearance of myelinated fibers were reported only in caudate-putamen of 20-week-old SHR. The same group also reported a marked WM increase in different areas (corpus callosum, anterior commissure, internal capsule, and caudate-putamen) of 20-week-old SHR-SP (52).

Recent studies of our group enclosed evidence of a microglial reaction in brain WM of 6-month-old SHR. Histochemistry of Bandeira simplicifolia isolectin BSI-B_4 revealed an increase of positive reaction in corpus callosum and fimbria fornix (Fig. 7) and no obvious changes in the striosomes. Quantitatively, this

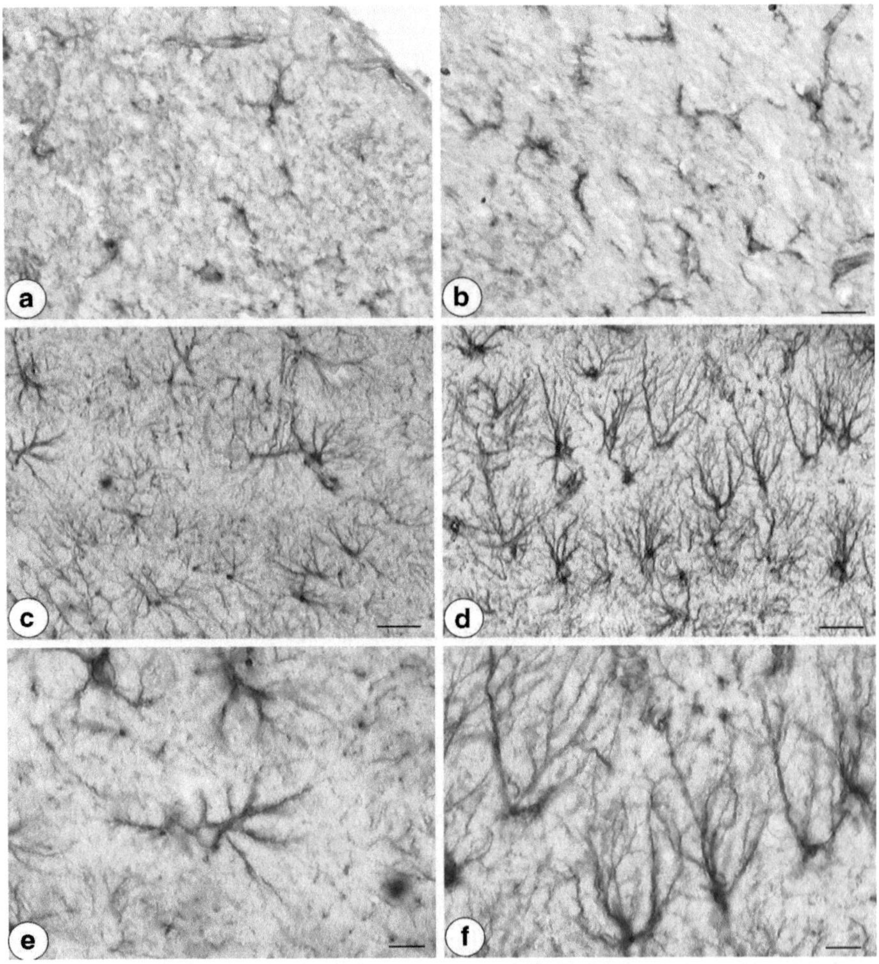

Fig. 7. (**a** and **b**) Microglial reaction in Fimbria Fornix from *Bandeira simplicifolia* isolectin BSI-B_4 binding. (**a**) 6-month-old Wistar-Kyoto (WKY) rat; (**b**) 6-month-old spontaneously hypertensive rat (SHR). Note in SHR the activation of microglia. Calibration bar: 25 μm. (**c–f**) Sections of rat dentate gyrus processed for glial fibrillary acidic protein (GFAP) immunohistochemistry. Pictures show parenchymal astrocytes. (**c**, **e**) 6-month-old WKY; (**d**, **f**) 6-month-old SHR. Note in SHR, the increase of dark-brown GFAP immunoreactive astrocytes in comparison with WKY rats. Calibration bar **c**, **d**: 25 μm. Calibration bar E, F: 10 μm.

phenomenon was more pronounced in brain areas undergoing volume reduction (84).

Major histocompatibility complex (MHC) class I antigen-immunoreactive microglia were found primarily in the WM of SHR-SP and to a lesser extent in that of SHR but not in normotensive WKY rats (52). These findings, which are consistent with human data reported, suggest an involvement of parenchymal microglia in hypertension. Hyperactivation of microglia may cause neuronal degeneration, due to release of proinflammatory cytokines such as interleukin-1, interleukin-6, and tumor necrosis factor-α (85). On the other hand, microglial activation results from the engagement of degenerative and phagocytic processes following the formation of necrotic or apoptotic tissue after neuronal injury (86).

Spontaneous hypertension after bilateral common carotid artery ligation (BCAL) in SHR may accelerate WM lesions with vacuolation of nerve fibers, microglial activation, and oligodendroglial DNA fragmentation (87). It has been suggested that apoptosis-like cell death of oligodendrocytes was induced in the WM of hypoperfused SHR brain. The development of oligodendroglial DNA fragmentation in the WM might be mainly due to cerebral hypoperfusion. Increase of infarct size and delayed recovery of local CBF have been reported in SHR after BCAL (88). Besides hypertension, acute reduction and delayed recovery of local CBF in SHR might affect oligodendroglial vulnerability (87).

4.2. Astroglia

Astroglia perform many functions, including cellular support during central nervous system development, ion homeostasis, and constitution of the BBB. They also provide neuronal nutrition, take up neurotransmitters, and contribute to neuroimmune function (89–91). The cytoskeletal protein glial fibrillary acidic protein (GFAP) represents a selective marker for mature astrocytes (92). Different studies have assessed the functional significance of this astrocyte protein using genetic knockout animals (93–95). Astrocytes in GFAP knockout animals show disturbed neuronal plasticity, loss of long-term depression (94), late-onset dysmyelination (93), and increased susceptibility to ischemia (96). Astrocytes and microglia respond to a great variety of lesions in the nervous tissue. In this sense, although astroglial cells react in a relatively constant manner, the intensity and time course of these changes largely depend on the type of lesion involved. Transformation of quiescent into reactive astrocytes is termed astrogliosis or reactive gliosis and may result in the formation of a glial scar. This phenomenon occurs primarily in the immediate vicinity of a lesion (89). Astrocytes distal from an injured area do not undergo extensive and permanent changes (97).

Several different central nervous system injuries are accompanied by proliferation and/or hypertrophy of astrocytes. An increased expression of brain GFAP and of GFAP mRNA was reported in association with aging or disease (98–101) and also specifically in brain neurodegenerative disorders (102, 103). A significant increase of GFAP mRNA and an increase of GFAP immunoreactivity were noticeable in various brain areas of SHR compared with those in normotensive WKY rats at 6 months, but not at 2 or 4 months of age (104) (Fig. 7). Immunohistochemistry revealed a numerical (hyperplasia) and size (hypertrophy) increase of GFAP-immunoreactive astrocytes in frontal cortex, striatum, and hippocampus of SHR (Fig. 7) (75, 76, 104, 105). The occurrence of astrogliosis in the brain of SHR with established hypertension suggests that hypertension induces a condition of brain suffering enough to increase biosynthesis and expression of GFAP, similarly as reported in several neurodegenerative disorders and in brain ischemia.

5. Neurochemistry and Chemical Neuroanatomy of SHR

Besides the gross and microanatomical changes described earlier, SHR present changes of some specific neurotransmitter systems that may have functional and behavioral relevance. Cholinergic, dopaminergic, and noradrenergic systems are most affected and extensively investigated.

5.1. Cholinergic System

Acetylcholine (ACh) plays an important role in central, peripheral, autonomic, and enteric nervous system neurotransmission. ACh is also involved in the endothelium-dependent vasodilatation in several vascular beds and is present in different nonneuronal cells (106). Brain cholinergic neurons form a global network characterized by nerve cell bodies located in selected areas from which ascending and, to a lesser extent, descending projections originate (Fig. 8). Ascending cholinergic pathways supplying cerebral cortex are involved in important higher brain functions such as attention, arousal, motivation, memory, and consciousness (106). The cholinergic system is altered in age-related neurodegenerative diseases, such as Alzheimer's disease, and is vulnerable to the effects of aging (106). A decline in the integrity of the cholinergic system characterized by a decrease of the ACh biosynthetic enzyme choline acetyltransferase (ChAT) and by a reduction of CSF ACh levels was also reported in patients with VaD (107). This, together with the observation of microanatomical changes typical of hypertensive brain damage in SHR, as well as of deficits in central nicotinic cholinergic receptors, has contributed to the validity of SHR as a possible model of age-related human neurological disorders characterized by impaired cholinergic function (108).

Cholinergic neurons localized in various brain regions are implicated in blood pressure regulation, and central ACh may

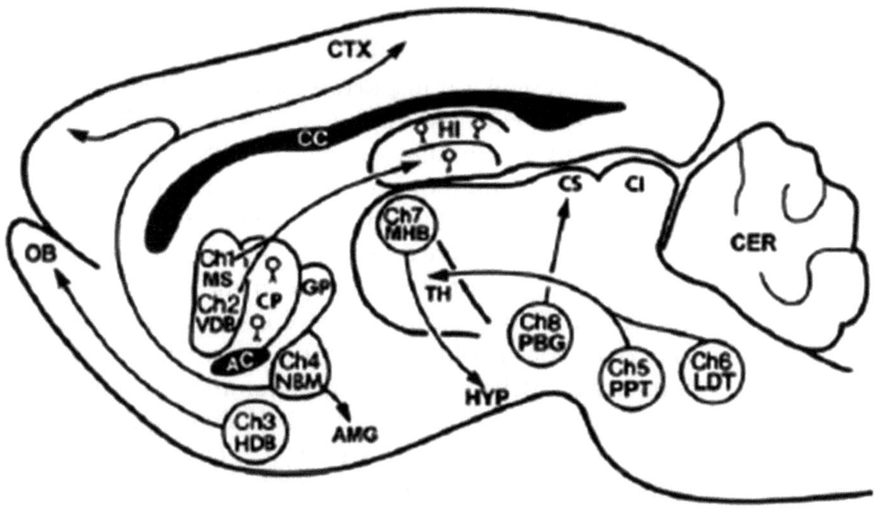

Fig. 8. Main cholinergic nuclei and their projections represented in rat brain. Ch1-Ch8: main cholinergic brain nuclei; AC: nucleus accumbens; AMG: amygdala; CC: corpus callosum; CTX: cerebral cortex; CER: cerebellum; CS: superior colliculus; CI: inferior colliculus; CP: caudate-putamen; GP: globus pallidus; HDB: horizontal diagonal band; HI: hippocampus; HYP: hypothalamus; MS: medial septum; NBM: nucleus basalis magnocellularis; OB: olfactory bulb; PBG: parabigeminal nucleus; TH: thalamus; VDB: ventral diagonal band; MHB: mesopontine tegmental nuclei; PPT: pedunculopontine tegmental nuclei; LTD: laterodorsal tegmental nuclei (Adapted from (106) with permission of publication from Bentham Science Publishers Ltd).

contribute to the pathogenesis of experimental hypertension. Brain regions expressing a cholinergic system related to hypertension are the rostral ventrolateral medulla (RVL), the posterior hypothalamus, and the lateral septal area (109, 110). The RVL is the major source of efferent sympathetic activity (111), and its cholinergic mechanisms are involved in the regulation of the cardiovascular system (110). Cholinergic function in RVL is enhanced in SHR, which could contribute to the development and/or maintenance of hypertension in this strain (110).

ChAT activity was reported to be enhanced in the brain stem of SHR (112). Intravenous administration of the ACh degradation inhibitor physostigmine, with or without coadministration of the peripheral muscarinic receptor antagonist methylatropine, evokes an exaggerated pressor response in unanesthetized SHR compared with normotensive WKY rats. In contrast, there is no difference in the hypertensive response to the cholinergic muscarinic receptor agonist oxotremorine in SHR or WKY rats pretreated with methylatropine (113).

The occurrence of an impaired cerebral cholinergic system, both in SHR and in SHR-SP, is documented. In SHR-SP, alterations of the cholinergic cerebral system consisting of a decrease of cerebrocortical and hippocampal levels of choline and ACh, as well as in cerebral spinal fluid, were found (114–116). Changes of ACh levels in cerebral cortex and hippocampus are related to impaired learning and memory functions (116, 117).

More recent studies have assessed the expression of vesicular ACh transporters (VAChT) in SHR (108, 118, 119). In nerve terminals, vesicular transporters pack neurotransmitters into synaptic vesicles, essential for neurotransmitter release. VAChT was investigated by immunochemical (108), and both immunochemical and immunohistochemical (118, 119) techniques in 4 and 15- (108) and 6-month-old SHR (118, 119). The studies reported a slight increase of the expression of the transporter in the hippocampus of the older age group investigated (108), and similarly to observed in frontal cortex, hippocampus and striatum of 6 months-old SHR (119). Treatment with the acetylcholinesterase (AChE) inhibitor galantamine countered the changes of VAChT expression in SHR (119). This suggests a greater sensitivity of the cholinergic system to pharmacological manipulation in this hypertensive strain. In humans, higher levels of VAChT expression were reported in early stages of Alzheimer's disease (120), as well in the hippocampus of individuals with mild cognitive impairment (MCI) (121, 122). These changes were considered as an up-regulation mechanism to counter cholinergic hypofunction in early phases of adult-onset cognitive dysfunctions. The same is probably true in 15-month-old SHR (108) in which the increased expression of VAChT may represent an attempt to compensate cholinergic impairment at this stage of hypertensive brain damage. Focal ischemia did not affect the expression of VAChT in SHR (123).

In SHR, learning and memory abilities explored with behavioral tests (spatial memory, water maze, conditioning tests, and radial arm maze) revealed some impairments (108, 124–127). These findings, combined with observations of deficits in nicotinic ACh receptors in various brain regions, including cerebral cortex, hippocampus, thalamus, and striatum in SHR compared with those in WKY rats (128–130), further support the view of impaired cholinergic neurotransmission in the brain of SHR. The reduced density of central nicotinic ACh receptors in prefrontal cortex, caudate-putamen, and enthorinal cortex but not in the hippocampus of 15-month-old SHR could, in part, explain the modest spatial learning decrements by SHR compared with that in WKY (130). Analysis of central muscarinic cholinergic receptors in the same animal model demonstrated a significant increase of M1 receptors in cingulate cortex, neostriatum, CA1 and CA3 subfields of hippocampus, posterior hypothalamic area, and substantia nigra (108, 126). An increased M2 receptor density was reported in olfactory tubercle, nucleus accumbens, and the basolateral amygdaloid nucleus in 12-week-old SHR (127) and in the neostriatum of 15-month-old SHR compared with age-matched WKY rats (108). The increased expression of muscarinic cholinergic receptors in SHR might represent a postsynaptic up-regulation consequent to reduced cholinergic cerebrocortical projections with age and/or a compensatory phenomenon following central nicotinic cholinergic receptor deficiency in older SHR (108).

In summary, similarly as reported in VaD, an impaired cholinergic neurotransmission characterizes both SHR and SHR-SP. The possibility that cholinergic neurotransmission mechanisms contribute to central nervous system plasticity or may improve brain circulation in this animal model is suggested by our recent studies detailed in the section 6.4. In these experiments, we have observed that treatment of SHR with the AChE inhibitor galantamine or with the cholinergic precursor choline alphoscerate (alpha-glyceryl-phosphoryl-choline), alone or in association, although not affecting blood pressure levels, exerts a neuroprotective effect countering typical hypertension-dependent microanatomical changes in SHR (131).

5.2. Dopaminergic System

SHR exhibit behavioral abnormalities, such as hyperactivity, lower attention, inhibition deficit, and hyperreactivity to stress, that resemble behavioral abnormalities of human attention-deficit with hyperactivity disorder (ADHD). Because dopamine has been implicated in the pathogenesis of ADHD, possible abnormalities of this system were extensively investigated in SHR (132).

In SHR, terminals of mesocortical, mesolimbic, and nigrostriatal dopaminergic neurons release less dopamine in response to electrical stimulation and/or depolarization as a result of exposure to high extracellular K^+ concentrations, than WKY (132, 133). Dopamine transporter (DAT) and vesicular storage of dopamine is impaired in SHR. DAT gene expression is transiently reduced in SHR midbrain during the first postnatal month and increased in adult SHR compared with controls (134, 135). Alterations in DAT gene expression can affect dopamine uptake and reutilization. This causes leakage of dopamine into the cytoplasm and increases d-amphetamine-induced transporter-mediated release. Release of dopamine evoked by electrical stimulation is reduced in prefrontal cortex of SHR, suggesting dopaminergic hypofunction in this cerebrocortical area. This hypofunction alters autoreceptor-mediated inhibition of noradrenaline release in SHR (136). From a receptor point of view, postsynaptic dopamine D1-like/D2-like receptors in the SHR are hyposensitive, while presynaptic dopamine D2-like receptors located particularly in the nucleus accumbens are hypersensitive compared with normotensive WKY rats (137). More recent studies have further characterized the dopaminergic system of SHR by analyzing genes coding for the five major subtypes of dopamine receptor (D1, D2, D3, D4, and D5), for enzymes responsible of catecholamine synthesis tyrosine hydroxylase and dopamine-beta-hydroxylase, and for DAT. Significantly lower expression of dopamine D4 receptor gene and protein synthesis was observed in the prefrontal cortex of SHR. No differences in the same genes were observed in the midbrain, prefrontal cortex, temporal cortex, striatum, or amygdala of SHR and WKY. These data, consistent with the situation seen in patients with ADHD, suggest a downregulation of dopamine D4 receptors, rather than a general downregulation of catecholamine synthesis in prefrontal cortex of SHR (73). In conclusion, the dopaminergic system is characterized by a hypofunction affecting primarily prefrontal cortex. DAT and dopamine D4 receptor genes are most probably involved in the dysfunctions of the dopaminergic system in specific brain areas, which resemble those identified in patients with ADHD.

5.3. Noradrenergic System

Analysis of SHR as a model of ADHD has also shown that, similarly as reported in the prefrontal cortex of children with ADHD, noradrenergic neurons of this cerebrocortical area are hyperfunctional in this animal model too.

Locus coeruleus is one of the main brain noradrenergic nuclei. Noradrenergic projections originating from this nucleus supply the entire cerebral cortex, various subcortical areas, cerebellum, and spinal cord. These nerve pathways are involved in attention, arousal, orienting, and vigilance (138). Noradrenergic projections to the prefrontal cortex modulate the transfer of information through neuronal circuits responsible for selective and sustained attention (138). In SHR, compared with normotensive WKY rats, noradrenaline levels are increased in several brain areas, including locus coeruleus, substantia nigra, and prefrontal cortex (139). This increase is probably because of an enhanced expression of the catecholamine biosynthetic enzyme tyrosine hydroxylase in the ventrolateral medulla oblongata (140, 141). Parallel to the increased noradrenaline levels, a downregulation of cerebrocortical alpha2-adrenoceptors, with decreased mRNA levels for the receptor and less-efficient alpha2-adrenoceptor-mediated inhibition of noradrenaline release, were found (132). In contrast, stimulus-evoked release of noradrenaline from prefrontal cortex slices of SHR was unchanged compared with that of WKY rats (142, 143). High firing of locus coeruleus neurons, overproduction of noradrenaline, and decreased alpha2-adrenoceptor-mediated regulation of the neurotransmitter, with its consequent spill over into the extracellular space are a trait of SHR. This may cause repeatedly increased release of noradrenaline from sympathetic nerve terminals and to consequent stress-dependent development of hypertension in SHR (144). In conclusion, the noradrenergic system is hyperactive in SHR, particularly in response to stress, thereby inducing an imbalance between noradrenergic hyperfunction and dopaminergic hypofunction similarly as occurs in ADHD.

5.4. Other Neurotransmitter Systems

Other neurotransmitter systems were investigated more sparsely. Some studies focused on the serotonergic system. Injection of serotonin into the nucleus tractus solitarius (NTS) elicits hypotension and bradycardia in rats. In SHR, this depressor response was markedly larger than in WKY rats. These results were considered suggestive that serotonergic system is altered in SHR rats, apparently acting tonically to reduce blood pressure, and that stimulation of 5HT-2 receptors in the NTS contributes to cardiovascular regulation independent of the baroreceptor reflex (145). In SHR-SP, repeated exposure to stress dramatically exacerbates neuronal death by transient ischemia. The 5-HT-4 receptor was suggested to be involved in this stress-induced exacerbating mechanism of the neuronal damage (146).

Amino acid neurotransmitter systems were investigated in SHR primarily in terms of their contribution to cardiovascular and blood pressure regulation but not for their possible interference with cognition and behavioural mechanisms; therefore, further investigations are required.

6. Preclinical Pharmacology and Pharmacotherapeutics

Risks of primary stroke and of secondary (recurrent) stroke are directly related to the blood pressure level. Clinical studies have clearly demonstrated that lowering blood pressure markedly reduces the incidence of stroke. Both SHR and SHR-SP were used in pharmacotherapy research to assess the effect of different neuroprotective treatments in case of stroke or hypertensive brain damage. Administration of antihypertensive doses of the α-adrenoceptor antagonist phentolamine or of the Ca^{2+} antagonist nicardipine decreased mean blood pressure during brain ischemia and suppressed the anomalous rise of arterial pressure occurring in ischemia. Although controlling blood pressure during an ischemic insult attenuates ischemia-induced damage of muscarinic cholinergic M1 receptors in the brain of SHR (147), lowering of blood pressure during stroke may exacerbate damage through reduction of blood supply to hypoperfused areas (steal effect).

A previous paper by our group has reviewed the influence of antihypertensive treatment on brain structure and function in animal models of hypertension with particular emphasis to SHR (22). Here we analyze the contribution of SHR and SHR-SP preclinical research in developing possible therapeutic strategies for protecting brain from arterial hypertension.

6.1. Calcium Antagonists

Arterial hypertension is characterized by altered Ca^{2+} metabolism not only in brain but also in other tissues, including vascular smooth muscle. In SHR, increased intracellular Ca^{2+} concentrations were attributed to genetic abnormalities in Ca^{2+}ATPase (132). Increased intracellular Ca^{2+} levels have several negative consequences on neurons, including decreased neurotransmitter release, altered Ca^{2+} signaling, impaired mitochondrial ATP synthesis and increased levels of reactive oxygen species. A better knowledge of altered Ca^{2+} metabolism in arterial hypertension has brought to the proposal of Ca^{2+} antagonists (Ca^{2+} channel blockers) as antihypertensive agents.

The dihydropyridine derivative nimodipine was investigated primarily in the treatment of brain ischemia. Protocols in general included intravenous administration of the drug or local application in the hippocampus. The compound was given before, during, and after ischemia in rabbit. Although nimodipine-sensitive Ca^{2+} channels (of the L-type) play a minor role in regulating Ca^{2+} influx of hippocampal neurons, the compound possesses a vasotropic action and may also protect brain neurons from neurotoxicity caused by Ca^{2+} overload (148). Studies on the effects of association between nimodipine and mannitol on infarct size and on the amount of apoptosis after transient focal cerebral ischemia have demonstrated that the two compounds provide neuroprotection by preventing both necrotic and apoptotic components of cell death (149). Another study has assessed the influence of a 12-week treatment with a dose of 15 mg kg^{-1} day^{-1} of nimodipine on hypertensive brain damage (71) and cerebrovascular changes (150) in SHR. Although the compound only partially lowered systolic blood pressure, it displayed cerebrovascular activity and was able to provide neuroprotection (71, 150).

Another series of studies have investigated the effects of isradipine on permanent and transient focal cerebral ischemia in SHR. Isradipine administered subcutaneously 30 min after permanent MCA occlusion significantly reduced the volume of ischemic brain damage in the cerebral hemisphere, cerebral cortex, and caudate nucleus when assessed at 24 h post-MCA occlusion. In SHR undergoing transient (2 h) MCA occlusion, isradipine administered 30-min post-MCA occlusion significantly reduced hemispheric infarct volume, whereas isradipine administered at the onset of reperfusion did not induce neuroprotection and did not counter functional deficit (measured using a subjective neurological scoring system) induced by MCA occlusion (151). The activity of AE0047, Ca^{2+} antagonist, in countering brain damage of SHR-SP with signs of stroke was compared with that of nicardipine and hydralazine. The drug (1 and 3 mg kg^{-1} day^{-1}) given daily to diseased SHR-SP prevented mortality and improved neurological symptoms. Nicardipine (10 mg kg^{-1} day^{-1}) and hydralazine (10 mg kg^{-1} day^{-1}) were less effective than AE0047 in a dosage equal to or higher than the hypotensive dose, respectively. AE0047 may be beneficial for treating the acute stage of stroke in humans because of its long-lasting hypotensive action and undefined direct actions on the cerebral vasculature (152).

Nicardipine is a second-generation dihydropyridine-type Ca^{2+} antagonist with high vascular selectivity and strong cerebral and coronary vasodilatory activity (153, 154). The compound significantly reduced systolic pressure and the thickness of the vascular tree and restored the number of neurons and the phosphorylated neurofilament in the frontal cortex and hippocampus, as well as decreased reactive astrogliosis in SHR (48, 154, 155). The effect of another Ca^{2+} channel antagonist NC1100 on CBF and metabolism in SHR was investigated. Pre- and postischemic administration of NC1100 attenuated cerebral ischemia and ameliorated the ischemia-related metabolic and circulatory derangement (156).

The positive effects of Ca^{2+} channel blockers on hypertension-dependent cerebrovascular and brain changes in SHR were confirmed by comparing lercanidipine, manidipine, nimodipine, and the nondihydropyridine-type vasodilatator hydralazine (71). Highest neuroprotective efficacy was exerted by lercanidipine and, to a lesser extent, by nimodipine, while nimodipine caused a smaller reduction of SBP compared with all other compounds (71). An MRI study has shown that reduced systolic blood pressure reversed cerebral edema and restored BBB integrity in SHR-SP treated with the long-lasting dihydropyridine-type Ca^{2+} antagonist amlodipine (157). In conclusion, although evidence for brain protection from dihydropyridine-type Ca^{2+} channel blockers in brain ischemia is not conclusive despite positive outcome of nimodipine, both nicardipine and lercanidipine effectively countered hypertensive brain damage and cerebrovascular impairment in SHR.

The nonselective Ca^{2+} channel blocker fluarizine has shown to possess neuroprotective activity and a dose-response neuroprotective morbidity–mortality relationship in the gerbil global cerebral ischemia model. The main limitation of the drug was apparently its narrow therapeutic range (158).

6.2. Angiotensin-Converting Enzyme Inhibitors and Angiotensin Receptor Antagonists

Alterations in CBF autoregulation have been reported in genetic and experimental animal model and in human hypertension. Chronic administration of ACE inhibitors seems to prevent and reverse these alterations. ACE inhibitors do not change CBF but normalize the lower limit of its autoregulation in hypertension (159). ACE inhibitors shift the lower and upper limits of cerebral autoregulation to the left and, in hypertensive animals, produce a shortening of the autoregulatory plateau (159). The effects of ACE inhibitors are mainly mediated through dilation of large arteries supplying the brain. ACE inhibitors have beneficial effects on the small resistance vessels of the brain, probably at least in part by decreasing angiotensinogen II formation. ACE inhibitors can reduce the diameter of pial arterioles in SHR, thereby protecting against focal cerebral ischemia and decreasing mortality in SHR-SP (160, 161). Moreover, drugs of this class attenuate ischemic brain metabolism in hypertensive animals (162) and improve neurologic outcome from incomplete cerebral ischemia in rats (163). They also decrease the medial thickness and inhibit the injury-induced proliferation of smooth muscle cells (164).

MRI was used to investigate the therapeutic effects of imidapril, an ACE inhibitor, on lesions developing in malignant SHR-SP (M-SHRSP). In the control group, the MRI score for cerebral lesions increased during the experiment, and seven out of eight control rats died within 17 days. Rats treated with imidapril showed less pronounced BBB damage and resolved brain edema within 1 week. At this time, neurological symptoms observed after stroke were also improved. None of the rats treated with imidapril showed recurrence of stroke, and their survival rate was better than untreated rats. Based on these preclinical data, imidapril was proposed to have therapeutic effect on stroke lesions, as well as prophylactic effects on the recurrence of stroke (165).

The use of the angiotensin II type 1 (AT1) receptor antagonists could also help to clarify the role of angiotensin II in the control of cerebral circulation. Acute administration of AT1 antagonists or angiotensin II type 2 (AT2) selective receptor ligands modulates the upper and lower limits of CBF autoregulation in rats (166, 167). Chronic treatment with the angiotensin II receptor antagonist candesartan cilexetil (0.1, 1 or 10 mg/kg) protects against stroke in SHR-SP and a low dose-reduced stroke incidence without affecting blood pressure (168). Acute administration of the potent and selective AT1 antagonist candesartan was reported to modify cerebrovascular autoregulatory response in normotensive and hypertensive rats (169). Treatment with candesartan (0.5 mg kg^{-1} day^{-1}) reduced the mean blood pressure in both WKY rats and SHR (159). In untreated adult SHR, the autoregulation curve significantly shifted to the right, toward higher blood pressure, when compared with the WKY rats. Treatment with candesartan significantly shifted the autoregulation toward the left in both WKY rats and SHR, and, after 14 days of treatment, differences between the autoregulation curves in SHR and WKY rats were partially reversed (159), thereby indicating that peripheral administration of candesartan affects cerebrovascular autoregulation. In general, AT1 receptor antagonists do not influence baseline CBF but shift the autoregulation curve toward lower blood pressure, with an effect similar to that of ACE inhibitors (159). A comparative analysis of the activity of the AT1 receptor antagonist losartan versus the ACE inhibitor enalapril on cerebral edema and proteinuria in SHR-SP showed that both losartan and enalapril decreased cerebral edema to baseline levels. Systolic blood pressure did not decrease after lorsartan treatment but decreased transiently after enalapril treatment. Cerebral edema and proteinuria were prevented and reduced in SHR-SP treated with either an AT1 receptor antagonist or an ACE inhibitor. In rats in which treatment was initiated after manifestation of cerebral edema, both cerebral edema and proteinuria reappeared despite continued treatment (170). Treatment of SHR from 3 to 6 months with telmisartan (2 mg kg^{-1} day^{-1}) or ramipril (1 mg kg^{-1} day^{-1}) normalized brain arteriolar pressure, preventing arteriolar wall thickening (171). Chronic increase in cerebral arteriolar blood pressure is accompanied by vascular remodeling characterized by blood vessel wall hypertrophy and diameter decrease. This increases cerebrovascular resistance which maintains CBF autoregulation at high systemic blood pressure levels but accentuates hypoperfusion at low arterial blood pressure levels. Available data emerging from the use of ACE inhibitors and AT1 receptor antagonist indicate that normalization of wall thickness represents a pressure-dependent phenomenon, whereas normalization of arteriolar diameter relates to a pressure-independent mechanism involving aldosterone (172). This suggests that intracellular sodium-dependent modulation of protein metabolism may represent a new drug target for arteriolar remodeling (172).

6.3. Other Antihypertensive Drugs

Carvedilol, a multiple-action antihypertensive agent, has been shown to scavenge free radicals and to inhibit lipid peroxidation in swine heart and rat brain homogenates. The neuroprotective effect of this drug was assessed on cultured cerebellar neurons and on CA1 hippocampal neurons of gerbils with experimental brain ischemia. Carvedilol provides neuroprotection both in *in vitro* and *in vivo* models of neuronal injury related to oxygen radicals. Hence, the compound may reduce the risk of cerebral ischemia and stroke due to its combined antihypertensive action and antioxidant properties (173).

Male SHR-SP treated with cicletanine, a furopyridine derivative, showed a reduced incidence of hypertensive cerebral damage measured in terms of occurrence of cerebral infarction and vascular alterations with fibrinoid necrosis. The prevention of hypertensive cerebrovascular damage by cicletanine probably depends on a direct action of the compound on the vascular wall (174).

Indapamide, a thiazide-related diuretic, effectively lowers blood pressure after both acute and chronic treatment in genetically and nongenetically determined forms of hypertension. Indapamide was shown to have excellent efficacy in protecting against target organ damage (heart, kidneys, and brain). It has been shown that indapamide lowers the response to sympathetic nerve stimulation, exhibits Ca^{2+} antagonist properties, enhances the production of prostacyclin, and limits the production of free radicals and of endothelium-dependent vasoconstrictor substances. Recent data suggest that low doses of indapamide exert synergistic effects in combination with other antihypertensive drugs such as ACE inhibitors (175). It cannot be excluded that positive effects on hypertensive brain damage elicited by indapamide are due to its multiple mechanisms of action. The influence of treatment with indapamide in bilateral carotid ligation-induced cerebral ischemia was also evaluated. In SHR, bilateral common carotid ligation induces an immediate and strong increase in arterial pressure values. Short-term treatment with 3 mg/kg intraperitoneal indapamide, a dose not inducing antihypertensive effects in SHR, decreased the bilateral carotid ligation-induced vasopressor responses (176).

6.4. Neuroprotection by Nonantihypertensive Drugs

Based on the assumption that SHR may represent a model of VBDs, they were also used to assess potential protective effects of nonhypotensive pharmacological treatments, including cholinergic neurotransmission enhancers, such as the choline precursor choline alphoscerate, the AChE inhibitor galantamine and the two drugs in association. Although the drugs under discussion did not change arterial blood pressure, treatment of male SHR of age 4 months for 8 weeks with choline alphoscerate countered nerve cell loss and glial reaction primarily in the CA1 subfields and in the dentate gyrus of the hippocampus (Fig. 9). The observation that treatment with choline alphoscerate attenuated the extent of glial reaction in the hippocampus of SHR was considered suggestive of a neuroprotective effect of this compound in this animal model of vascular brain damage (105). These results were extended by another investigation in which 32-week-old SHR and age-matched normotensive WKY rats were treated for 4 weeks with an oral dose of 3 mg kg^{-1} day^{-1} of galantamine, of 100 mg kg^{-1} day^{-1} of choline alphoscerate or their association in comparison with sham-treated groups. In SHR, the number of neurons in frontal cortex, in the CA1 subfield of the hippocampus and in the dentate gyrus was decreased compared with WKY rats. Astrogliosis, breakdown of phosphorylated neurofilament, unchanged microtubule associated protein-2 and altered AQP4 expression was found as well. Both galantamine and choline alphoscerate countered nerve cell loss. Choline alphoscerate, but not galantamine, decreased astrogliosis and restored expression of AQP4. Galantamine countered phosphorylated neurofilament breakdown to a greater extent than choline alphoscerate. Simultaneous administration resulted in a potentiated treatment effect. This study confirmed a neuroprotective effect of galantamine in SHR and has further highlighted a neuroprotective role of choline alphoscerate in the same model (131). This suggests that similarly as reported in adult-onset dementia disorders of vascular origin cholinergic neurotransmission enhancing drugs may have an activity in the hypertensive brain damage model represented by sensitivity of SHR (177). A cerebrovascular, neuronal, or combined activity on the two potential targets of these drugs may be at the basis of the neuroprotective effects mentioned earlier.

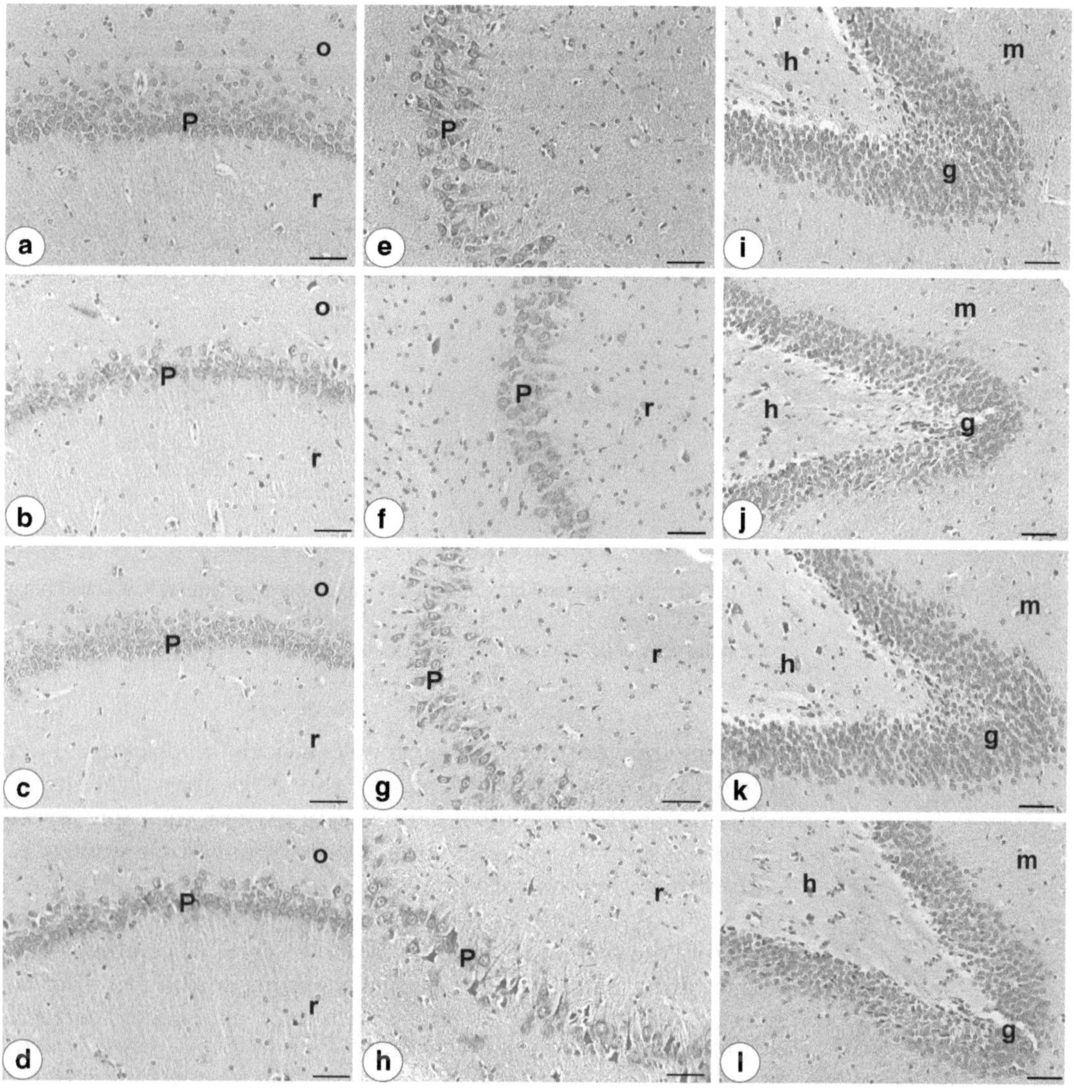

Fig. 9. Sections of the CA1 subfield (**a–d**), CA3 subfield (**e–h**) and dentate gyrus (**i–l**) of the hippocampus of normotensive Wistar-Kyoto (WKY) rats (**a**, **e** and **i**), SHR (**b**, **f** and **j**), spontaneously hypertensive rats (SHR) treated with α-glyceryl phosphorylcholine (GPC) (**c**, **g** and **k**), and SHR treated with phosphatidyl choline (PDC) (**d**, **h** and **l**). Cresyl violet staining. Note in SHR the reduced number of nerve cell profiles in the CA1 subfield and, to a lesser extent, in the dentate gyrus. Neuronal loss was countered by treatment with GPC and not by treatment with PDC. No changes in the number of nerve cell profiles were noticeable in the CA3 subfield of the different animal groups. P: Pyramidal neurons; o: stratum oriens; r: stratum radiatum: m: molecular layer; h: hilum; g: granule neurons. Calibration bar: 50 μm (Adapted from (105)).

7. Conclusion

As mentioned in the introduction, the evolution of the concept of cognitive complications of cerebrovascular disease bringing to the definition of multiinfarct dementia, VaD, VCI, and then VCD and the uncertainty in the classification of these pathologies (1, 2)

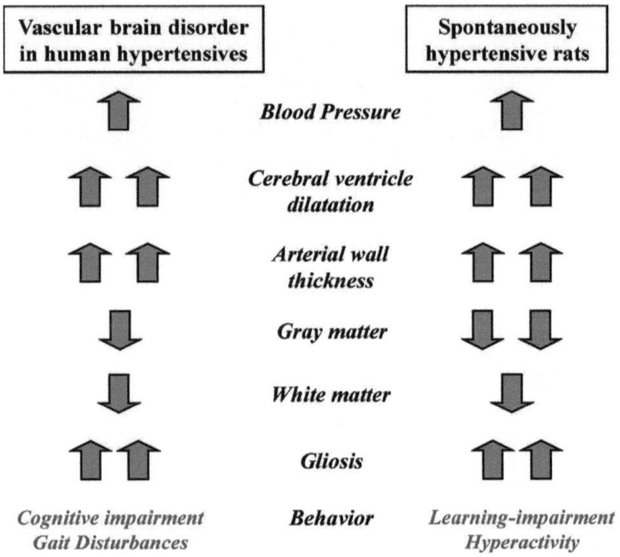

Fig. 10. Summary of general brain features in hypertensive patients with vascular brain disorder (VBD) and in SHR. Note the remarkable similarities in several parameters, supporting the view of usefulness of SHR as a model of VBD.

hampers the consideration of animal models of diseases under discussion. Dementia disorders, VCI and VCD, are typical human pathologies affecting higher brain functions, and, therefore, every model has several limitations and can provide only some similarities with cognate human pathology.

Brain alterations in SHR and in hypertensive patients are probably the consequence of structural changes of cerebral arteries, resulting in augmented arterial resistance leading to cerebral hypoperfusion (49, 150). This can induce changes of cerebral morphology with modified BBB osmotic fluid regulation, probable cytotoxic edema, gliosis, and nerve cell loss. As summarized in Fig. 10, these modifications are consistent with changes observed in the brain of hypertensive patients with cognitive impairment, suggesting that SHR may represent, from a morphological point of view, a model of VBD.

Cognitive deficits, including decrements in learning and memory, attention, perception, abstract reasoning, mental flexibility, psychomotor speed, constructional ability, and visuospatial task performance have been reported in hypertensive patients. The clinical significance of these findings is debated (178–181), although it is accepted that even "below average" performance on cognitive tasks may negatively affect daily activities (179). In SHR, hyperactivity, hyperreactivity, lower levels of anxiety, decrement of habituation capabilities (182–186), attention deficits, and decreased learning ability in behavioral tasks (125, 187) were reported. Impaired performance of memory tasks and of conditional avoidance tasks were also described (183, 188, 189). The same is true for operative conditioning functions (190), as well as for spatial learning tasks including water maze and radial arm maze procedures (125–127, 191, 192). In water maze experiments, 15-month-old SHR demonstrated inferior performance when compared with age-matched WKY rats (130). Impaired learning and memory were greater in older than in young SHR, suggesting that detrimental effects of hypertension may be additive with age (193). A decline of nicotinic ACh receptors and their binding sites was also found in SHR compared with control normotensive WKY rats (130). These findings are consisted with evidence of cholinergic hypofunction characterized by decreased ChAT activity in the cerebral cortex, hippocampus, and striatum of patients with VaD (1, 2, 107, 194). The same similarities are not shared by changes in catecholamine function, characterized by decreased dopaminergic tone and increased noradrenergic tone, which allowed considering SHR as a model of ADHD (132).

In summary, the majority of microanatomical, behavioral, and neurochemical data on SHR is in favor of the hypothesis that this strain is a suitable model of VBD. Changes in catecholaminergic transmission put forward SHR as a possible model of ADHD as well. Hence, at least in terms of brain involvement, SHR truly represent a multifaced model of two important groups of pathologies, VBD and ADHD. As for most models, researchers should always consider that SHR offer some similarities with corresponding human pathologies, but they do not suffer from the same disease.

Acknowledgements

Original studies summarizing microanatomical aspects of brain and BBB of SHR were supported in part by a grant from the Italian Ministry of Education, University and Scientific Research MIUR-COFIN No.2006060985_003.

References

1. Román GC, Royall DR (2004) 7 A diagnostic dilemma: is "Alzheimer's dementia" Alzheimer's disease, vascular dementia, or both? Lancet Neurol 3:141
2. Bowler JV (2007) Modern concept of vascular cognitive impairment. Br Med Bull 83:291–305
3. Prince MJ (1997) The treatment of hypertension in older people and its effect on cognitive function. Biomed Pharmacother 51: 208–212
4. Rigaud AS, Hanon O, Bouchacourt P, Forette F (2000) Cerebral complications of hypertension in the elderly. Rev Med Int 22: 959–968
5. Rigaud AS, Hanon O, Seux ML, Forette F (2001) Hypertension and dementia. Curr Hypertens Rep 3:454–457
6. In't Veld BA, Ruitenberg A, Hofman A, Stricker BH, Breteler MMB (2001) Antihypertensive drugs and incidence of dementia: the Rotterdam study. Neurobiol Aging 22:407–412

7. Seux ML, Forette F (1999) Effects of hypertension and its treatment on mental function. Curr Hypertens Rep 3:232–237
8. Weber MA (2003) The effects on dementia of long-term treatment of hypertension. Rev Cardiovasc Med 4:197–198
9. Mortel KF, Pavol MA, Wood S, et al (1994) Prospective studies of cerebral perfusion and cognitive testing among elderly normal volunteers and patients with ischemic vascular dementia and Alzheimer's disease. Angiology 45:1711–1718
10. Harrington F, Saxby BK, McKeith IG, Wesnes K, Ford GA (2000) Cognitive performance in hypertensive and normotensive older subjects. Hypertension 36:1079–1082
11. Takahashi N, Smithies O (2004) Human genetics, animal models and computer simulations for studying hypertension. Trends Genet 20:136–145
12. Lerman LO, Chade AR, Sica V, Napoli C (2005) Animal models of hypertension: an overview. J Lab Clin Med 146:160–173
13. White CJ, Ramee SR, Banks AK, Wiktor D, Price HL (1992) The Yucatan Miniature Swine: an atherogenic model to assess early patency rates of an endovascular stent. In: Swindle MM, Moody DC, Phillips LD (eds) Swine as models in biomedical research. Iowa State University Press, Ames, IA, pp 156–162
14. Sugiyama F, Yagami K, Paigen B (2001) Mouse models of blood pressure regulation and hypertension. Curr Hypertens Rep 3:41–48
15. Okamoto K, Aoki K (1963) Development of a strain of spontaneoulsy hypertensive rats. Jpn Circ J 27:282–293
16. Yamori Y (1994) Development of the spontaneously hypertensive rat (SHR), the stroke-prone SHR (SHRSP) and their various substrain models for hypertension-related cardiovascular disease. In: Ganten D, de Jong W (eds) Handbook ofhypertension. Experimental and genetic models of hypertension. Elsevier Science, England, pp 346–364
17. Okamoto K, Yamamoto K, Morita N, Ohta Y, Chikugo T, Higashizawa T, et al (1986) Establishment and use of the M strain of stroke-prone spontaneously hypertensive rat. J Hypertens 4:S21–S24
18. Folkow B (1982) Physiological aspects of primary hypertension. Physiol Rev 62:347–504
19. Hazama F, Haebara H, Amano S, Ozaki T (1977) Proceedings: Autoradiographic investigation of cell proliferation in the brain of spontaneously hypertensive rats. Acta Neuropathol 37:231–236
20. Johansson BB (1984) Cerebral vascular bed in hypertension and consequences for the brain. Hypertesion 6:81–86
21. Johansson BB, Fredriksson K (1985) Cerebral arteries in hypertension: structural and hemodynamic aspects. J Cardiovasc Pharmacol 7:S90–S93
22. Amenta F, Di Tullio MA, Tomassoni D (2003) Arterial hypertension and brain damage: evidence from animal models. Clin Exper Hypertens 25:359–380
23. Smith TL, Hutchins PM (1979) Central hemodynamics in the developmental stage of spontaneous hypertension in the unanesthetized rat. Hypertension 1:508–517
24. Watanabe K, Nishio T, Mori C, Kihara M, Yamori Y (1985) Changes in hemodynamics with advancing age in conscious spontaneously hypertensive rats. Jpn Circ J 49:446–450
25. Okamoto K, Yamori Y, Nagoaka A (1974) Establishment of the stroke-prone SHR. Circ Res 34/35:143–153
26. Dahl LK, Heine M, Tassinari L (1962) Effects of chronic excess salt ingestion. Evidence that genetic factors play an important role in susceptibility to experimental hypertension. J Exp Med 115:1173–1190
27. Ben-Ishay D, Saliternick R, Welner A (1972) Separation of two strains of rats inbred for dissimilar sensitivity to DOCA_salt hypertension. Experientia 28:1321–1322
28. Pinto YM, Paul M, Ganten D (1998) Lessons from rat models of hypertension: from Goldblatt to genetic engineering. Cardiovasc Res 39:77–88
29. Yamori Y (1999) Implication of hypertensive rat models for primordial nutritional prevention of cardiovascular diseases. Clin Exp Pharmacol Physiol 26:568–572
30. Fortepiani LA, Yanes L, Zhang H, Racusen LC, Reckelhoff JF (2003) Role of androgens in mediating renal injury in aging SHR. Hypertension 42:952–955
31. Reckelhoff JF, Granger JP (1999) Role of androgens in mediating hypertension and renal injury. Clin Exp Pharmacol Physiol 26:127–131
32. Fortepiani LA, Zhang H, Racusen L, Roberts LJ 2nd, Reckelhoff JF (2003) Characterization of an animal model of postmenopausal hypertension in spontaneously hypertensive rats. Hypertension 41:640–645
33. Hilbert P, Lindpaintner K, Beckmann JS, et al (1991) Chromosomal mapping of two genetic loci associated with blood pressure regulation in hereditary hypertensive rat. Nature 353:521–552

34. Jacob HJ, Lindpaintner K, Lincoln SE, et al (1991) Genetic mapping of a gene causing hypertension in the stroke-prone spontaneoulsy hypertensive rat. Cell 67:213–224
35. Nara Y, Nabika T, Ikeda K, Sawamura M, Endo J, Yamori Y (1991) Blood pressure cosegregates with a microsatellite of angiotensin I converting enzyme (ACE) in F2 generation from a cross between original normotensive Wistar-Kyoto rat (WKY) and stroke-prone spontaneously hypertensive rat (SHRSP). Biochem Biophys Res Commun 181:941–946
36. Okuda T, Sumiya T, Iwai N, Miyata T (2002) Difference of gene expression profiles in spontaneous hypertensive rats and Wistar-Kyoto rats from two sources. Biochem Biophys Res Commun 296:537–543
37. Lerman LO, Chade AR, Sica V, Napoli C (2005) Animal models of hypertension: an overview. J Lab Clin Med 146:160–173
38. Bing OH, Brooks WW, Robinson KG, et al (1995) The spontaneously hypertensive rat as a model of the transition from compensated left ventricular hypertrophy to failure. J Mol Cell Cardiol 27:383–396
39. Li Z, Bing OH, Long X, Robinson KG, Lakatta EG (1997) Increased cardiomyocyte apoptosis during the transition to heart failure in the spontaneously hypertensive rat. Am J Physiol 272:H2313–H2319
40. Mitchell GF, Pfeffer JM, Pfeffer MA (1997) The transition to failure in the spontaneously hypertensive rat. Am J Hypertens 10:120s–126s
41. Faraci FM, Heistad DD (1990) Regulation of large cerebral arteries and cerebral microvascular pressure. Circ Res 66:8–17
42. Welch KMA, Caplan LR, Reis DJ, Siesjo¨ DJ, Weir B (eds) (1997) Primer on cerebrovascular diseases. Academic Press, New York, pp 1–823
43. Edvinsson L, MacKenzie ET, McCulloch J (eds) (1993) Cerebral blood flow and metabolism. Raven Press, New York, pp 1–683
44. Fredriksson K, Nordborg C, Johansson BB (1984) The hemodynamic effect of bilateral carotid ligation and the morphometric characteristics of the main communicating circuit in normotensive and spontaneously hypertensive rats. Acta Physiol Scand 121:241–247
45. Johansson BB, Fredriksson K (1985) Cerebral arteries in hypertension: structural and hemodynamic aspects. J Cardiovasc Pharmacol 7:90–93
46. Heagerty AM, Aalkjaer C, Bund SJ, Korsgaard N, Mulvany MJ (1993) Small artery structure in hypertension. Dual processes of remodeling and growth. Hypertension 21:391–397
47. Mulvany MJ (1993) Resistance vessel structure and the pathogenesis of hypertension. J Hypertens Suppl 11:S7–S12
48. Amenta F, Strocchi P, Sabbatini M (1996) Vascular and neuronal hypertensive brain damage: protective effect of treatment with nicardipine. J Hypertens Suppl 14:S29–S35
49. Sabbatini M, Strocchi P, Vitaioli L, Amenta F (2001) Microanatomical Changes of Intracerebral Arteries in Spontaneously Hypertensive Rats: a Model of Cerebrovascular Disease of the Elderly. Mech Ageing Dev 122:1257–1268
50. Lee RM, Smeda JS (1985) Primary versus secondary structural changes of the blood vessels in hypertension. Can J Physiol Pharmacol 63:392–401
51. Folkow B, Hallbäck M, Lundgren Y, Sivertsson R, Weiss L (1973) Importance of adaptive changes in vascular design for establishment of primary hypertension, studied in man and in spontaneously hypertensive rats. Circ Res 32:2–16
52. Lin JX, Tomimoto H, Akiguchi I, Wakita H, Shibasaki H, Horie R (2001) White matter lesions and alteration of vascular cell composition in the brain of spontaneously hypertensive rats. Neuroreport 12:1835–1839
53. Abumiya T, Masuda J, Kawai J, Suzuki T, Ogata J (1996) Heterogeneity in the appearance and distribution of macrophage subsets and their possible involvement in hypertensive vascular lesions in rats. Lab Invest 75:125–136
54. Baumbach GL, Heistad DD (1983) Effects of sympathetic stimulation and changes in arterial pressure on segmental resistance of cerebral vessels in rabbits and cats. Circ Res 52:527–533
55. Schilling L, Wahl M (1997) Brain edema: pathogenesis and therapy. Kidney Int 59: S69–S75
56. Mueller SM, Heistad DD (1980) Effect of chronic hypertension on the blood-brain barrier. Hypertension 2:809–812
57. Mueller SM (1982) The blood-brain barrier in young spontaneously hypertensive rats. Acta Neurol Scand 65:623–628
58. Al-Sarraf H, Philip L (2003) Effect of hypertension on the integrity of blood brain and blood CSF barriers, cerebral blood flow and CSF secretion in the rat. Brain Res 975: 179–188
59. Nielsen S, Nagelhus EA, Amiry-Moghaddam M, Bourque C, Agre P, Ottersen OP (1997) Specialized membrane domains for water transport in glial cells: high resolution immunogold cytochemistry of aquaporin-4 in rat brain. J Neurosci 17:171–180

60. Rash JE, Yasumura T, Hudson CS, Agre P, Nielsen S (1998) Direct immunogold labeling of aquaporin-4 in square arrays of astrocyte and ependymocyte plasma membranes in rat brain and spinal cord. Proc Natl Acad Sci USA 95:11981–11986
61. Saadoun S, Papadopoulos M, Bell B, Krishna S, Davies D (2002) The aquaporin-4 water channel and brain tumour oedema. J Anat 200:528
62. Vizuete ML, Venero JL, Vargas C, Ilundáin AA, Echevarría M, Machado A, Cano J (1999) Differential upregulation of aquaporin-4 mRNA expression in reactive astrocytes after brain injury: potential role in brain edema. Neurobiol Dis 6:245–258
63. Ke C, Poon WS, Ng HK, Pang JC, Chan Y (2001) Heterogeneous responses of aquaporin-4 in oedema formation in a replicated severe traumatic brain injury model in rats. Neurosci Lett 301:21–24
64. Taniguchi M, Yamashita T, Kumura E, et al (2000) Induction of aquaporin-4 water channel mRNA after focal cerebral ischemia in rat. Brain Res Mol Brain Res 78:131–137
65. Vajda Z, Promeneur D, Dóczi T, et al (2000) Increased aquaporin-4 immunoreactivity in rat brain in response to sistemi hyponatremia. Biochem Biophys Res Commun 270:495–503
66. Ritter S, Dinh TT (1986) Progressive postnatal dilation of brain ventricles in spontaneously hypertensive rats. Brain Res 370:327–332
67. Tajima A, Hans FJ, Livingstone D, Wei L et al (1993) Smaller local brain volumes and cerebral atrophy in spontaneously hypertensive rats. Hypertension 21:105–111
68. Nelson DO, Boulant JA (1981) Altered CNS neuroanatomical organization of spontaneously hypertensive (SHR) rats. Brain Res 226:119–130
69. Sabbatini M, Baldoni E, Cadoni A, Vitaioli L, Zicca A, Amenta F (1999) Forebrain white matter in spontaneously hypertensive rats: a quantitative image analysis study. Neurosci Lett 265:5–8
70. Lehr RP Jr, Browning RA, Myers JH (1980) Gross morphological brain differences between Wistar-Kyoto and spontaneously hypertensive rats. Clin Exp Hypertens 2:123–127
71. Sabbatini M, Tomassoni D, Amenta F (2001) Hypertensive brain damage: comparative evaluation of protective effect to treatment with dihydropyridine derivatives in spontaneously hypertensive rats. Mech Ageing Dev 122:2085–2105
72. Lanari A, Silvestrelli G, De Dominicis P, Tomassoni D, Amenta F, Parnetti L (2007) Arterial hypertension and cognitive dysfunction in physiologic and pathologic aging of the brain. Am J Geriatr Cardiol 16:158–164
73. Li Q, Lu G, Antonio GE, Mak YT, Rudd JA, Fan M, Yew DT (2007) The usefulness of the spontaneously hypertensive rat to model attention-deficit/hyperactivity disorder (ADHD) may be explained by the differential expression of dopamine-related genes in the brain. Neurochem Int 50:848–857
74. Bendel P, Eilam R (1992) Quantitation of ventricular size in normal and spontaneously hypertensive rats by magnetic resonance imaging. Brain Res 574:224–228
75. Sabbatini M, Strocchi P, Vitaioli L, Amenta F (2000) The hippocampus in spontaneously hypertensive rats: a quantitative microanatomical study. Neuroscience 100:251–258
76. Sabbatini M, Catalani A, Consoli C, Marletta N, Tomassoni D, Avola R (2002) The hippocampus in spontaneously hypertensive rats: an animal model of vascular dementia? Mech Ageing Dev 123:547–559
77. Johansson BB (1986) Pentoxifylline: cerebral blood flow and glucose utilization in conscious spontaneously hypertensive rats. Stroke 17:744–747
78. Wei L, Lin SZ, Tajima A, Nakata H, Acuff V, Patlak C, Pettigrew K, Fenstermacher J (1992) Cerebral glucose utilization and blood flow in adult spontaneously hypertensive rats. Hypertension 20:501–510
79. Grabowski M, Nordborg C, Brundin P, Johansson BB (1988) Middle cerebral artery occlusion in the hypertensive and normotensive rat: a study of histopathology and behaviour. J Hypertens 6:405–411
80. Breteler MM, van Amerongen NM, van Swieten JC, et al (1994) Cognitive correlates of ventricular enlargement and cerebral white matter lesions on magnetic resonance imaging. The Rotterdam Study. Stroke 25:1109–1115
81. Pantoni L, Garcia JH (1997) Cognitive impairment and cellular/vascular changes in the cerebral white matter. Ann NY Acad Sci 826:92–102
82. Englund E (1998) Neuropathology of white matter changes in Alzheimer's disease and vascular dementia. Dement Geriatr Cogn Disord 9:6–12
83. Schmidt R, Mechtler L, Kinkel PR, Fazekas F, Kinkel WR, Freidl W (1993) Cognitive impairment after acute supratentorial stroke: a 6-month follow-up clinical and computed tomographic study. Eur Arch Psychiatry Clin Neurosci 243:11–15

84. Tomassoni D, Baldoni E, Di Tullio MA, Avola R, Vitaioli L (2002) Glial reaction in brain white matter of spontaneously hypertensive rats. Ital J Anat Embryol 109:146
85. Muñoz-Fernández MA, Fresno M (1998) The role of tumour necrosis factor, interleukin 6, interferon-gamma and inducible nitric oxide synthase in the development and pathology of the nervous system. Prog Neurobiol 56:307–340
86. Fu KY, Light AR, Matsushima GK, Maixner W (1999) Microglial reactions after subcutaneous formalin injection into the rat hind paw. Brain Res 825:59–67
87. Masumura M, Hata R, Nagai Y, Sawada T (2001) Oligodendroglial cell death with DNA fragmentation in the white matter under chronic cerebral hypoperfusion: comparison between normotensive and spontaneously hypertensive rats. Neurosci Res 39:401–412
88. Nishiyama Y, Katayama Y (1998) White matter changes of corpus callosum in normotensive and hypertensive rats following bilateral carotid artery stenosis. Nippon Ika Daigaku Zasshi 65:450–458
89. Ridet JL, Malhotra SK, Privat A, Gage FH (1997) Reactive astrocytes: cellular and molecular cues to biological function. Trends Neurosci 20:570–577
90. Holash JA, Noden DM, Stewart PA (1993) Re-evaluating the role of astrocytesin blood-brain barrier induction. Dev Dyn 197:14–25
91. Müller CM (1992) A role for glial cells in activity-dependent central nervous plasticity? Review and hypothesis. In: Bradley RJ (ed) International review of neurobiology. Academic Press, San Diego, CA, pp 215–281
92. Janeczko K (1993) Co-expression of GFAP and vimentin in astrocytes proliferating in response to injury in the mouse cerebral hemisphere. A combined autoradiographic and double immunocytochemical study. Int J Dev Neurosci 11:139–147
93. Liedtke W, Edelmann W, Bieri PL, et al (1996) GFAP is necessary for the integrity of CNS white matter architecture and long-term maintenance of myelination. Neuron 17:607–615
94. McCall MA, Gregg RG, Behringer RR, et al (1996) Targeted deletion in astrocyte intermediate filament (GFAP) alters neuronal physiology. Proc Natl Acad Sci USA 93: 6361–6366
95. Gimenez MA, Sim JE, Russell JH (2004) TNFR1-dependent VCAM-1 expression by astrocytes exposes the CNS to destructive inflammation. J Neuroimmunol 151: 116–125
96. Nawashiro H, Brenner M, Fukui S, Shima K, Hallenbeck JM (2000) High susceptibility to cerebral ischemia in GFAP-null mice. J Cereb Blood Flow Metab 20:1040–1044
97. Fernaud-Espinosa I, Nieto-Sampedro M, Bovolenta P (1993) Differential activation of microglia and astrocytes in aniso- and isomorphic gliotic tissue. Glia 8:277–291
98. Nichols NR, Day JR, Laping NJ, Johnson SA, Finch CE (1993) GFAP mRNA increases with age in rat and human brain. Neurobiol Aging 14:421–429
99. Linnemann D, Skarsfelt T (1994) Regional changes in expression of NCAM, GFAP, and S100 in aging rat brain. Neurobiol Aging 15:651–655
100. Cheung WM, Wang CK, Kuo JS, Lin TN (1999) Changes in the level of glial fibrillary acidic protein (GFAP) after mild and severe focal cerebral ischemia. Chin J Physiol 42:227–235
101. Catalani A, Sabbatini M, Consoli C, Cinque C, Tomassoni D, Azmitia E, Angelucci L, Amenta F (2002) Glial fibrillary acidic protein immunoreactive astrocytes in developing rat hippocampus. Mech Ageing Dev 123:481–490
102. Unger JW (1998) Glial reaction in aging and Alzheimer's disease. Microsc Res Tech 43: 24–28
103. Kristjánsdóttir R, Uvebrant P, Rosengren L (2001) Glial fibrillary acidic protein and neurofilament in children with cerebral white matter abnormalities. Neuropediatrics 32:307–312
104. Tomassoni D, Avola R, Di Tullio MA, Sabbatini M, Vitaioli L, Amenta F (2004) Increased expression of glial fibrillary acidic protein in the brain of spontaneously hypertensive rats. Clin Exp Hypertens 26:335–350
105. Tomassoni D, Avola R, Mignini F, Parnetti L, Amenta F (2006) Effect of treatment with choline alphoscerate on hippocampus microanatomy and glial reaction in spontaneously hypertensive rats. Brain Res 1120:183–190
106. Amenta F, Tayebati SK (2008) Pathways of acetylcholine synthesis, transport and release as targets for treatment of adult-onset cognitive dysfunction. Curr Med Chem 15:488–498
107. Gottfries CG, Blennow K, Karlsson I, Wallin A (1994) The neurochemistry of vascular dementia. Dementia 5:163–167
108. Hernandez CM, Høifødt H, Terry AV Jr (2003) Spontaneously hypertensive rats: further evaluation of age-related memory performance and cholinergic marker expression. J Psychiatry Neurosci 28:197–209

109. Buccafusco JJ (1996) The role of central cholinergic neurons in the regulation of blood pressure and in experimental hypertension. Pharmacol Rev 48:179–211
110. Kubo T (1998) Cholinergic mechanism and blood pressure regulation in the central nervous system. Brain Res Bull 46:475–481
111. Ross CA, Ruggiero DA, Joh TH, Park DH, Reis DJ (1984) Rostral ventrolateral medulla: selective projections to the thoracic autonomic cell column from the region containing C1 adrenaline neurons. J Comp Neurol 228:168–185
112. Helke CJ, Muth EA, Jacobowitz DM (1980) Changes in central cholinergic neurons in the spontaneously hypertensive rat. Brain Res 188:425–436
113. Kubo T, Hashimoto M (1979) Effects of intraventricular and intraspinal 6-hydroxydopamine on blood pressure of DOCA-saline hypertensive rats. Arch Int Pharmacodyn Ther 238:50–59
114. Togashi H, Matsumoto M, Yoshioka M, Hirokami M, Minami M, Saito H (1994) Neurochemical profiles in cerebrospinal fluid of stroke-prone spontaneously hypertensive rats. Neurosci Lett 166:117–120
115. Saito H, Togashi H, Yoshioka M, Nakamura N, Minami M, Parvez H (1995) Animal models of vascular dementia with emphasis on stroke-prone spontaneously hypertensive rats. Clin Exp Pharmacol Physiol Suppl 22: S257–S259
116. Kimura S, Saito H, Minami M, et al (2000) Pathogenesis of vascular dementia in stroke-prone spontaneously hypertensive rats. Toxicology 153:167–178
117. Togashi H, Kimura S, Matsumoto M, Yoshioka M, Minami M, Saito H (1996) Cholinergic changes in the hippocampus of stroke-prone spontaneously hypertensive rats. Stroke 27:520–525
118. Tayebati SK, Di Tullio MA, Amenta F (2004) Effect of treatment with the cholinesterase inhibitor rivastigmine on vesicular acetylcholine transporter and choline acetyltransferase in rat brain. Clin Exp Hypertens 26:363–373
119. Tayebati SK, Di Tullio MA, Amenta F (2008) Vesicular acetylcholine transporter (VAChT) in the brain of spontaneously hypertensive rats (SHR): effect of treatment with an acetylcholinesterase inhibitor. Clin Exp Hypertens 30(8):732–743
120. Hu L, Wong TP, Cote SL, Bell KF, Cuello AC (2003) The impact of Abeta-plaques on cortical cholinergic and non-cholinergic presynaptic boutons in Alzheimer's disease-like transgenic mice. Neuroscience 121:421–432
121. DeKosky ST, Ikonomovic MD, Styren SD, Beckett L, Wisniewski S, Bennett DA, Cochran EJ, Kordower JH, Mufson EJ (2002) Upregulation of choline acetyltransferase activity in hippocampus and frontal cortex of elderly subjects with mild cognitive impairment. Ann Neurol 51:145–155
122. DeKosky ST, Ikonomovic MD, Styren SD, et al (2002) Upregulation of choline acetyltransferase activity in hippocampus and frontal cortex of elderly subjects with mild cognitive impairment. Ann Neurol 51:143–144
123. Vemuganti R (2005) Decreased expression of vesicular GABA transporter, but not vesicular glutamate, acetylcholine and monoamine transporters in rat brain following focal ischemia. Neurochem Int 47:136–142
124. Knardahl S; Karlsen K (1984) Passive-avoidance behavior of spontaneously hypertensive rats. Behav Neural Biol 42:9–22
125. Wyss JM, Fisk G, van Groen T (1992) Impaired learning and memory in mature spontaneously hypertensive rats. Brain Res 592:135–140
126. Gattu M, Pauly JR, Boss KL, Summers JB, Buccafusco JJ (1997) Cognitive impairment in spontaneously hypertensive rats: role of central nicotinic receptors. I. Brain Res 771:89–103
127. Gattu M, Pauly JR, Urbanawiz S, Buccafusco JJ (1997) Autoradiographic comparison of muscarinic M1 and M2 binding sites in the CNS of spontaneously hypertensive and normotensive rats. Brain Res 771:173–183
128. Yamada S, Kagawa Y, Ushijima H, Takayanagi N, Tomita T, Hayashi E (1987) Brain nicotine cholinoceptor binding in spontaneous hypertension. Brain Res 410:212–218
129. Khan IM, Printz MP, Yaksh TL, Taylor P (1994) Augmented responses to intrathecal nicotinic agonists in spontaneous hypertension. Hypertension 24:611–619
130. Terry AVJ, Hernandez CM, Buccafusco JJ, Gattu M (2000) Deficits in spatial learning and nicotinic-acetylcholine receptors in older, spontaneouslyhypertensiverats. Neuroscience 101:357–368
131. Tayebati SK, Di Tullio MA, Tomassoni D, Amenta F (2008) Neuroprotective effect of treatment with galantamine and choline alphoscerate on brain microanatomy in spontaneously hypertensive rats. J Neurol Sci 283:187–194
132. Russell VA (2007) Reprint of "Neurobiology of animal models of attention-deficit hyperactivity disorder". J Neurosci Methods 166:I–XIV

133. Carboni E, Silvagni A, Valentini V, Di Chiara G (2003) Effect of amphetamine, cocaine and depolarization by high potassium on extracellular dopamine in the nucleus accumbens shell of SHR rats. An in vivo microdyalisis study. Neurosci Biobehav Rev 27:653–659
134. Leo D, Sorrentino E, Volpicelli F, Eyman M, Greco D, Viggiano D, di Porzio U, Perrone-Capano C (2003) Altered midbrain dopaminergic neurotransmission during development in an animal model of ADHD. Neurosci Biobehav Rev 27:661–669
135. Watanabe H, Kumon Y, Ohta S, Nakano K, Sakaki S, Matsuda S, Sakanaka M (1997) Protein synthesis inhibitor transiently reduces neuronal death in the thalamus of spontaneously hypertensive rats following cortical infarction. Neurosci Lett 233:25–28
136. Russell VA (2002) Hypodopaminergic and hypernoradrenergic activity in prefrontal cortex slices of an animal model for attention-deficit hyperactivity disorder – the spontaneously hypertensive rat. Behav Brain Res 130:191–196
137. Fujita S, Okutsu H, Yamaguchi H, et al (2003) Altered pre- and postsynaptic dopamine receptor functions in spontaneously hypertensive rat: an animal model of attention-deficit hyperactivity disorder. J Oral Sci 45:75–83
138. Solanto MV (1998) Neuropsychopharmacological mechanisms of stimulant drug action in attention-deficit hyperactivity disorder: a review and integration. Behav Brain Res 94:127–152
139. de Villiers AS, Russell VA, Sagvolden T, Searson A, Jaffer A, Taljaard JJ (1995) Alpha 2-adrenoceptor mediated inhibition of [^{3}H] dopamine release from nucleus accumbens slices and monoamine levels in a rat model for attention-deficit hyperactivity disorder. Neurochem Res 20:427–433
140. Reja V, Goodchild AK, Phillips JK, Pilowsky PM (2002) Tyrosine hydroxylase gene expression in ventrolateral medulla oblongata of WKY and SHR: a quantitative real-time polymerase chain reaction study. Auton Neurosci 98:79–84
141. Reja V, Goodchild AK, Pilowsky PM (2002) Catecholamine-related gene expression correlates with blood pressures in SHR. Hypertension 40:342–347
142. Russell VA, Wiggins TM (2000) Increased glutamate-stimulated norepinephrine release from prefrontal cortex slices of spontaneously hypertensive rats. Metab Brain Dis 15:297–304
143. Russell VA, de Villiers AS, Sagvolden T, Lamm MC, Taljaard JJ (2000) Methylphenidate affects striatal dopamine differently in an animal model for attention-deficit/hyperactivity disorder – the spontaneously hypertensive rat. Brain Res Bull 53:187–192
144. Printz MP, Jirout M, Jaworski R, Alemayehu A, Kren V (2003) Genetic Models in Applied Physiology. HXB/BXH rat recombinant inbred strain platform: a newly enhanced tool for cardiovascular, behavioral, and developmental genetics and genomics. J Appl Physiol 94:2510–2522
145. Tsukamoto K, Sved AF, Ito S, Komatsu K, Kanmatsuse K (2000) Enhanced serotonin-mediated responses in the nucleus tractus solitarius of spontaneously hypertensive rats. Brain Res 863:1–8
146. Sakurai-Yamashita Y, Yamashita K, Niwa M, Taniyama K (2003) Involvement of 5-hydroxytryptamine4 receptor in the exacerbation of neuronal loss by psychological stress in the hippocampus of SHRSP with a transient ischemia. Brain Res 973:92–98
147. Hirata H, Asanuma M, Tanaka K, Kondo Y, Ogawa N (1995) M1 receptors in blood pressure-controlled ischemic spontaneously hypertensive rats. Stroke 26:1268–1272
148. Lazarewicz JW, Pluta R, Puka M, Salinska E (1990) Diverse mechanisms of neuronal protection by nimodipine in experimental rabbit brain ischemia. Stroke 21:IV108–IV110
149. Korenkov AI, Pahnke J, Frei K, Warzok R, Schroeder HW, Frick R, Muljana L, Piek J, Yonekawa Y, Gaab MR (2000) Treatment with nimodipine or mannitol reduces programmed cell death and infarct size following focal cerebral ischemia. Neurosurg Rev 23:145–150
150. Sabbatini M, Tomassoni D, Amenta F (2001) Influence of treatment with Ca(2+) antagonists on cerebral vasculature of spontaneously hypertensive rats. Mech Ageing Dev 122:795–809
151. Campbell CA, Mackay KB, Patel S, King PD, Stretton JL, Hadingham SJ, Hamilton TC (1997) Effects of isradipine, an L-Type calcium channel blocker on permanent and transient focal cerebral ischemia in spontaneously hypertensive rats. Exp Neurol 148:45–50
152. Shinyama H, Nagai H, Kawamura T, Narita Y, Nakamura N, Kagitani Y (1998) Therapeutic effects of AE0047, a novel calcium antagonist, on progression of brain damage after stroke in stroke-prone spontaneously hypertensive rats. Gen Pharmacol 30:379–386
153. Sabbatini M, Strocchi P, Amenta F (1995) Nicardipine and treatment of cerebrovascular diseases with particular reference to hypertension-related disorders. Clin Exp Hypertens 17:719–750

154. Sabbatini M, Bellagamba G, Casado A, Tayebati SK, Venarucci D, Amenta F (2001) Protective effect of treatment with nicardipine on cerebrovascular tree of spontaneously hypertensive rats. Clin Exp Hypertens 23:143–155
155. Amenta F, Tomassoni D (2004) Treatment with nicardipine protects brain in an animal model of hypertension-induced damage. Clin Exp Hypertens 26:351–361
156. Sadoshima S, Ibayashi S, Nakane H, Okada Y, Ooboshi H, Fujishima M (1992) Attenuation of ischemic and postischemic damage to brain metabolism and circulation by a novel Ca2+ channel antagonist, NC-1100, in spontaneously hypertensive rats. Eur J Pharmacol 224:109–115
157. Blezer E, Nicolay K, Goldschmeding R, Koomans H, Joles J (2002) Reduction of cerebral injury in stroke-prone spontaneously hypertensive rats by amlodipine. Eur J Pharmacol 444:75–81
158. Zapater P, Moreno J, Horga JF (1997) Neuroprotection by the novel calcium antagonist PCA50938, nimodipine and flunarizine, in gerbil global brain ischemia. Brain Res 772:57–62
159. Saavedra JM, Nishimura Y (1999) Angiotensin and cerebral blood flow. Cell Mol Neurobiol 19:553–573
160. Hajdu MA, Heistad DD, Baumbach GL (1991) Effects of antihypertensive therapy on mechanics of cerebral arterioles in rats. Hypertension 17:308–316
161. Fujii K, Weno BL, Baumbach GL, Heistad DD (1992) Effect of antihypertensive treatment on focal cerebral infarction. Hypertension 19:713–716
162. Sadoshima S, Fujii K, Ooboshi H, Ibayashi S, Fujishima M (1993) Angiotensin converting enzyme inhibitors attenuate ischemic brain metabolism in hypertensive rats. Stroke 24:1561–1566
163. Werner C, Hoffman WE, Kochs E, Rabito SF, Miletich DJ (1991) Captopril improves neurologic outcome from incomplete cerebral ischemia in rats. Stroke 22:910–914
164. Véniant M, Clozel JP, Kuhn H, Clozel M (1992) Protective effect of cilazapril on the cerebral circulation. J Cardiovasc Pharmacol 19:S94–S99
165. Takahashi M, Fritz-Zieroth B, Ohta Y, Chikugo T (1994) Therapeutic effects of imidapril on cerebral lesions observed by magnetic resonance imaging in malignant stroke-prone spontaneously hypertensive rats. J Hypertens 12:761–768
166. Naveri L, Stromberg C, Saavedra JM (1994) Angiotensin II AT2 receptor stimulation extends the upper limit of cerebral blood flow autoregulation: agonist effects of CGP 42112 and PD 123319. J Cereb Blood Flow Metab 14:38–44
167. Stromberg C, Naveri L, Saavedra JM (1992) Angiotensin AT2 receptors regulate cerebral bood flow in rats. Neuroreport 3:703–704
168. Inada Y, Wada T, Ojima M, et al (1997) Protective effects of Candesartan cilexetil (TCV-116) against stroke, kidney dysfunction and cardiac hypertrophy in stroke prone spontaneously hypertensive rats. Clin Exp Hypertens 19:1079–1099
169. Vraamak T, Waldemar G, Stradgaard S, Paulson S (1995) Angiotensin II receptor antagonist CV-11974 and cerebral blood flow autoregulation. J Hypertens 13:755–761
170. Blezer EL, Nicolay K, Koomans HA, Joles JA (2001) Losartan versus enalapril on cerebral edema and proteinuria in stroke-prone hypertensive rats. Am J Hypertens 14:54–61
171. Dupuis F, Atkinson J, Liminana P, Chillon JM (2005) Comparative effects of the angiotensin I receptor blocker, telmisartan, and the angiotensin converting enzyme inhibitor, ramipril, on cerebrovascular structure in SHR. J Hypertens 23:1061–1066
172. Atkinson J, Dupuis F, Chillon JM (2007) Renin-angiotensin-aldosterone system. An old system offering new drug targets for the cerebral circulation. Ann Pharm Fr 65:195–202
173. Lysko PG, Lysko KA, Yue TL, Webb CL, Gu JL, Feuerstein G (1992) Neuroprotective effects of carvedilol, a new antihypertensive agent, in cultured rat cerebellar neurons and in gerbil global brain ischemia. Stroke 23:1630–1635
174. Ruchoux MM, Huguet F, Droy-Lefaix MT, Gelot A, Ruchoux P, Autret A (1989) The effect of cicletanine on cerebrovascular injury in stroke-prone spontaneously hypertensive rats. Am J Hypertens 2:683–689
175. Bataillard A, Schiavi P, Sassard J (1999) Pharmacological properties of indapamide. Rationale for use in hypertension. Clin Pharmacokinet 37:7–12
176. Delbarre B, Delbarre G (1988) Effect of indapamide on an experimental model of cerebral ischemia in hypertensive rats. Am J Med 84:20–25
177. Parnetti L, Mignini F, Tomassoni D, Traini E, Amenta F (2007) Cholinergic precursors in the treatment of cognitive impairment of vascular origin: ineffective approaches or need for re-evaluation? J Neurol Sci 257:264–269

178. Elias MF, Wolf PA, D'Agostino RB, Cobb J, White LR (1993) Untreated blood pressure level is inversely related to cognitive functioning: the Framingham Study. Am J Epidemiol 138:353–364
179. Elias MF, D'Agostino RB, Elias PK, Wolf PA (1995) Neuropsychological test performance, cognitive functioning, blood pressure, and age: the Framingham Heart Study. Exp Aging Res 21:369–391
180. Desmond DW, Tatemichi TK, Paik M, Stern Y (1993) Risk factors for cerebrovascular disease as correlates of cognitive function in a stroke-free cohort. Arch Neurol 50:162–166
181. Farmer ME, White LR, Abbott RD, et al (1987) Blood pressure and cognitive performance. The Framingham Study. Am J Epidemiol 126:1103–1114
182. Knardahl S, Sagvolden T (1979) Open-field behavior of spontaneously hypertensive rats. Behav Neural Biol 27:187–200
183. Knardahl S, Karlsel K (1984) Passive-avoidance behaviour of spontaneously hypertensive rats. Behav Neurol Biol 42:9–22
184. LeDoux JE, Sakaguchi A, Reis DJ (1982) Behaviorally selective cardiovascular hyperreactivity in spontaneously hypertensive rats. Evidence for hypoemotionality and enhanced appetitive motivation. Hypertension 4:853–863
185. Svensson L, Harthon C, Linder B (1991) Evidence for a dissociation between cardiovascular and behavioral reactivity in the spontaneously hypertensive rat. Physiol Behav 49:661–665
186. Skinner MH, Tan DX, Grossmann M, Pyne MT, Mahurin RK (1996) Effects of captopril and propranolol on cognitive function and cerebral blood flow in aged hypertensive rats. J Gerontol A Biol Sci Med Sci 51: B454–B460
187. Wultz B, Sagvolden T, Moser EI, Moser MB (1990) The spontaneously hypertensive rat as an animal model of attention-deficit hyperactivity disorder: effects of methylphenidate on exploratory behavior. Behav Neural Biol 53:88–102
188. Sutturer JR, Devito W (1980) Two-way shuttlebox and lever-press avoidance in the spontaneously hypertensive and normotensive rat. J Comp Physiol Psychol 94:155–163
189. Sutturer JR, Devito W, Rykaszewski I (1981) Developmental aspect of 2-way shuttlebox avoidance in the spontaneously hypertensive and normotensive rat. Dev Psychobiol 14:405–414
190. Meneses A, Castillo C, Ibarra M, Hong E (1996) Effects of aging and hypertension on learning, memory and activity in rats. Physiol Behav 60:341–345
191. Mori S, Kato M, Fujishima M (1995) Impaired maze learning and cerebral glucose utilization in aged hypertensive rats. Hypertension 25:545–553
192. Nakamura-Palacios EM, Caldas CK, Fiorini A, Chagas KD, Chagas KN, Vasquez EC (1996) Deficits and spatial learning and working memory in spontaneously hypertensive rats. Behav Brain Res 74:217–221
193. Meneses A, Hong E (1998) Spontaneously hypertensive rats: a potential model to identify drugs for treatment of learning disorders. Hypertension 31:968–972
194. Wallin A, Alafuzoff I, Carlsson A, Eckernas SA, Gottfries CG, Karlsson I, Svennerholm L, Winblad B (1989) Neurotransmitter deficits in a non-multi-infarct category of vascular dementia. Acta Neurol Scand 79:397–406

Part VII

Animal Models of Normal Pressure Hydrocephalus

Chapter 31

Animals Models of Normal Pressure Hydrocephalus

Petra M. Klinge

Abstract

Normal Pressure Hydrocephalus (NPH) is a syndrome of dementia, gait disturbances, and urinary incontinence affecting the elderly population. Neuroimaging shows ventriculomegaly, and from the 70s to the early 90s invasive testing of intracranial pressure (ICP) and cerebrospinal fluid (CSF) dynamics supported the concept of a disturbed CSF circulation in those patients. Implantation of ventricular shunts has shown considerable symptomatic improvement in individuals; however, age-related comorbidity and coexisting dementias, such Alzheimer's disease (AD) and subcortical arteriosclerotic encephalopathy (SAE), have conflicted both clinical diagnosis of the syndrome and outcome to treatment. Moreover, a pathological continuum co-existing in the dementias of SAE, AD, and NPH has been proposed more recently, as blood flow and CSF biomarker studies as well as brain biopsies taken at the time of shunt implantation have evidenced the role of chronic ischemia and white-matter disease in NPH patients and biopsies were classified as "Probable AD" according to the Consortium to establish a registry of Alzheimer's disease (CERAD). Aging has shown to be a main risk factor to NPH, but also to AD and SAE, and as the population is aging in both the developed and developing countries, the prevalence of NPH dementia is very likely to increase. As such, research aspects aim at an understanding of the comorbidity, and how it impacts the diagnosis, pathophysiology, and the outcome of NPH. Certainly, NPH has to be seen from the multi- and transdisciplinary perspective, and experimental models are needed to model the interaction of hydrocephalus with risk factors of aging, and the coexisting dementias. The kaolin-induced hydrocephalus model allows preselection of age, manipulation of the disease progress in animals, and modeling of a chronic and long-term hydrocephalus condition, as existing in NPH. Kaolin, an inert silica derivate, creates a CSF malabsorption through scar formation, when injected into the subarachnoidal cisterns. A progressive ventricular enlargement follows a transient increase in ICP, ventricles remain enlarged while ICP normalized, similar to the human condition. Kaolin hydrocephalus induced in the aging 12-months-old rat has supported both, AD- and SAE-pathological correlates the longer the hydrocephalus existed in the animals: a mild but chronic "sublethal" ischemia has been evidenced by C-14-Jodo-Autoradiographic blood flow studies, and intracerebral and perivascular accumulation of $A\beta_{1-40}$, $A\beta_{1-42}$ peptides and hyperphosphorylated TAU proteins were observed through qualitative and quantitative histology. An age-related CSF circulatory dysfunction and reduced CSF turnover with a subsequent failure to clear $A\beta_{1-40}$, $A\beta_{1-42}$, and other toxic metabolites out from the brain interstitial fluid was therefore suggested as a common pathological element in NPH and NPH coexisting with AD and SAE. Further evidence to a common element of metabolic failure existing in those dementias was supported by significant breakdown of blood-brain-barrier (BBB) receptors in the aging Kaolin-rat, the best characterized being the low-density-lipoprotein-related protein 1 (LRP-1), which allows Aβ to escape the brain through the capillary border. With aging, Aβ LRP-1 transport is reduced by 50%.

Finally, shunt treatment in Kaolin-hydrocephalus has shown that "irreversibility" in the hydrocephalic brain lies in the damage of "ischemia vulnerable" brain regions, such as the CA1 sector of the hippocampus.

Peter Paul De Deyn and Debby Van Dam (eds.), *Animal Models of Dementia*, Neuromethods, vol. 48,
DOI 10.1007/978-1-60761-898-0_31, © Springer Science+Business Media, LLC 2011

Changes in synaptophysin and neurofilament immunohistochemistry of CA1 have correlated with histological changes of "delayed neuronal death" and occurred early in the course of hydrocephalus. This warrants a timely shunting of patients. The studies in aging kaolin-induced hydrocephalus might allow insight into late-life disturbances in CSF circulation, CSF turnover, and BBB transport underlying several age-related dementias, including AD. This reconsiders the traditional concept of "shunts for dementia"; however, both clinical and experimental hydrocephalus research should embrace the multidisciplinary aspect as it might allow and bring further insight into the interaction of ICP, impaired CSF circulation, blood flow, and metabolic breakdown in the various dementias.

Key words: Normal pressure hydrocephalus, Kaolin-hydrocephalus, intracranial pressure, CSF circulation, shunt treatment

1. Introduction

1.1. Dementia in Normal Pressure Hydrocephalus

Normal pressure hydrocephalus (NPH) was first described by Hakim and Adams in the 1960s (1) as a syndrome consisting of dementia, gait disturbance, and urinary incontinence. Patients exhibited enlarged ventricles; however, there was absence of raised intracranial pressure (ICP). The initial series, which reported three cases, two posttraumatic and one idiopathic without known cause, encouraged publications in the 1970s and early 1980s on "Ventricular shunts for dementia" (2, 3) as Hakim went on to describe the "classic triad" of gait disturbance, incontinence, and dementia, which were improved with the removal of cerebrospinal fluid (CSF) (4, 5). However, in the 1980s and 1990s, the results on larger patient series treated accordingly were frustrating as the outcome was affected by variable improvement after shunt treatment. Despite considerable high percentages (20–30%) of surgical risks and persistent morbidity (6), long-term results were poor (7). Particularly in idiopathic NPH, long-term prognosis was affected by many comorbidities not related to the shunt procedure, such as cerebrovascular and cardiovascular disease and other age-related illnesses (7–13).

Although no definite epidemiological data exist (14), we are presently facing a situation of patient numbers increasing for both idiopathic NPH and secondary NPH patients in many neurological and neurosurgical services, in part, as a result from the improved healthcare and increased longevity. Calculations based on focused centers, such as memory or dementia clinics, estimated both incidence of NPH among the elderly population and prevalence of NPH in dementia ranging from 4% to 14% (15, 16). From own experiences, patients in their 80s are now seeking an improved quality of life, and despite the risks of surgery, are no longer willing to accept the age-related burden. As such, efforts are being made towards a more accurate diagnosis and a more individual "differential" treatment of those patients (17).

However, one of the main obstacles is that the pathophysiology of NPH is still poorly understood. The traditional concepts of a "CSF malabsorption" and the "Intracranial Pressure–Volume Relationship" (18–22) have failed to explain the whole spectrum of the disease, e.g., the origin and prognosis of the gait and cognitive symptoms, as well as the significance of the vascular pathology and cerebral atrophy as seen in imaging studies and autopsies (23–27) – Fig. 1. Also, measures of resistance to CSF outflow (Rout) and/or ICP-volume index, analysis of ICP oscillations ("B-waves") have shown invalid in the diagnosis of NPH (28), although those measures have shown an association with chronic hydrocephalus and CSF malabsorption in earlier studies (29). Moreover, neuropathological findings, such as arachnoid fibrosis or vascular pathology, obtained from biopsies taken at the time of shunt surgery in patients diagnosed with NPH, did not show any correlation of an increased Rout or disturbed pressure profile as evidenced by B-waves (30, 31).

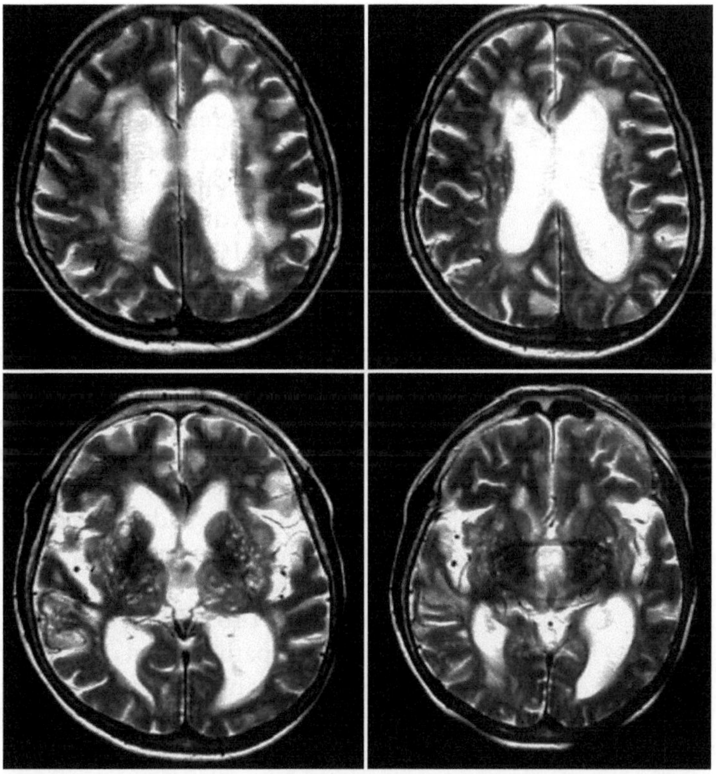

Fig. 1. Sagittal T2 MRI of idiopathic NPH. On axial T2-weighted magnetic resonance images deep white-matter and periventricular white-matter hyperintensities/lesions are frequently seen in idiopathic normal pressure hydrocephalus indicating cerebrovascular comorbidity. Around 80% of idiopathic NPH patients have vascular risk factors [8]. These lesions are not exclusion criteria for the diagnosis of NPH and decision-making for shunt treatment.

Neuroimaging, blood flow, brain metabolic studies, and analysis of CSF biomarkers have enlightened the presence and role of ischemia and metabolic failure in the NPH syndrome (32–41). Impairment of cerebrovascular reserve capacity (26, 27, 42), blood flow reductions, and reduced metabolism in frontal, supplementary motor, and anterior cingulated areas, as well as temporomesial, limbic areas have correlated with both the motor and cognitive dysfunction before and with changes after shunt treatment (32, 39, 43). Other PET-studies, using an anatomic region-of-interest analysis on coregistered MRI done in both secondary and idiopathic NPH, demonstrated cerebral blood flow reduction in the basal ganglia and the thalamus. This was suggested to correlate with the level of motor dysfunction in patients (33). Specific disturbances in regional brain metabolism have also been evidenced by 1H- and 31P-MRI Spectroscopy investigations in NPH patients (38): lactate peaks in periventricular areas and reduced cortical N-acetyl-aspartate/total creatine (NAA/tCr) ratios indicating neuronal disintegration and mitochondrial dysfunction have been shown and differed from controls and other forms of dementia. Microdialysis studies before and after high-volume CSF withdrawal have indicated an improvement in tissue oxygenation and increased glucose levels while glutamate levels decreased, comparable to a situation of postischemic recovery in patients (44).

Those measures have allowed access to the brain disease and the dementia in hydrocephalus, which is not simply a matter of ventriculomegaly, ICP, and "disturbed CSF-dynamics."

1.2. Comorbidity of Dementia in Idiopathic NPH

Of importance for an improved diagnosis and treatment of NPH is an understanding of the comorbidity, and how it impacts the pathophysiology and the outcome of the disease. Among the most important coexisting diseases are Binswanger's (Subcortical arteriosclerotic encephalopathy, SAE) and Alzheimer's dementia (AD). They exhibit clinical and imaging features in common with NPH (8, 23, 31, 45) and the above-mentioned series on neuropathological changes consistent with AD on cortical biopsies have further suggested that the two diseases may have some common pathophysiology (46, 47). Seven of 21 NPH patients (33%) in one series had neuropathological changes consistent with AD on cortical biopsy at the time of shunt placement (48). In a second series, one-third to one-half of NPH patients had Alzheimer-type neuropathology in cortical samples obtained at shunt implantation with dramatic clinical improvement (49) and, in a more recent series, 23 of 55 patients biopsies at the time of shunt placement for NPH were classified as "Probable AD" according to the CERAD (Consortium to establish a registry for Alzheimer's disease) criteria, including 75% of those with severe dementia (50). This coincidence of Alzheimer-type neuropathology among patients with NPH is greater than one would anticipate if the two diseases were unrelated. It was suggested, therefore, that the two diseases have a common basis in an age-related CSF circulatory dysfunction and a reduced CSF turnover with a subsequent failure to clear the various substances resulting from glial and neuronal metabolism out of the brain interstitial fluid (46, 51). These observations are supported by recent studies on elevated CSF contents of neuronal degrading products (hyperphosphorylated Tau (hpTau), amyloid-β (Aβ), and neurofilament) as well as biomarkers reflecting a disturbed neuronal metabolism, e.g., neuropeptides and transmitter-degrading products (41).

Among the neuronal CSF biomarkers, neurofilament protein has shown to be increased in idiopathic NPH, when compared to other dementias (52) or age-matched controls (53). Moreover, neurofilament levels correlated with the degree of dementia and cognitive impairment ("impaired wakefulness"), as well as improvement of cognition after shunt treatment (41, 54). Furthermore, reduction of neurofilament protein levels in both the lumbar and ventricular CSF after shunt treatment have correlated with a reduction of "irregular-type" periventricular white-matter hyperintensities seen in frontal T2-weighted turbo spin-echo sequences in patients with improving symptomatology (55). While neurofilament seems a marker of "reversible" damage in the human condition, sulfatide contents in NPH correlated with the degree of ischemia and "irreversible damage" in patients, as they were associated with poorer outcome to shunt treatment (40). However, there was also a clear overlap in CSF sulfatide measures comparing 43 patients with NPH and 19 patients with SAE, supporting the pathological continuum between both dementias (40).

2. Animal Models of Hydrocephalus and NPH

Different from human hydrocephalus and NPH, changes in the neuronal and glial metabolism have been largely described in animal models of hydrocephalus (56). Blood flow reductions in subcortical and cortical areas (57, 58), as well as mild but prolonged and chronic ischemia (59) have been evidenced. Here, the experimental results have much more straightforwardly evidenced that in the chronic hydrocephalic state cerebrovascular adaptive processes, e.g., angiogenesis and neuronal plasticity, might play a more important role in the disease process than ventricular enlargement and CSF dynamics (59, 60). Changes in various metabolic markers as well as cholinergic and dopaminergic transmitters have been also correlated with behavior and learning ability in chronic adult hydrocephalus (61–63).

In general, hydrocephalic models fall into the categories of obstructive, teratogenic, infectious, and genetically modified. Teratogenetic (e.g., vitamin-deficient, cytostatics), viral-induced, and genetic models, which are naturally occurring or induced, are most frequently prenatal or neonatal in onset (64–66). The most frequently used genetic models are the H-TX (67–69) and the LEW/Jms rat, both prenatal multigenic (70–72), and the Hy-3 mouse (73), where gene abnormality is present in hydin, a gene controlling cilia and flagella. However, all the infectious, genetic, and teratogenic models do not provide adequate models in the study of the significance of neurodegeneration and treatment options in the human chronic hydrocephalus condition (70). As to their pre- or neonatal onset and the developmentally acquired and toxic nature, as well as to the rapid development of ventriculomegaly, hydrocephalus, and ICP increases and detrimental effects, they do not apply to the generally mild and chronic onset of symptoms and disease duration and the mild to moderate ventriculomegaly, e.g., as observed in NPH, and also in other types of adult-or late-life-onset hydrocephalus (74, 75). Also, the coexisting complex brain malformations and systemic diseases and failures conflict the use of those models.

Even though some recent observations suggest a genetic linkage in a small subpopulation of patients with NPH (76, 77), the existing genetic-induced models are not considered to be of significance for the investigation of NPH-related dementia.

The most common method for creating hydrocephalus allowing a preselection of age of onset, the severity of ventriculomegaly and disease course are models of mechanical or chemical obstruction of the CSF pathways. Obstruction is created at the level of the aqueduct, the fourth ventricular outlets, the basal cisterns and/or venous system by injection of cyanoacrylate-gel (78, 79), Kaolin (80, 81), an inert silica derivate, or Silastic Oil (82–86). Ventricular dilatation is variable depending on the material and site of injection and obstruction. Furthermore, dilatation and its effects are more severe in younger animals (87, 88).

Those animal models mimic various forms of "obstructive" hydrocephalus; however, in light of the human condition, in idiopathic and secondary NPH, mostly a communicating type of hydrocephalus characterizes the chronic hydrocephalus disease (e.g., after trauma, hemorrhage, or meningitis), at least according to the classical definition of "communicating" versus "noncommunicating" hydrocephalus (89, 90).

Several attempts have been made in the past 20 years to induce communicating hydrocephalus in dogs, rats, and monkey by injecting kaolin, silicone, or Silastic oil into the subarachnoidal space (SAS) or the overlying cerebral hemispheres. However, the applied methods do not produce the disorder consistently and the site of obstruction has not been determined unequivocally. Nonmechanical induction methods and models, such as blood injection, bacterial inoculations, intrathecal injections of growth factors (e.g., transforming growth factor beta -TGF-β), and neurotoxins have also been used to produce "posthemorrhagic" and "postmeningitic" communicating hydrocephalus (91). More recently, renewed attempts to produce communicating-type hydrocephalus with kaolin injections into the cortical subarachnoid space or in the prepontine cisterns of adult rats have been reported, and the affected animals have been shown to survive for several months (91, 92).

However, to date there are no adequate "natural" NPH models available. All the above-mentioned procedures include the additional influence of the inducing chemical substances or implants on the pathophysiology, and thus are not true representations of the effects of the CSF obstruction or an absorptive defect alone on the brain.

Particularly when arguing for NPH, there is a lack of models to study across the age spectrum from brain development to senescent. Models using senescent animals might improve our understanding of the interaction of hydrocephalus with risk factors of aging, and the comorbidity of hypertensive vascular disease and AD, which are common in elderly patients with idiopathic NPH (Consensus from a NIH funded meeting held in Bethesda US, September 29–30, 2005: "Hydrocephalus: Myths, New Facts, and Clear Directions", Lit.) (93).

2.1. The Kaolin-Induced Model of the Adult and the Aging Rat

Hydrocephalus in the Kaolin models is generally induced by intracisternal injection of Kaolin suspension (aluminum silicate diluted with 0.9% saline; 1:4) under general anesthesia. In earlier studies of 3-month-old rats, 0.1 ml kaolin suspension was slowly injected through a 20-gauge needle inserted into the cisterna magna after microsurgical exposure of the atlanto-occipital membrane (59, 94, 95) – Fig. 2. The Kaolin-induced blockage of CSF circulation in the basal SAS results in hydrocephalus, increased resistance to CSF absorption, and consequent rises in ICP over a short-term period of 2 weeks. However, over a long-term period above 4–6 weeks, ICP normalizes, while resistance to CSF outflow remains above normal values (94). This is paralleled by a gradual progress of ventricular dilatation from short-term to long-term, and ventricles remain enlarged despite normalization of ICP, quite similar to the dynamics observed in NPH and adult chronic hydrocephalus (29)– Fig. 2.

Kaolin induces an unspecific inflammatory response in the basal cisterns. However, beyond the short-term period of 2 weeks, the inflammatory response was no longer observed, and scar formation dominated the basal SASs as shown by our group – Fig. 3. The findings on the regional and temporal progression and the onset of major histological changes in glial and neuronal markers in both cortical and subcortical regions, e.g. in the hippocampus (95), do no suggest a direct pathological effect of kaolin injection, as also described by others (91).

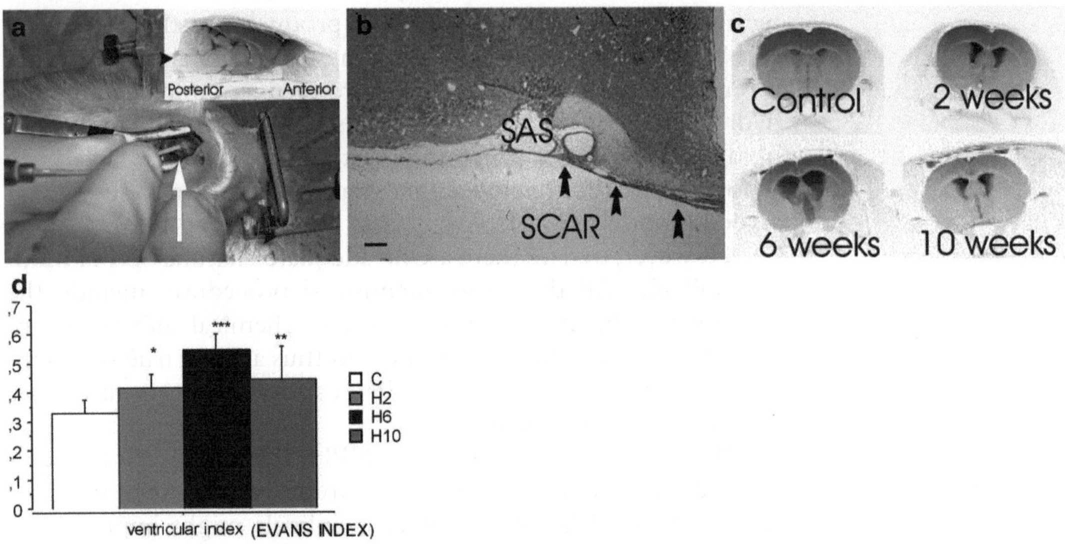

Fig. 2. Kaolin-induced hydrocephalus in the adult (3-month-old) rat. (**a**) Photograph of a Sprague–Dawley rat with the head mounted in a stereotaxic frame. *Arrow* points to catheter inserted into cisterna magna right below the exposed occipital bone. Inset shows lateral MRI view of rat brain (4.7 T BRUKER Biospect). *Arrowhead* points to cisternal site of kaolin injection. (**b**) H&E photomicrograph (bar = 500 μm) showing the scar formation at the site of kaolin injection in basal subarachnoidal space (SAS). The aluminum in kaolin is bound to silicate, and therefore not free to diffuse widely. As indicated in the figure, the kaolin suspension remains confined to basal SAS. Consequently, the aluminum does not distribute into parenchyma. (**c**) MRI (4.7 T BRUKER Biospect; RARE sequence) coronal section showing progressive ventricular dilatation over the course of hydrocephalus. Normal pressure hydrocephalus-like changes are consistently observed: moderate ventriculomegaly, reduced CSF formation rate, and increased resistance to CSF outflow. The induced hydrocephalus is well advanced at 6 weeks postinjection. This Sprague–Dawley model therefore mimics human NPH. (**d**) Evan's Index (as defined by the quotient of the largest ventricular diameter divided by the width of the brain at the level of the anterior commissure) measured in the MRI scans of the hydrocephalic rats at the assigned time points ($n = 16$ controls; $n = 5$ at 2 weeks, $n = 6$ at 6 weeks, $n = 7$ at 10 weeks; mean ± SD,* $p < 0.05$, ** $p < 0.01$, *** $p < 0.001$).

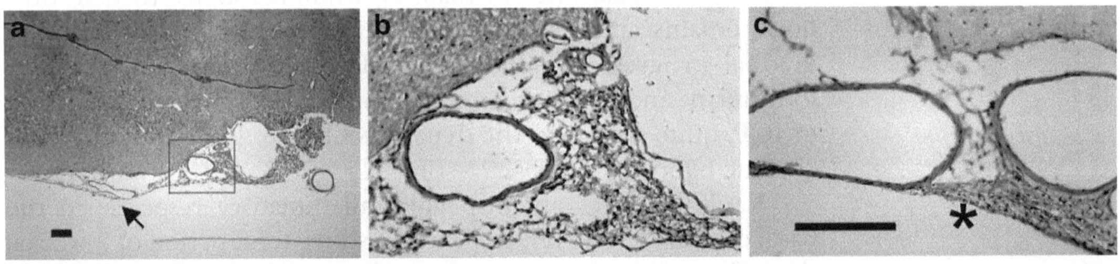

Fig. 3. Kaolin inflammatory reaction within the basal cisterns. Photomicrograph showing kaolin-related inflammation at 2 weeks (**a**, **b**) and at 6 weeks (**c**) hydrocephalus in the rat subarachnoidal space (SAS). The kaolin-related inflammatory reaction is confined to the medial parts of the SAS and does not extend significantly into the more lateral cortex, as indicated by the *arrow* (**a**, bar = 500 μm). The inset shows the acute inflammatory response at higher magnification, with lymphocytes, granulocytes, and macrophages within the arachnoid granulations (**b**). At 6 weeks, a chronic fibrosis is seen, again in a circumscribed region of the medial basal cisterns, as indicated by the asterisk (**c**, bar = 500 μm). The images also display the major arteries within the SAS, which appear morphologically intact.

Twelve-month-old rats were used to adopt to the aging NPH population, since they reflect human early senescence above the age of 50. In the aging rats, 0.03 ml kaolin suspension was sufficient to induce hydrocephalus as evidenced by measures of the Evan's index in the coronal histological specimen.

In both the adult and the aging rat model, studies on the vascular and AD-related comorbidity, as evidenced by the above described clinical observations, were performed by histological and immunohistological studies.

2.2. Histopathological Evidence of Vascular and Alzheimer-Related Comorbidity in Kaolin-Induced Hydrocephalus

Autoradiographic (C-14-Jodo-Antipyrin) studies in 3-month-old adult hydrocephalic rats have shown a prolonged decrease of cerebral blood flow in both cortical and subcortical regions ranging between 25% and 50% (59). This reflected a mild to moderate ischemia as seen in patients with cerebrovascular disease – Fig. 4. Those prolonged "sublethal, nonischemic" blood flow reductions are relevant to "silent stroke," the pathophysiological correlate of SAE, and have shown to induce a delayed selective neuronal injury in a rat model of four vessel occlusion (96). The relevance

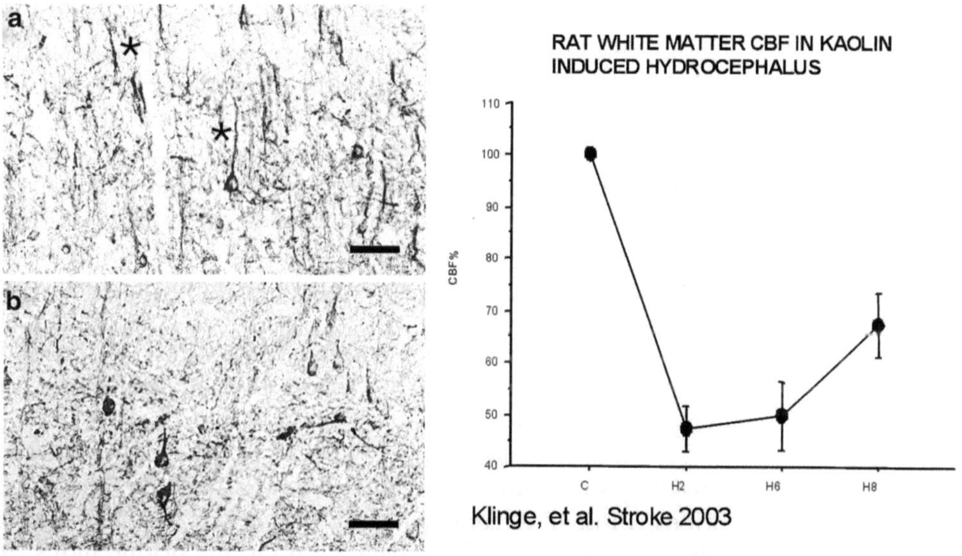

Fig. 4. Blood-flow autoradiographic studies in kaolin-induced hydrocephalus. (**a**, **b**) Neurofilament staining in the cortical pyramidal layer IV using specific antibody against the 68 kd neurofilament subunit. In controls (**a**), neurofilament 68 kD was predominately located in the dendrites and also in the perikaryon of the pyramidal-type cells; * dendrite staining. In the chronic hydrocephalus stage (**b**), at 8 weeks, the cytoarchitecture of the pyramidal layer was disturbed. There was a loss of the regular dendrite expression of neurofilament and some of the perikarya in the superficial cortical pyramidal layer displayed a strong increase in neurofilament-expression shifted towards the cell body; bar = 25 μm (**c**) Examples of [14C] iodoantipyrine autoradiography brain sections (kryosections 20 μm) for control and hydrocephalic animals, in which the optical densitometric measurement of C14-acitivity was done using C14 standards. Note the weaker density in the 2 and 6 weeks examples (59). The line graph displays [14C]iodoantipyrine autoradiography cerebral blood flow (CBF) in adult rats with kaolin hydrocephalus. CBF drops by 50% in 2 weeks after hydrocephalus induction and then recovers somewhat by 8 weeks, though never returning to normal despite normalization of intracranial pressure (ICP) (error bars = 1 SE).

of those prolonged "sublethal" blood flow reductions for the outcome of hydrocephalus are supported by findings of delayed neuronal injury and degeneration in ischemia-vulnerable areas in the kaolin-induced hydrocephalus model. Decreases in neurofilament and synaptophysin protein staining reactivity in the cortical pyramidal layer and the hippocampal CA-1 region have been observed, indicating an intrinsic susceptibility to cerebral hypoper-fu-sion (59).

In light of the human findings, the neurofilament protein staining in kaolin-induced hydrocephalus also supported the observations made in patients with NPH, namely that elevated neurofilament protein in the CSF suggest a "reversible" disease state (41, 53). Transient "reactive" increase in cortical neurofilament staining in the intermediate-term were seen in periventricular cortical areas (59) before chronic "breakdown" of neurofilament staining suggested neuronal damage in the long-term stages of hydrocephalus following the chronic ischemia – Fig. 4.

Aging is the most important risk factor for developing nonfamilial "sporadic" AD (46, 97, 98) and $A\beta_{1-40}$ and $A\beta_{1-42}$ peptide are the most important isoforms involved in AD plaque formation and in the AD-related pathological cascade, such as hpTau formation, oxidative stress, and microglial inflammatory responses (97). In the 12-month-old rats, both peptides have been immunohistochemically observed, both in the healthy control animals and in the hydrocephalic groups; however, not in the 3-month-old hydrocephalic and control rats. This supports the role of aging as a factor contributing to Aβ pathology, both in rodents and humans (99). Also, in the Brown–Norway Fisher aging rat model (100), $A\beta_{1-40}$ and $A\beta_{1-42}$ accumulations have been observed to dramatically increase during senescence at an age of 36 months, which equalled the three- to fourfold "load" in $A\beta_{1-42}$ in the cortex of the 12-month-old hydrocephalic rats, when counting the cortical $A\beta_{1-42}$-positively stained granules (101). $A\beta_{1-40}$, the more soluble of the two peptides, is usually present in the plaque periphery and stained in a diffuse, nonparticulate pattern, whereas $A\beta_{1-42}$-staining was characterized by small-, medium-, and large-size $A\beta_{1-42}$ granular accumulations, most likely reflecting the progressive self-aggregation of this less-soluble peptide, usually found in the plaque nidus (99) – Fig. 5. $A\beta_{1-42}$ accumulations were found in and near neurons; however, most strikingly at the microvessel border in both the cortex and the hippocampus the longer the hydrocephalus persisted. Those findings translate into the cerebral amyloidangiopathy seen in dementias. It was suggested that the amyloidangiopathy plays a role in white-matter pathology, chronic ischemia, and hypoperfusion as seen in patients with vascular dementia, and vascular dementia coexisting in AD and NPH (23, 102–104).

Much has been learned from the overexpression of Aβ as a result from the amyloid precursor protein (APP) cleavage induced

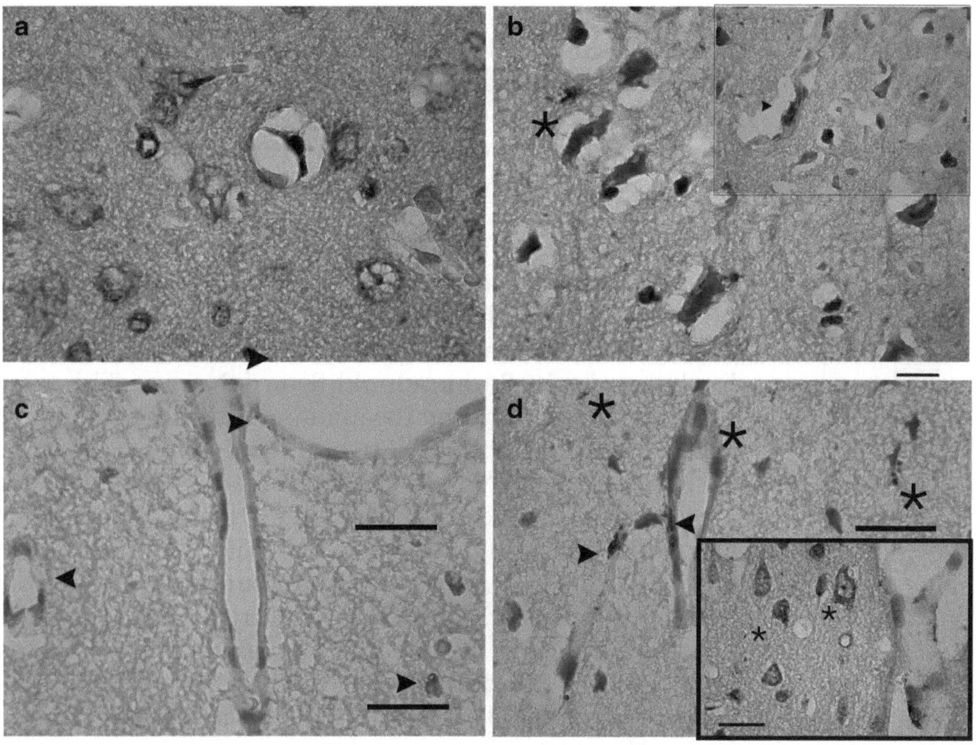

Fig. 5. Aβ_{1-40} and Aβ_{1-42} staining in the kaolin-induced hydrocephalus model of the aging rat. Photomicrograph showing cortical Aβ_{1-40}-positive staining in a 12-months nonhydrocephalic rat (**a**) and in a hydrocephalic rat 6 weeks after induction of hydrocephalus (**b**). In the control animals, a small amount of Aβ-positive staining is seen at a vessel border (*arrowhead*), whereas it is negative in the rest of the brain tissue. After 6 weeks of hydrocephalus, however, numerous Aβ_{1-40}-positively stained granules are seen both in the parenchyma and near neurons (*asterisks*). Larger and more numerous Aβ_{1-40}-positively stained deposits are also found at vessel borders (*inset*); bar= 25μm. Cortical Aβ_{1-42}-positive staining in a 12-month-old nonhydrocephalic rat (**c**, *arrowheads*) and in a hydrocephalic rat 6 weeks after kaolin induction (**d**) shows a small amount of age-related Aβ-positive staining around cortical vessels. After 6 weeks of hydrocephalus, amyloid positive granular staining is seen at vessels (*arrowheads*), in and near neurons (*inset*) and in the parenchyma (*asterisks*); bar = 25 μm.

by β- and γ-secretases, the so-called "amyloidogenic" pathway (105). On the other hand, there is Aβ metabolism characterized by a physiological clearance of Aβ peptides from the brain interstitial fluid via two main routes; (1) Clearance by diffusion and bulk flow from interstitial fluid to CSF, draining either into the lymphatic system or through the arachnoid granulations into the venous circulation (103), and (2) Aβ crossing the capillary endothelium of the blood-brain barrier (BBB) via specific receptors, the best characterized being low-density lipoprotein-related protein 1 (LRP-1) (106). Those mechanisms, however, have not been elaborately studied in relation to disease mechanisms and causes of neurodegeneration and dementia. In young mice, the clearance ratio of labeled Aβ across the BBB compared to the CSF route is nearly 9:1. However, with aging, LRP-1 transport is reduced by 50% (106). This would suggest a greater role for CSF production and turnover in Aβ clearance in aging brains, although CSF production and turnover, the secondary clearance route for Aβ, has also shown a two- to fourfold decrease in aging, hydrocephalus, and AD (107, 108).

BBB barrier dysfunction and leakage has been suggested in NPH as indicated by increase of CSF albumin ratios in patients (41, 109). In the kaolin-induced hydrocephalus of the aging rat, LRP-1 receptor staining dramatically decreased in both cortical and subcortical microvessels (100, 101). This resembled an accelerated effect of aging in hydrocephalus as the LRP-1 in 12-month-old hydrocephalus rats dropped down to levels seen in the 36-month-old Brown–Norway–Fisher senescent rats when comparing the immunohistochemical findings in kaolin-induced hydrocephalus with the findings of microvessel PCR in the Brown–Norway–Fisher aging rat (100). The findings suggest that Aβ accumulation and Alzheimer's pathology in chronic hydrocephalus of the elderly may be, at least in part, due to a failure of Aβ clearance, resulting from failure of the BBB transporters of Aβ (101). The decrease in LRP-1 receptor density has, in fact, been recently demonstrated in human AD brains (110). The dramatic change in BBB receptor ratios in both NPH and AD should lead to a net flux of Aβ into the brain (101, 110).

Neurofibrillary tangles (NFTs) are a primary pathological constituent of AD, and the number of NFTs present is well correlated to the severity of the dementia (111). A sequential phosphorylation of Tau protein has been reported as representing morphological "stages" of NFT formation: intraneuronal pre-NFT, intraneuronal NFT, and extraneuronal NFT are each phosphorylated at specific amino acids, constituting specific epitopes of hpTau (112). In the kaolin-model of the aging rat, an increased staining of both intraneuronal pre-NFT-hpTau (hp pT 231 and hp pT 262) and extraneuronal NFT-hpTau (AT 100) were investigated (113). At 6 and 10 weeks post hydrocephalic induction, pT231 staining showed neurons with focal somatic "condensed" increases of staining intensity, and loss of dendritic staining, attributable to hpTau migration from dendrites to the neuron soma ("excess Tau"), as seen in AD (113), while pT 262 did not display any differences. AT 100, the Tau variant that stains extracellular NFT's, containing substantial filamentous Tau – the last step in the progression of NFTs (113, 114) – showed increased staining in the hippocampus – Fig. 6. The prominent extracellular AT100 staining in the hippocampus raises further evidence towards a reflection of the stereotypical, sequential, and regional, hierarchical distribution of Tau pathology seen in AD (111), in the course of aging kaolin-induced hydrocephalus.

In NPH patients, reports differ regarding Tau levels in the lumbar CSF; both increases (115) and decreases (116, 117) have

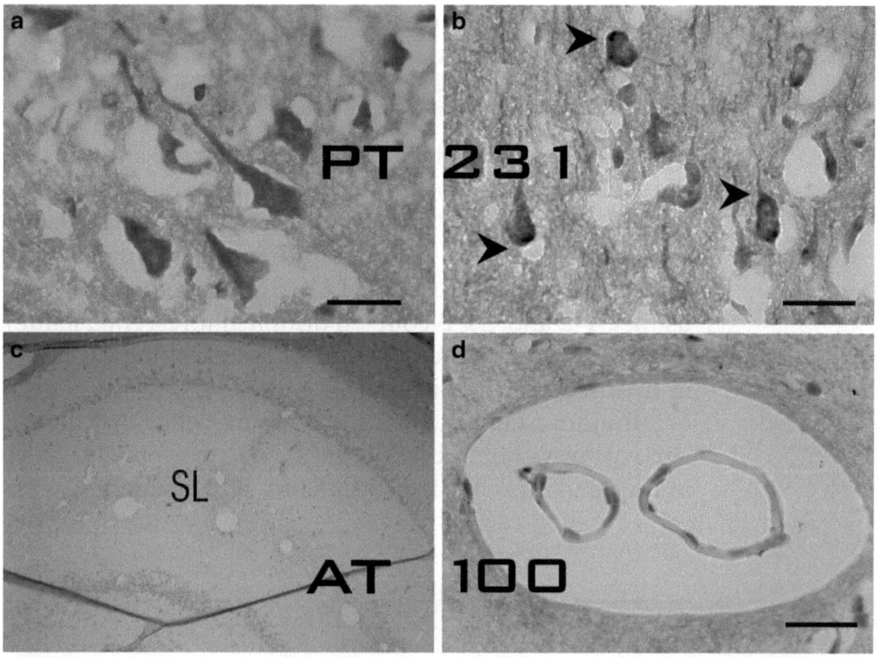

Fig. 6. Tau Pathology in the kaolin-induced hydrocephalus model of the aging rat. The photomicrograph shows cortical antihyperphosphorylated Tau pt 231 positive staining in a 12-month-old nonhydrocephalic rat (**a**) and in a hydrocephalic rat 10 weeks after kaolin induction (**b**). In the control animals, homogeneous age-related staining is seen throughout the soma and the dendrites of neurons, whereas, at 10 weeks, almost all neurons display a more irregular staining pattern with positively stained coarse granules in the soma periphery and at the axon hillock (*arrowheads*), areas of "excess" Tau accumulation in AD [113]. In the hippocampus of the hydrocephalic rats (**c**), extracellular AT100 staining was observed at 6 weeks. This antibody specifically recognizes paired helical filament Tau (PHF-Tau). The epitope of this antibody has been shown to contain phosphorylated Serine 121 and phosphorylated Threonine 214 residues (numbering according to human Tau 40) and prominently stains extracellular NFT's, containing substantial filamentous Tau, the last step in the progression of NFT's.The AT100-positive staining displayed a punctuate granular pattern in extracellular regions in proximity to hippocampal vessels, and was clearly associated with them (**d**); bar = 25 μm.

been observed. Neither study differentiated between regular and hpTau. Agren Wilson and coworkers (52) reported both low hpTau and Aβ levels in the lumbar CSF to be diagnostically relevant for NPH patients. In light of the findings on the Aβ and hpTau found in the kaolin-model, further evidence is needed on how reliably lumbar CSF protein content reflects CSF content as a result from cortical destruction or periventricular damage in dynamically altered CSF circulation of the hydrocephalic brain (100, 118). The different response in the hpTau variants in the kaolin model (119) warrants the study of the specific hpTau variants in human hydrocephalus.

Finally, the third component, the inflammatory oxidative stress response was evidenced by using a marker seen in and around dystrophic neurons and Aβ plaques in advanced AD, i.e., Lipoxygenase 12/15 (LOX12/15), probably being the major source of the oxidative stress (120, 121). LOX12/15 belongs to the family of lipid-peroxidizing enzymes, and has dual-specificity for forming both toxic arachidonic acid metabolites, 12 (S) and 15 (S) Hydroperoxy- eicosatetraenoic acids (12 (S) Hp ETE and 15 (S) Hp ETE) in a ratio of 3:1. In the kaolin-hydrocephalus of the aged rat, using the LOX 12/15-specific antibody as used in the AD-brains (120), immunostaining exhibited morphological findings in cortical dystrophic neurons and microglial-type cells similar to what has been found in AD samples (119, 122). Colocalization of the LOX 12/15 particles with hpTau and Aβ using CY-2 and CY-3 coupled double fluorescence staining added further support to the existence of the pathological cascade (123) in hydrocephalus – Fig. 7.

Immunostaining of APP did not evidence that the APP cleavage pathways are altered in chronic hydrocephalus as visualized in the aged rats with hydrocephalus. APP staining was prominent

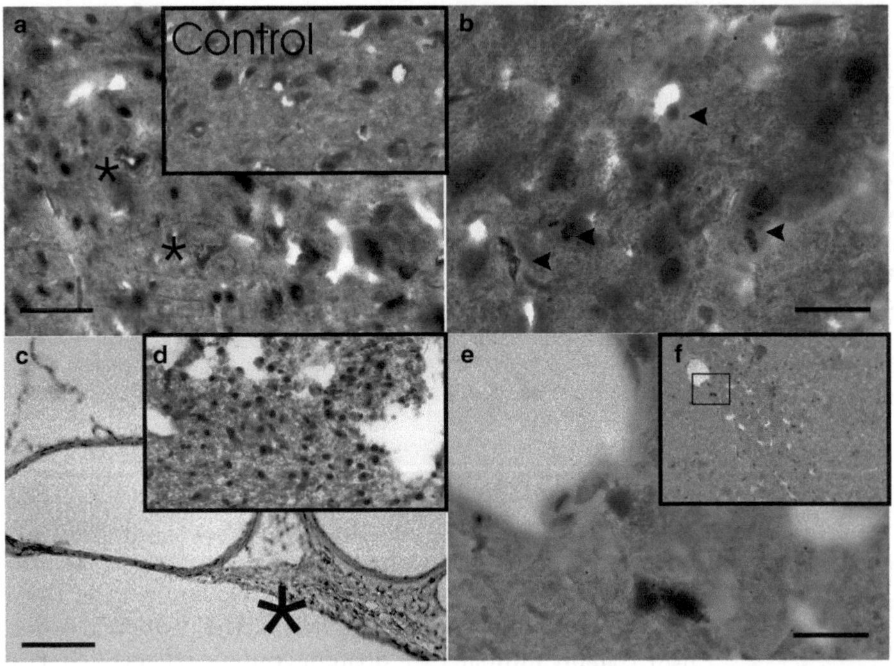

Fig. 7. Inflammatory Alzheimer-related response pathology in the kaolin-induced hydrocephalus model of the aging rat. Photomicrograph showing cortical 12/15 Lox positive staining in 12-month-old nonhydrocephalic rat (inset, ×40 magnification), and at 6 weeks after induction of kaolin hydrocephalus (**a**, bar = 50 μm). At 6 weeks of hydrocephalus, abundant positively stained "spots" are seen in and around neurons (*asterisks*). In controls, only rare and much smaller LOX positive spots are seen around neurons. At higher magnification (**b**, bar = 25 μm), the Lox-positive staining is intense within neuronal cell bodies and processes in the hydrocephalic rats (*arrowheads*), as described in Alzheimer's disease (AD) by Praticó et al. (AJP May 2004, Vol 164, No.5). In the subarachnoid space, however, where the kaolin-related chronic arachnoid inflammation is found (H&E; **c**, bar: 50 μm), no significant LOX-positive staining is observed (**d**, ×200 magnification). LOX-positive staining was also seen in cells that morphologically resemble microglia (**e**, bar = 25 μm). These cells were seen near vessel borders, predominantly observed in the hippocampus stratum lacunosum as seen in the overview (**f**, ×100 magnification).

in the fiber bundles crossing the caudate nucleus, but decreased markedly as hydrocephalus progressed.

In summary, experimental findings have evidenced that the comorbidities seen in NPH patients hallmark a common pathophysiological element existing in AD, vascular dementia, and in NPH. One has to acknowledge that the positive hpTau staining in the kaolin-induced rat model differs from the NFTs seen in NPH and AD patients. The pathological changes of human AD do not occur in rodents. However, the hpTau studies and the prior Aβ studies, as well as the observations on the AD-related inflammatory response, are likely to play a significant role in the high coincidence of AD and chronic ischemia in NPH, thereby allowing the study of the AD, SAE, and NPH continuum in relation to the reduced CSF absorption and turnover, to ischemia and BBB dysfunction.

2.3. Treatment of Kaolin-Induced Hydrocephalus

2.3.1. Shunt Treatment of Kaolin-Induced Hydrocephalus

It is perhaps the most important challenge of animal models in hydrocephalus to study the mechanism of recovery after shunt treatment. Much of it is still not entirely understood. In the majority of patients, ventricular width does not alter despite a dramatic clinical improvement, and yet, long-term outcome of shunt treatment in NPH is disappointing with recurrence occurring for both gait disturbances and dementia despite a functioning shunt. After 5 years of shunt treatment, significant improvement as rated by activity of daily living and/or quality of life measures is only found in 20% of the shunted patients (7).

Certainly, there is a need to identify "stages" of the disease and the "point-of-no-return." As such, both experimental and clinical studies have to target at an understanding of "reversible" versus "irreversible" disease signs (93). From NPH-related literature, there is evidence that a longer disease duration beyond a period of 2 years might be related to poorer outcome (6, 7, 124).

Shunt treatment in various animal models has been performed ever since, and incomplete reversibility of histological changes has been described with a preponderance of studies done in pediatric experimental hydrocephalus (125–134). Incomplete reversibility of neurotransmitter changes was correlated with an impaired learning ability after shunt treatment (62, 63).

Our studies in the adult kaolin-induced hydrocephalus model aimed at shunting the animals at two different time points to understand whether delaying of shunt treatment might be significant for "recovery" in the mature hydrocephalic brain. As the most vulnerable structure, changes in the hippocampus were investigated using immunohistochemical markers of oxidative stress and ischemia neuronal nitric oxide synthetase (nNOS) and heat-shock protein 70 and neuronal markers (synaptophysin 38kD and neurofilament 68kD) (59, 135).

Shunt placement was performed through a small midline longitudinal incision and a 2-mm-diameter burr-hole that was placed 2 mm posterior to the coronal suture and 2 mm lateral to the sagittal suture above the right hemisphere under a surgical microscope. A sterile transparent PE-catheter (PE 10) was inserted into the frontal horn of the right lateral ventricle. After CSF flow was established, the distal end of the catheter was tunneled under the skin of the neck and into a pocket that was created within the subcutaneous spaces of the neck. Animals were shunted at 1 (early) and at 3 weeks (late) after kaolin-induction and were reinvestigated at intervals of 1, 3, and 5 weeks after both shunt-placement times. After shunting, the course of ventricular width revealed no significant differences among hydrocephalic and shunted animals of both groups, which is in line with the findings of an unchanged ventricular width in human adult-type hydrocephalus – Fig. 8.

In the hydrocephalic rats, neuronal injury was evidenced by the immunohistochemical changes found in both the CA1 and CA3 hippocampal subfields. Early shunting at 1 week reversed most of the histological findings of increased nitric oxide synthetase staining in the CA1 sector and the "reactive" increases and changes seen for synaptophysin and neurofilament protein in both the CA1 and CA3 sectors. Late shunting, on the other hand, prompted a staining pattern comparable to the hydrocephalic stages, which virtually resembled those of the hydrocephalus animals – Fig. 8.

Certainly, these findings have important clinical implications for a timely shunting of hydrocephalic patients (136). They further demonstrate the "vulnerability" in the hydrocephalic brain, as evidenced by the hippocampal selective neuronal injury and the different responses and recovery in the neuronal markers. Different from neurofilament changes, immunohistochemical changes of synaptophysin did not recover at either time points of shunting. In animal models of ischemia, synaptophysin was suggested a sensitive marker of "reactive" synaptic changes related to impaired learning and memory processing (137). The minimal improvement seen in the histological findings at the level of the hippocampal CA1 sector after shunt treatment put further evidence to this hydrocephalic "vulnerability" (95, 136) – Fig. 7. Taking those findings into a broader perspective, and considering that the hippocampus would be affected early in the course of the chronic hydrocephalic disease, this might hold for the variability of cognitive deficits seen in NPH, and also for the poor and uncertain prognosis of memory processing function and dementia in NPH (7, 138).

Hippocampal injury in the course of chronic adult hydrocephalus was indicated in only a few studies (139, 140). An earlier study on Schaffer collateral responses in the CA1 pyramidal cell layer suggested disturbances in long-term potentiation of

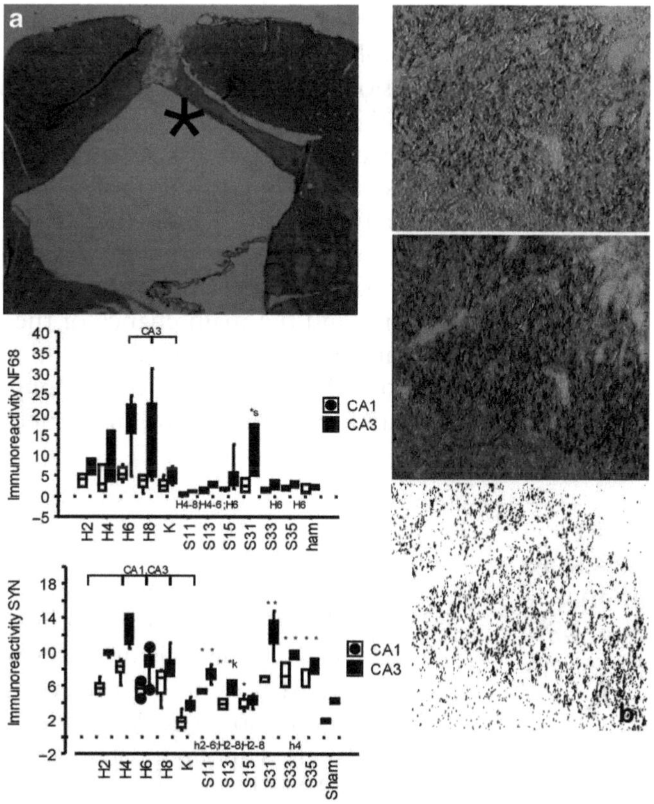

Fig. 8. Shunt treatment of kaolin-induced hydrocephalus and cellular recovery of neurofilament protein 68 kD and synaptophysin 38 kD. (**a**) Coronal brain slice evidencing the previous location of the ventricular catheter (*asterisks*) – brains were fixed with transcardial perfusion of 4% paraformaldehyde and postfixed in 30% sucrose solution for 72 h. (**b**) Synaptophysin staining in the hippocampal CA3 section shows focal increases in the apical pyramidal layer. The image sketches the black and white pixel analysis of immunoreactivity through an automatic image analysis process (Scion Image©). *Box plot* display the changes in "immunoreactivity" values in the hippocampal CA1 and CA3 subfields at 2 (H2), 4 (H4), 6 (H6), and 8 (H8) weeks following hydrocephalus induction, in unoperated controls (K), in sham-shunted animals (Sham), at 1 (S11), 3 (S13), 5 (S15) weeks of early shunt insertion (i.e. 1 week) and at 1 (S31), 3 (S33), 5 (S35) weeks of late shunt insertion (i.e. 3 weeks).* $p < 0.05$ (Fisher's PLSD) compared to control and sham group. Figures below the *box plots* indicate lower values of shunted groups compared to the hydrocephalus groups 2, 4, 6, and 8 weeks ($p < 0.05$; Fisher's PLSD); h = one region, H = both regions (CA1 and CA3).

population spikes in adult kaolin-induced hydrocephalus (141); however, the potential relevance of those findings for hydrocephalic pathophysiology has not been shown by studies that incorporate shunt treatment of the animals. Physicians tend to be conservative to avoid the surgical risks associated with shunting in NPH; withholding shunt treatment might impact the cognitive outcome of NPH.

Certainly, one has to acknowledge that shunting rodents cannot cover the whole spectrum of mechanisms relevant for human ventricular shunting, which is due to several technical and physiological factors. Probably, "siphoning" is an important physiological factor occurring in humans once in the upright position, mainly in the presence of differential pressure valve implants (6). It allows larger portions of CSF to drain into distal sites, mainly into the abdomen, and may lead to significantly negative ICPs (142). Paradoxically, it may have detrimental, as well as beneficial effects in human hydrocephalus (6). Technically, at least in smaller animals, shunt surgery has limitations regarding the choice and the maintenance of the distal placement sites. A subcutaneous pouch only allows a limited portion of the CSF to be absorbed at a time and over a long period of time since scarring around the catheter prevents function (126, 129).

3. Summary and Conclusions – Translational Aspects of NPH Basic Research

The studies in aging kaolin-induced hydrocephalus might allow insight into late-life disturbances in CSF circulation, CSF turnover, and BBB transport underlying several age-related dementias, including sporadic AD, which is not such a matter of genetic origin (such as apolipoprotein E4 and a2 macroglobulin dysmorphism) (107, 118), taken the pathological consequences of Aβ accumulations and the related toxic effects in both diseases, AD and NPH. This has implications for a common set of preventive and therapeutic modulations, utilizing the pathophysiological spectrum and continuum lying in metabolite and protein clearance defects resulting from CSF circulation in concert with BBB transport failures:

1. Restoration of CSF circulation and improved "clearance" of toxic brain metabolites, such as Aβ, e.g., by CSF drainage and shunt treatment
2. Pharmacological modulation of CSF circulation and BBB transport and neuroprotective treatment

This reconsiders the traditional concept of "shunts for dementia" (2) on the one hand, and opens the path for adjunct neuroprotective and neuromodulative therapeutic concepts in hydrocephalus. There is recent evidence existing for both:

1. Shunt treatment in AD patients has been reported from a prospective, randomised double-blinded, placebo-controlled trial performed in 2002 (143). A functioning-shunt group was compared to an occluded-shunt group using a novel, low-flow ventriculoperitoneal shunt system. Although initially

there was no differences in either group in the Global Deterioration Scale (GDS), a post hoc analysis on secondary activities of daily living measure (ADCS-ADL) as end-points and the Mathis Dementia Rating Scale showed a benefit of active shunting in early AD (144).

2. In the kaolin-induced model of the aging rat, the effect of human mesenchymal stem cells, secreting Glucagon-like-peptide-1, a neuroprotective peptide (145, 146), has shown reduced perivascular Aβ burden among a reduction of other glial and neuronal markers (GFAP, Heat-shock-protein 70, and nNOS) in the stem-cell treated group compared with untreated 6-week-old hydrocephalic rats. Alginate-encapsulated human mesenchymal stem cells were implanted into the right ventricle at the time of kaolin-induction and kept for a period of 5–6 weeks (147).

As such, both the clinical and experimental research activities should be aimed at alternative treatment facilities to augment the CSF circulation by improving BBB function, as well as the secretory and absorptive capacity of the choroid plexus, both of which contribute to the homeostasis of the brain interstitial fluid and neuronal metabolism (100). NPH has to be clearly seen from and inter- and multidisciplinary perspective, integrating both clinical and basic research elements. Shunt treatment has so far not sufficiently reduced the problem of dementia in NPH.

As a main conclusion from the Session on "Experimental Hydrocephalus," chaired by Dr. Hazel Jones, MD PhD (University of Florida) and Marc Del Bigio, MD PhD FRCPC (University of Manitoba, Winnipeg, Canada) at the international and interdisciplinary congress "Hydrocephalus 2008" (148), hydrocephalic models certainly mimic human hydrocephalus pathology to a high degree as they mirror ventricular enlargement, motor and cognitive changes, periventricular white matter and cortical damage, and respond to shunting. However, the details of the comparative development and neurobiology must be known, and to date there are no adequate "natural" NPH models available.

In the light of the human condition, behavior (motor and cognitive), imaging, CSF composition, tissue (histology and biochemistry), and response to drug and shunt therapy should be compared to bring further insight into the pathophysiology in the hydrocephalic brain. In NPH and hydrocephalus, both clinical and basic research needs a more comprehensive approach to the interaction of ICP, impaired CSF circulation and turnover, blood flow and brain metabolism, the impact of those parameter on the disease progress and on the different treatment potentials and modalities.

References

1. Hakim S, Adams RD (1965) The special clinical problem of symptomatic hydrocephalus with normal cerebrospinal fluid pressure. Observations on cerebrospinal fluid hydrodynamics. J Neurol Sci 2(4):307–327
2. Salmon JH, Armitage JL (1968) Surgical treatment of hydrocephalus ex-vacuo. Ventriculoatrial shunt for degenerative brain disease. Neurology 18(12):1223–1226
3. Petersen RC, Mokri B, Laws ER Jr (1985) Surgical treatment of idiopathic hydrocephalus in elderly patients. Neurology 35(3):307–311
4. Ojemann RG, Fisher CM, Adams RD, Sweet WH, New PF (1969) Further experience with the syndrome of "normal" pressure hydrocephalus. J Neurosurg 31(3):279–294
5. Fisher CM (1977) The clinical picture in occult hydrocephalus. Clin Neurosurg 24:270–284
6. Bergsneider M, Black PM, Klinge P, Marmarou A, Relkin N (2005) Surgical management of idiopathic normal-pressure hydrocephalus. Neurosurgery 57(3 Suppl):S29–S39
7. Klinge P, Marmarou A, Bergsneider M, Relkin N, Black PM (2005) Outcome of shunting in idiopathic normal-pressure hydrocephalus and the value of outcome assessment in shunted patients. Neurosurgery 57(3 Suppl):S40–S52
8. Krauss JK, Regel JP, Vach W, Droste DW, Borremans JJ, Mergner T (1996) Vascular risk factors and arteriosclerotic disease in idiopathic normal-pressure hydrocephalus of the elderly. Stroke 27(1):24–29
9. Malm J, Kristensen B, Stegmayr B, Fagerlund M, Koskinen LO (2000) Three-year survival and functional outcome of patients with idiopathic adult hydrocephalus syndrome. Neurology 55(4):576–578
10. Raftopoulos C, Massager N, Baleriaux D, Deleval J, Clarysse S, Brotchi J (1996) Prospective analysis by computed tomography and long-term outcome of 23 adult patients with chronic idiopathic hydrocephalus. Neurosurgery 38(1):51–59
11. Savolainen S, Hurskainen H, Paljarvi L, Alafuzoff I, Vapalahti M (2002) Five-year outcome of normal pressure hydrocephalus with or without a shunt: predictive value of the clinical signs, neuropsychological evaluation and infusion test. Acta Neurochir (Wien) 144(6):515–523
12. Aygok G, Marmarou A, Young HF (2005) Three-year outcome of shunted idiopathic NPH patients. Acta Neurochir Suppl 95:241–245
13. McGirt MJ, Williams MA, Rigamonti D (2007) Diagnosis, Treatment, and Analysis of Long-term outcomes in idiopathic normal-pressure hydrocephalus. Neurosurgery 60(1):E208
14. Marmarou A, Bergsneider M, Relkin N, Klinge P, Black PM (2005) Development of guidelines for idiopathic normal-pressure hydrocephalus: introduction. Neurosurgery 57(3 Suppl):S1–S3
15. Bech-Azeddine R, Waldemar G, Knudsen GM, Hogh P, Bruhn P, Wildschiodtz G, et al (2001) Idiopathic normal-pressure hydrocephalus: evaluation and findings in a multidisciplinary memory clinic. Eur J Neurol 8(6):601–611
16. Tisell M, Hoglund M, Wikkelso C (2005) National and regional incidence of surgery for adult hydrocephalus in Sweden. Acta Neurol Scand 112(2):72–75
17. Relkin N, Marmarou A, Klinge P, Bergsneider M, Black PM (2005) Diagnosing idiopathic normal-pressure hydrocephalus. Neurosurgery 57(3 Suppl):S4–S16
18. Shapiro K, Marmarou A, Shulman K (1980) Characterization of clinical CSF dynamics and neural axis compliance using the pressure-volume index: I. The normal pressure-volume index. Ann Neurol 7(6):508–514
19. Shulman K, Marmarou A (1968) Analysis of intracranial pressure in hydrocephalus. Dev Med Child Neurol Suppl-6
20. Borgesen SE, Gjerris F, Srensen SC (1978) The resistance to cerebrospinal fluid absorption in humans. A method of evaluation by lumbo-ventricular perfusion, with particular reference to normal pressure hydrocephalus. Acta Neurol Scand 57(1):88–96
21. Borgesen SE, Gjerris F (1987) Relationships between intracranial pressure, ventricular size, and resistance to CSF outflow. J Neurosurg 67(4):535–539
22. Borgesen SE (1984) Conductance to outflow of CSF in normal pressure hydrocephalus. Acta Neurochir (Wien) 71(1–2):1–45
23. Tullberg M, Hultin L, Ekholm S, Mansson JE, Fredman P, Wikkelso C (2002) White matter changes in normal pressure hydrocephalus and Binswanger disease: specificity, predictive value and correlations to axonal degeneration and demyelination. Acta Neurol Scand 105(6): 417–426
24. Bradley WG Jr, Whittemore AR, Watanabe AS, Davis SJ, Teresi LM, Homyak M (1991) Association of deep white matter infarction with chronic communicating hydrocephalus:

implications regarding the possible origin of normal-pressure hydrocephalus. AJNR Am J Neuroradiol 12(1):31–39

25. Jellinger K (1976) Neuropathological aspects of dementias resulting from abnormal blood and cerebrospinal fluid dynamics. Acta Neurol Belg 76(2):83–102
26. Klinge PM, Berding G, Brinker T, Knapp WH, Samii M (1999) A positron emission tomography study of cerebrovascular reserve before and after shunt surgery in patients with idiopathic chronic hydrocephalus. J Neurosurg 91(4):605–609
27. Klinge P, Berding G, Brinker T, Schuhmann M, Knapp WH, Samii M (2002) PET-studies in idiopathic chronic hydrocephalus before and after shunt-treatment: the role of risk factors for cerebrovascular disease (CVD) on cerebral hemodynamics. Acta Neurochir Suppl 81:43–45
28. Marmarou A, Bergsneider M, Klinge P, Relkin N, Black PM (2005) The value of supplemental prognostic tests for the preoperative assessment of idiopathic normal-pressure hydrocephalus. Neurosurgery 57(3 Suppl):S17–S28
29. Gjerris F, Borgesen SE, Sorensen PS, Boesen F, Schmidt K, Harmsen A, et al (1987) Resistance to cerebrospinal fluid outflow and intracranial pressure in patients with hydrocephalus after subarachnoid haemorrhage. Acta Neurochir (Wien) 88(3–4):79–86
30. Bech-Azeddine R, Hogh P, Juhler M, Gjerris F, Waldemar G (2007) Idiopathic normal-pressure hydrocephalus: clinical comorbidity correlated with cerebral biopsy findings and outcome of cerebrospinal fluid shunting. J Neurol Neurosurg Psychiatry 78(2):157–161
31. Savolainen S, Paljarvi L, Vapalahti M (1999) Prevalence of Alzheimer's disease in patients investigated for presumed normal pressure hydrocephalus: a clinical and neuropathological study. Acta Neurochir (Wien) 141(8):849–853
32. Klinge PM, Brooks DJ, Samii A, Weckesser E, van den HJ, Fricke H, et al (2008) Correlates of local cerebral blood flow (CBF) in normal pressure hydrocephalus patients before and after shunting – A retrospective analysis of [(15)O]H(2)O PET-CBF studies in 65 patients. Clin Neurol Neurosurg 110(4): 369–375
33. Owler BK, Momjian S, Czosnyka Z, Czosnyka M, Pena A, Harris NG, et al (2004) Normal pressure hydrocephalus and cerebral blood flow: a PET study of baseline values. J Cereb Blood Flow Metabol 24(1):17–23
34. Kondziella D, Sonnewald U, Tullberg M, Wikkelso C (2008) Brain metabolism in adult chronic hydrocephalus. J Neurochem 106(4):1515–1524
35. Agren-Wilsson A, Eklund A, Koskinen LO, Bergenheim AT, Malm J (2005) Brain energy metabolism and intracranial pressure in idiopathic adult hydrocephalus syndrome. J Neurol Neurosurg Psychiatry 76(8):1088–1093
36. Lenfeldt N, Hauksson J, Birgander R, Eklund A, Malm J (2008) Improvement after cerebrospinal fluid drainage is related to levels of N-acetyl-aspartate in idiopathic normal pressure hydrocephalus. Neurosurgery 62(1):135–141, discussion
37. Malm J, Kristensen B, Ekstedt J, Adolfsson R, Wester P (1991) CSF monoamine metabolites, cholinesterases and lactate in the adult hydrocephalus syndrome (normal pressure hydrocephalus) related to CSF hydrodynamic parameters. J Neurol Neurosurg Psychiatry 54(3):252–259
38. Braun KP, Vandertop WP, Gooskens RH, Tulleken KA, Nicolay K (2000) NMR spectroscopic evaluation of cerebral metabolism in hydrocephalus: a review. Neurol Res 22(1):51–64
39. Tullberg M, Hellstrom P, Piechnik SK, Starmark JE, Wikkelso C (2004) Impaired wakefulness is associated with reduced anterior cingulate CBF in patients with normal pressure hydrocephalus. Acta Neurol Scand 110(5):322–330
40. Tullberg M, Mansson JE, Fredman P, Lekman A, Blennow K, Ekman R, et al (2000) CSF sulfatide distinguishes between normal pressure hydrocephalus and subcortical arteriosclerotic encephalopathy. J Neurol Neurosurg Psychiatry 69(1):74–81
41. Tullberg M, Blennow K, Mansson JE, Fredman P, Tisell M, Wikkelso C (2008) Cerebrospinal fluid markers before and after shunting in patients with secondary and idiopathic normal pressure hydrocephalus. Cerebrospinal Fluid Res 5:9
42. Czosnyka ZH, Czosnyka M, Whitfield PC, Donovan T, Pickard JD (2002) Cerebral autoregulation among patients with symptoms of hydrocephalus. Neurosurgery 50(3):526–532
43. Lenfeldt N, Larsson A, Nyberg L, Andersson M, Birgander R, Eklund A, et al (2008) Idiopathic normal pressure hydrocephalus: increased supplementary motor activity accounts for improvement after CSF drainage. Brain 131:2904–2912
44. Agren-Wilsson A, Roslin M, Eklund A, Koskinen LO, Bergenheim AT, Malm J (2003) Intracerebral microdialysis and CSF hydrodynamics in idiopathic adult hydrocephalus

syndrome. J Neurol Neurosurg Psychiatry 74(2):217–221
45. Krauss JK, Regel JP, Vach W, Orszagh M, Jungling FD, Bohus M, et al (1997) White matter lesions in patients with idiopathic normal pressure hydrocephalus and in an age-matched control group: a comparative study. Neurosurgery 40(3):491–495
46. Silverberg GD, Mayo M, Saul T, Rubenstein E, McGuire D (2003) Alzheimer's disease, normal-pressure hydrocephalus, and senescent changes in CSF circulatory physiology: a hypothesis. Lancet Neurol 2(8):506–511
47. Silverberg GD (2004) Normal pressure hydrocephalus (NPH): ischaemia, CSF stagnation or both. Brain 127(Pt 5):947–948
48. Bech RA, Waldemar G, Gjerris F, Klinken L, Juhler M (1999) Shunting effects in patients with idiopathic normal pressure hydrocephalus; correlation with cerebral and leptomeningeal biopsy findings. Acta Neurochir (Wien) 141(6):633–639
49. Bech RA, Juhler M, Waldemar G, Klinken L, Gjerris F (1997) Frontal brain and leptomeningeal biopsy specimens correlated with cerebrospinal fluid outflow resistance and B-wave activity in patients suspected of normal-pressure hydrocephalus. Neurosurgery 40(3):497–502
50. Golomb J, Wisoff J, Miller DC, Boksay I, Kluger A, Weiner H, et al (2000) Alzheimer's disease comorbidity in normal pressure hydrocephalus: prevalence and shunt response. J Neurol Neurosurg Psychiatry 68(6):778–781
51. Weller RO, Massey A, Newman TA, Hutchings M, Kuo YM, Roher AE (1998) Cerebral amyloid angiopathy: amyloid beta accumulates in putative interstitial fluid drainage pathways in Alzheimer's disease. Am J Pathol 153(3):725–733
52. Agren-Wilsson A, Lekman A, Sjoberg W, Rosengren L, Blennow K, Bergenheim AT, et al (2007) CSF biomarkers in the evaluation of idiopathic normal pressure hydrocephalus. Acta Neurol Scand 116(5):333–339
53. Tullberg M, Rosengren L, Blomsterwall E, Karlsson JE, Wikkelso C (1998) CSF neurofilament and glial fibrillary acidic protein in normal pressure hydrocephalus. Neurology 50(4):1122–1127
54. Tarnaris A, Watkins LD, Kitchen ND (2006) Biomarkers in chronic adult hydrocephalus. Cerebrospinal Fluid Res 3:11
55. Tullberg M, Blennow K, Mansson JE, Fredman P, Tisell M, Wikkelso C (2007) Ventricular cerebrospinal fluid neurofilament protein levels decrease in parallel with white matter pathology after shunt surgery in normal pressure hydrocephalus. Eur J Neurol 14(3):248–254
56. Kondziella D, Ludemann W, Brinker T, Sletvold O, Sonnewald U (2002) Alterations in brain metabolism, CNS morphology and CSF dynamics in adult rats with kaolin-induced hydrocephalus. Brain Res 927(1):35–41
57. Richards HK, Bucknall RM, Jones HC, Pickard JD (1995) Uncoupling of LCBF and LCGU in two different models of hydrocephalus: a review. Childs Nerv Syst 11(5):288–292
58. da Silva MC, Michowicz S, Drake JM, Chumas PD, Tuor UI (1995) Reduced local cerebral blood flow in periventricular white matter in experimental neonatal hydrocephalus-restoration with CSF shunting. J Cereb Blood Flow Metab 15(6):1057–1065
59. Klinge P, Samii A, Mühlendyck A, Visnyei K, Meyer GJ, Walter GF, et al (2003) Cerebral hypoperfusion and delayed hippocampal response after induction of adult kaolin hydrocephalus. Stroke 34(1):193–199
60. Luciano MG, Skarupa DJ, Booth AM, Wood AS, Brant CL, Gdowski MJ (2001) Cerebrovascular adaptation in chronic hydrocephalus. J Cereb Blood Flow Metab 21(3):285–294
61. Tashiro Y, Chakrabortty S, Drake JM, Hattori T (1997) Progressive loss of glutamic acid decarboxylase, parvalbumin, and calbindin D28K immunoreactive neurons in the cerebral cortex and hippocampus of adult rat with experimental hydrocephalus. J Neurosurg 86(2):263–271
62. Tashiro Y, Drake JM, Chakrabortty S, Hattori T (1997) Functional injury of cholinergic, GABAergic and dopaminergic systems in the basal ganglia of adult rat with kaolin-induced hydrocephalus. Brain Res 770(1–2):45–52
63. Tashiro Y, Drake JM (1998) Reversibility of functionally injured neurotransmitter systems with shunt placement in hydrocephalic rats: implications for intellectual impairment in hydrocephalus. J Neurosurg 88(4):709–717
64. Jones HC, Chen GF, Yehia BR, Carter BJ, Akins EJ, Wolpin LC (2005) Single and multiple congenic strains for hydrocephalus in the H-Tx rat. Mamm Genome 16(4):251–261
65. Crews L, Wyss-Coray T, Masliah E (2004) Insights into the pathogenesis of hydrocephalus from transgenic and experimental animal models. Brain Pathol 14(3):312–316
66. Johanson C, Del Bigio M, Kinsman S, Miyan J, Pattisapu J, Robinson M, et al (2001) New models for analysing hydrocephalus and disorders of CSF volume transmission. Br J Neurosurg 15(3):281–283

67. Jones HC, Yehia B, Chen GF, Carter BJ (2004) Genetic analysis of inherited hydrocephalus in a rat model. Exp Neurol 190(1):79–90
68. Jones HC (1985) Cerebrospinal fluid pressure and resistance to absorption during development in normal and hydrocephalic mutant mice. Exp Neurol 90(1):162–172
69. Bucknall RM, Jones HC (1990) Electrocorticogram and sensory evoked potentials in the young hydrocephalic H-Tx rat. Z Kinderchir 45(Suppl 1):8–10
70. Oi S, Yamada H, Sato O, Matsumoto S (1996) Experimental models of congenital hydrocephalus and comparable clinical problems in the fetal and neonatal periods. Childs Nerv Syst 12(6):292–302
71. Jones HC, Totten CF, Mayorga DA, Yue M, Carter BJ (2005) Genetic loci for ventricular dilatation in the LEW/Jms rat with fetal-onset hydrocephalus are influenced by gender and genetic background. Cerebrospinal Fluid Res 2:2
72. Okuyama T, Hashi K, Sasaki S, Sudoh K (1985) Brain tissue damage in congenital hydrocephalus of the inbred rat, LEW/Jms – intracerebral cavity formation. No To Shinkei 37(11):1053–1057
73. McLone DG, Bondareff W, Raimondi AJ (1971) Brain edema in the hydrocephalic hy-3 mouse: submicroscopic morphology. J Neuropathol Exp Neurol 30(4):627–637
74. Oi S, Shimoda M, Shibata M, Honda Y, Togo K, Shinoda M, et al (2000) Pathophysiology of long-standing overt ventriculomegaly in adults. J Neurosurg 92(6): 933–940
75. Weller RO, Mitchell J (1980) Cerebrospinal fluid edema and its sequelae in hydrocephalus. Adv Neurol 28:111–123
76. Wilson RK, Williams MA (2007) Evidence that congenital hydrocephalus is a precursor to idiopathic normal pressure hydrocephalus (INPH) in only a subset of patients. J Neurol Neurosurg Psychiatry 78:508–511
77. Zhang J, Williams MA, Rigamonti D (2006) Genetics of human hydrocephalus. J Neurol 253(10):1255–1266
78. Johnson MJ, Ayzman I, Wood AS, Tkach JA, Klauschie J, Skarupa DJ, et al (1999) Development and characterization of an adult model of obstructive hydrocephalus. J Neurosci Methods 91(1–2):55–65
79. McAllister JP, Chovan P, Steiner CP, Johnson MJ, Ayzman I, Wood AS , et al (1998) Differential ventricular expansion in hydrocephalus. Eur J Pediatr Surg 8(Suppl 1):39–42
80. Del Bigio MR, Crook CR, Buist R (1997) Magnetic resonance imaging and behavioral analysis of immature rats with kaolin-induced hydrocephalus: pre- and postshunting observations. Exp Neurol 148(1):256–264
81. McAllister JP, Maugans TA, Shah MV, Truex RC Jr (1985) Neuronal effects of experimentally induced hydrocephalus in newborn rats. J Neurosurg 63(5):776–783
82. Del Bigio MR, Bruni JE (1987) Cerebral water content in silicone oil-induced hydrocephalic rabbits. Pediatr Neurosci 13(2):72–77
83 Sahar A, Hochwald GM, Ransohoff J (1970) Experimental hydrocephalus: cerebrospinal fluid formation and ventricular size as a function of intraventricular pressure. J Neurol Sci 11(1):81–91
84. Hochwald GM, Lux WE Jr, Sahar A, Ransohoff J (1972) Experimental hydrocephalus. Changes in cerebrospinal fluid dynamics as a function of time. Arch Neurol 26(2):120–129
85. Hochwald GM, Nakamura S, Camins MB (1981) The rat in experimental obstructive hydrocephalus. Z Kinderchir 34(4):403–410
86. Weller RO, Wisniewski H, Shulman K, Terry RD (1971) Experimental hydrocephalus in young dogs: histological and ultrastructural study of the brain tissue damage. J Neuropathol Exp Neurol 30(4):613–626
87. Cardoso ER, Del Bigio MR, Schroeder G (1989) Age-dependent changes of cerebral ventricular size. Part I: Review of intracranial fluid collections. Acta Neurochir (Wien) 97(1–2):40–46
88. Cardoso ER, Del Bigio MR (1989) Age-related changes of cerebral ventricular size. Part II: Normalization of ventricular size following shunting. Acta Neurochir (Wien) 97(3–4):135–138
89. Rekate HL (2008) The definition and classification of hydrocephalus: a personal recommendation to stimulate debate. Cerebrospinal Fluid Res 5:2
90. Rekate HL, Nadkarni TD, Wallace D (2008) The importance of the cortical subarachnoid space in understanding hydrocephalus. J Neurosurg Pediatrics 2(1):1–11
91. Li J, McAllister JP, Shen Y, Wagshul ME, Miller JM, Egnor MR, et al (2008) Communicating hydrocephalus in adult rats with kaolin obstruction of the basal cisterns or the cortical subarachnoid space. Exp Neurol 211(2):351–361
92. Slobodian I, Krassioukov-Enns D, Del Bigio MR (2007) Protein and synthetic polymer injection for induction of obstructive hydrocephalus in rats. Cerebrospinal Fluid Res 4:9

93. Williams MA, McAllister JP, Walker ML, Kranz DA, Bergsneider M, Del Bigio MR, et al (2007) Priorities for hydrocephalus research: report from a National Institutes of Health-sponsored workshop. J Neurosurg 107(5):345–357
94. Brinker T, Beck H, Klinge P, Kischnik B, Oi S, Samii M (1998) Sinusoidal intrathecal infusion for assessment of CSF dynamics in kaolin-induced hydrocephalus. Acta Neurochir (Wien) 140(10):1069–1075
95. Klinge P, Muhlendyck A, Lee S, Ludemann W, Groos S, Samii M, et al (2002) Temporal and regional profile of neuronal and glial cellular injury after induction of kaolin hydrocephalus. Acta Neurochir Suppl 81:275–277
96. Sekhon LH, Morgan MK, Spence I, Weber NC (1994) Chronic cerebral hypoperfusion and impaired neuronal function in rats. Stroke 25(5):1022–1027
97. Selkoe DJ (2000) Toward a comprehensive theory for Alzheimer's disease. Hypothesis: Alzheimer's disease is caused by the cerebral accumulation and cytotoxicity of amyloid beta-protein. Ann NY Acad Sci 924:17–25
98. Weller RO, Nicoll JA (2003) Cerebral amyloid angiopathy: pathogenesis and effects on the ageing and Alzheimer brain. Neurol Res 25(6):611–616
99. Cummings BJ (1997) Plaques and tangles: searching for primary events in a forest of data. Neurobiol Aging 18(4):358–362
100. Johanson CE, Duncan JA III, Klinge PM, Brinker T, Stopa EG, Silverberg GD (2008) Multiplicity of cerebrospinal fluid functions: New challenges in health and disease. Cerebrospinal Fluid Res 5:10
101. Klinge PM, Samii A, Niescken S, Brinker T, Silverberg GD (2006) Brain amyloid accumulates in aged rats with kaolin-induced hydrocephalus. Neuroreport 17(6):657–660
102. Weller RO, Cohen NR, Nicoll JA (2004) Cerebrovascular disease and the pathophysiology of Alzheimer's disease. Implications for therapy. Panminerva Med 46(4):239–251
103. Weller RO, Massey A, Kuo YM, Roher AE (2000) Cerebral amyloid angiopathy: accumulation of A beta in interstitial fluid drainage pathways in Alzheimer's disease. Ann NY Acad Sci 903:110–117
104. de la Torre JC (2004) Is Alzheimer's disease a neurodegenerative or a vascular disorder? Data, dogma, and dialectics. Lancet Neurol 3(3):184–190
105. Liu K, Solano I, Mann D, Lemere C, Mercken M, Trojanowski JQ, et al (2006) Characterization of Abeta11-40/42 peptide deposition in Alzheimer's disease and young Down's syndrome brains: implication of N-terminally truncated Abeta species in the pathogenesis of Alzheimer's disease. Acta Neuropathol 112(2):163–174
106. Shibata M, Yamada S, Kumar SR, Calero M, Bading J, Frangione B, et al (2000) Clearance of Alzheimer's amyloid-ss(1–40) peptide from brain by LDL receptor-related protein-1 at the blood-brain barrier. J Clin Invest 106(12):1489–1499
107. Silverberg GD, Heit G, Huhn S, Jaffe RA, Chang SD, Bronte-Stewart H, et al (2001) The cerebrospinal fluid production rate is reduced in dementia of the Alzheimer's type. Neurology 57(10):1763–1766
108. Silverberg GD, Huhn S, Jaffe RA (2002) Cerebrospinal fluid production is down-regulated in chronic hydrocephalus. J Neurosurg 97(6):1271–1275
109. Wikkelso C, Blomstrand C (1982) Cerebrospinal fluid proteins and cells in normal-pressure hydrocephalus. J Neurol 228(3):171–180
110. Donahue JE, Flaherty SL, Johanson CE, Duncan JA III, Silverberg GD, Miller MC, et al (2006) RAGE, LRP-1, and amyloid-beta protein in Alzheimer's disease. Acta Neuropathol 112(4):405–415
111. Braak H, Braak E (1997) Staging of Alzheimer-related cortical destruction. Int Psychogeriatr 9(Suppl 1):257–261
112. Kimura T, Ono T, Takamatsu J, Yamamoto H, Ikegami K, Kondo A, et al (1996) Sequential changes of tau-site-specific phosphorylation during development of paired helical filaments. Dementia 7(4):177–181
113. Augustinack JC, Schneider A, Mandelkow EM, Hyman BT (2002) Specific tau phosphorylation sites correlate with severity of neuronal cytopathology in Alzheimer's disease. Acta Neuropathol (Berl) 103(1):26–35
114. Gomez-Ramos P, Moran MA (1998) Ultrastructural aspects of neurofibrillary tangle formation in aging and Alzheimer's disease. Microsc Res Tech 43(1):49–58
115. Kudo T, Mima T, Hashimoto R, Nakao K, Morihara T, Tanimukai H, et al (2000) Tau protein is a potential biological marker for normal pressure hydrocephalus. Psychiatry Clin Neurosci 54(2):199–202
116. Tullberg M, Rosengren L, Blomsterwall E, Karlsson JE, Wikkelso C (1998) CSF neurofilament and glial fibrillary acidic protein in normal pressure hydrocephalus. Neurology 50(4):1122–1127

117. Lins H, Wichart I, Bancher C, Wallesch CW, Jellinger KA, Rosler N (2004) Immunoreactivities of amyloid beta peptide((1–42)) and total tau protein in lumbar cerebrospinal fluid of patients with normal pressure hydrocephalus. J Neural Transm 111(3):273–280
118. Johanson C, McMillan P, Tavares R, Spangenberger A, Duncan J, Silverberg G, et al (2004) Homeostatic capabilities of the choroid plexus epithelium in Alzheimer's disease. Cerebrospinal Fluid Res 1(1):3
119. Klinge PM, Heile A, Slone S, Johanson CE, Miles M, Duncan JA III et al (2008) Evidence of TAU pathology in kaolin-induced hydrocephalus model of the aged rat. Cerebrospinal Fluid Res 4(Suppl I):S 43
120. Pratico D, Zhukareva V, Yao Y, Uryu K, Funk CD, Lawson JA, et al (2004) 12/15-lipoxygenase is increased in Alzheimer's disease: possible involvement in brain oxidative stress. Am J Pathol 164(5):1655–1662
121. Pratico D, Yao Y, Rokach J, Mayo M, Silverberg GG, McGuire D (2004) Reduction of brain lipid peroxidation by CSF drainage in Alzheimer's disease patients. J Alzheimers Dis 6(4):385–389
122. Heile A, Knippenberg S, Miller M, Johanson CE, Silverberg GD, Klinge PM (2008) Pathology of AD-related inflammatory and oxidative stress mediators in Kaolin-induced hydrocephalus model of the aged rat. Clin Neurol Neurosurg 110(Suppl 1):S 17
123. Smith MA, Perry G (1996) Alzheimer disease: protein-protein interaction and oxidative stress. Bol Estud Med Biol 44(1–4):5–10
124. Larsson A, Wikkelso C, Bilting M, Stephensen H (1991) Clinical parameters in 74 consecutive patients shunt operated for normal pressure hydrocephalus. Acta Neurol Scand 84(6):475–482
125. Harris NG, McAllister JP, Jones HC (1993) The effect of untreated and shunt-treated hydrocephalus on cortical pyramidal neurone morphology in the H-Tx rat. Eur J Pediatr Surg 3(Suppl 1):31–32
126. Harris NG, Jones HC, Patel S (1994) Ventricle shunting in young H-Tx rats with inherited congenital hydrocephalus: a quantitative histological study of cortical grey matter. Childs Nerv Syst 10(5):293–301
127. Jones HC, Rivera KM, Harris NG (1995) Learning deficits in congenitally hydrocephalic rats and prevention by early shunt treatment. Childs Nerv Syst 11(11):655–660
128. Jones HC, Harris NG, Rocca JR, Andersohn RW (2000) Progressive tissue injury in infantile hydrocephalus and prevention/reversal with shunt treatment. Neurol Res 22(1):89–96
129. Lovely TJ, Miller DW, McAllister JP (1989) A technique for placing ventriculoperitoneal shunts in a neonatal model of hydrocephalus. J Neurosci Methods 29(3):201–206
130. McAllister JP, Cohen MI, O'Mara KA, Johnson MH (1991) Progression of experimental infantile hydrocephalus and effects of ventriculoperitoneal shunts: an analysis correlating magnetic resonance imaging with gross morphology. Neurosurgery 29(3):329–340
131. Eskandari R, McAllister JP, Miller JM, Ding Y, Ham SD, Shearer DM, et al (2004) Effects of hydrocephalus and ventriculoperitoneal shunt therapy on afferent and efferent connections in the feline sensorimotor cortex. J Neurosurg 101(2 Suppl):196–210
132. Miller JM, McAllister JP (2007) Reduction of astrogliosis and microgliosis by cerebrospinal fluid shunting in experimental hydrocephalus. Cerebrospinal Fluid Res 4:5
133. Del Bigio MR, Bruni JE (1988) Periventricular pathology in hydrocephalic rabbits before and after shunting. Acta Neuropathol 77(2):186–195
134. Del Bigio MR, Kanfer JN, Zhang YW (1997) Myelination delay in the cerebral white matter of immature rats with kaolin-induced hydrocephalus is reversible. J Neuropathol Exp Neurol 56(9):1053–1066
135. Klinge PM, Beck H, Brinker T, Walter GF, Samii M (1999) Induction of heat shock protein 70 in the rat brain following intracisternal infusion of autologous blood: evaluation of acute neuronal damage. J Neurosurg 91(5):843–850
136. Klinge PM, Gerkmann I, Samii A, Brinker T (2004) Incomplete reversibility of selective hippocampal response after early and delayed shunting in adult kaolin-hydrocephalus – implications for early shunt treatment. Cerebrospinal Fluid Res 1(Suppl 1):S 33
137. Martinez G, Di Giacomo C, Carnazza ML, Sorrenti V, Castana R, Barcellona ML, et al (1997) MAP2, synaptophysin immunostaining in rat brain and behavioral modifications after cerebral postischemic reperfusion. Dev Neurosci 19(6):457–464
138. Thomsen AM, Borgesen SE, Bruhn P, Gjerris F (1986) Prognosis of dementia in normal-pressure hydrocephalus after a shunt operation. Ann Neurol 20(3):304–310
139. Tashiro Y, Chakrabortty S, Drake JM, Hattori T (1997) Progressive loss of glutamic acid decarboxylase, parvalbumin, and calbindin D28K immunoreactive neurons in the cerebral cortex

and hippocampus of adult rat with experimental hydrocephalus. J Neurosurg 86(2):263–271
140. Shinoda M, Olson L (1997) Immunological aspects of kaolin-induced hydrocephalus. Int J Neurosci 92(1–2):9–28
141. Tsubokawa T, Katayama Y, Kawamata T (1988) Impaired hippocampal plasticity in experimental chronic hydrocephalus. Brain Inj 2(1):19–30
142. Bergsneider M, Yang I, Hu X, McArthur DL, Cook SW, Boscardin WJ (2004) Relationship between valve opening pressure, body position, and intracranial pressure in normal pressure hydrocephalus: paradigm for selection of programmable valve pressure setting. Neurosurgery 55(4):851–858
143. Silverberg GD, Levinthal E, Sullivan EV (2002) Assessment of low-flow CSF drainage as a treatment for AD: Results of a randomized pilot study. Neurology (in press)
144. Silverberg GD, Mayo M, Saul T, Fellmann J, Carvalho J, McGuire D (2008) Continuous CSF drainage in AD: results of a double-blind, randomized, placebo-controlled study. Neurology 71(3):202–209
145. Perry T, Greig NH (2002) The glucagon-like peptides: a new genre in therapeutic targets for intervention in Alzheimer's disease. J Alzheimers Dis 4(6):487–496
146. Perry T, Lahiri DK, Sambamurti K, Chen D, Mattson MP, Egan JM, et al (2003) Glucagon-like peptide-1 decreases endogenous amyloid-beta peptide (Abeta) levels and protects hippocampal neurons from death induced by Abeta and iron. J Neurosci Res 72(5):603–612
147. Heile A, Knippenberg S, Wallrap C, Klinge PM, Brinker T (2008) Ex vivo gene therapy of hydrocephalus: Intrathecal implantation of mesenchymal stem cells producing the neuroprotective peptide Glucagon like peptide-1. Clin Neurol Neurosurg 110(Suppl 1): S 30–S 31
148. Abstracts of the Congress "Hydrocephalus 2008" 17–20 September 2008, Hannover, Germany (2008). Clin Neurol Neurosurg 110(Suppl 1):1–46

Part VIII

Animal Models of Traumatic Dementia

Chapter 32

Animal Models of Traumatically-Induced Dementia

Jennifer E. Slemmer, Mohammad Z. Hossain, and John T. Weber

Abstract

Dementia in humans following traumatic brain injury (TBI) has been well documented in clinical populations, either after a single TBI or repeated mild TBI (rMTBI). In most cases, trauma-induced dementia follows a slow, chronic time-course, and in many cases mild injuries accumulate over time. Both single TBI and rMTBI have been modeled experimentally *in vivo*, most often in rodents, and various treatment strategies have been studied in order to reduce brain damage after injury. Here, we review the recent literature with regard to currently used *in vivo* models of TBI in animals, and studies conducted with them to investigate the link between TBI and the development of dementia-like pathology, such as that associated with Alzheimer's disease (AD). We also discuss the potential use of *in vitro* models of trauma for investigating a link between trauma and dementia.

Key words: Animal model, Alzheimer's disease, Dementia, Mouse, Rat, Repeated injury, TBI

1. Introduction

Traumatic brain injury (TBI) is generally defined as an injury to the brain caused by an external physical force, resulting in total or partial functional disability. Motor vehicle accidents are the primary cause of TBIs, while falls, sports, assaults, and gunshot wounds are also major causes of these injuries (1, 2). Globally, the incidence of TBI is reported as approximately 200 in 100,000 individuals with a mortality rate of about 20 per 100,000, making it one of the leading causes of death and disability worldwide (1, 2). In the United Kingdom, 200–300 per 100,000 people are hospitalized each year due to TBI (3), and the incidence is even higher in areas such as South Africa and southern Australia (4, 5). In the United States, TBI is associated with the death of approximately 51,000 people each year and causes long-term disability that affects an estimated 70,000–90,000 individuals

Peter Paul De Deyn and Debby Van Dam (eds.), *Animal Models of Dementia*, Neuromethods, vol. 48,
DOI 10.1007/978-1-60761-898-0_32, © Springer Science+Business Media, LLC 2011

annually (6). TBI is particularly prevalent in young individuals, and is the leading cause of mortality in infants (7). In the United Kingdom, TBI may represent as much as 20% of all deaths of individuals in the range of 5–45 years of age (5). In 1999, a National Institutes of Health Consensus Development Panel suggested that as many as 6.5 million individuals may be living with the consequences of TBI in the United States. In Europe, at least 11.5 million people suffer from disabilities related to a TBI (8). Although the total financial burden of TBI is difficult to determine, estimated costs of head injury per year in the United States alone are over $1 billion (9, 10). Given the global prevalence and enormous socioeconomic costs of TBI, it is important to understand the mechanisms that contribute to long-term dysfunction following trauma, so that we may devise appropriate treatment strategies to improve patient outcome.

The pathophysiology of TBI consists of two main phases: a primary (mechanical) phase of damage and a delayed secondary phase of damage. Primary damage occurs at the moment of sudden traumatic insult and includes contusion and laceration, intracranial hemorrhage, and diffuse axonal injury (10–13). Secondary damage includes processes that are initiated at the time of insult, but do not appear clinically for hours, or even days after the insult has occurred. Secondary processes include brain damage due to altered neurochemical mechanisms (e.g. increased glutamate release, generation of free radical species), activation of degradative enzymes (e.g. proteases), swelling (edema), and ischemia (for reviews, see (14–21)). TBI is also further defined by the severity of its symptoms, primarily based on the Glasgow Coma Scale (GCS) (22). Major classifications include severe TBI, which has a high fatality rate, moderate TBI, and mild TBI, which includes concussion and does not necessarily involve a loss of consciousness.

Studies and case reports from Europe in the 1950s and 1960s highlighted the coincidence of brain injury and cerebral deficits in specific human populations, such as dementia pugilistica (DP) in professional boxers (23–25), which results in severe motor and cognitive dysfunctions (26). These observations provided some of the first clues leading to the suggestion that mild, repetitive insults to the head could lead to dementia, most likely caused by neurodegeneration. Dementia is broadly defined as a clinical syndrome of cognitive and functional decline, frequently associated with behavior or personality changes (27). There are several different types of cortical and subcortical degenerative dementias (27, 28). Besides DP, the main cortical dementias include Alzheimer's disease (AD), vascular dementia, and Diffuse Lewy body dementia (DLBD), whereas subcortical dementias can result from diseases such as Huntington's, Parkinson's, syphilis, and AIDS. Some of these diseases are primarily associated with aging (e.g. AD, vascular dementia), some clearly are inherited (e.g. Huntington's), while others are due to environmental factors, such as infectious agents (syphilis, AIDS).

As with DP, there is growing evidence of a link between TBI and AD, suggesting that this type of neurodegenerative disease can be environmentally triggered. Initial evidence for this relationship was derived from a case report in which a 38-year-old man, who had experienced a TBI 16 years previously, developed early-onset AD (29). Since then, several epidemiological studies have been conducted to investigate this interaction. Although many studies have found no conclusive connection between TBI and AD, many others have demonstrated a definite relationship, and TBI is generally considered a robust risk factor for the future development of AD (for reviews, see (30, 31)).

As can be deduced from the literature, evidence for a link between head injury and the eventual development of cerebral dementia is correlative in nature. What is needed are in-depth studies with experimental models of TBI in order to demonstrate a true existence of a cause-and-effect relationship between trauma and dementia, as well as an understanding of the mechanisms that can lead to dementia following an injury, or repeated injuries, to the head. Although this area of research is largely in its infancy, several experimental approaches used to analyze trauma-induced dementia have been developed in recent years. This review provides an overview of the experimental methods that have been utilized to study the link between TBI and dementia.

2. *In Vivo* Models of Traumatically Induced Dementia: Single Injury

In addition to cognitive symptoms, dementias such as DP and AD are associated with specific types of neuropathological markers. In fact, AD in humans can only be fully confirmed postmortem via the presence of extracellular senile plaques, which are abnormal amyloid β (Aβ) protein deposits, and abnormal tau protein aggregation in specific brain regions (32). The tau protein is a functional component of the neuronal cytoskeleton of healthy neurons, and is a key component of the neurofibrillary plaques that are seen in both DP and AD (33). As with much basic research aimed at understanding neurodegenerative diseases, one is often shrinking down a large portion of a patient's lifetime into a few years or less. This can be challenging in one aspect, but beneficial if a disease state can be modeled properly. Any thorough study of the link between TBI and dementia should in theory encompass cognitive assessments of animals, as well as neuropathological evaluations at the conclusion of the study.

Early investigations into the effects of trauma on the development of dementia-like pathology in animals were largely descriptive in nature. In general, these studies utilized trauma devices, which delivered a head injury – usually to rats or mice – and measured cognitive impairment at various time points. These experiments also attempted to replicate clinical findings in human cadavers, in particular the existence of neuropathological markers, such as tau and Aβ protein, in animal brains after trauma (30, 31). Although moderate and severe TBI in humans have been fairly well characterized, satisfactory treatment options have been slow to develop. Recently, increased emphasis has been placed on patients who suffer from single or repeated mild TBI (rMTBI) as this population not only suffers from cognitive deficits, but may also respond better to treatment. Indeed, Sosin et al. (34) pointed out that over a million concussions occur annually in the United States, with a third of those cases resulting from sports-related injuries. Given the large number of new TBI cases worldwide per year, especially mild TBI, the coincidence of TBI and dementia can only increase. As the world population continues to age, however, especially in Western nations, the need for mechanistic studies has greatly increased. Overall, current aims have been to treat injured patients and to predict their recovery outcomes, the assumption being that there is a definite, causal link between TBI and dementia, such as AD.

Animal models of cognitive dysfunction, in particular those that use mammals, are based on the relative conservation of neuroanatomical structure–function relationships between humans and other mammals, such as rodents (35). To this end, multiple tests have been developed with which to discern cognitive deficits in rodents, including the Morris water maze (MWM), object recognition, fear conditioning, and conditioned taste aversion (35). These paradigms can be particularly useful after mild TBI, where impairment between groups may be more subtle. That being said, Nakayama et al. (36) believe that our current use of animal models for dementia may be hampered by the fact that they may die before the signs of neurodegenerative diseases are manifested. In addition, any deficits in cognitive tests in rodents subjected to experimental TBI may be due to dementia, or to the more immediate effects of the trauma itself. Therefore, the time it takes to observe cognitive deficits in animals, or the notion that cognitive deficits may worsen with time after injury, are important characteristics to take into account in these animal models.

Several experimental devices have been developed, which deliver reliable and reproducible brain injuries to rodents. The two most commonly used models in studies of traumatically induced dementia are lateral fluid percussion injury (LFPI) and controlled cortical impact (CCI). Injury severity can be easily

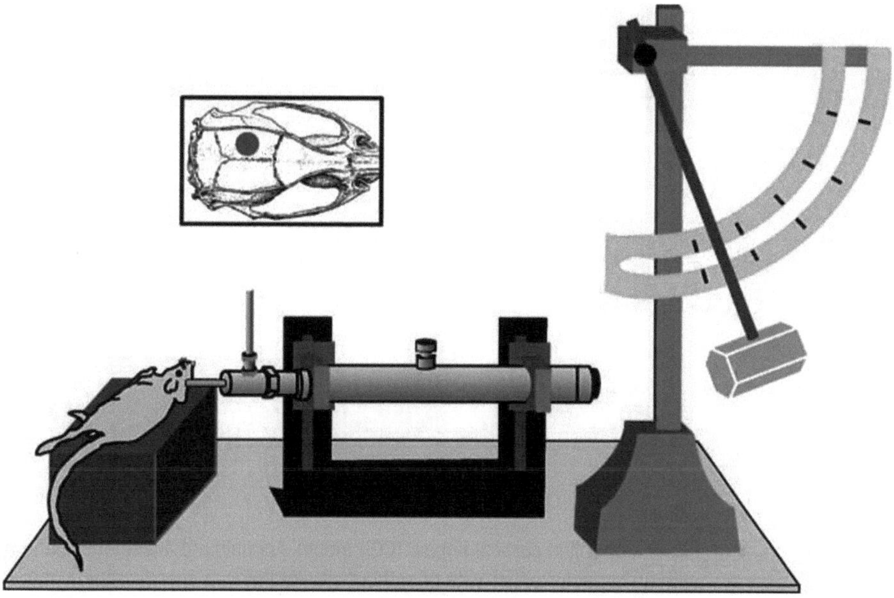

Fig. 1. Lateral fluid percussion injury (LFPI) model. Representative placement and site of originally described lateral (parasagittal) fluid percussion brain injury (*inset*) and schematic diagram of a fluid percussion brain injury device. A pendulum from a known height impacts the piston of a saline-filled reservoir, forcing a brief fluid bolus into the sealed cranial cavity (Reproduced from Thompson et al., 2005 with permission from Mary Ann Liebert, Inc.).

adjusted with these *in vivo* TBI models, in order to study mild, moderate, or severe levels of injury. To eliminate any confounding effects from surgery or anesthesia, "sham" animals are used as comparative controls, in that they undergo the same procedures as injured animals without the actual head injury.

LFPI is an experimental model of concussive TBI used in rodents (Fig. 1). In this model, a pendulum strikes a piston sending a brief percussion pulse of saline into the extradural space of the closed cranial cavity causing a quick displacement and deformation of the brain, mimicking concussive injury. The severity of the injury is determined by adjusting the height of the pendulum. LFPI results in focal as well as diffuse brain injury consisting of a cortical contusion at the impact site and cell death and axonal damage in the hippocampus, thalamus, and cortex (37–40). LFPI is widely used since it reproduces many pathophysiological features of human TBI including breakdown of the blood-brain barrier, neuronal cell loss, gliosis, altered ionic homeostasis, and long-term alterations in cognitive and motor function (37, 38, 40). LFPI has been extensively characterized and widely used in rats, and more recently in mice (39, 41). Although data generated in rats make for easier comparisons to the majority of previous studies in this animal, the use of mice to study TBI allows investigators to exploit transgenic and knockout mouse technology to directly test the role of specific genes or gene mutations in TBI.

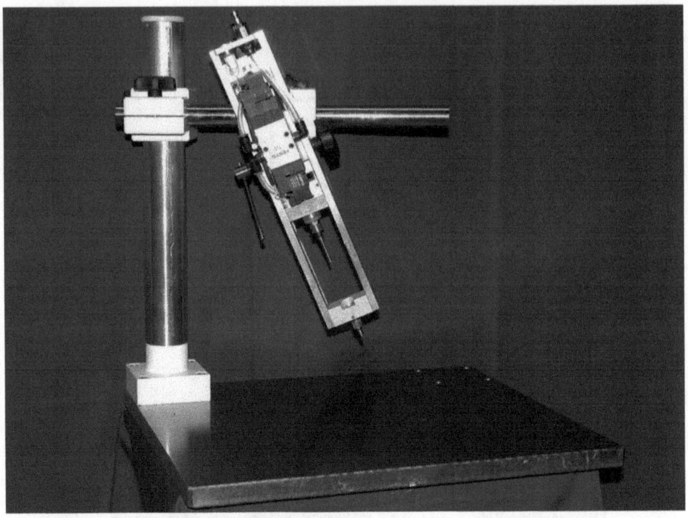

Fig. 2. Controlled cortical impact (CCI) model. Animals are anesthetized and are injured with an impactor either with or without craniotomy in a specific region of the skull. The severity of injury is determined by the impact velocity and duration of the impactor tip (photo supplied courtesy of Prof. Nikolaus Plesnila, Royal College of Surgeons in Ireland).

CCI is an *in vivo* model of TBI (Fig. 2), which produces a highly quantifiable impact to the brain or skull (13, 42). As with LFPI, animals generally undergo a craniotomy, and the severity of injury is determined by the impact velocity and duration of the impactor, which causes displacement of the brain of various levels. This model can also be used in the absence of craniotomy, which produces a closed head impact model, which more closely mimics TBI in the human population. Qualitatively, this model produces a similar pattern of injury throughout the brain when compared to LFPI. Some major advantages of the model are that the impact produced by CCI can be scaled across species due to its highly quantifiable nature (43), and it can be used to produce injury in several different areas of the cortex.

Using LFPI, Pierce et al. (44) showed increased amyloid precursor protein (APP), which is the protein from which Aβ protein is produced, in rats after injury. However, no increase in Aβ was noted. In a follow-up study, Pierce et al. (45) observed cognitive deficits for up to a year after severe LFPI. Masumura et al. (46) showed no changes in APP or Aβ after LFPI in the rat, whereas Murakami et al. (47) demonstrated increased APP in the cerebral cortex and hippocampus, but no increase in Aβ deposition. Ciallella et al. (48) also showed increased APP, but no Aβ increase using the CCI model in rats. In another study using CCI, Blasko et al. (49) found increases in the activity of β-secretase after injury, which is the first protease that cleaves APP, followed by the action of γ-secretase, which then leads to Aβ production. The activity of β-secretase increased transiently for 2 days after injury, but decreased back to basal levels by 7 days post-injury. Overall, these nontransgenic animal studies of TBI have failed to demonstrate a rapid increase in Aβ in the brain after injury.

The weight-drop model, which produces vertical acceleration movement of the head, has also been used to analyze cognitive performance after injury, with interesting results. This model consists of a mass free falling a known distance over a friction free guide rod and onto an impactor device, which is positioned on the head of the animal. Milman et al. (50) administered single, mild injuries to mice, and measured their cognitive abilities at 7, 30, 60, and 90 days post-injury using the swim T-maze and passive avoidance paradigms. They found that injured mice performed significantly worse than uninjured mice at day 30 only, indicating that the onset of cognitive impairment was slow, and that recovery from injury was possible. Slow onset of cognitive dysfunction could be indicative of dementia, but a recovery from such symptoms is not common in the human population.

To date, most of the previous studies have focused on the development of dementia, or specifically AD-related dementia, after moderate or severe levels of injury. Recent reviews have highlighted the advances being made in the generation of animal models of traumatically induced AD (31, 35). The focus on AD is understandable, as it accounts for more than half of all age-related dementias in modern society (51). However, the benefits of using transgenic animal models of dementia, such as AD (e.g. reduced time until disease onset, clear evidence of neuropathology) must be weighed against the relatively small percentage of cases with an obvious genetic link. For example, almost 90% of all AD cases are believed to be sporadic (52). Nonetheless, these studies provide insights into some of the potential pathological signs of AD that occur after TBI.

Recently, Brody and Holtzman (53) investigated the use of various MWM platform search strategies in wild-type mice, and in transgenic mice that overexpress a mutant human form of APP, before and after a single, moderate injury using the CCI device. Uninjured transgenic mice performed worse on the MWM (i.e. reduced reliance on spatial strategies) than wild-type animals on day 1, but equally as well by day 3. After TBI, however, the transgenic mice demonstrated poor strategy use as compared to injured wild-type mice, and the authors believe that this impairment provides direct comparisons to human TBI patients. Importantly, they did not find any Aβ deposition in transgenic mice, with or without injury; the question remains, however, as to whether Aβ does not play a role in traumatically induced dementia, or if this lack of deposition is attributed to the long-term progression of AD in humans.

Nakagawa et al. (54) also used a transgenic mouse model in which deposition of Aβ accumulates by 6 months of age. Both transgenic mice and wild-type controls were subjected to CCI, or remained uninjured, and brains were examined for Aβ deposition after 2, 5, or 8 months. Interestingly, TBI caused massive hippocampal neuron loss, which occurred concomitantly with a large, brief increase in Aβ, as well as decreased overall Aβ deposition, in older transgenic mice. The authors proposed a "two-hit" hypothesis in an attempt to explain their findings, and those of other groups who have found little or no evidence linking TBI and the development of dementia, particularly AD. In their hypothesis, TBI constitutes the second-hit, in which damage is manifested only after the first hit: high concentrations of Aβ.

In some of the earliest studies analyzing the effects of trauma on the development of AD-like pathology, Murai et al. (55) failed to demonstrate increased Aβ levels after injury induced by CCI in a transgenic mouse model that overexpresses human APP by twofold. In addition, injured animals showed similar motor and cognitive deficits as wild-type mice. However, Smith et al. (56) showed transient increased Aβ levels in the hippocampus of transgenic animals, which overexpress human APP by tenfold. This finding also correlated with a greater loss of neurons in the CA3 region of the hippocampus than wild-type mice. However, the formation of amyloid plaques occurred no faster than in injured wild-type mice. As with the studies in nontransgenic animals, rapid Aβ deposition has not been demonstrated as it has been in human studies (57–60). Also, an increase in the severity of injury does not seem to lead to an increase in Aβ deposits (30), and in some cases a higher level of injury may reduce Aβ deposition. This may be because more neurons containing Aβ are dying and being cleared from the brain after injury, but again, this observation does not correlate with findings in humans, where Aβ levels seem to increase with injury severity (58), as does the risk of developing AD (61).

DLBD in humans has been associated with an accumulation of neurofilament (NF)-rich inclusions. Although the mechanisms behind this formation are unclear, the fact that NFs are critical for proper neuronal function indicates that these inclusions could decrease cell survival. To this end, Nakamura et al. (62) subjected transgenic mice that develop Lewy-body-like inclusions, and their wild-type littermates, to single TBI using the CCI device, and evaluated their neuromotor function postinjury. Although injured wild-type mice showed dysfunction postinjury as compared to uninjured wild-type mice, injured transgenic mice exhibited motor difficulties for significantly longer periods of time posttrauma than all other groups. Therefore, TBI appeared to exacerbate the effects of dementia-like pathology in this model.

There is potential to analyze dementia-like pathology in nonrodent species. For example, using an ovine head impact model, Van Den Heuvel et al. (31) demonstrated a rapid increase in APP mRNA expression after injury. Smith et al. (64) also showed accumulation of Aβ in the brains of pigs subjected to rotational head injury, as early as 3 days after trauma. A study using rotational head injury in rabbits demonstrated a slight increase in Aβ after injury and a redistribution of NFs (65). Head et al. (66) found hallmarks of AD in aged, but not young cats, indicating that the cat may also be a good model for traumatically induced dementia. Drawbacks of using cats, however, include their long life-span – the "aged" cats used were 16 years of age or older – and the scarcity of studies using cats for acceleration–deceleration injury research. However, a longer life-span may also be a positive factor that would more closely mimic the human lifespan than rodents, which generally only live 1–2 years. In addition, you would need dedicated researchers and resources to carry out such long-term studies analyzing the mechanisms and possible therapeutic interventions to inhibit dementia after TBI.

3. *In Vivo* Models of Traumatically Induced Dementia: Repetitive Injury

Considerably less work has been conducted regarding the potential relationship between rMTBI and dementia. One of the potential problems with mimicking rMTBI is that, often, many years must pass before the onset of dementia in human patients. As a result, epidemiological data from human populations has been slow to accumulate; in this case, proper animal models could help decipher the mechanisms by which dementia may be triggered by repetitive brain injury.

Several lines of evidence suggest that rMTBI may have cumulative effects, leading to dementia-like neuropathology. In a recent review (67), rMTBI was defined as an initial mechanical insult to the head, followed by another mechanical insult to the head of the same or different magnitude. The development of animal models for the study of rMTBI has drawn largely on the findings after single moderate or severe injury. In most cases, rats or mice are subjected to one or multiple mild injuries, and both behavioral (e.g. MWM) and histopathological (e.g. silver stain) outcomes are measured. These studies often fall in one of two categories: the effect of rMTBI on normal mice (e.g. C57BL/6 strain) or the effect of rMTBI on transgenic mice (e.g. Tg2576 AD transgenic line).

Initial studies utilized normal mice and simple injury paradigms. Laurer et al. (68) administered zero, one, or two injuries (spaced 24 h apart) to adult mice, and found marked differences among the experimental groups. Although they did not find cognitive impairment in all three groups using the MWM, they did find exacerbated functional and behavioral impairment in the single MTBI mice. Moreover, whereas this impairment lasted for 3 days in the single MTBI animals, these deficits were found in rMTBI up to 7 days postinjury. Their work highlighted the importance of the interinjury interval, a parameter that can easily be adjusted both *in vivo* and *in vitro*. Similarly, Creeley et al. (69) utilized the weight-drop paradigm, in which 7-week-old mice were subjected to either three impacts, or remained uninjured. Interestingly, mice that experienced rMTBI were significantly slower to regain consciousness than uninjured mice. Spatial learning was also found to be markedly impaired in these mice using the MWM, possibly due to contrecoup brain injury, which is an injury occurring opposite to, or distant from, the sight of impact.

Later studies compared the effects of either multiple injuries or a variable interinjury interval. Longhi et al. (70) compared the cognitive functioning in mice receiving either zero, one, or two injuries (spaced either 3, 5, or 7 days apart), and found that, whereas mice injured twice after 3 or 5 days had significantly increased cognitive impairment compared to sham or single-injury mice, this impairment was not seen in mice injured twice 7 days apart. Their findings highlighted the fact that the interinjury "window" does not extend indefinitely after two injuries; what remains unknown is if that vulnerability would be increased after three or more injuries.

Similar to the numerous studies measuring the effects of a single trauma on the development of dementia, the progression of AD-like impairment has been a major focus of rMTBI research *in vivo*. As mentioned previously, the development of abnormal tau protein pathology is a potential molecular link between TBI and dementia. Interestingly, the human brain seems to be more resistant against tau pathology after injury. Yoshiyama et al. (71) used a robust injury paradigm in an attempt to model human DP in transgenic mice expressing the shortest human tau isoform (T44). Mice were subjected to four injuries a day, once a week, for 4 weeks, resulting in each mouse receiving a total of 16 injuries, and, surprisingly, they could find only one mouse that displayed pathology of DP at 9 months of age. Partly for this reason, the vast majority of animal studies have focused on the deposition of Aβ, or the intracellular processing of APP. Interestingly, deposition of Aβ has not been observed in the majority of nontransgenic animal studies after trauma (30, 68), and, as a result, many of the current models used to investigate traumatic dementia are derived from transgenic rodents that were originally created to investigate AD. For example, the transgenic mouse Tg2576, which is characterized by AD-like amyloidosis by 9 months of age, has been used in several investigations of rMTBI, and has become a popular animal model for traumatically induced dementia.

In contrast to the work of Brody and Holtzman (53) described previously for single injury, Uryu et al. (72) found that Tg2576

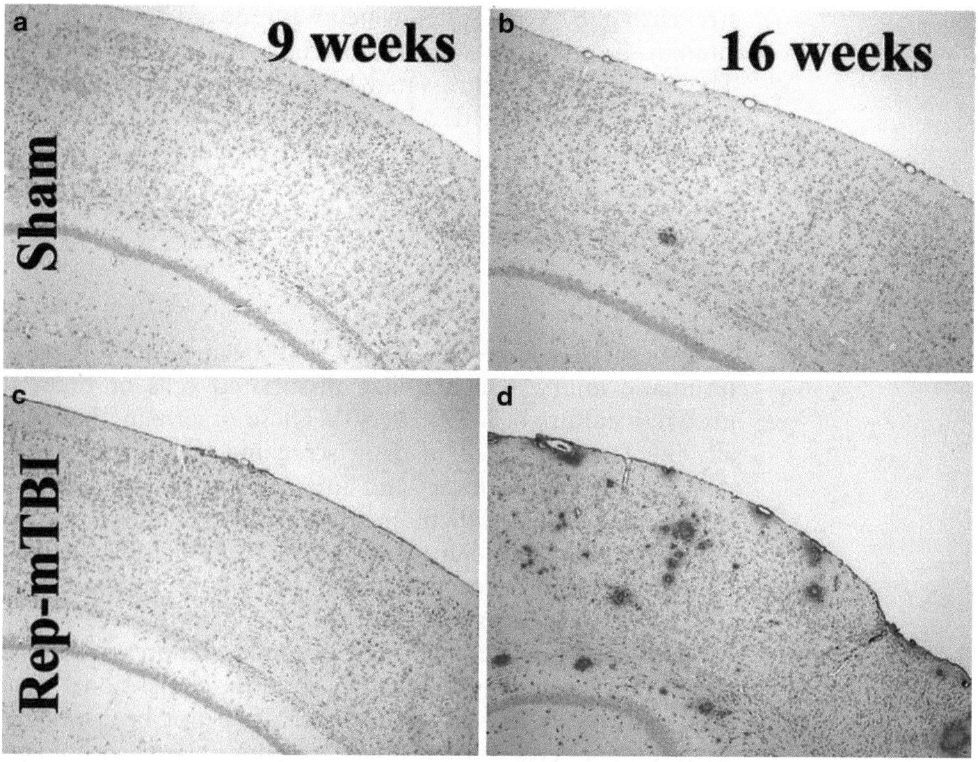

Fig. 3. Amyloid deposition in Tg2576 mice with sham or repeated mild traumatic brain injury (rMTBI) (**b**, **d**) with 4G8 immunohistochemistry at 9 (**a**, **c**) and 16 (**b**, **d**) weeks after mTBI. Senile plaques increased in an age-dependent manner in both sham and injured mice, but the largest number of Aβ-positive plaques are seen in the 16-week Rep-mTBI mice (**d**) (Modified from Uryu et al. (2002). Reprinted with permission from the Society for Neuroscience, 2002).

transgenic mice subjected to repeated, but not to single MTBI, displayed both cognitive deficits and Aβ deposition. As shown in Fig. 3, Aβ deposition did not occur in these mice at either 9 or 16 weeks post sham injury. In contrast, brain slices from Tg2576 mice that underwent rMTBI showed marked Aβ deposition (in the form of senile plaques) at 16 weeks postinjury. This finding highlights two key points. First, the transgenic background alone was not sufficient to induce Aβ deposition in these aged mice, thereby corroborating the "two-hit" hypothesis proposed by Nakagawa et al. (54). Second, the appearance of senile plaques followed a delayed time-scale, which is not surprising, as dementia is often manifested in humans long after TBI.

Increased incidence of dementia in humans has repeatedly been associated with increased age (73–76), and recent evidence links aging with the overproduction of free radicals via oxidative stress (21). TBI is also known to dramatically increase free radicals and reactive oxygen species (21, 77). Repetitive, but not single MTBI, has been previously shown to increase oxidative stress in Tg2576 mice (72), which was reduced by supplementing vitamin E, a known antioxidant, into the rodent chow (78). Therefore, oxidative stress may be one of the major contributing factors leading to the development of neurodegenerative disease following TBI.

4. *In Vitro* Models of Traumatically Induced Dementia?

Several *in vitro* approaches have now been developed to study traumatic injury, which utilize dissociated cells or brain slices grown in culture (13, 77, 79, 80). These *in vitro* methods generally increase the speed of drug screening and provide ease for deciphering altered inter- and intracellular mechanisms in real time, which may be difficult to study *in vivo*. Traumatic injury can be produced in slice cultures by a scratch/laceration method, by dropping a weight onto the slice, or by rolling a steel bar over the slice, which produces compression injury (for review, see (13)). Models in dissociated cultures produce compression or shear strain in two-dimensional (81), or three-dimensional systems (82). To date, these *in vitro* models have not been utilized to analyze the effects of trauma on dementia-related pathology. We have used a model of stretch-induced mechanical injury *in vitro* originally developed by Ellis et al. (83) (Fig. 4). This model has been extensively used by several research groups, with many different cell types derived from rats and mice (for review, see (77)). We have used this stretch injury model to study the response of cell cultures composed of neurons and glia from murine hippocampus (84, 85), cortex (86), and cerebellum (87) to trauma. Previous studies have characterized three levels of cell stretch (5.5, 6.5, and 7.5 mm deformations), defined as mild, moderate, and severe injury, which result in membrane deformation and biaxial strain or stretch of 31%, 38%, and 54%, respectively. This range of cell stretch has been shown to be relevant to what would occur in humans after rotational acceleration/deceleration injury (88).

In our initial studies of stretch-injured hippocampal cells, we began to separate the effects of injury to neurons and glia in mixed culture preparations using fluorescein diacetate retention and neuron-specific enolase release as markers of neuronal death and damage, and propidium iodide uptake and S-100B protein release as markers of glial damage. S-100B protein is a biomarker of injury commonly employed in the clinic to assess brain injury, as it is often found in the CSF of TBI patients (89, 90). S-100B is an intriguing compound as it acts as a microtubule disruptor (91) and thus an instigator of neuritic outgrowth, a necessary element in developing neurons (92). It has also been shown to be neuroprotective against excitotoxic

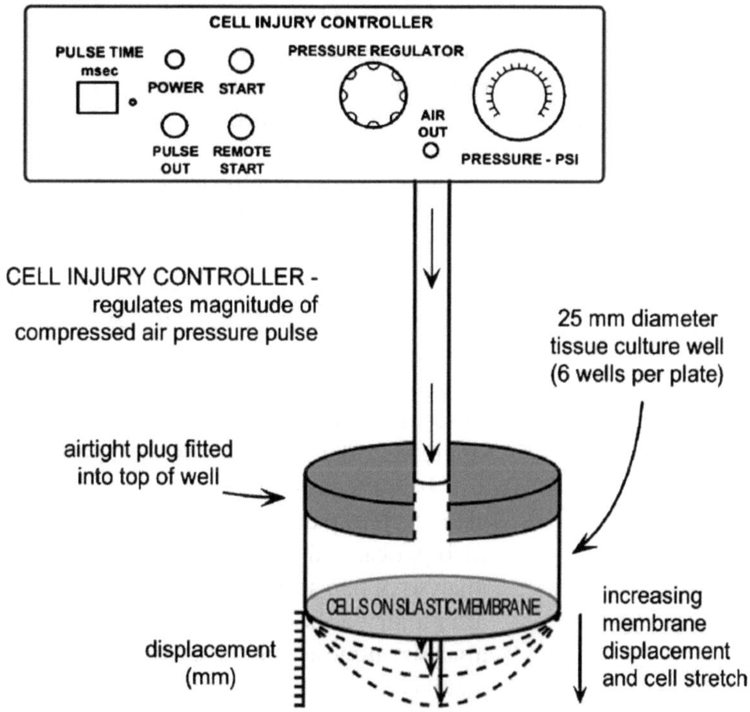

Fig. 4. *In vitro* model of mechanical injury. Cells are grown on deformable (Silastic) membranes and subjected to a controlled amount of stretch by delivery of an air pressure pulse of precise magnitude and duration. Variations in the pressure pulse produce varying deformations of the Silastic membrane, and therefore, stretch (Reproduced from Ellis et al. 1995 with permission from Mary Ann Liebert, Inc.).

damage and metabolic insults (93, 94). However, it may also be involved in neurodegeneration in AD and Down's syndrome (95, 96). As with TBI, S-100B is often elevated in AD patients (97), which suggests a possible mechanistic link between the two diseases. In addition, S-100B-expressing glial cells have been found in the vicinity of neurons expressing high APP and Aβ levels (98). We found elevated release of S-100B after injury in hippocampal (84, 85) and cerebellar cultures (86). Also, quantification revealed slightly elevated intracellular S-100B after single and repeated injury in hippocampal cells, suggesting it is actively produced after injury (85). We also found that if exogenous S-100B was added to cultures it was protective of hippocampal neurons against injury (85), but toxic against cerebellar Purkinje neurons (87), suggesting it has diverse effects.

S-100B has also been shown to stimulate large increases in intracellular free calcium in cultured neurons (99), similar to those produced by excessive glutamate levels. Recent evidence has suggested that altered calcium homeostasis may also be involved in AD progression (100). For example, cortical neurons from knock-in mice carrying an AD-linked mutation demonstrate enhanced released of calcium from inositol trisphosphate (IP_3)-sensitive internal stores (101). Also, activated APP releases calcium from internal stores and activates a store-operated calcium channel leading to further calcium influx (capacitative calcium influx), and neuronal death (102). Interestingly, remarkably similar alterations are observed in cortical neurons following traumatic injury, in that IP_3–mediated calcium signaling is perturbed and capacitative calcium influx is enhanced (103–105). These findings suggest altered calcium homeostasis as a contributing mechanism to dysfunction in AD and following trauma. This also raises the possibility of intracellular calcium stores and store-operated calcium influx as therapeutic targets for these disorders.

In some preliminary experiments, we have demonstrated detectable levels of pan-tau and phospho-tau (Fig. 5), as well as cleaved tau, in hippocampal cells after stretch injury. Quantitative Western blot analysis of pan and phospho-tau suggests a slight increase in tau levels after injury, although not statistically significant. Our initial qualitative analysis suggests higher cleaved tau levels after injury, although no significant differences between single and repeated injury were noted (unpublished data). However, we have not measured any protein levels later than 24 h after injury. It is possible that increases in proteins

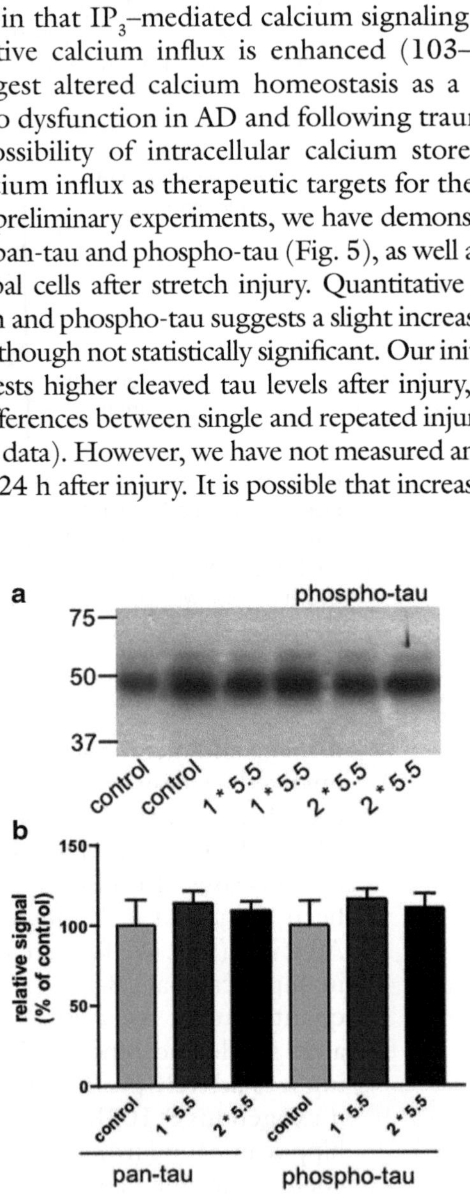

Fig. 5. (**a**) Representative Western blot of phospho-tau in hippocampal cultures (lanes 1–2, control; lanes 3–4, single 5.5 mm injury; lanes 5–6, two 5.5 mm injury administered within 1 h). (**b**) Quantitative assessment of protein levels of pan-tau and phospho-tau from hippocampal cultures (same conditions as **a**). All data are 24 h postinjury.

such as tau, APP, and Aβ may take longer than 24 h to be detected in this *in vitro* system. Obviously, *in vitro* models are limited in their ability to be able to test dementia-related pathology. Compared to *in vivo* models, the time-scale of injury and degeneration is being compressed to an even greater degree. However, such models could be useful for analyzing detailed intracellular pathways involved with injury progression, such as the relationship between S-100 and tau or Aβ pathology. As previously mentioned, they may also be useful for discerning potential therapeutic agents in the initial stages. However, these *in vitro* models would certainly only be meant to be used as complements to well-designed animal studies, or for providing the initial groundwork for hypothesis development and therapeutic potential before conducting *in vivo* experiments.

5. Future Directions

This review highlights the current findings, which link the development of dementia-like neuropathology and behavior in animals after single or repeated TBI. A clear relationship has been established with regard to moderate or severe injury in both human patients and animal models, but we strongly feel that more work must be conducted with regard to the cellular mechanisms, which may be responsible for the production of dementia after a traumatic insult. This is especially necessary for rMTBI studies, in which little is understood about the long-term prognosis of these patients, such as athletes. In conclusion, animal models of traumatically induced dementia need to be carefully developed, as any hope of treatment or prevention options rests on their proper use. Modeling single or repetitive TBI is relatively simple when compared to modeling the development of neurodegenerative diseases, such as AD. The continued use of transgenic rodent lines will greatly aid in our understanding of the relationship between dementia and TBI, as will the improvement of *in vitro* models of traumatic injury.

Acknowledgments

John T. Weber is currently supported by grants from the Canada Foundation for Innovation (CFI)-Leader's Opportunity Fund and the Natural Sciences and Engineering Research Council (NSERC). Jennifer E. Slemmer is currently supported by grants from the Atlantic Canada Opportunities Agency (ACOA). The authors would also like to thank Marva I. Sweeney (Department of Biology, University of Prince Edward Island) for her continued support; Dick Jaarsma and Eva Teuling (Department of Neuroscience, Erasmus Medical Centre, Rotterdam, The Netherlands) for conducting the Western blot analysis experiments; and Craig Malone for assistance with an earlier version of this manuscript.

References

1. Jennett B (1996) Epidemiology of head injury. J Neurol Neurosurg Psychiatry 60:362–369
2. Reilly P (2007) The impact of neurotrauma on society: an international perspective. Prog Brain Res 161:3–9
3. McGregor K, Pentland B (1997) Head injury rehabilitation in the U.K.: an economic perspective. Soc Sci Med 45:295–303
4. Nell V, Brown DS (1991) Epidemiology of traumatic brain injury in Johannesburg – II. Morbidity, mortality and etiology. Soc Sci Med 33:289–296
5. Hillier SL, Hiller JE, Metzer J (1997) Epidemiology of traumatic brain injury in South Australia. Brain Inj 11:649–659
6. Thurman D, Guerrero J (1999) Trends in hospitalization associated with traumatic brain injury. JAMA 282:954–957
7. Bayir H, Kochanek PM, Kagan VE (2006) Oxidative stress in immature brain after traumatic brain injury. Dev Neurosci 28:420–431
8. Schouten JW (2007) Neuroprotection in traumatic brain injury: a complex struggle against the biology of nature. Curr Opin Crit Care 13:134–142
9. Sosin DM, Sniezek JE, Waxweiler RJ (1995) Trends in death associated with traumatic brain injury, 1979 through 1992. Success and failure. JAMA 273:1778–1780
10. McIntosh TK, Smith DH, Meaney DF et al (1996) Neuropathological sequelae of traumatic brain injury: relationship to neurochemical and biomechanical mechanisms. Lab Invest 74:315–342
11. Patt S, Brodhun M (1999) Neuropathological sequelae of traumatic injury in the brain. An overview. Exp Toxicol.Pathol 51:119–123
12. LaPlaca MC, Simon CM, Prado GR et al (2007) CNS injury biomechanics and experimental models. Prog Brain Res 161:13–26
13. Spaethling JM, Geddes-Klein DM, Miller WJ et al (2007) Linking impact to cellular and molecular sequelae of CNS injury: modeling in vivo complexity with in vitro simplicity. Prog Brain Res 161:27–39
14. Bazan NG, Rodriguez de Turco EB, Allan G (1995) Mediators of injury in neurotrauma: intracellular signal transduction and gene expression. J Neurotrauma 12:791–814
15. Novack TA, Dillon MC, Jackson WT (1996) Neurochemical mechanisms in brain injury and treatment: a review. J Clin Exp Neuropsychol 18:685–706
16. Kermer P, Klocker N, Bahr M (1999) Neuronal death after brain injury. Models, mechanisms, and therapeutic strategies in vivo. Cell Tissue Res 298:383–395
17. Stelmasiak Z, Dudkowska-Konopa A, Rejdak K (2000) Head trauma and neuroprotection. Med Sci Monit 6:426–432
18. Leker RR, Shohami E (2002) Cerebral ischemia and trauma-different etiologies yet similar mechanisms: neuroprotective opportunities. Brain Res Brain Res Rev 39:55–73
19. Bramlett HM, Dietrich WD (2007) Progressive damage after brain and spinal cord injury: pathomechanisms and treatment strategies. Prog Brain Res 161:125–141
20. Farkas O, Povlishock JT (2007) Cellular and subcellular change evoked by diffuse traumatic brain injury: a complex web of change extending far beyond focal damage. Prog Brain Res 161:43–59
21. Slemmer JE, Shacka JJ, Sweeney MI et al (2008) Antioxidants and free radical scavengers for the treatment of stroke, traumatic brain injury and aging. Curr Med Chem 15:404–414
22. Woo BH, Thoidis G (2000) Epidemiology of traumatic brain injury. In: Woo BH, Nesathurai S (eds) The rehabilitation of people with traumatic brain injury. Blackwell Science, Malden, MA, pp 13–17
23. Brandenburg W, Hallervorden J (1954) Dementia pugilistica with anatomical findings. [Article in German). Virchows Arch 325:680–709
24. Grahmann H, Ule G (1957) Diagnosis of chronic cerebral symptoms in boxers (dementia pugilistica & traumatic encephalopathy of boxers). [Article in German]. Psychiatr Neurol (Basel) 134:261–283
25. Ryan AJ (1987) Intracranial injuries resulting from boxing: a review (1918–1985). Clin Sports Med 6:31–40

26. Jordan BD (2000) Chronic traumatic brain injury associated with boxing. Semin Neurol 20:179–185
27. Bouchard RW (2007) Diagnostic criteria of dementia. Can J Neurol Sci 34:S11–S18
28. Bonelli RM, Cummings JL (2008) Frontal-subcortical dementias. Neurologist 14:100–107
29. Rudelli R, Strom JO, Welch PT et al (1982) Posttraumatic premature Alzheimer's disease. Neuropathologic findings and pathogenetic considerations. Arch Neurol 39:570–575
30. Szczygielski J, Mautes A, Steudel WI et al (2005) Traumatic brain injury: cause or risk of Alzheimer's disease? A review of experimental findings. J Neural Transm 112:1547–1564
31. Van Den Heuvel C, Thornton E, Vink R (2007) Traumatic brain injury and Alzheimer's disease: a review. Prog Brain Res 161:303–316
32. Price JL, Davies PB, Morris JC, White DL (1991) The distribution of tangles, plaques and related immunohistochemical markers in healthy aging and Alzheimer's disease. Neurobiol Aging 12:295–312
33. Schmidt ML, Zhukareva V, Newell KL et al (2001) Tau isoform profile and phosphorylation state in dementia pugilistica recapitulate Alzheimer's disease. Acta Neuropathol 101:518–524
34. Sosin DM, Sniezek JE, Thurman DJ (1996) Incidence of mild and moderate brain injury in the United States, 1991. Brain Inj 10:47–54
35. Eriksen JL, Janus CG (2007) Plaques, tangles, and memory loss in mouse models of neurodegeneration. Behav Genet 37:79–100
36. Nakayama H, Uchida K, Doi K (2004) A comparative study of age-related brain pathology – are neurodegenerative diseases present in nonhuman animals? Med Hypotheses 63:198–202
37. Dixon CE, Lyeth BG, Povlishock JT et al (1987) A fluid percussion model of experimental brain injury in the rat. J Neurosurg 67:110–119
38. McIntosh TK, Vink R, Noble L et al (1989) Traumatic brain injury in the rat: characterization of a lateral fluid-percussion model. Neuroscience 28:233–244
39. Carbonell WS, Maris DO, McCall T et al (1998) Adaptation of the fluid percussion injury model to the mouse. J Neurotrauma 15:217–229
40. Thompson HJ, Lifshitz J, Marklund N et al (2005) Lateral fluid percussion brain injury: a 15-year review and evaluation. J Neurotrauma 22:42–75
41. Witgen BM, Lifshitz J, Smith ML et al (2005) Regional hippocampal alteration associated with cognitive deficit following experimental brain injury: a systems, network and cellular evaluation. Neuroscience 133:1–15
42. Lighthall JW (1988) Controlled cortical impact: a new experimental brain injury model. J Neurotrauma 5:1–15
43. Smith DH, Soares HD, Pierce JS et al (1995) A model of parasagittal controlled cortical impact in the mouse: cognitive and histopathologic effects. J Neurotrauma 12:169–178
44. Pierce JE, Trojanowski JQ, Graham DI et al (1996) Immunohistochemical characterization of alterations in the distribution of amyloid precursor proteins and β-amyloid peptide after experimental brain injury in the rat. J Neurosci 16:1083–1090
45. Pierce JE, Smith DH, Trojanowski JQ et al (1998) Enduring cognitive, neurobehavioral and histopathological changes persist for up to one year following severe experimental brain injury in rats. Neuroscience 87:359–369
46. Masumura M, Hata R, Uramoto H et al (2000) Altered expression of amyloid precursors proteins after traumatic brain injury in rats: in situ hybridization and immunohistochemical study. J Neurotrauma 17:123–134
47. Murakami N, Yamaki T, Iwamoto Y et al (1998) Experimental brain injury induces expression of amyloid precursor protein, which may be related to neuronal loss in the hippocampus. J Neurotrauma 15:993–1003
48. Ciallella JR, Ikonomovic MD, Paljug WR et al (2002) Changes in expression of amyloid precursor protein and interleukin-1beta after experimental traumatic brain injury in rats. J Neurotrauma 19:1555–1567
49. Blasko I, Beer R, Bigl M et al (2004) Experimental traumatic brain injury in rats stimulates the expression, production and activity of Alzheimer's disease beta-secretase (BACE-1). J Neural Transm 111: 523–536
50. Milman A, Rosenberg A, Weizman R et al (2005) Mild traumatic brain injury induces persistent cognitive deficits and behavioral disturbances in mice. J Neurotrauma 22: 1003–1010
51. Andersen OM, Schmidt V, Spoelgen R et al (2006) Molecular dissection of the interaction between amyloid precursor protein and its neuronal trafficking receptor SorLA/LR11. Biochemistry 45:2618–2628
52. Martin JB (1999) Molecular basis of the neurodegenerative disorders. N Engl J Med 340:1970–1980

53. Brody DL, Holtzman DM (2006) Morris water maze search strategy analysis in PDAPP mice before and after experimental traumatic brain injury. Exp Neurol 197:330–340
54. Nakagawa Y, Nakamura M, McIntosh TK et al (1999) Traumatic brain injury in young, amyloid-β peptide overexpressing transgenic mice induces marked ipsilateral hippocampal atrophy and diminished Aβ deposition during aging. J Comp Neurol 411:390–398
55. Murai H, Pierce JE, Raghupathi R et al (1998) Twofold overexpression of human β-amyloid precursor proteins in transgenic mice does not affect the neuromotor, cognitive, or neurodegenerative sequelae following experimental brain injury. J Comp Neurol 392:428–438
56. Smith DH, Nakamura M, McIntosh TK et al (1998) Brain trauma induces massive hippocampal neuron death linked to a surge in β-amyloid levels in mice overexpressing mutant amyloid precursor protein. Am J Pathol 153:1005–1010
57. Roberts GW, Gentleman SM, Lynch A et al (1991) βA4 amyloid protein deposition in brain after head trauma. Lancet 338:1422–1423
58. Roberts GW, Gentleman SM, Lynch A et al (1994) Beta amyloid protein deposition in the brain after severe head injury: implications for the pathogenesis of Alzheimer's disease. J Neurol Neurosurg Psychiat 57:419–425
59. Gentleman SM, Greenberg BD, Savage MJ et al (1997) AB42 is the predominant form of amyloid-b-protein in the brains of short-term survivors of head injury. Neuroreport 8:1519–1522
60. Ikonomovic MD, Uryu K, Abrahamson EE et al (2004) Alzheimer's pathology in human temporal cortex surgically excised after severe brain injury. Exp Neurol 190:192–203
61. Plassman BL, Havlik RJ, Steffens DC et al (2000) Documented head injury in early adulthood and risk of Alzheimer's disease and other dementias. Neurology 55:1158–1166
62. Nakamura M, Saatman KE, Galvin JE et al (1999) Increased vulnerability of NFH-LacZ transgenic mouse to traumatic brain injury-induced behavioral deficits and cortical damage. J Cereb Blood Flow Metab 19:762–770
63. Van den Heuvel C, Blumbergs PC et al (1999) Upregulation of amyloid precursor protein messenger RNA in response to traumatic brain injury: an ovine head impact model. Exp Neurol 159(2):441–450
64. Smith DH, Chen XH, Nonaka M et al (1999) Accumulation of amyloid beta and tau and the formation of neurofilament inclusions following diffuse brain injury in the pig. J Neuropathol Exp Neurol 58:982–992
65. Hamberger A, Huang YL, Zhu H et al (2003) Redistribution of neurofilaments and accumulation of β-amyloid protein after brain injury by rotational acceleration of the head. J Neurotrauma 20:169–178
66. Head E, Moffat K, Das P et al (2005) β-Amyloid deposition and tau phosphorylation in clinically characterized aged cats. Neurobiol Aging 26:749–763
67. Weber JT (2007) Experimental models of repetitive brain injuries. Prog Brain Res 161:253–261
68. Laurer HL, Bareyre FM, Lee VM et al (2001) Mild head injury increasing the brain's vulnerability to a second concussive impact. J Neurosurg 95:859–870
69. Creeley CE, Wozniak DF, Bayly PV et al (2004) Multiple episodes of mild traumatic brain injury result in impaired cognitive performance in mice. Acad Emerg Med 11:809–819
70. Longhi L, Saatman KE, Fujimoto S et al (2005) Temporal window of vulnerability to repetitive experimental concussive brain injury. Neurosurgery 56:364–374
71. Yoshiyama Y, Uryu K, Higuchi M et al (2005) Enhanced neurofibrillary tangle formation, cerebral atrophy, and cognitive deficits induced by repetitive mild brain injury in a transgenic tauopathy mouse model. J Neurotrauma 22:1134–1141
72. Uryu K, Laurer H, McIntosh T et al (2002) Repetitive mild brain trauma accelerates Aβ deposition, lipid peroxidation, and cognitive impairment in a transgenic mouse model of Alzheimer amyloidosis. J Neurosci 22:446–454
73. Morrison JH, Hof PR (2002) Selective vulnerability of corticocortical and hippocampal circuits in aging and Alzheimer's disease. Prog Brain Res 136:467–486
74. Burke SN, Barnes CA (2006) Neural plasticity in the ageing brain. Nat Rev Neurosci 7:30–40
75. Mahncke HW, Bronstone A, Merzenich MM (2006) Brain plasticity and functional losses in the aged: scientific bases for a novel intervention. Prog Brain Res 157:81–109
76. Persson J, Nyberg L (2006) Altered brain activity in healthy seniors: what does it mean? Prog Brain Res 157:45–56
77. Weber JT (2004) Calcium homeostasis following traumatic neuronal injury. Curr Neurovascular Res 1:151–171
78. Conte V, Uryu K, Fujimoto S et al (2004) Vitamin E reduces amyloidosis and improves cognitive function in Tg2576 mice following repetitive concussive brain injury. J Neurochem 90:758–764

79. Morrison B 3rd, Saatman KE, Meaney DF et al (1998) In vitro central nervous system models of mechanically induced trauma: a review. J Neurotrauma 15:911–928
80. Noraberg J, Poulsen FR, Blaabjerg M et al (2005) Organotypic hippocampal slice cultures for studies of brain damage, neuroprotection and neurorepair. Curr Drug Targets CNS Neurol Disord 4:435–452
81. LaPlaca MC, Lee VM, Thibault LE (1997) An in vitro model of traumatic neuronal injury: loading rate-dependent changes in acute cytosolic calcium and lactate dehydrogenase release. J Neurotrauma 14:355–368
82. LaPlaca MC, Cullen DK, McLoughlin JJ et al (2005) High rate shear strain of three-dimensional neural cell cultures: a new in vitro traumatic brain injury model. J Biomech 38:1093–1105
83. Ellis EF, McKinney JS, Willoughby KA et al (1995) A new model for rapid stretch-induced injury of cells in culture: characterization of the model using astrocytes. J Neurotrauma 12:325–339
84. Slemmer JE, Matser EJT, De Zeeuw CI et al (2002) Repeated mild injury causes cumulative damage to hippocampal cells. Brain 125:2699–2709
85. Slemmer JE, Weber JT (2005) The extent of cell damage following repeated injury to hippocampal cells is dependent on the severity of insult and inter-injury interval. Neurobiol Dis 18:421–431
86. Engel DC, Slemmer JE, Vlug AS et al (2005) Combined effects of mechanical and ischemic injury to cortical cells: secondary ischemia increases damage and decreases effects of neuroprotective agents. Neuropharmacology 49:985–995
87. Slemmer JE, Weber JT, De Zeeuw CI (2004) Cell death, glial protein alterations and elevated S-100β release in cerebellar cell cultures following mechanically induced trauma. Neurobiol Dis 15:563–572
88. Schreiber D, Gennarelli TA, Meaney DF (2005) Proceedings of the 1995 international research conference on biomechanics of impact (Brunen, Switzerland). In: Charpenne A (ed) International research council on biokinetics of impact (France), pp 233–244
89. Rothoerl RD, Brawanski A, Woertgen C (2000) S-100B protein serum levels after controlled cortical impact injury in the rat. Acta Neurochir 142:199–203
90. Herrmann M, Curio N, Jost S et al (2001) Release of biochemical markers of damage to neuronal and glial brain tissue is associated with short and long term neuropsychological outcome after traumatic brain injury. J Neurol Neurosurg Psychiatry 70:95–100
91. Donato R (1991) Perspectives in S-100 protein biology. Cell Calcium 12:713–726
92. Azmitia EC, Dolan K, Whitaker-Azmitia PM (1990) S-100β but not NGF, EGF, insulin or calmodulin is a CNS serotonergic growth factor. Brain Res 516:354–356
93. Barger SW, Van Eldik LJ, Mattson MP (1995) S-100β protects hippocampal neurons from damage induced by glucose deprivation. Brain Res 677:167–170
94. Ahlemeyer B, Beier H, Semkova I et al (2000) S-100β protects cultured neurons against glutamate- and staurosporine-induced damage and is involved in the antiapoptotic action of the 5 HT(1A)-receptor agonist, Bay x 3702. Brain Res 858:121–128
95. Whitaker-Azmitia PM, Wingate M, Borella A et al (1997) Transgenic mice overexpressing the neurotrophic factor S-100β show neuronal cytoskeletal and behavioral signs of altered aging processes: implications for Alzheimer's disease and Down's syndrome. Brain Res 776:51–60
96. Sheng JG, Mrak RE, Bales KR et al (2000) Overexpression of the neuritotrophic cytokine S100β precedes the appearance of neuritic β-amyloid plaques in APPV717F mice. J Neurochem 74:295–301
97. Petzold A, Jenkins R, Watt HC et al (2003) Cerebrospinal fluid S100B correlates with brain atrophy in Alzheimer's disease. Neurosci Lett 336:167–170
98. Mrak RE, Griffin WST (2001) Interleukin-1, neuroinflammation, and Alzheimer's disease. Neurobiol Aging 22:915–922
99 Barger SW, Van Eldik LJ (1992) S100β stimulates calcium fluxes in glial and neuronal cells. J Biol Chem 267:9689–9694
100. Ingram V (2003) Alzheimer's disease. Am Sci 91:312–321
101. Stutzmann GE, Caccamo A, LaFerla FM et al (2004) Dysregulated IP_3 signaling in cortical neurons of knock-in mice expressing an Alzheimer's-linked mutation in presenilin1 results in exaggerated Ca^{2+} signals and altered membrane excitability. J Neurosci 24:508–513
102. Bouron A, Mbebi C, Loeffler JP et al (2004) The beta-amyloid precursor protein controls a store-operated Ca^{2+} entry in cortical neurons. Eur J Neurosci 20:2071–2078
103. Weber JT, Rzigalinski BA, Willoughby KA et al (1999) Alterations in calcium-mediated signal transduction after traumatic injury of cortical neurons. Cell Calcium 26:289–299

104. Weber JT, Rzigalinski BA, Ellis EF (2001) Traumatic injury of cortical neurons causes changes in intracellular calcium stores and capacitative calcium influx. J Biol Chem 276:1800–1807

105. Weber JT, Rzigalinski BA, Ellis EF (2002) Calcium responses to caffeine and muscarinic receptor agonists are altered in traumatically injured neurons. J Neurotrauma 19: 1433–1443

Part IX

Animal Models of Toxic Dementia

Chapter 33

Animal Models of Alcohol-Induced Dementia

Angela Maria Ribeiro and Silvia R. Castanheira Pereira

Abstract

Alcoholic dementia is a disorder characterized by multiple cognitive deficits that include memory impairment associated with one or more cognitive disturbances listed in the present text. First, we characterize the disorder and describe aspects of using nonhuman models for studying specific and particular patterns of behavioral failures and biological dysfunctions found in cases of chronic alcohol consumption. Some kinds of animal models are depicted. Although there is no experimental model that displays all aspects considered as criteria for the diagnosis of alcoholic dementia, an animal model that is considered to be the most satisfactory to study behavioral and neurobiological aspects of this disease is that in which both high/chronic ethanol exposure and thiamine deficiency variables could be controlled. In this chapter, we show that animal models that manipulate only a single recognized etiological factor are less effective to elucidate the multiple influences that lead to alcoholic dementia. We conclude that even considering that only particular aspects of this disease could be approached using experimental animals; these studies can shed light on the biological processes, causing specific and particular patterns of cognitive failure.

Key words: Alcoholic dementia, Memory impairment, Chronic ethanol exposure, Thiamine deficiency

1. Alcoholic Dementia

The first step when using animals to study a cognitive dysfunction is to characterize the disorder. The one we are presently concerned with is "alcoholic dementia." Not everyone who drinks chronically develops dementia, and the major risk factors for disease development and the mechanisms by which this occurs have remained unclear. There are data indicating that abuse of alcohol alone does increase the risk of dementia, whereas the potential deleterious or beneficial effects of light to moderate drinking remain unclear (1–3). Alcohol abuse may contribute to cognitive impairment in several different ways. Acute use of alcohol (acute

Peter Paul De Deyn and Debby Van Dam (eds.), *Animal Models of Dementia*, Neuromethods, vol. 48,
DOI 10.1007/978-1-60761-898-0_33,

intoxication) impairs attention, memory, executive functions, and visuospatial skills (4–9), while chronic abuse causes neurocognitive deficits in memory, learning, visuospatial functions, psychomotor speed processing, executive functions, and decision-making and may lead to persistent amnesic disorder and alcoholic dementia (5, 10, 11). Among the several kinds of dementia, we cite substance-induced persisting dementia, which may be due to drug abuse, medication, or toxin exposure. A number of toxins have been reported to cause dementia, but the principal one in most societies is alcohol (ethanol). Alcoholic dementia is a disorder characterized by multiple cognitive deficits that include memory impairment associated with one or more cognitive disturbances listed in Table 1.

Cognitive impairment may range between severe dementia and subtle deficits, which may be reversible (at least in part) or permanent. Although chronic alcoholism is associated with cognitive impairments, there are still conflicting opinions about the pathogenesis of alcohol-related memory impairments (12). Much of this debate has revolved around the relative contributions of nutritional deficiency and alcohol to the neuropathological lesions (13, 14). One of the most important questions concerning alcoholic dementia is whether long-term alcohol use may have direct neurotoxic effects on the brain, leading to a characteristic dementia syndrome. Clinical and neuropsychological data point toward the existence of an alcoholic dementia without secondary complications. Improvements in neuroimaging technology, such

Table 1
Criteria for alcohol-induced persisting dementia in the Diagnostic and Statistical Manual of Mental Disorders (DSM-IV) (American Psychiatric Association (12))

A. The development of multiple cognitive deficits manifested by both 1. Memory impairment (impaired ability to learn new information or to recall previously learned information) (a) One (or more) of the following cognitive disturbances: (a) aphasia (language disturbance) (b) Apraxia (impaired ability to carry out motor activity despite intact motor function) (c) Agnosia (failure to recognize or identify objects despite intact sensory function) (d) Disturbance in executive functioning (i.e., planning, organizing, sequencing, abstracting)
B. The cognitive deficits in Criteria A1 and A2 each cause significant impairment in social or occupational functioning and represent a significant decline from a previous level or functioning
C. The deficits do not occur exclusively during the course of a delirium and persists beyond the usual duration of alcohol intoxication or withdrawal (code: 291.2)
D. There is evidence from the history, physical examination, or laboratory findings that the deficits are etiologically related to the persisting effects of alcohol use (code: 291.2)

as magnetic resonance imaging, magnetic resonance spectroscopy, and positron emission tomography have contributed significantly to unravel structural abnormalities in uncomplicated alcoholics (with no hepatic or thiamine deficiency problems) who are cognitively impaired (2, 15, 16). The existence of specific neurotoxic effects of alcohol on the central nervous system (CNS) has been approached in a number of studies (2, 17, 18).

To reduce subjective bias and standardize DSM-IV criteria (see Table 1) for alcohol-induced persisting dementia, Oslin et al. (19) proposed some criteria to improve the validity and reliability of the diagnosis of "Alcohol-related dementia" (ARD) (13). These authors reported that individuals with ARD are less cognitively impaired than patients with Alzheimer's disease and that both their cognitive and functional status frequently stabilizes. They consider ARD to differ from alcohol-induced dementia (DSM-IV) in that the presence of language impairment (aphasia) is less likely in ARD. This absence of aphasia in ARD is of clinical importance for the potential differentiation of the latter from Alzheimer's disease, frontotemporal dementia, and vascular dementia (20).

Brain neuropathological lesions observed in demented alcoholic patients are more extensive in patients who have additional vitamin B1 (thiamine) deficiency, i.e., Wernicke–Korsakoff Syndrome (WKS). In alcoholics, several factors may contribute to thiamine deficiency and to dysfunctions of the thiamine-dependent or -metabolizing enzymes: (i) nutritional thiamine deficiency due to poor eating habits often seen in alcoholics; (ii) impairment of thiamine absorption from the intestine (21); (iii) changes in thiamine-metabolizing enzyme activity in the brain, e.g., a significant decrease in thiamine pyrophosphokinase activity caused by excessive alcohol consumption (22); (iv) changes in activity of thiamine-dependent enzymes that participate in the metabolic pathway (pyruvate dehydrogenase, α-ketoglutarate dehydrogenase, and transketolase) (23).

The most characteristic symptoms of WKS are anterograde amnesia (i.e., the inability to learn and form new memories), retrograde amnesia (i.e., the loss of memories formed prior to the onset of WKS), and impairment of several cognitive processes (24). According to latter publication, these cognitive and memory deficits are accompanied by characteristic pathological damage to several brain structures. In these cases, alcohol may have a direct neurotoxic effect on cortical neurons, but much of the damage may be secondary to diencephalic pathology caused by thiamine deficiency, including loss of nerve cells in the thalamus, cell death and hemorrhages in the mammillary bodies, and enlargement of the third ventricle.

Results of a necropsy study by Harper (25) showed that 80% of 131 patients with WKS were not diagnosed as such during life. This is in accordance with some authors who have cast doubt on the existence of primary alcoholic dementia as a neuropathological entity and suggest that the majority of cases of alcoholic dementia are in fact WKS (26).

Wernicke's encephalopathy is a neuropsychiatric condition that can lead to dementia and occurs as a result of thiamine deficiency. It has been described as an acute neurological disorder characterized clinically by oculomotor abnormalities, cerebellar dysfunction, and altered mental state, and characterized pathologically by hemorrhagic lesions in the wall of the third and fourth ventricles (27). Korsakoff's syndrome is characterized by profound amnesia, disorientation, and frequent confabulation and is defined as a disproportionate impairment in memory, relative to other features of cognitive function, resulting from nutritional (thiamine) deficiency (28). Although the typical Korsakoff's syndrome presents as a pure amnesic state without significant intellectual decline, some patients are reported to show remarkable and overall dementia. Scanning studies have shown that patients with Korsakoff's syndrome have a more widespread cerebral and subcortical atrophy than alcoholic patients without amnesia (29). Because Korsakoff's syndrome often follows or accompanies Wernicke's encephalopathy, with the typical clinical pattern emerging when the acute global confusional state of the latter resolves, and both appear to share common pathological substrates, a number of researchers describe this continuum as WKS (26, 28, 30). The widely held view is that the cognitive deficits of varying types and severities found in alcoholics arise as a result of a coexisting thiamine deficiency syndrome (14). According to these authors, the fact that the degree of vulnerability of patients with Wernicke's encephalopathy to Korsakoff's syndrome appears to vary according to the history of alcoholism goes against the widely cited conclusion of Victor et al. (31) that Wernicke's encephalopathy and Korsakoff's syndrome are the acute and chronic phase of the same thiamine deficiency syndrome. Relating to the fact that alcohol consumption plays a relevant role in this syndrome, Martin et al. (32) consider that the most devastating neurological disorder caused by thiamine deficiency is WKS, which is seen most commonly in alcoholics. In short, clinical pathological issues regarding the so-called "alcoholic dementia" remain under debate. Although clinical observation favors the diagnosis of primary alcohol dementia, caused by direct alcohol neurotoxicity, further confirmation from the neuropathological and biochemical perspectives is required. Moriyama et al. (13) concluded that repeated episodes of subclinical WKS may partially account for the chronic state of primary alcoholic dementia, thus supporting the notion that primary alcoholic dementia exists in continuum with chronic and subclinical types of WKS. However, the role of thiamine deficiency remains undetermined in chronic "uncomplicated" alcoholism, but postmortem study indicates that Wernicke's encephalopathy is underdiagnosed in life (16, 32, 33). He et al. (34) consider that these unsolved possibilities can be addressed with larger *in vivo* and postmortem animal studies that manipulate thiamine levels in the context of bouts of high alcohol exposure aimed at paralleling human alcoholism. Uncomplicated alcoholics likely sustain repeated bouts of less profound nutritional deficiency, which may contribute to their milder *in vivo* neuroradiological findings when compared with those marking WKS (14). McIntosh and Chick (35) state that improvements in imaging technology and the continuous progresses in brain neurochemistry knowledge have started to unravel the association between alcohol, thiamine, and memory. It has been proposed that most organic brain syndromes in alcoholic patients are variants of the WKS, and that there is no need to consider a separate category of "alcoholic dementia."

2. Animal Models of Alcoholic Dementia

2.1. General Aspects on Using Nonhuman Models

Animal models may have either face validity (i.e., they mimic some aspect of the human condition) or predictive validity (i.e., results obtained with the animal model are predictive of actions of alcohol or of treatment efficacy in humans) (36). Basic research into the etiology of neurological disorders is often initiated in animal models. Nevertheless, most animal models are limited by the fact that animal models do not produce the plethora of behaviors that humans express. Many researchers study the relationships between cognitive and biological processes through comparative methods, using patients with cognitive impairments, mainly amnesic patients (37–39). However, the research techniques used in these studies are limited by ethical and/or practical considerations, and hence, the use of animal models is essential to obtain, in parallel with human experiments, the desired precision of manipulation and accuracy of measurement. As a first step to approach the questions regarding the biological basis of cognition, some laboratories have used simpler systems to study biological changes associated with learning. Using experimental models to study different aspects of cognitive deficits, researchers have obtained evidence that some behavioral modifications are associated with molecular neurobiological changes (40–42). Moreover, mammalian animal memory tasks analogous to human memory tests have been developed, and both yield results that support correspondence in mnemonic function (43, 44). The results obtained from experimental models may provide a starting point for an understanding of the relationship between organic and psychological phenomena and shed light on the hypothesis that each aspect of behavior is mediated by specific neural mechanisms in the brain. Considering that the basic biological substrate of mental processes by which a subject perceives, acts, learns, and remembers might be similar in different species, simpler systems (e.g., the rodent CNS) are useful models to understand some of the basic principles of interaction between mind and brain. As psychoactive drug effects develop gradually and can persist for a long time after cessation of chronic drug administration, this approach is useful to study the steps of a neurodegenerative process associated with cognitive performance impairment as in alcoholic dementia. In this case, we can follow each stage of the process at both the cognitive and biological level.

2.2. Animal Models of Alcoholism

Animal models have been developed to study various aspects of alcohol use and dependence, including alcohol-seeking behavior, alcohol-related organ damage, tolerance to alcohol, and physical dependence on alcohol. Additionally, because animal models can be genetically manipulated, they are valuable for research on the genetic determinants of alcoholism. Although no animal model replicates all aspects of this disorder, they can offer some advantages. For instance, they make it possible to dissect different components of the disorder. In this aspect, animal models can be useful to understand molecular mechanisms that play a relevant role in that disease and to study the correlations between specific behavioral deficits and neurobiological dysfunctions.

Factors potentially contributing to the behavioral performance of animals in different kinds of models for alcohol abuse include the species of choice, age of the animals, duration of alcohol treatment and abstinence, method of alcohol administration, and the model's face and predictive validity (45–47). It is unlikely that an animal model will be developed that completely mimics all the characteristic effects of acute or chronic alcohol ingestion in man. Thus, to describe some kinds of animal models that have been used to study aspects of chronic alcoholism and to subsequently highlight the relations between these and the criteria stated in DMS-IV for alcoholic dementia, the following paragraphs are divided into two main topics: (i) a short description of the types of animal models for studying chronic alcoholism and (ii) an animal model that is considered the most satisfactory to study some aspects of alcoholic dementia.

2.3. Some Animal Models for Studying Aspects of Chronic Alcoholism

The attention of researchers has been mostly directed to the following alcohol-associated phenomena: alcohol dependence, tolerance, withdrawal symptoms/signals, craving and relapse, alcohol-induced hepatic damage, and ethanol-reinforced-self-intoxication, each of which has been demonstrated in both man and animals. Many studies using animal models to approach behavioral aspects and biological mechanisms which are involved in these phenomena have been carried out (36). As mentioned before, the different effects of alcohol consumption depend on several factors, such as route and timing of alcohol administration (e.g., binge/intoxication or chronic consumption), strain, sex, and age of the animal. Many models have been validated in pharmacological studies and have provided some insight into the neurochemical and cellular changes underlying alcohol dependence, tolerance, and addictive behaviors. These topics are deeply approached in some extensive reviews describing different aspects of alcohol abuse (48–50). Models of relapse behavior have a high degree of predictive validity (51).

One of the main procedures used as a model of alcohol dependence to study many aspects of addictive behaviors is submitting rats to a prolonged exposure to alcohol vapor (52–54). This type of manipulation produces persistently increased alcohol intake in genetically nonselected rats. In addition, exposure to repeated cycles of intoxication and withdrawal, which mimics the course of the clinical condition, is extremely effective for inducing increased alcohol drinking (55, 56). Additionally, like intragastric administration, the dose of ethanol given to the animals can be controlled when exposure to alcohol vapor is used (57, 58).

Besides intragastric administration and exposure to alcohol vapor, prolonged oral administration of ethanol through either aqueous solutions (different concentrations have been used) or liquid diet is often used to study the chronic effects of the drug on brain and behavior (59–62). Using these administration routes, some authors have demonstrated that chronic alcohol consumption can induce cognitive deficits and neurobiological dysfunction (59, 63–65). It is important to emphasize that, unlike administration through vapor, in which the animal is forced to receive the ethanol treatment, when using oral administration the experimenter can give ethanol as the sole source of fluid (forced consumption) or give the animal the opportunity to choose between two solutions, with or without drug. In the latter case, the animal has voluntary access to ethanol.

According to Spanagel (48), when designing an animal model that covers several aspects of alcohol dependence and alcohol-related disease, a necessary precondition is that the laboratory animal has voluntary access to alcohol for a long time (at least for several months). The large variability in alcohol preference among individual animals and strains has allowed researchers to selectively breed rats for differential alcohol preference, generating pairs of animal strains that are characterized by particularly low or high alcohol consumption levels. For instance, two pairs of lines were generated in Finland and Sardinia. The Finnish model – called Alko Alcohol and Alko Nonalcohol rats – comprises two strains of albino rats that were selectively bred from 1963 onwards based on their preference or rejection of a 10% alcohol solution versus water (66, 67). The Sardinian alcohol-preferring rat line has also been selectively bred for high alcohol preference and consumption for over 20 years (68). Other important alcohol-preferring rat lines are the Indiana University alcohol preferring and the high-alcohol-drinking lines (69). Another model to study alcohol preference is that of knockout mice that can be used to assess the involvement of a particular gene in alcohol drinking behavior (70, 71).

In a review on animal models of alcohol abuse, Tabakoff and Hoffman (36) state that operant alcohol self-administration has not only been used to assess the reinforcing effect of alcohol but also to model the craving for alcohol experienced by abstinent alcoholics. When animals have been drinking alcohol regularly and are then subjected to a period of forced abstinence from alcohol, they show a reliable increase in alcohol intake when alcohol is again made available (i.e., the alcohol deprivation effect). Whether this apparently enhanced motivation to ingest alcohol is an accurate model of craving in humans is not clear. However, this model does have significant predictive validity; the drugs now used to reduce craving and relapse in humans (72, 73) can also block the increased responding associated with the alcohol deprivation effect in animal models (74). According to its aims, research addressing chronic ethanol treatment and alcohol self-administration with nonhuman models can be divided into: (i) demonstration that alcohol self-administration could be established; (ii) verification that ethanol self-administration can not only be induced but also maintained; (iii) establishment of a useful model for investigating potential pharmacotherapies; (iv) addressing consequences and risk factors for excessive alcohol drinking associated or not to a secondary cause (e.g., thiamine deficiency). Considering the latter, brain degeneration is among the numerous possible consequences of ethanol abuse, which can be related to a severe cognitive impairment. The relative contributions of a direct effect of alcohol and thiamine deficiency to biological and behavioral dysfunctions are still obscure. In this case, the use of animal models is useful because the neurochemical and molecular biological data can be determined and also its correlation to the specific behavioral aspects assessed.

As mentioned earlier, chronic alcohol consumption can lead to Korsakoff's syndrome, a memory deficiency attributed to diencephalic damage and/or to medial temporal or cortical dysfunction. A good number of animal models of Korsakoff's syndrome involve thiamine-deficient diet manipulation. Two kinds of treatments are described for inducing thiamine deficiency in laboratory animal models: (i) to submit them to a thiamine-deficient diet or (ii) to associate the thiamine-deficient diet treatment with pyrithiamine (an inhibitor of the kinase responsible for thiamine phosphorylation). Both types of treatment can be combined with chronic ethanol administration resulting in the development of multiple cognitive deficits.

Diencephalic dysfunctions have been described in thiamine-deficient animals (75). Neurochemical alterations along with the white matter loss that occurs in key fiber tracts from the diencephalon after thiamine deficiency (76) suggest that this vitamin deficiency likely causes system-level dysfunction, which probably underlies the deficits in learning and memory documented in this animal model. Moreover, animals treated with pyrithiamine associated with a thiamine-deficient diet showed severe neurological signs in relatively short periods of time (77). Morphological lesions in diencephalic regions, similar to those found in humans with WKS, have been observed in animals treated with pyrithiamine (78). Clinical signs include ataxia, lethargy, anorexia, seizures, and impaired righting reflexes and ultimately death after a 20-day treatment (77).

Pfefferbaum et al. (79) found that the brain regions clearly affected by thiamine deficiency in alcohol-preferring rats consuming high concentration of alcohol were the thalamus, mammillary nuclei, inferior colliculi, lateral and fourth ventricles, and hippocampus. Dixon and Harper (80) stated that the enduring macrostructural and neurochemical abnormalities involving critical nodes of the Papez circuit carry liabilities for development of amnesia and incomplete recovery of nonmnemonic cognitive and motor functions subserved by the affected neural systems.

2.4. The Most Satisfactory Animal Model to Study Aspects of Alcoholic Dementia

As explained earlier, there have been a large number of studies in which chronic ethanol is given to animals; however, relatively few have been specifically designed to provide information about the mechanisms underlying alcoholic dementia. There is no animal model that displays all aspects considered as criteria for the diagnosis of alcoholic dementia in DMS-IV. Moreover, even considering the use of an animal model to study only some aspects of this disease, the central question concerning the relative contribution of both alcohol- and thiamine-deficiency-related neurotoxicity to bring about a characteristic dementia syndrome should be taken into account. This is an essential concern because, as mentioned earlier, the role of thiamine deficiency remains undetermined in chronic "uncomplicated" alcoholism (32, 81). In addition to this, the widely held view is that the cognitive deficits of varying types and severities found in alcoholics arise as a result of a coexisting thiamine deficiency syndrome (14). Thus, the possibility of manipulating these two independent variables in a nonhuman model, namely high ethanol exposure and thiamine deficiency, might be crucial to address these unsolved questions and also contribute to our understanding of molecular and behavioral aspects of alcoholic dementia. Besides, the control of these variables also allows obtaining evidence about the etiology of the multiple persistent cognitive deficits.

Following, the DMS-IV criteria for alcoholic dementia are used to describe a rationale for the viability of using a model of chronic ethanol consumption associated to thiamine deficiency (a kind of WKS model) to study aspects of this dementia. These descriptions exemplify, both at the behavioral and biological levels, the effects of excessive alcohol drinking associated or not to severe or subclinical episodes of thiamine deficiency, which are related to some of the clinical signs that make up, according to DMS-IV criteria, an alcoholic dementia picture.

Next, the DMS-IV criteria are presented along with the exemplifications of data obtained using the WKS animal model and its aspects of face validity are pointed out. According to DMS-IV, the development of multiple cognitive deficits manifested by two clinical signs has to be considered for the diagnosis of alcoholic dementia: (i) memory impairment and (ii) one (or more) of the following cognitive disturbances: aphasia (language disturbance); apraxia (impaired ability to carry out motor activity despite intact motor function); agnosia (failure to recognize or identify objects despite intact sensory function); disturbance in executive functioning (i.e., planning, organizing, sequencing, abstracting). Except for aphasia, all the other clinical signs can be assessed in nonhuman models. Despite the existence of contradictory data, several pieces of evidence in the literature show memory impairments in nonhuman models similar to those displayed by patients with diagnosis of WKS (caused by the association of chronic alcohol consumption and thiamine deficiency). However, only in few of these studies, further experiments were carried out to assess other cognitive disturbances (e.g., apraxia, agnosia, and executive dysfunction).

Ciccia and Langlais (78) found learning and memory deficits in thiamine-deficient animals, whether or not submitted to chronic ethanol exposure, and verified that neurological symptoms were most associated with thiamine deficiency. These cognitive deficits appear to be caused by a synergistic interaction between chronic ethanol and thiamine deficiency. The authors also verified that ethanol alone affected short-term memory but did not alter long-term memory. Our group also showed that thiamine deficiency, whether or not associated to chronic ethanol treatment, induced a spatial learning and memory deficit. A significant interaction between the effects of chronic ethanol and thiamine deficiency was found (82). Furthermore, a thiamine deficiency episode induced a decrease in behavioral extinction (83). Some aspects of the observed cognitive deficits (memory and extinction) (82, 83) are correlated to neocortical and hippocampal cholinergic parameter changes. Recent lesion studies with rodents have suggested that the prefrontal cortex contributes to the temporal ordering of spatial and nonspatial events, to the level of attentional selection, as well as to the organization and planning of responses (84–86). Carvalho et al. (87) showed that, although the prefrontal cortex is considered by some authors as a nonvulnerable area to lesions caused by thiamine deficiency, this vitamin deficiency does cause neurochemical dysfunction in that region. In addition, cortical cholinergic projections are thought to play a fundamental role in cortical processing and to affect attentional and memory processes, including extinction (88, 89).

Considering that cholinergic neurotransmission in the basal forebrain plays a key role in attention, learning, and memory processes (90–92) and its dysfunction seems to be involved in the pathophysiology of dementias (93, 94), the biological data discussed here focus on this neurochemical system. Discussing all neurochemical systems that play a relevant role in the aforementioned cognitive processes would be beyond the scope of the present chapter.

Arendt et al. (95) reported extensive cholinergic cell loss in the nucleus basalis of Meynert (NbM) (origin of major cortical cholinergic projections) of patients with Korsakoff's disease, but not in chronic alcoholics without dementia. There is increasing evidence implicating cholinergic mechanisms in the neuropsychological deficits seen in chronic alcoholics. Other authors have also described the involvement of the cholinergic system in Korsakoff's disease (96, 97). Cholinergic dysfunction has been consistently linked to cognitive deficits in different neurodegenerative diseases, leading to the "cholinergic hypothesis," which states that a loss of cholinergic function in the CNS contributes to the cognitive decline associated with aging and dementia (98). A loss of neurons in the NbM complex is reflected by a decrease in the activity of choline acetyltransferase in cortex and hippocampus (95, 99), which are the target areas of these cholinergic neurons (100, 101).

Arendt et al. (102) suggested that rats chronically treated with ethanol represent a suitable animal model to test the cholinergic hypothesis of memory dysfunction, as well as to develop strategies for an amelioration of the impairment in memory and cognitive function in dementia disorders associated with degeneration of the NbM, such as alcoholic dementia and Alzheimer's disease. Floyd et al. (103) showed results indicating a neurotoxic effect of prolonged intake of ethanol on the basal forebrain cholinergic projection system, which may cause impairment of cholinergic innervation of target areas of the basal nucleus complex. Hodges et al. (59) verified that chronic alcohol (20% v/v in drinking water for 28 weeks) impaired acquisition of radial maze spatial and associative tasks by increasing both within-trial working and long-term reference memory errors. Alcohol-treated rats showed improvements in radial maze performance after treatment with cholinergic agonists (arecoline and nicotine) and disruption with antagonists (scopolamine and mecamylamine) at low doses, which did not affect controls. Transplants of cholinergic-rich basal forebrain and ventral mesencephalon fetal neural tissue in cortex and hippocampus of alcohol-treated rats improved radial maze performance to the control level. These results suggest that damage to the forebrain cholinergic projection system played an important part in the radial maze deficits displayed by alcohol-treated rats, since these animals were sensitive to cholinergic drug challenge and cholinergic-rich transplants, which restored cognitive function.

Research on memory enhancing effects of acetylcholinesterase (AChE) inhibitors was stimulated by the finding of diminished cholinergic markers in patients with Alzheimer's disease and the correlation of cognitive impairment to cholinergic deficits in these patients (104). The rationale behind the use of AChE inhibitors is their ability to prevent the breakdown of acetylcholine released from surviving nerve terminals. There is evidence that AChE inhibitors used to treat patients with Alzheimer's disease (105, 106), also ameliorate cognitive deficits of patients with diagnosis of ARD (107) and WKS (108). AChE inhibitors also improve performance in animal models of Alzheimer's disease (109) and in experimental animals with cognitive impairments induced by chronic alcohol consumption (110).

Using only chronic ethanol-treated rodents, several authors have observed memory and/or learning impairments, as well as other cognitive disturbances, such as apraxia, agnosia, and disturbance in executive functioning. Melcer et al. (111) submitted neonatal alcohol-preferring and nonpreferring rats to chronic ethanol treatment and tested them later when they were approximately 2 months old. These authors showed that rats from both lines exhibited motor impairments when tested for walking patterns (gait) and balance. Meyer et al. (112), using rats from nonselected lines, submitted the animals to a similar experimental design and also found abnormal gait when the animals were tested after 2 or 3 months. There is evidence that young animals submitted to an intermittent chronic ethanol treatment exhibited multiple cognitive and motor deficits during adult life, as manifested by performance impairments in conditional discrimination, object recognition, rotarod, and beam-walking tasks (113). Furthermore, Borlikova et al. (114) found that rats submitted intermittently to an ethanol-containing diet for 24 days with two intermediate 3-day withdrawal episodes showed a deficit in executive function as assessed through the performance in a negative patterning discrimination task. Santín et al. (115) also studied the effect of chronic ethanol treatment on executive function by testing rats in the Morris water maze task using place learning with multiple trials followed by a new place learning carried out in the same experimental context. These authors showed that chronically ethanol-treated rats displayed reduced behavioral flexibility. Slawecki (116) observed that adult rats exposed to ethanol vapor for 14 days are more sensitive than adolescent rats to the effects of ethanol on sustained attention. Borde et al. (117), using a delayed alternation task or a spatial reversal learning task (118), showed that chronic alcohol-treated mice displayed an impairment in the retrieval phase of memory processes or showed reversal deficits, respectively.

Beracochea et al. (119) found similar results and suggested that the memory deficits stem from a failure of some cholinergic-dependent retrieval processes. Using an eight arm-radial maze task, some authors showed that chronic alcohol-treated Sprague-Dawley rats presented a severe impairment in both spatial and nonspatial long- and short-term memory (47, 59, 120). Casamenti et al. (121), using an aversively motivated task, and Garcia- Moreno et al. (122), carrying out a spontaneous delayed-nonmatching-to-sample task, verified that the performance of chronic alcohol-treated Wistar rats was impaired. Our group observed retrograde, but not anterograde, amnesia in Fisher rats chronically treated with ethanol alone; however, the detected memory dysfunction was not correlated to central cholinergic parameter changes (123). When ethanol treatment was associated with thiamine deficiency, anterograde amnesia and neocortical/hippocampal cholinergic parameters changes were observed. Our findings have indicated that the association of chronic ethanol consumption with a severe thiamine deficiency episode causes learning (including extinction) and memory deficits and some of these cognitive dysfunction are significantly correlated to cholinergic changes (82, 83, 87).

Contradictory data about the effect of chronic ethanol alone on cholinergic system function can be found in the literature. Some authors have described cholinergic dysfunction in animals treated solely with chronic ethanol (47, 59, 119, 121), while others did not find this effect (123, 124). One of the possible explanations could be the existence of distinct susceptibility to ethanol effects on nutritional state of animals from different genetic patterns (distinct strains). In studies using chronic ethanol treatment alone, blood thiamine level was not measured, and thus secondary ethanol effects (thiamine deficiency) could not be discarded. However, using mice with normal blood thiamine levels, Farr et al. (125) verified that an 8-week ethanol treatment followed by 3 weeks of abstinence caused cognitive deficits in a T-maze foot-shock avoidance task. Irle and Markowitsch (126) showed that both long-term consumption of ethanol and a severe thiamine deficiency episode significantly impaired acquisition of an active two-way avoidance task. They showed that there was no difference between the effects of these two treatments on this behavioral parameter and that hippocampal and cerebellar damage was prominent in animals from both groups. Furthermore, the mammillary nuclei and a few cortical areas suffered from neuronal loss or damage. Moreover, White et al. (127) found that even a 4-day ethanol binge exposure in rats, after withdrawal, causes CNS damage and impairments in reversal learning assessed in the Morris water maze.

He et al. (34) showed that white matter lesions were more evident in the alcohol/thiamine-deficient group than in the water/thiamine-deficient group, which supports the hypothesis that chronic voluntary alcohol consumption exacerbates thiamine deficiency effects. This effect substantiates the potential existence of an alcohol/thiamine deficiency interaction in producing alcohol-related brain damage in human alcoholism. Uncomplicated chronic alcoholic patients show gray but not white matter cerebellar hemisphere volume deficits, whereas Korsakoff patients exhibit deficits in both tissue types. White matter lesion might be related to a system-level dysfunction, which probably underlies the multiple cognitive deficits documented in both chronic alcoholic patients (17) and related animal models (76).

3. Conclusions

Although there is no animal model that displays all aspects considered as criteria for the diagnosis of alcoholic dementia, an animal model that is considered to be the most satisfactory to study behavioral and neurobiological aspects of this disease in the present text is that in which both high/chronic ethanol exposure and thiamine deficiency variables could be controlled. Animal models that manipulate only a single recognized etiological factor are less effective to elucidate the multiple influences that lead to alcoholic dementia. Therefore, even considering that only particular aspects of this disease could be approached using experimental animals, these studies can shed light on the biological processes causing specific and particular patterns of cognitive failure.

References

1. Ruitenberg A, Van Swieten JC, Witteman JC, Mehta KM, van Duijn CM, Hofman A, Breteler MM (2002) Alcohol consumption and risk of dementia: the Rotterdam Study. Lancet 359(9303):281–286
2. Pfefferbaum A (2004) Alcoholism damages the brain, but does moderate alcohol use? Lancet Neurol 3(3):143–144
3. Robles N, Sabria J (2008) Effects of moderate chronic ethanol consumption on hippocampal nicotinic receptors and associative learning. Neurobiol Learn Mem 89(4):497–503
4. Balodis IM, Johnsrude IS, Olmstead MC (2007) Intact preference conditioning in acute intoxication despite deficient declarative knowledge and working memory. Alcohol Clin Exp Res 31(11):1800–1810
5. Weiss E, Marksteiner J (2007) Alcohol-related cognitive disorders with a focus on neuropsychology. Int J Disab Hum Dev 6(4):337–342
6. Hernandez OH, Vogel-Sprott M, Ke-Aznar VI (2007) Alcohol impairs the cognitive component of reaction time to an omitted stimulus: A replication and an extension. J Stud Alcohol Drugs 68(2):276–281
7. Bartholow BD, Pearson M, Sher KJ, Wieman LC, Fabiani M, Gratton G (2003) Effects of alcohol consumption and alcohol susceptibility on cognition: a psychophysiological examination. Biol Psychol 64(1–2):167–190
8. Wegner AJ, Fahle M (1999) Alcohol and visual performance. Prog Neuro-Psychopharm Biol Psychiatry 23(3):465–482
9. Finnigan F, Schulze D, Smalwood J (2007) Alcohol and the wandering mind: A new direction in the study of alcohol on attentional lapses. Int J Disabil Hum Dev 6(2):189–199

10. Hildebrandt H, Brokate B, Eling P, Lanz M (2004) Response shifting and inhibition, but not working memory, are impaired after long-term heavy alcohol consumption. Neuropsychology18(2):203–211
11. Flannery B, Fishbein D, Krupitsky E, Langevin D, Verbitskaya E, Bland C, Bolla K, Egorova V, Bushara N, Tsoy M, Zvartau E (2007) Gender differences in neurocognitive functioning among alcohol-dependent Russian patients. Alcohol Clin Exp Res 31(5):745–754
12. American Psychiatric Association (1994) Diagnostic criteria from DSM–IV. American Psychiatric Association, Washington, DC
13. Moriyama Y, Mimura M, Kato M, Kashima H (2006) Primary alcoholic dementia and alcohol-related dementia. Psychogeriatrics 6(3):114–118
14. Homewood J, Bond NW (1999) Thiamine deficiency and Korsakoff's syndrome: failure to find memory impairments following nonalcoholic Wernicke's encephalopathy. Alcohol 19(1):75–84
15. Butterworth RF (1995) Pathophysiology of alcoholic brain damage – Synergistic effects of alcohol, thiamine-deficiency and alcoholic liver-disease. J Neurochem 65:S202
16. Harper C (1979) Wernicke's encephalopathy: a more common disease than realised. A neuropathological study of 51 cases. J Neurol Neurosurg Psychiatry 42(3):226–231
17. Pfefferbaum A, Sullivan EV, Hedehus M, Adalsteinsson E, Lim KO, Moseley M (2000) In vivo detection and functional correlates of white matter microstructural disruption in chronic alcoholism. Alcohol Clin Exp Res 24(8):1214–1221
18. Harper C (1998) The neuropathology of alcohol-specific brain damage, or does alcohol damage the brain? J Neuropathol Exp Neurol 57(2):101–110
19. Oslin D, Atkinson RM, Smith DM, Hendrie H (1998) Alcohol related dementia: proposed clinical criteria. Int J Geriatr Psychol 13(4):203–212
20. Oslin DW, Cary MS (2003) Alcohol-related dementia: validation of diagnostic criteria. Am J Geriatr Psychiatry 11(4):441–447
21. Hoyumpa AM Jr (1980) Mechanisms of thiamine deficiency in chronic alcoholism. Am J Clin Nutr 33(12):2750–2761
22. Laforenza U, Patrini C, Gastaldi G, Rindi G (1990) Effects of acute and chronic ethanol administration on thiamine metabolizing enzymes in some brain areas and in other organs of the rat. Alcohol Alcohol 25(6):591–603
23. Pincus JH, Wells K (1972) Regional distribution of thiamine-dependent enzymes in normal and thiamine-deficient brain. Exp Neurol 37(3):495–501
24. Langlais PJ (1995) Alcohol-related thiamine deficiency impact on cognitive and memory functioning. Alcohol Health Res World 19(2):113–121
25. Harper C (1983) The incidence of Wernicke's encephalopathy in Australia – a neuropathological study of 131 cases. J Neurol Neurosurg Psychiatry 46(7):593–598
26. Victor M (1994) Alcoholic dementia. Can J Neurol Sci 21(2):88–99
27. Thomson AD, Cook CC, Guerrini I, Sheedy D, Harper C, Marshall EJ (2008) Wernicke's encephalopathy: 'Plus ca change, plus c'est la meme chose'. Alcohol Alcohol 43(2):180–186
28. Sechi G, Serra A (2007) Wernicke's encephalopathy: new clinical settings and recent advances in diagnosis and management. Lancet Neurol 6(5):442–455
29. Chick J (1997) Alcohol and the brain. Curr Opin Psychiatry 10(3):205–210
30. Kopelman MD (1985) Rates of forgetting in Alzheimer-type dementia and Korsakoffs syndrome. Neuropsychologia 23(5):623–638
31. Victor M, Adams RD, Collins GH (1989) The Wernicke-Korsakoff syndrome and related neurological disorders due to alcoholism and malnutrition, 2nd edn. F.A.Davis, Philadelphia, PA
32. Martin PR, Singleton CK, Hiller-Sturmhofel S (2003) The role of thiamine deficiency in alcoholic brain disease. Alcohol Res Health 27(2):134–142
33. Naidoo DP, Bramdev A, Cooper K (1996) Autopsy prevalence of Wernicke's encephalopathy in alcohol-related disease. S Afr Med J 86(9):1110–1112
34. He X, Sullivan EV, Stankovic RK, Harper CG, Pfefferbaum A (2007) Interaction of thiamine deficiency and voluntary alcohol consumption disrupts rat corpus callosum ultrastructure. Neuropsychopharmacology 32(10): 2207–2216
35. McIntosh C, Chick J (2004) Alcohol and the nervous system. J Neurol Neurosurg Psychiatry 75:16–21
36. Tabakoff B, Hoffman PL (2000) Animal models in alcohol research. Alcohol Res Health 24(2):77–84
37. Cohen NJ, Squire LR (1981) Retrograde-amnesia and remote memory impairment. Neuropsychologia 19(3):337–356
38. Becker JT, Butters N, Rivoira P, Miliotis P (1986) Asking the right questions: Problem

solving in male alcoholics and male alcoholics with Korsakoff's syndrome. Alcohol Clin Exp Res 10(6):641–646

39. Milner B, Corkin S, Teuber HL (1968) Further analysis of the hippocampal amnesic syndrome: 14-year follow-up study of H.M. Neuropsychologia 6(3):215–234
40. Pollonini G, Gao V, Rabe A, Palminiello S, Albertini G, Alberini CM (2008) Abnormal expression of synaptic proteins and neurotrophin-3 in the down syndrome mouse model Ts65Dn. Neuroscience 156(1):99–106
41. Savvaki M, Panagiotaropoulos T, Stamatakis A, Sargiannidou I, Karatzioula P, Watanabe K, Stylianopoulou F, Karagogeos D, Kleopa KA (2008) Impairment of learning and memory in TAG-1 deficient mice associated with shorter CNS internodes and disrupted juxtaparanodes. Mol Cell Neurosci 39(3):478–490
42. Wang Q, Liu Y, Zou X, Wang Q, An M, Guan X, He J, Tong Y, Ji J (2008) The hippocampal proteomic analysis of senescence-accelerated mouse: implications of Uchl3 and mitofilin in cognitive disorder and mitochondria dysfunction in SAMP8. Neurochem Res 33(9):1776–1782
43. Kesner RP, Novak JM (1982) Serial position curve in rats: role of the dorsal hippocampus. Science 218(4568):173–175
44. Squire LR, Spanis CW (1984) Long gradient of retrograde amnesia in mice: continuity with the findings in humans. Behav Neurosci 98(2):345–348
45. Getachew B, Hauser SR, Taylor RE, Tizabi Y (2008) Desipramine blocks alcohol-induced anxiety- and depressive-like behaviors in two rat strains. Pharmacol Biochem Behav 91(1):97–103
46. Oliveira-Silva IF, Pinto L, Pereira SR, Ferraz VP, Barbosa AJ, Coelho VA, Gualberto FF, Souza VF, Faleiro RR, Franco GC, Ribeiro AM (2007) Age-related deficit in behavioural extinction is counteracted by long-term ethanol consumption: correlation between 5-HIAA/5HT ratio in dorsal raphe nucleus and cognitive parameters. Behav Brain Res 180(2):226–234
47. Arendt T, Allen Y, Marchbanks RM, Schugens MM, Sinden J, Lantos PL, Gray JA (1989) Cholinergic system and memory in the rat – effects of chronic ethanol, embryonic basal forebrain brain transplants and excitotoxic lesions of cholinergic basal forebrain projection system. Neuroscience 33(3): 435–462
48. Spanagel R (2003) Alcohol addiction research: from animal models to clinics. Best Pract Res Clin Gastroenterol 17(4):507–518
49. Heilig M, Egli M (2008) Models for alcohol dependence: a clinical perspective. Drug Discov Today: Dis Models 2(4):313–318
50. Koob GE, Le Moal M (2008) Addiction and the brain antireward system. Annu Rev Psychol 59:29–53
51. Spanagel R, Holter SM (2000) Pharmacological validation of a new animal model of alcoholism. J Neural Transm 107(6):669–680
52. Walker BM, Koob GF (2008) Pharmacological evidence for a motivational role of kappa-opioid systems in ethanol dependence. Neuropsychopharmacology 33(3):643–652
53. Dhaher R, Finn D, Snelling C, Hitzemann R (2008) Lesions of the extended amygdala in C57BL/6J mice do not block the intermittent ethanol vapor-induced increase in ethanol consumption. Alcohol Clin Exp Res 32(2): 197–208
54. Zhao Y, Weiss F, Zorrilla EP (2007) Remission and resurgence of anxiety-like behavior across protracted withdrawal stages in ethanol-dependent rats. Alcohol Clin Exp Res 31(9):1505–1515
55. O'Dell LE, Roberts AJ, Smith RT, Koob GF (2004) Enhanced alcohol self-administration after intermittent versus continuous alcohol vapor exposure. Alcohol Clin Exp Res 28(11):1676–1682
56. Funk CK, Zorrilla EP, Lee MJ, Rice KC, Koob GF (2007) Corticotropin-releasing factor 1 antagonists selectively reduce ethanol self-administration in ethanol-dependent rats. Biol Psychiatry 61(1):78–86
57. Tomlinson D, Wilce P, Bedi KS (1998) Spatial learning ability of rats following differing levels of exposure to alcohol during early postnatal life. Physiol Behav 63(2):205–211
58. Ehlers CL, Somes C, Li TK, Lumeng L, Hwang BH, Jimenez P, Mathe AA (1999) Calcitonin gene-related peptide (CGRP) levels and alcohol. Int J Neuropsychopharmacol 2(3):173–179
59. Hodges H, Allen Y, Sinden J, Mitchell SN, Arendt T, Lantos PL, Gray JA (1991) The effects of cholinergic drugs and cholinergic-rich foetal neural transplants on alcohol-induced deficits in radial maze performance in rats. Behav Brain Res 41(1):7–28
60. Cadete-Leite A, Brandao F, Andrade JP, Ribeiro-da-Silva A, Paula-Barbosa MM (1997) The GABAergic system of the dentate gyrus after withdrawal from chronic alcohol consumption: effects of intracerebral grafting and putative neuroprotective agents. Alcohol Alcohol 32(4):471–484

61. Carroll MR, Rodd ZA, Murphy JM, Simon JR (2006) Chronic ethanol consumption increases dopamine uptake in the nucleus accumbens of high alcohol drinking rats. Alcohol 40(2):103–109
62. Christie BR, Swann SE, Fox CJ, Froc D, Lieblich SE, Redila V, Webber A (2005) Voluntary exercise rescues deficits in spatial memory and long-term potentiation in prenatal ethanol-exposed male rats. Eur J Neurosci 21(6):1719–1726
63. Baird TJ, Vanecek SA, Briscoe RJ, Vallett M, Carl KL, Gauvin DV (1998) Moderate, long-term, alcohol consumption potentiates normal, age-related spatial memory deficits in rats. Alcohol Clin Exp Res 22(3):628–636
64. Byrnes ML, Richardson DP, Brien JF, Reynolds JN, Dringenberg HC (2004) Spatial acquisition in the Morris water maze and hippocampal long-term potentiation in the adult guinea pig following brain growth spurt-prenatal ethanol exposure. Neurotoxicol Taratol 26(4):543–551
65. Pires RGW, Pereira SRC, Pittella JEH, Franco GC, Ferreira CLM, Fernandes PA, Ribeiro AM (2001) The contribution of mild thiamine deficiency and ethanol consumption to central cholinergic parameter dysfunction and rats' open-field performance impairment. Pharmacol Biochem Behav 70(2–3):227–235
66. Eriksson K (1968) Genetic selection for voluntary alcohol consumption in the albino rat. Science 159(3816):739–741
67. Sinclair JD, Le AD, Kiianmaa K (1989) The AA and ANA rat lines, selected for differences in voluntary alcohol consumption. Experientia 45(9):798–805
68. Colombo G, Agabio R, Diaz G, Fa M, Lobina C, Reali R, Gessa GL (1997) Sardinian alcohol-preferring rats prefer chocolate and sucrose over ethanol. Alcohol 14(6):611–615
69. Li TK, Lumeng L, Doolittle DP (1993) Selective breeding for alcohol preference and associated responses. Behav Genet 23(2):163–170
70. Ghozland S, Chu K, Kieffer BL, Roberts AJ (2005) Lack of stimulant and anxiolytic-like effects of ethanol and accelerated development of ethanol dependence in mu-opioid receptor knockout mice. Neuropharmacology 49(4): 493–501
71. Thiele TE, Naveilhan P, Ernfors P (2004) Assessment of ethanol consumption and water drinking by NPY Y(2) receptor knockout mice. Peptides 25(6):975–983
72. Han DH, Lyool IK, Sung YH, Lee SH, Renshaw PE (2008) The effect of acamprosate on alcohol and food craving in patients with alcohol dependence. Drug Alcohol Depend 93(3):279–283
73. Snyder JL, Bowers TG (2008) The efficacy of acamprosate and naltrexone in the treatment of alcohol dependence: a relative benefits analysis of randomized controlled trials. Am J Drug Alcohol Abuse 34(4):449–461
74. Zalewska-Kaszubska J, Gorska D, Dyr W, Czarnecka E (2008) Effect of chronic acamprosate treatment on voluntary alcohol intake and beta-endorphin plasma levels in rats selectively bred for high alcohol preference. Neurosci Lett 431(3):221–225
75. Savage LM, Chang Q, Gold PE (2003) Diencephalic damage decreases hippocampal acetylcholine release during spontaneous alternation testing. Learn Mem 10(4):242–246
76. Langlais PJ, Savage LM (1995) Thiamine deficiency in rats produces cognitive and memory deficits on spatial tasks that correlate with tissue loss in diencephalon, cortex and white matter. Behav Brain Res 68(1):75–89
77. Haas RH (1988) Thiamine and the brain. Annu Rev Nutr 8:483–515
78. Ciccia RM, Langlais PJ (2000) An examination of the synergistic interaction of ethanol and thiamine deficiency in the development of neurological signs and long-term cognitive and memory impairments. Alcohol Clin Exp Res 24(5):622–634
79. Pfefferbaum A, Adalsteinsson E, Bell RL, Sullivan EV (2007) Development and resolution of brain lesions caused by pyrithiamine- and dietary-induced thiamine deficiency and alcohol exposure in the alcohol-preferring rat: a longitudinal magnetic resonance imaging and spectroscopy study. Neuropsychopharmacology 32(5):1159–1177
80. Dixon G, Harper CG (2001) Quantitative analysis of glutamic acid decarboxylase-immunoreactive neurons in the anterior thalamus of the human brain. Brain Res 923(1–2):39–44
81. Harper C (1979) Wernicke's encephalopathy: a more common disease than realised. A neuropathological study of 51 cases. J Neurol Neurosurg Psychiatry 42(3):226–231
82. Pires RGW, Pereira SRC, Oliveira-Silva IF, Franco GC, Ribeiro AM (2005) Cholinergic parameters and the retrieval of learned and re-learned spatial information: a study using a model of Wernicke-Korsakoff Syndrome. Behav Brain Res 162(1):11–21
83. Pires RGW, Pereira SRC, Carvalho FM, Oliveira-Silva IF, Ferraz VP, Ribeiro AM

(2007) Correlation between phosphorylation level of a hippocampal 86 kDa protein and extinction of a behaviour in a model of Wernicke-Korsakoff syndrome. Behav Brain Res 180(1):102–106

84. Ragozzino ME, Wilcox C, Raso M, Kesner RP (1999) Involvement of rodent prefrontal cortex subregions in strategy switching. Behav Neurosci 113(1):32–41
85. Delatour B, Gisquet-Verrier P (2000) Functional role of rat prelimbic-infralimbic cortices in spatial memory: evidence for their involvement in attention and behavioural flexibility. Behav Brain Res 109(1):113–128
86. Dias R, Aggleton JP (2000) Effects of selective excitotoxic prefrontal lesions on acquisition of nonmatching- and matching-to-place in the T-maze in the rat: differential involvement of the prelimbic-infralimbic and anterior cingulate cortices in providing behavioural flexibility. Eur J Neurosci 12(12):4457–4466
87. Carvalho FM, Pereira SRC, Pires RGW, Ferraz VP, Romano-Silva MA, Oliveira-Silva LF, Ribeiro AM (2006) Thiamine deficiency decreases glutamate uptake in the prefrontal cortex and impairs spatial memory performance in a water maze test. Pharmacol Biochem Behav 83(4):481–489
88. Mason ST (1983) The neurochemistry and pharmacology of extinction behavior. Neurosci Biobehav Rev 7(3):325–347
89. Robbins TW (1997) Arousal systems and attentional processes. Biol Psychol 45(1–3): 57–71
90. Parikh V, Sarter M (2008) Cholinergic mediation of attention – contributions of phasic and tonic increases in prefrontal cholinergic activity. Mol Biophys Mech Arousal Alertness Atten 1129:225–235
91. Bainbridge NK, Koselke LR, Jeon J, Bailey KR, Wess J, Crawley JN, Wrenn CC (2008) Learning and memory impairments in a congenic C57BL/6 strain of mice that lacks the M-2 muscarinic acetylcholine receptor subtype. Behav Brain Res 190(1):50–58
92. Pauli WM, O'Reilly RC (2008) Attentional control of associative learning – a possible role of the central cholinergic system. Brain Res 1202:43–53
93. Keverne J, Ray M (2008) Neurochemistry of Alzheimer's disease. Psychiatry 7(1):6–8
94. Nardone R, Bergmann J, Tezzon F, Ladurner G, Golaszewski S (2008) Cholinergic dysfunction in subcortical ischaemic vascular dementia: a transcranial magnetic stimulation study. J Neural Transm 115(5):737–743
95. Arendt T, Bigl V, Arendt A, Tennstedt A (1983) Loss of neurons in the nucleus basalis of Meynert in Alzheimer's disease, paralysis agitans and Korsakoff's disease. Acta Neuropathol 61(2):101–108
96. Mair RG, Anderson CD, Langlais PJ, McEntee WJ (1988) Behavioral impairments, brain-lesions and monoaminergic activity in the rat following recovery from a bout of thiamine-deficiency. Behav Brain Res 27(3):223–239
97. Overstreet DH, Russell RW (1984) Selective breeding for differences in cholinergic function - sex-differences in the genetic-regulation of sensitivity to the anticholinesterase, Dfp. Behav Neural Biol 40(2):227–238
98. Bartus RT, Dean RL, Beer B, Lippa AS (1982) The cholinergic hypothesis of geriatric memory dysfunction. Science 217(4558):408–417
99. Gill SK, Ishak M, Dobransky T, Haroutunian V, Davis KL, Rylett RJ (2007) 82-kDa choline acetyltransferase is in nuclei of cholinergic neurons in human CNS and altered in aging and Alzheimer disease. Neurobiol Aging 28(7):1028–1040
100. Henny P, Jones BE (2008) Projections from basal forebrain to prefrontal cortex comprise cholinergic, GABAergic and glutamatergic inputs to pyramidal cells or interneurons. Eur J Neurosci 27(3):654–670
101. Garcia-Alloza M, Zaldua N, ez-Ariza M, Marcos B, Lasheras B, Javier Gil-Bea F, Ramirez MJ (2006) Effect of selective cholinergic denervation on the serotonergic system: implications for learning and memory. J Neuropathol Exp Neurol 65(11): 1074–1081
102. Arendt T, Henning D, Gray JA, Marchbanks R (1988) Loss of neurons in the rat basal forebrain cholinergic projection system after prolonged intake of ethanol. Brain Res Bull 21(4):563–569
103. Floyd EA, Young-Seigler AC, Ford BD, Reasor JD, Moore EL, Townsel JG, Rucker HK (1997) Chronic ethanol ingestion produces cholinergic hypofunction in rat brain. Alcohol 14(1):93–98
104. Sugimoto H (2008) The new approach in development of anti-Alzheimer's disease drugs via the cholinergic hypothesis. Chem Biol Interact 175(1–3):204–208
105. Tumiatti V, Bolognesi ML, Minarini A, Rosini M, Milelli A, Matera R, Melchiorre C (2008) Progress in acetylcholinesterase inhibitors for Alzheimer's disease: an update. Expert Opin Ther Pat 18(4):387–401

106. Brousseau G, Rourke BP, Burke B (2007) Acetylcholinesterase inhibitors, neuropsychiatric symptoms, and Alzheimer's disease subtypes: an alternate hypothesis to global cognitive enhancement. Exp Clin Psychopharmacol 15(6):546–554
107. Kim KY, Ke V, Adkins LM (2004) Donepezil for alcohol-related dementia: a case report. Pharmacotherapy 24(3):419–421
108. Cochrane M, Cochrane A, Jauhar P, Ashton E (2005) Acetylcholinesterase inhibitors for the treatment of Wernicke-Korsakoff syndrome – three further cases show response to donepezil. Alcohol Alcohol 40(2):151–154
109. Dong H, Csernansky CA, Martin MV, Bertchume A, Vallera D, Csernansky JG (2005) Acetylcholinesterase inhibitors ameliorate behavioral deficits in the Tg2576 mouse model of Alzheimer's disease. Psychopharmacology (Berl) 181(1):145–152
110. Mohammed AH (1993) Effects of cholinesterase inhibitors on learning and memory in rats: a brief review with special reference to THA. Acta Neurol Scand Suppl 149:13–15
111. Melcer T, Gonzalez D, Somes C, Riley EP (1995) Neonatal alcohol exposure and early development of motor-skills in alcohol-preferring and nonpreferring rats. Neurotoxicol Taratol 17(2):103–110
112. Meyer LS, Kotch LE, Riley EP (1990) Alterations in gait following ethanol exposure during the brain growth spurt in rats. Alcohol-Clin Exp Res 14(1):23–27
113. Pascual M, Blanco AM, Cauli O, Minarro J, Guerri C (2007) Intermittent ethanol exposure induces inflammatory brain damage and causes long-term behavioural alterations in adolescent rats. Eur J Neurosci 25(2): 541–550
114. Borlikova GG, Elbers NA, Stephens DN (2006) Repeated withdrawal from ethanol spares contextual fear conditioning and spatial learning but impairs negative patterning and induces over-responding: evidence for effect on frontal cortical but not hippocampal function? Eur J Neurosci 24(1):205–216
115. Santín LJ, Rubio S, Begega A, Arias JL (2000) Effects of chronic alcohol consumption on spatial reference and working memory tasks. Alcohol 20(2):149–159
116. Slawecki CJ (2006) Two-choice reaction time performance in Sprague-Dawley rats exposed to alcohol during adolescence or adulthood. Behav Pharmacol 17(7):605–614
117. Borde N, Jaffard R, Beracochea DJ (1996) Effects of methyl beta-carboline-3-carboxylate on memory impairments induced by chronic alcohol consumption in mice. Prog Neuropsychopharmacol Biol Psychiatry 20(8):1377–1387
118. Borde N, Beracochea DJ (1999) Effects of diazepam or chronic alcohol treatment on spatial reversal learning in mice. Pharmacol Biochem Behav 62(4):719–725
119. Beracochea D, Micheau J, Jaffard R (1992) Memory deficits following chronic alcohol consumption in mice: relationships with hippocampal and cortical cholinergic activities. Pharmacol Biochem Behav 42(4):749–753
120. Gal K, Bardos G (1994) The effect of chronic alcohol treatment on the radial maze performance of rats. Neuroreport 5(4):421–424
121. Casamenti F, Scali C, Vannucchi MG, Bartolini L, Pepeu G (1993) Long-term ethanol consumption by rats: effect on acetylcholine release in vivo, choline acetyltransferase activity, and behavior. Neuroscience 56(2):465–471
122. Garcia-Moreno LM, Conejo NM, Capilla A, Garcia-Sanchez O, Senderek K, Arias JL (2002) Chronic ethanol intake and object recognition in young and adult rats. Prog Neuropsychopharmacol Biol Psychiatry 26(5):831–837
123. Pereira SRC, Menezes GA, Franco GC, Costa AEB, Ribeiro AM (1998) Chronic ethanol consumption impairs spatial remote memory in rats but does not affect cortical cholinergic parameters. Pharmacol Biochem Behav 60(2):305–311
124. Fadda F, Cocco S, Stancampiano R, Rossetti ZL (1999) Long-term voluntary ethanol consumption affects neither spatial nor passive avoidance learning, nor hippocampal acetylcholine release in alcohol-preferring rats. Behav Brain Res 103(1):71–76
125. Farr SA, Scherrer JF, Banks WA, Flood JF, Morley JE (2005) Chronic ethanol consumption impairs learning and memory after cessation of ethanol. Alcohol Clin Exp Res 29(6):971–982
126. Irle E, Markowitsch HJ (1983) Widespread neuroanatomical damage and learning-deficits following chronic alcohol-consumption or vitamin-B1 (Thiamine) deficiency in rats. Behav Brain Res 9(3):277–294
127. White AM, Bae JG, Truesdale MC, Ahmad S, Wilson WA, Swartzwelder HS (2002) Chronic-intermittent ethanol exposure during adolescence prevents normal developmental changes in sensitivity to ethanol-induced motor impairments. Alcohol Clin Exp Res 26(7):960–968

Chapter 34

Animal Models of Metallic Dementia

Luigi F. Rodella

Abstract

Alzheimer's disease (AD) is nowadays the most widespread form of senile dementia in Western countries. AD is considered a multifactorial disease: in fact, it is probable that many concomitant factors (genetic and environmental) are relevant in the pathogenesis of this disease. In particular, among the environmental factors, a lot of recent interest has been raised concerning the possible involvement of metal dyshomeostasis. Several studies have indicated the imbalance of many trace elements in AD, including aluminum (Al), iron (Fe), lead (Pb), manganese (Mn), copper (Cu), and zinc (Zn). We have therefore analyzed the role of these metal ions in AD development.

Key words: Animal models, AD, Metals, Oxidative stress

1. Introduction

Several million people worldwide are affected by neurodegenerative diseases, a heterogeneous group of diseases affecting specific areas of the central nervous system (CNS). These can lead to gradual and progressive cognitive or movement impairment, depending on the type of neuronal cells undergoing selective degeneration (1). Alzheimer's disease (AD) is nowadays the most widespread form of senile dementia in Western countries. It is clinically characterized by gradual onset with progressive and irreversible cognitive decline. Memory impairment appears in the earliest stage of the disease while patients' motor and sensory functions are usually not affected until later stages (2). It is almost impossible to give an *in vivo* differential diagnosis between AD and normal aging. Indeed, only postmortem microscopic examination reveals the critical features of AD, i.e. accumulation of insoluble β-amyloid (Aβ) peptides in extracellular senile plaques

Peter Paul De Deyn and Debby Van Dam (eds.), *Animal Models of Dementia*, Neuromethods, vol. 48,
DOI 10.1007/978-1-60761-898-0_34, © Springer Science+Business Media, LLC 2011

and deposition of misfolded microtubule-associated tau protein in neurofibrillary tangles within neuronal bodies (3). Both these neuropathological hallmarks cause nerve cell death and so damage the functions of the CNS. Nowadays, the etiology of AD remains unknown. It is probable that many concomitant factors are relevant in AD pathogenesis. So we can consider AD as a multifactorial disease. AD is genetically complex: rare mutations in three genes (amyloid precursor protein (APP), and presenilins-1 and -2) are responsible for cases of familial early-onset (<65 years) autosomal dominant forms (1). Sporadic forms of AD are most probably due to the contribution of several susceptibility genes at multiple loci and their interaction with environmental factors. The major risk factor (other than aging) for the development of AD is a genetic polymorphism in the gene for ApoE, a lipoprotein involved in synaptic repair and in maintenance of neuronal structure (4). Individuals who carry one copy of the APOE-ε4 allele have a two- to threefold increased risk of AD, compared to the more common APOE-ε3 allele. However, the APOE-ε2 allele seems to give some protection (5). The finding that monozygotic twins do not necessarily both develop AD (6), clearly suggests that there is also an exogenous contribution related to diet, lifestyle, and the patient's anamnesis (7).

Among environmental factors, aging remains the major risk factor for AD in the general population. An interesting hypothesis states that free radicals produced during cellular respiration could play an important role both in the process of aging and in the development of AD (8). Another predisponent factor for AD is the occurrence of cerebral infarcts, whose incidence increases with aging (9). Moreover, several epidemiological studies have shown that head injury is also a risk factor (10) because it seems to be associated with the formation of neurofibrillary tangles (11) and Aβ deposition (12). Recent studies suggest that AD is an immune disorder from blood brain barrier (BBB) disruption (13), characterized by selective antibodies for cholinergic neurons. Even if there is no evidence due to the vicinity of rinencephalon and certain regions of the CNS involved in AD neurodegeneration, many authors have proposed a link between amyloid deposition and viral pathogens that enter the nervous system through olfactory receptors (14). At present, many authors accept that the immune system is involved in the etiology of AD. This premise is founded on the numerous alterations in AD brains that show the activation of immunological and inflammatory mechanisms. From an environmental point of view, a lot of recent interest has been raised concerning the possible involvement of metal dyshomeostasis in AD pathogenesis. We have therefore analyzed the role of metal ions in AD development.

2. Metals Dyshomeostasis and Oxidative Stress

The CNS appears to be particularly vulnerable to oxidative stress. A number of factors contribute to the relatively high vulnerability of the CNS to oxidative damage. These include low levels of the natural antioxidant glutathione present in neurons (15), a high proportion of polyunsaturated fatty acids contained in membranes (16), and a very high oxygen requirement due to the high metabolic activities of the brain (17). Oxidative stress is accompanied in varying degrees by a dyshomeostasis of metal ions, including redox-active transition metals (copper, iron, and manganese) as well as redox inactive metal ions such as zinc. Trace redox-active transition metals are essential for most biological reactions (e.g., in the synthesis of DNA, RNA, and proteins) and as cofactors for numerous enzymes, particularly those involved in respiration. Their deficiency can therefore lead to disturbances in CNS function. Accumulations of redox-active transition metals in tissues are cytotoxic due to their participation in many cellular disturbances characterized by increased free radical production (18). Mounting evidence indicates that dyshomeostasis of cerebral metals and APP/Aβ/metal redox interactions may contribute to the neuropathology of AD (19). Characteristic of AD is a dyshomeostasis of both redox-active and redox-inactive metal ions. Several studies have indicated the imbalance of many trace elements in AD, including aluminum (Al), lead (Pb), zinc (Zn), copper (Cu), manganese (Mn), and iron (Fe), with only the latter three being redox-active. Redox-inactive metal ions may be pathogenic by virtue of their ability to impair the metabolism of redox-active metal ions. Moreover, they may also contribute to neurodegeneration through their deleterious effects on protein and peptide structures, including for example pathological aggregation phenomena. Microparticle-induced X-ray emission data have revealed that insoluble Aβ plaques in postmortem AD brains have abnormally enriched Al, Cu, Fe, and Zn (20). Evidence for the imbalance of trace metal homeostasis in AD have led to the identification of possible interactions of metal ions with both APP and Aβ. Metals such as Al, Fe, Zn, and Cu *in vitro* promote Aβ aggregation in plaques (21–23). The finding that metal chelators can at least partially solubilize Aβ deposits and attenuate the cerebral Aβ burden in APP transgenic mouse models of AD (24, 25) suggests their possible therapeutical use (26, 27). Interestingly, a more recent study indicates that coprecipitant ions (such as Zn(II)) may be needed for *in vivo* Aβ aggregation since the $A\beta_{1-40}$ peptide is thermodynamically soluble at physiological concentrations without any factors (28). Moreover, the 50-untranslated region (50UTR) of APP mRNA has a functional iron-response element (IRE) for the bond with Fe (29), which is consistent with a role for APP as a redox-active metalloprotein (30). Some pilot studies also indicate that Fe, Cu, and Zn ions are able to promote APP expression linking its 50UTR in a dose-dependent manner. Cerebral biometal metabolism is probably intimately associated with APP processing. Again, the gene expression profile of AD brain implicates dysregulation of cerebral metal metabolism (31). Compared to controls, gene expression levels for metal regulatory proteins such as metallothionein III and metal regulatory factor-1 decreased more than fourfold in AD brain (32). In addition to oxidative damage to proteins, oxidative stress also causes oxidative damage to nucleic acids, in particular RNA. 8-Hydroxyguanosine is a nucleic acid oxidation marker commonly observed in the cytoplasm of those neurons particularly vulnerable to degeneration in AD (33). There is always the question of whether increased tissue burden of metals reflects at least partial dietary factors (e.g., there is mixed evidence for a role of dietary Cu in experimental models of AD). The brain normally concentrates Cu, Fe, and Zn in the neocortex (20). Environmental metal exposure has also been suggested to be a risk factor for AD. The most studied ones are Al and Fe (34, 35), while evidence for Pb, Mn, Cu, and Zn is sparse. Clearly, the complex biochemical relation between biometal metabolism and environmental metal exposure and AD pathophysiology warrants further study. All these findings are consistent with the refined amyloid cascade hypothesis (36). However, it is not known whether metal excess causes oxidative stress and neurodegeneration or whether metal excess is a consequence of neuronal cell loss.

3. Experimental Models of Metallotoxicity

Many experimental approaches can be used to study the potential metal involvement in the neuropathogenesis of AD. In particular, *in vivo* studies can be very useful to establish whether a metal is toxic for the CNS and if it can induce or facilitate the development of AD-like lesions. For the general design of an *in vivo* experiment, a large number of parameters have to be considered, including animal species, age of animals, and metal exposure routes (7). Many recent findings concerning metal-induced AD have been obtained using transgenic animals (e.g., AD mice or mice models with accelerated aging) combined with metal exposure. Studies on aged animals, or indeed studies with long-term exposure over months or years, are therefore better reflections of the human situation, where metals build-up (especially for nonacute toxicity ones) occurs over a long time. To date, a variety of *in vivo* procedures are available to assess the consequences of long-term exposure to metals. These comprise a range of behavioral tests, but also *in vivo* electrophysiological procedures (both in anesthetized and freely moving animals) and *in vivo* brain microdialysis. For example, more recently, the first animal MRI and near-infrared fluorescence studies were conducted in AD models (37), even though the main focus was general brain activity. Generally, systemic (e.g., oral and intraperitoneal) exposure routes resemble the human situation closer. However, it is very difficult to give an exact indication of the amount of metal-intake (especially for oral regimes) and there are many substantial species differences in metal bioavailability between humans and rodents. Therefore, injections of predefined metal solutions directly into the brain allow a better control over brain concentrations and can target specific brain regions. Both injections of metal into the ventricle (i.c.v.) or into the hippocampus have been conducted successfully, combined with controls using vehicle injections. This regime is less variable, since systemic exposure routes have large variations between animals due to peripheral metabolism such as renal clearance, liver metabolism, and accumulation of toxins in adipose tissue or bone. In this paper, we have focused on the most common *in vivo* experimental models to study metal involvement in neurodegeneration typical of AD.

3.1. Aluminum

3.1.1. Aluminum Toxicity

Al represents about 8% of the Earth's crust and is therefore the third most abundant element on Earth. Some researchers assert that in the "Aluminum Age" (38) in which we are living nowadays, we are subjected to continual chronic exposure from early embryonic development (39). An individual can be in contact with this metal from many environmental sources, through therapeutic treatments (e.g., antacid ingestion and vaccines) or through the use of certain cosmetic products. Moreover, Al is contained in many oven cooked foods, in certain other foods and beverages, in dust, and also in contaminants in potable tap water (40). In spite of its environmental abundance, no evidence exists to show a definite biological function for Al. On the contrary, it has been suggested that its chemical properties make it neutral or even incompatible with major physiological processes (41). Nevertheless, Al is not considered toxic in the same way as heavy metals, even if evidence of its toxicity when consumed in abundance does exist. Al toxicity, either in experimental animal models or in many different human clinical conditions, was the fulcrum of numerous scientific researches over the last century (42, 43). In 1886, Al was defined a neurotoxic metal for humans for the first time and the first research, which considered a relation between Al and certain alterations of the CNS, was published in 1937. From this moment, a wide ranging research has been carried out about the probable role of Al in dialysis syndrome (44), in cerebral functionality damages in some working categories (e.g., welders and foundry workers) (45), in amyotrophic lateral sclerosis (46), in Parkinson's disease (47), and in AD (48, 49). Even if the role of this metal in the etiopathogenesis of these diseases is still in discussion, it has been clearly demonstrated that Al can interfere with a great number of metabolic and cellular pathways either in the CNS or in other bodily systems. Even after decades of studies on the biological role of Al, the exact mechanism by which it generates its neurotoxicity is still unclear.

3.1.2. Entry of Aluminum into the Body

To face the abundance and toxicity of Al, living organisms have developed active and passive mechanisms to remove this metal. In humans, these mechanisms include: (1) the reduction of Al absorption mediated principally by the liver, and also by particular human intestinal formation, (2) the excretion of Al through the kidney and, in smaller quantities, through the bile, (3) the reduction of plasmatic levels of Al by its sequestration in bones and by its bond to transferrin (Tf), the most important plasmatic protein for metal transport in the blood (40). Even given its ubiquitous presence, only 0.06–0.1% of ingested Al is absorbed into the gastrointestinal (GI) tract (50). This limited uptake of Al can be explained by considering the capability of certain dietary components, such as citrate and fluoride, to form complexes with Al (51) and also due to the competition of Al with Ca, Mg, and Si for absorption (52). Finally, vitamin D and parathyroid hormone also seem to interfere with Al absorption (53). The factor which above all influences Al absorption is pH. Many epidemiological studies and some studies about the complex chemistry of Al have demonstrated that at pH values close to 7.9, Al tends to form aggregates, which are more difficult to absorb, and can reduce its toxicity (54, 55). For this reason, if it is true that drinking tap water is the principal Al source for humans, its absorption probably occurs mainly through the skin, nasal and pulmonary surfaces (for example taking a shower or a bath) rather than through GI tract surfaces that are more acid and could theoretically allow large absorption of Al. It should be noted, however, that this has never been verified (40). Nevertheless, low Al absorption levels, as well as small amounts that reach the blood, could – over sufficiently long periods – cause significant accumulations and alterations in certain target organs such as bones, the erythropoietic system, and the CNS (56). This is especially so if Al cannot be removed because of alterations to kidney function. This damage can be explained if we consider that Al is a very chemically active element. In biological systems, it can form more stable complexes than those formed by other elements (such as Ca and Mg) and so it can substitute these elements when constituting certain biological molecules (e.g., proteins, ATP, GTP, nucleic acids, phosphates) and change their functions (57, 58). The alteration of these molecules can in its turn upset the balance of certain human systems and apparatus, particularly those with major accumulations of Al (59).

3.1.3. Aluminum and the Central Nervous System

The neuronal microenvironment and particularly constant chemical–physical parameters in the cerebral interstitial fluid are necessary to optimize neuronal functionality (60). The maintenance of these constant parameters over long periods is due to both the kidney filtering blood and the isolation performed by the blood-brain barrier (BBB). The maintenance of these parameters over short periods, however, is due to an integration between BBB action and glial cell activity. In the CNS of healthy individuals, lower Al levels are found than in other tissues or organs (61). This seems to be due both to its partial exclusion from the brain because of BBB action and to an active mechanism that allows Al efflux from the CNS, probably as Al-citrate (62). In his/her life, an individual is subject to chronic exposure to Al. This prolonged exposure can exert pressure on the maintenance of the physical–chemical equilibrium of the interstitial cerebral fluid (63). Al, in fact, can modify BBB permeability influencing molecules and ions flux towards and from the CNS (64). However, Al uptake in the CNS is not dependent on BBB permeability alterations (65, 66). Modifying the short-term regulation of cerebral interstitial fluid equilibrium, Al can dilute transduction molecule concentrations and by favoring large neuronal intakes of neurotoxins, reduce neuronal vitality. Three probable mechanisms by which Al enters the CNS from blood circulation have been proposed: (1) bypassing the BBB, (2) through the nasal-olfactory pathway (Al inhaled through the nose goes directly into CNS through the olfactory bulb), and (3) from cerebrospinal fluid (CSF) (61, 67). The most rapid Al intake in the CNS is certainly through the BBB, because many carriers with high affinity for this metal have been identified in the BBB. The best mechanism is Tf-mediated Al transport (65). Florence and colleagues showed an increase in transferrin receptor (TfR) concentrations in rat brains subjected to chronic oral administration of Al-citrate (Table 1) (68). Another equally important system is the monocarbossilic acid carrier (MCT), a proton cotransporter located on either side of the BBB (69). The MCT-mediated transport of Al into the CNS seems also to be closely dependent on the presence of free-citrate molecules, because of the use of citrates as a vehicle. The maintenance of Al cerebral homeostasis is either by its expulsion out of the CNS (70), or by its uptake by neurons and glial cells that compete to internalize Al (71, 72). Moreover, neurons are long life cells, so they can accumulate Al insight themselves (48). In maintaining Al homeostasis, Al uptake by nervous cells is also very important. Even if the internalization of Al is still not completely described, it probably involves different mechanisms such as endocytosis, facilitated diffusion, and active transport that determine the region-specific distribution of Al in the CNS (48). It is still very difficult to establish the real distribution of Al in the CNS because the methods presently used have a lot of analytical problems due to the complexity of Al chemistry. In healthy subjects, Al concentrations vary between 50 and 196 μg/g bodyweight depending on the specific cerebral region. It has

Table 1
Animal models of aluminum-induced dementia

Reference	Experimental model	Exposure protocol	Effects
(73)	Male Wistar rats (22 months old)	Chronic (6 months) Al ($AlCl_3 \times 6H_2O$) administration (2 g/L) in drinking water	Increased contents of Al^{3+}, Cu^{2+}, Zn^{2+}, and Mn^{2+} prosencephalon, mesencephalon, and pons-medulla; No change in the cerebellum except a Cu^{2+} decrease Increased area occupied by the mossy fibers in the CA3 field (+32%)
(80)	Male Wistar rats	Chronic administration (3–5 weeks) of aluminum in drinking water (2.5% aluminum sulfate)	Increase in extracellular cGMP induced by nitric oxide generating agents (240%) Decreased cerebellar content of calmodulin (34%) and nitric oxide synthase (15%) Decreased basal activity of soluble guanylate cyclase (66%) Decreased basal cGMP in the cerebellar extracellular space (50%)
(108)	C57BL6 mice (5 weeks old)	Chronic exposure over 12 months to aluminum sulfate (2.5%) in drinking water	Deposition of Aß similar to that seen in congophilic amyloid angiopathy (CAA) in humans Reduction in neuronal expression of GRP78 similar to what has previously been observed in AD
(98)	Male Sprague–Dawley rats	Aluminum sulfate 2.5% in tap drinking water (corresponding to 0.2% Al) for 0.5, 1, 2, 3, 6, and 12 months	Reduced NADPH-d and nNOS positive neurons in the cerebral cortex time dependently, with the greatest effect appearing after 3 months Decrease in NADPH-d/NPY double-stained neurons with an apparent increase in NPY single-stained neurons between 1 and 3 months of exposition (these then decreased after 6–12 months)
(85)	Male albino rats	Aluminum sulfate ($Al_2(SO_4)_3 \cdot 18H_2O$) by gavage (21.2 mg/kg b.w.) for 35 days	Marked increase in glutamate and glutamine levels; significative decrease in GABA levels; spongioform changes in the neurons of hippocampus Nuclear deformity, and neurofibrillary degeneration, similar to NFT in AD

(continued)

Table 1 (continued)

Reference	Experimental model	Exposure protocol	Effects
(113)	Wistar rats	Al lactate (10 mg/kg of body weight/day) for 4 weeks through I.P. injections	Lower levels of acetylcholine Significant decrease in high-affinity choline uptake in the hippocampus; neurobehavioral deficits in terms of decreased active (52%) and passive (73.30%) avoidance tests
(110)	Tg2576 mice (5 months of age)	Diet supplemented with Al lactate (1 mg/g of diet) for 120 days	Impaired acquisition and retention of the water maze task Higher amounts of Aß fragments in brain Increase in the total number of proliferating cells in the dentate gyrus of hippocampus
(114)	Female Tg2576 mice and wild-type littermates	Diet supplemented with 1 mg Al/g for 6 months	In hippocampus, Cu concentrations decreased in nontreated mice, while Zn levels were higher in Al-treated mice (at 11 months) Not important metal changes
(107)	Old mice	2 µL of 0.25% aluminum chloride solution (i.c.v. injections once a day for 5 days)	Increased expression of HO-1 protein, and elevation of both HO activity and iron level in hippocampus
(68)	Male Wistar rats	Chronic oral administration (6 months) of Al citrate	Increased brain Al levels by 30-fold; Al tissue content significantly greater when the Al citrate was administered in an iron-free diet (distribution of Al in brain similar to that AD) Higher concentrations of transferrin receptors At the cellular level an extensive cytoplasmic vacuolation in astrocytes (especially) and neurons

been found in many cerebral areas (frontal, temporal, occipital, parietal and somatosensory cortex, hypothalamus, thalamus, hippocampus, and some cerebellar areas). The areas with the largest concentrations are certainly the temporal cortex and hippocampus, which are also those most involved in AD (41). An *in vivo* study by Fattoretti and colleagues reported an increase in Al^{3+} and other metals in the prosencephalon, mesencephalon, and pons-medulla of rats treated for 6 months with oral $AlCl_3$ (Table 1) (73). All Al that bypasses the BBB to the CNS is divided in dynamic equilibrium in the extracellular environment, intracellular environment, and membranes. In the blood, Al binds to proteins (like Tf) and to other ligands with a low molecular weight such as small peptides, nucleotides, nucleic acids, citrate, phosphate, and silicic acid. In the interstitial cerebral fluid, Al also binds to certain neurotransmitters, such as GABA and glutamate. Al is associated to BBB surfaces and to plasmatic membranes of glial cells and neurons. Al may be bound by phosphate groups in the hydrophilic heads of membrane phospholipids, which in their turn act as aggregation sites for Al. At an intracellular level, Al may bind ATP, even if most is confined to the endoplasmic reticulum and Golgi regions near the nucleus. Some studies claim that Al is also linked to nuclear chromatin. Between extracellular and intracellular environments, there are continuous exchanges of Al that could influence the physiology of the CNS.

3.1.4. The Relation Between Al and Nitroxidergic System: A Possible Toxicity Mechanism

Al is considered a neurotoxic metal (74, 75) and therefore could be an etiologic factor in the pathogenesis of various nervous system diseases such as AD (48, 49). Different mechanisms are understood by which Al manifests its toxicity in nonnervous organs (free radical deposition and damage of the lysosomal system and cytoskeleton) (76). The molecular mechanisms by which Al is toxic, however, are not completely understood. Some research suggest that Al interferes with glutamatergic neurotransmission (77, 78), thereby damaging the glutamate (Glu) -nitric oxide (NO)- monophosphate cyclic guanosine (cGMP) pathway *in vivo* (79). In a 1999 study, oral chronic aluminum sulfate administration (3–5 weeks) increased extracellular cGMP by 240%, induced by nitric oxide-generating agents in rat brains (Table 1) (80). In this pathway, the activation of ionotropic glutamate receptors (especially NMDA) increases intracellular Ca^{2+}, which, after binding with calmodulin, activates neuronal nitric oxide synthase (nNOS). NO is an atypical neurotransmitter because it does not accumulate in the synaptic vesicles, but is synthesized on request. In the CNS, NO is produced by nNOS using l-arginine as a substrate, and other synthesis factors (e.g., calmodulin, calcium, oxygen, NADPH, heme, and flavin). The neurons that produce NO are called nitroxidergic neurons. These represent about 1% of all neurons of the somatosensory cortex and are scattered among other neurons, which also act as interneurons (81). Once synthesized, NO spreads out of the presynaptic neuron that produced it, towards the postsynaptic neuron where it activates the soluble form of enzyme guanylate cyclases. This in turn increases cGMP concentrations (82). NO is a volumetric neurotransmitter, formed in one neuron; it can spread to nearby neurons and can activate their guanylate cyclases, which in their turn activate the Glu-NO-cGMP pathway in these neurons (79). This pathway is very important both under physiological and pathological conditions. Physiologically, it is fundamental because of its ability to modulate intercellular communication and important neurological processes, such as long-term potentiation, the basis of some forms of learning and memory (83, 84). Under pathological conditions, excessive activation of this pathway seems to be the basis for glutamatergic toxicity that is involved in the etiopathogenesis of many neurodegenerative diseases (80). A marked increase in Glu levels was observed in the hippocampus of albino rats treated for 35 days with aluminum sulfate (Table 1) (85). Neuronal NO was copurified with the enzyme NADPH-diaphorase (86–88). The role of Al in compromising the Glu-NO-cGMP pathway may be due to its ability to bypass the BBB and to form stable complexes with ATP, the principal transductional molecule for extracellular signals (89) and in the CNS, because it optimizes certain neurotransmitter activation (Glu, GABA, and acetylcholine) (90–92). Most ATP secreted in the interstitial cerebral fluid is combined with Mg, while only a little is free (93). When Al passes the BBB and penetrates into the CNS, it combines with ATP (both in cytosol and in interstitial cerebral fluid) more stably than with other metals (including Mg), thus exerting latent toxicity. If the Al–ATP complex replaces the Mg–ATP complex, it can potentiate glutamatergic neurotransmission (94). When, during glutamatergic neurotransmission, Glu and ATP are contemporarily released into the synaptic space, they act on their postsynaptic receptors to induce Ca^{2+} ion movements into nerve cells. In the presence of Al–ATP (rather than Mg–ATP), Ca channels remain open for too long and intracellular Ca^{2+} levels become too high. The consequent alteration in Ca^{2+} homeostasis compromises neuronal energy levels, causing neurotoxicity and neuronal death (95), in particular of nitroxidergic neurons. Therefore, some alterations in the Glu-NO-cGMP pathway can reduce nitroxidergic neurons and neuronal NO in certain cortical areas. These results were observed in AD patients' brains (96, 97). The same decrease of nitroxidergic transmission has also been observed in rat brains treated with chronic oral aluminum sulfate. Recent studies showed that in these brains, the number of nNOS-expressing neurons decreased proportionally to the duration of Al treatment (Table 1) (98, 99). This further strengthens the hypothesis that Al in drinking water can be an etiological factor in AD (100).

3.1.5. Aluminum and Oxidative Stress

Al biochemistry is very particular because, even if Al is not a redox metal, it has clear pro-oxidant activity (101). This metal can promote the oxidation of biological molecules in both *in vivo* and *in vitro* experimental models. Its ability to favor Fe-dependent (102) and Fe-independent (103, 104) lipid peroxidation, to induce Fe-independent NADH oxidation (105) and to form hydroxilic radicals (106) have all been demonstrated. Numerous possible mechanisms to explain how Al could have pro-oxidant effects have been proposed. All seem to have in common the formation of a very reactive Al species, i.e. the radical ion semi-reduced superoxide Al, which produces, due to its interaction with Fe, free radicals of oxygen both under pathological and physiological conditions (101). Al probably acts by stabling ferrous ions (Fe^{2+}), which are potent promoters of oxidative species generation as they catalyze Fenton's reaction, which forms ferric ions (Fe^{3+}) and OH^-. Moreover, Al is an activator of superoxide dismutase (SOD), the enzyme, which catalyzes hydrogen peroxide (H_2O_2) synthesis from superoxide ion, and an inhibitor of catalase, the enzyme which transforms H_2O_2 in less reactive species. Al can therefore increase H_2O_2 levels that, in turn, increase oxidative species production able to damage proteins, DNA, and membrane lipids causing oxidative stress inside cells. Numerous *in vivo* studies have confirmed Al's ability to induce oxidative stress. Yuan and colleagues showed increased expression and activity of HO-1 (a stress protein) in the hippocampus of elderly mice after an i.c.v. injection of aluminum chloride (Table 1) (107). Rodella and colleagues demonstrated a clearly reduced neuronal expression of GRP78 (another stress protein) after chronic exposure over 12 months to aluminum sulfate in mice (Table 1) (108).

3.1.6. Aluminum and β-Amyloid

Al, besides being a constituent of senile plaques, promotes Aβ aggregation (increasing it 100–1,000 times) and can increase its toxicity (109). This was confirmed by a recent study showing a higher amount of Aβ fragments in APP transgenic mice brains after Al administration, compared to nontreated transgenic mice (Table 1) (110). Another characteristic of Al is its ability to increase posttransductional modifications of APP and, therefore, to accelerate the production of Aβ after APP proteolysis. Al aids serine proteases (hampering its inhibitors) and so facilitates Aβ production and its accumulation and aggregation in senile plaques (111). Aβ is a metalloprotein able to bind ions of certain transition metals such as Zn, Cu, and Fe. This explains why high levels of these metals have been found in senile plaques (20). The bond between Aβ and the redox ions Cu and Fe overproduces H_2O_2, which causes oxidative damage and neuronal dysfunction by reducing them. Some researchers conversely believe that Aβ could be nontoxic itself, but toxic for neurons because of its ability to increase H_2O_2-mediated oxidative damage. The formation of plaques could compensate oxidative stress because, if Aβ binds metal ions in excess, it can chelate them from the action of H_2O_2 (112). Despite these contrasting results, most experimental evidence indicates that Al accumulation in brain tissue may have neuropathological relevance. In view of the great social impact of AD in which Al seems to be implicated, continued research is clearly justified (113, 114) (Table 1).

3.2. Iron

3.2.1. Iron Use in the Body and Brain

Fe is essential for life. Adequate Fe is important, because Fe deficiency causes anemia, which can lead to fatigue, decreased exercise tolerance, and general inanition. Thus, evolution has created reserves to balance periods of relative loss from such things as hemorrhage or bleeding. Intake of adequate bioavailable Fe, usually present in meat, is necessary to prevent anemia. Fe is found in the active centers of many enzymes and oxygen carrier proteins. It is a central element of the heme molecule, a critical part of hemoglobin essential for oxygen transport. Heme is also part of many essential enzymes, such as the cytochrome series, and Fe is found in cytochrome oxidase and in the iron–sulfur complexes of the oxidative chain important for producing ATP (115). Some enzymes involved in DNA synthesis, and succinate dehydrogenase of the tricarboxylic acid cycle, are also Fe-dependent (116). All these metabolic processes take place in the brain at a relatively high level. Indeed, the brain, which only takes up 2% of the total body weight, accounts for 25% of oxygen consumption, making the brain the highest oxygen-consuming organ in the body. Consequently, oxygen transportation and its associated metabolic activities are relatively high in the brain and create a high demand for Fe transportation to and use within the brain. Therefore, Fe is important to maintain many specific neurobiological processes. Additional demands of Fe in the brain come from myelogenesis and myelin maintenance. Oligodendrocytes, the cells responsible for myelin production and maintenance, are Fe-rich compared to other brain cells (117). The synthesis of neural transmitters is also Fe-dependent. Fe is needed as a cofactor to produce dopamine, norepinephrine, and serotonin, which play a significant role in certain chronic neurological disorders (118). There is also intense Fe staining in brain areas that receive GABA-ergic innervation (119) suggesting a possible relation between Fe and this inhibitory neurotransmitter.

3.2.2. Intra- and Extracellular Iron Transport

The dietary Fe pathway to the brain begins in the intestines where Fe^{3+} (ferric iron) is reduced by duodenal cytochrome-b to Fe^{2+} (ferrous iron). In this reduced form, the divalent metal transporter can carry Fe across the duodenal epithelium into the blood. In the blood, Fe^{2+} is oxidized to Fe^{3+} by ceruloplasmin to be coupled with Tf, which has two Fe-binding domains that bind the ferric form with very high affinity (120). Altered location of the Fe-transport protein Tf in senile plaques, rather than its regular location in the cytosol of nervous cells, may suggest that Tf becomes trapped within plaques when it attempts to transport Fe between cells (121). Studies examining Tf to Fe ratios in brain show that this ratio decreases in AD brains (122). This suggests a decrease in Fe mobility in AD patients consistent with the idea that the decrease in Tf in AD brains could be associated with a deficit in intracellular brain Fe mobilization or with a lower efflux of Fe from the brain. Such a decrease would increase Fe "labile pools" and would contribute to the increased susceptibility of the AD brain to Fe-induced oxidative damage (123). Furthermore, increased Tf subtype C2 frequency was reported in patients with AD (124). The presence of the C2 variant plus a hemochromatosis mutation increased the risk of AD fivefold, which was increased even more in the presence of APOE-ε4 allele (125). Fe homeostasis maintained in the brain is similar to systemic Fe homeostasis, although Fe circulating in the blood outside the CNS cannot directly cross the BBB. There are several pathways that can transfer Fe across the BBB. The first and probably most common pathway for Fe uptake in the brain is tightly controlled by the TfR expressed in neurons, in the endothelial cells of cerebral blood vessels, and in choroid plexus cells (126) that bind Fe circulating when bound to transferrin. The TfR-bound complex then enters the brain by endocytosis. In AD, significant decreases in TfR density were noted in the hippocampus and temporal and occipital cortices, but TfR density was unchanged in microvasculature, parietal and frontal cortices, and cerebellum (127). These data, even if difficult to understand, suggest that TfRs decrease in areas associated with memory loss (128). Several other transporter systems may also deliver Fe across the BBB, such as the divalent metal transporter and the lactoferrin receptor (35). A critical component of cellular Fe transport is the mechanism by which Fe is released into the cytoplasm. Removal of Fe from Tf within the endosome requires lower pH within the endosome itself. Decreased pH causes a lower affinity of Tf for Fe so that Fe dissociates from the carrier. Once Fe has been released into the cytosol from endosomes, it is either incorporated into Fe-requiring enzymes or exists at low concentrations (labile Fe pool).

3.2.3. Intracellular Iron Homeostasis

Labile Fe is potentially toxic and therefore, as a protection mechanism, cells have evolved an extensive and elegant control mechanism to regulate intracellular Fe bioavailability. Fe that is not immediately used in the cell is stored in ferritin. Ferritin is the most common Fe-storage protein in the brain and can store up to 4,500 Fe^{3+} atoms as an inorganic complex (129). Serum ferritin provides one measure of Fe storage, and a low ferritin blood level is a reliable indicator of Fe deficiency. It consists of two types of subunits, the heavy and light chains that work in a complementary way to store intracellular Fe. Heavy ferritin (21 kDa) efficiently sequestrates Fe and is found in organs with high Fe use and little Fe storage, while light ferritin (19 kDa) is associated with Fe storage (130). Studies by Connor and colleagues (131) found variations in ferritin levels. In elderly subjects, an increase in heavy and light ferritin coincides with an age-related increase in Fe. In AD, however, there is no concomitant increase in ferritin, despite Fe levels increase. In addition, ferritin isolated from the brains of AD patients has a higher Fe content than ferritin isolated from control brains (132). Together, these data suggest that Fe levels are increased in the brains of AD patients relative to ferritin levels. The rise in Fe levels without a simultaneous change in ferritin provides a source of labile Fe that generates free radicals. Finally, after the brain uses the Fe it has stored, it must leave the cells and the Cu-associated protein ceruloplasmin may facilitate cellular release of Fe (133). The importance of ceruloplasmin in brain Fe metabolism is highlighted by the extreme accumulations of Fe in patients with hereditary aceruloplasminemia (134). Fe leaves the body via bleeding or through the shedding of skin and other cells (135). Recently, it has become clear that the regulation/management of Fe at a cellular level, although primarily by Tf, TfR, and ferritin, is also controlled by lactotransferrin receptors, melanotransferrin, ceruloplasmin, and divalent cation transporters. So, disruption in these proteins can contribute to the altered brain Fe metabolism found in neurodegenerative disorders (20). Control at a posttranscriptional level of these proteins and therefore overall regulation of cellular Fe metabolism involves the action of two Fe regulatory proteins, IRP-1 and IRP-2. Iron responsive elements (IREs) found on certain mRNAs are targets for high-affinity interactions with the cytoplasmic protein IRP. The interaction of IRPs with IREs controls the translation or stability of mRNA (136). When intracellular Fe levels decrease, the affinity of IRP for IRE increases. IRPs bind to the IREs on 59UTR ferritin mRNA, preventing initiation of ferritin translation and decreasing cellular Fe storage capacity (137). At the same time, other IRPs can bind to IREs in the 39UTR of TfR mRNA, inhibiting degradation of TfR mRNAs and increasing TfR levels. Conversely, high intracellular Fe levels promote dissociation of IRPs from IREs on ferritin and TfR mRNA causing increased ferritin mRNA translation and decreased stability of the TfR mRNA (138). IRP-1, but not IRP-2, is rapidly activated by extracellular H_2O_2, establishing a regulatory connection between Fe metabolism control and oxidative stress response. IRPs are therefore critical to maintain cellular Fe homeostasis that could become dysfunctional in neurodegenerative disorders. Studies by Yamanaka and colleagues (139) showed that Al may stabilize IRP2, increasing its ability to bind to IREs. Further evidence for a disruption in the IRPs in AD comes from the immunocytochemical studies by Smith and colleagues who showed a selective association of IRP2, but not IRP1, with the pathological hallmarks of AD localizing to neurofibrillary tangles (NFT), senile plaques (SP), and neuropil threads (i.e., the neurofibrillary lesions) (140).

3.2.4. Iron and Oxidative Stress

The physiological, biochemical, and anatomical aspects of Fe homeostasis must be considered to understand the role of this metal in chronic neurological disorders. Although Fe's importance for normal neurological activity is clear, too high Fe concentration may have devastating effects. Fe is one of the most essential transition metals involved in the formation of oxygen-free radicals, owing to its interaction with H_2O_2 through Fenton chemistry. During the reduction of molecular oxygen, mitochondria produce superoxide (O_2^-). Enzymatic dismutation by SOD in turn yields H_2O_2. Although O_2^- and H_2O_2 by themselves are relatively nontoxic, H_2O_2, which is freely permeable in tissues, could produce highly toxic hydroxyl radical (OH^-) through a metal Fe-catalyzed Fenton reaction. This reaction is referred to as the superoxide-driven Fenton reaction. Superoxide may also produce H_2O_2 nonenzymatically. In its normal metabolic state, superoxide favors the oxidation of Fe^{2+} to Fe^{3+}. However, if the intracellular concentration of superoxide is high, the reaction helps reduce Fe^{3+} to Fe^{2+} and elaboration of hydroxyl radicals. In light of the high levels of oxygen consumption in the CNS, high levels of reactive oxygen species (ROS) are expected, as well as antioxidant defenses commensurate with free radical production. Under normal situations, oxidative balance is maintained and free radicals are detoxified. In disease, various modifications of macromolecules such as sugars, lipids, proteins, and nucleic acids, can lead to critical failure of biological functions and ultimately cell death (141, 142). It is now well established that neurodegenerative diseases are associated with oxidative imbalance and its consequences. Abnormally high levels of Fe have been found in a number of neurodegenerative disorders. Importantly, increased total Fe does not necessarily imply increased oxidative stress if there are concomitant increases in proteins that store Fe in redox inert forms. For example, ferritin (which may be upregulated in AD) contains a core ofinsoluble, unreactive ferrihydrate. Microglia are the major sites of ferritin bound Fe and are thought to be partly responsible for oxidative damage in neurodegenerative disorders. Interesting, oxidant damage is intimately involved with aging diseases, such as atherosclerosis, autoimmunity diseases, diabetes, fibrosis, inflammation, Parkinson's, and AD.

3.2.5. Iron Overaccumulation in Brain Regions Susceptible to AD

It has become apparent that Fe progressively accumulates in the affected brain areas of neurodegenerative disorders (143, 144). Overaccumulation of Fe in AD has been found within specific brain regions displaying selective vulnerability to neurodegeneration such as the hippocampus, cerebral cortex, and basal nucleus of Meynert (20, 145). Using an *in situ* Fe detection method, Sayre and colleagues (146) found redox-active Fe associated with both NFT and senile plaques. The association of Fe with NFT may be in part related to Fe binding to their primary protein constituent, tau. These results add to earlier studies reporting increased levels of Fe, Tf, and ferritin in AD. Further studies have suggested that accumulated Fe supports the AD pathology as a possible source of oxidative stress, demonstrating that neurons in AD brains have high oxidative loads (147, 148). Intrahippocampal unilateral infusion of a ferrous citrate solution in rats causes an increase in lipid peroxidation levels that extensively damage neurons of hippocampal regions (Table 2) (149). Moreover, de Lima and colleagues showed extensive oxidative damage (increased superoxide production and SOD activity) in rats' brains treated for a few days with oral Fe^{2+} (Table 2) (150). Postmortem analysis of Alzheimer patients' brains has revealed the activation of two enzymatic indicators of cellular oxidative stress: heme oxygenase (HO-1) (151) and NADPH oxidase (152). HO-1 was greatly increased in neurons and astrocytes of the hippocampus and cerebral cortex of Alzheimer subjects, colocalizing with senile plaques and NFT (153). HO-1 catalyzes the conversion of heme to Fe and biliverdin, which in turn, is reduced to bilirubin, an antioxidant. Since HO-1 is induced in proportion to the level of heme, HO-1 induction would suggest abnormal turnover of heme in AD. This may be connected to mitochondrial abnormalities in AD, which contain numerous heme proteins. Fe accumulation in AD has been shown to be particularly abundant in brain regions vulnerable to AD and at the microscopic level, it accumulates in senile plaques in AD (20). Increases in Fe accumulation

Table 2
Animal models of Iron induced dementia

Reference	Experimental model	Exposure protocol	Effects
(149)	Male Wistar rats	Intra hippocampal unilateral infusion of 1 μL Kreb's Ringer solution containing ferrous citrate (iron: 3.4 mmol/L) at a rate of 0.2 μL/min	Increase in lipid peroxidation levels; extensive damage in neurons from the CA1 and CA3 regions of the Fe^{2+}-treated animals
(150)	Male Wistar rats	10.0 mg/kg of oral Fe^{2+}at postnatal days 5–7, 12–14, 19–21, or 30–32	Impairments in long-term retention of object recognition memory Oxidative damage in the brain (increased superoxide production in submitochondrial particles) Increased superoxide dismutase activity in brain

and oxidative stress in AD brains are related to changes in the concentration of soluble and deposited Aβ protein. At a biochemical level, Aβ is defined as a redox-active metalloprotein that is precipitated by interactions with neocortical metal ions, especially Zn, Cu, and Fe resulting in its auto-aggregation and oligomerization (154, 155). Several studies have shown that oxidative stress increases Aβ production (156). On the other hand, Aβ itself may be able to generate ROS, driving a potential positive feedback loop whereby increased generation of ROS increases Aβ and vice versa. Three histidine and one tyrosine residue at the hydrophilic N-terminus of the peptide are crucial to bind metal to the peptide (157). Regarding Fe interaction, Fe^{3+} influences the formation of $A\beta_{1-42}$ fibrils (158). The interaction of redox-active Fe^{3+} with Aβ fibrils causes their chemical reduction and generation of reactive oxygen species, causing lipid membrane peroxidation, DNA breakdown, and protein oxidation (159). *In vitro*, Fe is needed for Aβ toxicity, since Fe deletion from the culture medium or the inclusion of adequate Fe chelators reduces or blocks toxicity, while overloading cells with Fe increases toxicity (160). Hence, abnormal Fe deposition, such as in AD plaques, may interact with Aβ to mediate free-radical-induced neurotoxicity and account in part, for the widespread oxidative damage in AD brains, accelerating the pathological process (161). Unexpectedly, the redox potential of Fe is significantly attenuated by Aβ, suggesting a neuroprotective chelating role for Aβ in disease pathogenesis (162). NFT are another hallmark lesion of AD and, interestingly, are another site of Fe accumulation. The presence of redox metals in NFT has been shown to induce oxidative stress via H_2O_2 (146). Fe^{3+} has the ability (similar to Al^{3+}) to bind with hyperphosphorylated tau and to induce its aggregation *in vitro*, leading to the formation of NFTs in AD. Although NFTs possibly act as a redox center, neurons lacking NFTs also contain oxidative modifications (163), suggesting that oxidative stress precedes NFT formation. Among the macromolecules modified by ROS, RNA may be an early target of oxidative damage in AD brains (164). Using 8-hydroxyguanosine as a marker of nucleic acid oxidation, oxidative damage can be found within the neuronal perikaryal cytoplasm. It is therefore possible that transition metals such as Fe play a pivotal role in the oxidation of RNA. Studies show that RNA oxidization is increased in AD brain (165) and subsequent protein translation is impaired (166) (Table 2).

3.3. Lead

3.3.1. General Aspects

Pb is a neurotoxin that continues to be considered a global environmental health hazard. In particular, the ability of Pb to interact in a flexible coordination with protein oxygen and sulfur atoms and form stable complexes with them increases its affinity for proteins when compared to other metals (167). Pb is a heavy metal with no known biological function in humans. Susceptibility to

Pb toxicity is influenced by several factors such as environmental exposure, age, and nutritional status. Human exposure to Pb occurs via food, water, air, and soil. It has been used for various applications in glass, pigments, makeup, water transport, wine, cooking, and more recently in antiknock fuel additives, electronic components, and batteries. People can also be exposed to Pb contamination from industrial sources, e.g., smelters and workers of Pb manufacturing industries (168). Thus, Pb is still a significant public health concern. Pb toxic effects depend on both the duration of exposure and the magnitude of the dose. Pb has well-characterized effects on every organ system, including cardiovascular (169), renal (170), immune (171), and reproductive (172) systems, as well as on bones and teeth (173). It has also been identified as a probable human carcinogen (174). However, the nervous system is especially sensitive to Pb effects. Additionally, it has been shown that Pb is stored in bone and released over time, especially during periods of bone demineralization such as pregnancy, lactation, and postmenopause. High Pb exposure in pregnant women can cause a low infant birth weight or prematurity. In addition, when the Pb level in maternal blood increases, Pb levels in breast milk also increase with additional risks to the neonate (175). Children are particularly sensitive to the deleterious effects of Pb. They can absorb 30–75% of Pb ingested (176), while adults absorb about 11% of Pb. In addition, pre- and perinatal exposure to Pb causes higher brain metal accumulation than later postnatal exposure due to an underdeveloped BBB in early life. The consequences of high blood Pb levels in children may be irreversible, including learning disabilities (177), behavioral problems, and mental retardation. At very high blood concentrations, Pb can cause convulsions, coma, and even death. Nutritional status is another significant risk factor for Pb intoxication. Fe, Zn, and Ca deficiencies increase the retention of ingested Pb, which can also increase Pb GI absorption (178), and affect Pb neurotoxicity susceptibility (179).

3.3.2. Lead Neurotoxicity and Its Possible Link with Alzheimer's Disease

The mechanisms underlying Pb neurotoxicity are still a matter of debate. So far, the effects of Pb on Ca fluxes and metabolism have been suggested as major mechanisms involved in Pb toxicity (180). Other potential mechanisms for Pb toxicity include its capacity to affect cell membrane biophysics, cause oxidative stress, and trigger certain transcription factors sensitive to oxidative status of the cell. Experimental studies have demonstrated that environmental factors (such as stress) can interact with Pb exposure. Bolin and colleagues found changes in the expression of copper/zinc SOD, manganese SOD, and reduced-form glutathion in rat brains after short-term or long-term lead acetate administration in drinking water (Table 3) (181). Tiffany-Castiglioni and colleagues showed that Pb binds to GRP78 during the Pb accumulation process in

Table 3
Animal models of Lead-induced dementia

Reference	Experimental model	Exposure protocol	Effects
(188)	Female monkeys (*Macaca fascicularis*); the monkeys were terminated 23 years later.	1.5 mg kg^{-1} day^{-1} of lead acetate from birth until 400 days of age via infant formula and vehicle after weaning	Elevated expression of AD-related genes (APP, BACE1 (beta-site APP cleaving enzyme 1)] and their transcriptional regulator (Sp1). Alterated levels, characteristics, and intracellular distribution of Abeta staining and amyloid plaques in the frontal association cortex. Decrease in DNA methyltransferase activity and higher levels of oxidative damage to DNA.
(186)	Long–Evans males hooded rats	200 ppm Pb (lead acetate) from day 1 after birth to day 20 through the drinking water or from18 to 20 months of age.	Formation of Abeta aggregates Developmental exposure to Pb did not alter the lifetime profiles of alpha-, beta-, and gamma-secretases in adult and aging animals.
(186)	Long–Evans males hooded rats.	200 ppm Pb (lead acetate) from day 1 after birth to day 20 through the drinking water or from 18 to 20 months of age	Transient induction of APP mRNA expression in neonates, but it exhibited a delayed overexpression 20 months after exposure to Pb had ceased. Rise in activity of the transcription factor Sp1 (a regulator of the APP gene).
(181)	Timed-pregnant Long–Evans hooded rats.	0.2% Lead acetate from postnatal day 1 through postnatal day 20 through the drinking water or from 18 to 20 months of age	20 months after Pb exposure stable induction of cerebral 8-hydroxy-2′-deoxyguanosine (Oxo8dG) Not altered DNA repair enzyme 8-oxoguanine DNA glycosylase (Ogg1) activity Changes in copper/zinc-superoxide dismutase (SOD1), manganese-SOD (SOD2), and reduced-form glutathion (GSH).

astroglial cells (182). As GRP78 is a stress protein, this binding may contribute to increased susceptibility of the brain to stress. Chaperone deficiency is implicated in a number of CNS diseases, including AD. The work published by Stewart and colleagues (183) revealed occupational exposure to Pb as a risk factor for AD. In this seminal work, the authors looked at the relation of Pb bone levels to the ApoE genotype. They concluded that the persistent CNS effects of Pb are more toxic in individuals with at least one APOE-ε4 allele (183). Momcilovic and colleagues found a tenfold increase of Pb in the white matter of AD patients (184). Moreover, Zawia and colleagues reported that Pb exposure in early life in both rodents and monkeys upregulates APP mRNA expression and increases the levels of its amyloidogenic cleavage product Aβ (Table 3) (185–187). These studies showed an association between developmental Pb exposure and amyloidogenesis during old age. They propose epigenetics as one of the potential mechanisms that mediate its involvement in AD etiopathogenesis and suggest that early Pb exposure may inhibit the methylation of CpG dinucleotides in the APP promoter, causing increased responsiveness of the APP gene later in life to transcriptional factors. This increased responsiveness could be the principal reason for upregulation of APP mRNAs in AD brains. In support of this, Wu and colleagues demonstrated decreased DNA methyltransferase activity and higher levels of oxidative damage to DNA in monkey brains treated with 1.5 mg kg^{-1} day^{-1} of lead acetate from birth until 400 days of age (Table 3) (188) (Table 3).

3.4. Manganese

3.4.1. General Aspects

Mn exists in a number of physical and chemical forms in the Earth's crust and it can assume different oxidation states. In living tissue, Mn has been found as Mn^{2+}, Mn^{3+}, and possibly as Mn^{4+} (189). Mn tends to form very tight complexes with various substances (190). As a result, its free plasma and tissue concentrations tend to be extremely low (191). In adults, dietary Mn concentrations influence the amount of Mn absorbed from the GI tract and the amount eliminated in bile. When dietary Mn levels are high, adaptive changes reduce GI absorption, enhance liver metabolism, and increase its biliary excretion (192). Thus, despite large fluctuations in oral Mn intake, brain and other tissue Mn levels remain relatively constant, at least in healthy adults. Recent studies suggest that high levels of Mn in drinking water (>300 mg/L) are associated with reduced intellectual function in children (193). High exposure to airborne Mn has also been associated with severe neurotoxic effects (194). Mn is an essential metal found in a variety of biological tissues and is necessary for a variety of physiological processes including amino acid, lipid, protein, and carbohydrate metabolism (195). In the brain, Mn is an important cofactor for a variety of enzymes, including the antioxidant enzyme SOD (196), as well as enzymes involved in neurotransmitter synthesis and metabolism (197). In excessive quantities, however, it can cause damage to the CNS. Mn enters the brain from the blood either across the cerebral capillaries and/or the cerebrospinal fluid. At normal plasma concentrations, Mn appears to enter the CNS primarily across the capillary endothelium, whereas at high plasma concentrations, transport across the choroid plexus appears to predominate (198), consistent with observations on the rapid appearance and high levels of Mn in this organ (199). It would appear that facilitated diffusion (198), active transport (198), divalent metal transporter 1-mediated transport (200), ZIP8- and Tf-dependent transport mechanisms (201) are all involved in shuttling Mn across the BBB.

3.4.2. Manganese and Manganism

Since the last century, the neurotoxicity of Mn has been well known (202) as manganism, and has been described and characterized by extrapyramidal dysfunction and neuropsychiatric symptoms. This syndrome has been observed throughout the world in hundreds of cases among miners and industrial workers who had been exposed to high levels of Mn (203). Cases of overt Mn poisoning were also reported due to environmental exposure in drinking water in Japan (204), or after drinking water low in magnesium but with a diet high in Mn content (205). Another possible source of Mn accumulation and clinical toxicity has been identified in excessive Mn levels in total parenteral solution, given intravenously to patients with chronic GI illnesses (206). In addition, since Mn is excreted almost completely via the biliaric tract, individuals with chronic liver failure are at the risk of developing hepatic encephalopathy, probably due to Mn accumulation in the brain (207).

3.4.3. Neurotoxicity of Mn: Parkinsonian Disturbances

Since Mn elimination from the CNS requires a long time, neurotoxic effects may occur later in life, increasing the frequency of Parkinsonian disturbances in elderly individuals. The signs and symptoms from relatively high levels of exposure include postural instability, mood and psychiatric changes (i.e., depression, agitation, hallucinations) (208), and Parkinsonian symptoms such as bradykinesia, rigidity, tremor, gait disturbance, postural instability, and dystonia and/or ataxia (209). Cognitive deficits such as memory impairment, reduced learning capacity, decreased mental flexibility, cognitive slowing (209), and difficulty with visuomotor and visuospatial information processing (210) have also been reported. The adverse effects of Mn on the nervous system are probably due to the failure of protective enzymes capable of detoxifying critical amounts of Mn or to alter its oxidation potential. On the one hand, the basis of Mn neurotoxicity hinges on the ability of the bivalent form Mn^{2+} to oxidize to the trivalent form Mn^{3+}, a powerful oxidizing species. On the other hand, Mn shows high affinity to neuromelanin-rich areas, such as the nigrostriatal tract (211). At this site, Mn favors the autoxidation of dopamine, generating toxic free radicals (212), which can cause substantial injury of the dopaminergic system. A multifactor hypothesis is more likely and could involve iron-induced oxidative stress and the direct interaction of Mn with dopaminergic terminal mitochondria, leading to selective mitochondrial dysfunction, subsequent excitotoxicity, and finally, neurodegeneration (202).

3.4.4. Manganese and Alzheimer's Disease

Although Mn has no reported direct interaction with Aβ, recent reports have highlighted the impact of the loss of Mn superoxide dismutase (MnSOD) on AD in APP transgenic models. MnSOD is found exclusively in mitochondria, and heterozygotic genetic ablation of MnSOD leads to increased oxidative stress. Mice lacking MnSOD die within the first week of life, and develop a complex heterogeneous phenotype arising from mitochondrial dysfunction and oxidative stress. Treatment of these mice with catalytic antioxidants increases their lifespan and rescues the peripheral phenotypes, while uncovering the CNS pathology. MnSOD null mice exhibit striking increases in tau phosphorylation, which can be remedied with an SOD-mimetic compound, previously reported to be sufficient to prevent neuropathology (Table 4) (213). Crossing MnSOD± mice with Tg2576 APP transgenic mice leads to increased amyloid burden and tau hyperphosphorylation (Table 4) (213). In similar studies, the genetic combination of MnSOD± with the Tg19959 model of AD increased amyloid burden (Table 4) (214). However, a more recent report observed decreased plaque burden in the parenchyma, with increased vascular amyloid, gliosis, and exaggerated neurocognitive deficits as a consequence of MnSOD heterozygosity in J20 APP transgenic mice (Table 4) (215). Another report showed gliosis and the development of pathological changes in glial morphology after intraperitoneal injection of manganese chloride in rats (Table 4) (216). Taken together, these reports indicate that MnSOD-induced mitochondrial oxidative stress affects AD-like pathologies, but clearly the effects vary according to the AD mouse model used. This raises the fact that Mn deficiency might increase the risk for AD. Mn nutritional status and homeostasis is an underexplored area and warrants more attention (Table 4).

3.5. Copper

3.5.1. Copper Metabolism

Cu is an essential component of countless enzymes and other proteins and it is critical for numerous vital reactions. Cu homeostasis is complex and the mechanisms of Cu trafficking, metabolism, and sequestration are currently being studied (217). A healthy human contains about 110 mg of Cu, with about 9 mg present in the brain (218, 219). There is a continual turnover of Cu, with most of it being recycled. Daily diets have around 0.6–1.6 mg of Cu

Table 4
Animal models of Manganese-induced dementia

Reference	Experimental model	Exposure protocol	Effects
(216)	Wistar rats	Manganese (II) chloride (50 mg/kg body weight, i.p.) once daily for 1 or 4 days	Increases in manganese levels of up to 232%, 523%, and 427% in the cerebral cortex, globus pallidus, and cerebellum; Development of pathological changes in glial morphology identified as Alzheimer type II astrocytosis in both cortical and subcortical structures
(213)	MnSOD null mice (a) and heterozygous MnSOD knockout mice combined with a well-characterized mouse model of AD (Tg2576) (b)	High and low doses of a catalytic antioxidant	(a) Striking elevations in the levels of tau phosphorylation (at Ser-396 and other phospho-epitopes of tau) in the low-dose antioxidant treated mice at AD-associated residues (b) Exacerbated amyloid burden and significantly reduced metal levels in brain Increased levels of Ser-396 phosphorylated tau
(215)	Human amyloid precursor protein (hAPP) transgenic mice	Inactivation of one MnSOD allele (MnSOD(±)) reducing MnSOD activity to approximately 50%	Decreased amyloid plaques in the brain parenchyma Development of cerebrovascular amyloidosis, gliosis, and plaque-independent neuritic dystrophy
(214)	Mice with a knockout of one allele of MnSOD crossed with Tg19959 mice (overexpression of a doubly mutated human APP)		Elevated oxidative stress and significantly increased brain Abeta levels and Abeta plaque burden

(contained especially in meat), but only about half of this ingested metal is absorbed into the small intestine (220) and then transported into blood. Cu transport is dependent on its own oxidation state: only Cu reduced by specific metalloreductases, i.e. Cu(I), can be transported. About 90% of blood Cu in humans is covalently bound to ceruloplasmin, while the remaining 10% is free or loosely bound to albumin and other molecules. Excess of free Cu (free or noncovalently bound Cu) is toxic to cells because of its ability to promote uncontrolled forms of reactive species through Fenton chemistry (Cu is a transition metal). It is therefore essential that mechanisms exist to sequestrate free intracellular Cu. Metallothionein is a cellular Cu-binding protein, which provides Cu storage in the liver, giving significant protection against high free Cu levels. It is a regulatory mechanism, which balances Cu import and export to ensure proper nutrient levels while minimizing toxic effects (221). Several neurodegenerative diseases are characterized by altered Cu homeostasis, which directly or indirectly increases oxidative stress, an important factor in neuronal toxicity (222). In particular, growing evidence supports a central role for altered Cu metabolism at many steps in development and progression of AD (155).

3.5.2. Copper and Amyloid Precursor Protein

APP and Aβ peptide are in fact strongly implicated in Cu metabolism. APP seems to have two Cu binding domains (CuBD) in the N-terminal region (223), including one in the Aβ sequence (224). Overexpression of APP in transgenic mice appears to reduce levels of Cu in the brain (225), while APP knockout mice have increased intracellular accumulation of Cu (226). These findings have led to suggestions that APP may be a Cu transporter (227). The interaction between Cu ions and CuBD can modulate Aβ production. In an *in vitro* model, Borchardt and colleagues showed that cells overexpressing APP reduced Aβ production after extracellular Cu(II) treatment (228). This reduction could be mediated by alterations in certain signaling mechanisms. For example, Scheuermann and colleagues showed that Aβ production increased when APP crosslinks to form dimers (229). CuBD is the most important site of dimerization of APP (230) and favors dimer formation in the absence of Cu ions: the effect of Cu(II) in lowering Aβ production may be elicited through a shift of monomer–dimer equilibrium to favor monomer forms of APP, which does not help Aβ production (231).

3.5.3. Copper Binds Aβ Peptide and Increases Oxidative Stress

The cores of senile plaques consist of aggregated Aβ peptides in which Cu levels are high (232). This accumulation of extracellular Cu is probably mediated by Cu binding to the Aβ peptide (Aβ has either high or low affinity Cu binding sites) (233). Cu binding to Aβ can increase amyloid peptide toxicity in two ways; first Maynard and colleagues showed that low levels of this metal can also induce an aggregation of synthetic peptide *in vitro* (234). Moreover, Cu binding to Aβ (Cu(II)–Aβ complex) catalyzes the production of hydrogen peroxide from O_2 through reduction of Cu(II) to Cu(I) (235, 236). This production might be significant for oxidative stress mediated neuronal dysfunction because H_2O_2 can diffuse through cell membranes, oxidizing intracellular proteins and lipids and generating highly toxic hydroxyl radical species (223, 237). Intracellular Cu appears less frequent in AD brains compared to controls (238). Oxidative stress due to Cu(II)–Aβ complex formation might deteriorate due to decreased or functionally deficient intracellular Cu. Cu bound in extracellular plaque deposits might lead to depleted intracellular Cu levels, so reducing the activity of several antioxidant cuproenzymes (superoxide dysmutase-1 (SOD-1), cytochrome C oxidase (COX), etc.) in the brain (239, 240). A 2003 study showed that Cu-supplemented sucrose-sweetened drinking water for 3 months restored superoxide dismutase-1 activity in transgenic APP23 mouse brains (Table 5) (241). There is probably a symbiotic process of increased oxidative stress coupled with weakened defense mechanisms that could contribute to neuronal damage. Therefore, Cu supplements could be a therapeutic approach to AD treatment. Alternatively, agents that redistribute Cu from extracellular space to the intracellular environment restoring Cu homeostasis also have potential as an AD therapy. Clioquinol (a Cu–Zn chelator) has received considerable attention because of its ability to cross the BBB, bind Cu(II), and rapidly dissolve Aβ peptides (242). However, a recent study showed that in young APP transgenic mice, clioquinol treatment was lethal and Cu administration through food caused clioquinol toxicity (Table 5) (243). In animal studies, Larry Sparks found that trace elements of Cu in drinking water in an AD rabbit model greatly enhanced Aβ accumulation in the brain and increased learning deficits (244). Moreover, Lee and colleagues found markedly reduced amyloid plaque burden and cerebral amyloid angiopathy in Tg2576 mice brains treated with daily DP-109, a metal (and in particular Cu) chelator (Table 5) (245). Contradicting these findings, recent data indicate that AD patients have higher Cu levels in the plasma and lower in the brain (238) and an early study did not find any differences between Cu levels between AD and control patients (246). In conclusion, Cu involvement in AD pathogenesis has not yet been definitively established (Table 5).

3.6. Zinc

3.6.1. Zinc Metabolism

Zn is the second most abundant trace element in the body after iron and considerably more abundant than Cu (247), but its presence in nature does not exceed 0.02% (248). Zn is a catalyst, structural and regulatory ion. It is essential for the activity of more than 300 enzymes influencing various organ functions. In addition, it is involved in DNA replication and transcription, and it is indispensable for the functionality of transcription factors that have "zinc finger domains." These Zn-binding domains are involved in most cellular functions, important in maintaining the homeostatic equilibrium of the organism (249). Therefore, a wide range of symptoms develops as a consequence of Zn deficiency (for example, distorted sensory function involving taste, smell, and vision) (250, 251). The body does not store Zn, so a constant dietary intake (especially from animal proteins and red meat) is necessary.

Table 5
Animal models of copper-induced dementia

Reference	Experimental model	Exposure protocol	Effects
(244)	New Zealand cholesterol-fed white rabbits	Distilled drinking water supplemented with 0.12 ppm copper ion (as copper sulfate)	Consistently increase in the numbers of damaged neurons and the intensity of neuronal Abeta immunoreactivity Copper-induced decrease in the clearance of overproduced Abeta resulting in memory deficits
(245)	Aged female hAbetaPP-transgenic Tg2576 mice	Daily administration with the lipophilic metal chelator DP-109 for 3 months	Markedly reduction of burden of amyloid plaques and degree of cerebral amyloid angiopathy in brains DP-109 treatment facilitates the transition of Abeta from insoluble to soluble forms in the cerebrum.
(243)	Young APP transgenic mice (a) and nontransgenics (b)	Animals fed until the age of 14 weeks with food pellets containing 356 mg/kg $CuSO_4 \times 5H_2O$ per day	Cu rescued the lethality of APP transgenics on clioquinol (CQ) treatment Cu treatment reduced plasma iron levels
(238)	Transgenic APP23 mice (12, 15, and 18 month of age)	Oral treatment for 3 months with Cu-supplemented sucrose-sweetened drinking water (treatment solution with 50 g/L sucrose and 0.25 g/L Cu–sulfate $\times 5H_2O$)	Cu restored the superoxide dismutase-1 activity in transgenic mouse brain Significant increase of brain Cu Cu treatment lowered endogenous CNS Abeta before a detectable reduction of amyloid plaques

In fact, Zn deficiency is widespread in the elderly who tend to have an inappropriate diet (252). On the contrary, toxicity associated with excessive exposure to Zn is less well known. Unlike other metals, homeostatic mechanisms regulating entry and excretion of Zn from cells are so efficient that no disorders are known to be associated with its excessive accumulation (253). Most (85–90%) body Zn is tightly bound within cells with metalloenzymes and Zn binding proteins and constituted the "fixed zinc pool." This pool turns over very slowly and is mainly responsible for housekeeping functions in cellular metabolism. The remaining 10–15% (labile Zn) comprises more dynamic pools that are readily depleted during Zn deficiency (254). Fixed Zn is uniformly distributed in the body, while labile Zn is concentrated in certain tissues and specific regions (255). The brain, for example, has the highest Zn content.

3.6.2. Zinc Distribution and Functions in Brain

Average total brain Zn is approximately 150 μmol/L (tenfold Zn serum levels) (256), with nanomolar concentrations in neuronal cytosol and brain extracellular fluids (257), and millimolar concentrations in the synaptic vesicles (258). The transportation of Zn into the brain parenchyma occurs via the BBB system (259). To maintain Zn homeostasis, brain capillary endothelial cells respond to changes in Zn status increasing or decreasing the uptake of this metal, respectively, in high or low Zn concentrations (260). Zn can be transferred freely through brain extracellular fluid compartments (259), but Zn uptake into neurons and glial cells involves transporters of the Zip Family (261). Zn distribution in the brain is not uniform. Higher Zn concentrations are found in the hippocampus, amygdala, and neocortex (262). In these regions, as in other regions of the brain, neurons (mainly inhibitory GABAergic neurons) are present that contain free Zn ions in the vesicles of their presynaptic boutons. These regions and neurons are generally termed "zinc-enriched" (263). The physiological role of synaptic Zn is not clear, but it may be involved in the modulation and release of glutamate, GABA, and glycine (264, 265) and it may influence the behavior of postsynaptic membrane receptors and ion channels. Zn exerts its modulatory action not only directly binding postsynaptic receptors, but also entering postsynaptic neurons and then exerting physiological functions like an intracellular signal similar to neurotransmitters (266). The largest source of neuropathological Zn comes from synaptic/vesicular Zn in the enriched regions of the brain (267). Current research suggests that Zn-binding metalloproteins (like metallothioneins) are a major source of the nonvesicular Zn pool (268). Some of these metalloproteins (like α2-macroglobulin) have been directly or indirectly associated with AD pathogenesis (269). Increased α2-macroglobulin binding to Aβ in the presence of Zn (270) and the abnormal expression of metallothionein in the AD brain (271) support this hypothesis. Increased levels of Zn bound to protein aggregates in AD brain might decrease Zn ion availability both in the brain and in the periphery system with subsequent impaired activity of the critical Zn-dependent enzymes.

3.6.3. Neurotoxicity of Zinc

Zn is relatively nontoxic compared to other transition metals such as Mn and Fe. However, direct evidence in mature cortical cell cultures exposed to several hundred micromolar concentrations of Zn shows that this metal might be a relatively potent, rapidly acting neurotoxin, and somewhat less potent gliotoxin (272). The influx of toxic Zn from presynaptic vesicles into postsynaptic degenerating neurons seems to be mainly responsible for the neurodegenerative process (273, 274), which in turn can be prevented by intraventricular injections of Zn-chelating agents (275). The mechanism by which the Zn ion exerts its potent neurotoxic effects involves many signaling pathways, including mitochondrial superoxide production (276), extramitochondrial generation of ROS, and disruption of metabolic enzyme activity. These ultimately activate apoptotic or necrotic processes (277). Research indicates that Zn metabolism is altered in AD (269). Previous studies that attempted to quantify cerebral Zn levels in AD have had highly variable results. Some researchers reported increased Zn concentrations in AD patients, especially in the hippocampus, the main brain region involved in AD neurodegeneration (278, 279), while others did not report any significant increase (280, 281), or even decreased Zn levels (282).

3.6.4. Zinc and Alzheimer's Disease

The role of Zn in AD pathogenesis was intensively studied when it was shown that APP had a specific and saturable Zn-binding site (283). Several Zn effects on APP have been observed, such as the modulation of APP interaction with extracellular matrix proteins (284, 285) and the inhibition of α-secretase cleavage of APP influencing Aβ peptide generation from APP (286). Moreover, Zn can increase the biological half-life of Aβ by protecting the peptide from proteolytic attack (22). Apart from these findings, there have been no reports to date which show that Zn binding to APP increases Aβ production *in vivo*. Several studies showed that Zn above 300 nM precipitates Aβ and induces Aβ aggregation in senile plaques (22). During synaptic transmission, the extracellular concentration of Zn rises to 300 μM with a possible contribution to Aβ deposition (287). Interestingly, Zn preserves the α-helical conformation of $A\beta_{1-40}$ and its aggregation is completely reversed by Zn chelation (288). Yoshiike and colleagues reported that Zn-induced Aβ precipitates are non-β-sheet and nonfibrillar in structure (289). The group reported that when Zn–Aβ is incubated in cultured cells, it confers cytoprotection. It has been shown that Zn-induced Aβ aggregates at physiological pH differ from Cu-induced Aβ aggregates in terms of their relative density and resolubilization state (Zn–Aβ aggregates are denser and harder to resolubilize than Cu–Aβ aggregates) (290). Zn appears to have a dual effect on Aβ cytotoxicity. In cultured primary hippocampal neurons, Zn is a double-edged sword, offering protection against Aβ toxicity at low concentrations and increasing toxicity at high concentrations (291). The exact mechanism of this protective effect is unclear, but these findings suggest that Zn in AD pathophysiology is not completely negative. Cuajungo and colleagues argued that Zn confers its protective effect by competing with Cu and/or Fe. Zn changes Aβ conformation to the extent that Cu and Fe ions cannot access their metal binding sites (292). These authors suggest that Zn-induced Aβ

Table 6
Animal models of zinc-induced dementia

Reference	Experimental model	Exposure protocol	Effects
(293)	APP/PS1 transgenic mouse model of AD	Animals fed with low dietary zinc for 3 months	Synaptic zinc distribution within cortical layers is paralleled by amyloid burden Prominent ZnT3 (a zinc transporter) immunoreactivity in dystrophic neurites within amyloid plaques Significant increase in total plaque volume (ca 25%) Level of oxidized proteins in brain tissue did not change

aggregation may be physiologically normal and beneficial for AD by protecting Aβ from oxidative interactions with Cu and Fe (292). Stoltenberg and his group showed a significant increase in total plaque volume (approx. 25%) in brain of APP/PS1 transgenic mice fed with low dietary zinc for 3 months (Table 6). Moreover, they showed prominent ZnT3 (a Zn transporter) immunoreactivity in dystrophic neurites within amyloid plaques (Table 6), indicating that Zn in amyloid plaques could be important (293). The biochemical relation between Aβ toxicity and oxidative stress in AD is quite complex and not fully understood. Metabolic signs of oxidative stress such as oxygen radical-mediated damage of brain proteins, lipids, and nucleic acids have been observed in AD (294, 295). Furthermore, synthetic Aβ peptides have been shown to induce lipid peroxidation of synaptosomes (296), and to be cytotoxic through mechanism that involve the generation of cellular superoxide radical and H_2O_2 (297) causing cell death influenced by transition metals (298). Zn is a redox inert metal and does not participate in oxidation–reduction chemistry. However, Zn does possess an indirect antioxidant property (299). Emerging evidence suggests that H_2O_2 production is central to Aβ-induced cytotoxicity (297). Recently, Huang and colleagues discovered that Aβ directly generates H_2O_2 through Cu(II)/Fe(II) reduction, and Zn(II) quenches H_2O_2 formation by Aβ. This may reflect a homeostatic mechanism to prevent excessive H_2O_2 production in the vicinity of Aβ accumulation (298). Although dietary Zn supplementation studies need further validation, some reports show that when taken daily (30 mg/kg for 1 year), Zn can delay cognitive decline found in AD (300). Nonetheless, further studies are needed in this area (Table 6).

4. Conclusions

Numerous experimental studies and clinical data suggest that brain biometal dysregulation (often due to environmental metal exposure) contributes to AD pathogenesis. AD is a neurodegenerative disease, in which clinical impairment correlates to the progressive loss of specific neuronal populations. Typical lesions include protein aggregates, which are vulnerable to oxidative stress and participate in free radical oxygen chemistry because of their high metal content. It has been shown that the dyshomeostasis of metal ions is pathogenetic in many neurodegenerative conditions. However, it is still unclear which types of early metabolic changes occur and which types are simply secondary signs of neuronal degeneration or effects of oxidative stress. The role of metal ions in mediating oxidative stress reflects the complex interaction between both redox-active and redox-inactive metals in competing with each other for available metal-binding proteins that may either enhance or reduce redox activity. So, when considering metal chelation as a possible therapy for AD, it is important to recognize the dual action of metals in catalyzing both prooxidant and antioxidant reactions. Certainly, improved animal and cellular model systems are needed that can be used to test pharmacological agents as prospective therapies.

Acknowledgments

I thank Dr. E. Foglio for her help in the preparation of the manuscript and Dr. R. Coates for the English revision.

References

1. Coppedè F, Mancuso M, Siciliano G et al (2006) Genes and the environment in neurodegeneration. Biosci Rep 26:341–367
2. Cummings JL, Vinters HV, Cole GM et al (1998) Alzheimer's disease: etiologies, pathophysiology, cognitive reserve, and treatment opportunities. Neurology 51:2–17; discussion 65–67
3. Masters CL, Simms G, Weinman NA et al (1985) Amyloid plaque core protein in Alzheimer disease and Down syndrome. Proc Natl Acad Sci USA 82:4245–4249
4. Poirier J, Delisle MC, Quirion R et al (1995) Apolipoprotein E4 allele as a predictor of cholinergic deficits and treatment outcome in Alzheimer disease. Proc Natl Acad Sci USA 92:12260–12264
5. Farrer LA, Cupples LA, Haines JL et al (1997) Effects of age, sex, and ethnicity on the association between apolipoprotein E genotype and Alzheimer disease. A meta-analysis. APOE and Alzheimer Disease Meta Analysis Consortium. JAMA 278: 1349–1356
6. Rapoport SI, Pettigrew KD, Schapiro MB (1991) Discordance and concordance of dementia of the Alzheimer type (DAT) in monozygotic twins indicate heritable and sporadic forms of Alzheimer's disease. Neurology 41:1549–1553
7. Platt B (2006) Experimental approaches to assess metallotoxicity and ageing in models of Alzheimer's disease. J Alzheimers Dis 10: 203–213

8. Smith MA, Sayre LM, Monnier VM et al (1995) Radical ageing in Alzheimer's disease. Trends Neurosci 18:172–176
9. Snowdon DA, Greiner LH, Mortimer JA et al (1997) Brain infarction and the clinical expression of Alzheimer disease. The Nun Study JAMA 277:813–817
10. Lye TC, Shores EA (2000) Traumatic brain injury as a risk factor for Alzheimer's disease: a review. Neuropsychol Rev 10:115–129
11. Smith CG, Veenhuis PE, Meyer RE (2001) Traumatic brain injury. North Carolina's challenge. N C Med J 62:328–334
12. Horsburgh K, McCarron MO, White F et al (2000) The role of apolipoprotein E in Alzheimer's disease, acute brain injury and cerebrovascular disease: evidence of common mechanisms and utility of animal models. Neurobiol Aging 21:245–255
13. Arshavsky YI (2006) Alzheimer's disease, brain immune privilege and memory: a hypothesis. J Neural Transm 113:1697–1707
14. Dobson CB, Wozniak MA, Itzhaki RF (2003) Do infectious agents play a role in dementia? Trends Microbiol 11:312–317
15. Cooper AJ, Kristal BS (1997) Multiple roles of glutathione in the central nervous system. Biol Chem 378:793–802
16. Hazel JR, Williams EE (1990) The role of alterations in membrane lipid composition in enabling physiological adaptation of organisms to their physical environment. Prog Lipid Res 29:167–227
17. Benzi G, Moretti A (1995) Are reactive oxygen species involved in Alzheimer's disease? Neurobiol Aging 16:661–674
18. Sayre LM, Perry G, Atwood CS et al (2000) The role of metals in neurodegenerative diseases. Cell Mol Biol (Noisy-le-grand) 46: 731–741
19. Huang X, Atwood CS, Moir RD et al (2004) Trace metal contamination initiates the apparent auto-aggregation, amyloidosis, and oligomerization of Alzheimer's Abeta peptides. J Biol Inorg Chem 9:954–960
20. Lovell MA, Robertson JD, Teesdale WJ et al (1998) Copper, iron and zinc in Alzheimer's disease senile plaques. J Neurol Sci 158:47–52
21. Atwood CS, Moir RD, Huang X et al (1998) Dramatic aggregation of Alzheimer abeta by Cu(II) is induced by conditions representing physiological acidosis. J Biol Chem 273: 12817–12826
22. Bush AI, Pettingell WH, Multhaup G et al (1994) Rapid induction of Alzheimer A beta amyloid formation by zinc. Science 265: 1464–1467
23. Huang X, Atwood CS, Moir RD et al (1997) Zinc-induced Alzheimer's Abeta1–40 aggregation is mediated by conformational factors. J Biol Chem 272:26464–26470
24. Cherny RA, Legg JT, McLean CA et al (1999) Aqueous dissolution of Alzheimer's disease Abeta amyloid deposits by biometal depletion. J Biol Chem 274:23223–23228
25. Cherny RA, Atwood CS, Xilinas ME et al (2001) Treatment with a copper-zinc chelator markedly and rapidly inhibits beta-amyloid accumulation in Alzheimer's disease transgenic mice. Neuron 30:665–676
26. Dong J, Atwood CS, Anderson VE et al (2003) Metal binding and oxidation of amyloid-beta within isolated senile plaque cores: Raman microscopic evidence. Biochemistry 42:2768–2773
27. Bishop GM, Robinson SR (2004) The amyloid paradox: amyloid-beta-metal complexes can be neurotoxic and neuroprotective. Brain Pathol 14:448–452
28. Sengupta P, Garai K, Sahoo B et al (2003) The amyloid beta peptide (Abeta(1–40)) is thermodynamically soluble at physiological concentrations. Biochemistry 42:10506–10513
29. Rogers JT, Randall JD, Cahill CM et al (2002) An iron-responsive element type II in the 5'-untranslated region of the Alzheimer's amyloid precursor protein transcript. J Biol Chem 277:45518–45528
30. Multhaup G, Schlicksupp A, Hesse L et al (1996) The amyloid precursor protein of Alzheimer's disease in the reduction of copper(II) to copper(I). Science 271:1406–1409
31. Loring JF, Wen X, Lee JM et al (2001) A gene expression profile of Alzheimer's disease. DNA Cell Biol 20:683–695
32. Colangelo V, Schurr J, Ball MJ et al (2002) Gene expression profiling of 12633 genes in Alzheimer hippocampal CA1: transcription and neurotrophic factor down-regulation and up-regulation of apoptotic and pro-inflammatory signaling. J Neurosci Res 70: 462–473
33. Nunomura A, Perry G, Aliev G et al (2001) Oxidative damage is the earliest event in Alzheimer disease. J Neuropathol Exp Neurol 60:759–767
34. Becaria A, Bondy SC, Campbell A (2003) Aluminum and copper interact in the promotion of oxidative but not inflammatory events: implications for Alzheimer's disease. J Alzheimers Dis 5:31–38
35. Ke Y, Qian ZM (2003) Iron misregulation in the brain: a primary cause of neurodegenerative disorders. Lancet Neurol 2:246–253

36. Hardy J, Selkoe DJ (2002) The amyloid hypothesis of Alzheimer's disease: progress and problems on the road to therapeutics. Science 297:353–356
37. Higuchi M, Iwata N, Matsuba Y et al (2005) 19F and 1H MRI detection of amyloid beta plaques in vivo. Nat Neurosci 8:527–533
38. Exley C (2004) The pro-oxidant activity of aluminium. Free Radic Biol Med 36:380–387
39. Golub MS, Domingo JL (1996) What we know and what we need to know about developmental aluminum toxicity. J Toxicol Environ Health 48:585–597
40. Jansson ET (2001) Aluminum exposure and Alzheimer's disease. J Alzheimers Dis 3:541–549
41. Gupta VB, Anitha S, Hedge ML et al (2004) Aluminium in Alzheimer's disease: are we still at a crossroad? Cell Mol Life Sci 62:143–158
42. Spafforth J (1921) Case of Aluminium poisoning. Lancet 1:1301
43. McLaughlin AIG, Kazantzis G, King E et al (1962) Pulmonary fibrosis and encephalopathy associated with the inhalation of aluminium dust. Br J Ind Med 19:253–263
44. Alfrey AC, Legendre GR, Kahny WD (1976) The dialysis encephalopathy syndrome, Possible aluminium intoxication. N Engl J Med 294: 184–188
45. Rifat SL, Eastwood MR, McLachlan DR et al (1990) Effect of exposure of miners to aluminium powder. Lancet 336:1162–1165
46. Wakayama I, Nerurkar VR, Strong MJ et al (1996) Comparative study of chronic aluminum-induced neurofilamentous aggregates with intracytoplasmic inclusions of amyotrophic lateral sclerosis. Acta Neuropathol (Berl) 92:545–554
47. Hirsch EC, Brandel JP, Galle P et al (1991) Iron and aluminium increase in the substantia nigra of patients with Parkinson's disease; an X-ray microanalysis. J Neurochem 56:446–451
48. Exley C (1999) A molecular mechanism of aluminium-induced Alzheimer's disease? Inorg Biochem 76:133–140
49. Good PF, Perl DP, Bierer LM et al (1992) Selective accumulation of aluminium and iron in the neurofibrillary tangles of Alzheimer's disease: A laser microprobe (LAMMA) study. Ann Neurol 31:286–292
50. Moore PB, Day JP, Taylor GA et al (2000) Absorption of aluminium-26 in Alzheimer's disease, measured using accelerator mass spectrometry. Dement Geriatr Cogn Disord 11: 66–69
51. Whitehead MW, Farrar G, Christie GL et al (1997) Mechanisms of aluminum absorption in rats. Am J Clin Nutr 65:1446–1452
52. Amstrong RA, Anderson J, Cowburn JD et al (1992) Aluminium administered in drinking water but not in the diet influences biopterin metabolism in the rodent. Biol Chem Hoppe Seyler 373:1075–1078
53. Crichton RR, Wilmet S, Legssyer R et al (2002) Molecular and cellular mechanisms of iron homeostasis and toxicity in mammalian cells. J Inorg Biochem 91:9–18
54. Bi S (1996) A model describing the complexing effect in the leaching of aluminum from cooking utensils. Environ Pollut 92:85–89
55. Birchall JD (1992) The interrelationship between silicon and aluminium in the biological effects of aluminium. Ciba Found Symp 169:50–61; discussion 61–68
56. Harrington CR, Wischik CM, McArthur FK et al (1994) Alzheimer's-disease-like changes in tau protein processing: association with aluminium accumulation in brains of renal dialysis patients. Lancet 343:993–997
57. Trapp GA (1986) Interactions of aluminum with cofactors, enzymes, and other proteins. Kidney Int Suppl 18:S12–S16
58. Macdonald TL, Martin RB (1988) Aluminum ion in biological systems. Trends BiochemSci 13:15–19
59. Dlugaszek M, Fiejka MA, Graczyk A et al (2000) Effects of various aluminium compounds given orally to mice on Al tissue distribution and tissue concentrations of essential elements. Pharmacol Toxicol 86:135–139
60. Cserr HF (1988) Role of secretion and bulk flow of brain interstitial fluid in brain volume regulation. Ann N Y Acad Sci 529:9–20
61. Yokel RA, Allen DD, Ackley DC (1999) The distribution of aluminium into and out of the brain. J Inorg Biochem 76:127–132
62. Yokel RA (2000) The toxicology of aluminum in the brain: a review. Neurotoxicology 21:813–828
63. Exley C, Burgess E, Day JP et al (1996) Aluminium toxicokinetics. J Toxicol Environ Health 48:569–584
64. Banks WA, Kastin AJ, Fasold MB (1998) Differential effect of aluminum on the blood-brain barrier transport of peptides, technetium and albumin. J Pharmacol Exp Ther 244:579–585
65. Roskams AJ, Connor JR (1990) Aluminum access to the brain: a role for transferrin and its receptor. Proc Natl Acad Sci U S A 87: 9024–9027
66. Walton J, Tuniz C, Fink D et al (1995) Uptake of trace amounts of aluminum into the brain from drinking water. Neurotoxicology 16:187–190

67. Perl DP, Good PF (1987) Uptake of aluminium into central nervous system along nasal-olfactory pathways. Lancet 1:1028
68. Florence AL, Gauthier A, Ponsar C et al (1994) An experimental animal model of aluminium overload. Neurodegeneration 3:315–323
69. Gerhart DZ, Enerson BE, Zhdankina OY et al (1997) Expression of monocarboxylate transporter MCT1 by brain endothelium and glia in adult and suckling rats. Am J Physiol 273: E207–E213
70. Ackley DC, Yokel RA (1998) Aluminum transport out of brain extracellular fluid is proton dependent and inhibited by mersalyl acid, suggesting mediation by the monocarboxylate transporter (MCT1). Toxicology 127:59–67
71. De Stasio G, Mercanti D, Ciotti MT et al (1994) Aluminium in rat cerebellar primary cultures: glial cells and GABAergic neurones. Neuoreport 5:1973–1976
72. Campbell A, Prasad KN, Bondy SC (1999) Aluminum-induced oxidative events in cell lines: glioma are more responsive than neuroblastoma. Free Radic Biol Med 26:1166–1167
73. Fattoretti P, Bertoni-Freddari C, Balietti M et al (2004) Chronic aluminum administration to old rats results in increased levels of brain metal ions and enlarged hippocampal mossy fibers. Ann N Y Acad Sci 1019:44–47
74. Boegman RJ, Bates LA (1984) Neurotoxicity of aluminium. Can J Physiol Pharmacol 62:1010–1014
75. Jope RS, Johnson GVW (1992) Neurotoxic effects of dietary aluminium. Ciba Found Symp 169:254–267
76. Becaria A, Campbell A, Bondy SC (2002) Aluminum as a toxicant. Toxicol Ind Health 18:309–320
77. Provan SD, Yokel RA (1992) Aluminium inhibits glutamate release from transverse rat hippocampal slices: role of G proteins, Ca channels and protein Kinase C. Neurotoxicology 13:413–420
78. Platt B, Haas H, Busselberg D (1994) Aluminium reduces glutamate-activated currents of rat hippocampal neurones. Neuroreport 5:2329–2332
79. Cucarella C, Hermenegildo C, Saez R et al (1998) Chronic exposure to aluminium impairs neuronal glutamate-nitric oxide-cyclic GMP pathway. J Neurochem 70:1609–1614
80. Hermenegildo C, Saez R, Minoia C et al (1999) Chronic exposure to aluminium impairs the glutamate-nitric oxide-cyclic GMP pathway in the rat in vivo. Neurochem Int 34:245–253
81. Bidmon HJ, Emde B, Kowalski T et al (2001) Nitric oxide synthase-I containing cortical interneurons co-express antioxidative enzymes and anti-apoptotic Bcl-2 following focal ischemia: evidence for direct and indirect mechanisms towards their resistance to neuropathology. J Chem Neuroanat 22:167–184
82. Lam HHD, Hanley DF, Trapp BD et al (1996) Induction of spinal cord neuronal nitric oxide synthase (NOS) after formalin injection in the rat hind paw. Neurosci Lett 210:201–204
83. Hawkins RD (1996) NO honey, I don't remember. Neuron 16:465–467
84. Boulton CL, Southam E, Garthwaite J (1995) Nitric oxide dependent long-term potentiation is blocked by a specific inhibitor of soluble guanylyl cyclase. Neuroscience 69:699–703
85. El-Rahman SS (2003) Neuropathology of aluminum toxicity in rats (glutamate and GABA impairment). Pharmacol Res 47:189–194
86. Hope BT, Micheal GJ, Knigge KM et al (1991) Neuronal NADPH-diaphorase is a nitric oxide synthase. Proc Natl Acad Sci USA 88: 2811–2820
87. Rodella L, Rezzani R, Agostini C et al (1995) Expression of NADPH-diaphorase and colocalization with Fos in the brain neurons of rat following visceral noxious stimulation. Brain Res 697:258–261
88. Rodella L, Rezzani R, Lanzi R et al (2001) Chronic exposure to aluminium decrease NADPH-diaphorase positive neurons in the rat cerebral cortex. Brain Research 889:229–233
89. Abbracchio MP, Brambilla R, Burnstock G et al (1999) Cyclo-oxygenase-2 mediates P2Y receptor-induced reactive astrogliosis. Br J Pharmacol 126:563–567
90. Barnard EA, Simon J, Webb TE (1997) Nucleotide receptors in the nervous system. An abundant component using diverse transduction mechanisms. Mol Neurobiol 15:103–129
91. Di Virgilio F, Chiozzi P, Falzoni S et al (1998) Cytolytic P2X purinoceptors. Cell Death Differ 5:191–199
92. Zinchuk VS, Okada T, Kobayashi T et al (1999) Ecto-ATPase activity in cerebellum: implication to the function of synaptic transmission. Brain Res 815:111–115
93. Wang X, Gruenstein EI (1997) Mechanism of synchronized Ca^{2+} oscillations in cortical neurons. Brain Res 767:239–249
94. Queiroz G, Gebicke-Haerter PJ, Schobert A et al (1997) Release of ATP from cultured rat astrocytes elicited by glutamate receptor activation. Neuroscience 78:1203–1208

95. Beal MF, Hyman BT, Koroshetz W (1993) Do defects in mitochondrial energy metabolism underlie the pathology of neurodegenerative diseases? Trends Neurosci 16:125–131
96. Tao Z, Van Gool D, Lammens M et al (1999) NADPH-diaphorase-containing neurons in cortex, subcortical white matter and neostriatum are selectively spared in Alzheimer's disease. Dement Geriatr Cogn Disord 10: 460–468
97. Gargiuolo L., Bermejo M, Liras A (2000) Reduced neuronal nitric oxide synthetase and c-protein kinase levels in Alzheimer's disease. Rev Neurol 30:301–303
98. Rodella LF, Ricci F, Borsani E et al (2006) Exposure to aluminium changes the NADPH-diaphorase/NPY pattern in the rat cerebral cortex. Arch Histol Cytol 69:13–21
99. Solfrizzi V, Panza F, Capurso A (2003) The role of diet in cognitive decline. J Neural Transm 110:95–110
100. Exley C, Esiri MM (2006) Severe cerebral congophilic angiopathy coincident with increased brain aluminium in a resident of Camelford, Cornwall, UK. J Neurol Neurosurg Psychiatry 77:877–879
101. Zatta P, Ibn-Lkhayat-Idrissi M, Zambenedetti P et al (2002) In vivo and in vitro effects of aluminum on the activity of mouse brain acetylcholinesterase. Brain Res Bull 59:41–45
102. Gutteridge JM, Quinlan GJ, Clark I et al (1985) Aluminium salts accelerate peroxidation of membrane lipids stimulated by iron salts. Biochim Biophys Acta 835:441–447
103. Verstraeten SV, Oteiza PI (2000) Effects of Al(3+) and related metals on membrane phase state and hydration: correlation with lipid oxidation. Arch Biochem Biophys 375:340–346
104. Meglio L, Oteiza PI (1999) Aluminum enhances melanin-induced lipid peroxidation. Neurochem Res 24:1001–1008
105. Kong S, Liochev S, Fridovich I (1992) Aluminum(III) facilitates the oxidation of NADH by the superoxide anion. Free Radic Biol Med 13:79–81
106. Méndez-Alvarez E, Soto-Otero R, Hermida-Ameijeiras A et al (2002) Effects of aluminum and zinc on the oxidative stress caused by 6-hydroxydopamine autoxidation: relevance for the pathogenesis of Parkinson's disease. Biochim Biophys Acta 1586:155–168
107. Yuan Y, Guo JZ, Zhou QX (2008) The homeostasis of iron and suppression of HO-1 involved in the protective effects of nimodipine on neurodegeneration induced by aluminum overloading in mice. Eur J Pharmacol 586:100–105
108. Rodella LF, Ricci F, Borsani E et al (2008) Aluminium exposure induces Alzheimer's disease-like histopathological alterations in mouse brain. Histol Histopathol 23: 433–439
109. Evans P, Harrington C (1998) Aluminosilicate particulate and beta-amyloid in vitro interactions: a model of Alzheimer plaque formation. Biochem Soc Trans 26:S251
110. Ribes D, Colomina MT, Vicens P et al (2008) Effects of oral aluminum exposure on behavior and neurogenesis in a transgenic mouse model of Alzheimer's disease. Exp Neurol 214:293–300
111. Clauberg M, Joshi JG (1993) Regulation of serine protease activity by aluminum: implications for Alzheimer's disease. Proc Natl Acad Sci U S A 90:1009–1012
112. Atwood CS, Huang X, Moir RD et al (2001) Neuroinflammatory responses in the Alzheimer's disease brain promote the oxidative post-translational modification of amyloid deposits. In: Iqbal K, Sisodia S, Winblad B (eds) Alzheimer's disease: advances in etiology, pathogenesis and therapeutics. Wiley, New York, pp 341–61
113. Julka D, Sandhir R, Gill KD (1995) Altered cholinergic metabolism in rat CNS following aluminum exposure: implications on learning performance. J Neurochem 65:2157–2164
114. Gómez M, Esparza JL, Cabré M et al (2008) Aluminum exposure through the diet: metal levels in AbetaPP transgenic mice, a model for Alzheimer's disease. Toxicology 249:214–219
115. Poss KD, Tonegawa S (1997) Heme oxygenase 1 is required for mammalian iron reutilization. Proc Natl Acad Sci USA 94:10919–10924
116. Wigglesworth JM, Baum H (1988) Iron-dependent enzymes in the brain. In: Youdim MBH (ed) Topics in neurochemistry and neuropharmacology. Taylor & Francis, London, pp 25–66
117. Connor JR, Menzies SL, St Martin SM et al (1990) Cellular distribution of transferrin, ferritin, and iron in normal and aged human brains. J Neurosci Res 27:595–611
118. Beard JL, Connor JR, Jones BC (1993) Iron in the brain. Nutr Rev 51:157–170
119. Hill JM (1985) Iron concentration reduced in ventral pallidum, globus pallidus, and substancia nigra by gaba-transaminase inhibitor, gamma- vinyl GABA. Brain Res 342:18–25

120. Ponka P (2004) Heredity causes of disturbed iron homeostasis in the central nervous system. Ann NY Acad Sci 1012:267–281
121. Connor JR, Menzies SL, St. Martin et al (1992) A histochemical study of iron, transferrin, and ferritin in Alzheimer's diseased brains. J Neurosci Res 31:75–83
122. Loeffler DA, Connor JR, Juneau PL et al (1995) Transferrin and iron in normal, Alzheimer's disease, and Parkinson's disease brain regions. J Neurochem 65:710–724
123. Smith MA, Sayre L, Perry G (1996) Is Alzheimer's a disease of oxidative stress? Alzheimer's Dis Rev 1:63–67
124. Zambenedetti P, De Bellis G, Biunno I et al (2003) Transferrin C2 variant does confer a risk for Alzheimer's disease in caucasians. J. Alzheimers Dis 5:423–427
125. Robson KJ, Lehmann DJ, Wimhurst VL et al (2004) Synergy between the C2 allele of transferrin and the C282Y allele of the haemochromatosis gene (HFE) as risk factors for developing Alzheimer's disease. J Med Genet 41:261–265
126. Moos T (1996) Immunohistochemical localization of intraneuronal transferrin receptor immunoreactivity in the adult mouse central nervous system. J Comp Neurol 375:675–692
127. Morris CM, Candy JM, Oakley AE et al (1989) Comparison of the regional distribution of transferrin receptors and aluminium in the forebrain of chronic renal dialysis patients. J Neurol Sci 94:295–306
128. Morris CM, Candy JM, Kerwin JM et al (1994) Transferrin receptors in the normal human hippocampus and in Alzheimer's disease. Neuropathol Appl Neurobiol 20: 473–477
129. Ford GC, Harrison PM, Rice DW et al (1984) Ferritin: design and formation of an ironstorage molecule. Philos Trans R Soc Lond B Biol Sci 304:551–565
130. Levi S, Yewdall SJ, Harrison PM et al (1992) Evidence that H- and L-chains have cooperative roles in the iron-uptake mechanism of human ferritin. J Biochem 288:591–596
131. Connor JR, Snyder BS, Arosio P et al (1995) A quantitative analysis of isoferritins in select regions of aged, Parkinsonian, and Alzheimer's diseased brains. J Neurochem 65:717–724
132. Fleming J, Joshi JG (1987) Ferritin: isolation of aluminum–ferritin complex from brain. Proc Natl Acad Sci USA 84:7866–7870
133. Floris G, Meddad R, Padiglia A et al (2000) The physiopathological significance of ceruloplasmin. Biochem Pharmacol 60: 1735–1741
134. Castellani RJ, Moreira PI, Liu G et al (2007) Iron: the Redox-active center of oxidative stress in Alzheimer disease. Neurochem Res 32(10):1640–1645
135. Pantopoulos K (2004) Iron metabolism and the IRE/IRP regulator system: an update. Ann NY Acad Sci 1012–1013
136. Hentze MW, Kuhn LC (1996) Molecular control of vertebrate iron metabolism: mRNA-based regulatory circuits operated by iron, nitric oxide, and oxidative stress. Proc Natl Acad Sci USA 93:8175–8182
137. Walden WE, Patino MM, Gaffield L (1989) Purification of a specific repressor of ferritin mRNA translation from rabbit liver. J Biol Chem 264:13765–13769
138. Chen OS, Blemings KP, Schalinske KL et al (1998) Dietary iron intake rapidly influences iron regulatory proteins, ferritin subunits and mitochondrial aconitase in rat liver. J Nutr 128:525–535
139. Yamanaka K, Minato N, Iwai K (1999) Stabilization of iron regulatory protein 2, IRP2, by aluminum. FEBS Lett 1999,462:216–220
140. Smith MA, Wehr K, Harris PLR et al (1998) Abnormal localization of iron regulatory protein in Alzheimer's disease. Brain Res 788:232–236
141. Halliwell B (2001) Role of free radicals in the neurodegenerative diseases: therapeutic implications for antioxidant treatment. Drugs Aging 18:685–716
142. Sayre LM, Smith MA, Perry G (2001) Chemistry and biochemistry of oxidative stress in neurodegenerative disease. Curr Med Chem 8:721–738
143. Bartzokis G, Tishler TA, Shin IS et al (2004) Brain ferritin iron as a risk factor for age at onset in neurodegenerative diseases. Ann NY Acad Sci 1012:224–236
144. Zecca L, Riederer P, Connor JA (2004) Iron, brain ageing and neurodegenerative disorders. Nat Rev Neurosci 5:863–873
145. Pinero DJ, Li N, Hu J et al (2001) The intracellular location of iron regulatory proteins is altered as a function of iron status in cell cultures and rat brain. J Nutr 131: 2831–2836
146. Sayre LM, Perry G, Harris PL et al (2000) In situ oxidative catalysis by neurofibrillary tangles and senile plaques in Alzheimer's disease: a central role for bound transition metals. J Neurochem 74:270–279
147. Castellani RJ, Honda K, Zhu X et al (2004) Contribution of redox-active iron and copper to oxidative damage in Alzheimer disease. Ageing Res Rev 3:319–326

148. Moreira PI, Siedlak SL, Aliev G et al (2005) Oxidative stress mechanisms and potential therapeutics in Alzheimer disease. J Neural Transm 112:921–932
149. Maharaj DS, Maharaj H, Daya S et al (2006) Melatonin and 6-hydroxymelatonin protect against iron-induced neurotoxicity. J Neurochem 96:78–81
150. de Lima MN, Polydoro M, Laranja DC et al (2005) Recognition memory impairment and brain oxidative stress induced by postnatal iron administration. Eur J Neurosci 21:2521–2528
151. Takeda A, Smith MA, Avila J et al (2000) In Alzheimer's disease, heme oxygenase is coincident with Alz50, an epitope of tau induced by 4-hydroxy-2-nonenal modification. J Neurochem 75:1234–1241
152. Shimohama S, Tanino H, Kawakami N et al (2000) Activation of NADPH oxidase in Alzheimer's disease brains. Biochem Biophys Res Commun 273:5–9
153. Schipper HM (2000) Heme oxygenase-1: role in brain aging and neurodegeneration. Exp Gerontol 35:821–830
154. Atwood CS, Obrenovich ME, Liu T et al (2003) Amyloid-beta: a chameleon walking in two worlds: a review of the trophic and toxic properties of amyloid-beta. Brain Res Brain Res Rev 43:1–16
155. Bush AI (2003) The metallobiology of Alzheimer's disease. Trends Neurosci 26: 207–214
156. Misonou H, Morishima-Kawashima M, Ihara Y (2000) Oxidative stress induces intracellular accumulation of amyloid beta-protein (Abeta) in human neuroblastoma cells. Biochemistry (Mosc) 39:6951–6959
157. Yang DS, McLaurin J, Qin K et al (2000) Examining the zinc binding site of the amyloid-beta peptide. Eur J Biochem 267: 6692–6698
158. House E, Collingwood J, Khan A et al (2004) Aluminium, iron, zinc and copper influence the in vitro formation of amyloid fibrils of Abeta42 in a manner which may have consequences for metal chelation therapy in Alzheimer's disease. J Alzheimers Dis 6: 291–301
159. Zhu X, Raina AK, Lee HG et al (2004) Oxidative stress signalling in Alzheimer's disease. Brain Res 1000:32–39
160. Rottkamp CA, Raina AK, Zhu X et al (2001) Redox-active iron mediates amyloid-beta toxicity. Free Radic Biol Med 30:447–450
161. Smith MA, Harris PL, Sayre LM et al (1997) Iron accumulation in Alzheimer disease is a source of redox-generated free radicals. Proc Natl Acad Sci USA 94:9866–9868
162. Rottkamp CA, Atwood CS, Joseph JA et al (2002) The state versus amyloid-beta: the trial of the most wanted criminal in Alzheimer disease. Peptides 23:1333–1341
163. Sayre LM, Zelasko DA, Harris PL et al (1997) 4-Hydroxynonenal- derived advanced lipid peroxidation end products are increased in Alzheimer's disease. J Neurochem 68: 2092–2097
164. Nunomura A, Perry G, Pappolla MA et al (1999) RNA oxidation is a prominent feature of vulnerable neurons in Alzheimer's disease. J Neurosci 19:1959–1964
165. Shan X, Tashiro H, Lin CL (2003) The identification and characterization of oxidized RNAs in Alzheimer's disease. J Neurosci 23: 4913–4921
166. Honda K, Smith MA, Zhu X et al (2005) Ribosomal RNA in Alzheimer disease is oxidized by bound redox-active iron. J Biol Chem 280:20978–20986
167. Garza A, Vega R, Soto E (2006) Cellular mechanisms of lead neurotoxicity. Med Sci Monit 12:57–65
168. White LD, Cory-Slechta DA, Gilbert ME et al (2007) New and evolving concepts in the neurotoxicology of lead. Toxicol Appl Pharmacol 225:1–27
169. Vaziri ND (2002) Pathogenesis of lead-induced hypertension: role of oxidative stress. J Hypertens Suppl 20:15–20
170. Gonick HC (2002) Lead, renal disease and hypertension. Am J Kidney Dis 40:202–204
171. Dietert RR, Piepenbrink MS (2006) Lead and immune function. Crit Rev Toxicol 36:359–385
172. Bellinger DC (2005) Teratogen update: lead and pregnancy. Birth Defects Res A Clin Mol Teratol 73:409–420
173. Hu H, Rabinowitz M, Smith D (1998) Bone lead as a biological marker in epidemiologic studies of chronic toxicity: conceptual paradigms. Environ Health Perspect 106:1–8
174. Silbergeld EK (2003) Facilitative mechanisms of lead as a carcinogen. Mutat Res 533:121–133
175. Li PJ, Sheng YZ, Wang QY et al (2000) Transfer of lead via placenta and breast milk in human. Biomed Environ Sci 13:85–89
176. Lidsky TI, Schneider JS (2003) Lead neurotoxicity in children: basic mechanisms and clinical correlates. Brain 126:5–19
177. White RF, Diamond R, Proctor S et al (1993) Residual cognitive deficits 50 years after lead

poisoning during childhood. Br J Ind Med 50:613–622

178. Ruff HA, Markowitz ME, Bijur PE et al (1996) Relationships among blood lead levels, iron deficiency, and cognitive development in two-year-old children. Environ Health Perspect 104:180–185
179. Aimo L, Oteiza PI (2006) Zinc deficiency increases the susceptibility of human neuroblastoma cells to lead-induced activator protein-1 activation. Toxicol Sci 91:184–191
180. Toscano CD, Guilarte TR (2005) Lead neurotoxicity: from exposure to molecular effects. Brain Res Brain Res Rev 49:529–554
181. Bolin CM, Basha R, Cox D et al (2006) Exposure to lead and the developmental origin of oxidative DNA damage in the aging brain. FASEB J 20:788–790
182. Tiffany-Castiglioni E, Zmudzki J, Wu JN et al (1987) Effects of lead treatment on intracellular iron and copper concentrations in cultured astroglia. Metab Brain Dis 2:61–79
183. Stewart WF, Schwartz BS, Simon D et al (2002) ApoE genotype, past adult lead exposure, and neurobehavioral function. Environ Health Perspect 110:501–505
184. Momcilovic B, Alkhatib HA, Duerre JA et al (2001) Environmental lead-210 and bismuth-210 accrue selectively in the brain proteins in Alzheimer disease and brain lipids in Parkinson disease. Alzheimer Dis Assoc Disord 15:106–115
185. Zawia NH, Basha MR (2005) Environmental risk factors and the developmental basis for Alzheimer's disease. Rev Neurosci 16:325–337
186. Basha MR, Murali M, Siddiqi HK et al (2005) Lead (Pb) exposure and its effect on APP proteolysis and Abeta aggregation. FASEB J 19:2083–2084
187. Basha MR, Wei W, Bakheet SA et al (2005) The fetal basis of amyloidogenesis: exposure to lead and latent overexpression of amyloid precursor protein and beta-amyloid in the aging brain. J Neurosci 25:823–829
188. Wu J, Basha MR, Brock B et al (2008) Alzheimer's disease (AD)-like pathology in aged monkeys after infantile exposure to environmental metal lead (Pb): evidence for a developmental origin and environmental link for AD. J Neurosci 28:3–9
189. Archibald FS, Tyree C (1987) Manganese poisoning and the attack of trivalent manganese upon catecholamines. Arch Biochem Biophys 256:638–650
190. Keen CL (1995) Overview of manganese toxicity. In: Velazquez S, Liaison EPA (eds) Proceedings of the workshop on the bioavailability and oral toxicity of manganese. Environmental Criteria and Assessment Office, US EPA, Washington D.C., pp 3–11
191. Cotzias, G.C., Horiuchi, K., Fuenzalida, S., Mena, I (1968) Chronic manganese poisoning: clearance of tissue manganese concentrations with persistence of the neurological picture. Neurology 18:376–382
192. Malecki AE, Devenyi AG, Beard JL et al (1999) Existing and emerging mechanisms for transport of iron and manganese to the brain. J Neurosci Res 56:113–122
193. Wasserman GA, Liu X, Parvez F et al (2006) Water manganese exposure and children's intellectual function in Araihazar, Bangladesh. Environ Health Perspect 114:124–129
194. Iregren A (1999) Manganese neurotoxicity in industrial exposures: proof of effects, critical exposure level, and sensitive tests. Neurotoxicology 20:315–323
195. Erikson KM, Syversen T, Aschner JL et al (2005) Interactions between excessive manganese exposures and dietary iron-deficiency in neurodegeneration. Environ Toxicol Pharmacol 19:415–421
196. Hurley LS, Keen CL (1987) Manganese. In: Underwood E, Mertz W (eds) Trace elements in human health and animal nutrition. Academic, New York, pp 185–225
197. Golub MS, Hogrefe CE, Germann SL et al (2005) Neurobehavioral evaluation of rhesus monkey infants fed cow's milk formula, sow formula or soy formula with added manganese. Neurotoxicol Teratol 27:615–627
198. Rabin O, Hegedus L, Bourre JM et al (1993) Rapid brain uptake of manganese(II) across the blood–brain barrier. J Neurochem 61: 509–517
199. London RE, Toney G, Gabel SA et al (1989) Magnetic resonance imaging studies of the brains of anesthetized rats treated with manganese chloride. Brain Res Bull 23:229–235
200. Garrick M, Dolan K, Horbinski C et al (2003) DMT1: a mammalian transporter for multiple metals. BioMetals 16:41–54
201. Aschner M, Gannon M (1994) Manganese (Mn) transport across the blood– brain barrier: saturable and transferrin-dependent transport mechanisms. Brain Res Bull 33:345–349
202. Verity A (1999) Manganese neurotoxicity: a mechanistic hypothesis, Neurotoxicology 20:489–498
203. Wang JD, Huang CC, Hwang YH et al (1989) Manganese induced parkinsonism: an outbreak due to an unrepaired ventilation control system in a ferromanganese smelter. Br J Ind. Med 46:856–859

204. Kawamura R, Ikuta H, Fukuzumi S et al (1941) Intoxication by manganese in well water. Kisato Arch Exp Med 18:145–171
205. Iwami O, Watanabe T, Moon CS et al (1994) Motor neuron disease on the Ku Peninsula of Japan: excess manganese intake from food coupled with low magnesium in drinking water as a risk factor. Sci Total Environ 149:121–135
206. Alves G, Thebot J, Tracqui A et al (1997) Neurological disorders due to brain manganese deposition in a jaundiced patient receiving long-term parenteral nutrition. JPEN 1:41–45
207. Krieger D, Krieger S, Jansen O et al (1995) Manganese and chronic hepatic encephalopathy. Lancet 346:270–274
208. Mena I, Marin O, Fuenzalida S et al (1967) Chronic manganese poisoning. Clinical picture and manganese turnover. Neurology 17:128–136
209. Josephs KA, Ahlskog JE, Klos KJ et al (2005) Neurologic manifestations in welders with pallidal MRI T1 hyperintensity. Neurology 64:2033–2039
210. Bowler RM, Gysens S, Diamond E et al (2003) Neuropsychological sequelae of exposure to welding fumes in a group of occupationally exposed men. Int J Hyg Environ Health 206:517–529
211. Lydén A, Larsson BS, Lindquist NG (1984) Melanin affinity of manganese. Acta Pharmacol Toxicol Copenhagen 55:133–138
212. Donaldson J, McGregor D, LaBella F (1982) Manganese neurotoxicity: a model for free radical mediated neurodegeneration? Can J Physiol Pharmacol 60:1398–1405
213. Melov S, Adlard PA, Morten K et al (2007) Mitochondrial oxidative stress causes hyperphosphorylation of tau. PLoS One 2:e536
214. Li F, Calingasan NY, Yu F et al (2004) Increased plaque burden in brains of APP mutant MnSOD heterozygous knockout mice. J Neurochem 89:1308–1312
215. Esposito L, Raber J, Kekonius L et al (2006) Reduction in mitochondrial superoxide dismutase modulates Alzheimer's disease-like pathology and accelerates the onset of behavioral changes in human amyloid precursor protein transgenic mice. J Neurosci 26: 5167–5179
216. Hazell AS, Normandin L, Norenberg MD et al (2006) Alzheimer type II astrocytic changes following sub-acute exposure to manganese and its prevention by antioxidant treatment. Neurosci Lett 396:167–171
217. Filiz G, Price KA, Caragounis A et al (2008) The role of metals in modulating metalloprotease activity in the AD brain. Eur Biophys J 37:315–321
218. Gaggelli E, Kozlowski H, Valensin D et al (2006) Copper homeostasis and neurodegenerative disorders (Alzheimer's, prion, and Parkinson's diseases and amyotrophic lateral sclerosis). Chem Rev 106:1995–2044
219. Macreadie IG (2008) Copper transport and Alzheimer's disease. Eur Biophys J37: 295–300
220. Linder MC, Hazegh-Azam M (1996) Copper biochemistry and molecular biology. Am J Clin Nutr 63:797–811
221. Peña MM, Lee J, Thiele DJ (1999) A delicate balance: homeostatic control of copper uptake and distribution. J Nutr 129: 1251–1260
222. Donnelly PS, Xiao Z, Wedd AG (2007) Copper and Alzheimer's disease. Curr Opin Chem Biol 11:128–133
223. Barnham KJ, McKinstry WJ, Multhaup G et al (2003) Structure of the Alzheimer's disease amyloid precursor protein copper binding domain. A regulator of neuronal copper homeostasis. J Biol Chem 278:17401–17407
224. Hesse L, Beher D, Masters CL et al (1994) The beta A4 amyloid precursor protein binding to copper. FEBS Lett 349:109–116
225. Phinney AL, Drisaldi B, Schmidt SD et al (2003) In vivo reduction of amyloid-beta by a mutant copper transporter. Proc Natl Acad Sci U S A 100:14193–14198
226. Bellingham SA, Ciccotosto GD, Needham BE et al (2004) Gene knockout of amyloid precursor protein and amyloid precursor-like protein-2 increases cellular copper levels in primary mouse cortical neurons and embryonic fibroblasts. J Neurochem 91:423–428
227. White AR, Zheng H, Galatis D et al (1998) Survival of cultured neurons from amyloid precursor protein knock-out mice against Alzheimer's amyloid-beta toxicity and oxidative stress. J Neurosci 18:6207–6217
228. Borchardt T, Camakaris J, Cappai R et al (1999) Copper inhibits beta-amyloid production and stimulates the non-amyloidogenic pathway of amyloid-precursor-protein secretion. Biochem J 344:461–467
229. Scheuermann S, Hambsch B, Hesse L et al (2001) Homodimerization of amyloid precursor protein and its implication in the amyloidogenic pathway of Alzheimer's disease. J Biol Chem 276:33923–33929
230. Soba P, Eggert S, Wagner K et al (2005) Homo- and heterodimerization of APP family members promotes intercellular adhesion. EMBO J 24:3624–3634

231. Kong GK, Miles LA, Crespi GA et al (2008) Copper binding to the Alzheimer's disease amyloid precursor protein. Eur Biophys J 37:269–279
232. Curtain CC, Ali F, Volitakis I et al (2001) Alzheimer's disease amyloid-beta binds copper and zinc to generate an allosterically ordered membrane-penetrating structure containing superoxide dismutase-like subunits. J Biol Chem 276:20466–20473
233. Atwood CS, Scarpa RC, Huang X et al (2000) Characterization of copper interactions with alzheimer amyloid beta peptides: identification of an attomolar-affinity copper binding site on amyloid beta1–42. J Neurochem 75:1219–1233
234. Maynard CJ, Bush AI, Masters CL et al (2005) Metals and amyloid-beta in Alzheimer's disease. Int J Exp Pathol 86:147–159
235. Huang X, Cuajungco MP, Atwood CS et al (1999) Cu(II) potentiation of alzheimer abeta neurotoxicity. Correlation with cell-free hydrogen peroxide production and metal reduction. J Biol Chem 274:37111–37116
236. Behl C, Davis JB, Lesley R et al (1994) Hydrogen peroxide mediates amyloid beta protein toxicity. Cell 77:817–827
237. White AR, Barnham KJ, Bush AI (2006) Metal homeostasis in Alzheimer's disease. Expert Rev Neurother 6:711–722
238. Bayer TA, Multhaup G (2005) Involvement of amyloid beta precursor protein (AbetaPP) modulated copper homeostasis in Alzheimer's disease. J Alzheimers Dis 8:201–206; discussion 209–215
239. Duara R, Grady C, Haxby J et al (1986) Positron emission tomography in Alzheimer's disease. Neurology 36:879–887
240. Cottrell DA, Blakely EL, Johnson MA et al (2001) Mitochondrial enzyme-deficient hippocampal neurons and choroidal cells in AD. Neurology 57:260–264
241. Bayer TA, Schäfer S, Simons A et al (2003) Dietary Cu stabilizes brain superoxide dismutase 1 activity and reduces amyloid Abeta production in APP23 transgenic mice. Proc Natl Acad Sci U S A 100:14187–14192
242. Cherny RA, Atwood CS, Xilinas ME et al (2001) Treatment with a copper-zinc chelator markedly and rapidly inhibits beta-amyloid accumulation in Alzheimer's disease transgenic mice. Neuron 30:665–676
243. Schäfer S, Pajonk FG, Multhaup G et al (2007) Copper and clioquinol treatment in young APP transgenic and wild-type mice: effects on life expectancy, body weight, and metal-ion levels. J Mol Med 85:405–413
244. Sparks LD (2004) Cholesterol, copper, and accumulation of thioflavine S-reactive Alzheimer's-like amyloid beta in rabbit brain. J Mol Neurosci 24:97–104
245. Lee JY, Friedman JE, Angel I et al (2004) The lipophilic metal chelator DP-109 reduces amyloid pathology in brains of human beta-amyloid precursor protein transgenic mice. Neurobiol Aging 25:1315–1321
246. Snaedal J, Kristinsson J, Gunnarsdóttir S et al (1998) Copper, ceruloplasmin and superoxide dismutase in patients with Alzheimer's disease: a case-control study. Dement Geriatr Cogn Disord 9:239–242
247. Choi DW, Koh JY (1998) Zinc and brain injury. Annu Rev Neurosci 21:347–375
248. Mills CF (1989) Zinc in human biology. Springer-Verlag, London
249. Coleman JE (1992) Zinc proteins: enzymes, storage proteins, transcription factors, and replication proteins. Annu Rev Biochem 61:897–946
250. Taylor A (1996) Detection and monitoring of disorders of essential trace elements. Ann Clin Biochem 33:486–510
251. Taylor A (1997) Measurement of zinc in clinical samples. Ann Clin Biochem 34:142–150
252. Martin CM (2006) Vitamins for the elderly: from A to zinc. Consult Pharm 21:450–464
253. Vallee BL, Falchuk KH (1993) The biochemical basis of zinc physiology. Physiol Rev 73:79–118
254. Vasto S, Candore G, Listì F et al (2008) Inflammation, genes and zinc in Alzheimer's disease. Brain Res Rev 58:96–105
255. Vasto S, Candore G, Aquino A et al (2006) The nACHR4 594C/T polymorphism in Alzheimer disease. Rejuvenation Res 9: 107–110
256. Takeda A (2000) Movement of zinc and its functional significance in the brain. Brain Res Brain Res Rev 34:137–148
257. Weiss JH, Sensi SL, Koh JY (2000) Zn(2+): a novel ionic mediator of neural injury in brain disease. Trends Pharmacol Sci 21:395–401
258. Frederickson CJ, Moncrieff DW (1994) Zinc-containing neurons. Biol Signals 3:127–139
259. Takeda A (2001) Zinc homeostasis and functions of zinc in the brain. Biometals 14:343–351
260. Lehmann HM, Brothwell BB, Volak LP et al (2002) Zinc status influences zinc transport by porcine brain capillary endothelial cells. J Nutr 132:2763–2768
261. Law W, Kelland EE, Sharp P et al (2003) Characterisation of zinc uptake into rat

cultured cerebrocortical oligodendrocyte progenitor cells. Neurosci Lett 4(352): 113–116
262. Frederickson CJ, Suh SW, Silva D et al (2000) Importance of zinc in the central nervous system: the zinc-containing neuron. J Nutr 130:1471–1483
263. Danscher G, Stoltenberg M, Juhl S (1994) How to detect gold, silver and mercury in human brain and other tissues by autometallographic silver amplification. Neuropathol Appl Neurobiol 20:454–467
264. Takeda A, Hirate M, Tamano H et al (2003) Release of glutamate and GABA in the hippocampus under zinc deficiency. J Neurosci Res 72:537–542
265. Laube B (2002) Potentiation of inhibitory glycinergic neurotransmission by Zn2+: a synergistic interplay between presynaptic P2X2 and postsynaptic glycine receptors. Eur J Neurosci 16:1025–1036
266. Burdette SC, Lippard SJ (2003) Meeting of the minds: metalloneurochemistry. Proc Natl Acad Sci U S A 100:3605–3610267. Frederickson CJ (1989) Neurobiology of zinc and zinc-containing neurons. Int Rev Neurobiol 31:145–238
268. St Croix CM, Wasserloos KJ, Dineley KE et al (2002) Nitric oxide-induced changes in intracellular zinc homeostasis are mediated by metallothionein/thionein. Am J Physiol Lung Cell Mol Physiol 282:85–92
269. Cuajungco MP, Lees GJ (1997) Zinc metabolism in the brain: relevance to human neurodegenerative disorders. Neurobiol Dis 4:137–169
270. Du Y, Ni B, Glinn M et al (1997) alpha2-Macroglobulin as a beta-amyloid peptide-binding plasma protein. J Neurochem 69:299–305
271. Carrasco J, Giralt M, Molinero A et al (1999) Metallothionein (MT)-III: generation of polyclonal antibodies, comparison with MT-I+II in the freeze lesioned rat brain and in a bioassay with astrocytes, and analysis of Alzheimer's disease brains. J Neurotrauma 16:1115–1129
272. Choi DW, Koh JY (1998) Zinc and brain injury. Annu Rev Neurosci 21:347–375
273. Sorensen JC, Mattsson B, Andreasen A et al (1998) Rapid disappearance of zinc positive terminals in focal brain ischemia. Brain Res 812:265–269
274. Tonder N, Johansen FF, Frederickson CJ et al (1990) Possible role of zinc in the selective degeneration of dentate hilar neurons after cerebral ischemia in the adult rat. Neurosci Lett 109:247–252
275. Suh SW, Chen JW, Motamedi M et al (2000) Evidence that synaptically-released zinc contributes to neuronal injury after traumatic brain injury. Brain Res 852:268–273
276. Weiss JH, Sensi SL, Koh JY (2000) Zn(2+): a novel ionic mediator of neural injury in brain disease. Trends Pharmacol Sci 21:395–401
277. Sensi SL, Jeng JM (2004) Rethinking the excitotoxic ionic milieu: the emerging role of Zn(2+) in ischemic neuronal injury. Curr Mol Med 4:87–111
278. González C, Martín T, Cacho J et al (1999) Serum zinc, copper, insulin and lipids in Alzheimer's disease epsilon 4 apolipoprotein E allele carriers. Eur J Clin Invest 29:637–642
279. Rulon LL, Robertson JD, Lovell MA et al (2000) Serum zinc levels and Alzheimer's disease. Biol Trace Elem Res 75:79–85
280. Basun H, Forssell LG, Wetterberg L et al (1991) Metals and trace elements in plasma and cerebrospinal fluid in normal aging and Alzheimer's disease. J Neural Transm Park Dis Dement Sect 3:231–258
281. Tully CL, Snowdon DA, Markesbery WR (1995) Serum zinc, senile plaques, and neurofibrillary tangles: findings from the Nun Study. Neuroreport 6(16):2105–2108
282. Panayi AE, Spyrou NM, Iversen BS et al (2002) Determination of cadmium and zinc in Alzheimer's brain tissue using inductively coupled plasma mass spectrometry. J Neurol Sci 195:1–10
283. Bush AI, Multhaup G, Moir RD et al (1993) A novel zinc(II) binding site modulates the function of the beta A4 amyloid protein precursor of Alzheimer's disease. J Biol Chem 268:16109–16112
284. Narindrasorasak S, Lowery DE, Altman RA et al (1992) Characterization of high affinity binding between laminin and Alzheimer's disease amyloid precursor proteins. Lab Invest 67:643–652
285. Narindrasorasak S, Altman RA, Gonzalez-DeWhitt P et al (1995) An interaction between basement membrane and Alzheimer amyloid precursor proteins suggests a role in the pathogenesis of Alzheimer's disease. Lab Invest 72:272–282
286. Roberts SB, Ripellino JA, Ingalls KM et al (1994) Non-amyloidogenic cleavage of the beta-amyloid precursor protein by an integral membrane metalloendopeptidase. J Biol Chem 269:3111–3116
287. Bush AI, Tanzi RE (2002) The galvanization of beta-amyloid in Alzheimer's disease. Proc Natl Acad Sci U S A 99:7317–7319

288. Huang X, Atwood CS, Moir RD et al (1997) Zinc-induced Alzheimer's Abeta1–40 aggregation is mediated by conformational factors. J Biol Chem 272:26464–26470
289. Yoshiike Y, Tanemura K, Murayama O et al (2001) New insights on how metals disrupt amyloid beta-aggregation and their effects on amyloid-beta cytotoxicity. J Biol Chem 276:32293–32299
290. Moir RD, Atwood CS, Romano DM et al (1999) Differential effects of apolipoprotein E isoforms on metal-induced aggregation of A beta using physiological concentrations. Biochemistry 38:4595–4603
291. Lovell MA, Xie C, Markesbery WR (1999) Protection against amyloid beta peptide toxicity by zinc. Brain Res 823:88–95
292. Cuajungco MP, Goldstein LE, Nunomura A et al (2000) Evidence that the beta-amyloid plaques of Alzheimer's disease represent the redox-silencing and entombment of abeta by zinc. J Biol Chem 275:19439–19442
293. Stoltenberg M, Bush AI, Bach G et al (2007) Amyloid plaques arise from zinc-enriched cortical layers in APP/PS1 transgenic mice and are paradoxically enlarged with dietary zinc deficiency. Neuroscience 150:357–369
294. Mecocci P, MacGarvey U, Beal MF (1994) Oxidative damage to mitochondrial DNA is increased in Alzheimer's disease. Ann Neurol 36:747–751
295. Smith MA, Sayre LM, Anderson VE et al (1998) Cytochemical demonstration of oxidative damage in Alzheimer disease by immunochemical enhancement of the carbonyl reaction with 2,4-dinitrophenylhydrazine. J Histochem Cytochem 46:731–735
296. Butterfield DA, Hensley K, Harris M et al (1994) beta-Amyloid peptide free radical fragments initiate synaptosomal lipoperoxidation in a sequence-specific fashion: implications to Alzheimer's disease. Biochem Biophys Res Commun 200:710–715
297. Behl C, Davis JB, Lesley R et al (1994) Hydrogen peroxide mediates amyloid beta protein toxicity. Cell 77:817–827
298. Huang X, Cuajungco MP, Atwood CS et al (1999) Cu(II) potentiation of alzheimer abeta neurotoxicity. Correlation with cell-free hydrogen peroxide production and metal reduction. J Biol Chem 274:37111–37116
299. Zago MP, Oteiza PI (2001) The antioxidant properties of zinc: interactions with iron and antioxidants. Free Radic Biol Med 31:266–274
300. Potocnik FC, van Rensburg SJ, Park C et al (1997) Zinc and platelet membrane microviscosity in Alzheimer's disease. The in vivo effect of zinc on platelet membranes and cognition. S Afr Med J 87:1116–1119

INDEX

A

B

Peter Paul De Deyn and Debby Van Dam (eds.), *Animal Models of Dementia*, Neuromethods, vol. 48, DOI 10.1007/978-1-60761-898-0, © Springer Science+Business Media, LLC 2011

E

F

G

H

I

K

L

M

N

O

P

R

S

T

U

V

W

X

Z

MIX
Papier aus verantwortungsvollen Quellen
Paper from responsible sources
FSC® C105338
FSC
www.fsc.org

If you have any concerns about our products,
you can contact us on
ProductSafety@springernature.com

In case Publisher is established outside the EU,
the EU authorized representative is:
Springer Nature Customer Service Center GmbH
Europaplatz 3, 69115 Heidelberg, Germany

Printed by Libri Plureos GmbH
in Hamburg, Germany